Chemical Thermodynamics of Materials

Chemical Thermodynamics of Materials

C. H. P. Lupis

North-Holland
New York • Amsterdam • London • Tokyo

Elsevier Science Publishing Co., Inc.
655 Avenue of the Americas, New York, NY 10010

Distributors outside the United States and Canada:

Elsevier Science Publishers B. V.
P.O. Box 211, 1000 AE Amsterdam, The Netherlands

Library of Congress Cataloging in Publication Data

Lupis, C. H. P.
Chemical Thermodynamics of Materials.

Includes bibliographies and index.
1. Thermodynamics. I. Title.
QD504.L87 541.3'69 82-2471
ISBN 0-444-00713-X AACR2
ISBN 0-444-00779-2 (pbk.)

Current printing (last digit):
10 9 8 7

Manufactured in the United States of America

To Marlyse, Alexis, and Jean-Claude

Contents

Preface

This book has been written for graduate students and senior students in metallurgy and materials science who have had a previous introductory course in thermodynamics. It is hoped that this work will also be useful to professional metallurgists, chemists, and chemical engineers.

The field of chemical thermodynamics covers a wide range of subjects. Those treated here include closed and open bulk systems, surfaces, and statistical models. Because of the length of this text, some other important topics have been omitted, such as electrochemical reactions, electrolytic solutions, and the effects of forces other than pressure forces. The thermodynamics of solids under stress is a particularly difficult subject to treat, and its complexity could not be handled adequately within the limited goals of this enterprise. Most of the applications in this text are for metals and alloys rather than aqueous solutions, ceramics, or polymers. However, the principles presented here apply to all materials, and it is hoped that this text will provide the reader with a sufficient basis for further studies in these other subjects. The development of practical applications has been sought while maintaining theoretical rigor. Elaborations on some topics of narrower interest are printed in smaller type for easier differentiation.

The material presented here was developed over a period of 15 years of teaching at Carnegie-Mellon University (formerly Carnegie Institute of Technology), and this author is grateful to the many students who contributed by their constructive comments. Part of this text was written during a sabbatical sponsored by Carnegie-Mellon University and the Alexander von Humboldt Foundation: it was spent at the Institut für Eisenhüttenkunde der Rheinisch-Westfälischen Technischen Hochschule in Aachen, West Germany. The hospitality of the Institute, and especially of Professor Tarek El Gammal, is gratefully acknowledged.

Any teacher is influenced by the way he himself was taught, and I wish to express here my deep gratitude to Professors John Chipman and John F. Elliott, both of the Massachusetts Institute of Technology, who awoke my interest in the subject of chemical thermodynamics. Many colleagues and friends offered valuable comments and suggestions. Although I cannot cite them all, I wish to acknowledge the contributions of Professor W. W. Mullins, Professor Gerard Lesoult, Dr. Philip Spencer, Professor Donald Sadoway, and Patrick Martin. Multiple drafts of this text were typed by Lillian Puher; her inexhaustible patience and cheerful good will have been sincerely appreciated. This book would not have been completed without the encouragement of my wife. Moreover, through her editorial work, the reader has been spared many ambiguities and poorly worded statements. For those which are left, I, alone, am responsible.

Chemical Thermodynamics of Materials

I Fundamental Principles and Equations for a Closed System

1. First Law

1.1. Enunciation of the First Law

The first law of thermodynamics states that *in any process energy is conserved.* Although historically the concept of energy has been slow to evolve, today it is almost intuitively accepted, and it no longer seems very peculiar that heat and work are only different measures of change in the same property of a system, its energy. The mathematical translation of this principle is

$$dE = \text{đ}q + \text{đ}w \tag{1}$$

where E, q, and w are, respectively, the energy, the heat, and the work *absorbed by the system* from the surroundings.

Applications of the first law are merely accounting processes. Generally one adds up in the work term all the nonthermal interactions. The most commonly encountered work term in thermochemistry is the work done against external pressures; it is equal to the external pressure P_{ext} multiplied by the resulting change of volume of the system:

$$w = -\int P_{\text{ext}}\, dV \tag{2}$$

We shall note that in a *reversible* process (defined below) the external pressure, P_{ext} is equal to the pressure P of the system; w is then $-\int P\,dV$.

Another kind of work is that done against the gravitational field. If a mass m is displaced by a height h, the corresponding work term is mgh, where g is the acceleration due to gravity. Other sources of energy such as kinetic energy, $\frac{1}{2}mv^2$ (where v is the velocity of the system), and electrical energy, VIt (voltage × current × time), must also be considered when relevant.

In the accounting of these energy terms, one should be careful to use compatible units. Heat terms are commonly expressed in calories, work terms in joules. The conversion is

$$1 \text{ cal} = 4.184 \text{ J} \tag{3}$$

When work against pressure is expressed in cm³ atm, the conversion factors are

$$1 \text{ cm}^3 \text{ atm} = 0.0242 \text{ cal} = 0.1013 \text{ J} \tag{4}$$

1.2. State Functions and Perfect Differentials

In equation (1) we have used different symbols for the increment of energy and the increments of heat and work. The symbol đ emphasizes that the infinitesimal increments $\text{đ}q$ and $\text{đ}w$ are not perfect differentials: in a transformation, their changes do not depend solely on the initial and final states of the system, but also on the transformation path linking these two states.

Let us elaborate on this further. If it is possible to identify (intrinsically) property A of a system, then in a state 1 A will have the value A_1, in a state 2 the value A_2. Regardless of the transformation linking these two states, the change in the property A will be $DA = A_2 - A_1$. A is said to be a *state function,* and its differential dA will be a perfect differential. The energy of a system may be identified as one of its properties, but not work or heat. Work and heat cannot be thought of as intrinsic properties of the system. That is, they cannot be associated with one given state of the system; they can only be defined in a transformation of the system. Consequently, the initial and final states of the system do not determine them unequivocally; the heat and work absorbed by the system between states 1 and 2 depend on the choice of the transformation path linking these two states.

As an example, let us suppose that the property A depends on the variables x and y. The function $A(x,y)$ is implicitly or explicitly known, and it is possible to write

$$dA = \left(\frac{\partial A}{\partial x}\right)_y dx + \left(\frac{\partial A}{\partial y}\right)_x dy \tag{5}$$

Also recall that

$$\frac{\partial^2 A}{\partial x\, \partial y} = \frac{\partial^2 A}{\partial y\, \partial x} \tag{6}$$

i.e., the second order derivative does not depend on the order of differentiation. Conversely, if we consider an infinitesimal change dA for which we only know that

$$dA = L(x,y)\, dx + M(x,y)\, dy \tag{7}$$

where L and M are functions of the independent variables x and y, it is not possible to assert a priori whether or not dA is a perfect differential, that is, whether or not there exists a function (or a property) $A(x,y)$. $L(x,y)$ is not necessarily equal to $(\partial A/\partial x)_y$ or $M(x,y)$ equal to $(\partial A/\partial y)_x$. However, a necessary and sufficient condition for dA to be a perfect differential is

$$\left(\frac{\partial L}{\partial y}\right)_x = \left(\frac{\partial M}{\partial x}\right)_y \tag{8}$$

which is the equivalent of equation (6). If this test on L and M is not successful, $\partial L/\partial y \neq \partial M/\partial x$, then dA is not a perfect differential, and no function $A(x,y)$ can be identified implicitly or explicitly. Also, equation (7) cannot be integrated between two states 1 and 2 unless one specifies a certain path, and the change DA will depend on that path. Alternatively, of course, if the test is successful [if condition (8) is obeyed], dA is a perfect differential and A is a property of the system; its change DA between states 1 and 2 is $A_2 - A_1$, regardless of any path chosen between the two states.

Let us discuss another aspect of the problem that at times may be confusing. Whereas it is clear, for example, that hardness is a property of a system, it may not be clear whether or not hardness is path dependent (or a state function). Consider a system with properties A and B that undergoes a transformation bringing it from an initial state 1 to a final state 2. Let us assume that the first property A is dependent on the variables x and y while the second B is dependent on the variables x, y, and z. If the states 1 and 2 are defined solely by the variables x and y, it is clear that while the change in A will be independent of the path followed, B will depend on the path, since different paths can lead to different final values z_{2a}, z_{2b} of the variable z (see Fig. 1) and different final states $2a$ and $2b$. Thus

$$[A(x,y)]_{2a} = [A(x,y)]_{2b}$$

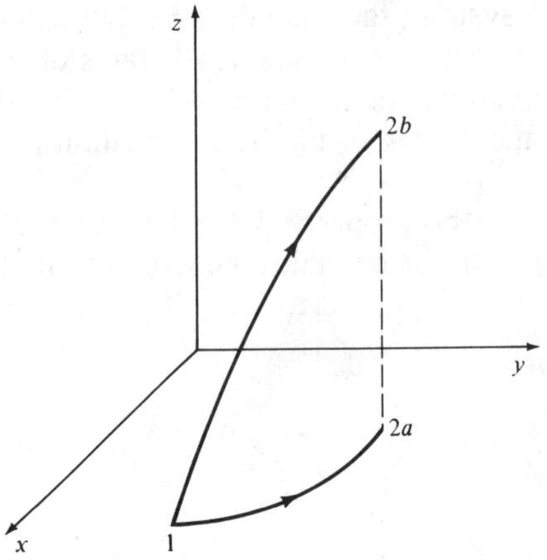

Figure 1. Importance of the variables for state functions. For the two paths 1–2*a* and 1–2*b*, the change in property $A(x,y)$ is the same; it is not the same for property $B(x,y,z)$.

leads to

$$(DA)_{\text{path } 1-2a} = (DA)_{\text{path } 1-2b}$$

whereas

$$[B(x,y,z)]_{2a} \neq [B(x,y,z)]_{2b}$$

leads to

$$(DB)_{\text{path } 1-2a} \neq (DB)_{\text{path } 1-2b}$$

In that sense, property B is often said to depend on the *history* of the system. However, if states 1 and 2 are now characterized by all three variables x, y, and z, then both A and B are now independent of the transformations linking the two states. The apparent dependence of B on the history of the system therefore resulted from an insufficient characterization of the initial and final states.

Hardness is thus a state function. Practically, however, the number of variables necessary to characterize the system (e.g., density and distribution pattern of dislocations, vacancies, and grain boundaries) may be too large to lend any practical value to that fact. For instance, the existence of a perfect differential for hardness is not very useful if we do not know how to express it in terms of the variables of the system.

Note that the difficulty in determining the dependence of a state function on the variables of the system depends on the precision we wish to bring to our measurements. For example, the energy of a system is, generally, readily accepted as a state function, and at the kilocalorie level it can usually be described by a small number of variables; at the calorie level, however, its description presents the same difficulty as in the case of hardness.

1.3. Heat Capacities and Enthalpy

When a substance absorbs heat its temperature is generally increased, and the relation between the quantity of heat absorbed $đq$ and the temperature increase dT is of considerable practical importance. Their ratio is called the *heat capacity* and is designated C:

$$C = đq/dT \tag{9}$$

Since $\text{đ}q$ is not a perfect differential, C will depend on the path of the transformation. The two most important kinds of transformations are those taking place at constant pressure and those at constant volume. They are designated C_p and C_v, respectively. Experimentally, it is usually much easier to carry and control processes at constant pressure than at constant volume, and experimental values of C_p are thus considerably more common than C_v. However, from a theoretical point of view it is easier to predict the values of C_v than of C_p. A comparison of theoretical predictions with experiment is often feasible because it is possible to calculate the difference $C_p - C_v$ by the relation (established in Section 2.6)

$$C_p - C_v = \alpha^2 VT/\beta \tag{10}$$

where α and β are the coefficients of expansion and compressibility:

$$\alpha = \frac{1}{V}\left(\frac{\partial V}{\partial T}\right)_P \tag{11}$$

$$\beta = \frac{-1}{V}\left(\frac{\partial V}{\partial P}\right)_T \tag{12}$$

If the work of a system is done only against external pressures, then the heat absorbed by this system at constant volume is identical to its increase in energy. Since

$$dE = \text{đ}q - P\,dV \tag{13}$$

and

$$(dE)_V = (\text{đ}q)_V \tag{14}$$

then

$$C_v = (\partial E/\partial T)_V \tag{15}$$

At constant pressure it is convenient to introduce the *enthalpy* function H, defined as

$$H \equiv E + PV \tag{16}$$

Then

$$dH = d(E + PV) = \text{đ}q + \text{đ}w + V\,dP + P\,dV \tag{17}$$

and when $\text{đ}w = -P\,dV$,

$$dH = \text{đ}q + V\,dP \tag{18}$$

Consequently,

$$(dH)_P = (\text{đ}q)_P \tag{19}$$

The heat absorbed (or released) by a system at constant pressure is equal to its enthalpy increase (or decrease). This explains why the enthalpy is often called *heat content*. The difference $H_T - H_{298}$ at constant pressure is termed *sensible heat*. A consequence of equation (19) is

$$C_p = (\partial H/\partial T)_P \tag{20}$$

The temperature dependence of the C_p of a substance is often expressed by an equation of the form

$$C_p = a + bT - (c/T^2) \tag{21}$$

which is empirical in nature and usually valid only above room temperature. Frequently the term c/T^2 is omitted, and when the data are scarce or the temperature range small, the term bT is also omitted (i.e., C_p is assumed to be constant). In certain cases, however, such as when a magnetic transformation occurs in the material, a simple equation such as (21) cannot represent the variation of C_p, even in a restricted range of temperature.

If a phase transformation occurs within the system, heat may be absorbed or released without resulting in a temperature change. The simplest example is that of a pure substance at its melting temperature. At constant pressure the heat absorbed merely transforms a portion of the substance from solid to liquid. The heat necessary for the transformation is the heat of fusion ΔH_f. It generally refers to one mole.

Figure 2 illustrates the temperature dependence of the enthalpy, at constant pressure, in the case of cadmium. The shape of the curve is typical of most metals; the magnitudes of the discontinuities generally increase with the transformation temperatures.

As another example, let us calculate the heat released by one pound (453.59 g) of copper when it is quenched under ordinary atmospheric pressure from the temperature of 1223°C (1500°K) where it is liquid, to room temperature (298°K). On the basis of 1 mol of copper (63.54 g), the data are as follows: $\Delta H_f = 3120$ cal at the temperature of fusion $T_f = 1357$°K, C_p(liquid) = 7.50 cal/°K, and C_p(solid) = $5.40 + 1.50 \times 10^{-3}T$ cal/°K. We may write

$$H_{1500} - H_{298} = (H_{1500} - H_{1357}) + (\Delta H_f)_{1357} + (H_{1357} - H_{298})$$

$$= \int_{1357}^{1500} 7.50\,dT + 3120 + \int_{298}^{1357} (5.40 + 1.50 \times 10^{-3}T)\,dT$$

$$= 1070 + 3120 + 7040 = 11230 \quad \text{cal/mol} \tag{22}$$

The heat released by 1 lb of copper is then equal to (453.59/63.54) × 11,230 = 80,180 cal.

If this pound of copper was quenched in 2 liter of water (2000 g) at room temperature (298°K), then the temperature of the water would be raised by approximately 40°C, since the *specific* heat capacity of water (calories per degree Kelvin *per gram*) is close to 1.[1]

1.4. Standard States

In thermodynamics one is rarely concerned with the absolute values of the energy, or the enthalpy, of a system; only changes (i.e., relative values) are recorded. It is possible, however, to assign a nonarbitrary absolute value to these functions through Einstein's discovery of the relation between mass and energy. Theoretically, the energy of a system is measurable by its mass and, for example, the "law" of conservation of mass is only a consequence of the law of conservation of energy.[2] However, this correspondence implies a change in the units which

[1] For an improved calculation of the temperature rise of the water, the change in the heat content of the copper from 298°K to the final temperature would have to be taken into account; this change is however quite small (approximately 230 vs 80,180 cal). For simplicity, the possible local formation and release of steam is neglected.

[2] Unlike energy, mass is not conserved. However, except when one deals with nuclear reactions, the changes are so small that the "law" of conservation of mass is, practically, very well respected.

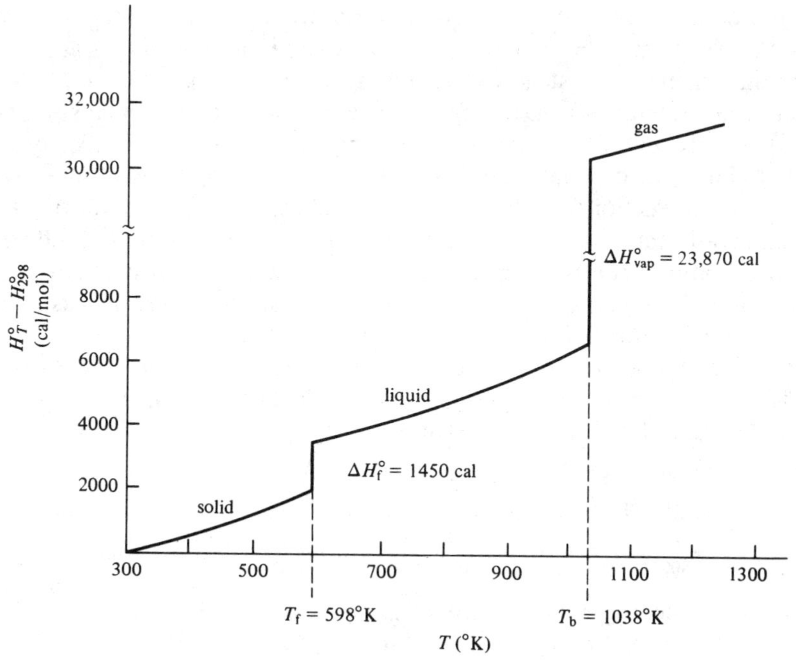

Figure 2. Standard enthalpy of cadmium.

introduces the square of the velocity of light, 1 g corresponding to 9×10^{20} erg or 20 billion kcal. Generally, chemical reactions do not involve more than a few kilocalories, and the corresponding changes in mass are thus well below the limits of detection. Consequently, unless one is specifically studying nuclear reactions, absolute values of the energy—and of the enthalpy—remain arbitrary for the elements of our material world. For example, the absolute values of the energies of hydrogen and helium can be chosen independently and arbitrarily, *unless* one considers the nuclear reaction: 4H = He, which releases nearly 600 billion cal (1 g-at. of helium, weighing 0.029 g, less than 4 g-at. of hydrogen), an amount well above the usual thermodynamic range. Although it is then possible in thermodynamics to choose an arbitrary value for the energy or the enthalpy of an element, generally no such value is chosen because there is no need for it.

Because only changes in the enthalpy of an element enter the calculations, it is very convenient, especially for purposes of tabulation, to choose standard reference states. *Unless otherwise specified, the standard state of an element is customarily chosen to be at a pressure of 1 atm and in the most stable structure of that element at the temperature at which it is investigated.* For gases, and to be rigorous, one should add the assumption of an ideal behavior (this point will be fully discussed in Chapter IV; for all practical purposes, it may now be ignored). For instance, at room temperature, the standard states of iron, mercury, and oxygen are, respectively, a bcc structure (α-iron) under 1 atm, a liquid under 1 atm, and a (perfect) diatomic gas at 1 atm. Thermodynamic functions evaluated at the standard state are usually identified by the superscript °; e.g., the standard enthalpies and heat capacities are designated H° and C°_p, respectively.

We must emphasize that it is possible to identify as a standard state one that does not correspond to the most stable form of the species under consideration. For example, it may be convenient to choose as the standard state of H_2O at 298°K that of the gas instead of the liquid, or one may choose at 298°K the fcc structure (austenite) of iron rather than the bcc one (ferrite). The standard state may also correspond to a *virtual state,* one that cannot be physically obtained but that can be theoretically defined and for which properties of interest can be calculated (see Section VII.3). Whatever the choice of structure for the definition of a standard state, we should recall that it corresponds to a fixed pressure, normally 1 atm. Summarizing then, the generally accepted convention is that, *unless otherwise specified,* the standard state of a species corresponds to the most stable form of that species under a pressure of 1 atm.

Standard enthalpies at various temperatures are usually tabulated as $H_T^\circ - H_{298}^\circ$. If we wish, for example, to calculate the enthalpy change of copper from the solid state at 1200°K under 100 atm to the liquid state at 1400°K under 1 atm, we may write

$$H_{1400}(\text{liq.}, P = 1) - H_{1200}(\text{sol.}, P = 100)$$

$$= (H_{1400}^\circ - H_{298}^\circ) - (H_{1200}^\circ - H_{298}^\circ) - [H_{1200}(\text{sol.}, P = 100) - H_{1200}^\circ] \quad (23)$$

Because the melting temperature of copper is 1357°K, H_{1400}° is identical to $H_{1400}(\text{liq.}, P = 1)$ and H_{1200}° is identical to $H_{1200}(\text{sol.}, P = 1)$. The first two groups of terms on the right-hand side of equation (23) can be found in various handbooks.[3] Equation (23) becomes

$$H_{1400}(\text{liq.}, P = 1) - H_{1200}(\text{sol.}, P = 100) = 10{,}480 - 5895 - \int_1^{100} (\partial H/\partial P)_{1200°\text{K}}\, dP \text{ cal} \quad (24)$$

We shall learn to evaluate the pressure dependence of H in Section 2, where we shall establish that $(\partial H/\partial P)_T = V(1 - \alpha T)$. The integral here is of the order of 20 cal, and the answer is thus 4565 cal.

1.5. Heats of Reactions

Let us assume that x mol of A react with y mol of B to form 1 mol of the compound A_xB_y. This transformation, or *reaction*, is described by the equation

$$xA + yB = A_xB_y \quad (25)$$

We have seen that at constant pressure the heat generated by any transformation is equal to the increase in enthalpy of the system. In the initial state, the enthalpy of the system is $xH(A) + yH(B)$, where $H(A)$ and $H(B)$ are, respectively, the enthalpies of 1 mol of A and 1 mol of B. In the final state the system has an enthalpy equal to the enthalpy of 1 mol of A_xB_y. The heat of the reaction is thus equal to

$$\Delta H = H(A_xB_y) - xH(A) - yH(B) \quad (26)$$

When ΔH is positive, heat is absorbed by the system and the reaction is said to

[3] See references on the compilation of data at the end of this chapter.

be *endothermic*. When ΔH is negative, heat is released by the system and the reaction is said to be *exothermic*.

We now consider the more general reaction

$$\nu_1 A_1 + \nu_2 A_2 + \cdots + \nu_k A_k = \nu_{k+1} A_{k+1} + \cdots + \nu_r A_r \tag{27}$$

where the A_is represent reactants ($A_1, \ldots, A_k$) or products ($A_{k+1}, \ldots, A_r$) and the ν_is balance the reaction stoichiometrically. It is slightly advantageous to rewrite the reaction (27)

$$\sum_{i=1}^{r} \nu_i A_i = 0 \tag{28}$$

and consider the numbers ν_i as algebraic: positive for the products and negative for the reactants. Thus, for reaction (25), $\nu_A = -x$, $\nu_B = -y$, and $\nu_{A_xB_y} = +1$.

The heat of the reaction (or, more tersely, its ΔH) is then defined as

$$\Delta H = \sum_{i=1}^{r} \nu_i H_i \tag{29}$$

where H_i is the enthalpy of the substance A_i. When A_i is not pure (e.g., when it is alloyed, such as carbon in iron), H_i represents its *partial molar enthalpy*. (The concept of partial molar properties will be presented and developed in subsequent chapters.)

When all the substances A_i are in their standard states, equation (29) becomes

$$\Delta H^\circ = \sum_{i=1}^{r} \nu_i H_i^\circ \tag{30}$$

ΔH (or ΔH°) represents the heat generated by the complete transformation of ν_i moles of the reactants A_i ($i = 1, \ldots, k$) into ν_i moles of the products A_i ($i = k+1, \ldots, r$) according to reaction (27) or (28). Actually, in a system containing n_i moles of the various components A_i, a reaction is rarely complete. Nevertheless, whether the reaction is complete or not, the heat generated may be calculated as follows.

Let us designate by dn_i the increment in the number of moles of the substance A_i produced by the reaction. All the increments dn_i are related to each other by the relation:

$$\frac{dn_1}{\nu_1} = \frac{dn_2}{\nu_2} = \cdots = \frac{dn_r}{\nu_r} = d\lambda \tag{31}$$

where λ is called the *progress variable* of the reaction. If $d\lambda$ is positive, reaction (27) proceeds forward (to the right; i.e., the numbers of moles of the products increase); if $d\lambda$ is negative, the reaction proceeds backward. The enthalpy of the system containing n_1 mol of component A_1, n_2 mol of component A_2, . . . , and n_r mol of component A_r is

$$H = \sum_{i=1}^{r} n_i H_i \tag{32}$$

At constant temperature and pressure, the increase in the enthalpy due to the progress of the reaction is

$$dH = \sum_{i=1}^{r} H_i \, dn_i \tag{33}$$

(This is evident when the enthalpies H_i are independent of the composition of the system, i.e., when the components A_i are pure. It also holds true, however, when the components A_i are not pure, in which case, the enthalpies H_i represent partial molar enthalpies.) Combining equations (33) and (31) yields

$$dH = \sum_{i=1}^{r} H_i \nu_i \, d\lambda = \left(\sum_{i=1}^{r} \nu_i H_i \right) d\lambda$$

or

$$dH = \Delta H \, d\lambda \tag{34}$$

The increase in the enthalpy of the system is the (algebraic) product of the ΔH of its reaction and the extent of its progress.

Consider for example a system containing 2 mol of calcium, 3 mol of carbon graphite, and 1 mol of calcium carbide CaC_2. The transformation may be characterized as follows:

	Ca	+	2C	=	CaC_2,	$\Delta H^\circ_{298} = -15{,}000$ cal.
initial state (mol):	2		3		1	
final state (mol):	$2 - \lambda$		$3 - 2\lambda$		$1 + \lambda$	

The reaction is complete and proceeds to the right. Consequently, $\lambda = 1.5$, and in its final state the system contains 0.5 mol of Ca and 2.5 mol of CaC_2. The reaction is exothermic ($\Delta H^\circ < 0$) and the heat released by the system is $\lambda \Delta H^\circ$, or $1.5 \times 15{,}000 = 22{,}500$ cal.

In Chapter V we shall learn to calculate accurately the direction and degree of advancement of an incomplete reaction. It may be noted however, that in the special case where all the substances A_i are pure and mutually immiscible solids and liquids, the reaction is generally complete; i.e., it proceeds until the supply of one of the components is exhausted. The formation of calcium carbide from calcium and graphite is such an example.

The ΔH of a reaction is a function of temperature, and to find its temperature dependence we note that

$$\left(\frac{\partial \Delta H}{\partial T} \right)_P = \left(\frac{\partial \left(\sum_{i=1}^{r} \nu_i H_i \right)}{\partial T} \right)_P = \sum_{i=1}^{r} \nu_i \left(\frac{\partial H_i}{\partial T} \right)_P = \sum_{i=1}^{r} \nu_i C_{\mathrm{p},i} \tag{35}$$

or

$$\left(\frac{\partial \Delta H}{\partial T} \right)_P = \Delta C_\mathrm{p} \tag{36a}$$

This expression is known as *Kirchhoff's law*. When all products and reactants are in their standard states, it becomes

$$\frac{d\Delta H^\circ}{dT} = \Delta C^\circ_\mathrm{p} \tag{36b}$$

1.6. Heats of Formation

The standard heat of the reaction corresponding to the formation of a substance from its elements in their standard states is defined as the standard heat of formation of that substance. For example, the standard heat of formation of carbon monoxide gas CO corresponds to the reaction

$$\text{C(graphite)} + \tfrac{1}{2}\text{O}_2\text{(gas)} = \text{CO(gas)} \tag{37}$$

and is equal to

$$H_f^\circ(\text{CO}) = \Delta H^\circ = H^\circ(\text{CO}) - H^\circ(\text{C(graphite)}) - \tfrac{1}{2}H^\circ(\text{O}_2) \tag{38}$$

It is obvious that the standard heat of formation of an element in its standard state is zero.

Knowledge of the heats of formations simplifies the calculation of the ΔHs of more complex reactions because these may be written as an algebraic sum of heats of formations:

$$\Delta H^\circ = \sum_{i=1}^{r} \nu_i H_{f,i}^\circ \tag{39}$$

and there are generally fewer nonzero terms in this equation than in the equation defining ΔH°:

$$\Delta H^\circ = \sum_{i=1}^{r} \nu_i H_i^\circ \tag{40}$$

Equation (39) may be demonstrated by recalling that since the difference ΔH° in the enthalpies of the final and initial states is independent of the path linking these two states, it is possible to decompose in a first step all the reactants into their elements (and in their standard states), and in a second step to recombine these elements into the reaction's products (see Fig. 3). In the first step, the variation

Figure 3. Heat of reaction as a function of heats of formation:

$$\Delta H = \sum_{j=k+1}^{r} \nu_j H_{f,j}(\text{products}) - \sum_{i=1}^{k} |\nu_i| H_{f,i}(\text{reactants})$$

$$= \sum_{i=1}^{r} \nu_i H_{f,i}$$

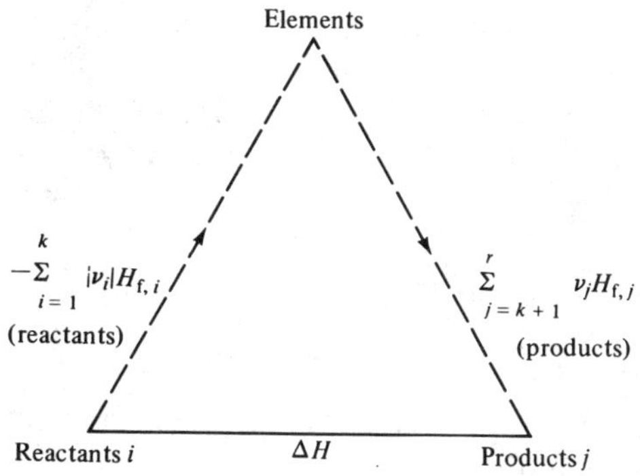

in the enthalpy of the system is equal to $-(\sum |\nu_i|\, H^\circ_{f,i})_{\text{reactants}}$ and in the second it is equal to $(\sum |\nu_i|\, H^\circ_{f,i})_{\text{products}}$. With the previous convention of algebraic signs for ν_i (positive for products, negative for reactants), we see that the total variation in the enthalpy ΔH° is indeed given by equation (39).

For example, let us consider the reaction

$$CO + H_2O = CO_2 + H_2 \tag{41}$$

Its ΔH° may be calculated as

$$\Delta H^\circ = H^\circ_f(CO_2) - H^\circ_f(CO) - H^\circ_f(H_2O) \tag{42}$$

The same result may be obtained through a series of algebraic manipulations. As previously, these are justified by the state function character of the enthalpy.

$$\begin{array}{lrcl|rcl}
\times(-1): & C + \tfrac{1}{2}O_2 & = & CO, & \Delta H^\circ_1 & = & H^\circ_f(CO) \\
\times(+1): & C + O_2 & = & CO_2, & \Delta H^\circ_2 & = & H^\circ_f(CO_2) \\
\times(-1): & H_2 + \tfrac{1}{2}O_2 & = & H_2O, & \Delta H^\circ_3 & = & H^\circ_f(H_2O) \\
\hline
 & CO + H_2O & = & CO_2 + H_2, & \Delta H^\circ & = & -\Delta H^\circ_1 + \Delta H^\circ_2 - \Delta H^\circ_3 \\
 & & & & & = & H^\circ_f(CO_2) - H^\circ_f(CO) - H^\circ_f(H_2O)
\end{array}$$

Alternatively, one may verify that

$$\begin{aligned}
H^\circ_f(CO_2) - H^\circ_f(CO) - H^\circ_f(H_2O) &= [H^\circ(CO_2) - H^\circ(C) - H^\circ(O_2)] \\
&\quad - [H^\circ(CO) - H^\circ(C) - \tfrac{1}{2}H^\circ(O_2)] \\
&\quad - [H^\circ(H_2O) - H^\circ(H_2) - \tfrac{1}{2}H^\circ(O_2)] \\
&= H^\circ(CO_2) + H^\circ(H_2) - H^\circ(CO) - H^\circ(H_2O) \\
&= \Delta H^\circ
\end{aligned}$$

2. Second Law

2.1. Enunciation of the Second Law

Under a given set of conditions, a system can be imagined to undergo several processes in which the energy is conserved (first law). However, it is common experience to observe that the only processes which occur are those which bring the system to a state of rest, i.e., to a state of equilibrium. By considering that this state of equilibrium is a property of the system—thus, that this state can be described by a function—the second principle (or law) of thermodynamics determines the direction and extent of such processes. *It affirms the existence of a state function, the entropy S, which for all reversible processes is defined by*

$$dS = đq_{\text{rev}}/T \tag{43}$$

and for all irreversible processes is such that

$$dS > đq/T \tag{44}$$

It may be noted that for an isolated system $đq = 0$ and thus, the entropy of an isolated system always tends to increase.

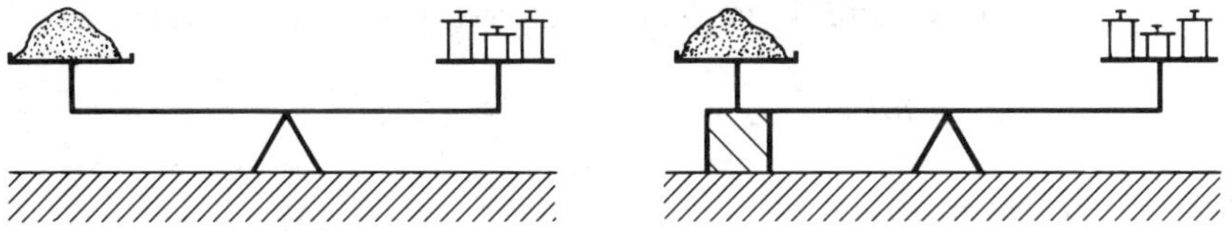

Figure 4. Examples of (a) reversible and (b) nonreversible equilibria.

We have used the words "reversible" and "irreversible," and their meaning should be recalled. A reversible equilibrium is characterized by a state of balance such that an infinitesimal change in the external conditions may cause a reversal in the direction of the process. For example, Fig. 4a illustrates a frictionless mechanical balance in a reversible equilibrium; if an infinitesimal weight is added or substracted to one of the scales, it will cause the balance to tip one way or the other. The same balance is shown in a nonreversible equilibrium in Fig. 4b where a block prevents the left scale from being lowered. Any decrease in the weight of the right scale will leave the balance in equilibrium, but an increase in the weight of the right scale may tip it.

Reversible processes are processes for which each stage is characterized by a state of reversible equilibrium. Actual processes are never reversible but they may be very nearly so, and certainly can be *imagined* to be so. This fact is important because thermodynamic calculations are much easier to perform for reversible processes. Changes in state functions do not depend on the individual path followed but only on the initial and final states. Thus, a reversible path is imagined linking these two states, along which the calculations are performed. A simple example will illustrate this procedure.

Consider a cylinder with two compartments. The first, of volume V_1, is filled with an ideal gas.[4] The second, of volume V_2, has been evacuated. By opening a suitable connection between the two compartments, the gas expands irreversibly and occupies the total volume $V_1 + V_2$. Since the system is isolated, its energy remains constant, and as the energy of an ideal gas depends solely on its temperature (see Section 2.5), the temperature of the system remains constant. To calculate the consequent increase in entropy, we imagine a frictionless piston which restores the gas to its initial state in a reversible fashion (and now through interactions with the surroundings). In this new transformation, the energy remains constant because it is a state function which depends only on the temperature. Consequently

$$q = -w = \int_{V_1+V_2}^{V_1} P\,dV = nRT \int_{V_1+V_2}^{V_1} \frac{dV}{V} = nRT \ln \frac{V_1}{V_1 + V_2} \tag{45}$$

and

$$S_1 - S_2 = \frac{q_{\text{rev}}}{T} = nR \ln \frac{V_1}{V_1 + V_2} \tag{46}$$

[4] The properties of an ideal gas are discussed in Section 2.5. We shall assume that the reader is already familiar with its equation of state, $PV = nRT$.

Thus, the entropy increase in the original irreversible process is

$$S_2 - S_1 = nR \ln \frac{V_1 + V_2}{V_1} \tag{47}$$

We verify that it is positive.

2.2. Entropy as a Measure of Irreversibility and Degradation

The fact that a reversible process can be considered as the limit toward which irreversible processes can approach indefinitely may be linked to the result that, for an isolated system, reversible processes correspond to no increase in the entropy and irreversible processes to a net gain of entropy. More generally and of fundamental importance, the entropy provides a measure of the degree of irreversibility of a transformation, or a measure of the degree of degradation of the system experiencing this irreversible transformation.

Let us elaborate qualitatively on this concept of entropy. (A rigorous presentation of the second law from this viewpoint is outside the scope of this text.) Two familiar and recurrent types of spontaneous or irreversible processes are the flow of heat from a high temperature to a low temperature and the conversion of work to heat (by friction for example). In the transformation of mechanical energy to thermal energy, we shall see that not only is the amount of energy exchanged important but also the temperature of the system.

Assume that a certain amount of work produces a heat q which is absorbed by a reservoir at the temperature T_h. The same amount of heat q may now be transferred from the reservoir at T_h to a second reservoir at a lower temperature T_l. If we now compare the conversion of work to heat received by the reservoir at T_h with the conversion of the same amount of work to the same amount of heat but for a reservoir at the lower temperature T_l, the second process must be more irreversible than the first. Indeed, the first process can be considered as an intermediate step to the fulfillment of the second, this fulfillment being achieved by an additional irreversible phenomenon, the flow of heat from T_h to T_l.

The degree of irreversibility of a transformation must obviously be proportional to the heat exchanged, and from what preceded it must also be a decreasing function of the temperature. The function q/T fulfills these prerequisites.

From considerations such as those above, it is thus possible to derive a statement of the second law based on the concept of entropy as a measure of the degree of irreversibility of a process, or of the degree of degradation of a system. For an excellent presentation of this concept, the reader is referred to the text by Lewis and Randall [1, Ch. 7].

2.3. Criteria of Equilibrium

The inequality (44) of the second law may take several forms. Its combination with equation (1) yields

$$dE - T\,dS - đw < 0 \tag{48}$$

and if—as we shall assume for simplicity in the rest of this treatment—the only work performed by the system is done against normal forces (pressure), then

$$dE - T\,dS + P\,dV < 0 \tag{49}$$

for all irreversible processes.

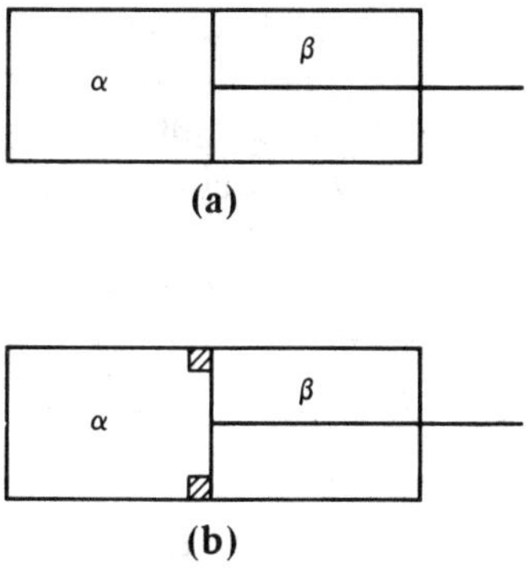

Figure 5. Illustration of an equilibrium criterion: $\delta V^{\alpha}(P^{\beta} - P^{\alpha}) \geq 0$. (a) Condition for equilibrium: $P^{\alpha} = P^{\beta}$. (b) Condition for equilibrium: $P^{\beta} \geqslant P^{\alpha}$.

An infinity of processes may be undergone or imagined to be undergone by the system. The consequent *virtual variations* of the thermodynamic functions will be designated by the symbol δ. Obviously, a class of such variations is in the field of equilibrium states if no irreversible change occurs, i.e., if the inequality (44) or (49) is violated. In other words, the thermodynamic criterion of equilibrium for a closed system may be expressed as

$$\delta S \leq \delta q / T \tag{50}$$

or

$$\delta E - T\,\delta S + P\,\delta V \geq 0 \tag{51}$$

For convenience, we shall denote by δZ the quantity

$$\delta Z = \delta E - T\,\delta S + P\,\delta V \tag{52}$$

The criterion of equilibrium is then

$$\delta Z \geq 0 \tag{53}$$

Let us clarify the meaning of the inequality (53). Figure 5a shows a piston separating two compartments α and β at the same pressures. If the piston is frictionless, it is in a state of reversible equilibrium. By contrast, in Fig. 5b small blocks in the cylinder prevent the piston from being displaced to the left; a change in the volume V^{α} of α can be only positive ($\delta V^{\alpha} \geq 0$). The condition then for equilibrium is that the pressure P^{β} in the right compartment be equal to or higher than that of the left compartment P^{α} (but not high enough to tear off the blocks!). The equilibrium is not reversible then, and any transformation in the state of the system which does not violate the criterion of equilibrium

$$\delta V^{\alpha}(P^{\beta} - P^{\alpha}) \geq 0 \tag{54}$$

is in the field of equilibrium states. Condition (54) is entirely analogous to criterion (53).[5]

When the process is reversible, the equality sign in condition (51) or (53) prevails on the inequality sign and yields

$$dE = T\,dS - P\,dV \tag{55}$$

---

[5] Although not a direct derivation of it (see Chapter II).

As discussed above, the importance of equation (55) lies in the fact that it can be applied to irreversible processes since the energy is a state function.

Depending on the restrictions to which a system is subjected, the introduction of certain thermodynamic functions is often very useful. For instance, if the process is carried out at constant entropy (e.g., $dq_{rev} = 0$, no heat is exchanged) and at constant pressure, the consideration of the enthalpy H [defined as $E + PV$ in equation (16)] simplifies considerably the calculations. Indeed,

$$dH = dE + P\,dV + V\,dP = T\,dS + V\,dP \tag{56}$$

or

$$(dH)_{S,P} = 0 \tag{57}$$

Also, since

$$(\delta Z)_{S,P} = (\delta E + P\,\delta V)_{S,P} = (\delta H)_{S,P} \tag{58}$$

the criterion of equilibrium becomes

$$(\delta H)_{S,P} \geq 0 \tag{59}$$

For similar reasons, at constant temperature and volume, it is natural to introduce the Helmholtz free energy A,

$$A \equiv E - TS \tag{60}$$

and, at constant temperature and pressure, the Gibbs free energy,

$$G \equiv E + PV - TS \tag{61}$$

since the combination of equation (55) with (60) and (61) yields

$$dA = -S\,dT - P\,dV \tag{62}$$

and

$$dG = -S\,dT + V\,dP \tag{63}$$

It is readily verified that

$$(\delta Z)_{V,T} = (\delta E - T\delta S)_{V,T} = (\delta A)_{V,T} \tag{64}$$

and

$$(\delta Z)_{P,T} = (\delta E - T\delta S + P\delta V)_{P,T} = (\delta G)_{P,T} \tag{65}$$

Consequently, the criterion of equilibrium may be expressed

$$(\delta A)_{T,V} \geq 0 \tag{66}$$

or

$$(\delta G)_{T,P} \geq 0 \tag{67}$$

The meaning of an inequality such as (67) is that at constant temperature and pressure the only possible variations which do not result in an irreversible process must increase the free energy of the system, i.e., at equilibrium the free energy is at a minimum. A spontaneous transformation can occur at constant temperature and pressure only if it is associated with a decrease in the Gibbs free energy of the system.

2.4. *Useful Relationships*

Equations (55), (56), (62), and (63) are the sources of many useful relations. For instance, dividing equation (56) by dT and keeping the pressure constant, we obtain

$$\left(\frac{\partial H}{\partial T}\right)_P = T\left(\frac{\partial S}{\partial T}\right)_P \tag{68a}$$

or

$$\left(\frac{\partial S}{\partial T}\right)_P = \frac{C_P}{T} \tag{68b}$$

where C_P is the specific heat at constant pressure. This equation may also be rewritten

$$(S)_{T_2} - (S)_{T_1} = \int_{T_1}^{T_2} C_P \, d \ln T \tag{69}$$

Similarly, the fact that dE, dH, dA, and dG are perfect differentials is expressed by *Maxwell's relations*. Equation (63), for example, yields

$$-\left(\frac{\partial S}{\partial P}\right)_T = \left(\frac{\partial V}{\partial T}\right)_P = \frac{\partial^2 G}{\partial P \, \partial T} \tag{70}$$

or

$$\left(\frac{\partial S}{\partial P}\right)_T = -\alpha V \tag{71}$$

where α is the coefficient of expansion defined in equation (11). These results are summarized in Table 1.

In addition, it may be noted that by dividing equation (56) by dP and keeping the temperature constant we obtain

$$\left(\frac{\partial H}{\partial P}\right)_T = T\left(\frac{\partial S}{\partial P}\right)_T + V = V(1 - \alpha T) \tag{72}$$

In deriving or in using thermodynamic equations such as those above, it is advisable to check that they are correct dimensionally, e.g., that a pressure × volume term is balanced by another energy term such as temperature × entropy, or temperature × heat capacity.

2.5. *Perfect Gases*

A "perfect gas" or an "ideal gas" is generally defined as a substance which obeys the following two conditions:

a. Its temperature, pressure, and volume obey the relation

$$PV = nRT \tag{73}$$

where n is the number of moles of that gas and R is a universal constant.

b. Its energy is a function of the temperature alone.

Table 1. Summary of Important Equations for the Thermodynamics of a Closed System

Thermodynamic restrictions	Relevant thermodynamic function	Differential	Criterion of equilibrium $\delta Z = (\delta E - T\,\delta S + P\,\delta V) \geq 0$	Maxwell's relations
E, V constant	S	$dS = đq_{rev}/T$	$(\delta S)_{E,V} \leq 0$	
S, V constant	E	$dE = T\,dS - P\,dV$	$(\delta E)_{S,V} \geq 0$	$\left(\frac{\partial T}{\partial V}\right)_S = -\left(\frac{\partial P}{\partial S}\right)_V = \frac{\partial^2 E}{\partial S\,\partial V}$
S, P constant	$H \equiv E + PV$	$dH = T\,dS + V\,dP$	$(\delta H)_{S,P} \geq 0$	$\left(\frac{\partial T}{\partial P}\right)_S = \left(\frac{\partial V}{\partial S}\right)_P = \frac{\partial^2 H}{\partial S\,\partial P}$
T, V constant	$A \equiv E - TS$	$dA = -S\,dT - P\,dV$	$(\delta A)_{T,V} \geq 0$	$\left(\frac{\partial S}{\partial V}\right)_T = \left(\frac{\partial P}{\partial T}\right)_V = -\frac{\partial^2 A}{\partial V\,\partial T}$
T, P constant	$G \equiv E + PV - TS$	$dG = -S\,dT + V\,dP$	$(\delta G)_{T,P} \geq 0$	$\left(\frac{\partial S}{\partial P}\right)_T = -\left(\frac{\partial V}{\partial T}\right)_P = -\frac{\partial^2 G}{\partial P\,\partial T}$

We will recall that these properties are approached by all gases when their pressure is indefinitely diminished. (In statistical thermodynamics, a perfect gas is characterized by the fact that there is no interaction between its molecules; for real gases when the pressure is diminished, the average distance between the molecules increases to the point where their interactions become negligible). Empirically (by extrapolation), it has been found that at 273.15°K (0°C), one mole of perfect gas under one atmosphere occupies a volume of 22414 cm^3. This corresponds to a value of R equal to 82.06 cm^3 atm/°K mol or 1.987 cal/°K mol.

The definition above (based on the conditions a and b) is usually given when the second law is ignored. Application of the second law makes condition b redundant: condition a is a sufficient condition for b.

To prove that a is sufficient, we need to prove that $(\partial E/\partial V)_T$ and $(\partial E/\partial P)_T$ are equal to zero when $PV = RT$ (for one mole of gas).

From the first law

$$dE = T\,dS - P\,dV \tag{74}$$

and consequently

$$\left(\frac{\partial E}{\partial V}\right)_T = T\left(\frac{\partial S}{\partial V}\right)_T - P \tag{75}$$

Using the Maxwell's relation corresponding to the second derivative of the Helmholtz free energy A with respect to V and T we may write

$$\left(\frac{\partial S}{\partial V}\right)_T = \left(\frac{\partial P}{\partial T}\right)_V = \frac{R}{V} \tag{76}$$

Equation (75) becomes

$$\left(\frac{\partial E}{\partial V}\right)_T = T\frac{R}{V} - P = 0 \tag{77}$$

since $PV = RT$.

To demonstrate that $(\partial E/\partial P)_T = 0$, it is preferable to use the equivalent statement $(\partial H/\partial P)_T = 0$; it is equivalent because

$$\left(\frac{\partial H}{\partial P}\right)_T = \left(\frac{\partial(E + PV)}{\partial P}\right)_T = \left(\frac{\partial(E + RT)}{\partial P}\right)_T = \left(\frac{\partial E}{\partial P}\right)_T \tag{78}$$

The procedure is then entirely similar to the previous one, using the expression of the first law in terms of dH, instead of dE, and Maxwell's relation corresponding to dG instead of dA.

Condition b alone would yield

$$\left(\frac{\partial E}{\partial V}\right)_T = T\left(\frac{\partial S}{\partial V}\right)_T - P = 0 \tag{79}$$

and using Maxwell's relation

$$T\left(\frac{\partial P}{\partial T}\right)_V = P \tag{80}$$

which, after integration, shows that at constant V, P is proportional to T:

$$P/T = f(V) \tag{81}$$

The function $f(V)$ cannot be defined without an additional condition. It may also be noted that because E is independent of V, its independence of P is also implicit (see Problem 11).

2.6. Relation Between C_p and C_v

In Section 1.3 we pointed out the significance of a relationship between C_p and C_v [equation (10)]. The second law allows the derivation of this relationship.

Considering energy as a function of the volume and the temperature, we have

$$dE = \left(\frac{\partial E}{\partial T}\right)_V dT + \left(\frac{\partial E}{\partial V}\right)_T dV = C_v\, dT + \left(\frac{\partial E}{\partial V}\right)_T dV \tag{82}$$

and

$$\left(\frac{\partial E}{\partial T}\right)_P = C_v + \left(\frac{\partial E}{\partial V}\right)_T \left(\frac{\partial V}{\partial T}\right)_P \tag{83}$$

or

$$C_p - C_v = \left(\frac{\partial E}{\partial V}\right)_T \left(\frac{\partial V}{\partial T}\right)_P + \left(\frac{\partial PV}{\partial T}\right)_P = \left(\frac{\partial V}{\partial T}\right)_P \left[P + \left(\frac{\partial E}{\partial V}\right)_T\right] \tag{84}$$

Equation (75) then leads to

$$C_p - C_v = T\left(\frac{\partial S}{\partial V}\right)_T \left(\frac{\partial V}{\partial T}\right)_P = T\left(\frac{\partial P}{\partial T}\right)_V \left(\frac{\partial V}{\partial T}\right)_P \tag{85}$$

But[6]

$$\left(\frac{\partial P}{\partial T}\right)_V \left(\frac{\partial T}{\partial V}\right)_P \left(\frac{\partial V}{\partial P}\right)_T = -1 \tag{86}$$

or

$$\left(\frac{\partial P}{\partial T}\right)_V = -\left(\frac{\partial V}{\partial T}\right)_P \Big/ \left(\frac{\partial V}{\partial P}\right)_T = \frac{\alpha}{\beta} \tag{87}$$

Consequently, equation (85) becomes

$$C_p - C_v = \alpha^2 VT/\beta \tag{88}$$

We note that C_p is always greater than C_v. In the case of perfect gases, it is easy to prove that the difference is equal to R.

[6] Quite generally, $(\partial z/\partial x)_y(\partial x/\partial y)_z(\partial y/\partial z)_x = -1$. Consider the function $z(x,y)$. We may write: $dz = (\partial z/\partial x)_y\, dx + (\partial z/\partial y)_x\, dy$. At constant z we obtain $(\partial z/\partial x)_y + (\partial z/\partial y)_x(\partial y/\partial x)_z = 0$, which is identical to the relation we wished to establish.

3. Third Law

3.1. Enunciation of the Third Law

In 1902 T. W. Richards found experimentally that, as the temperature decreases, the free energy of a reaction approaches asymptotically its enthalpy change. This result is not surprising since, for any transformation at constant temperature, the changes in the Gibbs free energy, enthalpy and entropy are related by the equation

$$DG = DH - TDS \tag{89}$$

and when $T \rightarrow 0°K$, $DG \rightarrow DH$. However, if we plot DG vs T (Fig. 6), we see that DG may approach DH with a vertical slope, a horizontal slope, or an oblique one. The value of the slope is the change in the entropy, since

$$\left(\frac{\partial DG}{\partial T}\right)_P = -DS \tag{90}$$

Examining Richard's data, Nernst found that the slope at the origin was never vertical and could be horizontal; in 1906, he suggested that at 0°K the entropy increment of reversible reactions among perfect crystalline solids is zero [2]. Planck attempted to generalize this principle [3], and stated that the entropies of all perfect crystalline solids at 0°K are zero. However, a better statement of the third law appears to be that of Lewis and Randall:

> If the entropy of each element in some [perfect] crystalline state be taken as zero at the absolute zero of temperature, every substance has a finite positive entropy; but at the absolute zero of temperature the entropy may become zero, and does so become in the case of perfect crystalline substances [4].

A number of remarks are in order. We stress first the difference between substances and elements: a substance is composed of elements. "Perfect" in the context above does not only mean without defects such as vacancies or dislocations, but also without disorder in the arrangement of the atoms (e.g., by substitution of two kinds of atoms). The choice of a zero reference value for the

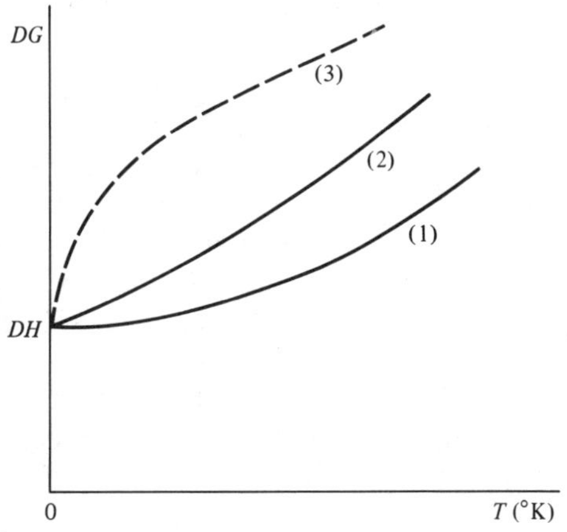

Figure 6. Change in the Gibbs free energy for a transformation occurring near 0°K. The vertical slope of the dashed curve (3) at 0°K is contrary to the third law. The horizontal slope of (1) is observed in the case of perfect substances, and the oblique slope of curve (2) is observed when the substances taking part in the reaction have retained some degree of disorder.

entropies of the elements of 0°K is conventional; its selection simplifies enormously the calculation of the entropies of substances but is not necessary. For example, the entropy of (perfect) α iron at 0°K could be chosen as 10 cal/°K mol, and that of carbon-graphite as 5 cal/°K mol. Then, at 0°K the entropies of γ iron and diamond would still be respectively 10 and 5 cal/°K mol, but the entropy of (perfect) cementite Fe_3C would be 35 cal/°K mol. Although the thermodynamic significance of the third law would remain unaltered ($DS = 0$ for the reaction corresponding to the formation of cementite), it is clear that a zero value for the entropies of all perfect substances at 0°K is much more practical.

To be more rigorous, the statement of Lewis and Randall above should be modified to include not only perfect crystalline substances at 0°K, but also substances in any state in true thermodynamic equilibrium at 0°K (and consequently without defects). A case in point would be liquid helium at 0°K.

An interesting alternate statement of the third law is that *the absolute zero of temperature can never be attained*. A demonstration of the equivalence of these two statements is given by Guggenheim [5] but is somewhat too lengthy to be reproduced here.

3.2. Experimental Verification

At first sight it would seem that an experimental verification of the third law would be very difficult: it applies to 0°K, a temperature which cannot be attained, and even if one were to wish to measure the DS at 0°K by extrapolation of entropy changes near 0°K, this would be extremely difficult since it is known that the kinetics of transformations at these temperatures are generally very slow. The difficulty, however, can be circumvented if it is remembered that the entropy is a state function.

The DS of a reaction (or transformation) at 0°K may be obtained as the sum of the entropy changes of three consecutive steps (see Fig. 7). In step 1, the reactants are brought to a temperature T; in step 2, the reaction takes place; and in step 3, the products are cooled to 0°K. In order to calculate the entropy changes of steps 1 and 3, it is necessary to know the heat capacities of the reactants and products [see equation (69)] at very low temperatures; however, if data are not available, the heat capacities may be estimated with a good degree of accuracy and the impact of an error in this estimate does not affect, in general, the precision of the overall result.

An example of such a procedure to verify the third law may be given for the transformation of rhombic sulfur to monoclinic sulfur at 0°K [1].

The transition temperature of rhombic sulfur to monoclinic sulfur is 368.5°K and the heat of transformation is 96.0 cal/mol. Since, at that temperature, the Gibbs free energies of the two structures in equilibrium must be the same, $DG = 0$ for the transformation and $DH = TDS$. Thus, for step 2 $DS_2 = 96.0/368.5 = 0.261$ cal/°K. For steps 1 and 3, the heat capacities of rhombic sulfur and metastable monoclinic sulfur have been measured by calorimetry in the interval 13–365°K. By extrapolation, in the interval 0–368.5°K, the entropy change of rhombic sulfur is [through equation (69)] $DS_1 = 8.810 \pm 0.05$ cal/°K; for monoclinic sulfur it is $DS_3 = -9.04 \pm 0.10$ cal/°K. Consequently, the entropy change

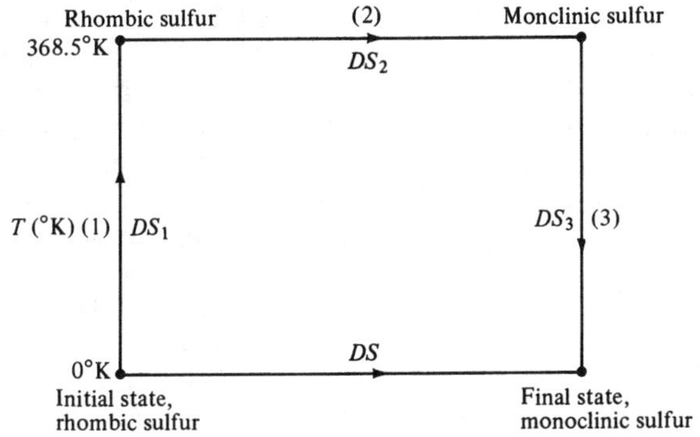

Figure 7. Experimental verification of the third law [1, p. 129]. At 0°K, the entropy change of the rhombic to monoclinic transformation of sulfur must be equal to zero. DS may be calculated as $DS = DS_1 + DS_2 + DS_3$ where

$$DS_1 = \int_0^{368.5} (C_p)_{\text{rhomb}}\, d \ln T$$

$$DS_2 = (DH)_{\text{transf}}/368.5$$

$$DS_3 = \int_{368.5}^{0} (C_p)_{\text{monocl}}\, d \ln T = -\int_0^{368.5} (C_p)_{\text{monocl}}\, d \ln T$$

(368.5°K is the temperature at which rhombic and monoclinic sulfur coexist under 1 atm.)

of the reaction S-rhombic to S-monoclinic at 0°K is

$$DS = DS_1 + DS_2 + DS_3 = 8.810 + 0.261 - 9.04 = 0.03 \quad \text{cal/°K} \tag{91}$$

This is equivalent to zero within the accuracy of the result (±0.15 cal/°K), and therefore confirms the third law.

3.3. *Interpretation*

Treatments of the second and third laws by statistical thermodynamics are beyond the scope of this text. We may indicate, however, that the entropy is related to a number W_{max} measuring the maximum number of quantum states for the system which are accessible under given conditions of energy, volume, etc. This relationship, called the Planck–Boltzmann equation, is

$$S = k \ln W_{max} \tag{92}$$

where k is Boltzmann's constant and is identical to the ratio of R, the gas constant, to N_0, Avogadro's number.

The second law was introduced by stating that although many transformations do conserve energy, the only observed transformations are those that bring a system to a state of rest. Restated in statistical terms, this means that a system will evolve in such a way as to adopt a state of maximum probability, this maximum probability being associated with the macroscopic state that has the highest number of microscopic quantum states under the applicable conditions of energy, volume, etc.

The significance of the third law is, then, that at 0°K the state of a system in true thermodynamic equilibrium corresponds to some single lowest quantum state. This may be achieved, for example, by a crystalline solid in which all the atoms are perfectly ordered. If this solid contains some disorder (e.g., a number of vacancies), it will have several quantum states corresponding to the same macroscopic state and its entropy will not be zero. A classical example is that of solid carbon monoxide at 0°K. Measurements have shown a residual entropy of about 1 cal/°K mol at 0°K. The state of lowest energy is that in which the CO molecules are all aligned in one direction. However, at some low temperatures many of the molecules are arranged in both forward and reverse positions (CO and OC), and as the crystal is cooled, a point is reached where the molecules no longer have the necessary activation energy to rotate and align themselves. Some disorder is therefore retained at very low temperatures, giving rise to a residual entropy. The maximum residual entropy due to this alignment disorder can be easily calculated. Since there are two states for each molecule (forward and reverse positions), for N_0 molecules: $W_{max} = 2^{N_0}$ and equation (92) yields

$$S = k \ln 2^{N_0} = R \ln 2 = 1.38 \quad \text{cal/°K mol} \tag{93}$$

The fact that the observed residual entropy is smaller than 1.38 indicates that the molecular orientation is not fully random.

3.4. Consequences

In equation (69) we derived that an entropy increase due to a temperature rise at constant pressure can be calculated through the heat capacity at constant pressure C_p. Because $S = 0$ at 0°K for all substances in internal equilibrium, this equation may be rewritten

$$S = \int_0^T \frac{C_p}{T}\, dT \tag{94}$$

Similarly, at constant volume

$$S = \int_0^T \frac{C_v}{T}\, dT \tag{95}$$

[It should be noted that the two entropies in equations (94) and (95) do not correspond to the same states.[7]] Moreover, because of the third law the entropy must remain finite. Consequently, it is seen that C_p and C_v must necessarily tend toward zero when $T \to 0$°K in order to prevent the divergence of the integrals at this limit. More specifically, C_p/T and C_v/T must remain finite for $T \to 0$°K. Experimentally, it is found that C_v is either proportional to T^3 or to T. Models of statistical thermodynamics establish that in the first case, the only atomic contribution to the value of C_v is of a vibrational nature, while in the second case, an electronic contribution is also present (see Section 3.5.2).

According to the third law, the entropy of a perfect substance at 0°K is equal

[7] For example if one starts with a material at 0°K, a pressure of P_0 and a volume of V_0, the entropy of equation (94) corresponds to $S(P_0,T)$ while that of equation (95) corresponds to $S(V_0,T)$. The latter two states are very different since to maintain a volume V_0 from 0 to T, a very high pressure may be required to balance the thermal expansion effect.

to zero regardless of the pressure. Consequently,

$$\left(\frac{\partial S}{\partial P}\right)_{T=0°K} = 0 \tag{96}$$

But because of Maxwell's relations

$$\left(\frac{\partial S}{\partial P}\right)_T = -\left(\frac{\partial V}{\partial T}\right)_P = -\alpha V \tag{97}$$

and thus

$$(\alpha)_{0°K} = 0 \tag{98}$$

Experimentally, the coefficient of thermal expansion has indeed been found to approach zero near 0°K.

The independence of the entropy on the volume at 0°K leads to

$$\left(\frac{\partial S}{\partial V}\right)_{T=0°K} = 0 \tag{99}$$

and since

$$\left(\frac{\partial S}{\partial V}\right)_T = \left(\frac{\partial P}{\partial T}\right)_V = \frac{\alpha}{\beta} \tag{100}$$

we obtain

$$\left(\frac{\alpha}{\beta}\right)_{0°K} = 0 \tag{101}$$

As suggested by equation (101), and contrary to the behavior of the coefficient of thermal expansion, there is no evidence that the coefficient of compressibility approaches zero at 0°K.

3.5. Estimates of Heat Capacities and Entropies

Because of equation (94) or (95), an estimate of the heat capacity yields an estimate of the entropy.

3.5.1. Heat Capacities of Gases

An elementary result of classical statistical mechanics is that if the energy of a system (such as a molecule) may be expressed under the form of a sum of squares (e.g., $\sum a_i q_i^2$ where q_i is a properly defined variable[8]), each term of the sum corresponding to a different variable or "degree of freedom," the mean value of the energy associated with each degree of freedom is $\frac{1}{2}kT$ (where k is Boltzmann's constant, a coefficient equal to R/N_0 where N_0 is Avogadro's number: 6.023×10^{23}). For example, the energy of an atom in a monatomic gas is essentially its kinetic energy: $\frac{1}{2}mv_x^2 + \frac{1}{2}mv_y^2 + \frac{1}{2}mv_z^2$, with m representing the mass of the atom and v_x,

---

[8] The conjugate coordinate of classical mechanics.

v_y, v_z, its velocities along three axes x, y, z. Consequently, the mean value of the energy of an atom is $\frac{3}{2}kT$, or for one mole of gas: $\frac{3}{2}RT$. The molar heat capacity at constant volume is thus

$$C_v = \tfrac{3}{2}R \qquad \text{(monatomic gases)} \tag{102}$$

For a diatomic molecule, there are additional contributions to consider: one vibrational contribution along the axis joining the atoms, and two rotational contributions around the two axes perpendicular to the common axis (and perpendicular to each other). Each rotational contribution yields one square term to the expression of the total energy and thus one $\frac{1}{2}R$ to C_v. The vibrational contribution yields two square terms: one representing a kinetic energy, the other a potential one. The theoretical total for C_v is then $\frac{7}{2}R$. Experimentally, however, this limit is approached only at relatively high temperatures; in addition, we have already established that at 0°K, $C_v = 0$. The consequent temperature dependence of C_v disagrees with the results of classical statistical mechanics and can only be resolved by the introduction of quantum mechanical principles.

For example, the vibrational contribution of a diatomic molecule may be analyzed through the model of an harmonic oscillator (Einstein's model) (e.g., [6]). The vibrational energy levels are then separated by the quantum $h\nu$, where h is Planck's constant (6.6252×10^{-27} erg sec) and ν is the characteristic vibrational frequency of the oscillator. It is convenient to define a characteristic vibrational temperature θ by the equation

$$h\nu = k\theta \tag{103}$$

With this parameter, it may be demonstrated that

$$C_v^{(\text{vib})} = R\left(\frac{\theta}{T}\right)^2 \frac{e^{\theta/T}}{(e^{\theta/T} - 1)^2} \tag{104}$$

The temperature dependence of this vibrational contribution to the total heat capacity is shown in Fig. 8. We note that for $T > \theta$, the value R which was predicted by classical statistical mechanics is rapidly approached. In the harmonic oscillator model,

$$\theta = \frac{h\nu}{k} = \frac{h}{2\pi k}\sqrt{\frac{f}{\mu}} \tag{105}$$

where f is the force binding the two atoms and μ is the *reduced mass* of the molecule [if m_1 and m_2 are the masses of the two atoms, $\mu = m_1 m_2/(m_1 + m_2)$]. For Cl_2, O_2, and CO, the vibrational temperatures are respectively 810, 2260, and 3120°K.

Rotational contributions may be treated in an analogous manner and yield similar results. The rotational energy is of the form

$$\epsilon_{\text{rot}} = n(n+1)\frac{h^2}{8\pi^2 I} = n(n+1)k\theta_{\text{rot}} \tag{106}$$

where I is the moment of inertia of the molecule and θ_{rot} is the characteristic temperature associated with the rotation. For most gases θ_{rot} is small; for Cl_2, O_2, and CO, it is respectively 0.35, 2.1, and 2.8°K. Thus, at ordinary temperatures

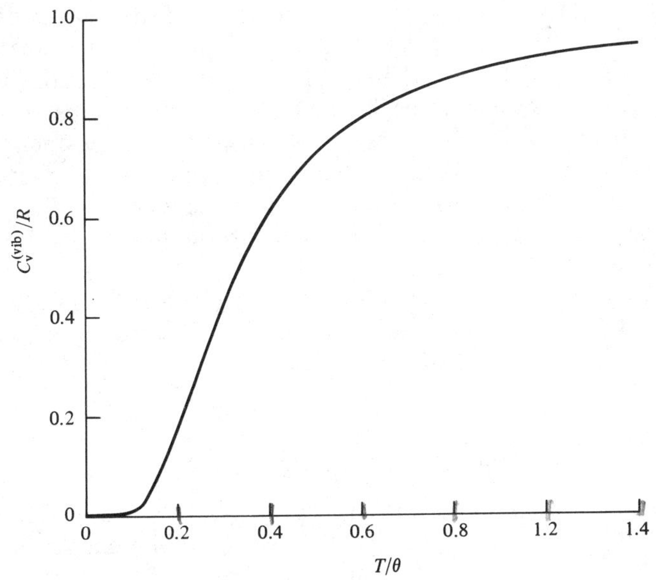

Figure 8. Temperature dependence of the vibrational contribution to the heat capacity according to Einstein's model (harmonic oscillator).

$\theta/T \ll 1$ and the results of classical statistical mechanics are satisfactory.[9] It also follows that for diatomic gases, the total heat capacity may be approximated by

$$C_V = R\left[\frac{5}{2} + \frac{u^2 e^u}{(e^u - 1)^2}\right] \quad \text{(diatomic gases)} \tag{107}$$

with

$$u = \frac{h\nu}{kT} = \frac{\theta}{T} \tag{108}$$

The principles discussed above also apply to gases with more complex molecules. In the case of a polyatomic molecule which has all its n atoms lying on a straight line, there are three translational degrees of freedom, two rotational degrees of freedom and $3n - 5$ vibrational contributions.[10] If the atoms do not lie on a straight line, there are three rotational contributions and $3n - 6$ vibrational ones. The consequent estimates of the heat capacity are, respectively,

$$C_V = R\left[\frac{5}{2} + \sum_{i=1}^{3n-5} \frac{u_i^2 e^{u_i}}{(e^{u_i} - 1)^2}\right] \quad \text{(gases with linear molecules)} \tag{109}$$

and

$$C_V = R\left[3 + \sum_{i=1}^{3n-6} \frac{u_i^2 e^{u_i}}{(e^{u_i} - 1)^2}\right] \quad \text{(gases with nonlinear molecules)} \tag{110}$$

[9] This is true for axes perpendicular to the axis joining the two atoms. For a rotation around the axis joining the atoms, the moment of inertia I is obviously extremely small and, consequently [equation (106)] θ_{rot} so large that the resulting rotational contribution is negligible.

[10] Note that the sum $3 + 2 + (3n - 5)$ is equal to the total number of independent coordinates necessary to describe the system.

The frequencies ν_i (corresponding to u_i) may be obtained from molecular spectroscopy. It is important, however, to note that the equations developed above are only estimates of a model and that in some cases many corrections are necessary. For example, the harmonic oscillator model assumes a restoring force strictly proportional to the displacement (Hooke's law). Real systems usually deviate slightly from this proportionality. There are also torsion effects, interactions between molecules and other contributions which can complicate significantly the analysis and are beyond the scope of this presentation (see Lewis and Randall [1, Chs. 6 and 27]).

Equation (88) may be used to estimate C_p from a given value of C_v; the assumption that the difference is equal to R (a property of perfect gases) is often justified.

3.5.2. Heat Capacities of Solids and Liquids

Each atom in a solid or liquid is surrounded by a shell of first (nearest) neighbors with additional shells of more distant neighbors. The central atom vibrates in its lattice site in a force field which is the sum of the separate forces exerted by all the neighbors on the central atom. In Einstein's model, each atom vibrates in its own cell independently of the vibrations of its neighbors and the motion is assumed isotropic about the average position. There are thus, for each atom, three independent and equivalent x, y, and z motions. If N is the number of atoms on the lattice, the system decomposes into a aggregate of $3N$ one-dimensional harmonic oscillators. The vibrational contribution being the only one assumed, the results of equation (104) are readily transposed:

$$C_v = 3R\left(\frac{\theta}{T}\right)^2 \frac{e^{\theta/T}}{(e^{\theta/T}-1)^2} \tag{111}$$

At high temperatures ($\theta/T \to 0$), C_v approaches the limiting value of $3R$. This is often referred to as the law of Dulong and Petit because they were first to discover that, ordinarily, $3R$ is a good estimate of C_v [7]. Equation (111) shows a fair degree of success, except at low temperatures where it predicts an exponential decrease; instead, a dependence on T^3 is generally observed.

The model has been refined by Debye who considered a range of frequencies (0–ν_D) instead of a single frequency [6]. The calculations yield

$$C_v = 3R\left[\frac{12}{u^3}\left(\int_0^u \frac{x^3\,dx}{e^x-1}\right) - \frac{3u}{e^u-1}\right] \tag{112}$$

with

$$u = \frac{h\nu_D}{kT} = \frac{\theta_D}{T} \tag{113}$$

Equation (112) is generally in excellent agreement with experimental results. In particular, it may be shown that, for $T \to 0$ (or $u \to \infty$), a dependence of C_v on T^3 is indeed followed (see Problem 16). Figure 9 illustrates the applicability of Debye's model and Table 2 lists typical values of the characteristic Debye temperature θ_D. For many of these solids, room temperature is above θ_D and the classical Dulong–Petit value of $3R$ is already practically attained at 298°K.

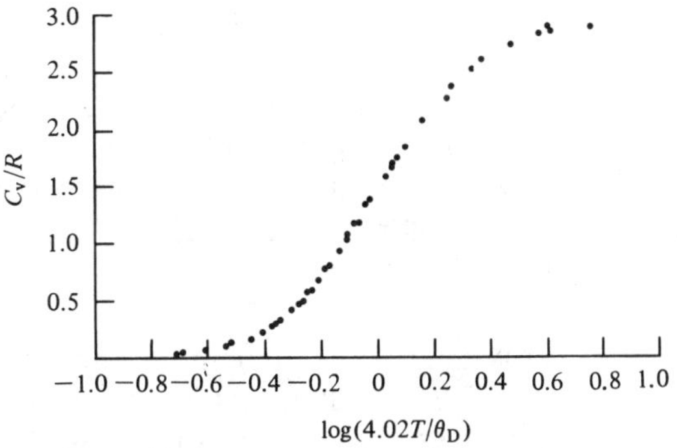

Figure 9. Superimposed experimental heat capacity points for Al, Cu, Pb, and C (diamond). (*From Hill* [*6*].)

In 1864 Kopp proposed that the heat capacity of a solid or liquid compound is equal to the sum of the heat capacities of its constituent elements in the solid or liquid states [9]. This rule is quite reliable at temperatures where the Dulong–Petit value of $3R$ per gram-atom applies, but it is only approximately valid at low temperatures. There are, moreover, many complicating factors which arise in solids at both low and high temperatures. One that arises for metals is the electronic contribution to the total heat capacity; it is predicted to increase at low temperatures with the first power of T:

$$C_V^{(\text{elec})} = \gamma T \tag{114}$$

This contribution is generally negligible relative to the vibrational one at ordinary temperatures, but for certain metals such as Cu and Al it becomes the predominant

Table 2. Typical Values of Debye temperatures (°K)

Li	Be											B	C	N	O	F	Ne
344	1440												2230				75
Na	Mg											Al	Si	P	S	Cl	Ar
158	400											428	645				92
K	Ca	Sc	Ti	V	Cr	Mn	Fe	Co	Ni	Cu	Zn	Ga	Ge	As	Se	Br	Kr
91	230	360	420	380	630	410	470	445	450	343	327	320	374	282	90		72
Rb	Sr	Y	Zr	Nb	Mo	Tc	Ru	Rh	Pd	Ag	Cd	In	Sn *w*	Sb	Te	I	Xe
56	147	280	291	275	450		600	480	274	225	209	108	200	211	153		64
Cs	Ba	La *β*	Hf	Ta	W	Re	Os	Ir	Pt	Au	Hg	Tl	Pb	Bi	Po	At	Rn
33	110	142	252	240	400	430	500	420	240	165	71.9	78.5	105	119			
Fr	Ra	Ac															

Ce	Pr	Nd	Pm	Sm	Eu	Gd	Tb	Dy	Ho	Er	Tm	Yb	Lu
						200		210				120	210
Th	Pa	U	Np	Pu	Am	Cm	Bk	Cf	Es	Fm	Md	No	Lw
163		207											

From Kittel [8].

one in the vicinity of 1°K ($\gamma_{Al} = 3.23 \times 10^{-4}$ and $\gamma_{Cu} = 1.65 \times 10^{-4}$ in cal/°K^2) [10]. Other types of contributions may originate from a reorientation of magnetic moments (e.g., in iron), from a change in the ordering of the atoms, from a rotation of a group of atoms, etc.

Generally, it is the value of C_p which is of practical interest. It may be deduced from C_v through the equation

$$C_p - C_v = \alpha^2 VT/\beta \tag{115}$$

provided the coefficients α and β are known. Although experimental values of α and β are often lacking, their orders of magnitude are generally sufficient to provide a reasonable estimate of the difference $C_p - C_v$. This difference is generally negligible at low temperatures but increases substantially with increasing temperatures. The typical case of copper is shown in Fig. 10.

Note that Kopp's empirical rule (also referred to as the Neumann–Kopp rule) applies equally well to heat capacities at constant pressure. In particular, it implies that the ΔC_p of a reaction involving only solid and liquid compounds is approximately equal to zero, and consequently, that the ΔH of this reaction is practically independent of the temperature.

3.5.3. Entropies

The temperature dependence of the entropy at constant pressure is illustrated in Fig. 11 in the case of magnesium at a pressure of 1 atm. Note that for any phase the curve must always increase because its slope is always positive:

$$\left(\frac{\partial S}{\partial T}\right)_P = \frac{C_p}{T} > 0 \tag{116}$$

There are discontinuities in the values of the function at temperatures of phase transformations such as melting or boiling.

For most metals, the entropy increase on fusion, ΔS_f°, is generally of the order of 2–3 cal/°K mol. This observation is sometimes referred to as Richard's rule.

Figure 10. Temperature dependence of the heat capacities of copper at constant pressure and at constant volume.

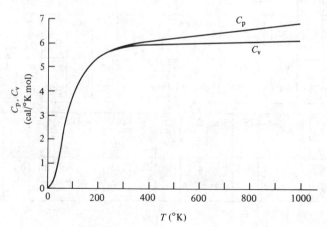

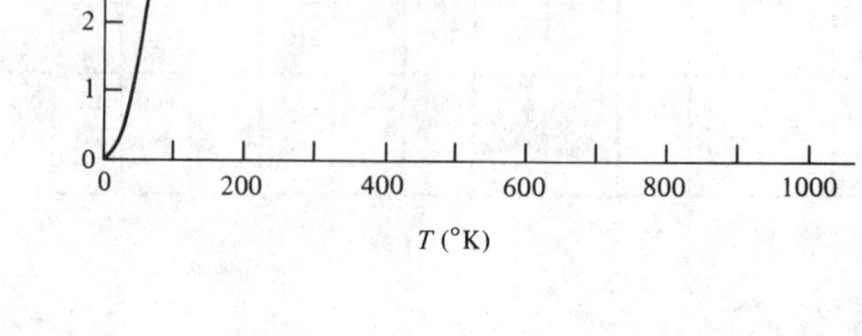

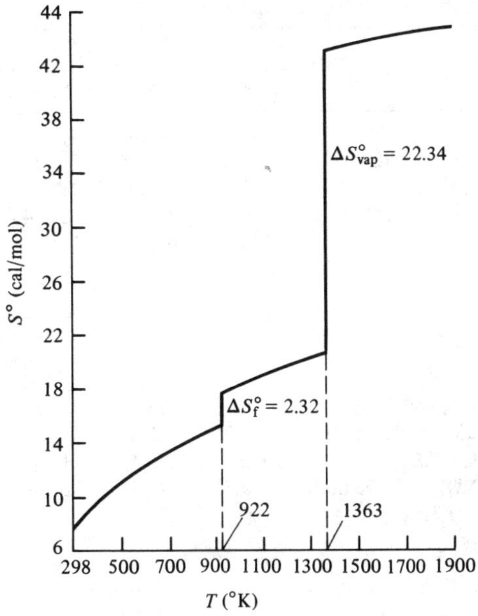

Figure 11. Temperature dependence of the entropy of magnesium at 1 atm.

Also, for most metals the entropy increment on boiling ΔS°_{vap} is of the order of 22 cal/°K mol. This is referred to as Trouton's rule. Table 3 lists some values of ΔS°_f and ΔS°_{vap}. It may be observed that there are occasional large deviations from both rules.

Better estimates of ΔS°_f are possible through consideration of a correlation between the entropy of fusion and the crystallographic structure of the solid phase [11]. For example, the hexagonal close-packed (hcp), face-centered cubic (fcc) and body-centered cubic (bcc) structures are characterized by entropies of fusion which differ from each other by about 0.25 cal/°K mol; i.e., $\Delta S^\circ_f(\text{hcp}) - \Delta S^\circ_f(\text{fcc}) \simeq \Delta S^\circ_f(\text{fcc}) - \Delta S^\circ_f(\text{bcc}) \simeq 0.25$ cal/°K mol. Other structures, such as rhombohedral, diamond cubic, or orthorhombic, are often associated with atomic forces which are not entirely metallic in character but are partly covalent or ionic. Generally, their entropies of fusion are significantly higher. For example, ΔS°_f (Sb,rh) = 5.26, ΔS°_f(Bi,rh) = 4.96, ΔS°_f(Ge,dc) = 7.30, and ΔS°_f(Ga,or) = 4.41 cal/°K mol.

Under atmospheric pressure and at room temperature, the entropy of a species may be obtained from

$$S^\circ_{298} = \int_0^{298} C^\circ_p \, d \ln T \tag{117}$$

At 298°K, C_p is generally very close to C_v [the difference, expressed in equation (115), is small] and Debye's model may be used to obtain an estimate of the entropy. It is readily seen (from Fig. 9, for example) that S will be relatively large when the vibrational characteristic temperature θ_D is small and vice-versa.

It is possible to demonstrate that Debye's temperature θ_D is equivalent to $\frac{3}{4}$ of Einstein's temperature θ_E. Thus, θ_D is proportional to $(f/M)^{1/2}$ [see equation (105)], where f is a measure of the strength of the bond between the atoms and M is the

Table 3. Entropy Values

	Elements								
	Na	K	Mg	Ca	Ba	Cr	W	Fe	Ni
S°_{298} (cal/°K mol)	12.23	15.46	7.81	9.9	14.92	5.65	7.80	6.52	7.14
ΔS°_f (cal/°K mol)	1.67	1.66	2.32	1.84	1.85	1.9	2.3	1.82	2.42
T_f (°K)	371.0	336.4	922	1112	1002	2130	3680	1809	1726
ΔS°_{vap} (cal/°K mol)	20.05	18.50	22.34	20.90	15.6	27.94	33.79	26.65	27.77
T_b (°K)	1156	1032	1363	1757	2171	2945	5828	3135	3187

	Elements								
	Pt	Cu	Ag	Au	Zn	Hg	C (graphite)	Si	Pb
S°_{298} (cal/°K mol)	9.95	7.92	10.20	11.35	9.95	18.14	1.37	4.50	15.65
ΔS°_f (cal/°K mol)	2.3	2.30	2.19	2.24	2.53	2.34	—	7.17	1.91
T_f (°K)	2042	1356.55	1234	1336.15	692.65	234.28	—	1685	600.6
ΔS°_{vap} (cal/°K mol)	29.72	25.31	24.59	25.53	23.36	22.47	—	26.4	20.99
T_b (°K)	4100	2836	2436	3130	1180	630	4100	3540	2023

	Gases								
	He	Mg (gas)	Pb (gas)	Ar	H_2	N_2	O_2	Cl_2	S_2
S°_{298} (cal/°K mol)	30.12	35.50	41.89	36.98	31.21	45.77	49.00	53.31	54.51

	Gases								
	CO	CO_2	H_2O (gas)	H_2S	COS	SO_2	SO_3	NH_3	CH_4
S°_{298} (cal/°K mol)	47.21	51.07	45.10	49.15	55.32	59.30	61.34	46.03	44.49

Data from Hultgren et al. [10] and Stull et al. [12]. ΔS°_f and ΔS°_{vap} are calculated at T_f and T_b, respectively.

atomic weight of the element considered. (Qualitative measures of the bond strength may originate from a comparison of melting or boiling temperatures, heats of vaporization, yield strengths or elastic constants, etc.) Thus a "hard" and light material such as carbon or silicon has a large θ and small S°_{298}, while a "soft" and heavy material such as mercury or lead has a small θ and large S°_{298}. Table 3 lists values of S°_{298} for several species.

It is important to consider the orders of magnitude of S°_{298}. Generally, solid elements have entropies of less than 15 cal/°K g-at., and liquids are nearly in the same range (as indicated by Richard's rule). Monoatomic gases have entropies of the order of 30–40 cal/°K mol while diatomic gases have larger entropies, between 45–55 cal/°K mol (with the notable exception of hydrogen, which because of its light weight, has a high characteristic temperature θ). Triatomic and larger molecules have even larger entropies in accordance with the preceding discussion of heat capacities [e.g., see equations (109) and (110)].

A consequence of Kopp's rule is that entropies of solid and liquid compounds may be roughly estimated from the additivity of the entropies of their constituent components in the solid or liquid states. Obviously then, a high number of gram atoms per mole of a compound correlates with a large value for its entropy. For example, the values of S°_{298} for Fe, $Fe_{0.95}O$, Fe_2O_3, and Fe_3O_4 are, respectively, 6.52, 13.76, 20.89, and 34.72 cal/°K mol (or cal/°K g-mol). Note that the values of S°_{298} would be 6.52, 7.05, 4.18, 4.96 cal/°K g-at.

Entropies of reactions are defined in the same way as enthalpies of reactions (see Section 1.5); i.e., the entropy of the reaction

$$\sum_{i=1}^{r} \nu_i A_i = 0 \tag{118}$$

is

$$\Delta S = \sum_{i=1}^{r} \nu_i S_i \tag{119}$$

where S_i is the molar entropy of the component A_i. Entropies of formation (S_f) have a definition which parallels that of the enthalpies of formation (Section 1.6).

In the study of chemical transformations, it is often very useful to know the size and order of magnitude of the $\Delta S°$ associated to a given reaction. These may often be simply deduced from the considerations above. Let us examine, for example, the following reactions:

$$Zr + O_2(g) = ZrO_2 \tag{120}$$

$$C + O_2(g) = CO_2(g) \tag{121}$$

$$C + CO_2(g) = 2\,CO(g) \tag{122}$$

$$C + MgO = Mg(g) + CO(g) \tag{123}$$

$$C + SiO_2 = SiO(g) + CO(g) \tag{124}$$

For reaction (120), it is clear that $\Delta S°_{298}$ is substantially negative since the entropy of the diatomic gas is much larger than that of the solid metal or oxide. Similarly, one should note for reaction (121) that $\Delta S°_{298}$ should be close to zero since both sides of the reaction contain the same number of gas moles, and that for reactions (122)–(124) the $\Delta S°_{298}$ are positive and in order of increasing values (e.g., because the entropy of the monatomic gas Mg is smaller than that of the diatomic gas SiO). Indeed, tabulated values of the entropies yield the values shown in Table 4.

In a first approximation, the temperature dependence of $\Delta S°$ may often be neglected. For reactions involving solid and liquid components, this is a consequence of Kopp's rule (see Section 3.5.2). However, even for reactions involving gases, it remains reasonably valid. Table 4 compares the $\Delta S°$ of reactions (120)–(124) at 298 and 1000°K.

Table 4. Values of $\Delta S°$

Reactions	$\Delta S°_{298}$ (cal/°K)	$\Delta S°_{1000}$ (cal/°K)
$Zr + O_2 = ZrO_2$	−46.3	−44.1
$C + O_2 = CO_2$	0.7	0.3
$C + CO_2 = 2\,CO$	42.0	41.9
$C + MgO = Mg(g) + CO$	74.9	72.1
$C + SiO_2 = SiO(g) + CO$	86.1	93.3

Data from Stull et al. [12].

4. Application to the Stability of Phases for One-Component Systems

4.1. Gibbs Free Energy Function

In our study of equilibrium criteria (Section 2.3) we introduced the Gibbs free energy function. We recall its definition:

$$G \equiv H - TS \tag{125}$$

We determined that at any given temperature T and pressure P a system is in equilibrium when it has reached the minimum value of its Gibbs free energy. Consequently, if at T and P a substance may be either in the α or β structure (or phase), the stable structure is that which corresponds to the lowest Gibbs free energy; i.e., α is more stable than β if $G^\alpha < G^\beta$.

It should be observed that if $G^\alpha < G^\beta$ at T and P, the inequality may be reversed at T' and P'. Consequently, it is important to know the temperature and pressure dependences of the Gibbs free energy of a phase.

Figure 12 illustrates the temperature dependence of the Gibbs free energies of various phases of zinc at 1 atm. Unlike the temperature dependence of the enthalpy and the entropy (Figs. 2 and 11), there are no discontinuities in the values of G for phase transformations: at the melting point $G^{(s)} = G^{(l)}$, and at the boiling point, $G^{(l)} = G^{(g)}$. The slopes of the curves are always negative since

$$\left(\frac{\partial G}{\partial T}\right)_P = -S < 0 \tag{126}$$

The curvatures are negative as well since

$$\left(\frac{\partial^2 G}{\partial T^2}\right)_P = -\left(\frac{\partial S}{\partial T}\right)_P = -\frac{C_P}{T} < 0 \tag{127}$$

The change in the slope of G at a transformation temperature corresponds to the ΔS of this transformation. Note in Fig. 12 that the discontinuity in the slope at T_f (equal to ΔS_f) is much smaller than that at T_b (and equal to ΔS_{vap}).

In Section 1.4, we saw that absolute values of the enthalpy are of no concern to the traditional field of thermochemistry; only relative values (i.e., changes) are measured and recorded. It is clear that this is also true of the Gibbs free energy since $G = H - TS$. This explains why, in Fig. 12, we have recorded values of $G^\circ - H^\circ_{298}$ rather than G°.

For a given pressure and at the transition temperature $T^{\alpha-\beta}$, that is, at the temperature where the phases α and β coexist, we must have

$$G^\alpha = G^\beta \tag{128a}$$

or

$$\Delta G^{\alpha\to\beta} = G^\beta - G^\alpha = 0 \tag{128b}$$

Therefore

$$H^\alpha - T^{\alpha-\beta}S^\alpha = H^\beta - T^{\alpha-\beta}S^\beta \tag{129a}$$

and

$$H^\beta - H^\alpha = T^{\alpha-\beta}(S^\beta - S^\alpha) \tag{129b}$$

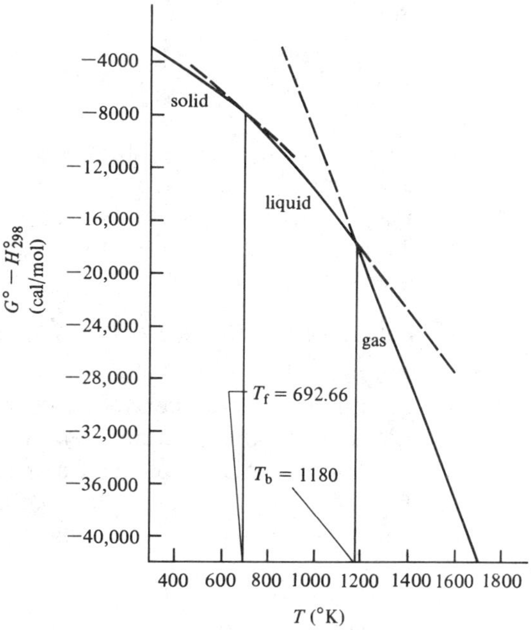

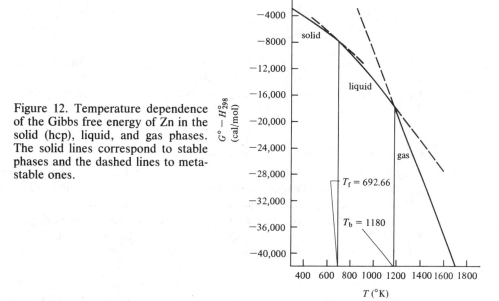

Figure 12. Temperature dependence of the Gibbs free energy of Zn in the solid (hcp), liquid, and gas phases. The solid lines correspond to stable phases and the dashed lines to metastable ones.

which we can rewrite

$$\Delta H^{\alpha \to \beta} = T^{\alpha - \beta} \Delta S^{\alpha \to \beta} \tag{129c}$$

This expression is useful in estimating heats of melting and heats of vaporization knowing the melting or boiling temperatures, since the entropies of melting and vaporization are often readily estimated (see Section 3.5.3).

To study the pressure dependence of the Gibbs free energy, we note that

$$\left(\frac{\partial G}{\partial P}\right)_T = V \tag{130}$$

Figure 13. Pressure dependence of the Gibbs free energies of carbon in graphite and diamond structures.

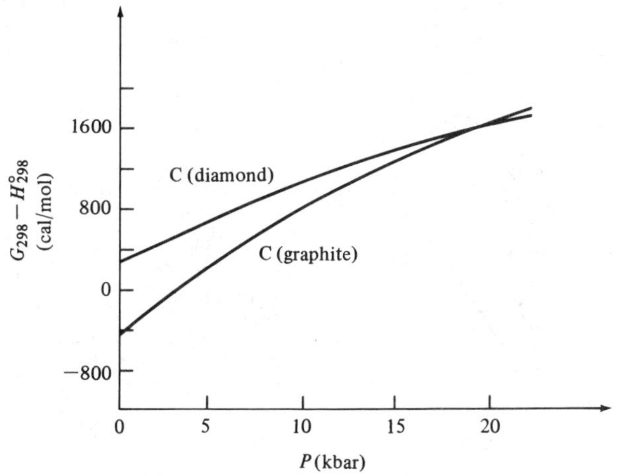

and

$$\left(\frac{\partial^2 G}{\partial P^2}\right)_T = \left(\frac{\partial V}{\partial P}\right)_T = -\beta V \tag{131}$$

Thus, a plot of G vs P shows curves of positive slopes but negative curvatures (see Fig. 13). High pressures favor phases of low molar volumes, i.e., of high density. For example, in the case of carbon at room temperature, high pressures stabilize the diamond structure with respect to the graphite structure.

4.2. Clausius–Clapeyron Equation

In a three-dimensional plot of G vs P and T, the Gibbs free energies of the phases α and β are represented by two surfaces (see Fig. 14). Their line of intersection yields the conditions of temperature and pressure under which they coexist. For example, the boiling point of a species (liquid–gas equilibrium) depends on the pressure. The equation of that line is of interest and may be obtained as follows. The phases α and β coexist when

$$G^\alpha = G^\beta \tag{132}$$

Along their coexistence line, any change in the value of G^α must be matched by an equal change in the value of G^β. Thus,

$$dG^\alpha = dG^\beta \tag{133}$$

or

$$-S^\alpha\, dT + V^\alpha\, dP = -S^\beta\, dT + V^\beta\, dP \tag{134}$$

Consequently,

$$\frac{dT}{dP} = \frac{V^\beta - V^\alpha}{S^\beta - S^\alpha} = \frac{\Delta V^{\alpha\to\beta}}{\Delta S^{\alpha\to\beta}} \tag{135}$$

This is the Clausius–Clapeyron equation. Using equation (129c), it may also be

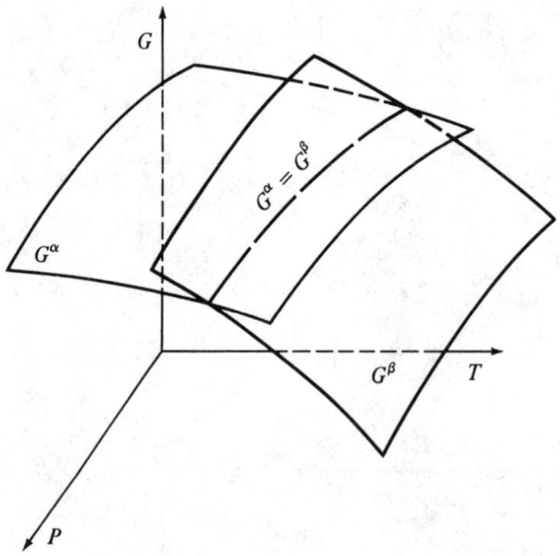

Figure 14. Illustration of the condition for the coexistence of two phases α and β. The intersection of the two free energy surfaces is a line $G^\alpha = G^\beta$ along which α and β coexist.

expressed

$$\frac{dT}{dP} = \frac{T^{\alpha\to\beta}\Delta V^{\alpha\to\beta}}{\Delta H^{\alpha\to\beta}} \tag{136}$$

Let us examine for example the value of dT/dP for the fusion of silver. With

$$T_f = 1234 \quad °K$$

$$\Delta V^{s\to 1} = 11.3 - 10.9 = 0.4 \quad cm^3/mol$$

$$\Delta H^{s\to 1} = 2855 \quad cal/mol$$

equation (136) yields

$$\frac{dT}{dP} = \frac{1234 \times 0.4}{2855} \times 0.0242 = 0.004 \quad °K/atm$$

The order of magnitude of this result is substantially the same for other elements, and we see therefore that melting temperatures are not substantially affected by minor variations in pressure: for example, changes in the fusion point of silver between high vacuum conditions and $P = 10$ atm are generally well below the limits of experimental detectability; they become noticeable only when the pressure attains a few hundred atmospheres. Figure 15 shows the stability domains of various phases of iron. We note that an increase in pressure favors the stability of the fcc phase (γ) with respect to that of the bcc phase (α or δ) and of the liquid phase. (V^γ is smaller than V^α, V^δ, and V^1.) ϵ is an hcp structure.

Changes in boiling temperatures are much larger than changes in melting temperatures since the corresponding ΔV is more than four orders of magnitude higher (V^g is 22,414 cm³/mol at 273 °K). For example, in the case of water at $P = 1$ atm,

$$\frac{dT}{dP} = \frac{373.15 \times (30{,}140 - 18.76)}{9717} \times 0.0242 = 28.01 \quad °K/atm$$

Noting that the molar volume of a gas is much larger than that of a solid or

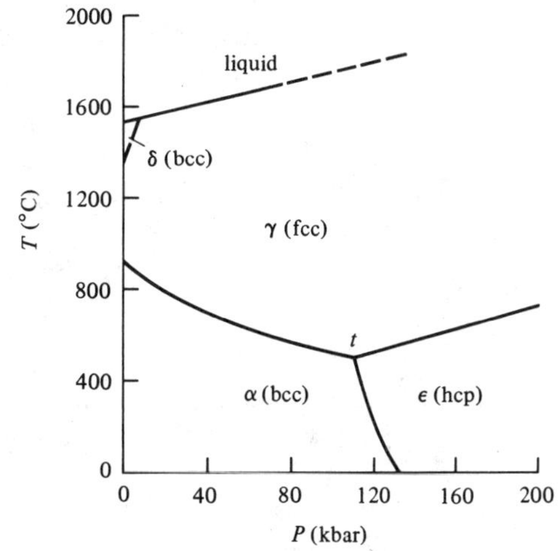

Figure 15. Pressure–temperature phase diagram of iron. (*After Takahashi and Bassett* [*13*] *and Bundy* [*14*].)

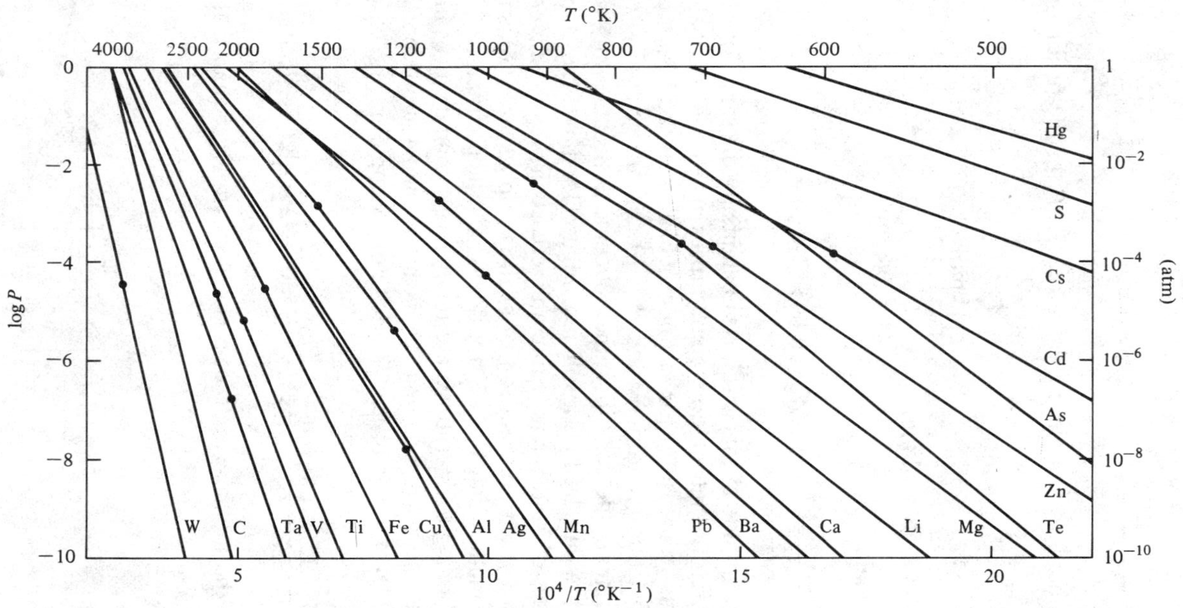

Figure 16. Vapor pressure of some elements. The black dots correspond to melting points. In the cases where there are several species of the same element in the gas phase (e.g., S_8, S_7, S_6, . . . for sulfur), the lines correspond to total vapor pressures. (*After Hultgten et al.* [*10*].)

liquid phase α

$$\Delta V^{\alpha\to g} = V^g - V^\alpha \simeq V^g \tag{137}$$

equation (136) may be rewritten

$$\frac{dT}{dP} = \frac{TV^g}{\Delta H_{vap}} \tag{138}$$

We shall assume that the gas behaves ideally. Consequently,

$$\frac{dT}{dP} \simeq \frac{RT^2}{P\Delta H_{vap}} \tag{139a}$$

or

$$\frac{dP}{P} \simeq \frac{\Delta H_{vap}}{RT^2}\,dT \tag{139b}$$

If we further assume that the heat of vaporization is not a strong function of pressure and temperature (see Problems 18 and 22), equation (139b) is easily integrated and yields

$$\ln P \simeq \frac{\Delta H_{vap}}{R}\left(\frac{1}{T_b} - \frac{1}{T}\right) \tag{140}$$

(the integration constant is determined by noting that $P = 1$ atm at $T = T_b$).

Figure 16 shows that in a plot of $\ln P$ vs $1/T$ the vapor pressures of the elements are expressed by lines which are very nearly straight, in accordance with equation (140).

4.3. Triple Points

In the preceding sections, we have established that two phases α and β of the same component coexist for conditions corresponding to the equality of their molar Gibbs free energies, i.e., for $G^\alpha = G^\beta$. If we now consider a third phase γ, the equilibria α–γ and β–γ are deduced by the same considerations. In a three-dimensional plot of G vs P and T (see Fig. 17), the three surfaces G^α, G^β, and G^γ intercept each other along three lines defined by the equations $G^\alpha = G^\beta$, $G^\alpha = G^\gamma$, and $G^\beta = G^\gamma$. It is clear that these three lines have a common point t

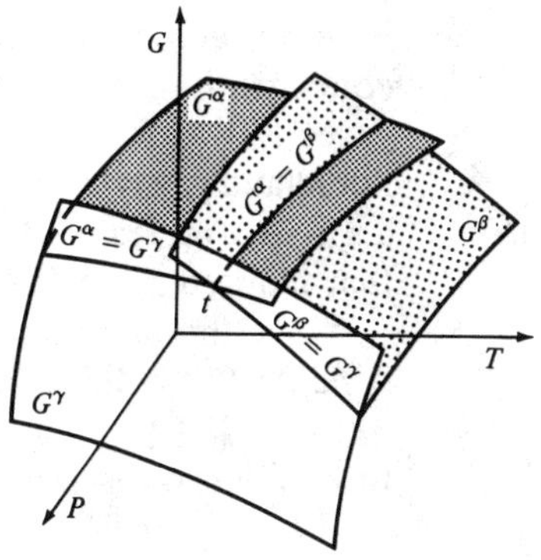

Figure 17. Illustration of the conditions for the coexistence of three phases α, β, and γ at the triple point t. At t, $G^\alpha = G^\beta = G^\gamma$.

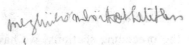

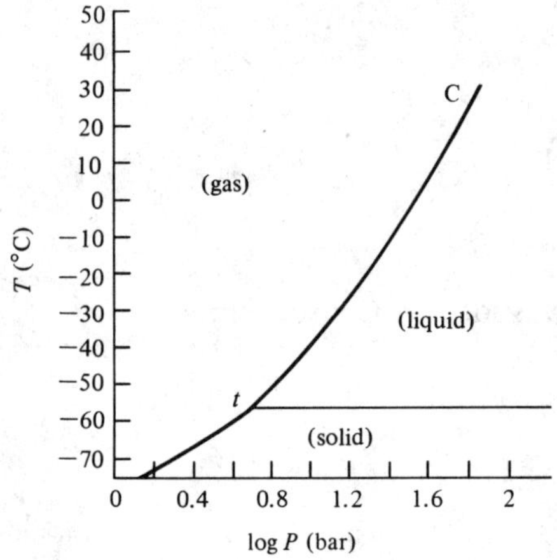

Figure 18. Pressure–temperature phase diagram for CO_2. (*From Gas Encyclopaedia* [*15*].)

at $G^\alpha = G^\beta = G^\gamma$. This point is called a *triple point* and corresponds to the equilibrium coexistence of all three phases.

In the pressure–temperature phase diagram of iron which is shown in Fig. 15, the point marked t is such a triple point: at roughly 110 kbar and 480°C, the bcc, fcc, and hcp phases coexist. Figure 18 shows another P–T phase diagram where the phases α, β, and γ correspond to the solid, liquid, and gas phases. Upon heating at a pressure above that of the triple point t the solid melts, whereas upon heating at a lower pressure, the solid sublimes. For most elements and compounds, P_t is well below atmospheric pressure (e.g., $P_t = 0.006$ atm for water) and, therefore, melting is generally observed. There are exceptions, one of the most notable being CO_2 for which $P_t = 5.1$ atm, $T_t = -56.6$°C. Thus, at ordinary pressures, "dry ice" (solid CO_2) sublimes.

4.4. Critical Points

In Section 4.1 and Fig. 12 we considered the liquid and gas phases of a substance as being quite distinct. However, under certain conditions of temperature and pressure, the liquid and gas phases are undistinguishable. Moreover, even under ordinary conditions where they are quite distinguishable, it is possible to pass from one to the other by processes in which the substance remains perfectly homogeneous.

These statements may be clarified by examining the P–V diagram of Fig. 19. At a temperature T_1 compressing the gas leads to condensation and a discontinuity is observed. At a higher temperature T_2 compressing the gas does not lead to condensation; the behavior of the gas resembles the hyperbolic behavior of perfect gases ($PV = RT$). T_c is the temperature at which the discontinuity between the liquid and gas phases vanishes. We also note that it is indeed possible to transform a gas A to a liquid B without any discontinuity by following a process such as that marked *AMNOB* in Fig. 19.

Since the liquid and gas phases correspond to a unique state of the matter, it is natural to believe that the two branches Av and uB of the discontinuous curve $AvuB$ at T_1 ($<T_c$) should be joined as a single curve by the segment vwu in order to exhibit no discontinuity. This is verified by equations of state for gases such as the van der Waals equation (see Problem 12).

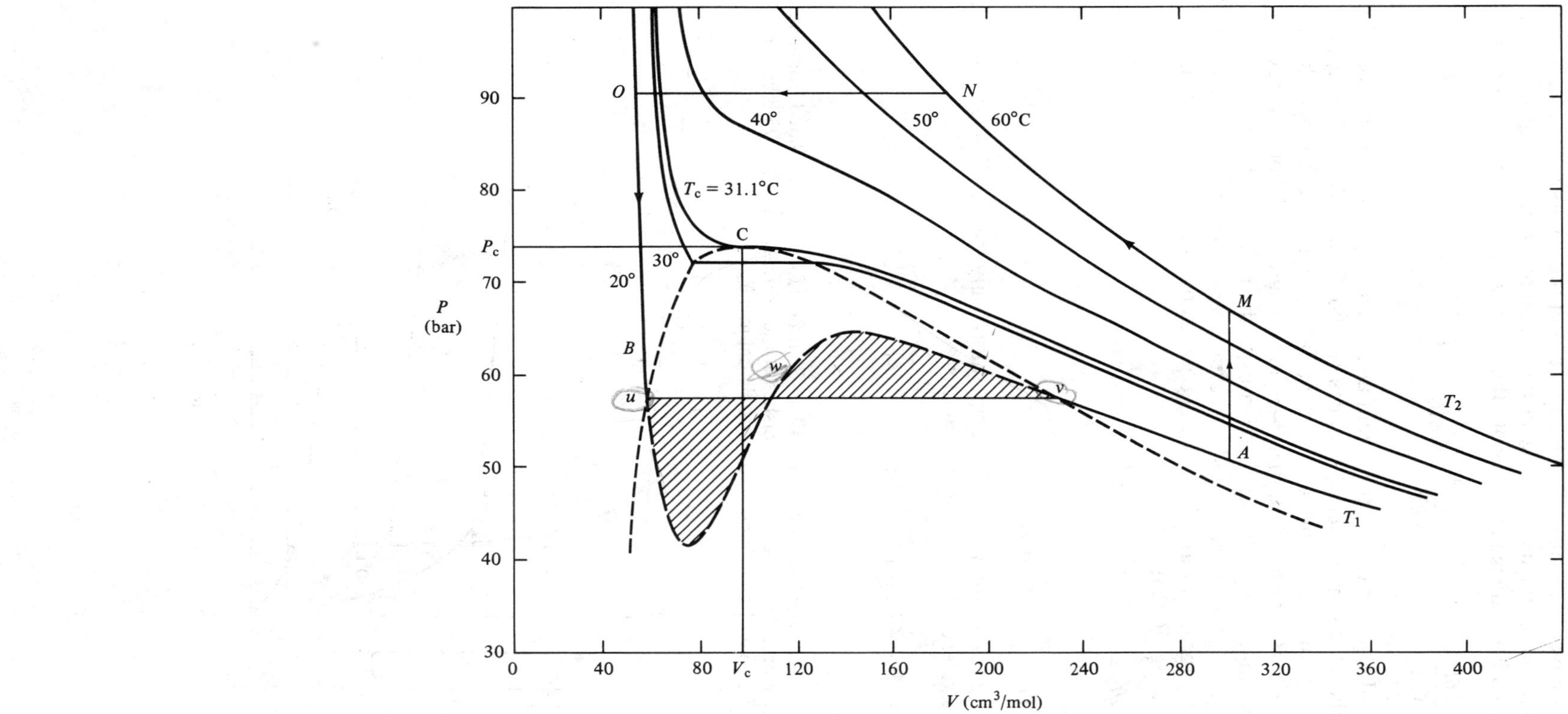

Figure 19. Pressure–volume diagram for CO_2. The coordinates of the critical point are $V_c = 94.9$ cm^3/mol and $P_c = 73.8$ bar [15] (1 bar = 0.987 atm). The liquid u being in equilibrium with the gas v, the two shaded areas must be equal. It is possible to pass from the gas A to the liquid B either by compression and condensation ($AvuB$), or by the process $AMNOB$ in which CO_2 remains always homogeneous.

The dome-shaped curve uCv is the locus of the points u and v and is called the *saturation curve*. Its apex C is the *critical point* at which the liquid and gas phases become undistinguishable. We note that it may be defined by the conditions

$$\left(\frac{\partial P}{\partial V}\right)_T = 0, \quad \left(\frac{\partial^2 P}{\partial V^2}\right)_T = 0, \quad \left(\frac{\partial^3 P}{\partial V^3}\right)_T > 0 \tag{141}$$

Since the properties of the liquid and gaseous phases can be joined continuously, the Gibbs free energy surfaces G^l and G^g in a G–P–T diagram must be portions of the same surface. That surface presents a fold which vanishes at $T = T_c$, $P = P_c$ (see Fig. 20). In a P–T phase diagram, the boundary corresponding to the liquid–gas equilibrium terminates at the point P_c, T_c.

In Fig. 19, the points u and v are positioned on the isothermal P–V curve $BuwvA$ by the condition $G^l = G^g$. The geometrical implication of that relation is the following. Recalling that

$$dG = -S\,dT + V\,dP \tag{142}$$

we have

$$G = \int V\,dP + \phi(T) \tag{143}$$

where $\phi(T)$ is a function of the temperature only. Since $G^l = G^g$, taking A as an arbitrary reference base, we may write

$$(G_u^l - G_A^g) - (G_v^g - G_A^g) = \int_A^u V\,dP - \int_A^v V\,dP = 0 \tag{144}$$

Figure 20. Isothermal sections of the G–P–T surface for CO_2 (calculated from data in [15]). The surface presents a fold which vanishes at the critical point $P_c = 73.8$ bar, $T_c = 31.1$°C. (*Note:* 1 kcal/kg of CO_2 is equivalent to 44 cal/mol.)

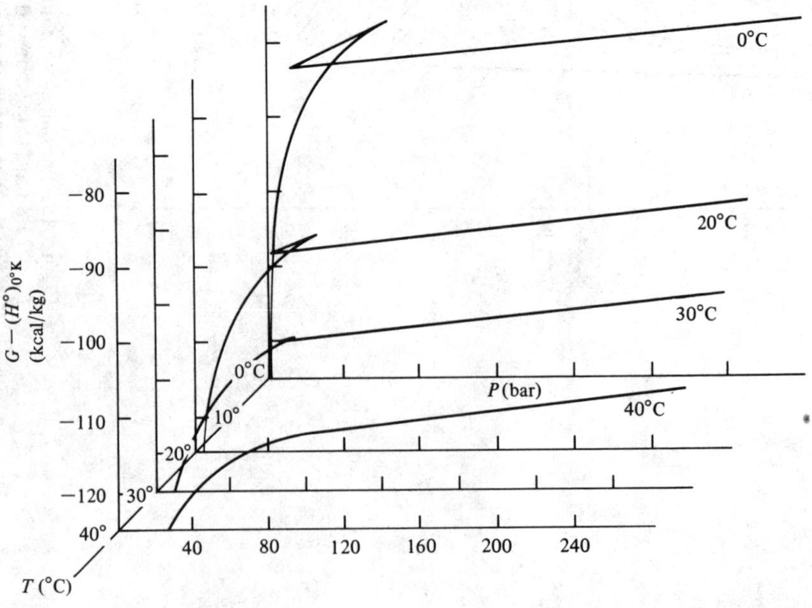

and

$$\int_u^v V\,dP = 0 \tag{145}$$

Consequently, the two shaded areas in Fig. 19 must be equal. This requirement fixes the level of the uv horizontal and thus determines the u, v points.

5. Summary

Let us briefly summarize some of the results seen in this chapter. The *first law* states that in any process the energy is conserved; e.g., any energy lost by a system is gained by its surroundings. Exchanges of energy are accounted for in terms of heat and work, and we may thus write

$$dE = đq + đw \tag{146}$$

The energy of a system is an intrinsic property (a state function), but heat and work are not. As a consequence, dE is a perfect differential but $đq$ and $đw$ are not; to integrate $đq$ and $đw$, a path must be specified. The work terms considered in this text are almost exclusively of the type $-P\,dV$.

The *second law* states the existence of another state function, the entropy S, defined for reversible processes by the equation

$$dS = đq_{\text{rev}}/T \tag{147}$$

and for irreversible processes by the inequality

$$dS > đq/T \tag{148}$$

Combining the first and second law we deduce that for any irreversible process

$$dE - T\,dS + P\,dV < 0 \tag{149}$$

Conversely, we may state that the criterion of equilibrium is

$$\delta E - T\,\delta S + P\,\delta V \geq 0 \tag{150}$$

Depending on the set of restrictions placed on the transformations envisaged (e.g., constant P and T), various thermodynamic functions become useful. These are the enthalpy ($H \equiv E + PV$), the Helmholtz free energy ($A \equiv E - TS$), and the Gibbs free energy ($G \equiv E + PV - TS$). Because they are state functions, their changes can always be calculated through reversible processes, and the expressions of their perfect differentials give rise to Maxwell's relationships.

Essentially, the *third law* enables us to assign a value of zero to the entropy of any substance at 0°K. At other temperatures, it is useful to know the order of magnitude of the entropy of various species and of the change in the entropy accompanying certain reactions.

Processes at constant temperature and pressure are often encountered, and under these conditions equilibrium is obtained when the Gibbs free energy reaches a minimum. This criterion of equilibrium enables us to determine which phase of a substance is the most stable and under what conditions of pressure and temperature two or three phases can coexist.

Problems

1. **a.** Calculate the energy absorbed (in cal) by a stationary wall when a car weighing one tonne impacts at 100 km/hr (62 mile/hr).
   b. A weight of 100 kg is dropped from a height of 3 m. Calculate the heat released (in cal) upon impact.
   c. A 500 W coil is immersed in a mug containing 500 cm^3 of water at 20°C. Calculate the temperature of the water after 5 min have passed, assuming that 20% of the energy available is lost to the surroundings.
   d. A 75 Ω resistance furnace is drawing a current of 5 A at steady state operation. What is the rate (in cal/sec) at which heat is lost?

2. The infinitesimal increment dz is defined by the following equation:

$$dz = \frac{x}{x^2 + y^4}\,dx - \frac{x^2 - y^4}{y(x^2 + y^4)}\,dy$$

Demonstrate that dz is a perfect differential. Find the function $z(x,y)$.

3. Consider the two differentials:

$$dz_1 = y(3x^2 + y^2)\,dx + x(x^2 + 2y^2)\,dy$$

$$dz_2 = y(3x^2 + y)\,dx + x(x^2 + 2y)\,dy$$

Integrate them between the points $x = 0$, $y = 0$ and $x = 1$, $y = 1$, along the paths $y = x$ and $y = x^2$.

4. Pure carbon is burnt with a 10% excess of air. Determine the adiabatic temperature of the flame. (It may be assumed that all the carbon is converted to CO_2; "adiabatic" means that there is no heat lost to the surroundings.)

5. **a.** Calculate the heat of reaction of methane CH_4 with air at 298°K, assuming that the products of reaction are CO_2 and H_2O(gas). (This heat of reaction is also called the *low calorific power* of CH_4.)
   b. Determine the low calorific power of ethane C_2H_6 in cal/mol and in kcal/m^3.
   c. Determine the low calorific power of natural gas analyzing 83% CH_4, 16% C_2H_6, and 1% N_2.

6. The natural gas described in Problem 5c is delivered at room temperature to a glass factory which operates a remelting furnace at 1600°K. The fuel is mixed with air, also at 298°K, and in 10% excess of the amount theoretically needed for complete combustion. The composition of air may be approximated as 21% O_2, 79% N_2.
   a. Calculate the composition of the flue gases and their sensible heat on the basis of one mole of natural gas. (All of the fuel's carbon is converted to CO_2, hydrogen to H_2O gas.)
   b. The furnace processes 2 tonnes of glass per hour and its heat losses to the surroundings average 100,000 kcal/hr. Calculate the fuel consumption at STP[11] (in m^3/hr) assuming that, for glass, $H^\circ_{1600} - H^\circ_{298} = 300$ kcal/kg.
   c. A heat exchanger is installed which transfers some of the sensible heat of the flue gases to the air of combustion. In this fashion, the air of combustion is preheated to 800°K. Calculate the decrease in fuel consumption.
   d. Let α be the cost of natural gas per cubic meter at STP. The cost of a preheater,

[11] STP is the conventional abbreviation for standard temperature and pressure, i.e., 0°C and 1 atm.

prorated on an hourly basis, is assumed to be equal to $15\alpha[(1 + r)/(1 - r)]$, where r is the *relative efficiency* of the preheater. r is defined as the ratio of the sensible heat of the air of combustion at the preheated temperature T to the sensible heat of that air at the temperature of the flue gases (1600°K). Calculate the optimum economic solution in terms of α. Estimate α and calculate the potential yearly savings assuming that the furnace operates 24 hr/day, 300 days/year. Determine the temperature T.

7. **a.** Demonstrate that, for given initial and final states at the same volume and temperature, the maximum work performed by a system along any path linking these two states is equal to the change in its Helmholtz free energy.
b. Show that if the initial and final states are at the same pressure and temperature, the maximum work performed by a system is equal to the change in its Gibbs free energy.

8. The difference $T\,dS - đq$ may be written $đq'$ or $T\,d_iS$. Clausius called $đq'$ the uncompensated heat and Prigogine and Defay refer to d_iS as the creation of entropy inside the system. Relate these quantities to the differential dZ in Section 2.3 and interpret their significance.

9. Evaluate $(\partial G/\partial T)_V$ for an element, in terms of functions measured or tabulated.

10. Show that for a perfect gas $đq$ is not a perfect differential but $đq/T$ is. ($1/T$ is then called an integrand; the existence of such an integrand forms the basis of Carath-eodory's presentation of the second law.)

11. Demonstrate that the assumption that E is independent of P at constant T leads to equation (81).

12. The van der Waals equation of state for an imperfect gas is

$$\left(P + \frac{a}{V^2}\right)(V - b) = RT$$

where a and b are two constants independent of the temperature.
a. Calculate the expressions of $(\partial P/\partial T)_V$, $(\partial E/\partial V)_T$, and of the coefficient of compressibility $\beta = -(1/V)(\partial V/\partial P)_T$.
b. Assume now that $a = 0$ (in this case the equation of state sometimes bears the name of Clausius) and calculate the coefficient of expansion α, $(\partial H/\partial P)_T$, and the difference $C_p - C_v = \alpha^2 VT/\beta$. For $b = 22$ cm^3, calculate the change in the molar enthalpy (in cal) of the gas when its pressure is changed from 1 to 100 atm.

13. Calculate the enthalpy and entropy changes of iron in the fcc structure when it is brought, without phase transformation, from a temperature of 1250°K and a pressure of 1 atm to 1600°K and 10,000 atm. Calculate also C_v at 1250°K and 1 atm. The data at 1250°K are as follows:

$$V^\circ_{Fe} = 7.31 \quad \text{cm}^3/\text{mol}$$

$$\alpha = 0.63 \times 10^{-4} \quad {}^\circ\text{K}^{-1}, \qquad \beta = 1.10 \times 10^{-6} \quad \text{atm}^{-1}$$

$$C^\circ_p = 5.80 + 1.98 \times 10^{-3}T \quad \text{cal/}^\circ\text{K mol}$$

14. Phase α of a species A transforms into phase β at 55°K and 1 atm. The heat capacities of A in the structures α and β are, respectively, $C_p^{\circ\alpha} = 2.1 \times 10^{-5}T^3$ and $C_p^{\circ\beta}$ =

$5.7 \times 10^{-5}T^3$ cal/°K mol. Calculate the enthalpy and entropy of transformation at 55°K and 0°K, at 1 atm.

15. Demonstrate that the slope of the coefficient of compressibility β vs T is horizontal at $T = 0$°K.

16. It may be established that

$$\int_0^{\infty} \frac{x^3}{e^x - 1}\,dx = \frac{\pi^4}{15}$$

Deduce from this result the asymptotic behavior of C_v for $T \to 0$ in the Debye model of a solid.

17. The temperature dependence of the heat capacity of anorthite, $CaAl_2Si_2O_8$, in the interval 298–1700°K, is represented by the equation

$$C_p^{\circ} = 66.42 + 13.70 \times 10^{-3}T - 16.89 \times 10^5 T^{-2} \quad \text{cal/°K mol}$$

Deduce the temperature dependence of $H_T^{\circ} - H_{298}^{\circ}$ and $S_T^{\circ} - S_{298}^{\circ}$, and calculate their values at $T = 1000$°K.

18. **a.** The heat of vaporization of water at 100°C is 9770 cal/mol. Plot $H_T^{\circ} - H_{298}^{\circ}$ and $S_T^{\circ} - S_{298}^{\circ}$ for H_2O between 273 and 400°K. Calculate the heat and entropy of vaporization of water at 298°K and indicate them on your graphs.

b. Calculate as accurately as you can with the data at your disposal the partial pressure of water vapor in equilibrium with liquid water at 25°C.

c. On a given sunny afternoon the temperature is 25°C and the humidity 80%. After sunset, the temperature falls rapidly. Calculate at what temperature the water vapor in the air condenses (dew point).

d. Plot the Gibbs free energy of water and water vapor as a function of pressure at 298°K. Calculate the difference $G^g - G^l$ in cal/mol at $P = 1$ atm, and deduce the pressure P at which the curves (G^g and G^l vs P) intersect. Compare this last result with that of part b. Data are

$$C_p(H_2O \text{ gas}) = 7.30 + 2.46 \times 10^{-3}\,T \quad \text{cal/°K mol}$$

$$C_p(H_2O \text{ liquid}) = 18.04 \quad \text{cal/°K mol}$$

19. The ice of an outdoor skating rink is at the ambient temperature of 27.5°F (−2.5°C). Calculate, in psi, the minimum pressure (applied for example by a skate) necessary to melt the ice.

Data: At 0°C, the specific volume of water is 1.000 cm³/g and that of ice is 1.090 cm³/g; the heat of fusion is 79.7 cal/g.

20. The boiling temperature of mercury is 630°K. Estimate the partial pressure of liquid mercury at room temperature (298°K).

21. At 400°C, liquid zinc has a vapor pressure of 10^{-4} atm. Estimate the boiling temperature of zinc knowing that its heat of vaporization is approximately equal to 28 kcal/mol.

22. In the integration of the Clausius–Clapeyron equation in the case of a gas–liquid equilibrium, it is usually assumed that ΔH_{vap} is constant. Is it equally proper to integrate the form of the Clausius–Clapeyron equation associated to ΔS_{vap} [equation (135)] assuming that ΔS_{vap} is constant? Discuss.

23. **a.** Assuming a van der Waals equation of state (see Problem 12), calculate the coordinates of the critical point in terms of the coefficients *a* and *b*.

b. In the *Handbook of Chemistry and Physics* for 1979 the coefficients *a* and *b* for CO_2 are listed as equal to 3.592 liter^2 atm/mol^2 and 4.267×10^{-2} liter/mol. Calculate the coordinates of the critical point and compare them to the data of Fig. 19.

References

1. G. N. Lewis and M. Randall, *Thermodynamics* (revised by K. S. Pitzer and L. Brewer). McGraw-Hill, New York, 1961.
2. W. Nernst, *Nachr. Akad. Wiss Göttingen Math. Phys.* **Kl,** 1 (1906).
3. M. Planck, *Thermodynamik,* 3rd ed. Veit, Leipzig, 1911, p. 279.
4. G. N. Lewis and M. Randall, *Thermodynamics and the Free Energy of Chemical Substances,* 1st ed. McGraw-Hill, New York, 1923, p. 448.
5. E. A. Guggenheim, *Thermodynamics, An advanced Treatment for Chemists and Physicists.* Interscience, New York, 1949, p. 161.
6. T. L. Hill, *An Introduction to Statistical Thermodynamics.* Addison-Wesley, Reading, MA, 1960, Ch. 5.
7. P. L. Dulong and A. T. Petit, *Ann. Chim. Phys.* **10,** 395 (1819).
8. C. Kittel, *Introduction to Solid State Physics,* 4th ed. Wiley, New York, 1971, p. 219.
9. H. Kopp, *Ann. Chem. Pharm. Suppl.* **3,** 1, 289 (1864).
10. R. Hultgren, P. D. Desai, D. T. Hawkins, M. Gleiser, K. K. Kelley, and D. D. Wagman, *Selected Values of the Thermodynamic Properties of the Elements.* Am. Soc. Metals, Metals Park, OH, 1973.
11. A. P. Miodownik, *Metallurgical Chemistry Symposium, 1971* (O. Kubaschewski, ed.). Her Majesty's Stationery Office, London, 1972, pp. 233–244.
12. D. R. Stull, H. Prophet, et al., *JANAF Thermochemical Tables,* NSRDS-NBS 37, 2nd ed. U.S. GPO, Washington, DC, 1971.
13. T. Takahashi and W. A. Bassett, *Science* **145,** 483–486 (1964).
14. F. P. Bundy, *J. Appl. Phys.* **36,** 616–620 (1965).
15. L'Air Liquide, *Gas Encyclopaedia.* Elsevier North Holland, New York, 1976.

Selected Bibliography

Textbooks

G. N. Lewis and M. Randall, *Thermodynamics,* 2nd ed. (revised by K. S. Pitzer and L. Brewer). McGraw-Hill, New York, 1961. Excellent presentation of the first, second, and third laws. Well-chosen examples.

D. R. Gaskell, *Introduction to Metallurgical Thermodynamics.* McGraw-Hill, New York, 1973. A text for a "first course," with numerous examples.

N. A. Gokcen, *Thermodynamics.* Techscience, Hawthorne, CA, 1975. Intermediate level, many problems, and a solutions manual (coauthored by L. R. Martin).

K. Denbigh, *The Principles of Chemical Equilibrium,* 2nd ed. Cambridge University Press, London, 1966.

Some more advanced texts are

H. B. Callen, *Thermodynamics.* Wiley, New York, 1960.

I. Prigogine and R. Defay, *Chemical Thermodynamics* (Transl. from the French: D. Everett), Longmans Green, London, 1954.

E. A. Guggenheim, *Thermodynamics, an Advanced Treatment for Chemists and Physicists,* 5th ed. North Holland, Amsterdam, 1967.

J. G. Kirkwood and I. Oppenheim, *Chemical Thermodynamics*. McGraw-Hill, New York, 1961. Few examples: concise, elegant, and rigorous treatment.

J. W. Gibbs, *Collected Works*. Yale University Press, New Haven, CT, 1948, Vol. I. A "classic," difficult to read, but worth the effort.

F. Fer, *Thermodynamique Macroscopique*, 2 vols. Gordon & Breach, Paris, 1970. General treatment, very advanced, extremely rigorous.

Compilations of Thermodynamic Data

D. R. Stull and H. Prophet, *JANAF Thermochemical Tables*, 2nd ed. NSRDS–NBS 37, U.S. GPO, Washington, DC, 1971; compilation continued by M. W. Chase, J. L. Cornutt, A. T. Hu, H. Prophet, R. A. McDonald, A. N. Syverud, and L. C. Walker, *J. Phys. Chem. Ref. Data* **3**, 311–480 (1974); **4**, 1–175 (1975); **7**, 793–940 (1978); reprints available from the American Chemical Society. The thermodynamic data listed in the compilations are illustrated in Appendix 4, which shows parts of the JANAF tables for CO, CO_2, H_2, H_2O, N_2, and O_2. The tabulated values are generally complemented by references and other data (not included in the appendix).

F. D. Rossini, D. D. Wagman, W. H. Evans, S. Levine, and I. Jaffee, Selected values of chemical thermodynamic properties, NBS Circ. 500, 1952; revised parts by D. D. Wagman, W. H. Evans, V. B. Parker, I. Harlow, S. M. Bailey, R. H. Schumm, and K. L. Churney, between 1965 and 1971.

I. Barin and O. Knacke, *Thermochemical Properties of Inorganic Substances*. Springer-Verlag, Berlin, 1973; supplement 1977.

J. F. Elliott and M. Gleiser, *Thermochemistry for Steelmaking*. Addison-Wesley, Reading, MA, Vol. 1, 1960.

C. E. Wicks and F. E. Block, Thermodynamic properties of 65 elements—their oxides, halides, carbides and nitrides, *Bur. of Mines Bull.* 605, U.S. GPO, Washington, DC, 1963.

K. K. Kelley, *Bull.* 601 (1962); K. K. Kelley and E. G. King, *Bull.* 592 (1961); K. K. Kelley, *Bull.* 584 (1960) (U.S. Bur. of Mines, U.S. GPO, Washington, DC).

R. Hultgren, P. D. Desai, D. T. Hawkins, M. Gleiser, K. K. Kelley, and D. D. Wagman, *Selected Values of the Thermodynamic Properties of the Elements*. Am. Soc. Metals, Metals Park, OH, 1973.

D. R. Stull, E. F. Westrum, and G. C. Sinke, *The Chemical Thermodynamics of Organic Compounds*. Wiley, New York, 1969.

R. C. Weast, ed., *Handbook of Chemistry and Physics*, 60th ed. Chemical Rubber Co., Cleveland, OH, 1979.

P. J. Spencer, *The Thermodynamic Properties of Silicates*, Nat. Phys. Lab., Rep. Chem. 21, England, February 1973.

K. K. Kelley, *Heats and Free Energies of Formation of Anhydrous Silicates*, Bur. of Mines R.I. 5901, U.S. Dept. of the Interior, Washington, DC, 1962.

II Fundamental Principles and Equations for an Open System

In the previous chapter, we studied principles and derived equations valid for any *closed* system, that is, for any system which allows heat and work transfers but no mass transfer across its boundaries. We shall now investigate the case of *open* systems. For simplicity we consider first the case of homogeneous open systems.

1. Introduction of the Chemical Potential

A homogeneous open system is a system which consists of a single phase and allows mass transfers across its boundaries. The energy of such a system will depend not only on the variables necessary to study a closed system (such as pressure and temperature), but also on the variables necessary to describe its size and composition. To analyze the system, it is convenient to choose as independent variables its entropy, its volume, and the numbers of moles n_i of its components:

$$E = E(S, V, n_i) \tag{1}$$

(The convenience of such a choice will be more apparent further in this section.) The perfect differential of E may be written

$$d\mathrm{E} = \left(\frac{\partial E}{\partial S}\right)_{V,n_i} dS + \left(\frac{\partial E}{\partial V}\right)_{S,n_i} dV + \sum_i \left(\frac{\partial E}{\partial n_i}\right)_{S,V,n_j} dn_i \tag{2}$$

The first two partial derivatives may be easily identified. Since the composition is kept constant (all n_i), they are identical to the partial derivatives $(\partial E/\partial S)_V$ and $(\partial E/\partial V)_S$ of a closed system, i.e., to T and $-P$, as can be readily seen in equation (I.55). Thus,

$$dE = T\,dS - P\,dV + \sum_i \left(\frac{\partial E}{\partial n_i}\right)_{S,V,n_j} dn_i \tag{3}$$

The last partial derivative is called the *chemical potential of component i* and is represented μ_i. Equation (3) then becomes

$$dE = T\,dS - P\,dV + \sum_i \mu_i\,dn_i \tag{4}$$

Recalling the definition of the enthalpy H [equation (I.16)], the Helmholtz free energy A [equation (I.60)] and the Gibbs free energy G [equation (I.61)], the following relations are derived:

$$dH = T\,dS + V\,dP + \sum_i \mu_i\,dn_i \tag{5}$$

$$dA = -S\,dT - P\,dV + \sum_i \mu_i\,dn_i \tag{6}$$

$$dG = -S\,dT + V\,dP + \sum_i \mu_i\,dn_i \tag{7}$$

The chemical potential μ_i may thus be defined by any of the following partial derivatives:

$$\mu_i = \left(\frac{\partial E}{\partial n_i}\right)_{S,V,n_j} = \left(\frac{\partial H}{\partial n_i}\right)_{S,P,n_j} = \left(\frac{\partial A}{\partial n_i}\right)_{T,V,n_j} = \left(\frac{\partial G}{\partial n_i}\right)_{T,P,n_j} \tag{8}$$

In particular, μ_i may be viewed as the rate of change of the total Gibbs free energy of the system when, holding the temperature and the pressure constant, an infinitesimal amount of the component i is added to the system without changing the number of moles of the other components j.[1]

To proceed further with this analytical study of open systems, it is helpful to recall the mathematical properties of Euler's homogeneous functions.

2. Extensive Properties

A function f of the variables x, y, z is said to be *homogeneous of degree n* if

$$f(\lambda x, \lambda y, \lambda z) = \lambda^n f(x, y, z) \tag{9}$$

where λ is an arbitrary parameter.

Let $u = \lambda x$, $v = \lambda y$, and $w = \lambda z$. Differentiation of equation (9) with respect to λ leads to

$$\frac{\partial f(u, v, w)}{\partial u}\frac{\partial u}{\partial \lambda} + \frac{\partial f(u, v, w)}{\partial v}\frac{\partial v}{\partial \lambda} + \frac{\partial f(u, v, w)}{\partial w}\frac{\partial w}{\partial \lambda} = n\lambda^{n-1} f(x, y, z) \tag{10}$$

or

$$x\frac{\partial f(\lambda x, \lambda y, \lambda z)}{\partial(\lambda x)} + y\frac{\partial f(\lambda x, \lambda y, \lambda z)}{\partial(\lambda y)} + z\frac{\partial f(\lambda x, \lambda y, \lambda z)}{\partial(\lambda z)} = n\lambda^{n-1} f(x, y, z) \tag{11}$$

Note that the operation performed for the derivative $\partial f/\partial u$ (or $\partial f/\partial(\lambda x)$) is identical to that performed for the derivative $\partial f/\partial x$, and for $\lambda = 1$, the two functions $\partial f(\lambda x, \lambda y, \lambda z)/\partial(\lambda x)$ and $\partial f(x, y, z)/\partial x$ are identical. Thus, equation (11) becomes

$$x\frac{\partial f(x, y, z)}{\partial x} + y\frac{\partial f(x, y, z)}{\partial y} + z\frac{\partial f(x, y, z)}{\partial z} = nf(x, y, z) \tag{12}$$

or, in abbreviated notation,

$$xf'_x + yf'_y + zf'_z = nf(x, y, z) \tag{13}$$

Let us now consider homogeneous functions of degree 1 ($n = 1$). Taking the total differential of equation (13) leads to

$$x\,df'_x + y\,df'_y + z\,df'_z + f'_x\,dx + f'_y\,dy + f'_z\,dz = f'_x\,dx + f'_y\,dy + f'_z\,dz \tag{14a}$$

and, after cancellation of common terms on both sides of the equation

$$x\,df'_x + y\,df'_y + z\,df'_z = 0 \tag{14b}$$

Equations (13) and (14) are two properties of homogeneous functions of degree 1 which play an important role in thermodynamics.

---

[1] One may also consider the addition of one mole of component i to a very large—mathematically, infinitely large—system.

It is an experimentally verified fact that the energy of a system is proportional to its size (assuming that the volume of the system is sufficiently large to make negligible the influence of its surfaces). For instance, if we double the number of moles of each component, the energy becomes twice as large as it was originally. The energy is said to be an *extensive property* of the system. Other functions such as the entropy, the volume, the enthalpy, and the free energy (Helmholtz or Gibbs) are also extensive properties, whereas temperature and pressure are unaffected by the size of the system; they are said to be *intensive properties*. Mathematically, the extensive property characteristic of a function may be translated by stating that it is a homogeneous function of degree 1. For instance, at a fixed temperature and pressure

$$E(\lambda n_1, \lambda n_2, \ldots, \lambda n_m) = \lambda E(n_1, n_2, \ldots, n_m) \tag{15}$$

$$S(\lambda n_1, \lambda n_2, \ldots, \lambda n_m) = \lambda S(n_1, n_2, \ldots, n_m) \tag{16}$$

$$V(\lambda n_1, \lambda n_2, \ldots, \lambda n_m) = \lambda V(n_1, n_2, \ldots, n_m) \tag{17}$$

Note that if we consider that the energy E is a function of the variables S, V, and n_i, it is obvious that E is also a homogeneous function of these variables:

$$E(\lambda S, \lambda V, \lambda n_i) = \lambda E(S, V, n_i) \tag{18}$$

Consequently, equations (13) and (14) are applicable. Recalling that the partial derivatives of E with regard to S, V, and n_i are respectively T, $-P$, and μ_i, equations (13) and (14) become

$$E = TS - PV + \sum_i n_i \mu_i \tag{19}$$

$$S\,dT - V\,dP + \sum_i n_i\,d\mu_i = 0 \tag{20}$$

Equation (20) is known as the *Gibbs–Duhem equation*. It is generally used under conditions of constant temperature and pressure, and has then the simpler expression

$$\sum_i n_i\,d\mu_i = 0 \tag{21}$$

Equation (19) leads to the following relations between integral thermodynamic functions:

$$H = E + PV = TS + \sum_i n_i \mu_i \tag{22}$$

$$A = E - TS = -PV + \sum_i n_i \mu_i \tag{23}$$

$$G = E + PV - TS = \sum_i n_i \mu_i \tag{24}$$

The last equation is important: the Gibbs free energy of a system is the sum of the chemical potentials of its constituents. It is not a mere consequence of the definition of the chemical potential μ_i as a partial derivative of G with respect to the number of moles n_i [equation (8)]. Note that for a system consisting of a single component the chemical potential of this component is identical to its molar Gibbs free energy (that is, to the Gibbs free energy of one mole of that substance).

3. Partial Molar Properties

3.1. Definitions and Relationships

The partial derivative of any extensive function Y with respect to the number of moles of a component i—keeping constant the number of moles of the other components j—is called a *partial molar*[2] property and is usually designated by the symbol $\overline{Y}_i$. For instance, the chemical potential of i (μ_i) is the partial molar Gibbs free energy of i:

$$\overline{G}_i = \left(\frac{\partial G}{\partial n_i}\right)_{P,T,n_j} = \mu_i \tag{25}$$

and the partial molar enthalpy, entropy, and volume of i, are, respectively,

$$\overline{H}_i = \left(\frac{\partial H}{\partial n_i}\right)_{P,T,n_j} \tag{26}$$

$$\overline{S}_i = \left(\frac{\partial S}{\partial n_i}\right)_{P,T,n_j} \tag{27}$$

$$\overline{V}_i = \left(\frac{\partial V}{\partial n_i}\right)_{P,T,n_j} \tag{28}$$

Partial molar derivatives need not be taken necessarily under conditions of constant temperature and pressure; they may be taken under different conditions such as constant temperature and volume. Practically, the first conditions are the most useful and unless otherwise specified, *partial molar properties* in this text (as in the literature) will refer to those defined at constant T and P. However, in Section 3.3 we shall examine some features of partial molar properties defined under conditions of constant volume, because these arise in the development of many statistical thermodynamic models.

Note that if the temperature and pressure are fixed, the extensive property Y depends only on the variables n_i and is thus a homogeneous function of these variables. Equation (13) then becomes

$$Y = \sum_i n_i \left(\frac{\partial Y}{\partial n_i}\right)_{P,T,n_j} = \sum_i n_i \overline{Y}_i \tag{29}$$

As in the derivation of equation (14), the total differential of Y and the differentiation of equation (29) leads to

$$-\left(\frac{\partial Y}{\partial T}\right)_{P,n_i} dT - \left(\frac{\partial Y}{\partial P}\right)_{T,n_i} dP + \sum_i n_i\, d\overline{Y}_i = 0 \tag{30}$$

or, at constant T and P,

$$\sum_i n_i\, d\overline{Y}_i = 0 \tag{31}$$

[2] The name *partial molal property* is also used, mostly by chemists.

Applied to the enthalpy, entropy, and volume functions, equation (30) (a more general form of the Gibbs–Duhem equation) yields

$$-C_p\,dT - V(1-\alpha T)\,dP + \sum_i n_i\,d\overline{H}_i = 0 \tag{32}$$

$$-\frac{C_p}{T}\,dT + \alpha V\,dP + \sum_i n_i\,d\overline{S}_i = 0 \tag{33}$$

$$-\alpha V\,dT + \beta V\,dP + \sum_i n_i\,d\overline{V}_i = 0 \tag{34}$$

where α and β are, respectively, the coefficients of expansion and compressibility of the phase under consideration. [The first two partial derivatives of equation (30) being taken at constant n_i, equations such as (I.20) and (I.72) derived for a closed system, are applicable.]

The variation of a partial molar property due to the change of another variable may often be obtained through Maxwell's relations (relations expressing the fact that the differentials of thermodynamic state functions are perfect differentials). For instance, equation (7) yields the following results on the variation of the chemical potential with temperature, pressure, and composition:

$$\left(\frac{\partial \mu_i}{\partial T}\right)_{P,n_i,n_j} = -\left(\frac{\partial S}{\partial n_i}\right)_{T,P,n_j} = -\overline{S}_i \tag{35}$$

$$\left(\frac{\partial \mu_i}{\partial P}\right)_{T,n_i,n_j} = \left(\frac{\partial V}{\partial n_i}\right)_{T,P,n_j} = \overline{V}_i \tag{36}$$

$$\left(\frac{\partial \mu_i}{\partial n_j}\right)_{T,P,n_i,n_k} = \left(\frac{\partial \mu_j}{\partial n_i}\right)_{T,P,n_j,n_k} \tag{37}$$

Also note that differentiation of the relation $G = H - TS$ with respect to n_i, at constant T, P, and n_j, leads to

$$\overline{G}_i = \mu_i = \overline{H}_i - T\overline{S}_i \tag{38}$$

or [using equation (35)]

$$\overline{H}_i = \mu_i - T\left(\frac{\partial \mu_i}{\partial T}\right)_{P,n_i,n_j} = \left(\frac{\partial(\mu_i/T)}{\partial(1/T)}\right)_{P,n_i,n_j} \tag{39}$$

A summary of the important equations analyzing the thermodynamics of homogeneous open systems is given in Table 1.

3.2. Graphical Representation: The Method of Intercepts

3.2.1. Binary Systems

Considering an extensive property such as the volume, the equality

$$V = \sum n_i\overline{V}_i \tag{40}$$

may be interpreted as the assignment of a volume $\overline{V}_i$ for each mole of component i, the volume of the system being then the sum of the volumes of the constituents.

Table 1. Summary of Important Equations for the Thermodynamics of Homogeneous Open Systems

Differential equations	Integral equations	Gibbs–Duhem equations at constant T and P
$dE = T\,dS - P\,dV + \sum_i \mu_i\,dn_i$	$E = TS - PV + \sum_i n_i\mu_i$	
$dH = T\,dS + VdP + \sum_i \mu_i\,dn_i$	$H = TS + \sum_i n_i\mu_i$	
$= C_P\,dT + V(1 - \alpha T)\,dP + \sum_i \overline{H}_i\,dn_i$	$= \sum_i n_i\overline{H}_i$	$\sum_i n_i\,d\overline{H}_i = 0$
$dA = -S\,dT - P\,dV + \sum_i \mu_i\,dn_i$	$A = -PV + \sum_i n_i\mu_i$	
$dG = -S\,dT + V\,dP + \sum_i \mu_i\,dn_i$	$G = \sum_i n_i\mu_i$	$\sum_i n_i\,d\mu_i = 0$
$dS = \frac{C_P}{T}\,dT - \alpha V\,dP + \sum_i \overline{S}_i\,dn_i$	$S = \sum_i n_i\overline{S}_i$	$\sum_i n_i\,d\overline{S}_i = 0$
$dV = \alpha V\,dT - \beta V\,dP + \sum_i \overline{V}_i\,dn_i$	$V = \sum_i n_i\overline{V}_i$	$\sum_i n_i\,d\overline{V}_i = 0$

Additional relations relevant to the study of the chemical potential:

$$\left(\frac{\partial\mu_i}{\partial T}\right)_{P,ni,nj} = -\overline{S}_i \qquad \left(\frac{\partial\mu_i}{\partial P}\right)_{T,ni,nj} = \overline{V}_i \qquad \left(\frac{\partial\mu_i}{\partial n_j}\right)_{P,T,ni,nk} = \left(\frac{\partial\mu_j}{\partial n_i}\right)_{P,T,nj,nk}$$

$$\mu_i = \overline{H}_i - T\overline{S}_i \qquad \overline{H}_i = \left(\frac{\partial(\mu_i/T)}{\partial(1/T)}\right)_{P,ni,nj}$$

This interpretation, however, presents some difficulty, since the partial molar volume may be found experimentally to be negative; upon further addition of i the volume of a solution may shrink instead of expand. As previously noted, a partial molar property measures a rate of change. Graphically, if we assume that the property Y of a binary A–B is known across the composition range, it is clear that $\overline{Y}_A$ and $\overline{Y}_B$ will depend on the position of the tangent at the composition under consideration. It is easy to demonstrate that in a plot of Y_m (Y per mole) versus the mole fraction X_B of B (Fig. 1), $\overline{Y}_A$ and $\overline{Y}_B$ are measured by the intercepts of the tangent with the axes of A and B. In Fig. 1,

$$BD = OM + OB\tan(\text{angle } PMD) \tag{41}$$

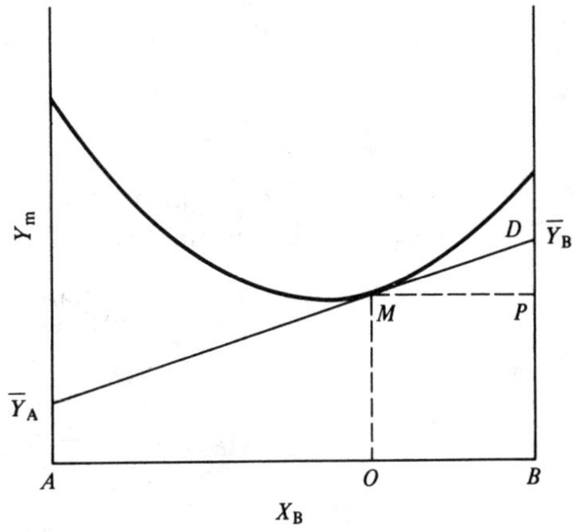

Figure 1. The graphical method of intercepts to measure the partial molar properties of a binary solution.

Hence, to demonstrate that $BD = \overline{Y}_B$, it is sufficient to show that

$$\overline{Y}_B = Y_m + (1 - X_B)\frac{\partial Y_m}{\partial X_B} \tag{42}$$

For a system of one mole

$$Y_m = X_A\overline{Y}_A + X_B\overline{Y}_B \tag{43}$$

and differentiating equation (43) with respect to X_B yields

$$\frac{\partial Y_m}{\partial X_B} = (\overline{Y}_B - \overline{Y}_A) + \left(X_A\frac{\partial \overline{Y}_A}{\partial X_B} + X_B\frac{\partial \overline{Y}_B}{\partial X_B}\right) \tag{44}$$

The second parenthesis is equal to zero because of the Gibbs–Duhem equation. Through equation (43), equation (44) may be rewritten

$$\frac{\partial Y_m}{\partial X_B} = \overline{Y}_B - \overline{Y}_A = \frac{\overline{Y}_B - Y_m}{X_A} \tag{45}$$

which is identical to equation (42).

The difference between the partial molar property of i and the partial derivative with respect to the mole fraction of i should be emphasized. It is a common mistake to consider them proportional. They are not because it is not possible to differentiate with respect to X_i and keep constant the number of moles of all the other components in the system.

The method of intercepts may be generalized to a multicomponent system. Let us derive the equivalent of equation (42) for a system of m components.

3.2.2. Multicomponent Systems

Although we could consider again the extensive property Y, we shall for illustrative purposes identify it now as the Gibbs free energy. For convenience, we continue to distinguish between G and G_m: these designate, respectively, the Gibbs free energy of the system and the molar Gibbs free energy of the system. If there is a total of n mol,

$$G = nG_m \tag{46}$$

To find $\overline{G}_i$ (i.e., μ_i), we need to differentiate G with respect to n_i; since G_m is generally analyzed in terms of mole fractions, we also need to consider the change of variables

$$(n_1, n_2, \ldots, n_j, \ldots, n_r) \rightarrow (n, X_2, \ldots, X_j, \ldots, X_r)$$

where n is the total number of moles ($= \sum n_j$) and X_j is the mole fraction of j ($= n_j/n$). Equation (46) yields

$$\begin{aligned}\overline{G}_i &= \left(\frac{\partial G}{\partial n_i}\right)_{n_k} = \left(\frac{\partial n}{\partial n_i}\right)_{n_k} G_m + n\left(\frac{\partial G_m}{\partial n_i}\right)_{n_k} \\ &= G_m + n\left(\frac{\partial G_m}{\partial n}\right)_{X_j}\left(\frac{\partial n}{\partial n_i}\right)_{n_k} + n\sum_{j=2}^{r}\left(\frac{\partial G_m}{\partial X_j}\right)_{n,X_k}\left(\frac{\partial X_j}{\partial n_i}\right)_{n_k}\end{aligned} \tag{47}$$

Because G_m depends only on the composition of the system and not on its size (see Problem 3),

$$\left(\frac{\partial G_m}{\partial n}\right)_{X_j} = 0 \tag{48}$$

Moreover,

$$\left(\frac{\partial n}{\partial n_i}\right)_{n_k} = 1 \tag{49}$$

and

$$\left(\frac{\partial X_j}{\partial n_i}\right)_{n_k} = \frac{\delta_{ij} - X_j}{n} \tag{50}$$

where δ_{ij} is Kronecker's symbol ($\delta_{ij} = 0$ for $i \neq j$ and $\delta_{ij} = 1$ for $i = j$). Thus, equation (47) becomes

$$\overline{G}_i = \mu_i = G_m + \sum_{j=2}^{r} (\delta_{ij} - X_j)\frac{\partial G_m}{\partial X_j} \tag{51}$$

For a ternary solution, it would yield

$$\mu_1 = G_m - X_2\frac{\partial G_m}{\partial X_2} - X_3\frac{\partial G_m}{\partial X_3} \tag{52a}$$

$$\mu_2 = G_m + (1 - X_2)\frac{\partial G_m}{\partial X_2} - X_3\frac{\partial G_m}{\partial X_3} \tag{52b}$$

$$\mu_3 = G_m - X_2\frac{\partial G_m}{\partial X_2} + (1 - X_3)\frac{\partial G_m}{\partial X_3} \tag{52c}$$

If we plot the Gibbs free energy surface of the phase under consideration vs X_2 and X_3 in a cartesian system of coordinates (see Fig. 2a), the equation of the plane (P) tangent to this surface at the point M of coordinates X_2, X_3, is

$$y - G_m = (x_2 - X_2)\frac{\partial G_m}{\partial X_2} + (x_3 - X_3)\frac{\partial G_m}{\partial X_3} \tag{52d}$$

where y, x_2, x_3 are the ordinate and abscissa of a point on (P). It is readily seen that equation (52a) expresses that (P) intercepts the vertical axis A, corresponding to the pure component 1 ($x_2 = x_3 = 0$), at an ordinate equal to μ_1. Equations (52b) and (52c) express similar results for μ_2 and μ_3.

In Chapter X, we shall see that to represent the composition of a ternary solution it is often more convenient to use an equilateral triangle rather than cartesian coordinates. An axis perpendicular to that triangle allows the representation of the molar Gibbs free energy (see Fig. 2b). Again equations (52) lead to the same result, namely that the intercept of the plane tangent to the surface G_m at the composition under consideration, with an axis corresponding to a pure component, yields the partial molar Gibbs free energy μ_i of that component.

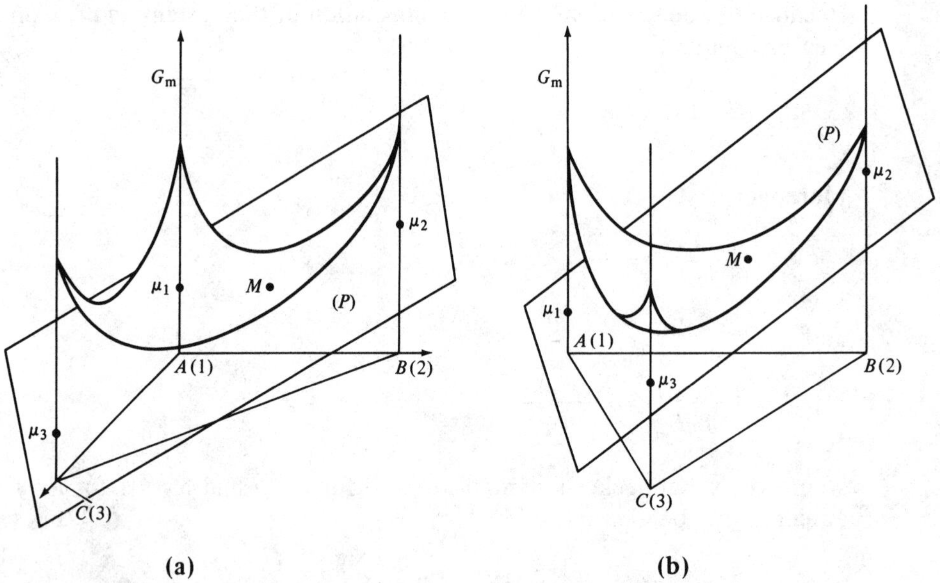

Figure 2. The graphical method of intercepts to measure the partial molar properties of a ternary system. The plane (P) is tangent to the Gibbs free energy surface at the point M and intercepts the vertical axes at μ_1, μ_2, and μ_3. (**a**) The cartesian system of coordinates. (**b**) Equilateral concentration triangle.

3.3. Constant Volume Conditions

We mentioned in Section 3.1 that although partial molar properties are usually defined under conditions of constant pressure, there are cases where theoretical considerations warrant conditions of constant volume.

It may be noted that the chemical potential which is usually defined as a partial molar Gibbs free energy at constant pressure, can also be defined as a partial molar Helmholtz free energy at constant volume (see Section 1):

$$\mu_i = \left(\frac{\partial G}{\partial n_i}\right)_{T,P,nj} = \left(\frac{\partial A}{\partial n_i}\right)_{T,V,nj} \tag{53}$$

The latter definition is often more useful than the former in the context of statistical thermodynamics models. Another function of some interest is the partial molar energy:

$$\tilde{E}_i = \left(\frac{\partial E}{\partial n_i}\right)_{T,V,nj} \tag{54}$$

where the symbol $\sim$ is used to recall that a condition of constant V rather than P is being used. In particular, equation (53) may be translated

$$\overline{G}_i = \tilde{A}_i \tag{55}$$

Finding the relation between $\overline{E}_i$ and $\tilde{E}_i$ necessitates an analytic derivation not unlike that leading to the relation between C_p and C_v. Consider E as a function of T, V, and n_j:

$$dE = \left(\frac{\partial E}{\partial T}\right)_{V,n_j} dT + \left(\frac{\partial E}{\partial V}\right)_{T,n_j} dV + \sum_{j=1}^{m} \tilde{E}_j\, dn_j \tag{56}$$

Dividing by dn_i at constant P and T we obtain

$$\overline{E}_i = \left(\frac{\partial E}{\partial V}\right)_{T,nj} \overline{V}_i + \tilde{E}_i \quad (57)$$

Moreover, from the second law, at constant n_j

$$dE = T\,dS - P\,dV \quad (58)$$

Consequently,

$$\left(\frac{\partial E}{\partial V}\right)_{T,nj} = T\left(\frac{\partial S}{\partial V}\right)_{T,nj} - P \quad (59)$$

and from Maxwell's relations

$$\left(\frac{\partial S}{\partial V}\right)_{T,nj} = \left(\frac{\partial P}{\partial T}\right)_{V,nj} = \frac{\alpha}{\beta} \quad (60)$$

Thus, combining equations (57), (59), and (60) yields

$$\overline{E}_i = \tilde{E}_i + (\alpha T\overline{V}_i/\beta) - P\overline{V}_i \quad (61)$$

We also note that since

$$H = E + PV \quad (62)$$

we have

$$\overline{H}_i = \overline{E}_i + P\overline{V}_i \quad (63)$$

and therefore

$$\overline{H}_i = \tilde{E}_i + (\alpha\overline{V}_i T/\beta) \quad (64)$$

Equation (64) links an experimental measure (the enthalpy increase of the system upon addition of i at constant P) to one which fits into a simpler theoretical framework (the energy increase of the system upon addition of i at constant volume). Equation (64) is of special interest in the case of interstitial solutions.

3.4. Independent Variables

In Section 2, we defined homogeneous functions and analyzed two of their properties [equations (12) and (14)]. These functions have two other properties of interest here (properties which can easily be derived):

a. If the function $f(x, y, z)$ is homogeneous of degree n with respect to x, y, z, a derivative such as $\partial f/\partial x$ is also homogeneous with respect to x, y, z, but of degree $n - 1$.
b. If a function $g(x, y, z)$ is homogeneous of degree 0 with respect to x, y, z, it depends on x, y, z, only through the variables x/X, y/X, z/X, where X is a linear combination of x, y, z.

We now recall that G is a homogeneous function of degree 1 with respect to the n_is, and A is a homogeneous function of degree 1 with respect to V and the n_is. From equation (53) and property a, we see therefore that μ_i must be a homogeneous function of degree 0 with respect to the n_is at constant T and P, and a homogeneous function of degree 0 with respect to V and the n_is at constant T.

In addition, from property b, we note that μ_i depends only on the variables T, P, n_i/X, or T, V/Y, n_i/Y, where X is a linear combination of the n_is, and Y is a linear combination of V and the n_is.

If we choose $X = Y = \sum_{i=1}^{m} n_i = n$, we find that μ_i depends only on the variables T, P, X_i, or T, V_m, X_i, where X_i is the mole fraction of i ($= n_i/n$) and V_m is the molar volume of the solution ($= V/n$). Moreover, since the X_is are not independent,

$$\sum_{i=1}^{m} X_i = 1 \tag{65}$$

μ_i is a function of the independent variables $T, P, X_2, \ldots, X_m$ or $T, V_m, X_2, \ldots, X_m$.

Other choices of X and Y are occasionally warranted. For example, in the case of a ternary solution 1–2–3, where 1 and 2 are substitutional components and 3 is an interstitial solute, it is convenient to consider the coordinates

$$Y_i = \frac{n_i}{n_1 + n_2} \qquad \text{for} \quad i = 1, 2, \text{ or } 3 \tag{66}$$

and the quantity

$$V_s = \frac{V}{n_1 + n_2} \tag{67}$$

which, in essence, represents the molar volume of the substitutional lattice. μ_i is then a function of the set of independent variables T, P, Y_2, Y_3, or T, V_s, Y_2, Y_3.

We emphasize that it would be incorrect to consider μ_i as a function of only the variables $T, V, X_2, \ldots, X_m$, or $T, V, Y_2, \ldots, Y_m$, for μ_i would then depend on the size of the system; it is not the total volume of the system which must be considered, but V_m, V_s, or any other such volume measure (V divided by any linear combination of the variables n_i and V).

Physically, this result may be easily understood. Consider a system at constant temperature, volume, and composition. Conceivably, by decreasing interatomic spacings, we could add a number of atoms n without altering the total volume, temperature, and composition of the solution. This would obviously cause a change in the chemical behavior of the components of the solution. More specifically, we calculate that

$$\left(\frac{\partial \mu_i}{\partial n}\right)_{T,V,X_j} = \left(\frac{\partial \mu_i}{\partial P}\right)_{T,X_j} \left(\frac{\partial P}{\partial n}\right)_{T,V,X_j} = \overline{V}_i \left(\frac{\partial P}{\partial n}\right)_{T,V,X_j}$$
$$= -\overline{V}_i \frac{(\partial V/\partial n)_{T,P,X_j}}{(\partial V/\partial P)_{T,n,X_j}} = \frac{\overline{V}_i}{\beta n} \neq 0 \tag{68}$$

To relate the composition dependences of μ_i at constant P and at constant V_m or V_s is an exercise which may be found in Section IV.6 and Ref. [1].

Considerations similar to those above for μ_i may also be applied to other homogeneous functions of degree 0, such as $\overline{H}_i$ and $\tilde{E}_i$ [2].

---

4. Derivation of the Conditions for Equilibrium in a Heterogeneous System

A heterogeneous system is a system which contains several phases, in contrast to a homogeneous system which consists of a single phase. In Chapter I, criteria of equilibrium were derived for closed systems (whether homogeneous or heterogeneous). We wish now to derive similar criteria for open homogeneous systems. The procedure consists in

a. grouping the phases exchanging heat and mass (homogeneous open systems) into a closed heterogeneous system;
b. applying, under one form or another, one of the criteria of equilibrium for a closed system; and
c. expressing the consequences of this criterion on the individual phases constituting the closed system.

Before attacking the general case, let us illustrate this procedure by applying it to a simple particular case.

4.1. Preliminary Example

A crucible of magnesium oxide containing iron, silver, and silicon is placed in a furnace at 1600°C under an argon atmosphere at a fixed pressure, e.g., 1 atm (see Fig. 3). Liquid iron and liquid silver are practically immiscible and we wish to study the equilibrium distribution of silicon between the two phases. MgO at 1600°C is quite "stable"; i.e., it does not react to any significant extent with the liquid phases. Similarly, the interaction with the gaseous phase may be neglected and for all practical purposes, the system of the two homogeneous phases (which we shall designate α and β: liquid silver–silicon alloy α and liquid iron–silicon alloy β) may be considered a closed heterogeneous system at constant temperature and pressure. The criterion of equilibrium under these conditions expresses that the Gibbs free energy must be minimum (see Chapter I), or that all the virtual variations δG in the field of equilibrium states must be positive:

$$(\delta G)_{P,T,n_i} \geq 0 \tag{69}$$

The free energy of the system being the sum of the free energies of the two phases

$$G = G^{\alpha} + G^{\beta} \tag{70}$$

In each phase, the Gibbs free energy may be simply expressed as a sum of

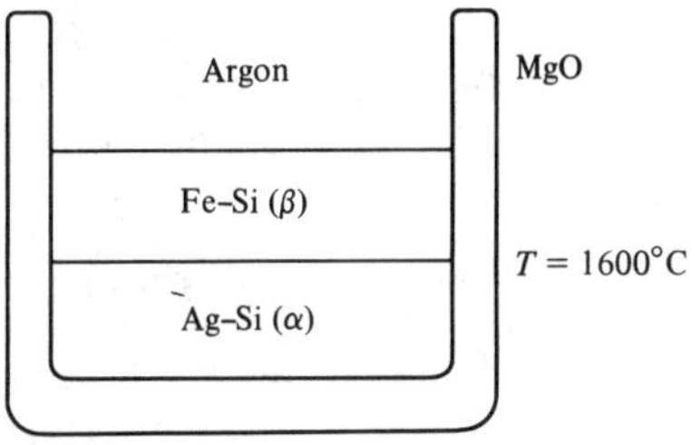

Figure 3. The silicon distributes itself between the two immiscible layers of silver and iron in such a way that, when equilibrium is reached, $\mu_{Si}^{\alpha} = \mu_{Si}^{\beta}$.

chemical potentials [equation (24)]:

$$G = (\mu_{Ag}^{\alpha} n_{Ag}^{\alpha} + \mu_{Si}^{\alpha} n_{Si}^{\alpha}) + (\mu_{Fe}^{\beta} n_{Fe}^{\beta} + \mu_{Si}^{\beta} n_{Si}^{\beta}) \tag{71}$$

If δn_{Si}^{α} moles of silicon are gained by the α phase at the expense of the β phase, the chemical potentials vary accordingly and the consequent variation of G is

$$\delta G = (\mu_{Si}^{\alpha}\, \delta n_{Si}^{\alpha} + \mu_{Si}^{\beta}\, \delta n_{Si}^{\beta}) + (n_{Ag}^{\alpha}\, \delta\mu_{Ag}^{\alpha} + n_{Si}^{\alpha}\, \delta\mu_{Si}^{\alpha}) + (n_{Fe}^{\beta}\, \delta\mu_{Fe}^{\beta} + n_{Si}^{\beta}\, \delta\mu_{Si}^{\beta}) \tag{72}$$

But because of the Gibbs–Duhem equation [equation (21)], the second and third sums of equation (72) are equal to zero, and

$$\delta G = \mu_{Si}^{\alpha}\, \delta n_{Si}^{\alpha} + \mu_{Si}^{\beta}\, \delta n_{Si}^{\beta} \tag{73}$$

[Note that this equation could have been immediately obtained by direct application of equation (7); however, the derivation above is natural and instructive.]

Since the total amount of silicon is constant, $\delta n_{Si}^{\alpha} = -\delta n_{Si}^{\beta}$, and consequently

$$\delta G = (\mu_{Si}^{\alpha} - \mu_{Si}^{\beta})\, \delta n_{Si}^{\alpha} \tag{74}$$

Thus, the criterion of equilibrium [condition (69)] becomes

$$(\mu_{Si}^{\alpha} - \mu_{Si}^{\beta})\, \delta n_{Si}^{\alpha} \geq 0 \tag{75}$$

If $\mu_{Si}^{\alpha} - \mu_{Si}^{\beta}$ were positive, a transfer of silicon from the α phase to the β phase ($\delta n_{Si}^{\alpha} < 0$) would violate condition (75) and the system would not be in equilibrium. Similarly, if $\mu_{Si}^{\alpha} - \mu_{Si}^{\beta}$ were negative, a transfer of silicon to the α phase from the β phase ($\delta n_{Si}^{\alpha} > 0$) would also violate the criterion of equilibrium. As there is no restriction on the values of δn_{Si}^{α}, the only way for condition (75) to be respected regardless of the values of δn_{Si}^{α} (i.e., no matter what process is envisaged) is to have $\mu_{Si}^{\alpha} - \mu_{Si}^{\beta} = 0$ or

$$\mu_{Si}^{\alpha} = \mu_{Si}^{\beta} \tag{76}$$

At equilibrium, the chemical potentials of silicon in the α and β phases must be equal.

Note that under conditions of nonequilibrium the irreversible process taking place must be such that

$$dG = (\mu_{Si}^{\alpha} - \mu_{Si}^{\beta})\, dn_{Si}^{\alpha} < 0 \tag{77}$$

If the chemical potential of silicon is higher in the phase β than in the phase α ($\mu_{Si}^{\beta} > \mu_{Si}^{\alpha}$), then dn_{Si}^{α} will be positive, that is, silicon will be transferred from β to α. This transfer of silicon from the phase of higher chemical potential to the phase of lower chemical potential will take place until the two chemical potentials of silicon attain the same value (equilibrium).

Essentially the same procedure will be now applied to the general case.

4.2. General Case

Let us consider a heterogeneous closed system consisting of the homogeneous open systems (phases) $\alpha, \beta, \ldots, \theta$. Each phase ν contains n_1^{ν} mol of component 1, n_2^{ν} mol of component 2, . . . , n_m^{ν} mol of component m. According to condition

(I.51), the criterion of equilibrium for this closed system is

$$(\delta E - T\,\delta S + P\,\delta V)_{n_i} \geq 0 \tag{78}$$

(the notation indicating that n_i remains constant emphasizes that the system is closed). Rather than impose on this system a restriction of constant temperature and pressure as in the preliminary example above, we shall impose the restriction that the entropy and volume of the total system remain unchanged. (This will allow us to derive more easily conditions on the temperature and pressure of each phase.) The criterion of equilibrium (78) then becomes

$$(\delta E)_{S,V,n_i} \geq 0 \tag{79}$$

Thus, any process which keeps constant the entropy, the volume, and the total number of moles of each component i and which, in addition, does not lower the energy of the total system, must be in the field of equilibrium states.

Our procedure will then consist in investigating the conditions to which the phases must be subjected, such that under the restrictions on S, V, and n_i noted above, no process can be found which violates the criterion of equilibrium (79). These conditions will then be the conditions of equilibrium for interacting homogeneous open systems.

The energy, entropy, and volume of the total system being the sums of the energies, entropies, and volumes of the individual phases, we may write

$$E = \sum_{\nu=\alpha}^{\theta} E^{\nu} \tag{80}$$

$$S = \sum_{\nu=\alpha}^{\theta} S^{\nu} \tag{81}$$

$$V = \sum_{\nu=\alpha}^{\theta} V^{\nu} \tag{82}$$

In addition

$$n_i = \sum_{\nu=\alpha}^{\theta} n_i^{\nu} \qquad \text{for} \quad i = 1, 2, \ldots, m \tag{83}$$

Consequently, the processes to be considered are such that

$$\delta E = \sum_{\nu=\alpha}^{\theta} \delta E^{\nu} \geq 0 \tag{84}$$

$$\delta S = \sum_{\nu=\alpha}^{\theta} \delta S^{\nu} = 0 \tag{85}$$

$$\delta V = \sum_{\nu=\alpha}^{\theta} \delta V^{\nu} = 0 \tag{86}$$

$$\delta n_i = \sum_{\nu=\alpha}^{\theta} \delta n_i^{\nu} = 0 \qquad \text{for} \quad i = 1, 2, \ldots, m \tag{87}$$

Using equation (4), the criterion of equilibrium (84) may be rewritten

$$\sum_{\nu=\alpha}^{\theta}\left(T^{\nu}\,\delta S^{\nu} - P^{\nu}\,\delta V^{\nu} + \sum_{i=1}^{m}\mu_i^{\nu}\,\delta n_i^{\nu}\right) \geq 0 \tag{88}$$

Thus, we shall now seek processes which obey condition (88) and satisfy the restrictions (85)–(87).

4.2.1. Pressure Conditions

A possible process may be characterized by the following variations in the entropy, volume, and composition of the phases:

$$\delta S^{\nu} = 0 \qquad \text{for} \quad \nu = \alpha, \beta, \ldots, \theta \tag{89}$$

$$\delta V^{\alpha} = -\delta V^{\beta}, \quad \delta V^{\nu} = 0 \qquad \text{for} \quad \nu = \gamma, \ldots, \theta \tag{90}$$

$$\delta n_i^{\nu} = 0 \qquad \text{for} \quad i = 1, 2, \ldots, m \quad \nu = \alpha, \beta, \ldots, \theta \tag{91}$$

Then, condition (88) becomes

$$-P^{\alpha}\,\delta V^{\alpha} - P^{\beta}\,\delta V^{\beta} \geq 0$$

or

$$(P^{\beta} - P^{\alpha})\delta V^{\alpha} \geq 0 \tag{92}$$

If there is no restriction on the change in volume of α or β, i.e., if δV^{α} can be either positive or negative, then at equilibrium we must have

$$P^{\alpha} = P^{\beta} \tag{93}$$

For if the pressures were not equal, it would be possible to choose a process with a δV^{α} of opposite sign to $P^{\beta} - P^{\alpha}$, which would violate condition (92) and thus be irreversible; consequently, the system would not be in equilibrium.

If, however, we were to assume that there are restrictions on the change of volume of the phase α which are such that δV^{α} is always positive (see Fig. I.4b), the equilibrium condition (92) would become

$$P^{\beta} \geq P^{\alpha} \tag{94}$$

In the case where the walls of the systems α and β are not deformable ($\delta V^{\alpha} = 0$), condition (92) is satisfied regardless of the values of the pressure P^{α} and P^{β}. Again the system is in equilibrium. It should be noted, however, that in both these examples where there are restrictions on the values of δV^{α}, the state of the system is not one of reversible equilibrium (see Section I.2).

The set of variational changes which led us to these results for the pressures of α and β can be readily modified to apply to the other phases of the total system. It is clear that if the walls of all the systems $\alpha, \beta, \ldots, \theta$ are totally deformable, the resulting equilibrium conditions are

$$P^{\alpha} = P^{\beta} = \cdots = P^{\theta} \tag{95}$$

The pressure is uniform throughout the entire heterogeneous system. But if certain walls are semideformable or nondeformable, the pressures of the corresponding phases can have different values, according to the preceding discussion.

4.2.2. Temperature Conditions

Another possible process may be characterized by the following variations:

$$\delta S^\alpha = -\delta S^\beta, \quad \delta S^\nu = 0 \qquad \text{for} \quad \nu = \gamma, \ldots, \theta \tag{96}$$

$$\delta V^\nu = 0 \qquad \text{for} \quad \nu = \alpha, \beta, \ldots, \theta \tag{97}$$

$$\delta n_i^\nu = 0 \qquad \text{for} \quad i = 1, 2, \ldots, m, \quad \nu = \alpha, \beta, \ldots, \theta \tag{98}$$

Applying the criterion of equilibrium (88) yields

$$T^\alpha \, \delta S^\alpha + T^\beta \, \delta S^\beta \geq 0$$

or

$$(T^\alpha - T^\beta) \, \delta S^\alpha \geq 0 \tag{99}$$

If there is no restriction on the flow of heat between α and β, δS^α may be either positive or negative, and condition (99) is satisfied only when

$$T^\alpha = T^\beta \tag{100}$$

If no heat flow is allowed between α and β (insulating walls), $\delta S^\alpha = 0$ and condition (99) is satisfied regardless of the values of T^α and T^β. This reasoning may be easily extended to the other phases of the system, and if there are no restrictions on the heat flow between the various phases, at equilibrium the temperature must be uniform throughout the entire heterogeneous system:

$$T^\alpha = T^\beta = \cdots = T^\theta \tag{101}$$

4.2.3. Conditions on the Chemical Potentials

We now consider the process characterized by the following equations:

$$\delta S^\nu = 0 \qquad \text{for} \quad \nu = \alpha, \beta, \ldots, \theta \tag{102}$$

$$\delta V^\nu = 0 \qquad \text{for} \quad \nu = \alpha, \beta, \ldots, \theta \tag{103}$$

$$\delta n_1^\alpha = -\delta n_1^\beta, \quad \delta n_1^\gamma = 0 \qquad \text{for} \quad \nu = \gamma, \ldots, \theta \tag{104}$$

$$\delta n_i^\nu = 0 \qquad \text{for} \quad i = 2, 3, \ldots, m, \quad \nu = \alpha, \beta, \ldots, \theta \tag{105}$$

and for which the criterion of equilibrium (88) becomes

$$\mu_1^\alpha \, \delta n_1^\alpha + \mu_1^\beta \, \delta n_1^\beta \geq 0$$

or

$$(\mu_1^\alpha - \mu_1^\beta) \, \delta n_1^\alpha \geq 0 \tag{106}$$

If there is no restriction on the transfer of species 1 between α and β, δn_1^α may be either positive or negative, and the equilibrium condition is that

$$\mu_1^\alpha = \mu_1^\beta \tag{107}$$

Note that if $\mu_1^\alpha > \mu_1^\beta$, the criterion of equilibrium is violated for $\delta n_1^\alpha < 0$; in other words, there will be a spontaneous (irreversible) transfer of species from the phase α where it has a high chemical potential, to the phase β where it has a lower chemical potential.

If no transfer of species 1 between the two phases is possible, then $\delta n_1^\alpha = 0$ and the criterion of equilibrium (106) is satisfied regardless of the values of μ_1^α and μ_1^β.

The result in equation (107) may be readily generalized to other species i and phases ν, and if there are no restrictions on the mass transfer of all these species among the various phases, the following equilibrium conditions are deduced:

$$\mu_i^\alpha = \mu_i^\beta = \cdots = \mu_i^\theta \qquad \text{for} \quad i = 1, 2, \ldots, m \tag{108}$$

To recapitulate, we have found that in the case where the walls of each phase are completely deformable and allow the transfer of both heat and mass, the selection of various simple processes impose the *necessary* equilibrium conditions

$$P^\alpha = P^\beta = \cdots = P^\theta = P \tag{109}$$

$$T^\alpha = T^\beta = \cdots = T^\theta = T \tag{110}$$

$$\mu_i^\alpha = \mu_i^\beta = \cdots = \mu_i^\theta = \mu_i \qquad \text{for} \quad i = 1, 2, \cdots, m \tag{111}$$

A question naturally arises: are these conditions also *sufficient?*

To demonstrate that they are, we note that given equations (109)–(111), the criterion of equilibrium (79) may be rewritten

$$(\delta E)_{S,V,n_i} = \left(\sum_{\nu=\alpha}^{\theta} T\,\delta S^\nu - P\,\delta V^\nu + \sum_{i=1}^{m} \mu_i\,\delta n_i^\nu \right)_{S,V,n_i} \geq 0 \tag{112}$$

or

$$\left(T \sum_{\nu=\alpha}^{\theta} \delta S^\nu - P \sum_{\nu=\alpha}^{\theta} \delta V^\nu + \sum_{i=1}^{m} \mu_i \sum_{\nu=\alpha}^{\theta} \delta n_i^\nu \right)_{S,V,n_i} \geq 0 \tag{113}$$

But since the total entropy, volume, and number of moles n_i must be kept constant, the summations on all the phases ν inside the brackets of condition (113) must be equal to zero [see equations (85)–(87)]. The left member of the inequality (113) is then equal to zero, and the criterion of equilibrium is indeed satisfied.

If there are restrictions on the interactions of the various phases in terms of entropy, volume, or mass, then all the conditions (109)–(111) are not necessary [e.g., some may be replaced by inequalities such as (94)], and it is not difficult to show that in this case the left member of the inequality (113) is positive instead of equal to zero.

5. Application of the Conditions for Equilibrium in a Heterogeneous System

5.1. Example

A quartz capsule containing two alloys, a binary Fe–C and a ternary Fe–C–V, is placed in a furnace at 1000°C (see Fig. 4). The atmosphere inside the capsule is a mixture of CO and CO_2 to which, at 1000°C, the quartz is practically nonreactive (Sections V.3 and V.4).

Carbon can be transferred from one alloy to the other through the gas phase by the reaction: CO_2(gas) + C(in alloy) = 2CO(gas). The partial pressures of iron

Figure 4. At equilibrium, the chemical potential of carbon in the Fe–C alloy is equal to that in the Fe–C–V alloy.

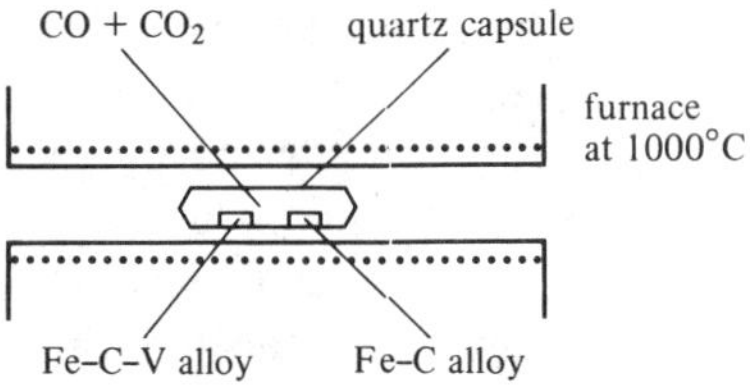

and vanadium are very low and since they do not form volatile compounds with the $CO–CO_2$ atmosphere, for all practical purposes there is no transfer of iron or vanadium from one alloy to the other. After a few hours at 1000°C, equilibrium with respect to carbon will have been achieved and the chemical potential of carbon in the binary alloy should be equal to that in the ternary alloy (although its concentration will remain different because of its interaction with vanadium). The chemical potentials of iron or vanadium, however, will not attain the same value in the two alloys.

Thus, given the restrictions of the experimental setup on the transfer of iron and vanadium, the system will have attained equilibrium. But it should be noted that these restrictions are time dependent. Indeed, if we were to wait a very long time, such as a few months, there would be a sufficient transfer of iron and vanadium (because of their vapor pressures) to also result in equal chemical potentials for each of these elements in the two alloys (the two alloys would then end up with the same composition). Thus, it may be preferable, in cases such as this, to speak of a *partial equilibrium* with respect to carbon. (The corresponding compositions of the alloys cannot be calculated before we reach Chapter IX.)

5.2. Common Tangent Construction

Let us assume that at a given temperature and pressure, two phases α and β (e.g., solid and liquid) of components 1 and 2 are interacting with no restriction on the transfers of mass. If the composition of the α phase is characterized by the mole fraction of 2, X_2^α, the chemical potentials of 1 and 2 can be graphically obtained by the method of intercepts (Section 3.2.1): the intercepts of the two axes by the

Figure 5. In case (**a**) the two phases α and β are not in equilibrium; they are in case (**b**).

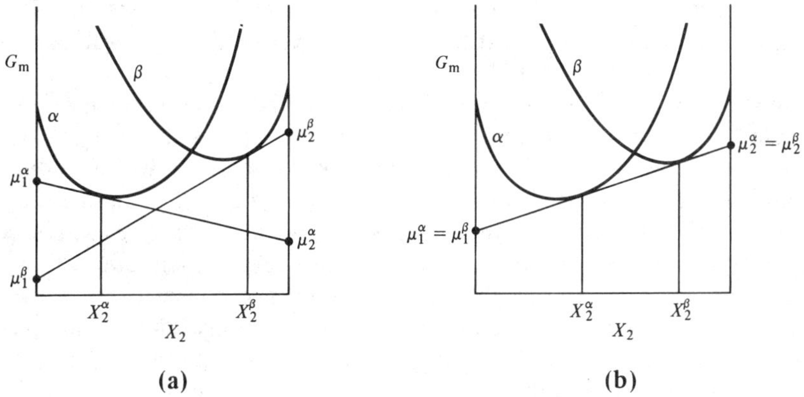

tangent to the Gibbs free energy curve of the α phase at the composition X_2^α represent μ_1^α and μ_2^α (see Fig. 5). A similar construction is possible for the phase β characterized by the composition X_2^β.

At equilibrium, we must have $\mu_1^\alpha = \mu_1^\beta$ and $\mu_2^\alpha = \mu_2^\beta$. It is apparent from Fig. 5 that the two curves G_m^α and G_m^β must then have a common tangent. This common tangent determines the composition of the two phases in equilibrium. This construction is very convenient and will be used extensively in our study of binary phase diagrams. It can also be extended to ternary systems (common tangent planes) and multicomponent ones (see Chapter X).

5.3. Phase Rule

The state of a phase is completely determined by its pressure, temperature, and composition. If it contains m components, then $m - 1$ variables must be specified for the composition, and 2 for P and T; this makes a total of $m + 1$ variables. In a system containing φ phases, there are $\varphi(m + 1)$ variables. However, if the system is in equilibrium, these variables are not independent: equations (109)–(111) for the temperatures, pressures, and chemical potentials are necessary if all the phases are free to interact (deformable walls and free transfer of heat and mass). Equations (109) and (110) on the equality of the temperatures and pressures yield $2(\varphi - 1)$ relations while the equality of the chemical potentials of each species for the φ phases yield $m(\varphi - 1)$ additional relations for a total $(m + 2)(\varphi - 1)$ relations. Consequently, there are only $\varphi(m + 1) - (m + 2)(\varphi - 1) = m + 2 - \varphi$ independent variables. This number of independent variables is called the *variance* or *degree of freedom* of the system; we shall designate it by the letter ϑ. Thus,

$$\vartheta = m + 2 - \varphi \tag{114}$$

For example, if the system contains only one component, $\vartheta = 3 - \varphi$. When only one phase is present, $\varphi = 1$ and $\vartheta = 2$; its pressure and temperature can then be chosen arbitrarily. If this component exists under two phases (e.g., solid and liquid) $\varphi = 2$, $\vartheta = 1$, and at any pressure there is only one temperature at which the two phases can coexist (e.g., to any given pressure corresponds only one temperature of fusion). The maximum number of coexisting phases is 3; they coexist at the *triple point*, which occurs at a determined temperature and pressure ($\vartheta = 0$).

In a binary system, if the pressure is fixed (e.g., $P = 1$ atm), then $\vartheta = 3 - \varphi$ and there will be at most three coexisting phases, this coexistence occurring at a single temperature (e.g., the eutectic or peritectic temperature). Additional examples may be found in Chapter X.

In the application of equation (114), the number of components m refer to the smallest number of constituents necessary to express the composition of each phase participating in the equilibrium. For example, if we consider the three phases CaO(s), CO_2(gas), and $CaCO_3$(s), the number of components is equal to 2: e.g., CaO and CO_2, since the composition of $CaCO_3$ may be expressed in terms of the concentrations of CaO and CO_2. (In this case, under equilibrium conditions and at a given temperature and pressure, only two phases can coexist.)

The phase rule is a very convenient tool to establish the soundness of a thermodynamic analysis [3]. Especially in complex situations, its application often pays substantial dividends.

Problems

1. Consider the following homogeneous function of degree 2:

$$z = (x + y)^2 e^{y/x} + xy$$

Calculate z'_x and z'_y and verify that they are homogeneous functions of degree 1. Calculate also the sum $xz'_x + yz'_y$ and check that it is equal to $2z$.

2. Consider the two following functions:

$$z_1 = \left(\frac{y - x}{y + x} \ln \frac{y}{x}\right) + \frac{y^2 - xy + x^2}{y^2 + x^2}$$

$$z_2 = \left(\frac{(y - x)^2}{y + x} \ln \frac{y}{x}\right) + \frac{y^2 - xy + x^2}{y + x}$$

Let $u = x/(x + y)$, $v = y/(x + y)$ and show that z_1 is a function of only one independent variable u but that z_2 is not. Would these results hold for $u = x/(ax + by)$, $v = y/(ax + by)$, where a and b are arbitrary constants?

3. **a.** Let Y be a homogeneous function of degree 1 with respect to the variables $n_1, n_2, \ldots, n_r$, and demonstrate that $Y_m = Y/n$ is independent of n at constant composition, i.e., at constant $X_1, X_2, \ldots, X_r$.
b. Assume that the functions Y_a, Y_b, and Y_c are homogeneous functions of degree 1 with respect to $n_1, n_2, \ldots, n_r$ and show that equation (51) is applicable to their sum.

4. Demonstrate that for a ternary solution

$$\mu_2 = G_m + (1 - X_2)\left(\frac{\partial G_m}{\partial X_2}\right)_{X_3/X_1}$$

Generalize this relation for a multicomponent solution.

5. The enthalpy of a given binary mixture may be expressed by the equation

$$H_m = X_1 H_1^\circ + X_2 H_2^\circ + (0.6 - 0.4\, X_2) X_1 X_2 \quad \text{kcal.}$$

Calculate dH_m/dX_2 and $\overline{H}_2$ at $X_2 = 0.5$.

6. The Gibbs free energy of the Al–Zn fcc phase at 1 atm pressure may be represented by the equation

$$G_m = X_{Al} G_{Al}^\circ + X_{Zn} G_{Zn}^\circ + RT(X_{Al} \ln X_{Al} + X_{Zn} \ln X_{Zn})$$
$$+ X_{Al} X_{Zn} (3150 X_{Al} + 2300 X_{Zn})(1 - T/4000) \quad \text{cal.}$$

where G_{Al}° and G_{Zn}° are the Gibbs free energies of pure Al and Zn (their numerical values may be left arbitrary).

Derive the analytical expressions of μ_{Zn}, $\overline{H}_{Zn}$, and $\overline{S}_{Zn}$ and plot these functions vs X_{Zn} at 600°K. Check your derivation of μ_{Zn} by the graphical method of intercepts. It is helpful to use the results of Problem 3b.

7. An iron–carbon alloy in the fcc structure is at a temperature of 1000°C and 1 atm. Its carbon content corresponds to $X_C = 0.05$. Calculate the change $D\mu_C$ in the chemical potential of carbon when the pressure is raised from 1 to 2000 atm at the same temperature and composition. It has been estimated that the molar volume of

such an alloy at 1000°C and 1 atm may be approximated by the equation $V_m = 7.32X_{Fe} + 3.73X_C + 0.15X_{Fe}X_C$ cm³/mol, and the coefficient of compressibility by $\beta = (1.09X_{Fe} + 1.92X_C)10^{-11}$ Pa^{-1}.

8. Let Y be a homogeneous function of degree 1 with respect to n_1, n_2, and n_3. Calculate $\overline{Y}_2$ at $X_2 = \frac{1}{2}$, $X_3 = \frac{1}{4}$ for $Y_m = \alpha X_1 X_2 X_3$.

9. The molar volume of a ternary solution may be expressed

$$V_m = 7X_A + 10X_B + 12X_C - 2X_AX_B + 3X_AX_BX_C \qquad \text{cm}^3/\text{mol}$$

Calculate the partial molar volumes $\overline{V}_A$, $\overline{V}_B$, $\overline{V}_C$ at $X_A = X_B = X_C = \frac{1}{3}$.

10. At 1560°C and $P = 1$ atm the chemical potentials of carbon and iron in the liquid binary phase may be expressed by the equations (see Section XVI.3.3)

$$\frac{\mu_{Fe} - G^\circ_{Fe}}{RT} = -\ln(1 + y_C) - 12 \ln \frac{1 + y_C e^{\Lambda}}{1 + y_C}$$

$$\frac{\mu_C - G^{\circ\,(gr)}_C}{RT} = -0.37 + \ln y_C + 24\Lambda$$

where G°_{Fe} and $G^{\circ\,(gr)}_C$ are the Gibbs free energies of pure liquid iron and carbon–graphite, y_C is equal to $X_C/(1 - 2\,X_C)$, and Λ is equal to

$$\Lambda = -\ln \tfrac{1}{2}[(1 + 1.15\, y_C + y_C^2)^{1/2} + (1 - y_C)]$$

Plot the Gibbs free energy of the liquid phase as a function of the mole fraction X_C. Since the Gibbs free energies are relative quantities, take $G^\circ_{Fe} = G^{\circ\,(gr)}_C = 0$. Deduce from that plot the solubility of graphite in liquid iron. Compare it to that given by the phase diagram (Fig. 19 of Chapter VIII).

11. **a.** Carbon–graphite, cementite (Fe_3C), and pure iron are placed in a furnace at 1000°C under 1 atm of argon. Apply the phase rule to the equilibrium of that system.

b. ZnS is reacted with pure oxygen to form $ZnSO_4$. What is the maximum number φ of phases coexisting at equilibrium at an arbitrary temperature and pressure? What is the value of φ if ZnS is reacted with air?

References

1. C. H. P. Lupis, *Acta Met.* **25,** 751–757 (1977).
2. C. H. P. Lupis, *Acta Met.* **26,** 211–215 (1978).
3. A. Findlay, *The Phase Rule and Its Applications*, revised by A. N. Campbell and N. O. Smith. Dover, New York, 1951.

Selected Bibliography

J. G. Kirkwood and I. Oppenheim, *Chemical Thermodynamics*. McGraw Hill, New York, 1961.

III Stability

1. Stable and Unstable Equilibria

We have previously derived a criterion for equilibrium but without examining the *stability* of this equilibrium. There are several kinds of stable or unstable equilibria and these are illustrated in Fig. 1 by a ball on rails of various shapes. Configurations 1a–c represent stable equilibria. In 1a the rail's curve presents no singularity; in 1b its slope has a discontinuity, and in 1c the ball cannot go to the right of the equilibrium position. These configurations are to be contrasted with the unstable equilibrium of 1d where a small displacement to the right or to the left removes the ball from its equilibrium position. In 1e the ball is unstable with respect to displacements toward the right, but stable with displacements toward the left (we overlook here forces due to acceleration). In 1f we have a *neutral* equilibrium and in 1g a *locally* stable equilibrium; i.e., the system is stable with respect to very small displacements but unstable with respect to large displacements. In the latter state, the system is also said to be *metastable*.

In Section 2 we shall be concerned with the study of equilibria which are stable with respect to all possible kinds of fluctuations but of infinitesimal values only. The study of stability with respect to large fluctuations is somewhat different because it is no longer restricted to the study of a local situation; we shall encounter it in our treatment of phase diagrams. In Section 3, stability criteria with respect to small composition fluctuations in binary solutions will be developed, and applied to the determination of the spinodal line and critical point in Section 4. In Section 5 a stability function ψ possessing certain convenient features will be defined. Its applications will be further developed in Chapter XI. Finally, in Section 6 we evaluate the free energy charges associated with the nucleation and growth of precipitates as illustrations of the principles developed in this chapter and as examples of calculations.

Figure 1. Equilibrium positions corresponding to various types of stability: **(a)–(c)** stable, **(d)** unstable, **(e)** stable for fluctuations to the left but unstable for fluctuations to the right, **(f)** neutral, **(g)** locally stable (metastable).

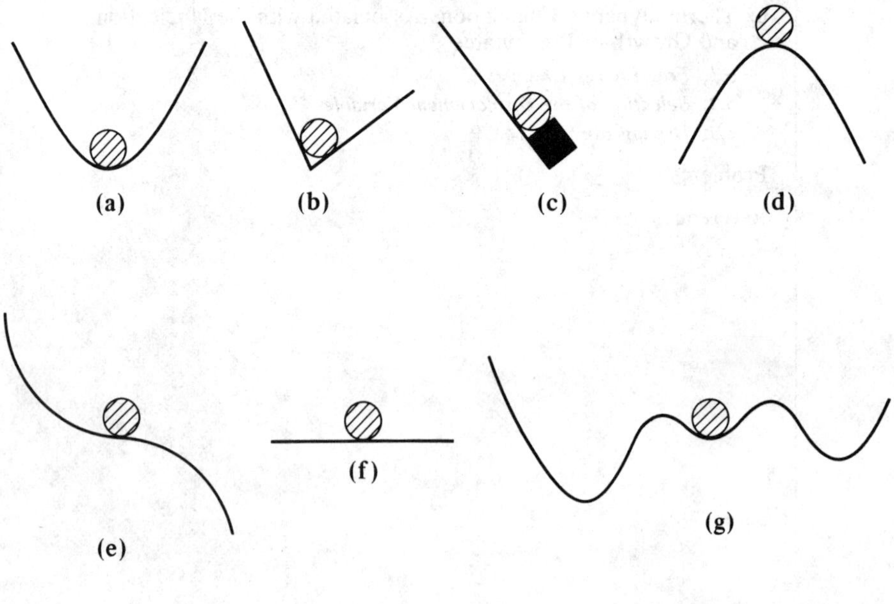

2. General Discussion of Stability Conditions with Respect to Infinitesimal Fluctuations

For a closed system at constant temperature and pressure the criterion of equilibrium is

$$(\delta G)_{P,T,n_i} \geq 0 \tag{1}$$

Pursuing the example of Fig. 1, if G is the gravitational potential of the ball and ξ measures the displacement from the equilibrium position, it may be seen that at any of the equilibrium positions of cases 1a,d–g $\delta G = (\partial G/\partial \xi)_{\xi=0}\, \delta\xi = 0$ because the tangent to the curve is horizontal. In case 1b the derivative has a singularity at the equilibrium point, but it may be seen that $\delta G > 0$ both to the right and to the left: to the right because both the slope of the tangent and ξ are positive, and to the left because they are both negative. In case 1c, δG is defined only to the left of the equilibrium position; there it is positive.

In general, the first derivative of G is sufficient to determine the points of equilibrium, but to examine the stability of these equilibria higher derivatives are needed.

Let DG represent the difference between the free energy of a state neighboring the equilibrium state and the free energy of that equilibrium state (characterized by $\xi = 0$). Regardless of the choice of the neighboring state,

$$\text{if} \quad DG > 0, \qquad \text{the equilibrium is stable} \tag{2a}$$

$$\text{if} \quad DG = 0, \qquad \text{the equilibrium is neutral} \tag{2b}$$

$$\text{if} \quad DG < 0, \qquad \text{the equilibrium is unstable} \tag{2c}$$

Let us now assume the existence of a Taylor series for the free energy at the equilibrium point; that is, we reject the cases where there is a discontinuity for the value of G or any of its derivatives at this point. Then

$$\begin{aligned} DG &= G - (G)_{\xi=0} \\ &= \left(\frac{\partial G}{\partial \xi}\right)_{\xi=0} \xi + \frac{1}{2}\left(\frac{\partial^2 G}{\partial \xi^2}\right)_{\xi=0} \xi^2 + \cdots + \frac{1}{n!}\left(\frac{\partial^n G}{\partial \xi^n}\right)_{\xi=0} \xi^n + \cdots \end{aligned} \tag{3}$$

which is of the form

$$DG = \delta G + \delta^2 G + \cdots + \delta^n G + \cdots \tag{4}$$

where δG is a variation of the first order, $\delta^2 G$ a variation of the second order, etc. As previously seen, at equilibrium $\delta G = 0$, and if $\delta^2 G \neq 0$, the sign of DG is the sign of $\delta^2 G$. Consequently, as ξ^2 is always positive, the equilibrium is stable if

$$(\partial^2 G/\partial \xi^2)_{\xi=0} > 0 \tag{5}$$

It is unstable if the second order derivative is negative.

If $(\partial^2 G/\partial \xi^2)_{\xi=0}$ is equal to zero, we have to examine higher order derivatives:

$$DG = \frac{1}{3!}\left(\frac{\partial^3 G}{\partial \xi^3}\right)_{\xi=0} \xi^3 + \frac{1}{4!}\left(\frac{\partial^4 G}{\partial \xi^4}\right)_{\xi=0} \xi^4 + \cdots \tag{6}$$

If $(\partial^3 G/\partial \xi^3)_{\xi=0}$ has a value which is different from zero, whether positive or

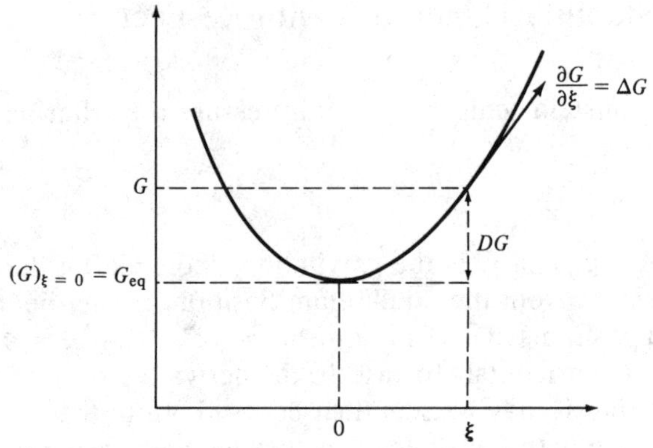

Figure 2. Gibbs free energy of a system undergoing a chemical reaction.

negative, it is possible to choose ξ (negative or positive) so as to make DG negative. Consequently, if $(\partial^3 G/\partial \xi^3)_{\xi=0}$ is different from zero, the equilibrium is unstable. If it is equal to zero, the equilibrium is stable if $(\partial^4 G/\partial \xi^4)_{\xi=0}$ is positive (since ξ^4 is always positive). If $(\partial^4 G/\partial \xi^4)_{\xi=0}$ is equal to zero, higher order derivatives again have to be considered (in an identical way).

It should be emphasized that ξ is any variable which affects the free energy of the system. For instance, in a system where a chemical reaction occurs, ξ may represent the progress variable of the reaction, $\partial G/\partial \xi$ will be recognized as the ΔG of the reaction, and at equilibrium we shall verify that $\Delta G = 0$ (see Fig. 2). Practically, the stability of this equilibrium is rarely studied, because if it has been attained spontaneously, it is necessarily stable.

Note that $\partial G/\partial \xi$ may be identified as a thermodynamic *driving force* toward an equilibrium state. Again, this is true whether ξ is the progress variable of a reaction, a geometric distance in a diffusion process, or any other variable measuring the displacement of a system from an equilibrium position.

The concepts discussed in this section will now be applied to an important case: that of stability with respect to infinitesimal composition fluctuations.

3. Stability Criteria for Infinitesimal Composition Fluctuations

Let us consider a homogeneous solution under conditions of constant temperature and pressure. Because of the movement of the atoms in the solution, there are local composition fluctuations of very small values. Through these fluctuations, the system may decompose into a mixture of "phases" characterized, at least initially, by compositions very close to that of the first homogeneous solution. If the Gibbs free energy of the mixture is lower than that of the homogeneous solution, then the latter is unstable and further decomposition may occur to produce states which have a still lower free energy. If the free energy of the homogeneous solution is lower than that of any mixture of phases, decomposition will not occur. Consequently, a necessary condition for the stability of a homogeneous solution is obtained by stating that the free energy of the initial state is lower than that of any mixture of states neighboring it. For simplicity, we shall first restrict the presentation to a binary system and consider the possibility of decomposing its homogeneous solution among only two phases. The generalization of the problem to any number of phases offers no difficulty.

3.1. First Method

The binary homogeneous solution consists of n mol, n_1 mol of component 1 and n_2 mol of component 2, corresponding to a concentration $X_2 = n_2/n$. The solution may decompose into two phases α and β of neighboring concentrations X_2^α and X_2^β, containing n^α and n^β mol characterized by the molar free energy values G_m^α and G_m^β. Let x be the relative proportion n^α/n of phase α in the mixture. We note that

$$X_2 = \frac{n_2}{n} = \frac{n_2^\alpha + n_2^\beta}{n} = \frac{n_2^\alpha}{n^\alpha}\frac{n^\alpha}{n} + \frac{n_2^\beta}{n^\beta}\frac{n^\beta}{n}$$

$$= X_2^\alpha x + X_2^\beta(1 - x) \tag{7a}$$

which also yields

$$x = \frac{X_2^\beta - X_2}{X_2^\beta - X_2^\alpha}, \qquad 1 - x = \frac{X_2 - X_2^\alpha}{X_2^\beta - X_2^\alpha} \tag{7b}$$

The molar free energy of the two-phase mixture is

$$G_m = \frac{n^\alpha}{n} G_m^\alpha + \frac{n^\beta}{n} G_m^\beta = xG_m^\alpha + (1 - x)G_m^\beta \tag{8a}$$

or

$$G_m = \frac{X_2^\beta - X_2}{X_2^\beta - X_2^\alpha} G_m^\alpha + \frac{X_2 - X_2^\alpha}{X_2^\beta - X_2^\alpha} G_m^\beta \tag{8b}$$

Equation (8b) is the equation of a straight line (see Fig. 3) passing by the two points A and B of coordinates (G_m^α, X_2^α) and (G_m^β, X_2^β). The free energy G_m of the mixture of overall composition X_2 is represented by the point M on this line.

We emphasize that the relative proportions of α and β, x and $1 - x$, are given by equation (7b); graphically, x is equal to QP/AP or MB/AB, and $1 - x$ is equal to AQ/AP or AM/AB. This result forms the basis of the *lever rule*.

In Fig. 4a it may be seen that for a free energy vs concentration curve which is concave upward, $\partial^2 G_m/\partial X_2^2 > 0$, the free energy of a two-phase mixture of neighboring compositions is always higher than the free energy of the homo-

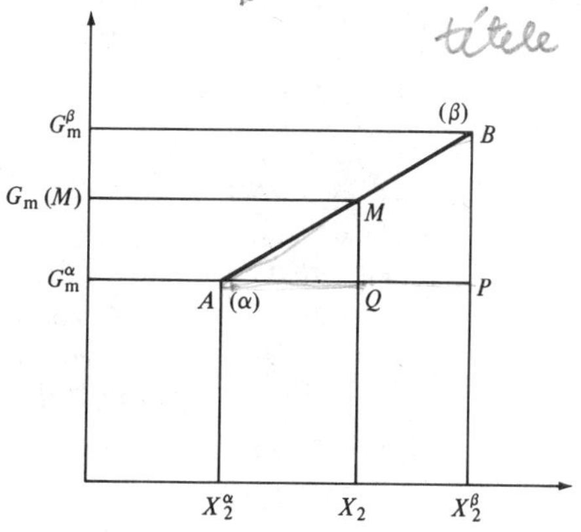

Figure 3. Graphical construction of the free energy of a mixture of phases α and β. The mixture contains QP/AP fraction of α and AQ/AP fraction of β and has the free energy $G_m(M)$.

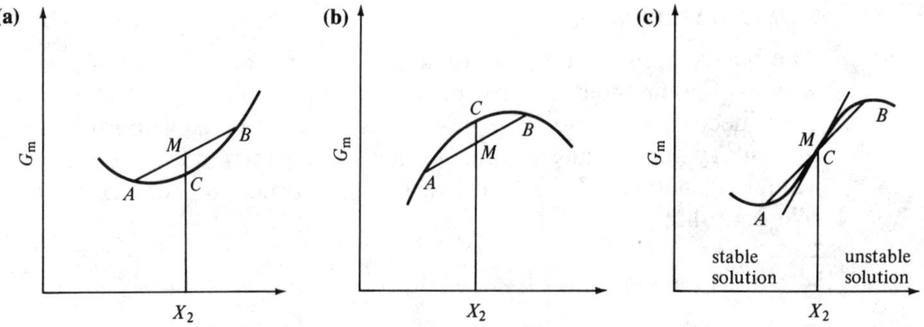

Figure 4. Criterion of stability for a homogeneous solution: the curvature of the free energy line must be positive. C is the representative point of the homogeneous solution, M that of a mixture of phases α and β.

(a) $\partial^2 G_m/\partial X_2^2 > 0$: stable solution.

(b) $\partial^2 G_m/\partial X_2^2 < 0$: unstable solution.

(c) $\partial^2 G_m/\partial X_2^2 = 0$: spinodal point.

geneous solution, since the point M on any chord joining two points A and B on each side of C is higher than C. Consequently,

$$\text{if} \quad \frac{\partial^2 G_m}{\partial X_2^2} > 0, \qquad \text{the solution is stable} \tag{9}$$

Similarly, if the curve is concave downward (Fig. 4b), M is lower than C. Thus,

$$\text{if} \quad \frac{\partial^2 G_m}{\partial X_2^2} < 0, \qquad \text{the solution is unstable} \tag{10}$$

The inflection point (Fig. 4c) at which $\partial^2 G_m/\partial X_2^2 = 0$ separates the regions of instability and stability (with respect to small fluctuations) and is called a *spinodal* point. The decomposition of a homogeneous solution resulting from infinitesimal fluctuations is also called *spinodal decomposition*.

The relation between the stability of the solution and the curvature of the free energy line may be established by an alternative method. Although more difficult, it is definitely more powerful and can easily be applied to the case of a multi-component system with any number of phases as well as to various other problems. It is also more in line with our discussion in Section 2.

3.2. Second Method

Again we consider the homogeneous solution of n_1 mol of component 1 and n_2 mol of component 2, which may decompose into the two phases α and β. Phase α would contain $\lambda_\alpha(n_1 + dn_1^\alpha)$ mol of 1 and $\lambda_\alpha(n_2 + dn_2^\alpha)$ mol of 2, and phase β would contain $\lambda_\beta(n_1 + dn_1^\beta)$ and $\lambda_\beta(n_2 + dn_2^\beta)$ mol of 1 and 2, respectively. Since the total number of moles of 1 or 2 remains constant, an elementary mass balance yields

$$\lambda_\alpha\, dn_1^\alpha + \lambda_\beta\, dn_1^\beta = 0 \tag{11a}$$

$$\lambda_\alpha\, dn_2^\alpha + \lambda_\beta\, dn_2^\beta = 0 \tag{11b}$$

and

$$\lambda_\alpha + \lambda_\beta = 1 \tag{12}$$

As the phases α and β are attained through continuous changes of the system (at least on the macroscopic level), it is valid to assume that the free energy is a single continuous function of the number of moles of each species. Thus, if the free energy of the homogeneous solution is

$$G = G(n_1, n_2) \tag{13}$$

those of the phases α and β are

$$G^\alpha = G[\lambda_\alpha(n_1 + dn_1^\alpha), \lambda_\alpha(n_2 + dn_2^\alpha)] \tag{14a}$$

$$G^\beta = G[\lambda_\beta(n_1 + dn_1^\beta), \lambda_\beta(n_2 + dn_2^\beta)] \tag{14b}$$

But since the Gibbs free energy is an extensive property (see Section II.2.2), equation (14) may be rewritten

$$G^\alpha = \lambda_\alpha G(n_1 + dn_1^\alpha, n_2 + dn_2^\alpha) \tag{15a}$$

$$G^\beta = \lambda_\beta G(n_1 + dn_1^\beta, n_2 + dn_2^\beta) \tag{15b}$$

To analyze the stability of the homogeneous solution, we must study the difference DG between the free energy of the mixture and that of the homogeneous solution (see Section 2):

$$DG = \lambda_\alpha G(n_1 + dn_1^\alpha, n_2 + dn_2^\alpha) + \lambda_\beta G(n_1 + dn_1^\beta, n_2 + dn_2^\beta) - G(n_1, n_2) \tag{16}$$

The dn_is measure the displacement from an equilibrium position (stable or unstable) and play the same role as ξ. By expanding into a Taylor series the first and second term on the right-hand member of equation (16), DG is expanded into a Taylor series with respect to the dn_is. For example, we obtain for the first term

$$\begin{aligned} G(n_1 + dn_1^\alpha, n_2 + dn_2^\alpha) = {} & G(n_1, n_2) + \left(\frac{\partial G}{\partial n_1}\right)_{n_2} dn_1^\alpha + \left(\frac{\partial G}{\partial n_2}\right)_{n_1} dn_2^\alpha \\ & + \frac{1}{2}\left(\frac{\partial^2 G}{\partial n_1^2}\right)(dn_1^\alpha)^2 + \frac{1}{2}\left(\frac{\partial^2 G}{\partial n_2^2}\right)(dn_2^\alpha)^2 + \left(\frac{\partial^2 G}{\partial n_1 \partial n_2}\right) dn_1^\alpha\, dn_2^\alpha \\ & + \text{higher order terms} \end{aligned} \tag{17}$$

or, in a more compact form,

$$\begin{aligned} G(n_1 + dn_1^\alpha, n_2 + dn_2^\alpha) = {} & G(n_1, n_2) + \sum_{i=1}^{2} \mu_i\, dn_i^\alpha + \tfrac{1}{2} \sum_{i,j=1}^{2} G_{ij}\, dn_i^\alpha\, dn_j^\alpha \\ & + \text{higher order terms} \end{aligned} \tag{18}$$

where μ_i is the chemical potential of i, $(\partial G/\partial n_i)_{n_j}$, and G_{ij} is the second derivative of G with respect to n_i and n_j. Consequently, equation (16) becomes

$$\begin{aligned} DG = {} & [(\lambda_\alpha + \lambda_\beta)G(n_1, n_2) - G(n_1, n_2)] + \left[\lambda_\alpha \sum_{i=1}^{2} \mu_i\, dn_i^\alpha + \lambda_\beta \sum_{i=1}^{2} \mu_i\, dn_i^\beta\right] \\ & + \left[\tfrac{1}{2}\lambda_\alpha \sum_{i,j=1}^{2} G_{ij}\, dn_i^\alpha\, dn_j^\alpha + \tfrac{1}{2}\lambda_\beta \sum_{i,j=1}^{2} G_{ij}\, dn_i^\beta\, dn_j^\beta\right] \\ & + \text{higher order terms} \end{aligned} \tag{19}$$

The first bracket is equal to zero because of equation (12). The second bracket is also equal to zero:

$$\lambda_\alpha \sum_{i=1}^{2} \mu_i \, dn_i^\alpha + \lambda_\beta \sum_{i=1}^{2} \mu_i \, dn_i^\beta = \sum_{i=1}^{2} \mu_i(\lambda_\alpha \, dn_i^\alpha + \lambda_\beta \, dn_i^\beta) = 0$$

because of equation (11). Thus, as expected DG is of the second order with respect to the dn_i variables and of the form

$$DG = \delta^2 G + \delta^3 G + \cdots \tag{20}$$

where

$$\delta^2 G = \tfrac{1}{2}\lambda_\alpha \sum_{i,j=1}^{2} G_{ij} \, dn_i^\alpha \, dn_j^\alpha + \tfrac{1}{2}\lambda_\beta \sum_{i,j=1}^{2} G_{ij} \, dn_i^\beta \, dn_j^\beta \tag{21}$$

As discussed in Section 2, the sign of $\delta^2 G$ yields the sign of DG and the criterion of stability for the homogeneous solution is

$$\delta^2 G \geq 0 \tag{22}$$

The equality sign corresponds to the spinodal point.

Using relations (11), equation (21) may be transformed into

$$\delta^2 G = \tfrac{1}{2}\lambda_\alpha(1 + \lambda_\alpha/\lambda_\beta) \sum_{i,j=1}^{2} G_{ij} \, dn_i^\alpha \, dn_j^\alpha \tag{23}$$

and since the λs are positive numbers, condition (22) becomes

$$\sum_{i,j=1}^{2} G_{ij} \, dn_i^\alpha \, dn_j^\alpha \geq 0 \tag{24}$$

It can be rewritten in matrix form

$$(dn_1^\alpha \quad dn_2^\alpha) \begin{pmatrix} G_{11} & G_{12} \\ G_{21} & G_{22} \end{pmatrix} \begin{pmatrix} dn_1^\alpha \\ dn_2^\alpha \end{pmatrix} \geq 0 \tag{25}$$

However, since dn_1^α and dn_2^α are arbitrary, condition (25) expresses that the quadratic form associated to the symmetric matrix G_{ij} (symmetric because $G_{ij} = G_{ji}$) must always be positive or null, regardless of the values of the dummy variables dn_i^α.

There are several equivalent necessary and sufficient conditions which express that a quadratic form is invariably positive or null. One of them is to write that the eigenvalues e of the G_{ij} matrix are all positive or null, or that the coefficients of the characteristic equation are alternately positive and negative. The characteristic equation is

$$e^2 - e(G_{11} + G_{22}) + (G_{11}G_{22} - G_{12}^2) = 0 \tag{26}$$

The last parenthesis, equal to the determinant of the G_{ij} matrix, is null because of the Gibbs–Duhem relation:

$$n_1 \, d\mu_1 + n_2 \, d\mu_2 = 0 \tag{27}$$

Indeed, dividing by dn_1 and by dn_2, we obtain, respectively,

$$n_1 G_{11} + n_2 G_{21} = 0 \tag{28}$$

$$n_1 G_{12} + n_2 G_{22} = 0 \tag{29}$$

and since these two equations must be compatible for values of n_1 and n_2 different from zero,

$$G_{11}/G_{12} = G_{21}/G_{22}$$

or

$$G_{11}G_{22} = G_{12}^2 \tag{30}$$

From equation (26) we deduce, then, that one eigenvalue is zero and the second is

$$e = G_{11} + G_{22} \tag{31}$$

Since it must be positive or null,

$$G_{11} + G_{22} \geq 0 \tag{32}$$

However, because of equation (30) G_{11} and G_{22} must have the same sign. Consequently, an equivalent condition is

$$G_{22} = \partial^2 G/\partial n_2^2 \geq 0 \tag{33}$$

The second derivative of the free energy with respect to the number of moles of 2 (G_{22}) may be easily related to the second derivative of the molar free energy with respect to the mole fraction of 2. The result is

$$G_{22} = \frac{X_1^2}{n}\frac{\partial^2 G_m}{\partial X_2^2} \tag{34}$$

These derivatives are of the same sign. Consequently, our stability criterion (33) for infinitesimal composition fluctuations becomes

$$\partial^2 G_m/\partial X_2^2 \geq 0 \tag{35}$$

It is now identical to that found by the first method. It is also possible to show that

$$G_{22} = \frac{X_1}{n}\frac{\partial \mu_2}{\partial X_2} \tag{36}$$

and it thus follows that, for a stable solution, the chemical potential of a species always increases with its concentration.

There is another equivalent, but less familiar, way of expressing that the quadratic form associated to the G_{ij} matrix is invariably positive or null. Hancock [1] demonstrated that a necessary and sufficient condition is that each of the following "diagonal" determinants be always positive or null:

$$G_{11} \geq 0 \tag{37}$$

$$\begin{vmatrix} G_{11} & G_{12} \\ G_{21} & G_{22} \end{vmatrix} \geq 0 \tag{38}$$

The second condition is trivial; the determinant associated to the G_{ij} matrix is null because of the Gibbs–Duhem equation [see equation (30)]. Thus the only remaining condition is $G_{11} \geq 0$, which, as we have seen, is equivalent to $G_{22} \geq 0$.

Presenting three different techniques to demonstrate the same result may appear superfluous, but is quite useful. The first technique is the most intuitive, but it is difficult to extend to a multicomponent system. The second (quadratic form and eigenvalues) is easier to generalize but leads to cumbersome expressions. The third (quadratic form and diagonal determinants) is closely related to the second but uses a theorem which requires a more intimate familiarity with the algebra of determinants; it is also the easiest to apply in the case of a multicomponent system.

4. Spinodal Line and Critical Point

A spinodal point is a point at which the second derivative of the Gibbs free energy with respect to composition is equal to zero. At this point the quadratic form $\delta^2 G$ is also equal to zero and

$$DG = \frac{1}{3!}\left(\frac{\partial^3 G}{\partial \xi^3}\right)_{\xi=0} \xi^3 + \frac{1}{4!}\left(\frac{\partial^4 G}{\partial \xi^4}\right)_{\xi=0} \xi^4 + O(\xi^5) \tag{39}$$

If the third order derivative is not zero, we see that DG may be either positive or negative according to the value of ξ; e.g., if $(\partial^3 G/\partial \xi^3)_{\xi=0}$ is negative, then $DG<0$ if $\xi>0$, and $DG>0$ if $\xi<0$. Since the small composition fluctuations to which a system is subject can be of any sign, there are fluctuations for which $DG<0$ and the system is unstable; that is, it will decompose.

We shall see that at each end of the composition range ($X_1 \to 0$, $X_2 \to 0$), the tangent $\partial G_m/\partial X_2$ is $-\infty$ and the curvature is necessarily positive. Thus, if there is one spinodal point (separating a region of positive curvature from a region of negative curvature), there must necessarily be a second one. The number of spinodal points is always even at a given temperature.

Let us consider, at a temperature T, a system which presents a phase α unstable with respect to small fluctuations in the composition range PQ (the spinodal points) and unstable with respect to large fluctuations in the range AB (see Fig. 5). This instability is usually due to a positive enthalpy of mixing H. The free

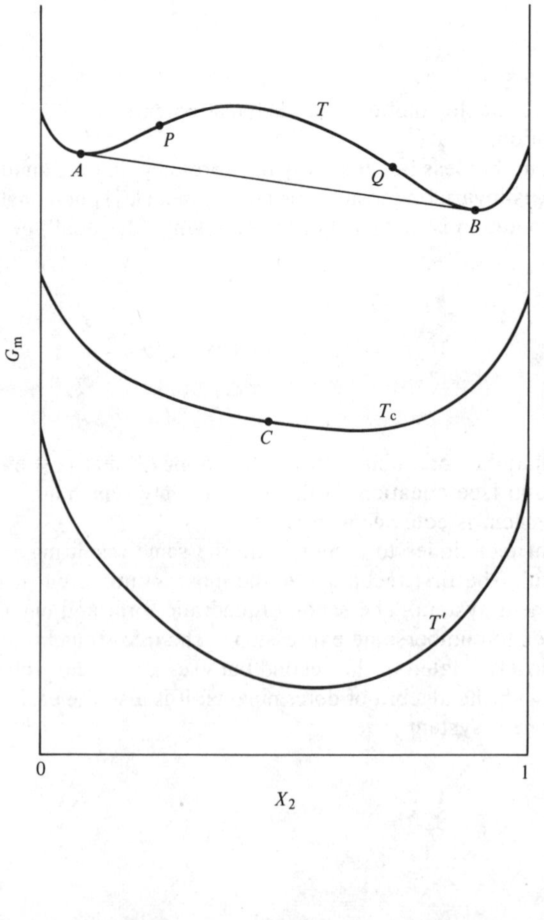

Figure 5. Typical temperature dependence of the free energy of a system presenting a miscibility gap. ($T<T_C<T'$.)

energy of the phase is made of two contributions, the enthalpy H and the entropy contribution $-TS$. H plays the role of a segregation force whereas the entropy plays the opposing role of a "randomizing" force. At a low temperature T, the term TS is less important than H and there is a region of instability where segregation occurs. At a high temperature T', the term $T'S$ is sufficiently large to offset the effect of H and the phase is stable. As the temperature is raised from T to T', the region of instability shrinks until it vanishes at a temperature which is called the *critical temperature* T_C (see Fig. 5). At this temperature, all four points A, B, P, and Q have merged into a single critical point C. We shall demonstrate that at this point $\partial^3 G_m/\partial X_2^3$, the third order derivative of G_m with respect to X_2, is necessarily equal to zero.

This may be seen as follows. Let us imagine a straight line intersecting a curve at the points O and M. If we let the point M move toward O the line changes its position until ultimately it becomes tangent to the curve when M has merged with O. Thus, an ordinary tangent to a curve intersects this curve at two points which are infinitely close to each other. In the same way, it may be seen that at an inflexion point, the tangent intersects the curve at three points infinitely close to each other (contact of the third order). The tangent at the critical point may be viewed as the ultimate position of the common tangent AB which was intersecting the curve at four points: two points at A and two points at B. When A and B merge into C, this contact point is then of the fourth order (it counts for four). Now, if $y = f(x)$ represents the equation of the curve and $y = ax + b$ the equation of the line, the roots of the equation

$$z = f(x) - ax - b = 0$$

yield the abscissa x_0 of the intersection points. If two of these roots have the same value (contact of order 2)

$$\left(\frac{\partial z}{\partial x}\right)_{x=x_0} = 0 \quad \text{or} \quad \left(\frac{\partial f}{\partial x}\right)_{x=x_0} = a \tag{40}$$

and we find the well-known result that the slope of the line is equal to the derivative of f at this point. If the root is of the third order,

$$\left(\frac{\partial^2 z}{\partial x^2}\right)_{x=x_0} = 0 \quad \text{or} \quad \left(\frac{\partial^2 f}{\partial x^2}\right)_{x=x_0} = 0 \tag{41}$$

Thus, at an inflexion point the second derivative is equal to zero (and at the spinodal point $\partial^2 G_m/\partial X_2^2 = 0$) and equation (40) remains also valid. Finally, if the root is of the fourth order, we must have in addition to equations (40) and (41)

$$\left(\frac{\partial^3 z}{\partial x^3}\right)_{x=x_0} = 0 \quad \text{or} \quad \left(\frac{\partial^3 f}{\partial x^3}\right)_{x=x_0} = 0 \tag{42}$$

Consequently, at the critical point,

$$\left(\frac{\partial^2 G_m}{\partial X_2^2}\right)_{X_2 = X_2^C} = 0 \tag{43}$$

and

$$\left(\frac{\partial^3 G_m}{\partial X_2^3}\right)_{X_2 = X_2^C} = 0 \tag{44}$$

[The equation corresponding to (40) is of little interest.] For completeness, one should add the inequality

$$\left(\frac{\partial^4 G_m}{\partial X_2^4}\right)_{X_2 = X_2^C} \geq 0 \tag{45}$$

4.1. Case of a Regular Solution

The Gibbs free energy of a *regular* solution may be described by the equation

$$G_m = X_1 G_1^\circ + X_2 G_2^\circ + RT(X_1 \ln X_1 + X_2 \ln X_2) + \Omega X_1 X_2 \tag{46}$$

where G_1° and G_2° are the free energies of the pure components 1 and 2 and Ω is a parameter characterizing the interaction of the atoms 1 and 2 in the solution. Differentiating equation (46) with respect to X_2 yields

$$\frac{\partial G_m}{\partial X_2} = (G_2^\circ - G_1^\circ) + RT(-\ln X_1 + \ln X_2) + \Omega(1 - 2X_2) \tag{47}$$

$$\frac{\partial^2 G_m}{\partial X_2^2} = RT\left(\frac{1}{X_1} + \frac{1}{X_2}\right) - 2\Omega \tag{48}$$

$$\frac{\partial^3 G_m}{\partial X_2^3} = RT\left(\frac{1}{X_1^2} - \frac{1}{X_2^2}\right) \tag{49}$$

The spinodal line is determined by equation (43) or

$$X_1 X_2 = RT/2\Omega \tag{50}$$

This is the line of a symmetrical parabola (see Fig. 6). The critical point is determined by equations (43) and (44), which yield

$$X_1 = X_2 = 0.5, \qquad T_C = \Omega/2R \tag{51}$$

Figure 6. Example of a miscibility gap and a spinodal line in the case of a regular solution (Ω = 4000 cal).

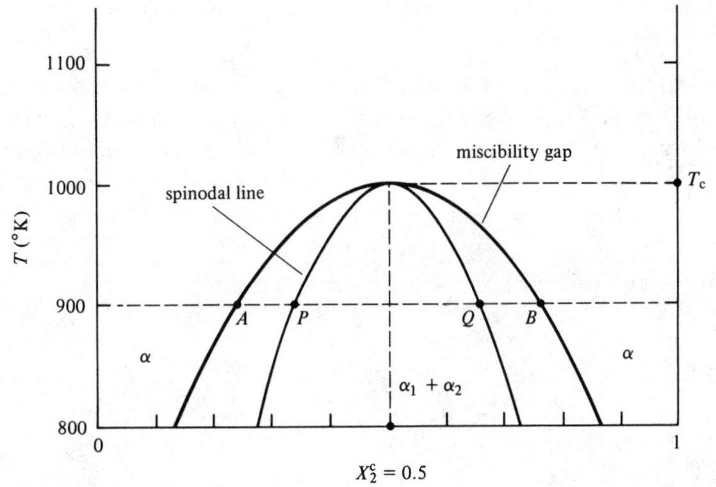

It is easy to show that the equation of the miscibility gap is

$$\ln \frac{X_2}{X_1} = \frac{\Omega}{RT}(X_2 - X_1) \tag{52a}$$

or

$$\ln \frac{X_2}{X_1} = \frac{2T_C}{T}(X_2 - X_1) \tag{52b}$$

It is also symmetric with respect to the axis $X_2 = 0.5$.

Let AB represent the width of the miscibility gap (Fig. 6) and PQ the width of the spinodal domain. It is possible to demonstrate (see Problem 2) that the limit of the ratio AB/PQ for $T \to T_C$ is equal to $\sqrt{3}$. Consequently, in the vicinity of the critical temperature,

$$AB \simeq \sqrt{3}\, PQ \tag{53}$$

4.2. General Case

The study of the miscibility gap and spinodal domains may be generalized without consideration of a particular type of solution such as the regular solution above.

Let us designate by X_2^α and X_2^β the two compositions of the spinodal at a temperature T. It will be advantageous to enter the following variables:

$$\Delta X^\alpha = X_2^\alpha - X_2^C, \qquad \Delta X^\beta = X_2^\beta - X_2^C \tag{54}$$

$$x = X_2^\beta - X_2^\alpha = \Delta X^\beta - \Delta X^\alpha \tag{55}$$

$$y = \tfrac{1}{2}(X_2^\alpha + X_2^\beta) - X_2^C = \tfrac{1}{2}(\Delta X^\alpha + \Delta X^\beta) \tag{56}$$

The geometrical significance of these variables is illustrated in Fig. 7. x is the width of the spinodal domain at a given temperature and y a measure of its asymmetry.

Figure 7. Geometrical significance of the variables x and y chosen to study the shape of the spinodal line, or of the miscibility gap, near the critical point.

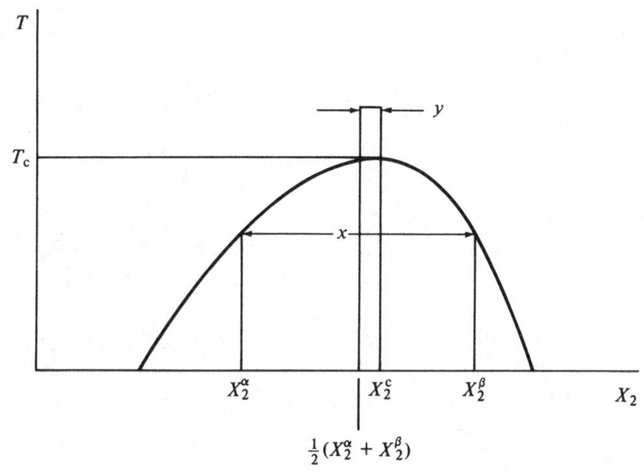

We now assume that the Gibbs free energy of the solution can be developed in a Taylor series at the critical point X_2^C, T_C:

$$G_m(X_2, T) = G_m(X_2^C, T_C) + G_{1,0}\Delta X + G_{0,1}\Delta T$$
$$+ \tfrac{1}{2}[G_{2,0}(\Delta X)^2 + 2G_{1,1}\Delta X\Delta T + G_{0,2}(\Delta T)^2] + \cdots$$
$$+ \frac{1}{n!}\Bigg[G_{n,0}(\Delta X)^n + \cdots + \frac{n!}{(n-i)!i!}G_{n-i,i}(\Delta X)^{n-i}(\Delta T)^i$$
$$+ \cdots + G_{0,n}(\Delta T)^n\Bigg] + \cdots \tag{57}$$

where

$$\Delta X = X_2 - X_2^C, \qquad \Delta T = T - T_C \tag{58}$$

and

$$G_{i,j} = \left(\frac{\partial^{i+j}G_m}{\partial X_2^i \partial T^j}\right)_{X_2^C, T_C} \tag{59}$$

Since the derivatives $G_{i,j}$ are calculated at the critical point, we note that $G_{2,0} = G_{3,0} = 0$ and $G_{4,0} \geq 0$.

Neglecting terms of the fifth order and above in the expansion series, a composition X_2^α of the spinodal line is obtained by writing that at that point

$$\frac{\partial^2 G_m(X_2^\alpha, T)}{\partial X_2^2} = 0 \tag{60}$$

or, through equation (57),

$$G_{2,1}\Delta T + \tfrac{1}{2}G_{4,0}(\Delta X^\alpha)^2 + G_{3,1}\Delta X^\alpha\,\Delta T + \tfrac{1}{2}G_{2,2}(\Delta T)^2 = 0 \tag{61}$$

At the same temperature T the other composition X_2^β of the spinodal is similarly obtained:

$$G_{2,1}\Delta T + \tfrac{1}{2}G_{4,0}(\Delta X^\beta)^2 + G_{3,1}\Delta X^\beta\,\Delta T + \tfrac{1}{2}G_{2,2}(\Delta T)^2 = 0 \tag{62}$$

Substracting equation (61) from equation (62) yields

$$G_{3,1}\Delta T(\Delta X^\beta - \Delta X^\alpha) + \tfrac{1}{2}G_{4,0}[(\Delta X^\beta)^2 - (\Delta X^\alpha)^2] = 0 \tag{63}$$

or, through equation (56),

$$y = -\frac{G_{3,1}\Delta T}{G_{4,0}} = -\frac{(\partial^3 S_m/\partial X_2^3)_{X_2^C, T_C}}{(\partial^4 G_m/\partial X_2^4)_{X_2^C, T_C}}(T_C - T) \tag{64}$$

An analysis of this expression will be given later.

Adding equations (61) and (62) and rearranging the terms yield

$$x^2 = 4y^2 - \frac{8\Delta T(G_{2,1} + \tfrac{1}{2}G_{2,2}\Delta T)}{G_{4,0}} \tag{65}$$

(Note that $\Delta T = T - T_C$ is, in general, negative for any point of the spinodal.)

At temperatures close to the critical temperature, the terms in $(\Delta T)^2$ in equation (65) are negligible when compared to the term in ΔT, and thus x^2 is then proportional to ΔT, or x is then proportional to $(\Delta T)^{1/2}$.

To compare the shape of the spinodal line to that of the miscibility gap, we must find the compositions X_2^α and X_2^β of the points on the Gibbs free energy curve which, at a given temperature, have a common tangent. This may be done by writing

$$\mu_1^\alpha = \mu_1^\beta \tag{66}$$

$$\mu_2^\alpha = \mu_2^\beta \tag{67}$$

Recalling that

$$\mu_i = G_m + (1 - X_i)\frac{\partial G_m}{\partial X_i} \qquad (i = 1 \text{ or } 2) \tag{68}$$

and using for G_m the Taylor series of equation (57) where we neglect terms of the fifth order and higher, equation (66) becomes

$$(y - X_1^C)\,\Delta T\,(G_{2,1} + \tfrac{1}{2}G_{2,2}\Delta T) - G_{3,1}\Delta T[X_1^C\,y - \tfrac{1}{12}(12y^2 + x^2)]$$
$$- \tfrac{1}{24}G_{4,0}[X_1^C\,(12y^2 + x^2) - 3y(4y^2 + x^2)] = 0 \tag{69}$$

where x and y now refer to the characteristics of the miscibility gap line. Similarly, equation (67) becomes

$$(y + X_2^C)\Delta T(G_{2,1} + \tfrac{1}{2}G_{2,2}\Delta T) + G_{3,1}\Delta T[X_2^C\,y + \tfrac{1}{12}(12y^2 + x^2)]$$
$$+ \tfrac{1}{24}G_{4,0}[X_2^C\,(12y^2 + x^2) + 3y(4y^2 + x^2)] = 0 \tag{70}$$

Subtracting equation (69) from (70) yields after rearrangement

$$x^2 = -\frac{24\,\Delta T}{G_{4,0}}[(G_{2,1} + \tfrac{1}{2}G_{2,2}\Delta T) + G_{3,1}y] - 12y^2 \tag{71}$$

Adding equation (69) to (70) yields

$$(2y + 2X_2^C - 1)\Delta T(G_{2,1} + \tfrac{1}{2}G_{2,2}\Delta T)$$
$$+ G_{3,1}\Delta T[y(2X_2^C - 1) + \tfrac{1}{6}\,(12y^2 + x^2)]$$
$$+ \tfrac{1}{24}G_{4,0}[(12y^2 + x^2)\,(2X_2^C - 1) + 6y(4y^2 + x^2)] = 0 \tag{72}$$

Replacing x^2 in equation (72) by its expression in equation (71) yields after elimination of second order terms (i.e., near the critical point)

$$y = -G_{3,1}\Delta T/G_{4,0} \tag{73}$$

which is identical to equation (64). The asymmetry of the spinodal line is then identical to that of the miscibility gap line near the critical point, or, in other words, at any temperature near the critical temperature the center of the diameter of the spinodal line coincides with the center of the diameter of the miscibility gap. Combining equations (71) and (73) yields

$$x^2 = 12y^2 - 24\Delta T(G_{2,1} + \tfrac{1}{2}G_{2,2}\Delta T)/G_{4,0} \tag{74}$$

Comparing equations (74) and (65) shows that

$$x^2_{\text{misc. gap}} = 3x^2_{\text{spinodal}}$$

or

$$x_{\text{misc. gap}} = \sqrt{3}x_{\text{spinodal}} \tag{75}$$

Thus, the square root of three rule that was found for the regular solution model of equation (53) is a much more general result.

As was the case for the spinodal line, we also note in equation (74) that near the critical point $x_{\text{misc. gap}}$ will be proportional to the square root of $T_C - T$. This is often referred to as the *parabolic rate*.

Empirically, however, for systems where the data are sufficiently accurate to provide for a good analysis of the dependence of x on $T_C - T$ near the critical point, it turns out that x is proportional to $(T_C - T)^n$ where n is close to $\frac{1}{3}$ and not $\frac{1}{2}$. Geometrically, this means that the miscibility gap and spinodal lines will appear somewhat flatter near their apex than predicted through the preceding calculations. Obviously, these results invalidate the assumption of the existence of a Taylor series for the Gibbs free energy near the critical point. A theoretical analysis based on the Ising model [2] also indicates that the unstable region in the miscibility gap is larger than the ratio of $1/\sqrt{3}$ predicted by equation (75) and is equal to 1/1.18 [3]. For further reading on the general subject of critical phenomena, Ref. [4] is recommended.

We note that since in most cases of metallurgical interest, the parabolic rate and $\sqrt{3}$ rule predict results which are within the scatter of the experimental results [5], the analyticity of the Gibbs free energy function near critical points will be accepted here as a matter of convenience.

We now turn to the expression of y in equation (64) to note that since at the critical point $\partial^3 G_m/\partial X_2^3 = 0$, it also follows that

$$\left(\frac{\partial^3 H_m}{\partial X_2^3}\right)_{X_2^C, T_C} = T_C \left(\frac{\partial^3 S_m}{\partial X_2^3}\right)_{X_2^C, T_C} \tag{76}$$

Consequently, equation (64) may be rewritten

$$y = -\frac{T_C - T}{T_C}\frac{(\partial^3 H_m/\partial X_2^3)_{X_2^C, T_C}}{(\partial^4 G_m/\partial X_2^4)_{X_2^C, T_C}} \tag{77}$$

Thus, for solutions the enthalpy of which can be expressed as a polynomial of the second order only (e.g., *regular* or *quasi-regular* solutions) $y = 0$. It may also be noted that y is of the sign opposite to $(\partial^3 H_m/\partial X_2^3)_{X_2^C, T_C}$ since $(\partial^4 G_m/\partial X_2^4)_{X_2^C, T_C}$ must be positive.

5. Stability Function ψ

In 1971 Lupis and Gaye [6] defined a stability function ψ to study the thermodynamic properties of multicomponent solutions (see Chapter XI). In the case of a binary solution, the defining equation of ψ reduces to

$$\psi = X_1 X_2\, d^2(G_m/RT)/dX_2^2 \tag{78}$$

For $\psi > 0$ the solution is stable with respect to small composition fluctations, and for $\psi < 0$ it is unstable (see Fig. 8). At the spinodal compositions $\psi = 0$. Equation

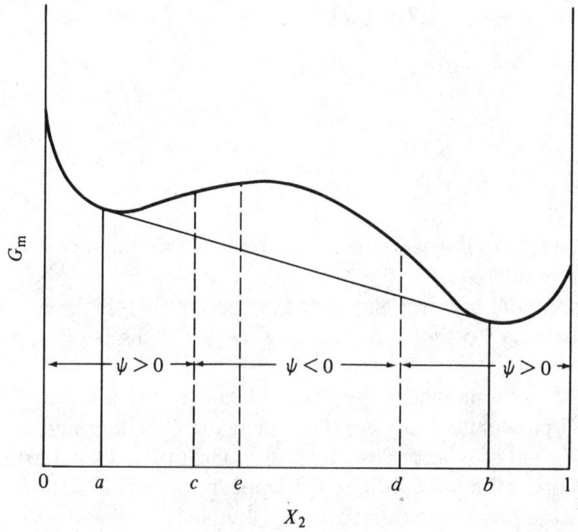

Figure 8. Illustration of the ranges of composition where ψ is positive for a system exhibiting a miscibility gap.

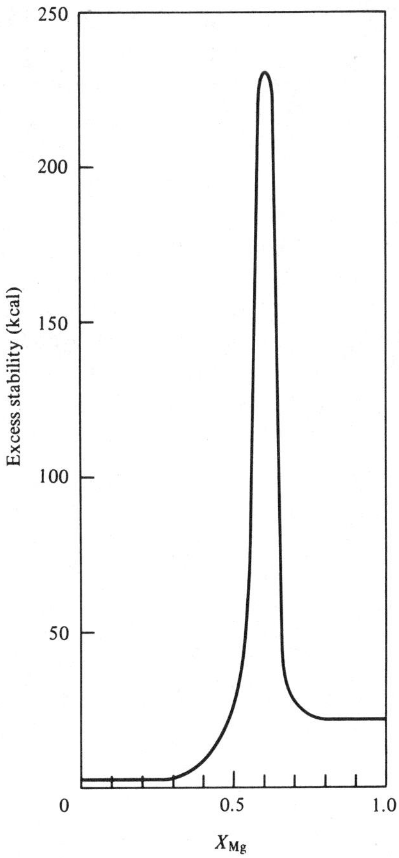

Figure 9. Excess stability α^E of the Mg–Bi liquid phase at 700°C [7].

(48) shows that for a regular solution

$$\psi = 1 - (2\Omega/RT)X_1X_2 \tag{79}$$

and for an ideal solution (see Section IV.4 or VI.2) $\psi = 1$.

The function ψ is closely related to other stability functions, α and α^E, introduced by Darken in 1967 [7]:

$$\alpha = d^2G_m/dX_2^2 \tag{80}$$

and

$$\alpha^E = \alpha - (RT/X_1X_2) \tag{81}$$

Clearly,

$$\psi = \alpha X_1X_2/RT = 1 + (\alpha^E X_1X_2/RT) \tag{82}$$

The advantages of the ψ function over the α and α^E functions originate in the cases of ternary and higher order systems.

In Chapter XI we shall see that the search for negative values of ψ can provide upper and lower bounds for the values of parameters in the formalism associated with the thermodynamic properties of solutions. Strong positive values of ψ are also of interest in the study of the structure of a solution.

For example, Darken [7] showed that in liquid solutions α^E exhibits large peaks at stoichiometric compositions which may be anticipated either from valence considerations or from the presence of intermediate solid state phases at lower temperatures. The Mg–Bi system in Fig. 9 shows a peak which is particularly pronounced.

Equation (82) shows that if a large peak is observed for α^E, it should also be observed for ψ and α. It is difficult to assess how abundant these peaks are in metallic solutions. Often, the data lack the accuracy needed for a definite result, because the stability functions α^E and ψ are very sensitive to small experimental errors. Nevertheless, the significance of the peaks may be somewhat elucidated through statistical thermodynamic studies.

Such an investigation was attempted by Lupis et al. [8] through the central atoms model (see Chapters XV and XVI). The basic entity used to describe the solution A–B is the cluster composed of a central atom (A or B) and its nearest

Figure 10. Values of the free energy of mixing, activity coefficients, and stability function as predicted by the central atoms model. The dotted lines correspond to a linear case, and the solid lines to a superposed preferred configuration of the type AB_2.

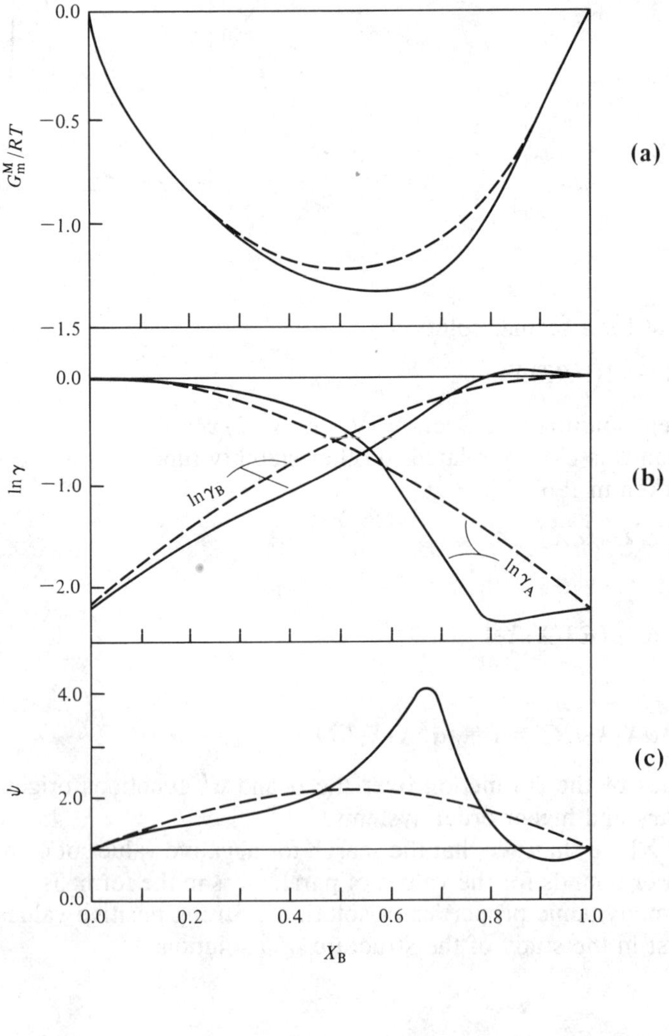

neighbors' shell of Z atoms. To the central atom is associated an energy which varies with the number i of B atoms in the nearest neighbor shell. Figure 10a shows two cases: the first corresponds to a variation of this energy which is linear with i, the second to the superposition of a peak for configurations corresponding to the stoichiometry AB_2. The magnitude of the peak is about 10% of the value of the linear contribution. We note in Fig. 10b that the perturbations of the excess Gibbs free energy and of the activity coefficients affect a substantial part of the composition range. In contrast, in Fig. 10c the perturbation of the stability function ψ is much more localized. In addition, the peak of that function does correspond to the favored stoichiometry. Clearly, however, much remains to be done for an adequate understanding of the stability of clusters in homogeneous solutions.

6. Thermodynamic Calculations Associated with the Nucleation and Growth of Precipitates

6.1. Free Energy Changes

In the study of the nucleation and growth of precipitates, it is often important to evaluate the *driving free energy* of the transformation, that is, the difference DG in the values of the Gibbs free energy at time t and at equilibrium ($t = \infty$). If ξ is a variable measuring the displacement of the system from its equilibrium position ($\xi = 0$), we may write

$$DG = (G)_{\xi=0} - G = -\frac{1}{2}\left(\frac{\partial^2 G}{\partial \xi^2}\right)_{\xi=0} \xi^2 + O(\xi^3) \tag{83}$$

where $O(\xi^3)$ designates terms of order three and higher. Comparing equations (83) and (3), it may be noted that we changed the sign definition of DG. This is because in Section 2 we were interested in departures from the equilibrium state $\xi = 0$ to a nonequilibrium state ξ, whereas in the present case we are interested in the reverse process, i.e., a departure from the nonequilibrium state ξ to the equilibrium state $\xi = 0$.

The evaluation of DG is very similar in both cases of nucleation and growth. Initially, the binary system 1–2 is in equilibrium at a temperature T', in the one-phase field α in the case of nucleation (state A in Fig. 11a) and in the two-phase field α–β in the case of growth (state A in Fig. 11b). The number of moles of components 1 and 2 in each of the phases α and β are $n_1'^{\alpha}$, $n_2'^{\alpha}$ and $n_1'^{\beta}$, $n_2'^{\beta}$ (in the nucleation case $n_1'^{\beta} = n_2'^{\beta} = 0$). The system is then abruptly brought to a neighboring temperature T (state B) and, with time, proceeds to an equilibrium (state C) corresponding to the concentrations n_1^{α}, n_2^{α} and n_1^{β}, n_2^{β}. The transformation B to C is isothermal and isobaric, and the driving energy is then $DG = G_C - G_B$.

It is convenient to introduce the following notation. Let λ_α and λ_β designate the fractions of the solution in the phases α and β:

$$\lambda_\alpha = \frac{n^\alpha}{n^\circ}, \qquad \lambda_\beta = \frac{n^\beta}{n^\circ} \tag{84}$$

where n^α and n^β are the total number of moles in the phases α and β and $n^\circ = n^\alpha + n^\beta$. Also, let

$$Dn_i^\alpha = n_i^\alpha - n_i'^{\alpha}, \qquad Dn_i^\beta = n_i^\beta - n_i'^{\beta} \tag{85}$$

where $i = 1$ or 2. An elementary mass balance yields

$$n_i'^{\alpha} + n_i'^{\beta} = n_i^\alpha + n_i^\beta \tag{86a}$$

or

$$Dn_i^\alpha = -Dn_i^\beta \tag{86b}$$

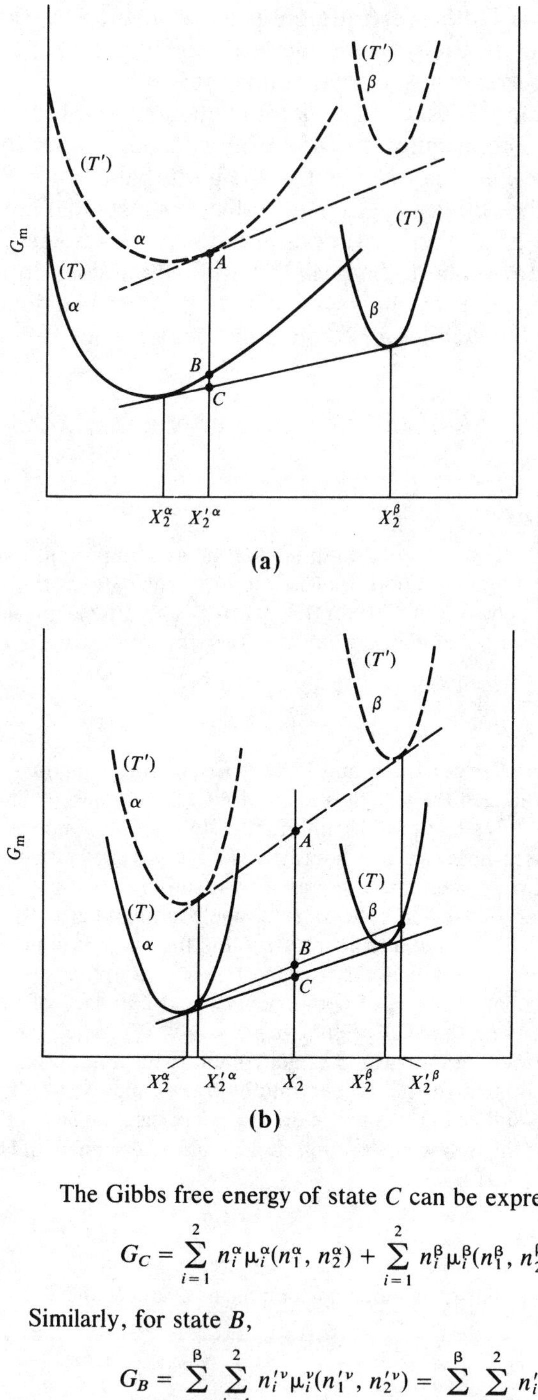

Figure 11. Graphical representation of the free energy changes associated with **(a)** nucleation and **(b)** growth. Note that n_i^ν ($\nu = \alpha$ or β, $i = 1$ or 2) is equal to $n^\nu X_i^\nu$ or $n^\circ \lambda_\nu X_i^\nu$.

The Gibbs free energy of state C can be expressed as a sum of chemical potentials:

$$G_C = \sum_{i=1}^{2} n_i^\alpha \mu_i^\alpha(n_1^\alpha, n_2^\alpha) + \sum_{i=1}^{2} n_i^\beta \mu_i^\beta(n_1^\beta, n_2^\beta) = \sum_{\nu=\alpha}^{\beta} \sum_{i=1}^{2} n_i^\nu \mu_i^\nu \tag{87}$$

Similarly, for state B,

$$G_B = \sum_{\nu=\alpha}^{\beta} \sum_{i=1}^{2} n_i'^\nu \mu_i^\nu(n_1'^\nu, n_2'^\nu) = \sum_{\nu=\alpha}^{\beta} \sum_{i=1}^{2} n_i'^\nu \mu_i'^\nu \tag{88}$$

We note that in the case of nucleation, $n_i'^\beta = 0$ and equation (88) reduces accordingly. The difference $DG = G_C - G_B$ may be written

$$DG = -\sum_{\nu=\alpha}^{\beta} \sum_{i=1}^{2} (n_i'^\nu \mu_i'^\nu - n_i^\nu \mu_i^\nu) = -\sum_{\nu=\alpha}^{\beta} \sum_{i=1}^{2} n_i'(\mu_i'^\nu - \mu_i^\nu) - Dn_i^\nu \mu_i^\nu \tag{89}$$

Recalling equation (86b) and the fact that $\mu_i^\alpha = \mu_i^\beta$ at equilibrium, equation (89) becomes

$$DG = -\sum_{\nu=\alpha}^{\beta} \sum_{i=1}^{2} n_i'^\nu(\mu_i'^\nu - \mu_i^\nu) \tag{90}$$

In the case of nucleation, equation (90) is further simplified:

$$DG = -\sum_{i=1}^{2} n_i'^\alpha(\mu_i'^\alpha - \mu_i^\alpha) \tag{91}$$

If we only consider small deviations from equilibrium, the chemical potentials may be developed into a Taylor series with respect to the Dn_is. Neglecting terms of the third order and higher, the calculations yield the following expressions of DG. In the case of nucleation,

$$DG = -n°RT \frac{\lambda_\beta^2}{\lambda_\alpha} \frac{(X_2^\beta - X_2^\alpha)^2}{2X_1^\alpha X_2^\alpha} \psi^\alpha \tag{92}$$

where ψ^α is the stability function of the α phase, and in the case of growth,

$$DG = -n°RT \left[\frac{\lambda_\alpha'^2}{\lambda_\alpha} \frac{(X_2'^\alpha - X_2^\alpha)^2}{2X_1^\alpha X_2^\alpha} \psi^\alpha + \frac{\lambda_\beta'^2}{\lambda_\beta} \frac{(X_2'^\beta - X_2^\beta)^2}{2X_1^\beta X_2^\beta} \psi^\beta \right] \tag{93}$$

The relative amounts λ of each phase and the mole fractions X appearing in equations (92) and (93) are illustrated graphically in Fig. 12.

Figure 12. Graphical analysis of some terms appearing in the expressions of DG [equations (92) and (93)].

$$\lambda_\alpha = \frac{CN}{MN} = \frac{X_2^\beta - X_2}{X_2^\beta - X_2^\alpha}, \qquad \lambda_\beta = \frac{MC}{MN} = \frac{X_2 - X_2^\alpha}{X_2^\beta - X_2^\alpha}$$

$$\lambda_\alpha' = \frac{AN'}{M'N'} = \frac{X_2'^\beta - X_2}{X_2'^\beta - X_2'^\alpha}, \qquad \lambda_\beta' = \frac{M'A}{M'N'} = \frac{X_2 - X_2'^\alpha}{X_2'^\beta - X_2'^\alpha}$$

Note that in the case of nucleation the temperature T' of the initial state is above T^*.

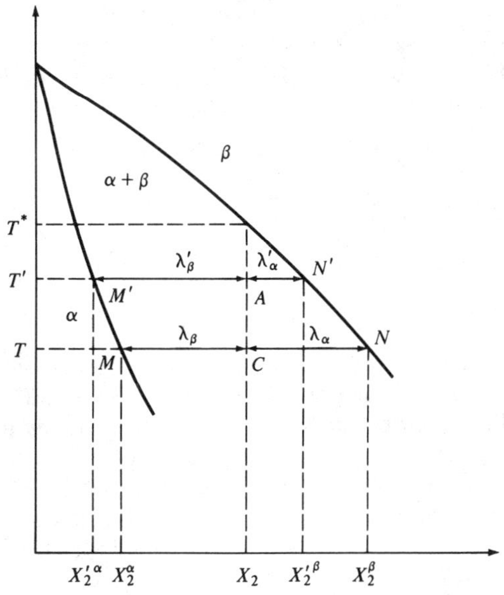

In the case of growth of a stoichiometric compound, we observe that in equations (90) and (93) the terms in β vanish.

6.2. Selection of the Displacement Variable

The displacement variable ξ may be measured in several different ways. ξ may be equal to Dn^β, the increase in the number of moles of β, to the change DV^β in the volume V^β of the β phase, or to the change Dx in a morphological characteristic x of the precipitates (thickness of a plate, radius of a sphere). If A represents the total area of all the precipitates, and V_m^β their molar volume,

$$Dn^\beta = \frac{DV^\beta}{V_m^\beta} = \frac{A}{V_m^\beta} Dx \tag{94}$$

Although the transformation BC is isothermal (at the temperature T), in the case of precipitate growth, state B is uniquely determined by the state of equilibrium A at T', and it is often convenient to measure the displacement ξ from the equilibrium C at T by the temperature difference $DT = T - T'$. However, in the case of precipitate nucleation, the state B does not uniquely correspond to the state A at the temperature T', but to any state of the same composition in the temperature interval $[T', T^*]$ where T^* is the temperature at which β starts precipitating when there is no nucleation barrier (see Fig. 12). The temperature displacement variable in this case is then $T - T^*$. We shall also designate it by DT.

It is obvious then that the preceding calculations of DG relative to the nucleation of precipitates may be considered as a special case of the calculations relative to the growth of precipitates, where the initial state corresponds to the temperature T^*. Similarly, the calculations relating Dn_1^β and Dn_2^β to DT and DG to DT are valid for both cases of nucleation and growth. A possible procedure for these calculations is as follows.

In any equilibrium state at T, T', or T^*, the chemical potentials of i ($i = 1$ or 2) must be equal in the phases α and β. Consequently,

$$\mu_i^\alpha(T' \text{ or } T^*, n_1'^\alpha, n_2'^\alpha) - \mu_i^\alpha(T, n_1^\alpha, n_2^\alpha) = \mu_i^\beta(T' \text{ or } T^*, n_1'^\beta, n_2'^\beta) - \mu_i^\beta(T, n_1^\beta, n_2^\beta) \tag{95}$$

Upon expanding μ_i^ν into a Taylor series with respect to Dn_1^ν, Dn_2^ν, and DT and noting that $Dn_i^\alpha = -Dn_i^\beta$, equation (95) yields a relation between Dn_1^β, Dn_2^β, and DT. Since there are *two* such equations (for $i = 1$ and $i = 2$), Dn_1^β and Dn_2^β may be solved in terms of DT. Inserting their expressions into the Taylor series expansion of DG at T [in equation (90)] with respect to Dn_1^β, Dn_2^β yields DG as a function of DT. The calculations are long and need not be reproduced here. Neglecting terms of order 3 and higher, the resulting expression of DG is

$$\frac{DG}{n^\circ} = -\frac{(DT)^2}{2(X_2^\beta - X_2^\alpha)^2}\left[\frac{\lambda^\alpha(\Delta S^{\alpha\to\beta})^2}{(\partial^2 G_m^\alpha/\partial(X_2^\alpha)^2)} + \frac{\lambda^\beta(\Delta S^{\beta\to\alpha})^2}{(\partial^2 G_m^\beta/\partial(X_2^\beta)^2)}\right] + O(DT^3) \tag{96}$$

where

$$\Delta S^{\alpha\to\beta} = S_m^\beta - X_1^\beta S_1^\alpha - X_2^\beta S_2^\alpha \tag{97}$$

and represents the entropy change of the reaction corresponding to the formation of 1 mol of β from X_1^β mol of 1 in the phase α and X_2^β mol of 2 also in the phase α. $\Delta S^{\beta\to\alpha}$ has a similar definition and may be obtained from equation (97) by interchanging the indices α and β.

In the case where β is a stoichiometric compound, the curvature of the Gibbs free energy of the β phase is infinite and the second term in the brackets of equation (96) vanishes.

6.3. Driving Forces

In atomistic analyses of the processes of nucleation and growth, one needs to evaluate the change in the Gibbs free energy of the system as a function of the progress of the transformation. In the preceding sections we calculated only the differences between the final and initial values of the Gibbs free energy $[G_C - G_B = G(\xi_{eq}) - G(\xi_0)]$ and not the function itself $G(\xi)$. Obtaining this function necessitates knowledge of the transformation path. Consequently, the driving force $\Delta G = \partial G/\partial \xi$ is itself dependent on the transformation path. Generally, this path can be estimated only through certain assumptions.

We recall that state B is at temperature T of the final equilibrium and is characterized by compositions of the phases α and β which correspond to the equilibrium state at temperature T'. Let us assume that the transformation is achieved through a series of such states at temperature T, but corresponding to equilibrium states at temperatures between T' and T. Then the expressions of DG we found above also represent the function $G(\xi)$ (they differ only by a constant). In the special case where the precipitate β is stoichiometric, this assumption is automatically satisfied since the transformation path is unique (we disregard concentration gradients in the α phase). If the β phase is not stoichiometric, the assumption is reasonable but only atomistic and kinetic considerations can ascertain its value. Nevertheless, we shall adopt it here.

In the transformations considered above, G is an analytic function which can be expanded into a Taylor series in the vicinity of its equilibrium value:

$$G(\xi) = G(\xi_{eq}) + \tfrac{1}{2}(\partial^2 G/\partial \xi^2)_{\xi_{eq}}(\xi - \xi_{eq})^2 + O[(\xi - \xi_{eq})^3] \tag{98}$$

Consequently,

$$\Delta G = \frac{\partial G}{\partial \xi} = -\frac{\partial DG}{\partial \xi} = \left(\frac{\partial^2 G}{\partial \xi^2}\right)_{\xi_{eq}} (\xi - \xi_{eq}) + O[(\xi - \xi_{eq})^2] \tag{99}$$

and

$$\Delta G = 2DG/(\xi - \xi_{eq}) + O[(\xi - \xi_{eq})^2] \tag{100}$$

Thus, the expressions above for DG readily provide associated expressions for ΔG. We note that ΔG is of the first order with respect to the displacement variable.

The driving free energies or driving forces we calculated are sometimes called *chemical* driving free energies or forces. If we consider atomistic fluctuations which provoke the formation of a nucleus of surface dA, work must be provided to create the new surface. This work is equal to $\sigma\, dA$, where σ is the interfacial tension between the phase β of the nucleus and the phase α of the matrix (see Chapter XIII). The change in the Gibbs free energy of the system for a small nucleus is

$$dG = \frac{\partial G}{\partial n^\beta} dn^\beta + \sigma\, dA \tag{101}$$

If the nucleus is spherical and of radius r,

$$dn^\beta = \tfrac{4}{3}\pi r^3 / V_m^\beta \tag{102}$$

and

$$dA = 4\pi r^2 \tag{103}$$

Thus,

$$G(r) - G(r = 0) = \tfrac{4}{3}\pi r^3 (\Delta G / V_m^\beta) + 4\pi r^2 \sigma \tag{104}$$

ΔG is negative whereas σ is positive. The function $G(r)$ has a maximum (see Fig. 13) at

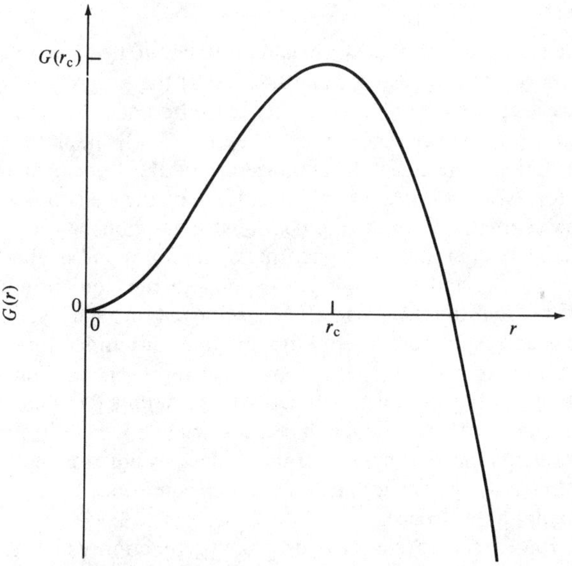

Figure 13. Gibbs free energy of the system as a function of the radius of its nucleus. The curve represents the equation

$$\frac{G(r) - G(0)}{G(r_c) - G(0)} = 3\left(\frac{r}{r_c}\right)^2 - 2\left(\frac{r}{r_c}\right)^3$$

where r_c and $G(r_c) - G(0)$ are identified in equations (106) and (107) in terms of the chemical driving force ΔG and the interfacial tension σ.

$r = r_c$. Below the critical radius r_c the nucleus is unstable (G increases); beyond r_c it is stable and grows. r_c is calculated by

$$\left(\frac{\partial G(r)}{\partial r}\right)_{r_c} = 0 \tag{105}$$

which yields

$$r_c = -\frac{2\sigma}{\Delta G / V_m^\beta} \tag{106}$$

At the critical radius, the *activation energy* $G(r_c) - G(0)$ is then equal to

$$G(r_c) - G(0) = \tfrac{16}{3}\pi \frac{\sigma^3}{(\Delta G / V_m^\beta)^2} \tag{107}$$

Since ΔG is of the first order with respect to the displacement variable, the activation energy is proportional to the cube of the interfacial tension and inversely proportional to the square of the displacement variable (e.g., DT)..

The study of nucleation and growth does not lie within the scope of this text. The preceding considerations were offered as examples of thermodynamic calculations rather than as a thorough treatment of the subject. For further information on the problems of nucleation and growth, the reader is referred to other textbooks [9–11].

Problems

1. Consider a one-component fluid at constant V and T. Imagine that it decomposes into two nearly identical phases of molar volumes V'_m and V''_m. Demonstrate that the criterion of stability with respect to such density fluctuations is that the compressibility coefficient β must remain positive.

2. Demonstrate in the case of a regular solution that the limit of the ratio AB/PQ (see Fig. 6) for $T \rightarrow T_C$ is equal to $\sqrt{3}$. Use the analytic equations (50) and (52) describing the spinodal line and miscibility boundary. [*Hint*: Both curves being symmetric with respect to the axis $X_2 = 0.5$, the easiest procedure is to write that a point on either curve has the composition $X_1 = 0.5 - \frac{1}{2}x$, $X_2 = 0.5 + \frac{1}{2}x$, where x is the width of the associated domain, and to develop the necessary equations as Taylor series with respect to x.]

3. The fcc phase of the Al–Zn system exhibits a miscibility gap. The molar Gibbs free energy of that phase may be described as

$$G_m = X_1 G_1^\circ + X_2 G_2^\circ + RT(X_1 \ln X_1 + X_2 \ln X_2)$$
$$+ X_1 X_2 (3150 X_1 + 2300 X_2)\,[1 - (T/4000)] \quad \text{cal}$$

for which Al $\equiv$ 1 and Zn $\equiv$ 2. Calculate the composition and temperature of the miscibility gap's critical point.

4. Using the results of Problem 3, calculate the asymmetry of the miscibility gap $y/(T_c - T)$ (see Fig. 7) in the case of the Al–Zn fcc phase through equation (77).

5. The molar Gibbs free energy of a binary (regular) solution is expressed

$$G_m = X_1 G_1^\circ + X_2 G_2^\circ + RT(X_1 \ln X_1 + X_2 \ln X_2) + \Omega X_1 X_2$$

where Ω is a constant. Calculate G_{11}, G_{22}, and G_{12}, where G_{ij} is the second derivative of G with respect to n_i and n_j. Verify that $G_{11}G_{22} = G_{12}^2$.

6. Calculate and plot the stability function ψ for the liquid Hg–K solution at 600°K on the basis of the data compiled by Hultgren et al. [12].

7. **a.** Demonstrate that, neglecting third order terms, the driving free energy DG for nucleation (see Section 6.1) may be written

$$DG = -\tfrac{1}{2}[(G_{11}^\alpha)^{1/2} n_1^\beta - (G_{22}^\alpha)^{1/2} n_2^\beta]^2$$

b. Demonstrate that, neglecting third order terms, the driving free energy DG for growth may be written

$$DG = -\tfrac{1}{2}[(G_{11}^\alpha)^{1/2} Dn_1^\alpha - (G_{22}^\alpha)^{1/2} Dn_2^\alpha]^2 - \tfrac{1}{2}[(G_{11}^\beta)^{1/2} Dn_1^\beta - (G_{22}^\beta)^{1/2} Dn_2^\beta]^2$$

8. **a.** Derive equation (92) from the equation given in Problem 7a.
b. Derive equation (93) from the equation given in Problem 7b.

9. Calculate DG for the nucleation and growth of a stoichiometric compound β from a regular solution α in terms of the initial composition $X_2'^\alpha$ and of the final composition X_2^α in the α phase.

References

1. H. Hancock, *Theory of Maxima and Minima*. Ginn, Boston, 1917, p. 91.
2. D. S. Gaunt and G. A. Baker, *Phys. Rev.* **B1,** 1184 (1970).
3. J. W. Cahn, *Proceedings of the Darken Conference, 1976* (R. M. Fischer, R. A. Oriani, and E. T. Turkdogan, eds.). U.S. Steel, pp. 399–404.
4. H. E. Stanley, *Introduction to Phase Transitions and Critical Phenomena*. Oxford University Press, Oxford, 1971.
5. H. E. Cook and J. E. Hilliard, *Trans. Met. Soc. AIME* **233,** 142–146 (1965).
6. C. H. P. Lupis and H. Gaye, *Metallurgical Chemistry, Proceedings of a Symposium held at Brunel University and the National Physical Laboratory, July 1971* (O. Kubaschewski, ed.). Her Majesty's Stationery Office, London, 1972, pp. 469–482.
7. L. S. Darken, *Trans. Met. Soc. AIME* **239,** 80–89 (1967).
8. C. H. P. Lupis, H. Gaye, and G. Bernard, *Scr. Met.* **4,** 497–502 (1970).
9. P. G. Shewmon, *Transformation in Metals*. McGraw-Hill, New York, 1969.
10. J. Burke, *The Kinetics of Phase Transformations in Metals*. Pergamon, Oxford, 1965.
11. J. W. Christian, *The Theory of Transformations in Metals and Alloys*. Pergamon, Oxford, 1965.
12. R. Hultgren, P. D. Desai, D. T. Hawkins, M. Gleiser, and K. K. Kelley, *Selected Values of the Thermodynamic Properties of Binary Alloys*. Am. Soc. Metals, Metals Park, OH, 1973.

IV Chemical Potentials, Fugacities, and Activities

So far, our thermodynamic driving forces and conditions for equilibrium have been expressed in terms of chemical potentials, and knowledge of the functional dependence of the chemical potentials on composition has been implicitly assumed. Actually, the chemical potential of a species i is a property which is affected by all the environmental conditions to which i is subjected, and a detailed knowledge of the atomic structure of the system of which i is a component is usually necessary. Thus, the functional dependence of a chemical potential on composition is by no means a trivial problem; indeed, many of the remaining chapters are devoted to its study.

In the course of this treatment several auxiliary functions need to be introduced. Their definitions often present some problems of clarity, and it is advantageous to study first the chemical potential of a single (pure) component.

1. Chemical Potential of a Single Component

1.1. Perfect Gas

The chemical potential μ_i of a pure species i is identical to its molar Gibbs free energy, G_i [see equation (II.24)]. Thus, at constant temperature, we may write

$$(d\mu_i)_T = (dG_i)_T = (V_i\,dP)_T \tag{1}$$

where V_i is the molar volume of i.

In the case where i is a perfect (or ideal) gas, its equation of state is

$$PV_i = RT \tag{2}$$

and equation (1) yields

$$(d\mu_i)_T = RT(d\ln P)_T \tag{3}$$

Upon integration between two states at pressures P and $P°$ (and the same temperature T), we obtain

$$\mu_i(T, P) = \mu_i(T, P°) + RT\ln(P/P°) \tag{4}$$

The state characterized by T and $P°$ is referred to as a *standard state*. The quantity $RT\ln(P/P°)$ measures the deviation from that standard state. Generally, standard states are defined at a pressure $P° = 1$ atm. *With that selection and the convention that P will now be measured in atmospheres*, equation (4) becomes

$$\mu_i(T, P) = \mu_i^\circ(T) + RT\ln P \tag{5}$$

We emphasize the fact that $\mu_i^\circ(T)$ is a function of the temperature alone since the pressure is fixed.

The characterization of the gas by equation (5) is equivalent to that given by its equation of state. Indeed, since

$$\left(\frac{\partial\mu_i}{\partial P}\right)_T = \overline{V}_i = V_i \tag{6}$$

and because of equation (5)

$$\left(\frac{\partial \mu_i}{\partial P}\right)_T = \frac{RT}{P} \tag{7}$$

a comparison of equations (6) and (7) leads directly to equation (2).

1.2. Real Gases; the Fugacity Function

In the general case where a gas is imperfect, the relation between the chemical potential and the pressure cannot be derived without additional knowledge of the state of the system. Nevertheless, the deviations from ideality are usually small enough so that it is mathematically convenient to substitute for the chemical potential another function which is related to the pressure in a simpler way than the nearly logarithmic one suggested by equation (5). (Observe that when P tends to zero μ_i tends to $-\infty$ and the practical description of μ_i becomes very difficult.) The *fugacity* function is thus introduced. It is defined by the isothermal equation

$$(d\mu_i)_T = RT(d \ln f_i)_T \tag{8}$$

Upon integration between two states differing only by their pressures P and $P°$, equation (8) becomes

$$\mu_i(T, P) = \mu_i(T, P°) + RT \ln(f_i/f°) \tag{9}$$

Again, it is convenient to consider the state T, $P°$ as a standard state relative to which we measure $\mu_i(T, P)$. In the definition of this standard state, the pressure is fixed but the temperature is not. Thus, we may write

$$\mu_i(T, P) = \mu_i^\circ(T) + RT \ln(f_i/f_i^\circ) = \mu_i^\circ(T) + RT \ln f_i \tag{10}$$

where f_i°, the value of the fugacity in the standard state of the gas, is chosen to be equal to 1. We note that since f_i/f_i° and f_i° are both dimensionless, f_i is also dimensionless.

The standard state is arbitrary and may be real or hypothetical. It is convenient to choose it as the *state of the gas under 1 atm pressure if it were to behave as a perfect gas.*

Let us assess the consequence and significance of this choice. The consequence is that, if the gas is perfect, the fugacity is equal to $P/(1 \text{ atm})$ and, if we agree to measure the pressure in atmospheres, the fugacity is equal to the numerical value of the pressure. The significance of this selection for the standard state rests on the experimental evidence that the behavior of a real gas approaches that of a perfect one when its pressure is reduced. (Atomistically, this is easily understood since the interactions between the molecules decrease when their distances increase.) Thus, the fugacity of a real gas approaches asymptotically its pressure when the latter tends to zero (see Fig. 1). The standard state is therefore based on the extrapolation of a behavior observed in the limit case of a state of infinite dilution (i.e., zero density).

Most metallurgical processes are carried out under pressures which do not exceed appreciably one atmosphere, and in these cases the difference between fugacity and pressure (i.e., its numerical value when measured in atmospheres)

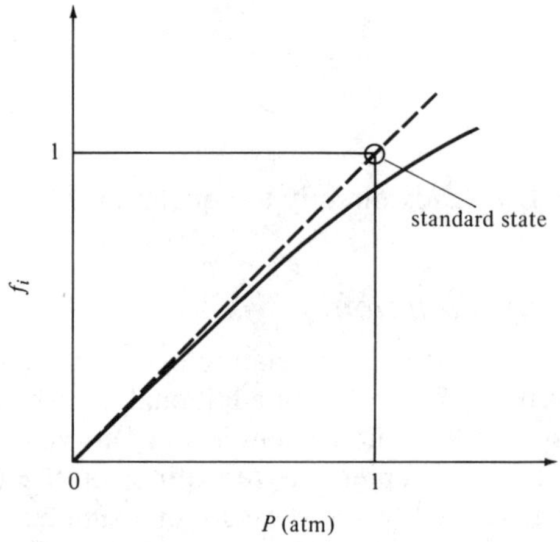

Figure 1. Standard state commonly adopted for a gas. The deviation between the two lines is exaggerated in order to present a clearer illustration.

is small enough to be neglected. In subsequent sections we shall usually assume the perfect gas law to be adequate:

$$f_i \simeq P \tag{11}$$

Nonetheless, a more accurate description of the relation between fugacity and pressure is necessary for the assessment of the approximation (11) and is briefly given below.

1.2.1. Example of Calculation of the Fugacity Function

The equation of state of any sufficiently dilute gas is $PV_i = RT$. At higher concentrations, the equation of state deviates from this formula and an expansion series with respect to the density $1/V_i$ yields

$$\frac{PV_i}{RT} = 1 + \frac{B_{i,2}}{V_i} + \frac{B_{i,3}}{V_i^2} + \cdots \tag{12}$$

The coefficients of this expansion are called *virial coefficients* and depend only on the temperature. The series always converges when the temperature T is above the critical temperature T_c. Below T_c the series converges only for values of the density up to that of the saturated vapor, at which point the gas condenses.

Alternatively, it is possible to write an expansion series in terms of the pressure:

$$PV_i/RT = 1 + C_{i,2}P + C_{i,3}P^2 + \cdots \tag{13}$$

where the C coefficients also depend only on the temperature. It may easily be seen, for example, that

$$B_{i,2} = C_{i,2}RT \tag{14}$$

The relationships between higher order coefficients of the two series are somewhat more complex. The convenience of one series relative to the other depends on the problem at hand.

Equations (6) and (8) yield

$$\left(\frac{\partial \ln f_i}{\partial P}\right)_T = \frac{V_i}{RT} \tag{15}$$

and using the equation of state (13) at constant temperature,

$$d \ln f_i = \frac{1}{P}(1 + C_{i,2}P + C_{i,3}P^2 + \cdots)\, dP$$

$$= \frac{dP}{P} + (C_{i,2} + C_{i,3}P + \cdots)\, dP \tag{16}$$

or

$$d \ln(f_i/P) = (C_{i,2} + C_{i,3}P + \cdots)\, dP \tag{17}$$

Integrating yields

$$\ln(f_i/P) = C_{i,2}P + \tfrac{1}{2}C_{i,3}P^2 + \cdots \tag{18a}$$

or

$$f_i/P = 1 + C_{i,2}P + \tfrac{1}{2}(C_{i,2}^2 + C_{i,3})P^2 + \cdots \tag{18b}$$

The integration constant in equation (18a) is equal to zero since, for $P \to 0$, $f_i \to P$ and $\ln(f_i/P) \to 0$.

There are many experimental measurements and theoretical estimates of the virial coefficients. For example, the following expression for B_2 (for convenience, the subscript i has been dropped) has been derived in statistical mechanics [1]:

$$B_2(T) = -\tfrac{1}{2}N_0 \int_0^\infty (e^{-u(r)/kT} - 1)4\pi r^2\, dr \tag{19}$$

where N_0 is Avogadro's number and $u(r)$ is the interaction energy between two molecules of the gas at a distance r. In the case of a Lennard–Jones potential

$$u(r) = -2u^*\left(\frac{r^*}{r}\right)^6 + u^*\left(\frac{r^*}{r}\right)^{12} \tag{20}$$

the function $B_2(T)$ is illustrated in Fig. 2. Another estimate of B_2 in terms of the critical temperature and pressure is due to Berthelot [2]:

$$B_2 = 0.07\frac{RT_c}{P_c}\left(1 - 6\frac{T_c^2}{T^2}\right) \tag{21}$$

It shows that B_2 is generally low for gases with a low critical temperature and a high critical pressure. The formula has proved to be satisfactory for the evaluation of small deviations from the perfect gas law [3, Ch. 16].

Table 1 further illustrates the effect of density or pressure on the deviation of f/P from 1 in the case of nitrogen at 0°C [3, Ch. 16].

Table 1. Fugacity of Nitrogen at 0°C [3]

P (atm)	f/P	P (atm)	f/P
1	0.99955	300	1.0055
10	0.9956	400	1.062
50	0.9812	600	1.239
100	0.9703	800	1.495
150	0.9672	1000	1.839
200	0.9721		

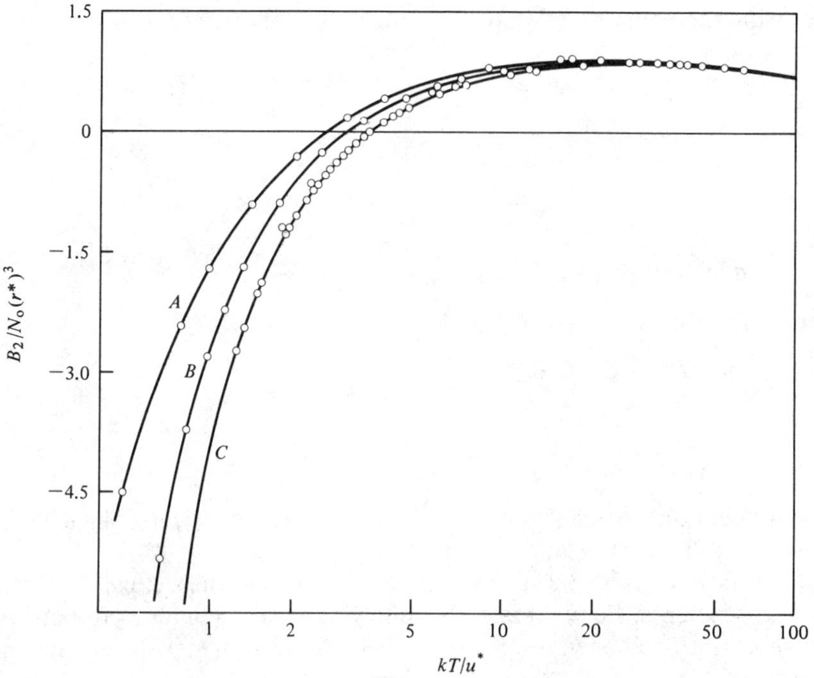

Figure 2. Reduced second virial coefficient, $B_2/N_0(r^*)^3$ as a function of the reduced temperature, kT/u^*. Curve C is calculated from equations (19) and (20). The experimental points on C are a mixture of data for Ar, Ne, N_2, and CH_4. Curves A and B incorporate quantum effects for He and H_2, respectively. (*After Hill* [*1*].)

1.3. Solids and Liquids

For a pure solid or liquid we still have the equation

$$(\partial \mu_i / \partial P)_T = V_i \tag{22}$$

However, the molar volume of a species in its condensed state is generally much lower than the molar volume of this species in the gaseous state (roughly 0.5×10^{-3} smaller). Consequently, the effect of pressure on the chemical potential of a solid or liquid is much smaller and often negligible. Nevertheless, to study this effect it is possible to introduce the fugacity function by the same isothermal equation:

$$(d\mu_i)_T = RT(d \ln f_i)_T \tag{23}$$

Upon integration,

$$\mu_i(T, P) = \mu_i^\circ(T) + RT \ln f_i \tag{24}$$

where $\mu_i^\circ(T)$ represents the standard state of the species (generally at $P = 1$ atm) and $f_i = 1$ (dimensionless number) when the species is in this standard state.

To evaluate the order of magnitude of f_i, let us assume that at a certain temperature T the pressure dependence of the volume in a given pressure range can be expressed by

$$V_i = V_i^\circ[1 - \beta(P - 1)] \tag{25}$$

Equations (22), (23), and (25) then yield

$$\ln f_i = \frac{1}{RT}\int_1^P V_i\,dP = \frac{1}{RT}\int_1^P V_i^\circ [1 - \beta(P - 1)]\,dP$$

$$= \frac{V_i^\circ}{RT}(P - 1) - \frac{V_i^\circ\beta}{2\,RT}(P - 1)^2 \tag{26}$$

Assuming that $V_i^\circ = 10\ \text{cm}^3$, $\beta = 10^{-6}\ \text{atm}^{-1}$, for $T = 1000°\text{K}$ and $P = 10$ atm,

$$\ln f_i = 0.001 - 6 \times 10^{-9} = 0.001 \tag{27a}$$

or

$$f_i = 1.001 \tag{27b}$$

and for the chemical potential

$$\mu_i(1000°\text{K, 10 atm}) - \mu_i(1000°\text{K, 1 atm}) = 2\ \text{cal} \tag{28}$$

The effect of pressure is thus seen to be negligible at modest pressures (at which most metallurgical processes occur). It does become significant when pressures of the order of kilobars are considered (1 bar = 0.98692 atm).

2. Mixture of Ideal Gases

As noted in the introduction, knowledge of the functional dependence of the chemical potential on composition is essential to many problems of thermochemistry. The results established in Section 1 on pure components allow us to approach now the subject of mixtures on a better basis. We start with the study of a mixture of ideal gases.

2.1. Definition

A mixture of m ideal gases may be defined by the condition that for each constituent i of the gaseous solution

$$P\overline{V}_i = RT \qquad (i = 1, 2, \ldots m) \tag{29}$$

Recalling that

$$V = \sum_{i=1}^{m} n_i\overline{V}_i \tag{30}$$

equation (29) yields

$$PV = \sum_{i=1}^{m} n_iRT = nRT \tag{31}$$

The gaseous mixture itself follows then the equation of state of a perfect gas. Moreover, a comparison of equations (29) and (31) yields immediately

$$\overline{V}_i = V/n \tag{32}$$

The partial molar volume of any component is therefore identical to the molar volume of a perfect gas.

It may be observed that equation (31) is an alternative possible definition of a mixture of ideal gases since the differentiation of equation (31) with respect to n_i at constant T, P, and n_j yields equation (29).

Regardless of whether the gases are ideal or not, the *partial pressure* of a component i is defined by the equation

$$p_i = (n_i/n)P \tag{33}$$

We note that

$$\sum_{i=1}^{m} p_i = P \tag{34}$$

Combining equations (31) and (33), we obtain

$$p_i V = n_i RT \tag{35}$$

p_i is therefore the pressure that the component i would exert if it were the only one exerting a pressure in the same volume V.

2.2. Interpretation

Through statistical thermodynamics methods it may be shown that if the molecules of a pure gas are assumed not to interact with each other and to have a negligible size, the equation of state of a perfect gas can be derived. These assumptions remain the same for a mixture of perfect gases. The close resemblance of the equations of state (29) or (35) and (31) may then be more easily understood.

Indeed, if the n_i molecules of component i ignore the presence of the molecules of the other components (because of their lack of interaction), the pressure due to these n_i molecules (p_i) must be identical to the pressure that these n_i molecules would have if the gas i were pure, that is, n_iRT/V. Consequently, $p_iV = n_iRT$ and by definition of p_i [equation (33)] the equation of state (31) $PV = nRT$ follows. As observed earlier, it is a sufficient condition for equation (29) $P\overline{V}_i = RT$.

When the pressure P of a real gaseous mixture decreases, its density also decreases (V increases). The interactions among the molecules also decrease since they depend on the mutual distances of these molecules. Moreover, the effect of their size becomes less important since it becomes a smaller fraction of the total volume available to the molecules. Consequently, when the pressure P tends to zero, a mixture of real gases should tend to behave as a mixture of ideal gases (as in the case of a pure gas). This conclusion is supported by experimental observations.

2.3. Fugacities

The fugacity function continues to be defined by the same isothermal equation:

$$(d\mu_i)_T = RT(d \ln f_i)_T \tag{36}$$

Upon integration it yields

$$\mu_i(T, P, X_j) = \mu_i^\circ(T) + RT \ln f_i \tag{37}$$

where X_j is the mole fraction of any component j. It is important to note that the fugacity f_i is a function of temperature, pressure, and composition, whereas the integration constant μ_i° is a function of T alone. μ_i° may be identified as the chemical potential of i in its standard state, generally corresponding to a pressure of 1 atm and the composition $X_i = 1$ (pure i).

Noting that

$$\left(\frac{\partial \mu_i}{\partial P}\right)_{T,X_j} = RT\left(\frac{\partial \ln f_i}{\partial P}\right)_{T,X_j} = \overline{V}_i \tag{38}$$

we obtain

$$\ln\frac{f_i}{f_i'} = \int_{P'}^{P} \frac{\overline{V}_i}{RT}\,dP = \int_{P'}^{P}\left(\frac{\overline{V}_i}{RT} - \frac{1}{P}\right) dP + \ln\frac{P}{P'} \tag{39}$$

or

$$\ln\frac{f_i}{P} = \ln\frac{f_i'}{P'} + \int_{P'}^{P}\left(\frac{\overline{V}_i}{RT} - \frac{1}{P}\right) dP \tag{40}$$

Let us now choose $P' = 0$ and establish that

$$(f_i/P)_{P\to 0} = X_i \tag{41a}$$

or

$$(f_i/p_i)_{P\to 0} = 1 \tag{41b}$$

Consider two compartments separated by a membrane permeable only to the molecule i of a gaseous mixture (Fig. 3). The first compartment α contains only the species i, while the second compartment β contains a mixture of real gases. At equilibrium $\mu_i^\alpha = \mu_i^\beta$, and if we choose identical standard states for i in both compartments, $f_i^\alpha = f_i^\beta$. In the case where the pressures in the system are small, the pure gas in α will behave as a perfect gas and consequently $f_i^\alpha = P^\alpha$. In the β compartment, the gaseous mixture will behave as a mixture of ideal gases and the molecules of the component i will not "see" the molecules of the other gaseous components. They will also ignore the presence of the semipermeable wall and for these molecules the entire space $\alpha + \beta$ will look uniform. Thus, the density of the molecules in the α compartment should be identical to the density of the molecules i in the β compartment; in other words, $p_i^\beta = P^\alpha$. It thus follows that $f_i^\beta = p_i^\beta$. Since this result is valid only at low pressures, it is identical to equation (41).

Equation (41) has thus been justified. It has not been "demonstrated," in the sense that it is additional information which does not follow from the definitions of fugacity and mixture of ideal gases.

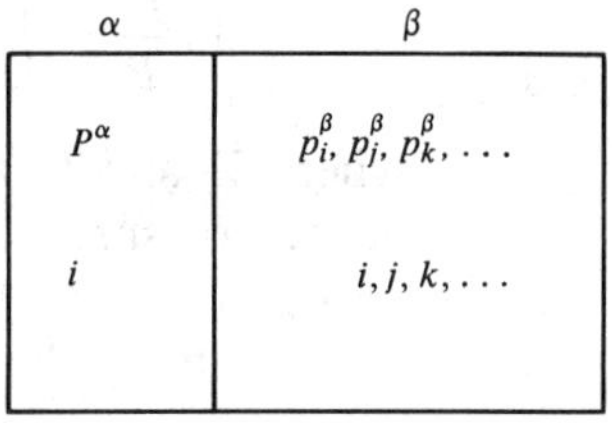

Figure 3. The wall separating the two compartments α and β is permeable only to the i molecules. At equilibrium $\mu_i^\alpha = \mu_i^\beta$; at low pressure, this condition yields $P^\alpha = p_i^\beta$.

Equation (40) becomes

$$\ln \frac{f_i}{P} = \ln X_i + \int_0^P \left(\frac{\overline{V}_i}{RT} - \frac{1}{P} \right) dP \tag{42a}$$

or

$$\ln f_i = \ln p_i + \int_0^P \left(\frac{\overline{V}_i}{RT} - \frac{1}{P} \right) dP \tag{42b}$$

In the case of perfect gases, the term in parenthesis is identical to zero regardless of the total pressure of the system. Consequently, the fugacity becomes identical to the partial pressure (or more rigorously, to the numerical value of the partial pressure when measured in atmospheres):

$$f_i = p_i \tag{43}$$

3. Fugacities in a Mixture of Real Gases

In the case of a mixture of real gases, the fugacity of a component is still defined by the same equation:

$$\mu_i = \mu_i^\circ(T) + RT \ln f_i \tag{44}$$

and the standard state is generally chosen to correspond to the state in which i is pure and behaves as a perfect gas under a pressure of one atmosphere. Equation (42b) still holds:

$$\ln \frac{f_i}{p_i} = \int_0^P \left(\frac{\overline{V}_i}{RT} - \frac{1}{P} \right) dP \tag{45}$$

At very low pressures, the gas mixture behaves as a mixture of ideal gases and the integrand tends toward zero (with the result that $f_i \rightarrow p_i$). At other pressures, experimental values of $\overline{V}_i$ must be available over the whole range of integration.

For example, at a fixed pressure and temperature, the volume of the mixture must be measured as a function of composition to deduce the partial molar volume of i at a given composition. The same measurements can then be repeated at other pressures. The partial molar volume $\overline{V}_i$ at a given composition is then obtained as a function of pressure and the integration in equation (45) is carried out at this composition. The procedure requires a large amount of experimental data and is not frequently performed.

3.1. Ideal Solution of Imperfect Gases

When component i is pure, equation (42a) becomes

$$\ln \frac{f_i^\dagger}{P} = \int_0^P \left(\frac{V_i}{RT} - \frac{1}{P} \right) dP \tag{46}$$

where V_i and $f_i^\dagger$ are, respectively, the molar volume and the fugacity of pure i at the pressure P. Subtracting equation (46) from equation (42a) yields

$$\ln f_i = \ln(X_i f_i^\dagger) + \frac{1}{RT} \int_0^P (\overline{V}_i - V_i) \, dP \tag{47}$$

An ideal solution of (perfect or imperfect) gases is defined by

$$f_i = X_i f_i^{\dagger} \tag{48}$$

Equation (47) shows that for an ideal solution

$$\overline{V}_i = V_i \tag{49}$$

which is an equivalent definition of an ideal solution. For example, if at all pressures the volume of a binary gaseous mixture is linearly dependent on the mole fraction of one of the two components, the mixture is ideal.

Equation (48) offers a convenient way of estimating the fugacity of an imperfect gas in a mixture, since rather extensive tabulations of single gas fugacities are available in the literature. The ideal solution approximation in equation (48) is also referred to as the Lewis and Randall rule [4, Ch. 19].

It may be demonstrated that if, at a given pressure and temperature, one component i follows equation (48) at all compositions of the multicomponent gaseous mixture, then the fugacities of all the other components j of the mixture are also proportional to their mole fractions X_j. For example, in a binary mixture, at constant temperature and pressure, the Gibbs–Duhem equation yields

$$X_1\, d\mu_1 + X_2\, d\mu_2 = 0 \tag{50a}$$

or

$$X_1\, d\ln f_1 + X_2\, d\ln f_2 = 0 \tag{50b}$$

If $f_1 = X_1 f_1^{\dagger}$, equation (50b) becomes

$$X_2\, d\ln f_2 = -X_1\, d\ln X_1 = -dX_1 = dX_2 \tag{51a}$$

or

$$d\ln f_2 = d\ln X_2 \tag{51b}$$

Consequently,

$$f_2 = X_2\, f_2^{\dagger} \tag{52}$$

4. Solid and Liquid Solutions; the Activity Function

In Sections 2 and 3, the fugacity function was introduced to simplify the functional relation between the chemical potential of a gaseous species i and its concentration (or the partial pressure p_i). A similar study may be conducted for condensed phases.

The fugacity function is again defined by consideration of changes in the chemical potential at constant temperature:

$$(d\mu_i)_T = RT(d\ln f_i)_T \tag{53}$$

or

$$\mu_i = \mu_i^{\circ}(T) + RT\ln f_i \tag{54}$$

Again, the standard state is generally defined at a pressure of 1 atm. However, the choice of the composition and the structure of i in that state depends on the problem at hand.

The *activity function* a_i is introduced to study changes in the chemical potential due to composition at constant temperature and pressure. It is defined by

$$(d\mu_i)_{T,P} = RT(d \ln a_i)_{T,P} \tag{55}$$

or

$$\mu_i = \mu_i^*(T, P) + RT \ln a_i \tag{56}$$

μ_i^* differs from μ_i° in that it depends on both temperature and pressure whereas μ_i° depends only on the temperature. μ_i^* is the chemical potential of i in a *reference state*, with the same composition and structure as in the standard state but at a variable pressure P.

The fugacity and activity functions may be related through the coefficient Γ_i,

$$f_i = \Gamma_i a_i \tag{57}$$

which analyzes the pressure dependence of the reference state:

$$\ln \Gamma_i = \frac{\mu_i^*(T, P) - \mu_i^\circ(T, P = 1 \text{ atm})}{RT} = \int_1^P \frac{\overline{V}_i^*}{RT} dP \tag{58}$$

$\overline{V}_i^*$ is the partial molar volume of i at the composition of the reference state (which often corresponds to pure i). In Section 3.1, we saw that the effect of pressure on the chemical potentials of solids and liquids is generally small. Consequently, at moderate pressures Γ_i may generally be taken as equal to 1 and the activity as equal to the fugacity.

A detailed discussion of possible reference states will be given in Chapter VII. In the interim, we shall adopt as reference state for the chemical potential of i in any given phase (e.g., a bcc phase) the state in which i is pure and in the structure of that phase (e.g., bcc). For convenience, we shall refer to that state as a *Raoultian reference state*.

With this choice $a_i = 1$ for $X_i = 1$. To deduce the value of a_i when $X_i \to 0$, we must know the value of μ_i for $X_i \to 0$. Imagine that we place a dilute solution in an evacuated closed reaction chamber. All elements being at least partially volatile, a gas phase is formed and equilibrium is attained when the chemical potential of each element is the same in the gas phase and in the solution. It is natural to expect that if the solution is very dilute in i, the resulting partial pressure p_i will also be very small. But we have seen that when $p_i \to 0$, $\mu_i \to -\infty$. Consequently, also in the solution, when $X_i \to 0$, $\mu_i \to -\infty$. Considering equation (56), it is immediately seen then that for $X_i \to 0$, $a_i \to 0$.

We have established the result that $a_i = 0$ for $X_i = 0$, and $a_i = 1$ for $X_i = 1$. The special solutions for which

$$a_i = X_i \tag{59}$$

at all compositions are called *ideal solutions*.

Real solutions tend to differ from that strict linearity. However there are so many solutions for which this is a reasonable approximation that, indeed, we shall generally focus our attention on the deviations from this ideality. Essentially, this is done by introducing a correction factor to the strict proportionality between a_i and X_i. The correction factor is called the *activity coefficient* and is designated by the symbol γ_i. It is defined by

$$a_i = \gamma_i X_i \tag{60}$$

In the case of an ideal solution, $\gamma_i = 1$ at all concentrations. In real solutions, it is a function of composition, temperature, and pressure and its deviation from the value of 1 will measure the deviation of the solution's behavior from the ideal model.

Note that so far all that has been accomplished is to transfer our ignorance of the composition dependence of μ_i to the activity a_i, or to the activity coefficient γ_i. This substitution is justified by reasons of mathematical convenience which find their basis both in experimental results and in theoretical models. Clearly, this is the ultimate justification of any formalism.

5. Partial Vapor Pressure of a Solute

Let us consider a solute i in a solid or liquid phase α in equilibrium with respect to its partial pressure in the gas phase g. Since the chemical potentials of i in the two phases must be equal, we may write

$$\mu_i^{\circ(\alpha)}(T) + RT \ln \Gamma_i^\alpha a_i^\alpha = \mu_i^{\circ(g)}(T) + RT \ln f_i^g \tag{61}$$

or

$$\Gamma_i^\alpha a_i^\alpha = f_i^g \exp \frac{\mu_i^{\circ(g)}(T) - \mu_i^{\circ(\alpha)}(T)}{RT} = f_i^g K(T) \tag{62}$$

where $K(T)$ is a function of the temperature alone. At moderate pressures, Γ_i^α may be taken as equal to 1 and equation (62) becomes

$$a_i^\alpha = f_i^g K \tag{63}$$

If the standard state of i in the α phase corresponds to pure i in the structure of α, equation (63) yields

$$a_i^\alpha = f_i^g / f_i^{\dagger(g)} \tag{64}$$

where $f_i^{\dagger(g)}$ is the fugacity of pure i. At moderate pressures we have seen that f_i^g may be replaced by the partial pressure p_i and $f_i^{\dagger(g)}$ may be identified as the vapor pressure P_i of pure i in the α phase at the temperature T. Consequently,

$$a_i = p_i / P_i \tag{65}$$

(The superscript α has been dropped for convenience, but may be reinstated in the case of possible confusion—see Problem 6.) Equation (65) is very useful and is the basis of many experimental methods in the thermodynamic study of solutions.

We note that in the case of a solid or liquid ideal solution

$$a_i = X_i = p_i / P_i \tag{66}$$

i.e., the partial pressure of a solute is proportional to its mole fraction in the condensed solution.

6. Composition Dependence of the Activity Under Conditions of Constant Volume

Most experimental conditions correspond to conditions of constant pressure rather than constant volume. However, in the development of many theoretical models, conditions of constant volume prevail. Consequently, it is of interest to relate the composition dependences of the activity function under conditions of constant pressure and constant molar volume V_{m}. The calculations below follow closely the results established in Section II.3.

The chemical potential of a species i in a solution of m components 1, 2, . . . , m, is a function of the independent variables P, T, and mole fractions $X_2, X_3, \ldots, X_m$. We may then write

$$d\mu_i = \left(\frac{\partial \mu_i}{\partial T}\right)_{P,X_j} dT + \left(\frac{\partial \mu_i}{\partial P}\right)_{T,X_j} dP + \sum_{j=2}^{m} \left(\frac{\partial \mu_i}{\partial X_j}\right)_{T,P,X_k} dX_j \tag{67}$$

Consequently,

$$\left(\frac{\partial \mu_i}{\partial X_j}\right)_{T,V_{\mathrm{m}},X_k} = \left(\frac{\partial \mu_i}{\partial X_j}\right)_{T,P,X_k} + \left(\frac{\partial \mu_i}{\partial P}\right)_{T,X_j} \left(\frac{\partial P}{\partial X_j}\right)_{T,V_{\mathrm{m}},X_k} \tag{68}$$

Moreover,

$$\left(\frac{\partial P}{\partial X_j}\right)_{T,V_{\mathrm{m}},X_k} = -\frac{(\partial V_{\mathrm{m}}/\partial X_j)_{T,P,X_k}}{(\partial V_{\mathrm{m}}/\partial P)_{T,X_j,X_k}} = \frac{\overline{V}_j - \overline{V}_1}{\beta V_{\mathrm{m}}} \tag{69}$$

where β is the isothermal compressibility of the solution. Equation (69) is an immediate consequence of the mathematical relationship

$$\left(\frac{\partial y}{\partial x}\right)_z \left(\frac{\partial x}{\partial z}\right)_y \left(\frac{\partial z}{\partial y}\right)_x = -1 \tag{70}$$

(see Section I.2.6). Combining equations (68) and (69) yields

$$\left(\frac{\partial \mu_i}{\partial X_j}\right)_{T,P,X_k} - \left(\frac{\partial \mu_i}{\partial X_j}\right)_{T,V_{\mathrm{m}},X_k} = \frac{\overline{V}_i(\overline{V}_1 - \overline{V}_j)}{\beta V_{\mathrm{m}}} \tag{71}$$

The definition of the activity function [equation (56)] may be rewritten

$$(d\mu_i)_T = \overline{V}_i^* \, dP + RT \, d\ln a_i \tag{72}$$

It yields

$$\left(\frac{\partial \mu_i}{\partial X_j}\right)_{T,P,X_k} = RT\left(\frac{\partial \ln a_i}{\partial X_j}\right)_{T,P,X_k} \tag{73}$$

and

$$\left(\frac{\partial \mu_i}{\partial X_j}\right)_{T,V_{\mathrm{m}},X_k} = RT\left(\frac{\partial \ln a_i}{\partial X_j}\right)_{T,V_{\mathrm{m}},X_k} + \overline{V}_i^* \left(\frac{\partial P}{\partial X_j}\right)_{T,V_{\mathrm{m}},X_k} \tag{74}$$

Combining equations (71), (73), (74), and (69), we obtain

$$\left(\frac{\partial \ln a_i}{\partial X_j}\right)_{T,P,X_k} - \left(\frac{\partial \ln a_i}{\partial X_j}\right)_{T,V_{\mathrm{m}},X_k} = \frac{(\overline{V}_i - \overline{V}_i^*)(\overline{V}_1 - \overline{V}_j)}{RT\beta V_{\mathrm{m}}} \tag{75}$$

When the reference state of i corresponds to pure i, $\overline{V}_i^* = V_i$. If the partial molar volume of i, $\overline{V}_i$, is practically independent of composition (i.e., $\overline{V}_i = V_i$; the solution obeys *Vegard's law*), then the composition dependence of the activity is the same at constant pressure and at constant molar volume.

Lupis has further elaborated on the determination of thermodynamic properties under conditions of constant pressure and constant volume [5].

Problems

1. Calculate the virial coefficient B_2 for a hypothetical gas of hard spheres:

$$u(r) = \begin{cases} \infty & \text{for} \quad r < a \\ 0 & \text{for} \quad r \geq a \end{cases}$$

2. The parameters of the Lennard–Jones potential for oxygen (O_2) are $u^*/k = 118°K$ and $r^* = 3.88$ Å. Estimate (from Fig. 2) the second virial coefficient B_2 at 273°K and 1800°K and the corresponding values of the fugacity at $P = 1000$ atm.

3. Assuming that a given gas follows the Van der Waals equation of state,

$$\left(P + \frac{a}{V^2}\right)(V - b) = RT$$

Find an analytical expression of $\ln(f/P)$.

The behavior of oxygen may be represented by the coefficients to $a = 1.35$ liter^2 atm/mol and $b = 3.15 \times 10^{-2}$ liter/mol. Estimate the fugacity of oxygen at 1000 atm and 273°K.

4. The Van der Waals coefficients a (in liter^2 atm/mol) and b (in liter/mol) for nitrogen are listed as $a = 1.390$ and $b = 0.0391$. For methane CH_4 they are given the values $a = 2.253$ and $b = 0.0428$. Estimate the fugacities of these gases in a mixture of 30% CH_4, 70% N_2 at room temperature and a pressure of 1000 atm.

5. Assume that methane follows the Van der Waals equation of state with the values of a and b given in Problem 4 and calculate its critical temperature, pressure, and volume.

6. A liquid alloy A–B of composition $X_B^l = 0.65$ is in equilibrium with a solid alloy A–B of composition $X_B^s = 0.40$ at $T = 1200°K$. At that temperature, pure liquid B has a vapor pressure equal to 2.5×10^{-3} atm. Calculate the vapor pressure of metastable pure solid B at the same temperature, assuming that B behaves ideally in both the solid and liquid phases.

7. Consider the binary solution 1–2. Show that if it is stable with respect to small composition fluctuations, the activity of component 2 must necessarily increase with X_2. Write the thermodynamic conditions that must be satisfied at the critical point in terms of the composition dependence of $\ln \gamma_2$.

8. A mixture of Ag and Cu is placed in a reaction chamber which is evacuated and sealed. The mixture is then heated to 1500°K and forms a liquid alloy of composition X_{Ag}. Calculate and plot the composition of the gas phase, $Y_{Ag} = p_{Ag}/(p_{Ag} + p_{Cu})$, as a function of X_{Ag}. Calculate the slope dY_{Ag}/dX_{Ag} at $X_{Ag} = 0$ and at $X_{Ag} = 1$.
Data: At 1500°K, the vapor pressures of pure Ag and Cu are, respectively, 3.6×10^{-4} and 8.6×10^{-6} atm. The behaviors of the solutes may be approximated by the equations $(\ln \gamma_{Ag})/X_{Cu}^2 = (\ln \gamma_{Cu})/X_{Ag}^2 = 1.14$.

9. Element B boils at 2500°K with a standard enthalpy of vaporization equal to 55,000 cal/mol. At 1000°K and 1 atm it forms with component A a liquid solution characterized by the following properties: $\ln \gamma_B = -3.2X_A^2$ and $V_m = 7X_A + 10X_B - 4X_AX_B$ cm^3/mol. The liquid solution is equilibrated with a gaseous phase containing A, B, and argon. It may be assumed that the gases behave ideally and that argon has no solubility in the liquid solution. Determine the activity of B in the solution and its partial pressure in the gas phase at $T = 1000$°K, $X_B = 0.25$, and $P = 1$ atm as well as at $P = 8000$ atm. Specify any other assumption you find it necessary to make.

References

1. T. L. Hill, *An Introduction to Statistical Thermodynamics*. Addison-Wesley, Reading, MA, 1960, p. 267.
2. D. Berthelot, *Trav. mem. bur. intern. poids mesures*, No. 13 (1907).
3. G. N. Lewis and M. Randall, *Thermodynamics* (revised by K. S. Pitzer and L. Brewer). McGraw-Hill, New York, 1961.
4. G. N. Lewis and M. Randall, *Thermodynamics*. McGraw-Hill, New York, 1923.
5. C. H. P. Lupis, *Acta Met.* **25,** 751–757 (1977); **26,** 211–215 (1978).

V Chemical Reactions

1. Case of a Single Chemical Reaction

1.1. General Treatment

Let us consider a system containing n_1 mol of species A_1, n_2 mol of species A_2, . . . , n_r mol of species A_r. The system is maintained at constant temperature and pressure but undergoes a transformation for which n_1 varies by dn_1, n_2 by dn_2, . . . , n_r by dn_r. We shall assume that this transformation may be represented by a single chemical reaction. With the convention described in Section I.1.5 this reaction may be written

$$\sum_{i=1}^{r} \nu_i A_i = 0 \tag{1}$$

The ν_i are the numbers necessary to balance stoichiometrically the reaction; we recall that they are positive when associated with "products" and negative when associated with "reactants." The increments dn_i are related to each other by the relations

$$dn_1/\nu_1 = dn_2/\nu_2 = \cdots = dn_r/\nu_r = d\lambda \tag{2}$$

where λ is a parameter designated the *progress variable* of the reaction.

Assuming that the transformation may be described by a single chemical reaction is equivalent to assuming that the transformation depends on a single independent variable. The number of moles n_i of any particular component A_i may be chosen as that independent variable. Generally, however, no component A_i plays a very special role in the transformation. For reasons of symmetry and elegance in the thermodynamic treatment of that transformation, it is then preferable to introduce the progress variable λ and select it as the independent variable.

The Gibbs free energy of a system varies with temperature, pressure, and the number of moles of its components according to the formula (see Section II.1)

$$dG = -S\,dT + V\,dP + \sum_{i=1}^{r} \mu_i\,dn_i \tag{3}$$

For the transformation corresponding to the chemical reaction (1), the dn_is are not independent but vary according to equation (2). Consequently, equation (3) becomes

$$dG = -S\,dT + V\,dP + \left(\sum_{i=1}^{r} \nu_i\mu_i\right) d\lambda \tag{4}$$

At constant temperature and pressure, the criterion of equilibrium for a closed system is

$$(\delta G)_{P,T} \geq 0 \tag{5}$$

Consequently,

$$\left(\sum_{i=1}^{r} \nu_i\mu_i\right) \delta\lambda \geq 0 \tag{6}$$

Let, by definition,

$$\Delta G = \sum_{i=1}^{r} \nu_i \mu_i \tag{7}$$

We note that because of equation (4) ΔG can also be defined

$$\Delta G = \left(\frac{\partial G}{\partial \lambda}\right)_{P,T} \tag{8}$$

The criterion of equilibrium expressed by (6) becomes

$$\Delta G\, \delta\lambda \geq 0 \tag{9}$$

$\delta\lambda$ is any conceivable virtual variation of λ in the field of equilibrium states and may be either positive or negative. Since the sign of ΔG is not arbitrary, the only instance in which the criterion of equilibrium (9) is obeyed is when

$$\Delta G = \sum_{i=1}^{r} \nu_i \mu_i = 0 \tag{10}$$

Consequently, equation (10) is our equilibrium condition.

Let us now designate by $d\lambda$ all *spontaneous* (irreversible) variations of λ, thus outside the field of equilibrium states. Obviously, these variations must be such that

$$(dG)_{P,T} = \Delta G\, d\lambda < 0 \tag{11}$$

If ΔG is negative, $d\lambda$ must be positive and the reaction will proceed "forward" (i.e., more of the products will be formed). Conversely, if ΔG is positive, $d\lambda$ must be negative and the reaction will proceed "backward" (more of the reactants will be formed). At equilibrium, the reaction has an equivalent tendency to move forward or backward, and the condition of equilibrium is given by equation (10).

Figure 1 illustrates the change in the Gibbs free energy corresponding to the progress of the reaction. It is important to note that ΔG is different from the total change DG in the Gibbs free energy of the system. ΔG can be represented by a slope [equation (8)] and is of the nature of a force, i.e., the derivative of a potential with respect to a displacement variable (in this case λ).

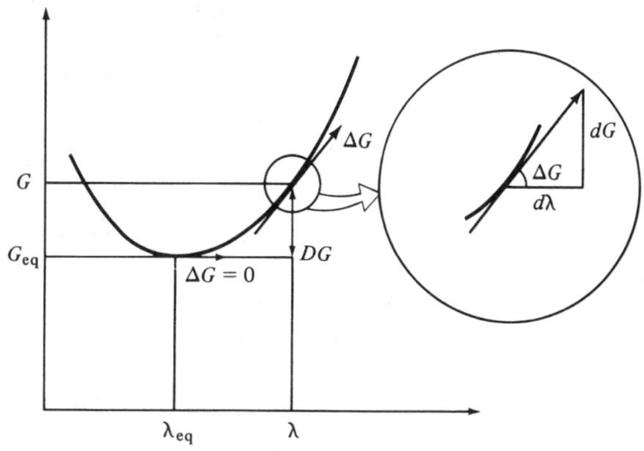

Figure 1. Gibbs free energy of a system undergoing a chemical reaction as a function of the progress variable at constant temperature and pressure. We see that if the slope ΔG is positive, λ must decrease ($d\lambda < 0$) in order to reach its equilibrium value. Conversely, for $\Delta G < 0$, λ must increase ($d\lambda > 0$).

A mechanical analogy for ΔG, perceived as a thermodynamic driving force F, would be a frictionless piston separating two compartments at pressures p' and p'' (see Fig. 2). If λ defines the position of the piston and F is of the same sign as $p'' - p'$ we see that when $F > 0$, $d\lambda < 0$ and when $F < 0$, $d\lambda > 0$. Equilibrium is obtained when $F = 0$.

In equation (7) the ΔG of a reaction has been defined as a weighted sum of chemical potentials. To relate it to the composition of the system under study, it is more practical to analyze it in terms of fugacities. Recalling that

$$\mu_i = \mu_i^\circ(T) + RT \ln f_i \tag{12}$$

it may be seen that it is convenient to consider the ΔG° of the reaction:

$$\Delta G^\circ = \sum_{i=1}^{r} \nu_i \mu_i^\circ \tag{13}$$

which is the value of ΔG when all the substances A_i are in their standard states, since substracting equation (13) from equation (7) yields

$$\Delta G - \Delta G^\circ = \sum_{i=1}^{r} \nu_i(\mu_i - \mu_i^\circ) = RT \sum_{i=1}^{r} \ln f_i^{\nu_i} \tag{14a}$$

If we now differentiate between the products ($\nu_i > 0$, $i = 1, 2, \ldots, k$) and the reactants ($\nu_i < 0$, $i = k + 1, k + 2, \ldots, r$), equation (14a) may be rewritten

$$\Delta G - \Delta G^\circ = RT \ln \frac{(f_1)^{\nu_1}(f_2)^{\nu_2} \cdots (f_k)^{\nu_k}}{(f_{k+1})^{|\nu_{k+1}|} \cdots (f_r)^{|\nu_r|}} \tag{14b}$$

The argument of the logarithm in this equation is often referred to as the *reaction's quotient* and is designated by the letter Q:

$$\Delta G - \Delta G^\circ = RT \ln Q \tag{15}$$

At moderate pressures, it should be noted that for gases

$$f_i \simeq p_i \tag{16a}$$

and for solids and liquids

$$f_j = a_j \Gamma_j \simeq a_j \tag{16b}$$

Consequently, in the expression of Q the fugacities may often be replaced by partial pressures and activities. For example, if components 1 and $k + 1$ are

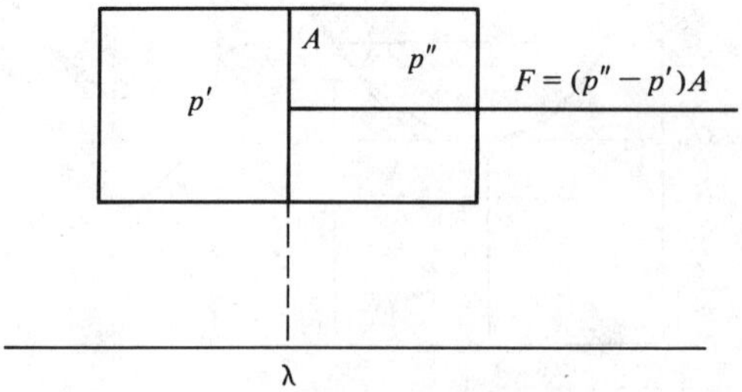

Figure 2. Analogy of a frictionless piston. The piston is in equilibrium when $F = 0$; if $F > 0$, $d\lambda < 0$, and if $F < 0$, $d\lambda > 0$.

gaseous and the others are either solid or liquid,

$$Q \simeq \frac{(p_1)^{\nu_1}(a_2)^{\nu_2} \cdots (a_k)^{\nu_k}}{(p_{k+1})^{|\nu_{k+1}|} \cdots (a_r)^{|\nu_r|}} \tag{17}$$

At equilibrium

$$\Delta G = 0 \tag{18}$$

and Q takes the equilibrium value designated by K:

$$(Q)_{\mathrm{eq}} = K \tag{19}$$

Equation (15) then yields

$$\Delta G^\circ = -RT \ln K \tag{20}$$

K is called the *equilibrium constant* and depends solely on the choice of the standard states. As such, it is dependent on temperature (see Section 1.3) but independent of the actual composition of the system (i.e., the individual concentrations of the species).

Rewriting equation (15) as

$$\Delta G = RT \ln Q/K \tag{21}$$

it is seen that if $Q < K$ the reaction proceeds forward, and if $Q > K$ the reaction proceeds backward.

1.2. Example

Consider a gaseous mixture of 20% CO, 20% CO_2, 50% H_2, and 10% H_2O, which is brought to a temperature T and at 1 atm. We wish to calculate the equilibrium composition of the gaseous phase at $T = 700$, 1130, and 1500°K. The reaction between the species and the states of the system may be characterized as follows:

	CO_2	+	H_2	=	CO	+	H_2O
initial state:	0.2		0.5		0.2		0.1
final state:	$0.2 - \lambda$		$0.5 - \lambda$		$0.2 + \lambda$		$0.1 + \lambda$

(Note that there is a total of 1 mol of gas in each state. Since we are only seeking relative percentages, this choice is merely for convenience.) The fugacities of the substances may be replaced by partial pressures without any noticeable error and

$$K = \left(\frac{p_{\mathrm{CO}}p_{\mathrm{H_2O}}}{p_{\mathrm{CO_2}}p_{\mathrm{H_2}}}\right)_{\mathrm{eq}} \tag{22a}$$

But $p_i = (n_i/n_{\mathrm{total}})P$, and consequently, if λ corresponds to the value of the progress variable at equilibrium,[1]

$$K = \frac{(0.2 + \lambda)(0.1 + \lambda)}{(0.2 - \lambda)(0.5 - \lambda)} \tag{22b}$$

[1] In equation (22b), n_{total} and P have canceled. This is not always the case.

Table 1. The Gas Phase

Gas	Initial state (%)	Equilibrium concentration (%)		
		700°K	1130°K	1500°K
CO	20	16.7	28	31.7
CO_2	20	23.3	12	8.3
H_2	50	53.3	42	38.3
H_2O	10	6.7	18	21.7

Given the value of K this equation is easily solved. At $T = 700$, 1130, and 1500°K, tables of standard free energies of formation allow the calculation of ΔG° in a manner identical to that described for the calculation of ΔH° in Section I.1.6. The following values are found:

$$(\Delta G^\circ)_{700^\circ\mathrm{K}} = 3350 \text{ cal}, \quad (\Delta G^\circ)_{1130^\circ\mathrm{K}} = 0 \text{ cal}, \quad (\Delta G^\circ)_{1500^\circ\mathrm{K}} = -2300 \text{ cal}$$

Since

$$\log K = -(\Delta G^\circ/4.575T) \tag{23}$$

we find

$$(K)_{700^\circ\mathrm{K}} = 0.09 \quad (K)_{1130^\circ\mathrm{K}} = 1, \quad (K)_{1500^\circ\mathrm{K}} = 2.16$$

These values of K may be compared to the value of Q for the initial composition

$$Q = \frac{0.2 \times 0.1}{0.2 \times 0.5} = 0.2$$

We see that at 700°K the reaction is proceeding backward ($Q > K$), while at 1130 and 1500°K it is proceeding forward ($Q < K$).

Solving equation (22) yields the values of the progress variable λ:

$$(\lambda)_{700^\circ\mathrm{K}} = -0.033, \quad (\lambda)_{1130^\circ\mathrm{K}} = 0.08, \quad (\lambda)_{1500^\circ\mathrm{K}} = 0.117$$

The equilibrium compositions of the gas phase are summarized in Table 1.

1.3. Effect of Temperature and Pressure

We recall that because of the definition of the Gibbs free energy $G = H - TS$ we have $\mu_i = \overline{H}_i - T\overline{S}_i$ where $\overline{H}_i$ and $\overline{S}_i$ are, respectively, the partial molar enthalpy and entropy of i. Consequently,

$$\Delta G = \sum_{i=1}^{r} \nu_i \mu_i = \sum_{i=1}^{r} \nu_i(\overline{H}_i - T\overline{S}_i) = \left(\sum_{i=1}^{r} \nu_i \overline{H}_i\right) - T\left(\sum_{i=1}^{r} \nu_i \overline{S}_i\right) \tag{24}$$

or

$$\Delta G = \Delta H - T\Delta S \tag{25}$$

ΔH and ΔS are respectively the heat and entropy of the reaction. At equilibrium $\Delta G = 0$ and $\Delta H = T\Delta S$.

When all the elements are in their standard states

$$\Delta G^\circ = \Delta H^\circ - T\Delta S^\circ \tag{26}$$

The standard heat of the reaction ΔH° being equal to $\sum_{i=1}^{r} \nu_i \overline{H}_i^\circ$, and the standard entropy of the reaction ΔS° being equal to $\sum_{i=1}^{r} \nu_i \overline{S}_i^\circ$, combining equations (20) and (26) yields

$$\ln K = -\frac{\Delta H^\circ}{RT} + \frac{\Delta S^\circ}{R} \tag{27}$$

To study the effect of temperature, we note that

$$\left(\frac{\partial G}{\partial T}\right)_P = -S \quad \text{and} \quad \left(\frac{\partial (G/T)}{\partial (1/T)}\right)_P = H \tag{28}$$

Thus,

$$\left(\frac{\partial \Delta G}{\partial T}\right)_P = -\Delta S, \qquad \frac{d\Delta G^\circ}{dT} = -\Delta S^\circ \tag{29}$$

and

$$\frac{d(\Delta G^\circ/T)}{d(1/T)} = -\frac{d(R \ln K)}{d(1/T)} = \Delta H^\circ \tag{30a}$$

or

$$\frac{d \ln K}{d(1/T)} = -\frac{\Delta H^\circ}{R} \tag{30b}$$

This equation is often referred to as the *Van't Hoff equation.* It may also be deduced from equation (27) and does not necessitate the assumption that ΔH° is independent of temperature.

It may be observed that graphically ΔG° should be plotted vs T whereas $\ln K$ should be plotted vs $1/T$. The plots are linear to the extent that the ΔC_p° of the reaction is small (and thus has a negligible effect on the curvature in the range of temperature investigated).

The effect of pressure is measured by

$$\left(\frac{\partial \Delta G}{\partial P}\right)_T = \sum_{i=1}^{r} \nu_i \left(\frac{\partial \mu_i}{\partial P}\right) = \sum_{i=1}^{r} \nu_i \overline{V}_i = \Delta V \tag{31}$$

When the component i is in the physical state referred to as its standard state, $(\partial\mu_i/\partial P)_{\text{at std. st.}}$ has the value $\overline{V}_i^\circ$, the partial molar volume of i in its standard state, in the same way that $(\partial\mu_i/\partial T)_{\text{at std. st.}}$ has the value $\overline{S}_i^\circ$. However, whereas $(\partial\mu_i^\circ/\partial T)_{P=P^\circ} = S_i^\circ \neq 0$, $(\partial\mu_i^\circ/\partial P)_{T=T^\circ} = 0$. These seemingly contradictory statements are a result of the convention that while the standard state is defined at the temperature of interest for the problem at hand, it is defined at a fixed pressure (generally $P^\circ = 1$ atm), which is independent of the pressure actually exerted on the system. Consequently, the standard values μ_i° are temperature dependent (T° varies) and pressure independent (P° is fixed). Other conventions which would assign similar roles to P and T are possible but less convenient. This explains why in equations (29) and (30) we have used total differentials (d) instead of partial derivatives (∂) for the functions ΔG° and $\ln K$. Alternatively, of course, one could write

$$\left(\frac{\partial \Delta G^\circ}{\partial P}\right)_T = 0 \quad \text{and} \quad \left(\frac{\partial \ln K}{\partial P}\right)_T = 0 \tag{32}$$

It may also be noted that equations (15) and (31) yield

$$\left(\frac{\partial \ln Q}{\partial P}\right)_T = \frac{\Delta V}{RT} \tag{33}$$

2. Le Chatelier–Braun Principle

The fugacity of a substance increases with its concentration. From the definition of K as a ratio of fugacities, it may thus be seen that an increase in the value of K results in the formation of more products and less reactants; i.e., the reaction proceeds forward. Similarly, a decrease in the value of K is translated for the reaction by a move backward. Equation (30) shows that an increase in temperature moves the equilibrium forward if it is endothermic, backward if it is exothermic, or in other words, moves the equilibrium in the direction in which the system absorbs heat. Similarly, a decrease in temperature moves the equilibrium in the direction where the system produces heat (becomes exothermic). Thus, the system reacts in a way which opposes the change in temperature and tends to nullify its effect.

It is interesting to consider where the contrary proposition would lead. If an increase in temperature moves the reaction in an exothermic direction, the system would then produce heat which would increase the temperature of the system, produce more heat, again increase the temperature, . . .—a chain reaction which may end in disaster. Obviously then, the system would be unstable and one should expect the results of Chapter III on the stability of a system to be incorporated in the description of the temperature effect on the equilibrium of the system.

Actually, this temperature effect is just an illustration of a very general principle, called the *Le Chatelier–Braun principle: If a system in equilibrium is subjected to constraints which displace the equilibrium, the reaction proceeds in such a direction as to accommodate the constraints and partially nullify their effects.*

In Section 1 we established

$$dG = -S\,dT + V\,dP + \Delta G\,d\lambda \tag{34}$$

where

$$\Delta G = \sum \nu_i \mu_i = (\partial G/\partial \lambda)_{P,T} \tag{35}$$

The perfect differential of ΔG may be written

$$d(\Delta G) = d\left(\frac{\partial G}{\partial \lambda}\right)_{P,T} = \frac{\partial}{\partial T}\left(\frac{\partial G}{\partial \lambda}\right)_{P,T} dT + \frac{\partial}{\partial P}\left(\frac{\partial G}{\partial \lambda}\right)_{P,T} dP + \left(\frac{\partial^2 G}{\partial \lambda^2}\right)_{P,T} d\lambda \tag{36a}$$

or

$$d(\Delta G) = \frac{\partial}{\partial \lambda}\left(\frac{\partial G}{\partial T}\right)_P dT + \frac{\partial}{\partial \lambda}\left(\frac{\partial G}{\partial P}\right)_T dP + \left(\frac{\partial^2 G}{\partial \lambda^2}\right)_{P,T} d\lambda \tag{36b}$$

$$d(\Delta G) = -\left(\frac{\partial S}{\partial \lambda}\right)_{P,T} dT + \left(\frac{\partial V}{\partial \lambda}\right)_{P,T} dP + \left(\frac{\partial^2 G}{\partial \lambda^2}\right)_{P,T} d\lambda \tag{36c}$$

In the field of equilibrium states

$$\Delta G = 0 \qquad \text{and} \qquad d(\Delta G) = 0 \tag{37}$$

Consequently, equation (36c) yields

$$\left(\frac{\partial \lambda_{\text{eq}}}{\partial P}\right)_T = -\frac{(\partial V/\partial \lambda)_{P,T}}{(\partial^2 G/\partial \lambda^2)_{P,T}} \tag{38}$$

and

$$\left(\frac{\partial \lambda_{\text{eq}}}{\partial T}\right)_P = \frac{(\partial S/\partial \lambda)_{P,T}}{(\partial^2 G/\partial \lambda^2)_{P,T}} \tag{39}$$

But

$$(dV)_{P,T} = \sum_{i=1}^{r} dn_i\, \overline{V}_i = \sum_{i=1}^{r} \nu_i \overline{V}_i\, d\lambda = \Delta V\, d\lambda \tag{40}$$

or

$$(\partial V/\partial \lambda)_{P,T} = \Delta V \tag{41}$$

and similarly

$$(dS)_{P,T} = \sum_{i=1}^{r} dn_i\, \overline{S}_i = \sum_{i=1}^{r} \nu_i \overline{S}_i\, d\lambda = \Delta S\, d\lambda \tag{42}$$

or

$$(\partial S/\partial \lambda)_{P,T} = \Delta S \tag{43}$$

Therefore,

$$\left(\frac{\partial \lambda_{\text{eq}}}{\partial P}\right)_T = \frac{-\Delta V}{(\partial^2 G/\partial \lambda^2)_{P,T}} \tag{44}$$

$$\left(\frac{\partial \lambda_{\text{eq}}}{\partial T}\right)_P = \frac{\Delta S}{(\partial^2 G/\partial \lambda^2)_{P,T}} = \frac{\Delta H}{T(\partial^2 G/\partial \lambda^2)_{P,T}} \tag{45}$$

In Chapter III we saw that for a stable system the second derivative of the Gibbs free energy with respect to the displacement variable $(\partial^2 G/\partial \lambda^2)$ must be positive. Consequently, equation (44) shows that an increase in pressure causes the reaction to proceed in the direction in which the volume of the system is decreased. Similarly, equation (45) shows that an increase in temperature causes the reaction to proceed in the direction in which heat is absorbed, forward for an endothermic reaction ($\Delta H > 0$), backward for an exothermic one ($\Delta H < 0$). These results have already been obtained in our discussion of the signs of the derivatives of ln K with respect to T.

The preceding demonstration in terms of λ may be generalized easily to include other forces. Assuming that the system is a function of the variable r, the total differential of the free energy of the system undergoing a chemical reaction becomes

$$dG = -S\,dT + V\,dP + \Delta G\,d\lambda + R\,dr \tag{46}$$

where

$$R = (\partial G/\partial r)_{P,T,n_i \text{ or } \lambda} \tag{47}$$

The equivalent of equation (36) is now

$$d\left(\frac{\partial G}{\partial \lambda}\right)_{P,T,r} = -\left(\frac{\partial S}{\partial \lambda}\right)_{P,T,r} dT + \left(\frac{\partial V}{\partial \lambda}\right)_{P,T,r} dP + \left(\frac{\partial R}{\partial \lambda}\right)_{P,T,r} dr + \left(\frac{\partial^2 G}{\partial \lambda^2}\right)_{P,T,r} d\lambda \tag{48}$$

and in the field of equilibrium states

$$\left(\frac{\partial \lambda_e}{\partial r}\right)_{P,T} = -\frac{(\partial R/\partial \lambda)_{P,T,r}}{(\partial^2 G/\partial \lambda^2)_{P,T,r}} \tag{49}$$

An increase in r at constant T and P will displace the equilibrium in the direction in which R is decreased. Consequently, the work term increment $R\,dr$ of the free energy (corresponding to this increase dr) is reduced by the displacement of the equilibrium. A good general discussion of Le Chatelier–Braun principle has been given by Bever and Rocca [1].

3. Case of Simultaneous Reactions

3.1. General Treatment

Again, let us consider a system undergoing transformations which result in changes of the number of moles of the r components of that system. We shall now assume that a single chemical reaction is unable to describe these changes and that a minimum of p chemical reactions is necessary:

$$\sum_{i=1}^{r} \nu_i^s \mathrm{A}_i = 0 \qquad (s = 1, 2, \ldots, p) \tag{50}$$

Each reaction is characterized by a progress variable λ_s:

$$\frac{dn_1^s}{\nu_1^s} = \frac{dn_2^s}{\nu_2^s} = \cdots = \frac{dn_r^s}{\nu_r^s} = d\lambda_s \tag{51}$$

and the total change in the number of moles of species i is

$$dn_i = \sum_{s=1}^{p} dn_i^s = \sum_{s=1}^{p} \nu_i^s\,d\lambda_s \tag{52}$$

Proceeding as for the case of a single reaction, the criterion of equilibrium becomes

$$(\delta G)_{P,T} = \sum_{i=1}^{r} \mu_i \left(\sum_{s=1}^{p} \nu_i^s\,\delta\lambda_s\right) = \sum_{s=1}^{p}\left(\sum_{i=1}^{r} \nu_i^s \mu_i\right) d\lambda_s \geq 0 \tag{53}$$

or with

$$\Delta G_s = \sum \nu_i^s \mu_i \tag{54}$$

$$(\delta G)_{P,T} = \sum_{s=1}^{p} \Delta G_s \, \delta\lambda_s \geq 0 \tag{55}$$

Alternatively, the system is subject to an irreversible transformation when

$$(dG)_{P,T} = \sum_{s=1}^{p} \Delta G_s \, d\lambda_s < 0 \tag{56}$$

If all the $d\lambda_s$ are totally arbitrary (independent and uncoupled), then for a given sign of ΔG_s, $d\lambda_s$ must have the opposite sign to lower the total free energy of the system; i.e., for each reaction s,

$$\Delta G_s \, d\lambda_s < 0 \qquad (s = 1, 2, \ldots, p) \tag{57}$$

The condition for equilibrium becomes

$$\Delta G_s = 0 \qquad (s = 1, 2, \ldots, p) \tag{58}$$

If the $d\lambda_s$ are not totally arbitrary (i.e., if they are independent but coupled), the inequalities (57) are no longer necessary. Only inequality (56) remains valid. A mechanical analogy would be two wagons on the opposite sides of a hill (see Fig. 3). If they are coupled by a spring, the acceleration of the heavier wagon in rolling downhill may force the lighter wagon uphill. The concept of coupled reactions will be presented more explicitly in Section 6.2. We note here that although coupled reactions are frequently encountered in biology, they seem far less common in metallurgy and materials science. Consequently, in this text we shall treat the cases of simultaneous independent reactions with the assumption that these reactions are uncoupled. This assumption has no bearing on the determination of the final (equilibrium) state of the system, only on its evolution.

The formulas developed for a single reaction can be readily extended to the case of several reactions. Assuming that there are p reactions,

$$\nu_1^{(1)} A_1 + \nu_2^{(1)} A_2 + \cdots + \nu_r^{(1)} A_r = 0$$

$$\nu_1^{(2)} A_1 + \nu_2^{(2)} A_2 + \cdots + \nu_r^{(2)} A_r = 0$$

$$\vdots$$

$$\nu_1^{(p)} A_1 + \nu_2^{(p)} A_2 + \cdots + \nu_r^{(p)} A_r = 0$$

or, in matrix and vector notation,[2]

$$(\nu)\boldsymbol{A} = \boldsymbol{0} \tag{59}$$

By definition

$$\boldsymbol{\Delta G} = (\nu)\boldsymbol{\mu} \tag{60}$$

and

$$\boldsymbol{\Delta G}^\circ = (\nu)\boldsymbol{\mu}^\circ \tag{61}$$

[2] Parentheses identify matrices, bold type vectors.

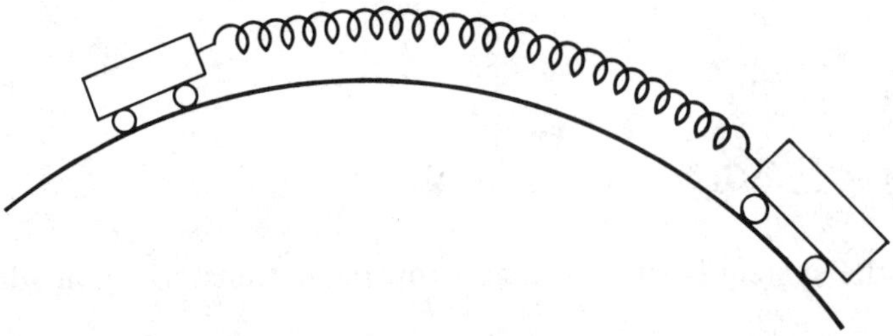

Figure 3. Analogy for coupled reactions. If the wagons are coupled by a spring, the acceleration of the heavier wagon in rolling downhill may force the lighter wagon uphill. The positions of the two wagons can be fixed arbitrarily but not their movements.

Consequently,

$$\mathbf{\Delta G} - \mathbf{\Delta G}^\circ = (\nu)(\boldsymbol{\mu} - \boldsymbol{\mu}^\circ) = RT(\nu) \ln \boldsymbol{f} = RT \ln \boldsymbol{Q} \tag{62}$$

At equilibrium, condition (58) yields

$$\mathbf{\Delta G} = \mathbf{0} \tag{63}$$

and if $\boldsymbol{K} = \boldsymbol{Q}_{\text{eq}}$, equation (62) yields

$$\mathbf{\Delta G}^\circ = -RT \ln \boldsymbol{K} \tag{64}$$

Note that the enthalpy terms may be obtained as follows:

$$\mathbf{\Delta H} = \left(\frac{\partial(\mathbf{\Delta G}/T)}{\partial(1/T)}\right)_P = \left(\frac{d(\mathbf{\Delta G}^\circ/T)}{d(1/T)}\right) + R\left(\frac{\partial \ln \boldsymbol{Q}}{\partial(1/T)}\right)_P \tag{65}$$

or

$$\mathbf{\Delta H} = \mathbf{\Delta H}^\circ + R\left(\frac{\partial \ln \boldsymbol{Q}}{\partial(1/T)}\right)_P \tag{66}$$

In the case where the fugacities of the substances A_i are assumed to be replaceable by partial pressures and concentrations, $\ln \boldsymbol{Q}$ is independent of temperature and $\mathbf{\Delta H}$ is thus identical to $\mathbf{\Delta H}^\circ$. We also note that

$$\mathbf{\Delta S} = -\left(\frac{\partial \mathbf{\Delta G}}{\partial T}\right)_P = \mathbf{\Delta S}^\circ - \left(\frac{\partial}{\partial T}(RT \ln \boldsymbol{Q})\right)_P \tag{67a}$$

or

$$\mathbf{\Delta S} = \mathbf{\Delta S}^\circ - R \ln \boldsymbol{Q} - RT\left(\frac{\partial \ln \boldsymbol{Q}}{\partial T}\right)_P \tag{67b}$$

The free energy change of the system is

$$dG = \mathbf{\Delta G} \cdot \boldsymbol{d\lambda} \tag{68}$$

Similarly, the heat generated by the reactions in the system is

$$dH = \mathbf{\Delta H} \cdot \boldsymbol{d\lambda} \tag{69}$$

The general problem of calculating the equilibrium composition of a system—

and thus, the progress variables of its reactions—may be analyzed as follows. Let us designate by n_i the initial number of moles and by $n_i^* = n_i + dn_i$ the final (equilibrium) number of moles of species i. In vector notation, $\boldsymbol{n}$ is known and $\boldsymbol{n}^*$ or $\boldsymbol{dn}$ is the unknown.

$$\boldsymbol{dn} = \boldsymbol{n}^* - \boldsymbol{n} \tag{70}$$

Among the r reacting species, we select a maximum of p *independent* reactions (such that no line of the ν matrix can be obtained by a linear combination of the other lines of the matrix):

$$\boldsymbol{\Delta G}^\circ = -RT \ln \boldsymbol{K} \tag{71}$$

Equation (71) yields p relations to calculate the r unknowns. But from mass balance considerations [equation (52)]

$$\boldsymbol{dn} = (\nu)^{\mathrm{t}}\, \boldsymbol{d\lambda} \tag{72}$$

where $(\nu)^{\mathrm{t}}$ is the transposed matrix of (ν). Equation (72) yields r additional relations with p unknowns (the $d\lambda_s$), or $r - p$ additional relations to calculate the dn_i. Thus, we have a total of $(r - p) + p = r$ independent relations to calculate the r unknowns, and the problem has a solution.

The $r - p$ mass balance equations need not take the form of equation (72). They may be written without explicitly introducing progress variables (see the following examples). Their form is a matter of convenience which depends on the nature of the problem at hand.

Note that equations (70) and (72) are valid whether the increments $\boldsymbol{dn}$ and $\boldsymbol{d\lambda}$ are infinitesimal or finite. However, equations (68) and (69) are valid only for infinitesimal increments of G and H. When these increments are finite, the equations must be rewritten

$$DG = G_{\text{final}} - G_{\text{initial}} = \int_{\text{initial}}^{\text{final}} \boldsymbol{\Delta G}\cdot\boldsymbol{d\lambda} \tag{73}$$

and

$$DH = H_{\text{final}} - H_{\text{initial}} = \int_{\text{initial}}^{\text{final}} \boldsymbol{\Delta H}\cdot\boldsymbol{d\lambda} \tag{74}$$

In general, it is difficult to evaluate these integrals directly and it is easier to calculate directly the initial and final values of G and H. However, when the final state is the equilibrium state, the calculation can be simplified:

$$DG = G^* - G = \boldsymbol{n}^*\cdot\boldsymbol{\mu}^* - \boldsymbol{n}\cdot\boldsymbol{\mu} \tag{75}$$

$$= (\boldsymbol{n}^* - \boldsymbol{n})\cdot\boldsymbol{\mu}^* + \boldsymbol{n}\cdot(\boldsymbol{\mu}^* - \boldsymbol{\mu}) \tag{76}$$

and through equations (70) and (72)

$$DG = (\nu)^{\mathrm{t}}\boldsymbol{D\lambda}\cdot\boldsymbol{\mu}^* + \boldsymbol{n}\cdot(\boldsymbol{\mu}^* - \boldsymbol{\mu})$$

$$= \boldsymbol{D\lambda}\cdot(\nu)\boldsymbol{\mu}^* + \boldsymbol{n}\cdot(\boldsymbol{\mu}^* - \boldsymbol{\mu}) \tag{77}$$

But at equilibrium

$$\boldsymbol{\Delta G}^* = (\nu)\boldsymbol{\mu}^* = 0 \tag{78}$$

and consequently,

$$DG = \boldsymbol{n}\cdot(\boldsymbol{\mu}^* - \boldsymbol{\mu}) = -RT\boldsymbol{n}\cdot\mathbf{ln}(\boldsymbol{f}/\boldsymbol{f}^*) \tag{79}$$

where $\mathbf{ln}(\boldsymbol{f}/\boldsymbol{f}^*)$ is the vector with components $\ln(f_i/f_i^*)$. Similar notation for ratios will appear several times in this section. An alternative expression of DG is also possible:

$$DG = (\boldsymbol{n}^* - \boldsymbol{n})\cdot\boldsymbol{\mu} + \boldsymbol{n}^*\cdot(\boldsymbol{\mu}^* - \boldsymbol{\mu}) \tag{80}$$

and after transformation

$$DG = \boldsymbol{\Delta G}\cdot\boldsymbol{D\lambda} - RT\boldsymbol{n}^*\cdot\mathbf{ln}(\boldsymbol{f}/\boldsymbol{f}^*) \tag{81}$$

This equation emphasizes the fact that although $dG = \boldsymbol{\Delta G}\cdot\boldsymbol{d\lambda}$, the integrated quantity DG is different from $\boldsymbol{\Delta G}\cdot\boldsymbol{D\lambda}$ [by the nonnegligible term $RT\boldsymbol{n}\cdot\mathbf{ln}(\boldsymbol{f}/\boldsymbol{f}^*)$]. From a practical standpoint, however, equation (79) is generally more convenient than equation (81).

Expressions for DH may be derived in a similar way. In the special case where fugacities can be replaced by mole fractions and partial pressures, we have seen that

$$\boldsymbol{\Delta H} = \boldsymbol{\Delta H}^\circ \tag{82}$$

and since $\boldsymbol{\Delta H}^\circ$ is independent of the progress variables, equation (74) becomes

$$DH = \boldsymbol{\Delta H}^\circ\cdot\boldsymbol{D\lambda} \tag{83}$$

If the deviations from equilibrium are small (without being infinitesimal), the expression for DG may be further simplified. We note that

$$\left(\frac{\partial \Delta G_i}{\partial \lambda_j}\right)_{\lambda_i,\lambda_k} = \left(\frac{\partial^2 G}{\partial \lambda_i\, \partial \lambda_j}\right)_{\lambda_k} = \left(\frac{\partial \Delta G_j}{\partial \lambda_i}\right)_{\lambda_j,\lambda_k} \tag{84}$$

A Taylor series expansion around the equilibrium state yields

$$G(\ldots, \lambda_i, \ldots) = G(\ldots, \lambda_i^*, \ldots) + \frac{1}{2}\sum_{i=1}^{p}\sum_{j=1}^{p}\left(\frac{\partial^2 G}{\partial \lambda_i\, \partial \lambda_j}\right)_{\text{eq}}(\lambda_i - \lambda_i^*)(\lambda_j - \lambda_j^*) \tag{85}$$

The first order derivatives do not appear, since

$$(\partial G/\partial \lambda_i)_{\text{eq}} = \Delta G_i^* = 0 \tag{86}$$

and the third order terms are assumed negligible (since the $\lambda_i - \lambda_i^*$ are small for small deviations from equilibrium). Thus, the Gibbs free energy surface is assimilated to a paraboloid in the vicinity of its minimum (the equilibrium state). Equation (85) yields

$$\Delta G_i = \left(\frac{\partial G}{\partial \lambda_i}\right)_{\lambda_j} = \sum_{j=1}^{p}\left(\frac{\partial^2 G}{\partial \lambda_i\, \partial \lambda_j}\right)_{\text{eq}}(\lambda_j - \lambda_j^*) \tag{87}$$

[the coefficient $\frac{1}{2}$ in equation (85) disappears because, in the double summation,

i appears in the first summation as well as in the second.] Consequently,

$$DG = \tfrac{1}{2} \sum_{i=1}^{p} \Delta G_i(\lambda_i - \lambda_i^*) \tag{88}$$

$$= \tfrac{1}{2}\Delta \mathbf{G} \cdot D\boldsymbol{\lambda} = (RT/2)\, \mathbf{ln}\, \boldsymbol{Q/K} \cdot D\boldsymbol{\lambda} \tag{89}$$

In Chapter III, we analyzed in detail the free energy changes associated with the nucleation and growth of precipitates. These transformations may be considered as chemical reactions, and the results of this section apply to these transformations as well. It may easily be verified for example that equation (III.90) is identical to equation (79).

It should be emphasized here that the formalism of chemical reactions is just an expression of results which may equally well be studied through free energy curves and phase diagrams. The choice of the approach is a matter of convenience. The formalism of chemical reactions is often simpler when the transformation involves stoichiometric compounds, i.e., free energy curves of negligible width.

3.2. Example: Composition of the Gas Phase of a Furnace

The gas phase of a heat-treating furnace at 1000°K and a pressure of 1.1 atm is assumed in equilibrium. A gas sample is taken and analyzed for its main constituents. On a volume percent basis the results are as follows: 20.5% CO, 18.5% CO_2, 5.5% CH_4, 14% H_2, 2.3% H_2O, and 39.2% N_2. However, of necessity the analysis is not done at the furnace temperature, and therefore the percentages analyzed do not represent the percentages in the furnace. Nevertheless, it is possible to deduce these by imagining that the gas is replaced in the furnace and by calculating its composition when equilibrium is achieved. We shall proceed to do this by two different methods.

3.2.1. First Method

Since volume percentages in the gas are proportional to numbers of moles, it is convenient to consider a total number of moles equal to 1. Mass balances on the carbon, oxygen, hydrogen, and nitrogen elements give the following relations between the relative concentrations of the species in the sample analyzed (n_i') and in the furnace (n_i):

$$n_C = n'_{CO} + n'_{CO_2} + n'_{CH_4} = n_{CO} + n_{CO_2} + n_{CH_4} \tag{90a}$$

$$n_O = n'_{CO} + 2n'_{CO_2} + n'_{H_2O} = n_{CO} + 2n_{CO_2} + n_{H_2O} \tag{91a}$$

$$n_H = 2n'_{H_2} + 2n'_{H_2O} + 4n'_{CH_4} = 2n_{H_2} + 2n_{H_2O} + 4n_{CH_4} \tag{92a}$$

$$n'_{N_2} = n_{N_2} \tag{93a}$$

or

$$n_{CO} + n_{CO_2} + n_{CH_4} = 0.205 + 0.185 + 0.055 = 0.445 \tag{90b}$$

$$n_{CO} + 2n_{CO_2} + n_{H_2O} = 0.205 + 0.370 + 0.023 = 0.598 \tag{91b}$$

$$2n_{H_2} + 2n_{H_2O} + 4n_{CH_4} = 0.280 + 0.046 + 0.220 = 0.546 \tag{92b}$$

$$n_{N_2} = 0.392 \tag{93b}$$

Table 2

Species	Gas analysis (vol. %)	Actual gas composition (vol. %)	Actual partial pressures $P = 1.1$ atm
CO	20.5	30.2	0.332
CO_2	18.5	9.8	0.108
CH_4	5.5	0.23	0.0026
H_2	14.0	20.0	0.220
H_2O	2.3	4.3	0.047
N_2	39.2	35.5	0.390

Moreover, two independent chemical reactions may be written

$$CO_2 + H_2 = CO + H_2O, \qquad (\Delta G_1^\circ)_{1000^\circ K} = 850 \quad \text{cal} \tag{94}$$

$$2CO + 2H_2 = CH_4 + CO_2, \qquad (\Delta G_2^\circ)_{1000^\circ K} = 5910 \quad \text{cal} \tag{95}$$

They correspond to two equilibrium constants K_1 and K_2:

$$\log K_1 = 850/4575 = -0.186 \qquad \text{or} \qquad K_1 = 0.65 \tag{96}$$

$$\log K_2 = -5910/4575 = -1.292 \qquad \text{or} \qquad K_2 = 0.051 \tag{97}$$

For convenience let

$$n_g = n_{CO} + n_{CO_2} + n_{CH_4} + n_{H_2} + n_{H_2O} + n_{N_2} \tag{98}$$

Then

$$K_1 = \frac{p_{CO}p_{H_2O}}{p_{CO_2}p_{H_2}} = \frac{n_{CO}n_{H_2O}}{n_{CO_2}n_{H_2}} = 0.65 \tag{99}$$

and

$$K_2 = \frac{p_{CO_2}p_{CH_4}}{p_{CO}^2 p_{H_2}^2} = \frac{n_{CO_2}n_{CH_4}}{n_{CO}^2 n_{H_2}^2} \times \frac{n_g^2}{P^2} = 0.051 \tag{100}$$

We thus have a system of seven equations [(90)–(93) and (98)–(100)] with seven unknowns (n_{CO}, n_{CO_2}, n_{CH_4}, n_{H_2}, n_{H_2O}, n_{N_2}, and n_g), which may be solved numerically. The results are summarized in Table 2. An alternative solution by the method of progress variables is also possible, and generally simpler to apply.

3.2.2. Second Method

Five unknowns are identified: n_{CO}, n_{CO_2}, n_{CH_4}, n_{H_2}, and n_{H_2O} (n_{N_2} remains unchanged since nitrogen participates in no reaction). Since it is possible to write three mass balances on the elements C, O, and H, two additional equations are needed and they must correspond to two independent chemical reactions. Having established that there are only two independent reactions, we select arbitrarily the reactions (94) and (95):

$$CO_2 + H_2 = CO + H_2O \tag{94}$$

$$2CO + 2H_2 = CH_4 + CO_2 \tag{95}$$

If λ_1 and λ_2 are the two progress variables associated with these reactions, on the basis of 1 mol of sample we may write

$$n_{CO} = 0.205 + \lambda_1 - 2\lambda_2 \tag{101}$$

$$n_{CO_2} = 0.185 - \lambda_1 + \lambda_2 \tag{102}$$

$$n_{CH_4} = 0.055 + \lambda_2 \tag{103}$$

$$n_{H_2} = 0.14 - \lambda_1 - 2\lambda_2 \tag{104}$$

$$n_{H_2O} = 0.023 + \lambda_1 \tag{105}$$

$$n_{N_2} = 0.392 \tag{106}$$

$$n_g = 1 - 2\lambda_2 \tag{107}$$

At equilibrium, equations (94) and (95) yield

$$K_1 = \frac{(0.205 + \lambda_1 - 2\lambda_2)(0.023 + \lambda_1)}{(0.185 - \lambda_1 + \lambda_2)(0.14 - \lambda_1 - 2\lambda_2)} = 0.65 \tag{108}$$

$$K_2 = \frac{(0.055 + \lambda_2)(0.185 - \lambda_1 + \lambda_2)(1 - 2\lambda_2)^2}{(0.205 + \lambda_1 - 2\lambda_2)^2(0.14 - \lambda_1 - 2\lambda_2)^2(1.1)^2} = 0.051 \tag{109}$$

This system of two equations [which are equivalent to (96) and (97)] can now be solved for the two unknowns λ_1 and λ_2. Through equations (101)–(105) these progress variables yield the number of moles of each species and therefore the composition of the gas.

Several numerical techniques may be used to solve for λ_1 and λ_2. The following iteration technique is one of the simplest.

The system characterized by the analysis of the gas is not in equilibrium at 1000°K, and to determine whether the reactions (94) and (95) would proceed to the right or to the left (i.e., to determine the signs of λ_1 and λ_2) we calculate the reaction quotients Q_1 and Q_2:

$$Q_1 = \frac{p_{CO}p_{H_2O}}{p_{CO_2}p_{H_2}} = \frac{0.205 \times 0.023}{0.185 \times 0.14} = 0.182 \tag{110}$$

$$Q_2 = \frac{p_{CH_4}p_{CO_2}}{p_{CO}^2 p_{H_2}^2} = \frac{0.055 \times 0.185}{(0.205)^2 \times (0.14)^2} \times \frac{1}{(1.1)^2} = 10.2 \tag{111}$$

We note that $Q_1 < K_1$ and $Q_2 > K_2$. Thus $\Delta G_1 < 0$ and $\Delta G_2 > 0$. Reaction (94) must then proceed to the right ($\lambda_1 > 0$) and reaction (95) to the left ($\lambda_2 < 0$).

Since λ_2 must also be larger than -0.055 in order for n_{CH_4} to remain positive [equation (103)], we may assume for λ_2 a starting value of -0.030. Equation (108) may then be solved for λ_1 and yields $\lambda_1 = 0.027$. [Note that we do not solve equation (109) for λ_1 since it is too sensitive to the assumed value of λ_2.] With this value of λ_1 we solve equation (109) for λ_2 and obtain $\lambda_2 = -0.052$. With this improved value of λ_2 we solve again equation (108) and find $\lambda_1 = 0.024$. This value in equation (109) leads to $\lambda_2 = -0.052$. The iteration has thus converged and the values of the two progress variables at equilibrium are $\lambda_1 = 0.024$ and $\lambda_2 = -0.052$. They yield the results already listed in Table 2.

3.2.3. Remarks

Considering the reaction

$$CH_4 + 2H_2O = CO_2 + 4H_2 \tag{112}$$

it may not be obvious whether it is independent of reactions (94) and (95). If we let

$$A_1 \equiv CO, \quad A_2 \equiv CO_2, \quad A_3 \equiv CH_4, \quad A_4 \equiv H_2, \quad A_5 \equiv H_2O \tag{113}$$

a general method consists in calculating the rank of the matrix (ν) associated with reactions (94), (95), and (112):

$$(\nu) = \begin{pmatrix} 1 & -1 & 0 & -1 & 1 \\ -2 & 1 & 1 & -2 & 0 \\ 0 & 1 & -1 & 4 & -2 \end{pmatrix} \tag{114}$$

It may be verified here that any determinant of size 3 is equal to zero and that the rank of the matrix is therefore equal to 2. The consequent linear combination between the three rows of the matrix [i.e., between the chemical reactions (94), (95), and (112)] is

$$\boldsymbol{R}_{112} = -2\boldsymbol{R}_{94} - \boldsymbol{R}_{95} \tag{115}$$

where $\boldsymbol{R}_s$ corresponds to the reaction s.

We have stressed that often the easiest way of finding the number p of independent reactions is to subtract from the number of unknowns the number of mass balance equations. However, even after the number p is known, it may be necessary to verify that the p reactions selected are indeed independent.

When the numbers of unknowns and reactions are large, computer techniques become most convenient. Many software programs are now available that use a large array of search techniques [2].

We shall now present another example, typical of many cases, where the calculations are easy because the concentrations of the components have different orders of magnitude. By neglecting certain small terms, one may take into account the mass balances in much simpler ways.

3.3. Example: Decomposition of Silica

A small tube of pure silica (SiO_2) with a capacity of 20 cm^3 is sealed at room temperature (298°K). It is then placed in a furnace at 1800°K. We wish to determine the total pressure and the composition of the gas phase inside the tube due to the decomposition of SiO_2 in the following cases:

a. the tube has been sealed under vacuum,
b. the tube has been sealed under air, and
c. the tube has been sealed under vacuum but contains a few grams of graphite.

This example is of practical interest because, at high temperatures, one is often faced with the problem of possible contamination of a sample by the walls of a reaction chamber through the gaseous atmosphere of the chamber.

At 1800°K the following data may be found:

$$Si + \tfrac{1}{2}O_2(g) = SiO(g), \quad \Delta G^\circ = -57{,}400 \text{ cal} \tag{116}$$

$$Si + O_2(g) = SiO_2, \quad \Delta G^\circ = -140{,}200 \text{ cal} \tag{117}$$

$$C + \tfrac{1}{2}O_2(g) = CO(g), \quad \Delta G^\circ = -64{,}480 \text{ cal} \tag{118}$$

$$C + O_2(g) = CO_2(g), \quad \Delta G^\circ = -94{,}720 \text{ cal} \tag{119}$$

To treat this problem we first note that the standard free energy change of the reaction

$$SiO_2 = SiO(g) + \tfrac{1}{2}O_2(g) \tag{120}$$

may be calculated as

$$\Delta G^\circ = 140{,}200 - 57{,}400 = 82{,}800 \quad \text{cal} \tag{121}$$

Consequently,

$$\log K = -\frac{82{,}800}{4.575 \times 1800} = -10.0 \tag{122}$$

or

$$p_{SiO} p_{O_2}^{1/2} = K = 10^{-10} \tag{123}$$

a. In the case where the tube has been sealed under vacuum the decomposition of SiO_2 into SiO(g) and O_2(g) yields, for every mole of SiO, $\frac{1}{2}$ mol O_2. Consequently,

$$p_{O_2} = \tfrac{1}{2} p_{SiO} \tag{124}$$

and

$$(\tfrac{1}{2})^{1/2} (p_{SiO})^{3/2} = 10^{-10} \tag{125}$$

or

$$p_{SiO} = 2.7 \times 10^{-7} \quad \text{atm}$$

$$p_{O_2} = 1.3 \times 10^{-7} \quad \text{atm}$$

$$P = p_{SiO} + p_{O_2} = 4.0 \times 10^{-7} \quad \text{atm}$$

b. If the tube has been sealed under air, the atmosphere is necessarily more oxidizing than in the previous case and the equilibrium (120) will be displaced to the left ($p_{SiO} < 2.7 \times 10^{-7}$).

At 298°K the partial pressure of oxygen was 0.21 atm. At 1800°K, if there were no decomposition of the silica, it would be

$$p_{O_2} = 0.21 \times 1800/298 = 1.27 \quad \text{atm}$$

Because of the decomposition of SiO_2, p_{O_2} is higher, but by a very negligible amount; to the value of 1.27 must be added a value smaller than the previous 1.3×10^{-7} (since the equilibrium is displaced now to the left). Consequently, in the conditions of the experiment

$$p_{O_2} = 1.27 \quad \text{atm}$$

p_{SiO} may be calculated through equation (123):

$$p_{SiO} = 10^{-10}/(1.27)^{1/2} = 9 \times 10^{-11} \quad \text{atm}$$

The partial pressure of nitrogen is

$$p_{N_2} = 0.79 \times 1800/298 = 4.77 \quad \text{atm}$$

and

$$P = p_{O_2} + p_{N_2} + p_{SiO} = 6.0 \quad \text{atm}$$

c. Because of the presence of carbon, the atmosphere is now more reducing than in case a (p_{O_2} will be smaller than 1.3×10^{-7}) and the decomposition of SiO_2 will be more pronounced.

Through examination of possible reactions such as

$$C + \tfrac{1}{2}O_2(g) = CO(g), \qquad p_{CO}/(p_{O_2})^{1/2} = K = 6.3 \times 10^{7} \tag{126}$$

and

$$2CO(g) = C + CO_2(g), \qquad p_{CO_2}/(p_{CO})^2 = K = 7 \times 10^{-5} \tag{127}$$

it is clear that CO may be present in a significant amount, O_2 and CO_2 in very small amounts. Thus, the only two species present in any significant amount in the gas phase will be CO and SiO.

The transformation in the system may then be described by the single overall reaction

$$C + SiO_2 = CO(g) + SiO(g), \qquad p_{CO}p_{SiO} = K = 6 \times 10^{-3} \tag{128}$$

for which

$$p_{CO} = p_{SiO} \tag{129}$$

(and which is true only if the amount of CO transformed into CO_2, or formed by C plus O_2, is negligible). Consequently,

$$p_{CO} = p_{SiO} = (6 \times 10^{-3})^{1/2} = 0.08 \quad \text{atm} \tag{130}$$

It must be emphasized that the overall reaction (128) need not represent the actual mechanism of the decomposition of SiO_2 (for instance, the graphite powder may not be in contact with the silica).

Our assumptions concerning the smallness of p_{O_2} and p_{CO_2} may be justified quantitatively. Even if the results of equation (130) were inaccurate, they would still be sufficiently precise to calculate the orders of magnitude of p_{O_2} and p_{CO_2} using equations (126) and (127). The calculations yield

$$p_{O_2} = 1.5 \times 10^{-18} \quad \text{atm}, \qquad p_{CO_2} = 4.2 \times 10^{-7} \quad \text{atm}$$

Our assumptions are thus justified and the results of equations (124) and (125) are accurate. If we had found p_{O_2} or p_{CO_2} of the same order of magnitude as p_{CO} and p_{SiO}, we should have restarted the calculations introducing the whole set of mass balance equations.

The total pressure is

$$P = p_{CO} + p_{SiO} + p_{O_2} + p_{CO_2} \simeq p_{CO} + p_{SiO} = 0.16 \quad \text{atm}$$

Note that silica may be present either in the vitreous or the crystalline state. The value of ΔG° in equation (117) corresponds to cristobalite (a crystalline phase). Actually, the temperature of 1800°K (1527°C) is too high for vitreous silica; it would soften below this temperature or, given time, would transform into cristobalite, which is the stable phase at high temperatures.

4. Application to Oxygen and Sulfur Potential Diagrams

Let us consider the equilibrium between a metal M, its oxide M_xO_2, and a gaseous phase of oxygen potential p_{O_2}:

$$xM + O_2 = M_xO_2 \tag{131}$$

(when $x = 2$, the right member of the reaction may be written 2MO). Assuming that the metal and its oxide are in their standard states, the equilibrium constant corresponding to this reaction is just the inverse of p_{O_2}:

$$K = 1/p_{O_2} \tag{132}$$

and

$$\Delta G^\circ = RT \ln p_{O_2} \tag{133}$$

If the oxygen potential is lower than $1/K$, the metal is not oxidized (or its oxide decomposes), and if it is higher, all the metal is oxidized.

A plot of the standard free energies of formation of various oxides vs the temperature is given in Fig. 4. This method of representation originated from Ellingham [3] and was further popularized by Richardson and Jeffes [4]. Its features will now be described.

The slopes of all the representative lines are equal to $-\Delta S°$ since $d\Delta G°/dT = -\Delta S°$. Moreover, since for all the metal–metal oxide equilibria $\Delta S° = S°(M_xO_2) - xS°(M) - S°(O_2)$ and the entropy of 1 mol of oxygen is much higher than the entropy difference for a metal and its oxide, $\Delta S°$ is negative and the slopes are positive. The lines are practically straight, showing that the $\Delta C_p°$ have a negligible influence on the temperature dependence of the corresponding $\Delta G°$. The breaks in the slopes are associated with a phase transformation of the metal or its oxide, such as melting or boiling.

The line representing the reaction $C + O_2 = CO_2$ is almost horizontal because the entropy of 1 mol of carbon dioxide is nearly equal to the sum of the entropies of 1 mol of oxygen and 1 mol of carbon. (The contribution of the latter is small; at 298°K, we have 1.4 cal/°K per mole of graphite vs 49 cal/°K per mole of O_2 and 51 cal/°K per mole of CO_2.) The line representing the equilibrium $2C + O_2 = 2CO$ has a negative slope for similar reasons. Note that for both these reactions the $\Delta G°$s are calculated considering CO and CO_2 to be in their standard states: each is at a pressure (or, more rigorously, a fugacity) of 1 atm.

If on the line representing the equilibrium (131) we locate the point corresponding to the temperature T and join it by a straight line to the point O on the upper left of the diagram, the intercept of this line with the scale of p_{O_2} on the right of the diagram yields the equilibrium value of p_{O_2} at this temperature T. For instance, at 1200°C the partial pressure of oxygen in equilibrium with nickel and its oxide NiO is found to be slightly higher than 10^{-8} atm. The basis for this construction is understood if one observes that the point O corresponds to 0°K and $RT \ln p_{O_2} = 0$; consequently, the slope of the constructed line is $RT \ln p_{O_2}/T$ or $R \ln p_{O_2}$, a measure of p_{O_2} independent of T. Thus, the scale on the right is just a calibration of this slope.

Similarly, two points H and C on the axis corresponding to 0°K permit by the same construction a direct reading of the ratios p_{H_2}/p_{H_2O} and p_{CO}/p_{CO_2} in equilibrium with the metal and its oxide. (See Problem 15 for a justification of this construction.) For example, at 1200°C nickel and nickel oxide are in equilibrium with an H_2–H_2O atmosphere corresponding to a ratio of 10^{-2} and with a CO–CO_2 atmosphere corresponding to a ratio of about $10^{-1.5}$ (~0.03).

The metals which have the lowest lines on the diagram correspond to the most stable oxides, that is, to those which best resist decomposition. Indeed, a very low oxygen potential is needed to reduce these oxides to the metallic state.

In practice, one often has to deal with situations where the metals and their oxides are not in their standard states. Then their activities are not equal to 1 and the results of the diagram have to be modified. However, even in these cases, the diagram generally indicates the orders of magnitude involved, and in this respect it has proved to be very useful.

It must also be stressed that the diagram yields equilibrium data only. For example, in experimental setups the choice of copper at 500°C to purify gaseous streams of their oxygen content may appear questionable. At 500°C the corre-

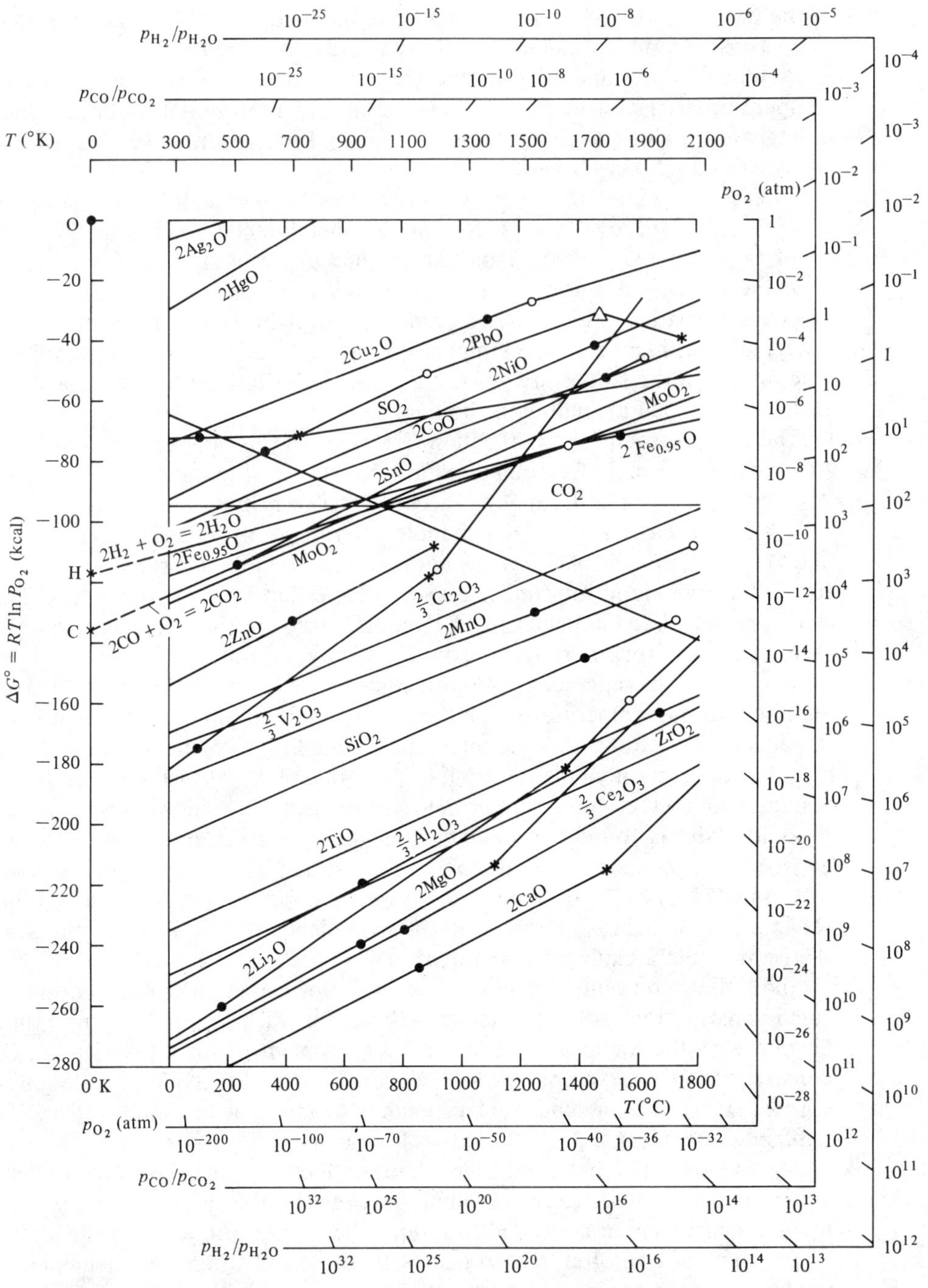

Figure 4. Standard free energy of formation of some selected oxides [4]. The $\Delta G°$ corresponds to the reaction $(2x/y)M + O_2 = (2/y)M_xO_y$. The symbols ● and * identify the melting and boiling points, respectively, of the metals; ○ and △ identify the melting and boiling points, respectively, of the oxides.

sponding equilibrium oxygen pressure is 10^{-16} atm. Lower temperatures would yield lower potentials, but unless one uses a much slower flow rate (or catalytic copper), the gas composition would have a larger departure from its equilibrium one and higher oxygen potentials would be likely. Temperatures higher than 500°C would favor the kinetics of the reaction but at the expense of a higher oxygen equilibrium potential. Experimentally, 500°C appears to offer a reasonable compromise. It may also be noted that copper is often preferred to other metals, such

Figure 5. Standard free energies of the reaction $(2x/y)M + S_2(g) = (2/y)M_xS_y$ for selected sulfides. The symbols ● and * identify the melting and boiling points, respectively, of the metals; ○ and △ identify the melting and boiling points, respectively, of the sulfides.

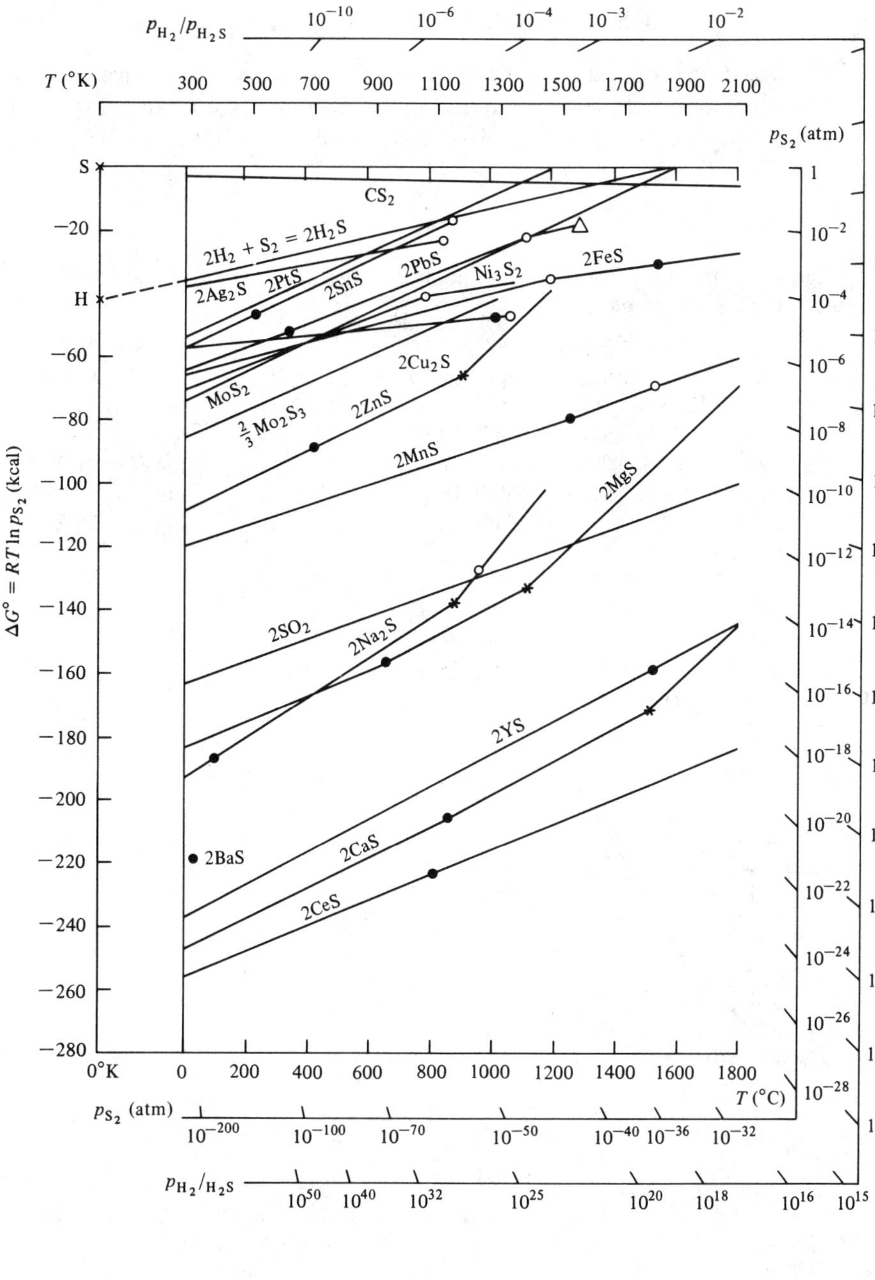

as silicon or magnesium, for other practical reasons. It is readily available and easy to rejuvenate (i.e., reduce its oxide) by passing hydrogen through it; moreover, it is likely that much lower oxygen potentials could not be retained in the gas flow because of all the unavoidable leaks in an apparatus.

A diagram for the standard free energies of formation of the sulfides is given in Fig. 5 and may be used in an entirely similar way. The points S and H on the 0°K axis yield, by the same construction as previously, the partial pressure p_{S_2} and the ratio p_{H_2}/p_{H_2S} corresponding to the reactions

$$xM + S_2(g) = M_xS_2 \qquad \text{with} \qquad K = 1/p_{S_2} \tag{134}$$

and

$$\tfrac{1}{2}xM + H_2S = \tfrac{1}{2}M_xS_2 + H_2 \qquad \text{with} \qquad K = p_{H_2}/p_{H_2S} \tag{135}$$

"Roasting" operations are common in extractive metallurgy and consist in transforming a sulfide to an oxide or to a sulfate. Generally, the oxide can be reduced by carbon (from coal or coke) and the sulfate can be leached by an

Figure 6. Stability domains of compounds in the Zn–S–O system at 900°K [6]. The lines describe the reactions

(a)	$ZnS + 2O_2 = ZnSO_4$,	$\log K = 26.61$
(b)	$3ZnSO_4 = ZnO{\cdot}2ZnSO_4 + SO_2 + \tfrac{1}{2}O_2$,	$\log K = -3.98$
(c)	$3ZnS + \tfrac{11}{2}O_2 = ZnO{\cdot}2ZnSO_4 + SO_2$,	$\log K = 75.84$
(d)	$ZnO{\cdot}2ZnSO_4 = 3ZnO + 2SO_2 + O_2$,	$\log K = -10.52$
(e)	$ZnS + \tfrac{3}{2}O_2 = ZnO + SO_2$,	$\log K = 21.77$
(f)	$Zn + SO_2 = ZnS + O_2$,	$\log K = -6.85$
(g)	$2Zn + O_2 = 2ZnO$,	$\log K = 29.84$

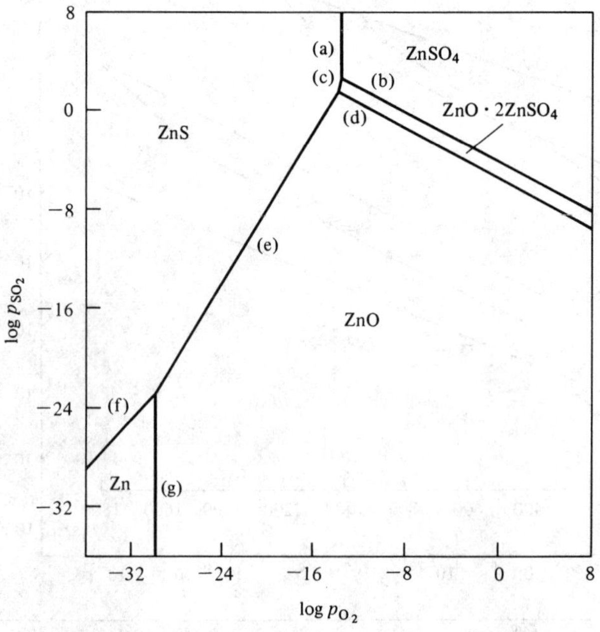

aqueous solution. Typical roasting reactions are

$$M_xS + \tfrac{3}{2}O_2 = M_xO + SO_2 \tag{136}$$

$$M_xS + 2O_2 = M_xSO_4 \tag{137}$$

At any given temperature, the phase rule shows that the variance ϑ of the system is equal to

$$\vartheta = c + 1 - \varphi \tag{138}$$

where c is the number of components and φ is the number of phases (see Section II.5.3). If the partial pressures of SO_2 and O_2 are arbitrarily fixed a maximum of only two phases can coexist, the gaseous phase and a condensed phase. The coexistence of two condensed phases corresponds to a relation between p_{SO_2} and p_{O_2} (since only one of the two partial pressures remains arbitrary). For example, in a plot of log p_{SO_2} vs log p_{O_2}, reaction (136) describing the equilibrium between the oxide and the sulfide yields a straight line. The stability domains of various oxides, sulfides, and sulfates are conveniently described by such plots, which are often referred to as Kellogg's diagrams [5]. Figure 6 illustrates the case of zinc.

5. Application to Some Important Metallurgical Equilibria

5.1. Boudouard Reaction

Coal or its derivative coke is one of the most important metallurgical fuels, and some of the equilibria to which it gives rise will be briefly considered as an illustration of the preceding thermodynamic treatment.

Let us consider a furnace at temperature T and 1 atm in which coke and air are reacted. To be specific, let us assume that 100 metric tonnes/day of coke are reacted with 10,000 m^3(STP)/hr of air.[3] We wish to calculate the composition of the flue gases and the heat evolved by the reactions.

The feed rate of coke corresponds to $10^8/(24 \times 12) = 3.47 \times 10^5$ (mol C)/hr (or $3.47 \times 10^5 \times$ %C/100 if we wish to take into account the fact that coke is not 100% carbon). The flow rate of air corresponds to $10^{10}/22{,}414 = 4.46 \times 10^5$ (mol air)/hr, or 0.94×10^5 mol of O_2 and 3.52×10^5 (mol of N_2)/hr. It is clear that the ratio of oxygen to carbon supplies is such that some of the carbon will remain unburned.

For the basis of our calculations we shall arbitrarily consider one mole of air. This corresponds to 3.47/4.46 or 0.78 mol C. The reactions pertaining to the direct formation of carbon monoxide and carbon dioxide are

$$C + \tfrac{1}{2}O_2 = CO \tag{139}$$

$$C + O_2 = CO_2 \tag{140}$$

After its formation, CO_2 further reacts with carbon to form CO:

$$CO_2 + C = 2CO \tag{141}$$

[3] Recall that STP is the conventional abbreviation for "standard conditions of temperature and pressure": $T = 273°K$, $P = 1$ atm.

To calculate the composition of the flue gases, knowledge of the actual physical mechanisms by which the gaseous species are formed is unnecessary and we note that the reactions (139)–(141) are not independent.

$$\boldsymbol{R}_{141} = 2\boldsymbol{R}_{139} - \boldsymbol{R}_{140} \tag{142}$$

A solution to the problem can be obtained by the general method followed in the example of Section 3.2.1. The mass balance equations for carbon, oxygen, and nitrogen are

$$n_C + n_{CO} + n_{CO_2} = 0.78 \tag{143}$$

$$n_{CO} + 2n_{CO_2} + 2n_{O_2} = 0.42 \tag{144}$$

$$n_{N_2} = 0.79 \tag{145}$$

Let

$$n_{CO} + n_{CO_2} + n_{O_2} + n_{N_2} = n_g \tag{146}$$

The two reactions (139) and (140) yield

$$K_{139} = \frac{p_{CO}}{p_{O_2}^{1/2}} = \frac{n_{CO}}{n_{O_2}^{1/2} n_g^{1/2}} \tag{147}$$

$$K_{140} = \frac{p_{CO_2}}{p_{O_2}} = \frac{n_{CO_2}}{n_{O_2}} \tag{148}$$

We have six equations [(143)–(148)] with six unknowns, and the problem can thus be solved numerically.

Alternatively, the problem can be solved through the use of two progress variables associated with reactions (139) and (140) (see Section 3.2.2). There is however a much simpler method if one makes use of the obvious fact that the number of oxygen moles left in the gaseous phase is bound to be very small. Let us choose reactions (140) and (141) as our independent basis. It is clear that in this case the progress variable associated to the reaction (140) will be equal to 0.21 since (practically) all the oxygen must be consumed by reaction (140) alone.

We are thus left with the determination of only one progress variable:

$$CO_2 + C = 2CO \tag{141}$$

$$0.21 \qquad 0: \qquad n_g = 0.79 + 0.21 = 1$$

$$0.21 - \lambda \qquad 2\lambda: \qquad n_g = 0.79 + 0.21 + \lambda = 1 + \lambda$$

Consequently,

$$K_{141} = \frac{4\lambda^2}{(0.21 - \lambda)(1 + \lambda)} \tag{149}$$

or

$$\lambda^2[(4/K_{141}) + 1] + 0.79\lambda - 0.21 = 0 \tag{150}$$

a quadratic equation easily solved for λ.

At $T = 1000°K$, we have $\Delta G° = -1300$ cal, $\log K_{141} = 0.284$, and $K_{141} = 1.92$. λ is then found to be equal to 0.162 and the composition of the flue gases is

$p_{CO} = 0.28$, $p_{CO_2} = 0.041$, and $p_{N_2} = 0.68$. To calculate the actual pressure of oxygen, one can use either reaction (139) or (140). With reaction (140)

$$(K_{140})_{1000°K} = p_{CO_2}/p_{O_2} \tag{151}$$

or

$$5 \times 10^{20} = 0.041/p_{O_2}$$

and

$$p_{O_2} = 0.8 \times 10^{-22}$$

Our previous assumption that diatomic oxygen in the flue gases is essentially depleted is thus amply justified.

The heat released by 1 mol of air can be calculated by equation (83):

$$\begin{aligned} DH &= \Delta H^\circ_{140} D\lambda_{140} + \Delta H^\circ_{141} D\lambda_{141} \\ &= -94{,}320 \times 0.21 + 40{,}780 \times 0.162 \\ &\simeq -13{,}200 \quad \text{cal/(mol air)} \end{aligned} \tag{152}$$

Note that if one were to choose reactions (139) and (140) as our basis of independent reactions, then it is obvious from the values of p_{CO} and p_{CO_2} that the progress variables would have the values of 0.28 and 0.04 on the basis of 1 mol of flue gases. Consequently,

$$\begin{aligned} DH &= \Delta H^\circ_{139} D\lambda_{139} + \Delta H^\circ_{140} D\lambda_{140} \\ &= -26{,}770 \times 0.28 - 94{,}320 \times 0.041 \\ &\simeq -11{,}360 \quad \text{cal/(mol flue gas)} \end{aligned} \tag{153a}$$

or, since 1 mol of air generates 1.162 (i.e., $1 + \lambda$) mol of flue gases,

$$DH = -11{,}360 \times 1.162 \simeq -13{,}200 \quad \text{cal/(mol air)} \tag{153b}$$

which as expected, is identical to the result of equation (152).

Coke consumption is $0.21 + 0.162 = 0.372$ (mol C)/(mol air), or $100 \times 0.372/0.78 = 47.8$ tonne/day. It should be noted, however, that the numerical values of the flow rates do not affect the composition of the flue gases and the heat evolved DH as long as there remains an excess of carbon. It should also be noted that the numerical value of the progress variable associated with any given reaction [e.g., reaction (140)] depends on the other reactions chosen to constitute a basis of independent reactions.

The calculations we made at 1000°K can be repeated at any other temperature. The results are presented in Fig. 7 as a plot of $p_{CO}/(p_{CO} + p_{CO_2})$ vs T. At low temperature carbon dioxide predominates, whereas at high temperatures carbon monoxide predominates. This result can also be seen as a consequence of the Le Chatelier–Braun principle, since reaction (141) is endothermic. Note that another consequence of the same principle is that a decrease in pressure favors the formation of more carbon monoxide since the ΔV of reaction (141) is positive. Similar calculations can be made in the case of pure oxygen instead of air; they yield, qualitatively, the same results (see Fig. 7).

The reduction of carbon dioxide by carbon [equation (141)] is often called the

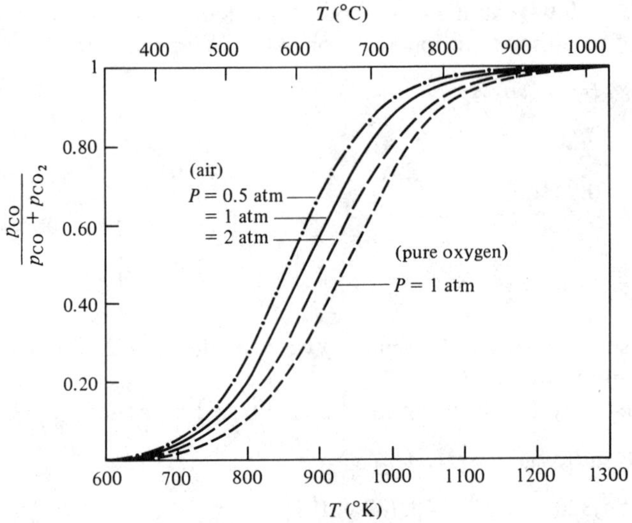

Figure 7. Equilibrium corresponding to the Boudouard reaction, $CO_2 + C = 2CO$. The three upper curves correspond to the reaction of carbon with air at a total pressure of 0.5, 1, and 2 atm. The lowest curve corresponds to the reaction of carbon with pure oxygen at a pressure of 1 atm.

Boudouard reaction and the characteristic S-shaped curves of Fig. 7 are referred to as *Boudouard curves*.

5.2. Dissociation of Carbon Dioxide

Carbon monoxide, carbon dioxide, and oxygen can be in equilibrium with or without the presence of carbon as a separate phase. In both cases, their relative partial pressures are related by the reaction corresponding to the dissociation of CO_2 into CO and O_2:

$$CO_2 = CO + \tfrac{1}{2}O_2 \tag{154}$$

which can be obtained as a linear combination of reactions (139) and (140):

$$\boldsymbol{R}_{154} = \boldsymbol{R}_{139} - \boldsymbol{R}_{140} \tag{155}$$

If carbon is present as a separate phase, then the ratio of p_{CO} to p_{CO_2} is fixed by the Boudouard equilibrium. If carbon is not present as a separate phase (e.g., when there is an excess of oxygen with respect to carbon), the ratio of p_{CO} to p_{CO_2} is arbitrary. Whatever its value, it determines the oxygen potential, since

$$K_{154} = \frac{p_{CO}p_{O_2}^{1/2}}{p_{CO_2}} \tag{156a}$$

or

$$p_{O_2} = [K_{154}(p_{CO_2}/p_{CO})]^2 \tag{156b}$$

For instance at 1500°K where $K_{154} = 5 \times 10^{-6}$, a ratio $p_{CO}/p_{CO_2} = \frac{1}{2}$ yields an oxygen potential

$$p_{O_2} = (5 \times 10^{-6} \times 2)^2 = 10^{-10}$$

To maintain a given (low) potential of oxygen in a reaction chamber at temperature T, it is often convenient to pass a flow of CO and CO_2 adjusted to the ratio corresponding to the desired oxygen potential. The mixture can be prepared at any temperature (e.g., room temperature) because reaction (154) occurring at T does not proceed to an extent which would significantly alter the values of p_{CO} and p_{CO_2}, since the values of p_{O_2} [and therefore of the progress variable associated with reaction (154)] remain very small. This is not true, however, when carbon is precipitated, because the values of p_{CO} and p_{CO_2} can be then drastically altered by the progress of reaction (141). Thus, experimentally, it may often be important to check on the presence of carbon deposits.

As an example, let us consider the following case. The flue gases of the reactor previously considered and containing, at 1000°K, 28% CO, 4% CO_2, and 68% N_2, are passed through another reaction chamber at a lower temperature T. Since the Boudouard reaction is endothermic, the equilibrium should be shifted backward; i.e., more of the reactants CO_2 and C should be formed if there is no nucleation barrier. The driving free energy DG associated with the precipitation of carbon may be calculated by equation (73). For instance, if $T = 800$°K, on the basis of 1 mol of flue gases, the progress variable associated to the Boudouard reaction is found to be equal to -0.12 and results in the following equilibrium partial pressures: $p_{CO} = 0.045$, $p_{CO_2} = 0.182$, and $p_{N_2} = 0.773$. Equation (76) then yields

$$DG = -RT\left\{n_{CO}\ln\left[\frac{p_{CO}}{(p_{CO})_{eq}}\right] + n_{CO_2}\ln\left[\frac{p_{CO_2}}{(p_{CO_2})_{eq}}\right] + n_{N_2}\ln\left[\frac{p_{N_2}}{(p_{N_2})_{eq}}\right]\right\}$$

$$DG = -4.575 \times 800\left\{0.28\log\frac{0.28}{0.045} + 0.04\log\frac{0.04}{0.182} + 0.68\log\frac{0.68}{0.773}\right\} \qquad (157)$$

$$= -579 \quad \text{cal/(mol flue gas)}$$

or

$$DG = -579/0.12 = -4823 \quad \text{cal/(mol C)}$$

Application of equation (156), with $K_{154} = 1.3 \times 10^{-14}$ at 800°K, shows that p_{O_2} is equal to 3.4×10^{-30} if carbon does not precipitate, to 2.7×10^{-27} if carbon precipitates.

5.3. Reduction of Iron Oxides

The reactions between iron, carbon, and their oxides are of paramount importance in the extraction of iron from its ore; they also provide another good illustration of a type of calculation often encountered in metallurgy.

Hematite (Fe_2O_3), magnetite (Fe_3O_4), and wüstite[4] (FeO) give rise to the

[4] Wüstite is a nonstoichiometric phase more accurately described as Fe_xO, with x close to 1 (about 0.95). See Chapter XII.

following equilibria:

$$3Fe_2O_3 = 2Fe_3O_4 + \tfrac{1}{2}O_2 \tag{158}$$

$$Fe_3O_4 = 3FeO + \tfrac{1}{2}O_2 \tag{159}$$

$$FeO = Fe + \tfrac{1}{2}O_2 \tag{160}$$

At any given temperature, there is only one oxygen potential allowing the equilibrium coexistence of two oxides, or of one oxide and the metal. This oxygen potential may be fixed by a CO–CO_2 gas mixture, according to reaction (154) or equation (156). This may also be seen by combining reaction (154) with reactions (158)–(160).

$$3Fe_2O_3 + CO = 2Fe_3O_4 + CO_2 \tag{161}$$

$$Fe_3O_4 + CO = 3FeO + CO_2 \tag{162}$$

$$FeO + CO = Fe + CO_2 \tag{163}$$

Figure 8 illustrates the values of $p_{CO}/(p_{CO} + p_{CO_2})$ corresponding to the coexistence of two solid phases as a function of temperature. Pressure has no significant effect on these equilibria since the ΔV of these reactions is negligible. These curves determine the domains of stability of the metal and its oxides. For example, at $T = 727°C$, Fe_2O_3 is reduced to Fe_3O_4 by mixtures even very dilute in CO and Fe_3O_4 is stable for mixtures containing a ratio of p_{CO} to $p_{CO} + p_{CO_2}$ up to 0.34; for values of this ratio between 0.34 and 0.62 wüstite Fe_xO is stable, and above 0.62, pure iron is stable.

It is interesting now to consider the case where the reducing atmosphere containing CO and CO_2 is also in contact with carbon. This is the case of the blast furnace, where air burnt at the tuyeres and thus transformed to a mixture of CO, CO_2, and N_2 sweeps through iron ore and coke. Superimposing the Boudouard curve corresponding to $P = 1$ atm to the iron oxide reduction diagram (Fig. 8), one can deduce the following results.

Figure 8. Reduction of iron oxides. The Boudouard curve corresponds to the reaction of air with carbon at a pressure of 1 atm.

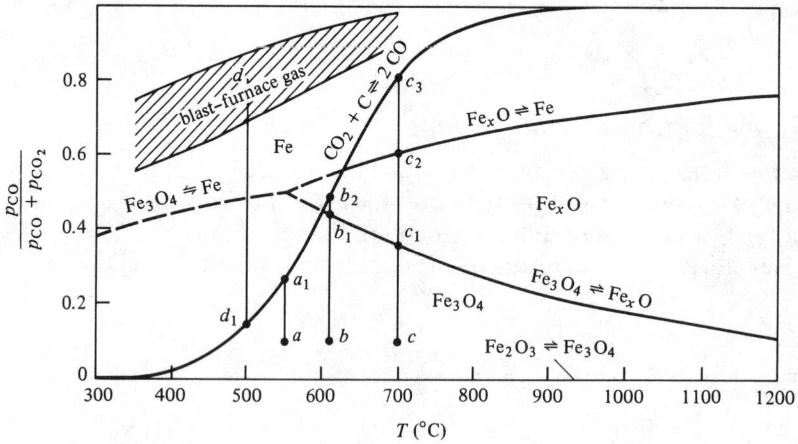

Let us consider a gaseous mixture corresponding to point a in Fig. 8 at 550°C. Fe_3O_4 is then stable. Because of the presence of carbon, some carbon dioxide will be dissociated until the carbon monoxide concentration of the gas builds up to point a_1 on the Boudouard curve. But since a_1 is still in the stability domain of Fe_3O_4, Fe_3O_4 is not reduced.

If, however, we start with the same gaseous mixture but at the slightly higher temperature of 610°C (point b in Fig. 8), we see that to reach the Boudouard curve the representative point of the gaseous composition crosses the stability boundary of wüstite. Because of the dissociation of CO_2, the concentration of CO is increased, and when it reaches point b_1, Fe_3O_4 is reduced to FeO. According to equation (162) this produces CO_2, which is consumed by carbon to form CO. Until all the carbon or all the Fe_3O_4 is used up, the composition of the gas phase remains at level b_1. If there is an excess of carbon, eventually all the magnetite will be reduced to wüstite. The concentration of carbon monoxide will then reach level b_2, corresponding to the Boudouard equilibrium.

With a gaseous composition represented by point c at a temperature higher than 625°C, magnetite is reduced first to wüstite (c_1) and then to pure iron (c_2). Finally, the composition of the gas reaches level c_3 (if there is sufficient carbon).

If the composition of the gas phase lies above the Boudouard curve (e.g., at point d), precipitation of carbon should occur and the composition of the gas phase should be shifted from d to d_1. The boundaries corresponding to the oxidation and reduction of iron and its oxides are metastable and can be observed only if the nucleation of carbon does not occur, or if its nucleation and growth are very slow. These metastable boundaries are represented as dotted lines in Fig. 8.

In the iron-making blast furnace, the composition of the gas phase is such that the reduction of iron oxides is theoretically possible at all temperatures (see the shaded area of Fig. 8).

6. Remarks on the Progress Variables

6.1. Change of Basis

In Section 3.1, we observed that although many chemical reactions may be written, only p among them are linearly independent. The criterion of independence is that the corresponding matrix (ν) is not singular (i.e., at least one determinant of size p is different from zero, ensuring a rank p for the matrix). The choice of the p independent reactions is arbitrary, but we have seen in studying the Boudouard reaction that proper selection of these reactions could simplify the calculations considerably, since for the chosen basis of reactions the value of at least one progress variable became evident. The problem we now address ourselves to is the following: given two possible bases of independent reactions, if the progress variables corresponding to one basis are found, what are the values of the progress variables on the other basis?

Let us designate by $\boldsymbol{R}'_1, \boldsymbol{R}'_2, \ldots, \boldsymbol{R}'_p$, the first system of p independent reactions, and by $\boldsymbol{R}''_1, \boldsymbol{R}''_2, \ldots, \boldsymbol{R}''_p$, the second system of p independent reactions. (A vector $\boldsymbol{R}$ corresponds to a linear vector of the ν matrix.) Each $\boldsymbol{R}''_j$ may be expressed as a linear combination of the $\boldsymbol{R}'_i$ vectors:

$$\boldsymbol{R}''_j = \alpha_{j,1}\boldsymbol{R}'_1 + \alpha_{j,2}\boldsymbol{R}'_2 + \cdots + \alpha_{j,p}\boldsymbol{R}'_p \tag{164a}$$

or

$$(R'') = (\alpha)(R') \tag{164b}$$

where (R'') and (R') are the arrays of vectors $\boldsymbol{R}''$ and $\boldsymbol{R}'$ and are by definition identical to the matrices (ν'') and (ν'):

$$(\nu'') = (\alpha)(\nu') \tag{164c}$$

In addition, equation (72) yields

$$\boldsymbol{dn} = (\nu')^{\mathrm{t}}\, \boldsymbol{d\lambda}' = (\nu'')^{\mathrm{t}}\, \boldsymbol{d\lambda}'' \tag{165}$$

Consequently,

$$(\nu')^{\mathrm{t}}\, \boldsymbol{d\lambda}' = [(\alpha)(\nu')]^{\mathrm{t}}\, \boldsymbol{d\lambda}'' \tag{166}$$

or

$$(\nu')^{\mathrm{t}}\, \boldsymbol{d\lambda}' = (\nu')^{\mathrm{t}}[(\alpha)^{\mathrm{t}}\, \boldsymbol{d\lambda}''] \tag{167}$$

Because (ν') [and thus $(\nu')^{\mathrm{t}}$] is not singular, it follows from equation (167) that

$$\boldsymbol{d\lambda}' = (\alpha)^{\mathrm{t}}\, \boldsymbol{d\lambda}'' \tag{168}$$

It should be emphasized that this relation is true whether the increments considered are infinitesimal (dn) or finite (Dn):

$$\boldsymbol{D\lambda}' = (\alpha)^{\mathrm{t}}\, \boldsymbol{D\lambda}'' \tag{169}$$

It expresses the correspondence we sought.

It may be noted that because of equation (164)

$$\Delta G_j'' = \alpha_{j,1}\Delta G_1' + \alpha_{j,2}\Delta G_2' + \cdots + \alpha_{j,p}\Delta G_p' \tag{170a}$$

or

$$\boldsymbol{\Delta G}'' = (\alpha)\boldsymbol{\Delta G}' \tag{170b}$$

and

$$dG = \boldsymbol{\Delta G}' \cdot \boldsymbol{d\lambda}' = \boldsymbol{\Delta G}'' \cdot \boldsymbol{d\lambda}'' = (\alpha)\boldsymbol{\Delta G}' \cdot \boldsymbol{d\lambda}'' = \boldsymbol{\Delta G}' \cdot (\alpha)^{\mathrm{t}}\, \boldsymbol{d\lambda}'' \tag{171}$$

which could also be considered to lead to equation (168). It is obvious that ΔG in equation (170) may be replaced by ΔH, ΔS, or ΔV.

For an illustration of these equations, let us turn to the study of Section 5.1 on the oxidation of carbon. The first basis of independent reactions we selected was

$$C + \tfrac{1}{2}O_2 = CO \qquad (\boldsymbol{R}_1', \Delta G_1') \tag{172}$$

$$C + O_2 = CO_2 \qquad (\boldsymbol{R}_2', \Delta G_2') \tag{173}$$

to which we preferred the second basis:

$$C + O_2 = CO_2 \qquad (\boldsymbol{R}_1'', \Delta G_1'') \tag{174}$$

$$C + CO_2 = 2CO \qquad (\boldsymbol{R}_2'', \Delta G_2'') \tag{175}$$

We note that

$$\boldsymbol{R}_1'' = \boldsymbol{R}_2', \qquad \Delta G_1'' = \Delta G_2' \tag{176}$$

$$\boldsymbol{R}_2'' = 2\boldsymbol{R}_1' - \boldsymbol{R}_2', \qquad \Delta G_2'' = 2\Delta G_1' - \Delta G_2' \tag{177}$$

which yields

$$(\alpha) = \begin{pmatrix} 0 & 1 \\ 2 & -1 \end{pmatrix} \tag{178}$$

Thus, equation (169) becomes

$$\boldsymbol{\lambda}' = \begin{pmatrix} 0 & 2 \\ 1 & -1 \end{pmatrix} \boldsymbol{\lambda}'' \tag{179}$$

or

$$\lambda_1' = 2\lambda_2'' \tag{180a}$$

$$\lambda_2' = \lambda_1'' - \lambda_2'' \tag{180b}$$

(The initial values of the λ variables are taken as zero, so that $D\lambda = \lambda$.) Since we found that the values of λ_1'' and λ_2'' were respectively 0.21 and 0.162,

$$\lambda_1' = 2 \times 0.162 = 0.324$$

$$\lambda_2' = 0.21 - 0.162 = 0.048$$

(these values correspond to 0.28×1.162 and 0.41×1.162 in Section 5.1). We can verify for example that DH may be calculated either as

$$DH = -26{,}770 \times 0.324 - 94{,}320 \times 0.048 \simeq -13{,}200 \quad \text{cal/(mol air)}$$

or as

$$DH = -94{,}320 \times 0.21 + 40{,}780 \times 0.162 \simeq -13{,}200 \quad \text{cal/(mol air)}$$

Of course, after a basis of reactions has been chosen and the progress variables found, it is not necessary to determine the values of the progress variables associated to another basis in order to deduce all the thermodynamic properties of the transformation. Nevertheless, the preceding relationship [equation (169)] is interesting on two accounts. First, it emphasizes the fact that the numerical value of a progress variable associated with a given reaction is meaningless unless the other reactions of the basis have been specified. Second, when studying the kinetics of a transformation and its possible rate controlling steps, one may wish to evaluate the progress of several different reactions in various bases.

6.2. Coupled Reactions

Consider a system undergoing a spontaneous transformation that may be described by p independent chemical reactions. In Section 3.1, we stated that if the reactions are not coupled, then for each reaction we must have

$$\Delta G_s \, d\lambda_s \le 0 \qquad (s = 1, 2, \ldots, p) \tag{181}$$

but these conditions are no longer necessary if the reactions are coupled.

To clarify the concept of coupled reactions, let us restrict our study to the case of just two simultaneous and independent reactions, characterized by the driving forces ΔG_1, ΔG_2, and the progress variables λ_1, λ_2. For any spontaneous evolution of the system,

$$(dG)_{P,T} = \Delta G_1 \, d\lambda_1 + \Delta G_2 \, d\lambda_2 < 0 \tag{182}$$

As a function of time, the decrease in the Gibbs free energy of the system is expressed

$$(dG/dt)_{P,T} = \Delta G_1 \dot{\lambda}_1 + \Delta G_2 \dot{\lambda}_2 < 0 \tag{183}$$

where t is the time and $\dot{\lambda}_1$, $\dot{\lambda}_2$ are, respectively, the speeds $d\lambda_1/dt$, $d\lambda_2/dt$ at which reactions 1 and 2 progress. The reactions are said to be coupled if $\dot{\lambda}_1$, $\dot{\lambda}_2$ are functions of both ΔG_1 and ΔG_2:

$$\dot{\lambda}_1 = f(\Delta G_1, \Delta G_2) \tag{184a}$$

$$\dot{\lambda}_2 = g(\Delta G_1, \Delta G_2) \tag{184b}$$

If $\dot{\lambda}_1$ is a function of only ΔG_1 and $\dot{\lambda}_2$ is a function of only ΔG_2, the reactions are uncoupled.

In the case of small deviations from equilibrium and slow progress, $\dot{\lambda}_1$ and $\dot{\lambda}_2$ may be considered to be linear functions of ΔG_1 and ΔG_2:

$$\dot{\lambda}_1 = L_{11}(-\Delta G_1) + L_{12}(-\Delta G_2) \tag{185a}$$

$$\dot{\lambda}_2 = L_{21}(-\Delta G_1) + L_{22}(-\Delta G_2) \tag{185b}$$

The expression dG/dT becomes

$$(dG/dt)_{P,T} = -[L_{11}(\Delta G_1)^2 + (L_{12} + L_{21})\Delta G_1 \Delta G_2 + L_{22}(\Delta G_2)^2] \tag{186}$$

The necessary and sufficient conditions for this quadratic function to be negative definite are

$$L_{11} \geq 0, \qquad L_{22} \geq 0 \tag{187a}$$

$$4L_{11}L_{22} - (L_{12} + L_{21})^2 \geq 0 \tag{187b}$$

We observe that it is possible that the system decreases its Gibbs free energy while violating one of the conditions (181). For example, it is possible to have simultaneously

$$\Delta G_1 \dot{\lambda}_1 = -L_{11}(\Delta G_1)^2 + L_{12}\Delta G_1 \Delta G_2 < 0 \tag{188a}$$

$$\Delta G_2 \dot{\lambda}_2 = -L_{21}\Delta G_1 \Delta G_2 + L_{22}(\Delta G_2)^2 > 0 \tag{188b}$$

and

$$dG/dT = \Delta G_1 \dot{\lambda}_1 + \Delta G_2 \dot{\lambda}_2 < 0 \tag{188c}$$

In this case, reaction (2) may proceed forward even though its driving force is positive.

If the reactions are uncoupled

$$L_{12} = L_{21} = 0 \tag{189}$$

and the inequalities (187) reduce to (187a). In this case, conditions (181) are respected for both reactions:

$$\Delta G_1 \dot{\lambda}_1 = -L_{11}(\Delta G_1)^2 < 0 \tag{190a}$$

$$\Delta G_2 \dot{\lambda}_2 = -L_{22}(\Delta G_2)^2 < 0 \tag{190b}$$

Unlike in the case of coupled reactions, it is not possible for one of the reactions to progress forward if its ΔG is positive or backward if its ΔG is negative. These results may be generalized to the case in which the λ_i are not linear functions of the driving forces ΔG_j.

We stress the difference between the concept of independent reactions and that of coupled reactions. The concept of dependent or independent reactions arises essentially from mass balances; the time element does not intervene. The concept of coupled or uncoupled reactions can only be considered in conjunction with the kinetics of the reactions. Returning to the analogy of Fig. 3, we see that the position of the two wagons can be fixed independently of each other, but their movements cannot; e.g., the acceleration of one has an impact on the other through the spring.

For further elaboration on the matter, the texts of Fer [7] and Glansdorf and Prigogine [8] are recommended. We also note that through statistical mechanics and the principle of microscopic reversibility it may be demonstrated that

$$L_{12} = L_{21} \tag{191}$$

which is known as Onsager's reciprocal relation.

Problems

Note: the best way, perhaps the only way, of mastering the contents of this chapter is to work out *many* problems involving chemical reactions. The texts cited at the end of this chapter should be helpful in this regard.

1. Exposed to air at room temperature, silver oxidizes. When heated, the oxide film disappears. Calculate the temperature at which the oxide dissociates knowing that the Gibbs free energy of formation of Ag_2O is given as $-7000 + 15.25T$ (cal/mol).

2. A flow of 90% CH_4, 10% H_2 is passed in a furnace at 900°K. Calculate the composition of the gaseous mixture at this temperature, assuming equilibrium and a pressure of 1.2 atm. The standard Gibbs free energy of formation of methane (CH_4) at 900°K is +2030 cal.

3. In the carbothermic reduction of magnesium oxide, briquettes of MgO and C are heated at high temperatures in a nonoxidizing atmosphere. Calculate the temperature at which the sum of the pressures of Mg(gas) and CO reach 1 atm. In the vicinity of 2000°K, the Gibbs free energies of formation of MgO(solid) and CO may be represented, respectively, by the expressions $-174{,}000 + 48.7T$ and $-28{,}000 - 20.2T$ (cal/mol).

4. The following data have been obtained for the oxygen potentials corresponding to the coexistence of the two oxides CuO and Cu_2O:

T (°K):	1173	1223	1273	1303	1350
p_{O_2} (atm):	0.0208	0.0498	0.1303	0.225	0.504

Estimate $\Delta H°$ and $\Delta S°$ of the following reaction:

$$2CuO = Cu_2O + \tfrac{1}{2}O_2$$

and express its $\Delta G°$ as a function of temperature.

5. In a reactor at 600°K and 1 atm pressure, 10,000 m^3(STP)/hr of sulfur dioxide is mixed with 15,000 m^3(STP)/hr of hydrogen sulfide. Calculate, assuming equilibrium, the composition of the flue gases, the production rate of liquid sulfur, and the heat evolved per hour. At 600°K, the standard free energies and enthalpies of formation of SO_2, H_2S, and H_2O are

	SO_2	H_2S	H_2O	
$\Delta G°$	−71,790	−10,130	−51,160	(cal/mol)
$\Delta H°$	−72,820	−7,260	−58,500	(cal/mol)

6. An investigator wishes to mix two flows of SO_3 and SO_2 in a ratio x such that the resulting partial pressure of oxygen is equal to 0.05 atm in a reactor at 1100°K and under a total atmosphere of 1.2 atm (assuming equilibrium). Calculate x given the data

$$SO_2(g) + \tfrac{1}{2}O_2(g) = SO_3(g), \qquad \Delta G° = -22{,}600 + 21.36T \text{ cal}$$

7. A specimen of titanium is to be heat treated in a furnace at 1600°K in a flow of hydrogen gas. How dry should be the hydrogen in order to prevent oxidation (i.e., what partial pressure of H_2O can you tolerate)? At 1600°K, the Gibbs free energies

of formation of TiO, TiO_2, and H_2O are, respectively, −94,600, −156,300 and −37,930 cal.

8. An investigator wishes to reduce 100 g of finely divided powder of MnO to metallic Mn by passing a flow of hydrogen, at the rate of 100 cm^3 (STP)/min on the charge in a furnace at 1000°K and 1 atm. Assume that the kinetics of the reduction reaction are very fast and estimate the time it would take to reduce all the oxide powder. At 1000°K, the Gibbs free energies of formation of MnO and H_2O are, respectively, −74,550 and −46,040 cal.

9. A flow of nitrogen of 0.2 mol/hr bubbles through liquid $SnCl_4$ at 308°K and is assumed saturated with $SnCl_4$(gas). A second flow of nitrogen of 0.4 mol/hr bubbles through water at 319.3°K and is saturated with water vapor at that temperature. The two flows mix in a reaction chamber at 700°K in order to deposit SnO_2 by chemical reaction. The pressure in the reaction chamber is maintained at 1 atm.

a. The boiling temperature of $SnCl_4$ is 386°K and its heat of vaporization is 8325 cal/mol. Estimate its vapor pressure at 308°K.

b. Calculate the composition of the gas phase resulting from the mixture of the two nitrogen flows *before* $SnCl_4$ reacts with H_2O vapor.

c. Assume that 1 mol of gas of the composition found in question **b** is isolated and reaches equilibrium at 700°K and 1 atm. Calculate the corresponding number of moles of SnO_2 deposited.

d. Estimate the steady state deposition rate of SnO_2 in the actual reaction chamber. Discuss your estimate.

e. Would an increase in total pressure increase the deposition rate of SnO_2? Why?

Data: At 319.3°K, the vapor pressure of water is 0.1 atm. At 700°K, the standard Gibbs free energies of formation of $SnCl_4$, H_2O, SnO_2, and HCl are, respectively, −95.11, −49.93, −104.19, and −23.58 kcal/mol.

10. Carbon graphite is reacted with air at 2500°K. The ratio of carbon to air, in number of moles, is $\frac{1}{6}$. Calculate the composition of the flue gases and the heat evolved per mole of carbon. The total pressure is maintained at 1 atm.

11. The oxides AO and BO form an ideal binary solution at 1800°K. At this temperature, it is also known that the metals A and B form an ideal liquid solution. An ore containing an equimolar solution of AO and BO is placed in a furnace at 1800°K in the presence of graphite in order to reduce the oxides to their metallic components. The furnace is evacuated and a mechanical pump maintains a pressure of 10^{-4} atm.

a. Calculate the equilibrium mole fractions of A and B in the metallic solution and of AO and BO in the oxide solution. Deduce the metallic yields of A and B (e.g., the number of moles of A in the metallic solution divided by the initial number of moles of A in the ore).

b. What practical suggestions could you offer to increase the metallic yields?

c. A different ore now contains AO and BO as pure separate phases. Indicate the consequences of this morphology (contrast with part a).

The data at 1800°K are

$$AO + C = A + CO, \qquad K_1 = 2.17 \times 10^{-5}$$

$$BO + C = B + CO, \qquad K_2 = 1.0 \times 10^{-3}$$

The solubilities of C in the metallic solution and in the oxide solution are negligible.

12. A mix of 2 lb of graphite powder and 2 lb of quartz powder is placed in a furnace, the volume of which is 10 liter. The furnace gases are evacuated and the furnace is closed; its temperature is increased to and maintained at 1500°C. The graphite reacts with the quartz to form silicon carbide whiskers.

Find at equilibrium the partial pressures of CO and SiO and the quantities of SiC produced and graphite reacted. At 1773°K, the following free energies of formation (in cal/mol) are known:

SiO_2(c)	SiO(g)	SiC(c)	CO(g)
−142,420	−58,810	−13,620	−63,790

13. One mole each of ammonia, nitrogen, and carbon dioxide is fed into a furnace at 400°K, where the pressure is maintained at 1 atm. Calculate the composition of the gas phase at equilibrium assuming that the only species present in any significant amount are N_2, NH_3, H_2, H_2O, CO, and CO_2. What is the change in the Gibbs free energy of the system at 400°K before and after equilibrium? At 400°K, the free energies of formation of NH_3, H_2O, CO, and CO_2 are, respectively, −1600, −53,350, −35,000, and −94,300 cal/mol.

14. **a.** In a reactor maintained at 1000°K and a pressure of 1 atm, a flow of dry air ($\%O_2 = 0.21$, $\%N_2 = 0.79$) is reacted with an excess of coke. In Section 5.1, we calculated that on the basis of 1 mol air, the coke consumption is 0.372 mol C. Assuming equilibrium, we found that the composition of the flue gases to be $p_{CO} = 0.28$, $p_{CO_2} = 0.041$, and $p_{N_2} = 0.68$, and that the heat evolved was −13,200 cal/mol air.

b. The dry air is now replaced by wet air containing 3% H_2O. To find the new composition of the flue gases, the carbon consumption, and the heat evolved, proceed as follows:

1. Find the new composition of the air.
2. Determine the unknowns and how many independent reactions have to be considered.
3. Select a basis of independent reactions and express the partial pressures of CO, CO_2, H_2, and H_2O in terms of the progress variables associated to these reactions.
4. Solve for the progress variables (e.g., using an iteration technique).
5. Deduce the composition of the flue gases, the coke consumption, and the heat evolved per mole of air.

c. The wet air is now enriched with oxygen and its composition is $\%O_2 = 25$, $\%N_2 = 72$, $\%H_2O = 3$. Find the composition of the flue gases, the coke consumption, and the heat evolved by proceeding as in part b.

d. The flue gases are maintained at 1000°K and pass in another reactor at 1000°K where they react with wüstite. Compare the number of moles of wüstite reduced to iron when the flue gases originate from 1 mol of dry air (part a), wet air (part b), or wet enriched air (part c). Also compare in the three cases the coke consumption and the net heat evolved per mole of iron.

At 1000°K, the following Gibbs free energies and enthalpies of formation are known:

	CO	CO_2	H_2O	$Fe_{0.95}O$
$\Delta G°$ (cal/mol)	−47,860	−94,630	−46,040	−47,690
$\Delta H°$ (cal/mol)	−26,770	−94,320	−59,250	−62,940

15. Justify the graphical construction relative to the points C and H on the Ellingham diagram (Fig. 4).

16. **a.** 500 g of ZnO and 100 g of C are mixed and placed in a reaction chamber of 0.1 m^3 capacity. The chamber is closed, evacuated, and heated to 1200°K. Consider the two reactions ZnO + C = Zn(g) + CO and 2CO = CO_2 + C. Find the progress variables associated with these reactions, the weight of zinc dissociated from ZnO and present in the gas phase, and the pressure in the reaction chamber.

b. Apply the results of Section 6.1 to find the progress variables associated with the reactions ZnO + C = Zn(g) + CO and ZnO + CO = Zn(g) + CO.

At 1200°K the Gibbs free energies of formation of ZnO, CO, and CO_2 are −53,350, −52,050, and −94,680 cal/mol, respectively.

References

1. M. B. Bever and R. Rocca, *Rev. Met. (Paris)* **48**(5), 363–368 (1951).
2. T. M. Besmann, *SOLGASMIX-PV, a Computer Program to Calculate Equilibrium Relationships in Complex Chemical Systems,* Oak Ridge Nat. Lab. Rep. TM-5775, April 1977.
3. H. T. T. Ellingham, *J. Soc. Chem. Ind. (London)* **63,** 125 (1944).
4. F. D. Richardson and J. H. E. Jeffes, *J. Iron Steel Inst. (London)* **160,** 261 (1948); **161,** 229 (1949); **163,** 397 (1949); with G. Withers, **166,** 213 (1950); **171,** 165 (1952); F.D.R. alone, **175,** 53 (1953).
5. H. H. Kellogg and S. K. Basu, *Trans. AIME* **218,** 70–81 (1960).
6. K. Natesan, Ph.D. thesis, Carnegie-Mellon University, 1968.
7. F. Fer, *Thermodynamique Macroscopique.* Gordon and Breach, New York, 1971, Vol. II.
8. P. Glandsorf and I. Prigogine, *Thermodynamic Theory of Structure, Stability and Fluctuations.* Wiley Interscience, London, 1971.

Selected Bibliography

G. N. Lewis and M. Randall, *Thermodynamics,* 2nd ed. (revised by K. S. Pitzer and L. Brewer). McGraw-Hill, New York, 1961.

D. R. Gaskell, *Introduction to Metallurgical Thermodynamics.* McGraw-Hill, New York, 1973.

N. A. Gokcen, *Thermodynamics* (and solutions manual by N. A. Gokcen and L. R. Martin). Techscience, Hawthorne, CA, 1975.

G. S. Upadhyaya and R. K. Dube, *Problems in Metallurgical Thermodynamics and Kinetics.* Pergamon, London, 1977.

A. E. Morris, *Problem Manual for Metallurgical Thermodynamics.* University of Missouri-Rolla, 1973.

J. Hilsenrath, *Summary of on-line or interactive physico-chemical numerical data systems,* NBS Tech. Note 1122, Washington, DC, 1980.

VI Binary Solutions

In most of the examples in the previous chapters we assumed that the materials were pure. Usually, in practical applications, we do not dispose of pure elements and pure compounds but of mixtures of metals (alloys) and compounds (slags, mattes, glasses, etc.). The properties of a component in solution may differ considerably from the properties of that component when it is pure. The dependence of the activity and other thermodynamic functions on composition is an important subject of the thermochemistry of materials. It is not a simple subject either, and we shall restrict the introductory treatment of this chapter to binary solutions.

1. Thermodynamic Functions of Mixing

Let us consider a system of n_A mol of pure A and n_B mol of pure B (see Fig. 1) in the same structure α, at the temperature T and pressure P. Focusing our attention on the thermodynamic extensive property Y, the value of Y for the system of A and B before mixing is

$$Y_{b.m} = n_A Y_A + n_B Y_B \tag{1}$$

where Y_A and Y_B are, respectively, the molar values of Y for pure A and pure B. Maintaining the same temperature and pressure, we mix A and B to obtain a solution of $n_A + n_B$ mol of A and B in the same structure α. In this new system Y takes the value

$$Y_{a.m} = n_A \overline{Y}_A + n_B \overline{Y}_B \tag{2}$$

where $\overline{Y}_A$ and $\overline{Y}_B$ represent the partial molar properties of A and B, and the subscript a.m means "after mixing." The difference in the values of Y after mixing and before mixing is designated Y^M (Y of mixing):

$$Y^M = Y_{a.m} - Y_{b.m} = n_A(\overline{Y}_A - Y_A) + n_B(\overline{Y}_B - Y_B) \tag{3}$$

It may be rewritten[1]

$$Y^M = n_A Y_A^M + n_B Y_B^M \tag{4}$$

Applying these equations to the Gibbs free energy, we obtain

$$G^M = n_A(\mu_A - G_A) + n_B(\mu_B - G_B) = n_A G_A^M + n_B G_B^M \tag{5}$$

The activity function was defined in Section IV.4 by the equation

$$\mu_i = \mu_i^*(T, P) + RT \ln a_i \tag{6}$$

If the reference state is selected to be Raoultian, that is, if it corresponds to the pure component in the same structure as that of the solution, then equation (5) becomes

$$G^M = RT(n_A \ln a_A + n_B \ln a_B) \tag{7}$$

The molar Gibbs free energy of mixing is obtained by dividing equation (7) by $n_A + n_B$:

$$G_m^M = RT(X_A \ln a_A + X_B \ln a_B) \tag{8}$$

[1] The partial molar property of mixing Y_A^M could be written $\overline{Y}_A^M$. We choose to drop the bar for reasons of convenience. The superscript M already indicates that A is in solution; Y_A^M must therefore refer to a *partial* molar property.

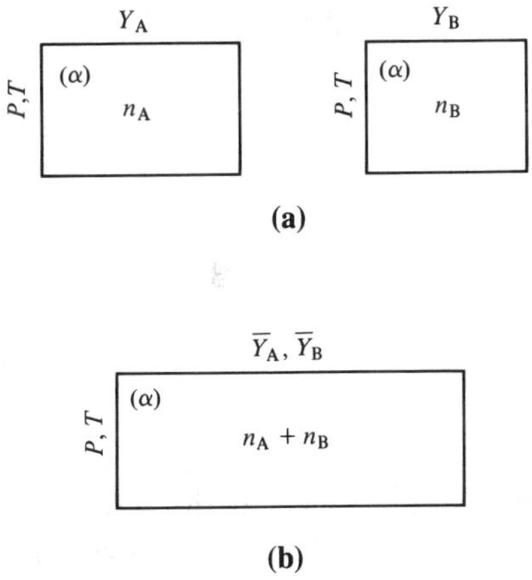

Figure 1. Illustration of a mixing property: $Y^{M} = Y_{a.m} - Y_{b.m} = n_A(\overline{Y}_A - Y_A) + n_B(\overline{Y}_B - Y_B)$. **(a)** System before mixing: $Y_{b.m} = n_A Y_A + n_B Y_B$. **(b)** System after mixing: $Y_{a.m} = n_A \overline{Y}_A + n_B \overline{Y}_B$.

Similarly, the molar entropy, enthalpy, and volume of mixing are

$$S_m^M = X_A(\overline{S}_A - S_A) + X_B(\overline{S}_B - S_B) = X_A S_A^M + X_B S_B^M \tag{9}$$

$$H_m^M = X_A(\overline{H}_A - H_A) + X_B(\overline{H}_B - H_B) = X_A H_A^M + X_B H_B^M \tag{10}$$

$$V_m^M = X_A(\overline{V}_A - V_A) + X_B(\overline{V}_B - V_B) = X_A V_A^M + X_B V_B^M \tag{11}$$

Since the Gibbs free energy, entropy, enthalpy, and volume are interrelated by the equations

$$G = H - TS \tag{12}$$

$$S = -(\partial G/\partial T)_P \tag{13}$$

$$H = \left[\frac{\partial(G/T)}{\partial(1/T)}\right]_P \tag{14}$$

$$V = (\partial G/\partial P)_T \tag{15}$$

it is clear that we also have

$$G^M = H^M - TS^M \tag{16}$$

$$S^M = -(\partial G^M/\partial T)_P \tag{17}$$

$$H^M = \left[\frac{\partial(G^M/T)}{\partial(1/T)}\right]_P \tag{18}$$

$$V^M = (\partial G^M/\partial P)_T \tag{19}$$

In addition, differentiation of equations (16)–(19) with respect to n_A at constant n_B, T, and P yields the following relations between the partial molar properties of mixing of A:

$$G_A^M = RT \ln a_A = H_A^M - TS_A^M \tag{20}$$

$$S_A^M = -\left(\frac{\partial G_A^M}{\partial T}\right)_P = -R \ln a_A - RT\left(\frac{\partial \ln a_A}{\partial T}\right)_P \tag{21}$$

$$H_A^M = R\left(\frac{\partial \ln a_A}{\partial 1/T}\right)_P \tag{22}$$

$$V_A^M = \left(\frac{\partial G_A^M}{\partial P}\right)_T = RT\left(\frac{\partial \ln a_A}{\partial P}\right)_T \tag{23}$$

Similar equations hold for the partial molar properties of B.

2. Ideal Solution

A solution is called *ideal* if the activity of one of its constituents is proportional to its mole fraction over the entire composition range. If the reference state is Raoultian ($a_A = 1$ for $X_A = 1$), then the proportionality becomes an equality:

$$a_A = X_A \tag{24}$$

Through the Gibbs–Duhem equation, it may easily be demonstrated (see Section 5) that if one of the components behaves ideally, the other component must also behave ideally. In other words, equation (24) implies

$$a_B = X_B \tag{25}$$

(assuming again a Raoultian reference state for B).

Equation (8) shows that the molar Gibbs free energy of mixing of an ideal solution is equal to

$$G_m^{M(id)} = RT(X_A \ln X_A + X_B \ln X_B) \tag{26}$$

The molar entropy of mixing may be obtained by differentiating equation (26) with respect to T [equation (17)]:

$$S_m^{M(id)} = -R(X_A \ln X_A + X_B \ln X_B) \tag{27}$$

The sign of $G^{M(id)}$ is negative while that of $S^{M(id)}$ is positive: upon mixing, the Gibbs free energy is decreased and the entropy (a measure of the disorder of the system) is increased. In addition, equations (18) and (19) show that for an ideal solution there is no heat evolved in the mixing process and there is no change of volume:

$$H^{M(id)} = 0 \tag{28}$$

$$V^{M(id)} = 0 \tag{29}$$

The partial molar properties of a component i (A or B) are readily obtained:

$$G_i^{M(id)} = RT \ln X_i \tag{30}$$

$$S_i^{M(id)} = -R \ln X_i \tag{31}$$

$$H_i^{M(id)} = 0 \tag{32}$$

$$V_i^{M(id)} = 0 \tag{33}$$

These results are illustrated in Fig. 2.

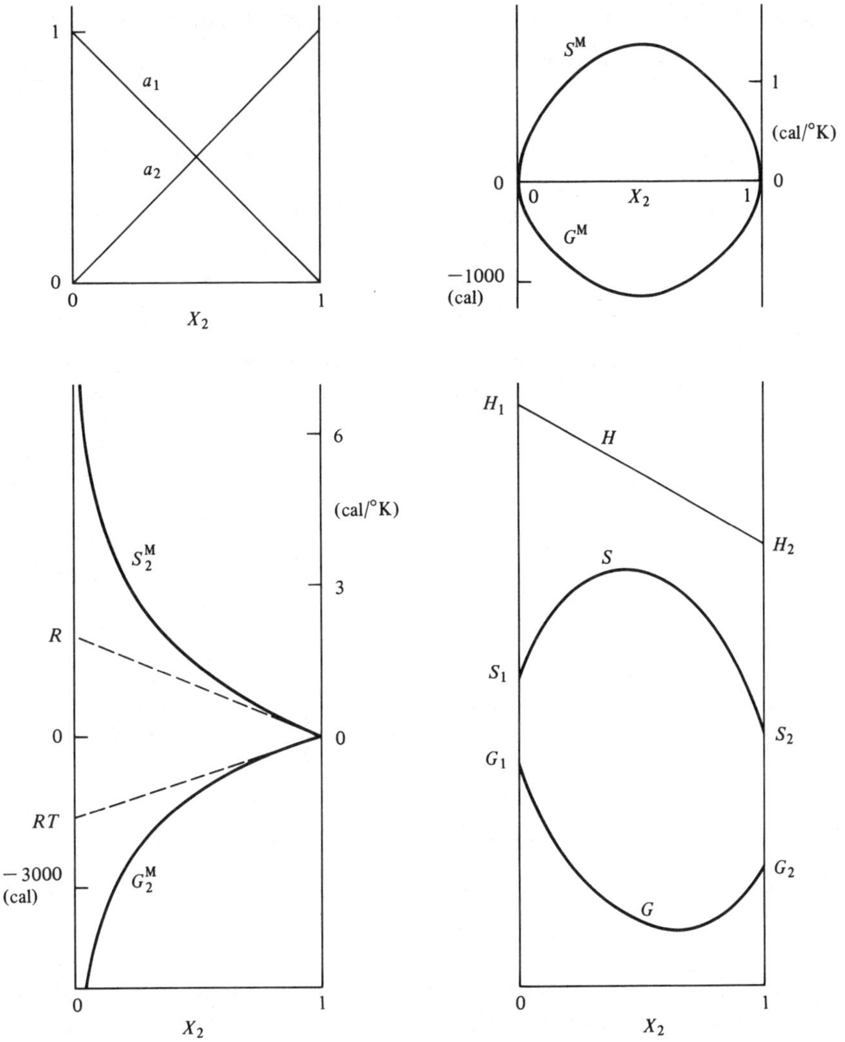

Figure 2. Composition dependence of some thermodynamic properties in the case of an ideal solution. (The Gibbs free energy values are calculated at $T = 800°K$.)

3. Excess Properties

The properties of real solutions differ from the properties of ideal ones. However, in most cases, the approximation of the ideal model is sufficiently close that it becomes convenient to consider the deviation from that model. Thus, any mixing property Y^M will be calculated as the sum of an ideal contribution $Y^{M(id)}$ and of an *excess contribution* Y^E:

$$Y^M = Y^{M(id)} + Y^E \tag{34a}$$

or

$$Y^E = Y^M - Y^{M(id)} \tag{34b}$$

It is clear that this definition applies also to partial molar properties:

$$Y_i^{\mathrm{E}} = Y_i^{\mathrm{M}} - Y_i^{\mathrm{M(id)}} \tag{35}$$

and that

$$Y_{\mathrm{m}}^{\mathrm{E}} = X_{\mathrm{A}} Y_{\mathrm{A}}^{\mathrm{E}} + X_{\mathrm{B}} Y_{\mathrm{B}}^{\mathrm{E}} \tag{36}$$

Substituting in these equations the Gibbs free energy for Y we obtain

$$G_i^{\mathrm{E}} = RT \ln a_i - RT \ln X_i = RT \ln \gamma_i \tag{37}$$

$$G_{\mathrm{m}}^{\mathrm{E}} = RT(X_{\mathrm{A}} \ln \gamma_{\mathrm{A}} + X_{\mathrm{B}} \ln \gamma_{\mathrm{B}}) \tag{38}$$

Recalling the general relationships between the Gibbs free energy, entropy, enthalpy, and volume [equations (16)–(19)], we readily establish that

$$G^{\mathrm{E}} = H^{\mathrm{E}} - TS^{\mathrm{E}} \tag{39}$$

$$G_i^{\mathrm{E}} = H_i^{\mathrm{E}} - TS_i^{\mathrm{E}} \tag{40}$$

which combined with equation (37) yields

$$\ln \gamma_i = \frac{H_i^{\mathrm{E}}}{RT} - \frac{S_i^{\mathrm{E}}}{R} \tag{41}$$

Moreover,

$$S_i^{\mathrm{E}} = S_i^{\mathrm{M}} + R \ln X_i = -R \ln \gamma_i - RT\left(\frac{\partial \ln \gamma_i}{\partial T}\right)_P \tag{42}$$

$$H_i^{\mathrm{E}} = H_i^{\mathrm{M}} = R\left(\frac{\partial \ln \gamma_i}{\partial (1/T)}\right)_P \tag{43}$$

$$V_i^{\mathrm{E}} = V_i^{\mathrm{M}} = RT\left(\frac{\partial \ln \gamma_i}{\partial P}\right)_T \tag{44}$$

All these results remain identical in the case of a multicomponent system. They are summarized in Tables 1 and 2.

Table 1. Mixing and Excess Thermodynamic Functions: Integral Properties

Functions of mixing	Ideal solution properties	Excess functions
$Y^{\mathrm{M}} = Y_{\mathrm{a.m}} - Y_{\mathrm{b.m}}$	$Y^{\mathrm{M(id)}}$	$Y^{\mathrm{E}} = Y^{\mathrm{M}} - Y^{\mathrm{M(id)}}$
$G_{\mathrm{m}}^{\mathrm{M}} = RT \sum X_i \ln a_i$	$G_{\mathrm{m}}^{\mathrm{M(id)}} = RT \sum X_i \ln X_i$	$G_{\mathrm{m}}^{\mathrm{E}} = RT \sum X_i \ln \gamma_i$
$H_{\mathrm{m}}^{\mathrm{M}} = \left(\frac{\partial G_{\mathrm{m}}^{\mathrm{M}}/T}{\partial 1/T}\right)_P = R \sum X_i \left(\frac{\partial \ln \gamma_i}{\partial 1/T}\right)_P$	$H^{\mathrm{M(id)}} = 0$	$H^{\mathrm{E}} = H^{\mathrm{M}}$
$S^{\mathrm{M}} = -\left(\frac{\partial G^{\mathrm{M}}}{\partial T}\right)_P$	$S_{\mathrm{m}}^{\mathrm{M}} = -R \sum X_i \ln X_i$	$S_{\mathrm{m}}^{\mathrm{E}} = S_{\mathrm{m}}^{\mathrm{M}} + R \sum X_i \ln X_i$
$G^{\mathrm{M}} = H^{\mathrm{M}} - TS^{\mathrm{M}}$	$G^{\mathrm{M}} = -TS^{\mathrm{M}}$	$G^{\mathrm{E}} = H^{\mathrm{E}} - TS^{\mathrm{E}}$
$V_{\mathrm{m}}^{\mathrm{M}} = \left(\frac{\partial G_{\mathrm{m}}^{\mathrm{M}}}{\partial P}\right)_T = RT \sum X_i \left(\frac{\partial \ln \gamma_i}{\partial P}\right)_T$	$V^{\mathrm{M}} = 0$	$V^{\mathrm{E}} = \mathrm{V}^{\mathrm{M}}$

The subscripts a.m and b.m stand for "after mixing" and "before mixing."

Table 2. Mixing and Excess Thermodynamic Functions: Partial Molar Properties

Functions of mixing	Ideal solution properties	Excess functions
$Y_i^M = \bar{Y}_{i\,(a.m)} - Y_{i\,(b.m)}$	$Y_i^{M(id)}$	$Y_i^E = Y_i^M - Y_i^{M(id)}$
$G_i^M = RT \ln a_i$	$G_i^{M(id)} = RT \ln X_i$	$G_i^E = RT \ln \gamma_i$
$H_i^M = \left(\frac{\partial G_i^M/T}{\partial 1/T}\right)_P = R\left(\frac{\partial \ln \gamma_i}{\partial 1/T}\right)_P$	$H_i^{M(id)} = 0$	$H_i^E = H_i^M$
$S_i^M = -\left(\frac{\partial G_i^M}{\partial T}\right)_P$	$S_i^{M(id)} = R \ln X_i$	$S_i^E = S_i^M + R \ln X_i$
$G_i^M = H_i^M - TS_i^M$	$G_i^{M(id)} = TS_i^{M(id)}$	$G_i^E = H_i^E - TS_i^E$
$V_i^M = \left(\frac{\partial G_i^M}{\partial P}\right)_T = RT\left(\frac{\partial \ln \gamma_i}{\partial P}\right)_T$	$V_i^{M(id)} = 0$	$V_i^E = V_i^M$

The subscripts a.m and b.m stand for "after mixing" and "before mixing."

As emphasized by these expressions of the excess properties, the activity coefficient is a natural choice to study departures from ideality. Note, however, that we do not write

$$a_i = a_i^{(id)} + a_i^E = X_i + a_i^E$$

but

$$\ln a_i = \ln a_i^{(id)} + \ln a_i^E = \ln X_i + \ln \gamma_i \tag{45}$$

Deviations from ideality are said to be positive when $\ln \gamma_i$ is positive ($\gamma_i > 1$)

Figure 3. Activities in the Cd–Mg system at 650°C [2]. At $X_{Mg} = 0.6$ the activity coefficient $\gamma_{Mg} = QM/0Q = QM/QP = 0.72$.

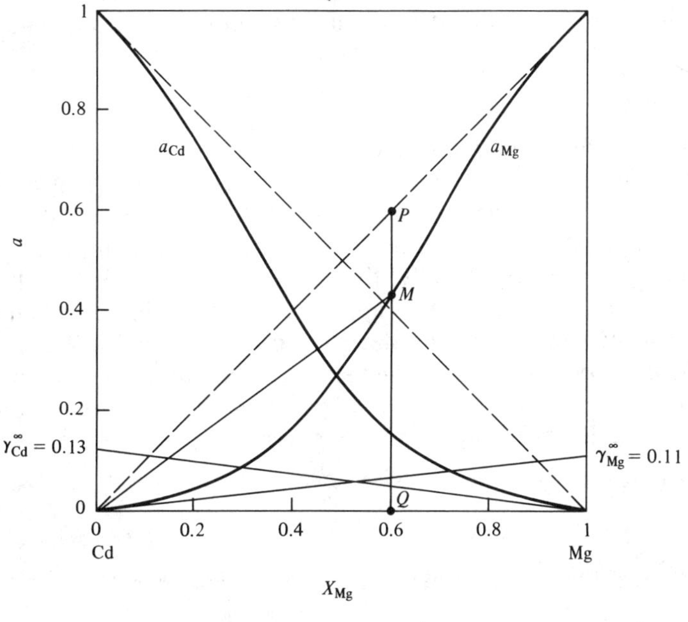

and negative when $\ln \gamma_i$ is negative ($\gamma_i < 1$).[2] The activity curves of the liquid Cd–Mg system at 650°C, which are shown in Fig. 3, are typical of many systems. In this example, the deviations from ideality are negative. We note that the activity coefficient $\gamma_i = a_i/X_i$ is represented graphically by the ratio $QM/0Q$ and therefore by the slope of the line $0M$; it is also equal to the ratio QM/QP since $QP = 0Q$. At the point of infinite dilution, $X_i = 0$, the activity coefficient takes the value γ_i^∞. This is also the value of the slope of the activity curve at $X_i = 0$ since that slope is the limit position of the line $0M$. Actually, this result is the graphical illustration of L'Hospital's rule: the ratio of two continuous functions, each tending towards zero, is equal to the ratio of their derivatives. Thus,

$$\gamma_i^\infty = \left(\frac{a_i}{X_i}\right)_{X_i \to 0} = \left(\frac{da_i}{dX_i}\right)_{X_i \to 0} \tag{46}$$

4. Raoult's and Henry's Laws

The study of dilute solutions has long been of major importance to physical chemists. This is not surprising since most solutions are dilute with respect to certain solutes (generally termed impurities when undesirable). Two laws are relevant to this study. The first, Henry's law, results from the experimental findings of Henry in 1803 [3]. The second, Raoult's law, dates from 1887 [4]. The concept of an activity function was of course unknown then, and these laws were enunciated in a way which would be inadequate today if we wish to consider them as laws. Often in the literature, even modern versions of their historical formulations are imprecise and confuse their significance. This is unfortunate since these laws are relevant to the foundation of the formalisms of solutions and to the possible atomistic models supporting them. Consequently, these laws will be presented and discussed in this section at some length.

Before proceeding further, we shall adopt here a convention that is fairly widely adopted in the technical literature. Wherever we deal with binary or multicomponent dilute solutions, component 1 will always represent the solvent.

4.1. Definitions

Consider the binary dilute solution 1–2. It is often stated that the solvent 1 follows Raoult's law in a given range of concentration (which includes that of the pure element 1) if in that range the activity curve of 1 is the line of ideal mixing, $a_1 = X_1$. This statement is unsatisfactory since it only yields a different name for the behavior of ideal mixing. Actually, the interesting feature of Raoult's law is that the solvent of a nonideal solution approaches ideal behavior when its concentration approaches unity, i.e., the state of purity (see Fig. 4). We shall make this feature the basis of our definition and state that *Raoult's law is obeyed when the slope of the solvent's activity curve at the point* $X_1 = 1$ *is the line of ideal mixing*. Mathematically, this statement is expressed

$$(da_1/dX_1)_{X_1 \to 1} = 1 \tag{47}$$

---

[2] It is possible, however, to have $\ln \gamma_A > 0$ and $\ln \gamma_B < 0$. This is the case, for example, of the Co–Fe solution at 1600°C [1].

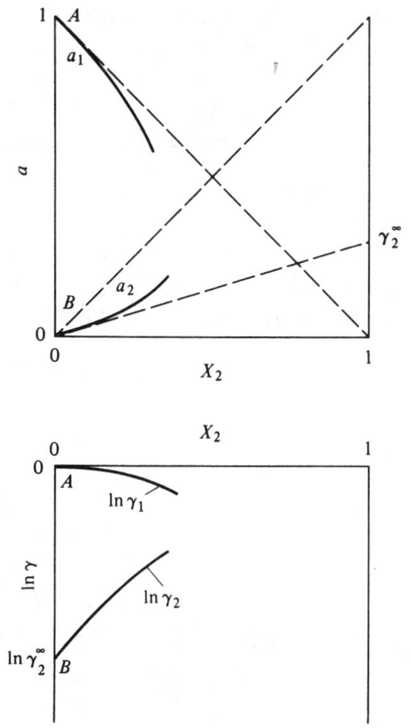

Figure 4. Raoult's law applies at the point A, Henry's law at the point B. At A the activity curve is tangent to the line of ideal mixing and the curve $\ln \gamma_1$ has a tangent which is horizontal. At B the slope of the activity curve is neither vertical nor horizontal; i.e., $\ln \gamma_2^\infty$ is finite (Henry's zeroth order law).

(When the reference state is not Raoultian, the value of the solvent's activity at $X_1 = 1$ replaces the value of 1 in the right member of this equation.)

In terms of the activity coefficient, equation (47) becomes

$$\left(\frac{d\,(\gamma_1 X_1)}{dX_1}\right)_{X_1 \to 1} = \left(\gamma_1 + X_1 \frac{d\gamma_1}{dX_1}\right)_{X_1 \to 1} = 1 \tag{48}$$

and since $(\gamma_1)_{X_1 \to 1} = 1$,

$$(d\gamma_1/dX_1)_{X_1 \to 1} = 0 \tag{49}$$

It will be advantageous to transform equation (49) into

$$\left(\frac{d \ln \gamma_1}{dX_2}\right)_{X_1 \to 1} = 0 \tag{50}$$

Equation (50) is not only a necessary consequence of Raoult's law; it is also a sufficient condition for it.

Considering the solution in the same range of composition, Henry's law applies to the solute 2. A common definition of Henry's law is that the activity of the solute 2 is proportional to its concentration when it is very dilute. Clearly, this proportionality should exist only as an approximation, i.e., it is only within the experimental scatter that an activity curve may be replaced by its tangent at the point of infinite dilution. The argument that both Raoult's and Henry's laws should properly be statements about limiting slopes may seem obvious. That it should be made explicit rather than implicit will become more apparent below.

A solute 2 obeys Henry's law if, at the point of infinite dilution of 2 where a_2 is zero, *the slope of the activity curve a_2 vs X_2 has a nonzero finite value:*

$$(da_2/dX_2)_{X_2\to 0} = \text{nonzero finite value} \tag{51}$$

Through the definition of the activity coefficient and L'Hospital's rule, we note that

$$\gamma_2^\infty \equiv (\gamma_2)_{X_2\to 0} = \left(\frac{a_2}{X_2}\right)_{X_2\to 0} = \left(\frac{da_2}{dX_2}\right)_{X_2\to 0} \tag{52}$$

Consequently, γ_2^∞ is finite and different from zero. Moreover, since

$$\left(\frac{da_2}{dX_2}\right)_{X_2\to 0} = \left(\gamma_2 + X_2\frac{d\gamma_2}{dX_2}\right)_{X_2\to 0} \tag{53}$$

a comparison of equations (52) and (53) yields

$$\left(X_2\frac{d\ln\gamma_2}{dX_2}\right)_{X_2\to 0} = 0 \tag{54}$$

Equation (54) is a *necessary* consequence of Henry's law (in the form proposed here), but it is not a *sufficient* condition for it [i.e., equation (51) cannot be derived from equation (54)].

4.2. *Raoult's Law as a Consequence of Henry's Law*

We recall that, at constant temperature and pressure, the Gibbs–Duhem equation yields

$$n_1\, d\mu_1 + n_2\, d\mu_2 = 0 \tag{55}$$

or in terms of activities and mole fractions

$$X_1\, d\ln a_1 + X_2\, d\ln a_2 = 0 \tag{56}$$

Moreover, since

$$X_1\, d\ln X_1 + X_2\, d\ln X_2 = X_1\frac{dX_1}{X_1} + X_2\frac{dX_2}{X_2} = dX_1 + dX_2 = 0 \tag{57}$$

equation (56) may be rewritten

$$X_1\, d\ln\gamma_1 + X_2\, d\ln\gamma_2 = 0 \tag{58}$$

Thus, at the point of infinite dilution

$$1\cdot\left(\frac{d\ln\gamma_1}{dX_2}\right)_{X_1\to 1} + \left(X_2\frac{d\ln\gamma_2}{dX_2}\right)_{X_2\to 0} = 0 \tag{59}$$

When Henry's law is obeyed, equation (54) is satisfied and, consequently, equation (59) shows that equation (50) is also necessarily satisfied, i.e., Raoult's law is necessarily obeyed. But, if Raoult's law is obeyed, we can only deduce that equation (54) is satisfied and not that Henry's law is obeyed [since equation (54) is necessary but not sufficient].

Thus, Raoult's law is a consequence of Henry's law, but Henry's law is not a consequence of Raoult's law.

4.3. Henry's Zeroth Order and First Order Laws

To metallurgists, the applicability of Henry's law is commonly considered to mean that in a small concentration range $(0, X_2)$ the activity coefficient γ_2 is approximately constant, and thus that $(d \ln \gamma_2/dX_2)_{X_2\to 0}$ is not infinite. This condition is more restrictive than the definition of Henry's law adopted above. While it is supported by experimental evidence for nonelectrolytic solutions (and metallic solutions in particular), it cannot be applied to electrolytic solutions (see Fig. 5).

The Debye–Hückel theory [7a] shows that in dilute solutions the logarithm of the activity coefficient of an ionic species i varies proportionally to the square root of the ionic strength:

$$\ln \gamma_i \propto \Big(\sum_j z_j^2 X_j\Big)^{1/2} \tag{60}$$

where z_j is the charge of the ionic species j. Since the activity and activity coefficient of an ionic species cannot be measured, it is usually the electrolyte's mean activity $a_\pm$ and activity coefficient $\gamma_\pm$ which are considered [5]. From equation (60) it follows that $\ln \gamma_\pm$ varies as $X_2^{1/2}$ in dilute solutions, a result in good agreement with experimental observations [5]. We also note that the mathematical consequences of this result are that $(d \ln \gamma_\pm/dX_2)_{X_2\to 0}$ is infinite, while $\ln \gamma_\pm^\infty$ is

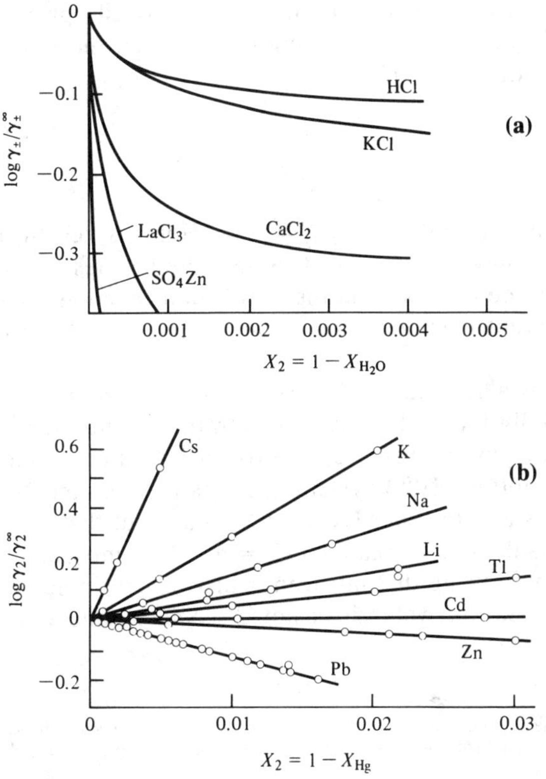

Figure 5. **(a)** Aqueous solutions at 25°C. (*After data compiled by Glasstone* [5].) **(b)** Liquid amalgams at 25°C. (*After data compiled by Darken and Turkdogan* [6].) Henry's zeroth order law is obeyed in both cases **(a)** and **(b)**, whereas Henry's first order law is not obeyed in **(a)** but obeyed in **(b)**. [In **(a)** the zeroth law is obeyed by the ions, not the electrolyte.]

finite and $(X_2\, d \ln \gamma_\pm / dX_2)_{X_2 \to 0} = 0$. In equation (53), $(da_\pm / dX_2)_{X_2 \to 0}$ is then seen to be finite. (It may be checked that the solute 2 itself does not obey Henry's law, as defined in equation (51), because $\gamma_2 \to 0$ as $X_2 \to 0$.)

For nonelectrolytic solutions, the fact that $(d \ln \gamma_2 / dX_2)_{X_2 \to 0}$ is finite has a range of applicability which is wide enough to warrant being called a law. Thus, it appears desirable to distinguish between Henry's law as previously defined—or *Henry's zeroth order law*—which states the finiteness of $\ln \gamma_2^\infty$, and *Henry's first order law* which states the finiteness of both $\ln \gamma_2^\infty$ and $(d \ln \gamma_2 / dX_2)_{X_2 \to 0}$.

Perhaps the best justification for such a distinction is that obedience to the zeroth order law may be interpreted in terms of the selection of the species describing the system, and to the first order law in terms of the short range effectiveness of the forces in the solution.

As a simple illustration, let us consider the case of nitrogen solubility in liquid iron. Numerous experimental investigations have shown that at moderate pressures, the concentration of nitrogen in iron is proportional to the square root of the partial pressure of nitrogen. The binary system may be described by the couple of species Fe and N_2, or Fe and N. If we select the species Fe and N_2 we can write the absorption reaction

$$N_2(g) = N_2(\text{in iron}) \tag{61}$$

and the equilibrium constant

$$K = \frac{a_{N_2}}{p_{N_2}} = \frac{\gamma_{N_2} X_{N_2}}{p_{N_2}} \tag{62}$$

At high dilutions, all measures of concentrations are practically proportional to each other (e.g., $X_{N_2} \simeq \frac{1}{2} X_N$) and it is equivalent to write that the square root of p_{N_2} is proportional to X_{N_2} or to X_N. Thus,

$$\sqrt{p_{N_2}} = h X_{N_2} \tag{63}$$

$$\gamma_{N_2} = K h^2 X_{N_2} \tag{64}$$

For $X_{N_2} \to 0$, we see that $\gamma_{N_2}^\infty = 0$ or $\ln \gamma_{N_2}^\infty = -\infty$; Henry's zeroth order law is not obeyed. If, on the contrary, we were to select the species Fe and N, we would find by a similar procedure [writing the absorption reaction $\frac{1}{2}N_2(g) = N$], that γ_N^∞ is finite and different from zero, i.e., that Henry's zeroth order law is obeyed.

A converse example may also be given. If we consider the solubility of nitrogen in water, it is experimentally found that the concentration of nitrogen is approximately proportional to the partial pressure of nitrogen.[3] As a consequence, it is easy to show that a description of the system in terms of H_2O and N_2 leads to an applicability of Henry's zeroth order law ($\ln \gamma_{N_2}^\infty$ finite) but that a description in terms of H_2O and N results in its violation ($\gamma_N^\infty = \infty$). The importance of these considerations is that they indicate that nitrogen is dissociated in liquid iron and is present as atoms, whereas in water it is not dissociated and is present as

[3] A similar result is observed for the solubility of nitrogen in blood. It is well known to scuba divers (as nitrogen narcosis) and engenders a feeling akin to alcoholic intoxication. A martini law has been enunciated: each 50 feet of depth is equivalent to one martini.

diatomic molecules. Thus, Henry's zeroth order law is relevant to a study of the structure of a solution.

Henry's first order law may give us some information on the interactions in solution. Through various statistical models (similar to those of the free volume [7]) it is possible to see that the derivatives of ln γ_i with respect to composition involve integrals which converge when the net interaction potential between atoms or molecules decreases rapidly with the distance and which diverge when it decreases slowly with the distance (e.g., as $1/r$ between electrical charges). Thus, it is not surprising to find that Henry's first order law is obeyed by metallic solutions and is not by electrolytic solutions. It should be emphasized that this does not prove that the forces between solute atoms in metallic solutions are necessarily short range; it merely establishes that in most cases the thermodynamic properties of the solutions can be adequately described through short range forces.

5. Integration of the Gibbs–Duhem Equation

In the experimental investigation of the thermodynamic properties of a solution, it is often much easier to measure the behavior of a solute rather than that of the solution itself. Moreover, if a partial molar property of one component has been measured over a given concentration range, it is not necessary to measure the partial molar property of the other component in that range: it can be deduced through the Gibbs–Duhem equation.

Let us assume that the activity a_1 of component 1 has been measured over the entire range of concentration. In equation (56), we have seen the Gibbs–Duhem equation in the form

$$X_1\, d \ln a_1 + X_2\, d \ln a_2 = 0 \tag{65}$$

Consequently,

$$\ln a_2 - (\ln a_2)_{X_2=1} = -\int_{X_2=1}^{X_2} \frac{X_1}{X_2}\, d \ln a_1 \tag{66}$$

or, if the reference state of 2 is Raoultian,

$$\ln a_2 = -\int_{X_2=1}^{X_2} \frac{X_1}{X_2}\, d \ln a_1 \tag{67}$$

This integral leads however to some difficulties. For, if we were to plot X_1/X_2 vs $-\ln a_1$, the resulting curve would have two asymptotic branches ($-\ln a_1 = \infty$ for $X_1/X_2 = 0$, and $X_1/X_2 \to \infty$ for $-\ln a_1 = 0$; see Fig. 6), and would make the measure of the area under it quite difficult. It is possible to rewrite the Gibbs–Duhem equation in terms of the activity coefficient γ [see equation (58)]:

$$X_1\, d \ln \gamma_1 + X_2\, d \ln \gamma_2 = 0 \tag{68}$$

so that

$$\ln \gamma_2 = -\int_{X_2=1}^{X_2} \frac{X_1}{X_2}\, d \ln \gamma_1 \tag{69}$$

In a plot of X_1/X_2 vs $-\ln \gamma_1$, we see now that the resulting curve has only one

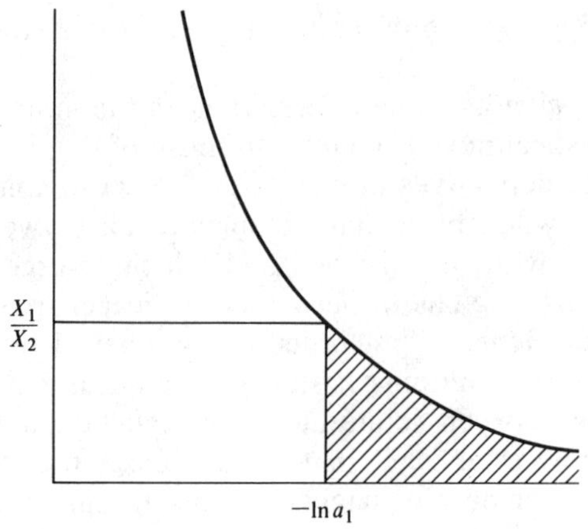

Figure 6. Integration of the Gibbs–Duhem equation under the form of equation (67). The shaded area measures $\ln a_2$ at the composition X_2.

asymptotic branch (for $X_2 \to 0$, $-\ln \gamma_1 = 0$ and $X_1/X_2 \to \infty$; see Fig. 7). Still, it is desirable to transform the Gibbs–Duhem equation in such a way as to remove this last difficulty.

If $\ln \gamma_1$ can be expanded into a Taylor series with respect to X_2 near the point of infinite dilution for component 2,

$$\ln \gamma_1 = (\ln \gamma_1)_{X_2 \to 0} + \left(\frac{d \ln \gamma_1}{dX_2}\right)_{X_2 \to 0} X_2 + \frac{1}{2}\left(\frac{d^2 \ln \gamma_1}{dX_2^2}\right)_{X_2 \to 0} X_2^2 + \cdots \quad (70)$$

We note that $(\ln \gamma_1)_{X_2 \to 0} = 0$ because of the Raoultian reference state ($\gamma_1 = 1$ at $X_1 = 1$). Moreover, $(d \ln \gamma_1/dX_2)_{X_2 \to 0} = 0$ because of Raoult's law [equation (50)]. Consequently, the term of lowest order in this expansion is of the second order with respect to X_2, and X_2^2 can be factored out:

$$\ln \gamma_1 = X_2^2(A_0 + A_1X_2 + A_2X_2^2 + \cdots) \quad (71)$$

The function $\ln \gamma_1/X_2^2$ is thus well behaved for $X_2 \to 0$ and its introduction in equation (68) eliminates the denominator which yields the asymptotic branch.

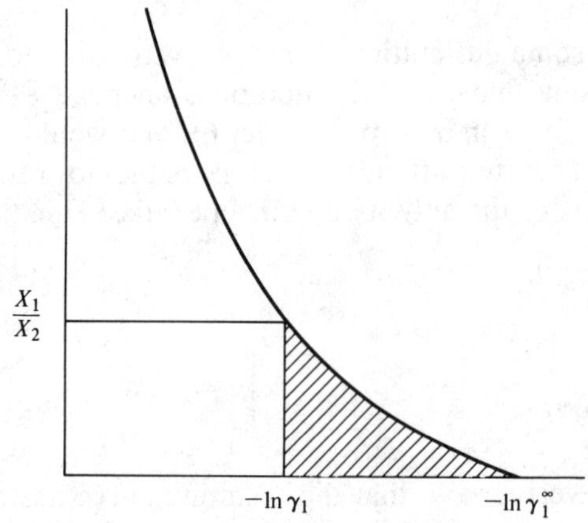

Figure 7. Integration of the Gibbs–Duhem equation under the form of equation (69). The shaded area measures $\ln \gamma_2$ at the composition X_2.

Darken and Gurry [8] were the first to use this technique through the introduction of the α function:

$$\alpha_i = \frac{\ln \gamma_i}{(1 - X_i)^2} \qquad (i = 1 \text{ or } 2) \tag{72}$$

Since

$$d \ln \gamma_1 = d(\alpha_1 X_2^2) = 2\alpha_1 X_2\, dX_2 + X_2^2\, d\alpha_1 \tag{73}$$

equation (69) becomes

$$\ln \gamma_2 = -\int_{X_2=1}^{X_2} 2\alpha_1 X_1\, dX_2 - \int_{X_2=1}^{X_2} X_1 X_2\, d\alpha_1 \tag{74}$$

Integrating by parts the second integral, we obtain

$$\int_{X_2=1}^{X_2} X_1 X_2\, d\alpha_1 = [\alpha_1 X_1 X_2]_{X_2=1}^{X_2} - \int_{X_2=1}^{X_2} \alpha_1(1 - 2X_2)\, dX_2 \tag{75}$$

Equation (74) then yields

$$\ln \gamma_2 = -\alpha_1 X_1 X_2 - \int_{X_2=1}^{X_2} \alpha_1\, dX_2 \tag{76}$$

The function α_1 is generally well behaved and the integral in equation (76) offers no difficulty.

It may be noted that equation (76) yields the following expression for $\ln \gamma_2^\infty$:

$$\ln \gamma_2^\infty = \int_{X_2=0}^{X_2=1} \alpha_1\, dX_2 \tag{77}$$

It shows that $\ln \gamma_2^\infty$ is the average value of α_1 over the entire composition range.

In the case of a regular solution [equation (III.46) or see Section VII.2] it is relatively straightforward to demonstrate that

$$\alpha_1 = \alpha_2 = \text{constant} \tag{78}$$

[The value of the constant is Ω/RT in equation (III.46)].

Other integration techniques are also used. Often, they depend on the form in which the data are obtained. For example, in the Kundsen cell-mass spectroscopy method of studying the thermodynamic properties of a binary solution, the experimentally measured quantity is proportional to the ratio of the activities of the two components. Knowing their concentrations, we deduce a quantity proportional to the ratio of the activity coefficients of the two components. The Gibbs–Duhem equation (68) may then be rewritten

$$d \ln \gamma_2 + X_1\, d \ln(\gamma_1/\gamma_2) = 0 \tag{79}$$

and yields

$$\ln \gamma_2 = -\int_{X_2=1}^{X_2} X_1\, d \ln(\gamma_1/\gamma_2) \tag{80}$$

$\ln \gamma_2$ is obtained through a plot of X_1 vs $-\ln(\gamma_1/\gamma_2)$ (see Problem 6). $\ln \gamma_1$ is similarly obtained by inverting in equation (80) the subscripts 1 and 2. This integration technique is due to Belton and Fruehan [1].

We note that instead of integrating the Gibbs–Duhem graphically, it is often more expedient to do it analytically. Analytic expressions of the thermodynamic properties of binary solutions constitute the subject of the next chapter.

Problems

1. The activity of B in a solution A–B at $T = 1000°K$ is found to be equal to 0.1 at a mole fraction $X_B = 0.01$. In a calorimeter with a reservoir of A at 1000°K, it is measured that very small additions of B absorb 7000 cal/mol. Estimate the activity of B at 1500°K for $X_B = 0.02$. Justify any assumption you make.

2. The activity coefficient of a solvent 1 may be represented by the following equation:

$$\ln \gamma_1 = X_2^2 (\alpha + \beta X_2)$$

Calculate the corresponding expressions of $\ln \gamma_2$ and G_m^M.

3. **a.** The excess enthalpy and entropy of liquid Ag–Cu solutions at 1423°K may be represented by the equations

$$H_m^E = (5500X_{Cu} + 3900X_{Ag})X_{Ag}X_{Cu} \quad \text{cal}$$
$$S_m^E = (1.430X_{Cu} + 0.323X_{Ag})X_{Ag}X_{Cu} \quad \text{cal/°K}$$

Determine the partial molar enthalpy and entropy of copper as a function of concentration. Calculate the activity of copper at $X_{Cu} = 0.5$.

b. A crucible containing 1 mol Ag, 2 mol Fe, and 0.26 mol Cu, is placed in a furnace at 1823°K under neutral atmosphere. Ag and Fe are practically immiscible and form two liquid layers. After equilibrium is achieved, a sample is taken in the iron layer and analysis shows a mole fraction of copper equal to 0.03. From the data estimate the activity coefficient of copper in iron at that concentration.

4. Demonstrate that $G_m^E/RT = (1 - X_2) \int_{X_2=0}^{X_2} \alpha_2 \, dX_2$.

5. The following data have been obtained for Pb in liquid Ag at 1000°C:

X_{Pb} =	0.096	0.143	0.182	0.261	0.293	0.322	0.361	0.403	0.445	0.478
a_{Pb} =	0.169	0.254	0.337	0.473	0.508	0.551	0.589	0.604	0.648	0.685
X_{Pb} =	0.522	0.590	0.643	0.682	0.742	0.797	0.853	0.891	0.956	
a_{Pb} =	0.691	0.730	0.747	0.768	0.811	0.829	0.884	0.924	0.983	

Draw a_{Pb}, $\ln \gamma_{Pb}$, and α_{Pb} vs X_{Pb}. Integrate graphically the Gibbs–Duhem equation and plot a_{Ag} and $\ln \gamma_{Ag}$ vs X_{Pb}.

6. Analysis by mass spectroscopy of the effusion of a Knudsen cell containing a liquid Cu–Sn solution at 1400°C yields the ratio I_{Cu}^+/I_{Sn}^+ of the relative ion intensities of isotopes of Cu and Sn. The ratio I_{Cu}^+/I_{Sn}^+ is proportional to the ratio a_{Cu}/a_{Sn} of the activities of Cu and Sn. The results are

X_{Cu} =	0.041	0.200	0.358	0.523	0.645	0.761	0.863	0.940
$\ln(I_{Cu}^+/I_{Sn}^+)$ =	−1.84	−1.07	−0.68	−0.32	0.03	0.57	1.25	1.97

Deduce from these data the activity and activity coefficient of Cu. Plot a_{Cu} and $\ln \gamma_{Cu}$ vs X_{Cu}.

References

1. G. R. Belton and R. J. Fruehan, *J. Phys. Chem.* **71,** 1403 (1967).
2. R. Hultgren, P. D. Desai, D. T. Hawkins, M. Gleiser, and K. K. Kelley, *Selected Values of the Thermodynamic Properties of Binary Alloys*. Am. Soc. for Metals, Metals Park, OH, 1973.
3. W. Henry, *Phil. Trans. R. Soc. London Part I*, 29–43 and 274–277 (1803).
4. F. M. Raoult, *C. R. Acad. Sci. Ser. C* **104,** 1430 (1887); *Z. Phys. Chem.* **2,** 353 (1888).
5. S. Glasstone, *Introduction to Electrochemistry*. Van Nostrand, New York, 1942, p. 139.
6. L. S. Darken and E. T. Turkdogan, U.S. Steel Edgard Bain Laboratory, Monroeville, PA, private communication, 1968.
7. T. L. Hill, *An Introduction to Statistical Thermodynamics*. Addison-Wesley, Reading, MA, 1960, pp. (a) 321–327, (b) 382–386.
8. L. S. Darken and R. W. Gurry, *Physical Chemistry of Metals*. McGraw-Hill, New York, 1953, p. 264.

VII Thermodynamic Formalisms Associated with Binary Metallic Solutions

The problem of information storage and retrieval is becoming increasingly important, and to provide a convenient system of classification, analytic forms of the composition dependence of thermodynamic properties need to be found. The usefulness of these forms is especially obvious when the number of components is larger than two, that is, when the use of two-dimensional diagrams becomes cumbersome and inadequate. A case in point is that of dilute multicomponent metallic solutions; it will be analyzed in detail in Chapter IX. The lengthy presentation of dilute binary metallic solutions which follows in Section 1 would be somewhat unwarranted were it not for the fact that the principles brought forth will be quite helpful in the presentation of the multicomponent case.

Section 2 considers some simple analytical functions which span the entire composition range, and Section 3 focuses on the often delicate problem of defining and selecting adequate reference states and composition coordinates. Sections 4 and 5 examine some consequences of these selections.

1. Dilute Solutions

1.1. Approximation of a Series by a Polynomial

The most widely used forms of representation of an integral or partial molar property have been polynomials. (Other functions, such as Fourier series, have also been used but have met with less success.) The variable is usually the mole fraction of the solute (X_2) or of the solvent (X_1):

$$Y(X_2) = A_0 + A_1X_2 + \cdots + A_nX_2^n = \sum_{i=0}^{n} A_iX_2^i \tag{1}$$

Sometimes it is a polynomial combination of both:

$$Y = X_1X_2 \sum_{i=0}^{n} B_i(X_1 - X_2)^i \tag{2}$$

$$Y = C_1X_1^nX_2 + C_2X_1^{n-1}X_2^2 + \cdots + C_nX_1X_2^n \tag{3}$$

It may be noted that forms (2) and (3), which were proposed by Guggenheim [1] and Borelius [2], respectively, may be generated by the general polynomial (1). Depending on the original form adopted, the number n of parameters varies, and there may be several relationships between the A_is.

The polynomial expression for Y in equation (1) is itself a particular case of the considerably more general Taylor series representation of Y:

$$Y = Y(X_2^\circ) + (X_2 - X_2^\circ)\, Y'(X_2^\circ) + \frac{(X_2 - X_2^\circ)^2}{2!}\, Y''(X_2^\circ) + \cdots + \frac{(X_2 - X_2^\circ)^i}{i!}\, Y^{(i)}(X_2^\circ) + \cdots \tag{4}$$

where $Y^{(i)}(X_2^\circ)$ is the ith derivative of Y at the point X_2° ($0 \le X_2^\circ \le 1$). It is readily seen that equation (4) is equivalent to (1) when, for instance,

$$X_2^\circ = 0 \tag{5a}$$

and

$$\frac{Y^{(i)}(X_2^\circ)}{i!} = \begin{cases} A_i & \text{for } i = 0, 1, 2, \ldots, n \\ 0 & \text{for } i = n+1, n+2, \ldots, \infty \end{cases} \tag{5b}$$

It should be emphasized that from a classical thermodynamics viewpoint the existence of the Taylor series (i.e., the fact that all the partial derivatives at the point X_2° are finite) is an assumption. However, the assumption is not very restrictive—e.g., it is obviously less restrictive than the assumption of a polynomial of finite degree—and is in good agreement with most of our data on metallic solutions. It is also justified by models of statistical thermodynamics, which relate it to the short range effectiveness of the atomic forces in the solution.

Note that the existence of a Taylor series for the partial molar excess free energy of the solute 2,

$$\frac{G_2^E}{RT} = \ln\gamma_2 = \ln\gamma_2^\infty + \left(\frac{\partial \ln\gamma_2}{\partial X_2}\right)_{X_2\to 0} X_2 + \frac{1}{2}\left(\frac{\partial^2 \ln\gamma_2}{\partial X_2^2}\right)_{X_2\to 0} X_2^2 + \cdots \tag{6}$$

implies obedience to Henry's zeroth order law (since $\ln\gamma_2^\infty$ is finite) and to Henry's first order law [since $(\partial \ln \gamma_2/\partial X_2)_{X_2\to 0}$ is finite]. By extension, it may be stated that Henry's law of any order n is also implied since the derivative of any order n must exist for the existence of the Taylor series.

It is not necessary to choose the point of infinite dilution $X_2 = 0$ as the point at which the series is expanded. Such a choice, however, is natural as well as convenient for the study of dilute solutions, and in no way affects the generality of the subsequent analysis.

Equation (4) may be rewritten

$$Y(X_2) = Y_n(X_2) + R_n(X_2) \tag{7}$$

where Y_n is the polynomial of degree n

$$Y_n(X_2) = Y(0) + X_2 Y'(0) + \cdots + \frac{X_2^n}{n!} Y^{(n)}(0) \tag{8a}$$

and R_n is the remainder

$$R_n = \frac{X_2^{n+1}}{(n+1)!} Y^{(n+1)}(0) + \frac{X_2^{n+2}}{(n+2)!} Y^{(n+2)}(0) + \cdots \tag{8b}$$

R_n may be also expressed[1]

$$R_n = \frac{X_2^{n+1}}{n+1} Y^{(n+1)}(\xi) \tag{9}$$

where ξ represents a particular value of the interval $(0, X_2)$, which generally varies with X_2.

In dilute solutions, X_2 being much smaller than 1, the absolute value of R_n decreases as the degree n of the polynomial increases, as may be seen from the expression for R_n in equation (9). If the experimental error in Y is ΔY, it is then possible to select the

[1] This result is demonstrated in several mathematics texts [e.g., 3].

minimum value of n such that

$$|R_n| < \Delta Y \tag{10}$$

and for all practical purposes the infinite series representation of Y is replaced by the polynomial Y_n containing $n + 1$ parameters.

Obviously, the difficulty generally lies in the estimation of the order of magnitude of the remainder R_n (or of the possible values of the derivatives of Y). However, such an estimate often may be attained by comparison of the system under investigation with other systems of similar behavior.

1.2. Application to ln γ_2; *Free Energy Interaction Coefficients*

Let us apply these principles to the function $Y = G_2^E/RT = \ln \gamma_2$, which is often under more direct experimental investigation than the logarithm of the activity coefficient of the solvent $\ln \gamma_1$, or than the integral excess free energy $G^E = RT \times (X_1 \ln \gamma_1 + X_2 \ln \gamma_2)$.

The Taylor series expansion of $\ln \gamma_2$ at the point of infinite dilution $X_2 \to 0$ (or $X_1 \to 1$) yields

$$\ln \gamma_2 = \ln \gamma_2^\infty + \left(\frac{\partial \ln \gamma_2}{\partial X_2}\right)_{X_2 \to 0} X_2 + \frac{1}{2}\left(\frac{\partial^2 \ln \gamma_2}{\partial X_2^2}\right)_{X_2 \to 0} X_2^2 + \cdots + \frac{1}{n!}\left(\frac{\partial^n \ln \gamma_2}{\partial X_2^n}\right)_{X_2 \to 0} X_2^n + \cdots \tag{11}$$

or

$$\ln \gamma_2 = \sum_{i=0}^{\infty} J_i^{(2)} X_2^i \tag{12}$$

with

$$J_i^{(2)} = \frac{1}{i!}\left(\frac{\partial^i \ln \gamma_2}{\partial X_2^i}\right)_{X_2 \to 0} \tag{13}$$

The coefficient $J_i^{(2)}$ has received a special name: *interaction coefficient* or *self-interaction coefficient* of order i. The self-interaction coefficient of zeroth order is just the value of $\ln \gamma_2$ at infinite solution: $\ln \gamma_2^\infty$. The self-interaction coefficient of order 1 is the most widely used of all the $J_i^{(2)}$ coefficients and is generally designated by $\epsilon_2^{(2)}$ (read "epsilon 2 on 2"). It was introduced in the literature by the work of Wagner [4] and Chipman [5]. It is a measure of how an increase in the concentration of 2 affects the activity coefficient of 2 (hence the name "self-interaction"), and is the coefficient of prime interest.

For higher concentrations, Lupis and Elliott [6] have used the second order interaction coefficient $J_2^{(i)}$ and designated it by the symbol $\rho_2^{(2)}$. The expansion of $\ln \gamma_2$ may then be written

$$\ln \gamma_2 = \ln \gamma_2^\infty + \epsilon_2^{(2)} X_2 + \rho_2^{(2)} X_2^2 + \text{higher order terms} \tag{14}$$

Figure 1 shows a plot of $\ln \gamma_2$ vs X_2 and illustrates the significance of the interaction coefficients. $\ln \gamma_2^\infty$ is the limiting value of the function when $X_2 \to 0$, $\epsilon_2^{(2)}$ is the slope of the curve at this point, and $\rho_2^{(2)}$ is its curvature.

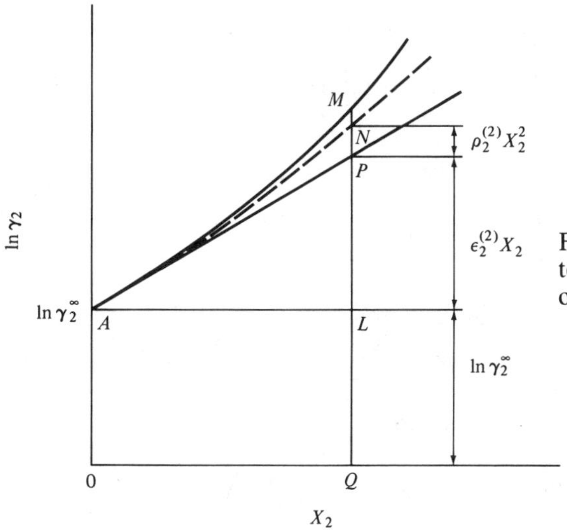

Figure 1. Illustration of low order terms in the Taylor series expansion of $\ln \gamma_2$.

The degree n of the polynomial to be retained in the expansion of $\ln \gamma_2$ [equations (11) and (6)] depends on the error $\Delta \ln \gamma_2$ which must be tolerated when measuring or calculating $\ln \gamma_2$. n can be determined according to the inequality (10). If n is chosen to be 1, this implies that at high dilutions the curve $\ln \gamma_2$ vs X_2 is replaced by the tangent AP (Fig. 1). If n is chosen to be 2, then the curve is replaced by the parabola AN, which has the tangent AP at A. If n is chosen to be 3, then the curve is replaced by another which has the tangent AP, the curvature of the parabola AN at A, and the same third order derivative, and so forth. It may be seen in Fig. 1 that the value of $\ln \gamma_2$ is approached as a sum of decreasingly small contributions:

$$QM = QL + LP + PN + \cdots$$

where $QM = \ln \gamma_2$, $QL = \ln \gamma_2^\infty$, $LP = \epsilon_2^{(2)}X_2$, $PN = \rho_2^{(2)}X_2^2$, etc.

Although interaction coefficients are defined at infinite dilution, by necessity they are measured and used at finite concentrations. This does not invalidate the approach, but its limitations should be clearly understood. The following examples should be helpful.

Let us suppose that measurements of $\ln \gamma_2$ have been obtained near the point X_2'. The difference between the measured slope $(\partial \ln \gamma_2/\partial X_2)_{X_2=X_2'}$ at X_2' and the value of $\epsilon_2^{(2)}$ or $(\partial \ln \gamma_2/\partial X_2)_{X_2\to 0}$ may be expressed

$$\left(\frac{\partial \ln \gamma_2}{\partial X_2}\right)_{X_2=X_2'} - \epsilon_2^{(2)} = X_2'\left(\frac{\partial^2 \ln \gamma_2}{\partial X_2^2}\right)_{X_2=\xi} \qquad (0 \le \xi \le X_2') \tag{15}$$

(as shown in Fig. 2.) Thus if we assume that the absolute value of the second derivative of $\ln \gamma_2$ with respect to X_2 over the interval studied is, for instance, less than 10, at $X_2' = 0.01$ the maximum error on $\epsilon_2^{(2)}$ is less than 0.1 (and negligible in most cases).

Assuming now that the interaction coefficients are accurately known, let us examine the error entailed by replacing the infinite series of equation (11) by a polynomial formalism of finite degree. Retaining a polynomial of degree 1, the value of $\ln \gamma_2$ is calculated:

$$\ln \gamma_2 = \ln \gamma_2^\infty + \epsilon_2^{(2)}X_2 \tag{16}$$

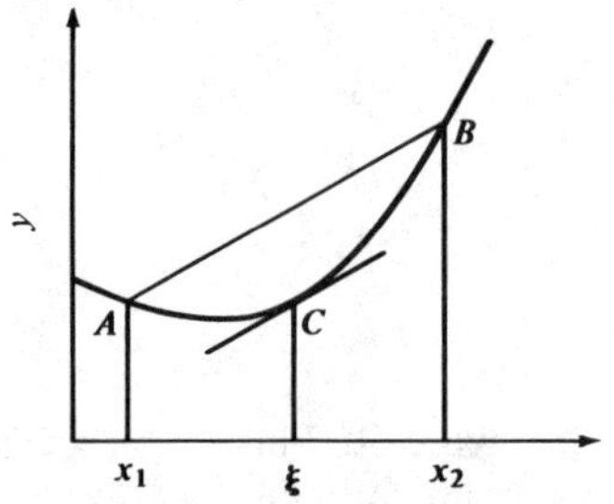

Figure 2. For a curve without singularity, to any straight line joining two points A and B there corresponds at least one point C in the interval AB at which the tangent is parallel to AB:

$$\frac{y(x_2) - y(x_1)}{x_2 - x_1} = y'(\xi) \qquad \text{with} \quad x_1 \leq \xi \leq x_2$$

This result applies to equation (15) with $y = \partial \ln \gamma_2/\partial X_2$, $x = X_2$, $x_1 = 0$, $x_2 = X_2'$.

instead of

$$\ln \gamma_2 = \ln \gamma_2^\infty + \epsilon_2^{(2)} X_2 + \tfrac{1}{2}(\partial^2 \ln \gamma_2/\partial X_2^2)_{X_2=\xi} X_2^2 \qquad (0 \leq \xi \leq X_2) \tag{17}$$

At a concentration X_2' the error is equal to $\frac{1}{2}(\partial^2 \ln \gamma_2/\partial X_2^2)_{X_2=\xi}(X_2')^2$. With the previous assumption that the second derivative of ln γ_2 is less than 10, at $X_2' = 0.01$ the error is less than 0.5×10^{-3} (i.e., the relative error on γ_2 is less than 1/2000). At $X_2' = 0.1$ the error is less than 0.05. If at that higher concentration the possible error is considered to be too large, a polynomial of higher degree (at least 2) must be retained.

The orders of magnitude of the successive derivatives of ln γ_2 (i.e., of the interaction coefficients) depend very much on the system studied. Generally, the stronger the atomic interactions between the two components the larger the values of these interaction coefficients are likely to be. As we have mentioned, the order of magnitude of these coefficients may be estimated with a fair degree of confidence by comparison with data on similar systems.

1.3. Enthalpy and Entropy Interaction Coefficients

The same series or polynomial formalism may be applied to the partial molar excess enthalpy and entropy of the solute 2, H_2^E and S_2^E.

$$H_2^E = \sum_{i=0}^{\infty} L_i^{(2)} X_2^i = H_2^{E\infty} + \eta_2^{(2)} X_2 + \cdots \tag{18}$$

$$S_2^E = \sum_{i=0}^{\infty} K_i^{(2)} X_2^i = S_2^{E\infty} + \sigma_2^{(2)} X_2 + \cdots \tag{19}$$

The coefficients of these series are called enthalpy and entropy interaction coefficients. Note that since

$$G_2^E = RT \ln \gamma_2 = H_2^E - TS_2^E \tag{20}$$

independently of the concentration X_2, it follows that

$$J_i^{(2)} = \frac{L_i^{(2)}}{RT} - \frac{K_i^{(2)}}{R} \tag{21}$$

or

$$\ln \gamma_2^\infty = \frac{H_2^{E\infty}}{RT} - \frac{S_2^{E\infty}}{R} \tag{22}$$

and

$$\epsilon_2^{(2)} = \frac{\eta_2^{(2)}}{RT} - \frac{\sigma_2^{(2)}}{R} \tag{23}$$

Obviously, series similar to those defined in equations (12), (18), and (19) may be written for the partial molar properties of the solvent:

$$G_1^E = RT \ln \gamma_1 = RT \sum_{i=0}^{\infty} J_i^{(1)} X_2^i \tag{24}$$

$$H_1^E = \sum_{i=0}^{\infty} L_i^{(1)} X_2^i \tag{25}$$

$$S_1^E = \sum_{i=0}^{\infty} K_i^{(1)} X_2^i \tag{26}$$

or for integral molar properties such as the excess free energy:

$$G^E = RT \sum_{i=0}^{\infty} \Phi_i X_2^i \tag{27}$$

The coefficients of series (24)–(27) do not receive the name "interaction coefficients." This is reserved for the coefficients of the Taylor series expansion of a partial molar property of the *solute* with respect to its mole fraction. The coefficients of the series (24)–(27) may, however, be easily calculated in terms of interaction coefficients through the use of the Gibbs–Duhem equation.

1.4. Application of the Gibbs–Duhem Equation

We recall that at constant temperature and pressure the Gibbs–Duhem equation yields (see Section VI.5)

$$X_1\, d \ln \gamma_1 + X_2\, d \ln \gamma_2 = 0 \tag{28}$$

or

$$(1 - X_2) \frac{\partial \ln \gamma_1}{\partial X_2} + X_2 \frac{\partial \ln \gamma_2}{\partial X_2} = 0 \tag{29}$$

$\partial \ln \gamma_1/\partial X_2$ and $\partial \ln \gamma_2/\partial X_2$ are easily calculated from equations (24) and (12), and equation (29) yields

$$(1 - X_2)\,(J_1^{(1)} + 2J_2^{(1)} X_2 + \cdots) + X_2(\epsilon_2^{(2)} + 2\rho_2^{(2)} X_2 + \cdots) = 0 \tag{30a}$$

or

$$J_1^{(1)} + X_2(-J_1^{(1)} + 2J_2^{(1)} + \epsilon_2^{(2)}) + X_2^2(\cdots) + \cdots = 0 \tag{30b}$$

Equation (30b) is an identity; the Gibbs–Duhem equation, which it represents, is a result that stands independently of the value of the concentration X_2. Consequently, each coefficient of the polynomial in the left member of equation (30b) must be equal to zero. For example,

$$J_1^{(1)} = 0 \tag{31}$$

$$J_2^{(1)} = -\tfrac{1}{2}\epsilon_2^{(2)} \tag{32}$$

Note that $J_1^{(1)}$ is the first derivative of $\ln \gamma_1$ at $X_2 \to 0$ and that equation (31) is then equivalent to a statement of Raoult's law (see Section VI.4.1):

$$(\partial \ln \gamma_1/\partial X_2)_{X_2\to 0} = 0 \tag{33}$$

It is not surprising that Raoult's law is obeyed. It is obtained as a consequence of Henry's law since, as previously emphasized, the assumption of a Taylor series for $\ln \gamma_2$ implies that Henry's law is obeyed.

The adoption of a Raoultian standard state for the activity of 1 entails that at $X_1 = 1$ (or $X_2 = 0$), a_1 and γ_1 are equal to 1. Thus $(\ln \gamma_1)_{X_2=0} = J_0^{(1)} = 0$ and the series expansion of $\ln \gamma_1$ starts with a second order term:

$$\ln \gamma_1 = -\tfrac{1}{2}\epsilon_2^{(2)}X_2^2 + \text{higher order terms} \tag{34}$$

a result which we already found in our introduction of the α function (Section VI.5).

The coefficients of the Taylor series for G^E may be easily obtained by recalling that

$$G^E = RT[(1 - X_2)\ln \gamma_1 + X_2 \ln \gamma_2] \tag{35}$$

Consequently,

$$G^E = RT(X_2 \ln \gamma_2^\infty + \tfrac{1}{2}\epsilon_2^{(2)}X_2^2 + \cdots) \tag{36}$$

The relationships among the general coefficients $J_i^{(1)}$, $J_i^{(2)}$, and Φ_i may be obtained in similar ways (see Problem 2).

2. Use of Polynomials Across the Composition Range

The use of polynomials to represent free energy functions across the composition range is based essentially on the same premises as those presented for dilute solutions. In the dilute ranges, Raoult's and Henry's laws must generally be obeyed and the numbers of coefficients or parameters in the polynomials depend on the complexity of the systems investigated and the accuracy of our information.

The formalism of a regular solution has already been encountered; the excess Gibbs free energy is then expressed

$$G^E = X_1X_2\Omega \tag{37}$$

and as a consequence,

$$G_1^E/RT = \ln \gamma_1 = (\Omega/RT)X_2^2 \tag{38}$$

$$G_2^E/RT = \ln \gamma_2 = (\Omega/RT)X_1^2 \tag{39}$$

Equation (37) is that of a parabola with a vertical axis of symmetry at $X_1 = X_2 = 0.5$, while equation (38) [or (39)] is that of a parabola which has its summit at X_1 (or X_2) = 0 (see Fig. 3). The formulas contain one single parameter and thus are very practical for interpolation or extrapolation when little information on the system is available (a much too frequent occurrence). Moreover, the parameter Ω has a simple atomistic interpretation: it compares the energy u_{12} of the bond between the dissimilar atoms 1 and 2 with the arithmetic mean of the bonds between like atoms:

$$\Omega = Z\,[u_{12} - \tfrac{1}{2}(u_{11} + u_{22})] \tag{40}$$

where Z is the number of nearest neighbors to an atom (e.g., 12 in the case of an fcc lattice). Ω is negative for a net attraction between the atoms 1 and 2 and positive for a net repulsion between the atoms 1 and 2 (see Section XV.3 for a detailed treatment of the model).

Often, however, the regular solution is unable to represent the properties of real solutions and a second parameter is introduced. Equation (37) then becomes

$$G^E = X_1X_2(A_{21}X_1 + A_{12}X_2) \tag{41}$$

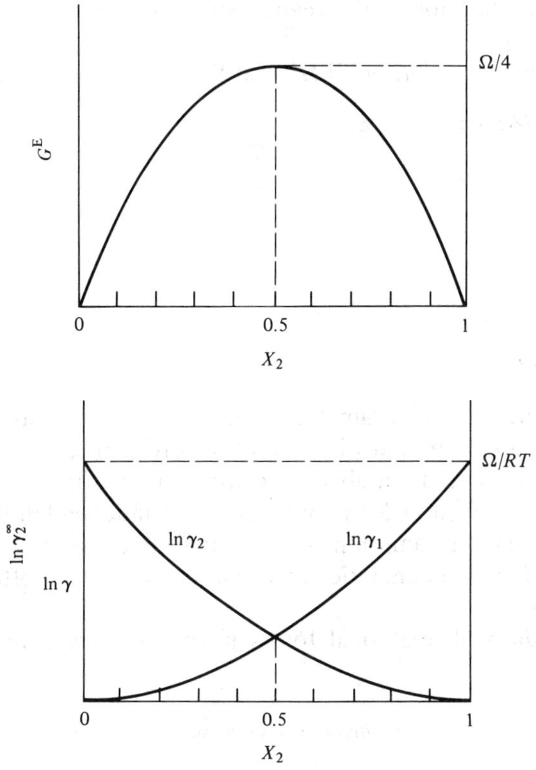

Figure 3. Graphical representation of free energy functions for a regular solution.

Solutions which obey this type of representation have been termed *subregular* [7]. Of course, if two parameters are inadequate to represent the data, a third may be introduced:

$$G^E = X_1X_2(A_{21}X_1 + A_{12}X_2 + A_{22}X_1X_2) \tag{42}$$

and so forth:

$$G^E = \sum_{i=1}^{m} \sum_{j=1}^{n} X_1^i X_2^j A_{ij} \tag{43}$$

In most metallurgical solutions three parameters or fewer represent the data well within the experimental scatter.

Note that equation (42) simplifies to (41) when $A_{22} = 0$; it further simplifies to the regular solution equation (37) when in addition $A_{12} = A_{21}$. The expressions for $\ln \gamma_1$ and $\ln \gamma_2$ can be easily deduced. In particular, equation (42) leads to

$$RT \ln \gamma_1 = X_2^2[2X_1A_{21} + (1 - 2X_1)A_{12} + X_1(2 - 3X_1)A_{22}] \tag{44}$$

and

$$RT \ln \gamma_2 = X_1^2[2X_2A_{12} + (1 - 2X_2)A_{21} + X_2(2 - 3X_2)A_{22}] \tag{45}$$

Obviously, the same procedure is applicable to the excess enthalpy and the excess entropy. In general, however, there is much more uncertainty on the values of the excess entropy.[2] It is often assumed to be equal to 0:

$$S^E = 0 \qquad \text{or} \qquad S^M = -R(X_1 \ln X_1 + X_2 \ln X_2) \tag{46}$$

This is, in particular, the assumption of the regular solution. [In equation (37) Ω is temperature independent.]

In the case of a *quasi-regular solution* [8] H^E and S^E have the same form (37):

$$H^E = X_1X_2\Omega = X_1X_2B_{11} \tag{47}$$

$$S^E = X_1X_2C_{11} \tag{48}$$

If we let

$$\tau = B_{11}/C_{11} = H^E/S^E \tag{49}$$

then G^E may be rewritten

$$G^E = X_1X_2\Omega[1 - (T/\tau)] \tag{50}$$

This form and the significance of the parameter τ will be discussed in some detail in Chapter XV. Let us just note here that τ is often of the order of 2500 ± 1000°K, but not necessarily in this range and not always positive and that the assumption of a proportionality between H^E and S^E may be retained independently of the parabolic form [(47)–(48)]. The assumption is convenient because it introduces only one parameter for S^E (given an analytic representation of H^E); it often leads to very satisfactory results.

A good number of additional analytical forms have been presented in the

[2] Excess entropy functions are generally determined as *differences* between excess enthalpy and excess Gibbs free energy functions [e.g., $S^E = (H^E - G^E)/T$].

literature. Darken, for instance, has recommended the use of a quadratic formalism [9]. More promising are the methods based on orthogonal functions. Williams, for instance, noted that a power series is not the most adequate mathematical way of fitting data, because the values of the parameters are very sensitive to the number of terms retained, and introduced a series of orthogonal functions which seems to yield fairly satisfactory results [10]; in the work of Bale and Pelton, these orthogonal functions are Legendre polynomials [11]. Generally, however, unless the system presents some anomaly, simple polynomials of one to three terms are fairly adequate.

Ideally, of course, if a statistical model were to predict accurately the properties of a solution, the model would then automatically provide the desired formalism. Although no model yet has been found that can be applied to the majority of metallic solutions, some models can indeed provide the basis for formalisms different from the ones presented above, and which apply to certain classes of solutions (e.g., interstitial solutions).

3. Composition Coordinates and Standard States for the Measure of the Activity Function

3.1. Change of Reference State

As emphasized in Chapter IV, the study of the chemical potential of a species i is essentially a study of its composition dependence, and we found it convenient to introduce the activity function a_i defined by the differential equation

$$(d\mu_i)_{T,P} = RT\, d \ln a_i \tag{51}$$

or by its integral form

$$\mu_i = \mu_i^*(T, P) + RT \ln a_i \tag{52}$$

μ_i^* is an arbitrary function of the pressure and the temperature (independent of concentration) which represents the value of the chemical potential of i in a reference state.

A change in the choice of μ_i^* changes the measure of the activity a_i but only by a proportionality factor. This may be seen as follows. Changing reference states $\mu_i^{*\prime}$ to $\mu_i^{*\prime\prime}$ leads to

$$\mu_i = \mu_i^{*\prime} + RT \ln a_i' = \mu_i^{*\prime\prime} + RT \ln a_i'' \tag{53}$$

and

$$a_i'/a_i'' = \exp[-(\mu_i^{*\prime} - \mu_i^{*\prime\prime})/RT] \tag{54}$$

which is a function of only T and P. Consequently, the ratio of two activities defined by different reference states is independent of composition.

When i is in its reference state, its activity is equal to 1.

To define adequately the activity a_i, the reference state must be defined unequivocally. At the temperature and pressure considered all the characteristics of the solution necessary to calculate μ_i^* must be known, including the composition of the solution and its structure (liquid, solid fcc, or solid bcc). In general, the selected phase is that which is stable under the conditions considered; in case of possible ambiguity, the selection of the phase must be made explicit. It must be emphasized that the definition of the reference state is purely arbitrary and as such can be entirely hypothetical, i.e., correspond to no physical reality.

A reference state becomes a standard state when the pressure is conventionally fixed, usually at 1 atm (see Chapter IV). Recalling that the effect of pressure on the chemical potential of a component in a solid or liquid phase is generally of little significance at moderate pressures, the difference between reference and standard states can be neglected in most metallurgical applications. As a result, reference and standard states are often used interchangeably.

3.2. Composition Coordinates

Another element of arbitrariness entering the study of the activity function is the choice of the composition coordinate. Quite generally, the activity is written as the product of a composition coordinate and an *activity coefficient*. Thus, the definition of the activity coefficient depends on the composition scale adopted as well as on the standard state. So far we have used only the mole fraction coordinate X_i:

$$X_i = n_i \Big/ \sum_{j=1}^{m} n_j \tag{55}$$

and in theoretical applications it is by far the most commonly used. However, in industrial practice weights are of paramount importance and concentrations are generally expressed in weight percents. If m_i represents the mass of component i in the solution and M_i its atomic or molecular weight (per mole), then

$$\%i/100 = m_i \Big/ \sum_{j=1}^{m} m_j \tag{56}$$

and

$$\%i/100 = X_i M_i \Big/ \sum_{j=1}^{m} X_j M_j \tag{57}$$

or

$$X_i = \frac{\%i/M_i}{\sum_{j=1}^{m} (\%j/M_j)} \tag{58}$$

Other composition coordinates, generally suggested by models of statistical thermodynamics, are occasionally introduced. For example, in the case of a binary interstitial solution 1–2, where 1 is the solvent and 2 the interstitial solute, if we assume for simplicity that the number n_i of interstitial sites is equal to the number n_S of substitutional sites (as in an fcc lattice), then the coordinate

$$y_2 = \frac{n_2}{n_\mathrm{i} - n_2} = \frac{n_2}{n_\mathrm{S} - n_2} = \frac{n_2}{n_1 - n_2} \tag{59a}$$

or

$$y_2 = \frac{X_2}{1 - 2X_2} \tag{59b}$$

enters quite naturally the calculations and its use simplifies the expressions of the results of the model.

In principle, any coordinate may be used for data reduction or theoretical models, but from a practical point of view a multiplicity of coordinates entails a multiplicity of thermodynamic functions (e.g., activity coefficients) and reference states, which can be very cumbersome and confusing to the nonspecialist; in our opinion, it is important to convert the final results of an investigation into one or two (at most) reasonably unified systems of coordinates to allow simple tabulations and comparisons.

In this treatment, we shall continue to use mole fractions, but in addition we shall introduce weight percents. The most common choices of reference (or standard) states based on these two coordinates will be analyzed now in some detail. The presentation will be restricted to the case of binary solutions, although it can be readily extended to the case of multicomponent ones. Also *for convenience, the pressure will be fixed (to 1 atm) so that reference states become standard states.*

3.3. Raoultian Standard State and Mole Fraction Composition Coordinate

The standard state for the component 2 is chosen now to correspond to the state of the pure substance 2. Thus

$$\mu_2 = \mu_2^\circ \qquad \text{or} \qquad a_2^\circ = 1 \quad \text{for } X_2 = 1 \tag{60}$$

where a_2° is the activity of 2 in its standard state. The standard state is identified in Fig. 4. Note that

$$(a_2)_{X_2=1} = (a_2^\circ)_{X_2=1} \tag{61}$$

The name of Raoultian standard state is adopted here to emphasize the fact that the standard state is chosen at the composition where Raoult's law applies ($X_2 = 1$)—although the choice of this standard state does not necessarily imply obedience to Raoult's law. For convenience, to distinguish this activity scale from the subsequent ones we shall identify it by the notation a_2^R. The symbol reserved to the activity coefficient is γ_2. By definition,

$$\gamma_2 = a_2^R/X_2 \tag{62}$$

From equations (60)–(62), it follows that

$$(\gamma_2)_{X_2=1} = 1 \tag{63}$$

3.4. Henrian Standard State and Mole Fraction Composition Coordinate

The choice of the previous standard state is unnatural for the study of the behavior of species in very dilute concentrations (such as nitrogen in iron). In such cases it is more meaningful to define the standard state according to what is happening in very dilute solutions. Assuming obedience to Henry's law, the *Henrian* standard state is defined as the hypothetical state obtained by extrapolating the Henrian

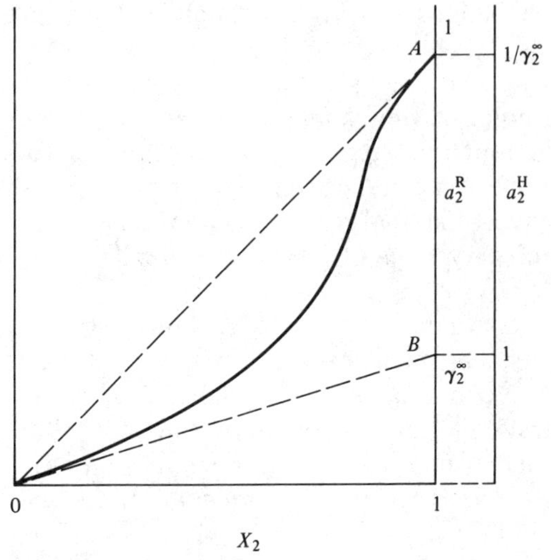

Figure 4. A and B are, respectively, the representative points of the Raoultian and Henrian standard states.

behavior of 2 to the concentration where 2 is pure. In other words, to obtain the standard state, Henry's law is assumed to apply not only at infinite dilution but across the composition range to the point $X_2 = 1$ (see Fig. 4).

Note that equation (60) is still valid:

$$\mu_2 = \mu_2^\circ \qquad \text{or} \qquad a_2^\circ = 1 \quad \text{for } X_2 = 1$$

The important difference is that the state of pure 2 is now an extrapolated one and usually quite different from the one which could be observed at this concentration. Thus, contrary to equation (61),

$$(a_2)_{X_2=1} \neq (a_2^\circ)_{X_2=1} \tag{64}$$

The activity function so defined will be designated a_2^H. By definition, the corresponding activity coefficient will be

$$\varphi_2 = a_2^H/X_2 \tag{65}$$

In the standard state, the value of φ_2 is obviously 1. At infinite dilution, the activity coefficient is equal to the slope of the activity curve at $X_2 = 0$ (see Section VI.3), and consequently φ_2 at infinite dilution is also equal to 1:

$$(\varphi_2)_{X_2\to 0} = 1 \tag{66}$$

(Note that it is possible to select other Henrian standard states on the line of Henry's law; however, at a composition of, let us say, $X_2 = 0.5$, the limit of the activity coefficient at $X_2 = 0$ would be 2 instead of 1.)

It is of interest to find the conversion relationship between the activities a_2^R and a_2^H. Recalling that the ratio of two activities defined on the basis of different standard states is a constant [equation (54)], the value of this constant may be obtained at the point of infinite dilution. Thus,

$$\frac{a_2^R}{a_2^H} = \frac{\gamma_2 X_2}{\varphi_2 X_2} = \frac{\gamma_2}{\varphi_2} = \left(\frac{\gamma_2}{\varphi_2}\right)_{X_2\to 0} = \frac{\gamma_2^\infty}{1} \tag{67a}$$

in which we note that

$$\varphi_2 = \gamma_2/\gamma_2^\infty \tag{67b}$$

and where γ_2^∞ is the value of γ_2 at infinite dilution.

3.5. Henrian Standard State and Weight Percent Composition Coordinate

The standard state is again selected to lie on Henry's law line but at the concentration of 2 equal to 1% (see Fig. 5). On this scale, the activity may be designated a_2^{h} and the activity coefficient f_2:

$$f_2 = a_2^{\mathrm{h}}/\%2 \tag{68}$$

In the standard state f_2 is equal to 1, and since at infinite dilution the value of f_2 becomes equal to the slope of the Henrian line, we also have

$$(f_2)_{\%2\to 0} = 1 \tag{69}$$

The conversion relation between a_2^{R} and a_2^{h} may be found by the same procedure as that followed in equation (67):

$$\frac{a_2^{\mathrm{h}}}{a_2^{\mathrm{R}}} = \frac{f_2\%2}{\gamma_2 X_2} = \left(\frac{f_2\%2}{\gamma_2 X_2}\right)_{X_2\to 0} = \frac{1}{\gamma_2^\infty}\left(\frac{\%2}{X_2}\right)_{X_2\to 0} \tag{70}$$

and recalling that

$$\frac{\%2}{100} = \frac{X_2 M_2}{X_1 M_1 + X_2 M_2} \tag{71}$$

where M_1 and M_2 are the atomic weights of the solvent 1 and the solute 2,

$$\left(\frac{\%2}{X_2}\right)_{X_2\to 0} = \frac{100 M_2}{M_1} \tag{72}$$

Thus

$$a_2^{\mathrm{h}}/a_2^{\mathrm{R}} = 100 M_2/\gamma_2^\infty M_1 \tag{73}$$

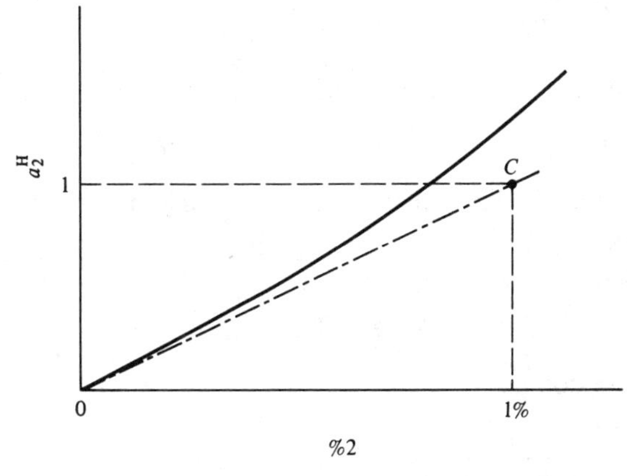

Figure 5. C is the representative point of the Henrian standard state on a weight percent basis.

and in another form

$$2.3 \log f_2 = \ln\gamma_2 + \ln(X_2/\%2) + \ln(100M_2/\gamma_2^\infty M_1) \tag{74}$$

or, through equation (71),

$$2.3 \log f_2 = \ln(\gamma_2/\gamma_2^\infty) + \ln[X_1 + X_2(M_2/M_1)] \tag{75}$$

It should be noted that whereas a natural logarithm is customarily used for γ_2, a logarithm to the base 10 is generally used for f_2.

4. Interaction Coefficients Based on a Weight Percent Composition Coordinate

4.1. Free Energy Interaction Coefficients

In Section 1.2 we developed into a series the natural logarithm of the activity coefficient of the solute 2 with respect to the mole fraction of 2 at the point of infinite dilution of 2. A similar Taylor series may be written for the decimal logarithm of the activity coefficient f_2:

$$\log f_2 = (\log f_2)_{\%2\to 0} + \left(\frac{\partial \log f_2}{\partial \%2}\right)_{\%2\to 0} \%2$$

$$+ \frac{1}{2}\left(\frac{\partial^2 \log f_2}{\partial (\%2)^2}\right)_{\%2\to 0} (\%2)^2 + \text{higher order terms} \tag{76}$$

Because of the Henrian standard state associated with the use of f_2 we have seen that $f_2 = 1$ at $\%2 \to 0$. Thus, the first term of the series is equal to zero. As in the analysis of the series expansion of $\ln \gamma_2$, the coefficient of the second term may be called a *self-interaction coefficient of the first order*. It is denoted $e_2^{(2)}$:

$$e_2^{(2)} = \left(\frac{\partial \log f_2}{\partial \%2}\right)_{\%2\to 0} \tag{77}$$

Similarly

$$r_2^{(2)} = \frac{1}{2}\left(\frac{\partial^2 \log f_2}{\partial (\%2)^2}\right)_{\%2\to 0} \tag{78}$$

and so forth. Equation (76) then becomes

$$\log f_2 = e_2^{(2)}\%2 + r_2^{(2)}(\%2)^2 + \text{higher order terms} \tag{79}$$

which may be compared with its parallel on a mole fraction scale:

$$\ln \varphi_2 = \ln(\gamma_2/\gamma_2^\infty) = \epsilon_2^{(2)} X_2 + \rho_2^{(2)} X_2^2 + \text{higher order terms} \tag{80}$$

A conversion formula relating $e_2^{(2)}$ and $\epsilon_2^{(2)}$ may be readily obtained by differentiation of equation (75):

$$2.3 \frac{\partial \log f_2}{\partial \%2} = \frac{\partial \ln \gamma_2}{\partial X_2} + \frac{M_2 - M_1}{X_1 M_1 + X_2 M_2} \frac{\partial X_2}{\partial \%2} \tag{81}$$

At infinite dilution of 2 it becomes

$$2.3 e_2^{(2)} = \left(\epsilon_2^{(2)} - \frac{M_1 - M_2}{M_1}\right) \frac{M_1}{100 M_2}$$

or

$$e_2^{(2)} = 0.434 \times 10^{-2}\left(\frac{M_1}{M_2}\epsilon_2^{(2)} - \frac{M_1 - M_2}{M_2}\right) \tag{82}$$

Conversely,

$$\epsilon_2^{(2)} = 230\frac{M_2}{M_1}e_2^{(2)} + \frac{M_1 - M_2}{M_1} \tag{83}$$

The reader should note that the last term of equation (83) has, mistakenly, often been omitted in the literature [12].

A differentiation of equation (81) leads similarly to the conversion relationship between $r_2^{(2)}$ and $\rho_2^{(2)}$ [6]:

$$r_2^{(2)} = \frac{0.434 \times 10^{-2}}{M_2^2}[M_1^2\rho_2^{(2)} - M_1(M_1 - M_2)\epsilon_2^{(2)} + \tfrac{1}{2}(M_1 - M_2)^2] \tag{84}$$

or

$$\rho_2^{(2)} = \frac{230\,M_2}{M_1^2}[10^2 M_2 r_2^{(2)} + (M_1 - M_2)e_2^{(2)}] + \frac{1}{2}\left(\frac{M_1 - M_2}{M_1}\right)^2 \tag{85}$$

4.2. Enthalpy and Entropy Interaction Coefficients

$\ln \gamma_2$ being equal to G_2^E/RT, writing it as the sum of an excess enthalpy contribution (H_2^E/RT) and entropy contribution $(-S_2^E/R)$ provides simple relations between free energy, enthalpy, and entropy interaction coefficients on a mole fraction basis [see equations (21)–(23)]. On a weight percent basis, no simple relation exists between $\log f_2$ and G_2^E/RT, and therefore between $\log f_2$, H_2^E, and S_2^E. However, it is advantageous to distinguish enthalpy and entropy contributions in $\log f_2$ and its associated interaction coefficients, and consequently the introduction of two new functions $\mathcal{H}_2^E$ and $\mathcal{S}_2^E$ are found to be warranted [6,13].

We shall define these functions by the relation

$$2.3RT \log f_2 = \mathcal{H}_2^E - T\mathcal{S}_2^E \tag{86}$$

in obvious parallel to equation (20). To relate $\mathcal{H}_2^E$ and $\mathcal{S}_2^E$ to H_2^E and S_2^E, equation (75) may be used. It yields

$$RT \ln(\gamma_2/\gamma_2^\infty) + RT \ln[X_1 + X_2(M_2/M_1)] = \mathcal{H}_2^E - T\mathcal{S}_2^E \tag{87}$$

or

$$(H_2^E - H_2^{E\infty}) - T\left[(S_2^E - S_2^{E\infty}) - R\ln\left(X_1 + X_2\frac{M_2}{M_1}\right)\right] = \mathcal{H}_2^E - T\mathcal{S}_2^E \tag{88}$$

Since equation (86) is valid at all temperatures, it follows that

$$\mathcal{H}_2^E = H_2^E - H_2^{E\infty} \tag{89}$$

and

$$\mathcal{S}_2^E = (S_2^E - S_2^{E\infty}) - R\ln[X_1 + X_2(M_2/M_1)] \tag{90}$$

Thus, $\mathcal{H}_2^E$ is simply related to H_2^E, but $\mathcal{S}_2^E$ to S_2^E is not.

Expanding $\mathcal{H}_2^E$ and $\mathcal{S}_2^E$ in Taylor series with respect to %2 yields

$$\mathcal{H}_2^E = \left(\frac{\partial \mathcal{H}_2^E}{\partial \%2}\right)_{\%2\to 0} \%2 + \frac{1}{2}\left(\frac{\partial^2 \mathcal{H}_2^E}{\partial (\%2)^2}\right)_{\%2\to 0} (\%2)^2 + \cdots \tag{91}$$

$$\mathcal{S}_2^E = \left(\frac{\partial \mathcal{S}_2^E}{\partial \%2}\right)_{\%2\to 0} \%2 + \frac{1}{2}\left(\frac{\partial^2 \mathcal{S}_2^E}{\partial (\%2)^2}\right)_{\%2\to 0} (\%2)^2 + \cdots \tag{92}$$

The coefficients of these series are, respectively, the enthalpy and entropy interaction coefficients on a weight percent basis. The first order enthalpy interaction coefficient is generally denoted $h_2^{(2)}$:

$$h_2^{(2)} = (\partial \mathcal{H}_2^E / \partial \%2)_{\%2\to 0} \tag{93}$$

and the first order entropy interaction coefficient $s_2^{(2)}$:

$$s_2^{(2)} = (\partial \mathcal{S}_2^E / \partial \%2)_{\%2\to 0} \tag{94}$$

The conversion formula between $\eta_2^{(2)}$ and $h_2^{(2)}$ is readily obtained through equation (89):

$$\frac{\partial H_2^E}{\partial X_2} = \frac{\partial \mathcal{H}_2^E}{\partial \%2}\frac{\partial \%2}{\partial X_2} \tag{95}$$

and for $X_2 \to 0$

$$\eta_2^{(2)} = 100(M_2/M_1)h_2^{(2)} \tag{96}$$

Similarly, equation (90) leads to

$$\sigma_2^{(2)} = 100\frac{M_2}{M_1} s_2^{(2)} - R\frac{M_1 - M_2}{M_1} \tag{97}$$

The relationship among the free energy, enthalpy, and entropy interaction

Table 1. Self-Interaction Coefficients

Mole fraction composition coordinate	*Weight percent composition coordinate*
$G_2^E/RT = \ln\gamma_2 = \ln\gamma_2^\infty + \epsilon_2^{(2)}X_2 + \rho_2^{(2)}X_2^2 + \cdots$	$\log f_2 = e_2^{(2)}\%2 + r_2^{(2)}(\%2)^2 + \cdots$
$H_2^E = H_2^{E\infty} + \eta_2^{(2)}X_2 + \cdots$	$\mathcal{H}_2^E = h_2^{(2)}\%2 + \cdots$
$S_2^E = S_2^{E\infty} + \sigma_2^{(2)}X_2 + \cdots$	$\mathcal{S}_2^E = s_2^{(2)}\%2 + \cdots$
$\ln\gamma_2 = H_2^E/RT - S_2^E/R$	$\log f_2 = \mathcal{H}_2^E/2.3RT - \mathcal{S}_2^E/2.3R$
$\epsilon_2^{(2)} = \eta_2^{(2)}/RT - \sigma_2^{(2)}/R$	$e_2^{(2)} = h_2^{(2)}/2.3RT - s_2^{(2)}/2.3R$
Conversion formulas	
$2.3\log f_2 = \ln(\gamma_2/\gamma_2^\infty) + \ln[X_1 + X_2(M_2/M_1)]$	$\epsilon_2^{(2)} = 230\,(M_2/M_1)e_2^{(2)} + (M_1 - M_2)/M_1$
$\mathcal{H}_2^E = H_2^E - H_2^{E\infty}$	$\eta_2^{(2)} = 100\,(M_2/M_1)h_2^{(2)}$
$\mathcal{S}_2^E = (S_2^E - S_2^{E\infty}) - R\ln[X_1 + X_2(M_2/M_1)]$	$\sigma_2^{(2)} = 100\,(M_2/M_1)s_2^{(2)} - R\,(M_1 - M_2)/M_1$

Based on Taylor's series: $f(x) = f(0) + (df/dx)_{x=0}x + \frac{1}{2}(d^2f/dx)_{x=0}x^2 + \cdots$.

coefficients is immediately obtained by differentiation of equation (86):

$$e_2^{(2)} = (h_2^{(2)}/2.3RT) - (s_2^{(2)}/2.3R) \tag{98}$$

and yields the temperature dependence of the coefficient $e_2^{(2)}$.

Table 1 summarizes many of these formulas.

Orders of magnitude for the interaction coefficients should be kept in mind. The ϵ coefficients usually vary from -10 to $+10$, the η coefficients are of the order of a few kilocalories, and the σ coefficients of a few calories per degree Kelvin. More extreme values are of course possible and known to occur for some systems. The corresponding coefficients e, h, and s are about 10^2 times smaller, as may be seen from the conversion formulas. The ρ coefficients are of the same order of magnitude as ϵ (a few units), while the r coefficients are of the order of 10^{-4} (a hundred times smaller than e).

5. Application to Chemical Reactions

5.1. Notation

In the thermodynamic analysis of chemical reactions, the choice of the standard states for the various reactants and products is very important. Any given value of $\Delta G°$ corresponding to the reaction

$$\sum_{i=1}^{r} \nu_i A_i = 0 \tag{99}$$

is meaningless without a clear specification of the standard state of each species A_i since

$$\Delta G° = \sum_{i=1}^{r} \nu_i \mu_i^\circ \tag{100}$$

For brevity's sake, the following convention is often adopted.

If, in reaction (99), a Henrian standard state based on a mole fraction composition coordinate is chosen for A_i, then A_i appears underlined in the reaction. If a Henrian standard state based on a weight percent coordinate is chosen instead, the symbol becomes $\underline{A_i}(\%)$. If a Raoultian standard state is chosen, A_i remains as written. In case of possible ambiguity, additional information is written between parentheses following A_i. For instance, "Fe(γ)" specifies a Raoultian standard state for iron in the fcc lattice structure; "$\underline{O}$(% in liquid iron)" indicates that the standard state is taken as 1% of oxygen in liquid iron (not solid iron, and not an iron alloy).

5.2. Solubility of Gases

Let us consider the dissolution of nitrogen in liquid iron:

$$\tfrac{1}{2}N_2(g) = \underline{N}(\%) \tag{101}$$

At 1600°C, the $\Delta G°$ of this reaction equals 11,560 cal [14] and consequently,

$$\log K = \frac{-11{,}560}{4.575 \times 1873} = -1.35 \qquad \text{or} \qquad K = 0.045 \tag{102}$$

Thus,

$$\frac{\%\mathrm{N} f_{\mathrm{N}}}{p_{\mathrm{N}_2}^{1/2}} = 0.045 \tag{103}$$

In dilute solutions ($\%\mathrm{N} \rightarrow 0$), $f_{\mathrm{N}} \rightarrow 1$. Thus, unless f_{N} becomes significantly different from 1 as %N increases, the solubility of nitrogen under a partial pressure of nitrogen equal to one atmosphere[3] is approximately equal to 0.045. To analyze the correction to this solubility value because of the composition dependence of f_{N} we note that

$$\log f_{\mathrm{N}} = e_{\mathrm{N}}^{\mathrm{N}} \%\mathrm{N} + r_{\mathrm{N}}^{\mathrm{N}} (\%\mathrm{N})^2 + \cdots \tag{104}$$

Since %N is of the order of 0.045 or below (for $p_{\mathrm{N}_2} \leq 1$ atm), the term $r_{\mathrm{N}}^{\mathrm{N}}(\%\mathrm{N})^2$ is certainly negligible in front of $e_{\mathrm{N}}^{\mathrm{N}}\%\mathrm{N}$ (the ratio is likely to be of the order $10^{-2} \times 0.04 = 4 \times 10^{-4}$). Thus, equation (104) becomes

$$\log f_{\mathrm{N}} \simeq e_{\mathrm{N}}^{\mathrm{N}} \%\mathrm{N} \tag{105}$$

$e_{\mathrm{N}}^{\mathrm{N}}$ is likely to be of the order of 0.1 at most, and thus $e_{\mathrm{N}}^{\mathrm{N}}\%\mathrm{N}$ is probably smaller than 4×10^{-3}. Consequently, the correction to be made to the value of 1 for f_{N} is less than 0.01 (1% of its value) and well within the general experimental scatter of nitrogen solubility measurements. For the usual range of partial pressures (0–1 atm) a value of $e_{\mathrm{N}}^{\mathrm{N}}$ is then unlikely to be obtained experimentally, and the approximation $f_{\mathrm{N}} = 1$ is justified. It yields

$$\%\mathrm{N} = 0.045 p_{\mathrm{N}_2}^{1/2} \tag{106}$$

The proportionality of the solubility of a diatomic gas to the square root of its partial pressure is known as *Sieverts's law*. Any deviation from this proportionality, arising from the composition dependence of the activity coefficient, is referred to as a deviation from Sieverts's law. It should be noted that Sieverts's law is in essence the application of Henry's first order law, and that the range of validity of Sieverts's law (i.e., the range where there is no deviation) is determined by the approximation corresponding to the replacement of the activity curve by its tangent at the point of infinite dilution.

The solubility of nitrogen in liquid chromium is much larger than in liquid iron, and at 1800°C it is equal to 5.3% [15]. The correction for f_{N} [equation (105)] becomes quite significant and yields large deviations from Sieverts's law (see Fig. 6), which have been experimentally observed. They have led to a value of $e_{\mathrm{N}}^{\mathrm{N}}$ equal to approximately 0.1 (and thus to $f_{\mathrm{N}} = 3.4$ at $\%\mathrm{N} = 5.3$ or $p_{\mathrm{N}_2} = 1$ atm).

5.3. Henrian Standard States and the Calculation of $\Delta G°$

Chemical reactions involving Henrian standard states instead of Raoultian standard states are often simpler to use, and the calculation of the corresponding $\Delta G°$s should be understood. The following examples are fairly typical of some ordinary procedures.

---

[3] Solubility values are often quoted without explicit specification of the corresponding partial pressures; in these cases, it is generally understood that these values correspond to partial pressures equal to 1 atm.

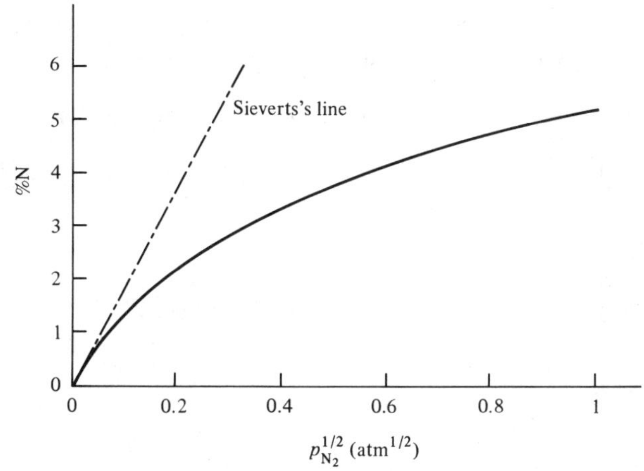

Figure 6. Solubility of nitrogen in liquid chromium at 1800°C (case of unusually large solubilities and deviations from Sieverts's law).

5.3.1. First Example

Let us assume that we are interested in the reaction of chromium in liquid iron with oxygen gas which forms chromium oxide Cr_2O_3, at temperatures in the vicinity of 1900°K. We choose to write the reaction

$$2\underline{Cr}(\%) + \tfrac{3}{2}O_2(g) = Cr_2O_3(s) \tag{107}$$

and wish to find the corresponding $\Delta G°$ as a function of temperature. The information available is the following. The enthalpy and free energy of formation of $Cr_2O_3(s)$ at 1900°K are $-270{,}600$ and $-154{,}100$ cal, respectively. Cr melts at 2171°K, its heat of fusion is equal to 5000 cal, and it forms with liquid iron a nearly ideal solution.

Reaction (107) may be obtained as a combination of the reaction corresponding to the formation of the oxide from pure chromium,

$$2Cr(s) + \tfrac{3}{2}O_2(g) = Cr_2O_3(s) \tag{108}$$

and the dissolution reaction:

$$Cr(s) = \underline{Cr}(\%) \tag{109}$$

The $\Delta G°$ of reaction (108) is known, but that of reaction (109) is not immediately apparent. It is again advantageous to consider reaction (109) as the sum of the two following reactions:

$$Cr(s) = Cr(l) \tag{110}$$

and

$$Cr(l) = \underline{Cr}(\%) \tag{111}$$

Reaction (110) represents the melting of pure chromium, and reaction (111) describes the change in the free energy of chromium upon mixing.

At the temperature of fusion of pure Cr, $\Delta G^\circ = 0$ and $\Delta H_f^\circ = T_f \Delta S_f^\circ$. Neglecting the difference in the heat capacities of solid and liquid chromium,

$$\Delta G_{110}^\circ = \Delta H_f^\circ - T\Delta S_f^\circ = 5000 - (5000/2171)T = 5000 - 2.30T \tag{112}$$

(the subscript 110 identifying the reaction to which this ΔG° corresponds).
By definition,

$$\Delta G_{111}^\circ = \mu_{Cr}^{oh} - \mu_{Cr}^{oR} \tag{113}$$

and we have previously seen [equations (54) and (73)] that

$$a_{Cr}^{h}/a_{Cr}^{R} = \exp[-(\mu_{Cr}^{oh} - \mu_{Cr}^{oR})/RT] = 100M_{Cr}/\gamma_{Cr}^{\infty}M_{Fe} \tag{114}$$

Thus,

$$\Delta G_{111}^\circ = -RT\ln(100\,M_{Cr}/\gamma_{Cr}^{\infty}M_{Fe}) \tag{115}$$

Since the Fe–Cr solution is nearly ideal, $\gamma_{Cr}^{\infty} \simeq 1$ and

$$\Delta G_{111}^\circ = -RT\ln(100M_{Cr}/M_{Fe}) = -9.01T \tag{116}$$

For reaction (108), we note that at 1900°K the entropy of formation of Cr_2O_3(s) is

$$\Delta S^\circ = -(\Delta G^\circ - \Delta H^\circ)/T = -(154{,}100 + 270{,}600)/1900 = -61.3 \quad \text{cal/°K} \tag{117}$$

Consequently,

$$\Delta G_{108}^\circ = -270{,}600 + 61.3T \tag{118}$$

The ΔG° of reaction (107) may now be obtained as an algebraic combination of the ΔG°s of reactions (108), (110), and (111) in equations (112), (116), and (118):

$$\Delta G_{107}^\circ = \Delta G_{108}^\circ - 2\Delta G_{110}^\circ - 2\Delta G_{111}^\circ = -280{,}600 + 83.92T \tag{119}$$

5.3.2. Second Example

We now turn to a second example of calculations in which the determination of a ΔG° associated with a Henrian state is based upon an extrapolation. Let us assume that an investigator wishes to determine the activity of sulfur in a metal A at various concentrations of sulfur and a temperature T, e.g., 1000°K. The A–S solution may be equilibrated with a gas flow in which the sulfur potential is controlled by fixing the ratio of H_2S to H_2. The equilibrium may be represented by the reaction

$$H_2 + \underline{S}(\%) = H_2S, \qquad \Delta G^\circ = -RT\ln K \tag{120}$$

with

$$K = \frac{p_{H_2S}}{p_{H_2}}\frac{1}{a_S} = \frac{p_{H_2S}}{p_{H_2}}\frac{1}{(f_S\%S)} \tag{121}$$

Although a_S is fixed by the ratio of H_2S to H_2, its value is not known since the value of K is unknown. What can be experimentally determined is the value of

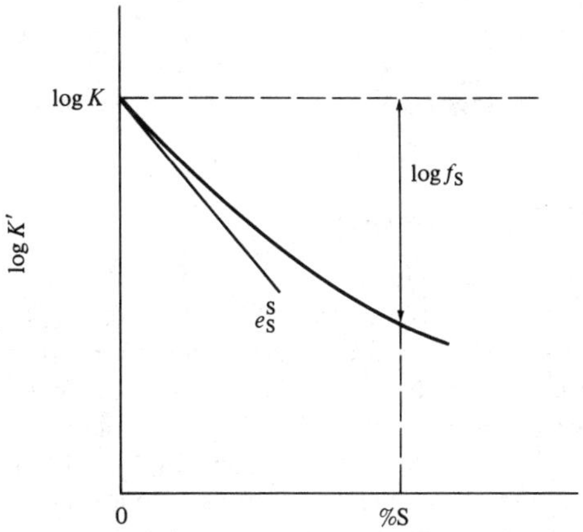

Figure 7. Determination of the value of log K as the limit value of log K' or log $p_{H_2S}/(p_{H_2}\%S)$. Equation (123) shows that log f_S may be obtained as the difference between log K' and log K. The slope of log K' at the origin yields e_S^S.

K', where K' is defined

$$K' = \frac{p_{H_2S}}{p_{H_2}} \frac{1}{\%S} \tag{122}$$

Comparing equations (121) and (122), we may write

$$\log K = \log K' - \log f_S \tag{123}$$

In a plot of log K' vs %S (see Fig. 7), we see that the extrapolated value of log K' at %S = 0 yields the value of log K (and consequently of $\Delta G°$), since log $f_S \rightarrow 0$ for %S $\rightarrow$ 0. We also note that the slope of log K' vs %S at the origin yields the value of e_S^S.

Once the value of the $\Delta G°$ of reaction (120) has been thus determined, it may be observed that the value of $\Delta G°$ for the reaction

$$\tfrac{1}{2}S_2(g) = \underline{S}(\%) \tag{124}$$

is readily obtained by combining reaction (120) with that corresponding to the formation of H_2S:

$$H_2 + \tfrac{1}{2}S_2(g) = H_2S \tag{125}$$

Most practical applications of chemical thermodynamics involve standard states or reference states. Serious errors can be made without proper attention to the identification of such states and to the calculations permitting conversion from one state to another. Although the theoretical principles associated with such exercises are fairly straightforward, the subject proves to be a difficult one for the unprepared student.

Problems

1. Expand as a Taylor series the function $y = 1 + [\ln(1 + x)]/(1 + x)$. How many terms in the series must be retained to calculate y at $x = 0.05$, at $x = 0.1$ and at $x = 0.2$ with an error smaller than 0.005? With an error smaller than 0.001? Compare the values of y obtained by the truncated series with those obtained directly on the closed form of y. Plot y and the parabola approximating y (the series truncated at the second order term) in the interval $x = 0$ to 0.2.

2. Establish the relations

$$J_i^{(2)} = J_i^{(1)} - [(i+1)/i]J_{i+1}^{(1)} \quad \text{and} \quad \phi_i = J_i^{(1)}/(i-1)$$

3. **a.** Demonstrate that

$$(\partial^2 a_1/\partial X_2^2)_{X_1\to 1} = -\epsilon_2^{(2)} \quad \text{and} \quad (\partial^2 a_2/\partial X_2^2)_{X_2\to 0} = 2\gamma_2^\infty \epsilon_2^{(2)}$$

b. Given that $a_1 = 0.87$ at $X_2 = 0.1$, estimate $\epsilon_2^{(2)}$. Assume that $\gamma_2^\infty = 0.05$ and plot carefully the two curves a_1 and a_2 in the interval $X_2 = 0$ to 0.1. As the result on the curvature of a_2 shows, note that for negative deviations from ideality (γ_2^∞ small), the curve is closest to its tangent when its slope is lowest. This is contrary to what would be expected by simple drawing interpolation between a tangent of low slope γ_2^∞ at $X_2 = 0$ and a tangent of slope 1 at $X_2 = 1$. The reverse is true for high positive deviations from ideality (strong curvatures will be observed). At low concentrations, the shapes of the activity curves on each side of the ideal line are thus, generally, not mirrorlike to one another.

4. The activity of thallium in liquid cadmium at 400°C has been measured to be equal to 0.12 at $T_{Tl} = 0.03$, and 0.34 at $X_{Tl} = 0.15$. At 550°C, it is found that $a_{Tl} = 0.085$ at $X_{Tl} = 0.03$ and $a_{Tl} = 0.28$ at $X_{Tl} = 0.15$. From these data, estimate ϵ_{Tl}^{Tl}, η_{Tl}^{Tl}, and σ_{Tl}^{Tl} at 400°C.

5. A study of the liquid Sn–Zn solution at 750°K determines that $\gamma_{Sn}^\infty = 4.58$ and $\gamma_{Zn}^\infty = 1.96$. Find an analytical expression of G^E compatible with these data and calculate the activities of Sn and Zn at $X_{Sn} = X_{Zn} = 0.5$.

6. Estimate the minimum amount of gold in a Ag–Au alloy which would prevent oxidation when exposed to air at room temperature. At 298°K, $\ln \gamma_{Ag}^\infty \simeq -5$ and the Gibbs free energy of formation of Ag_2O is -2500 cal/mol.

7. Consider the following oxidation reaction:

$$Mn + \tfrac{1}{2}O_2(g) = MnO$$

Mn melts at 1517°K with a heat of fusion of 3500 cal/g-at., while MnO melts at 2058°K with a heat of fusion of 13,000 cal/mol.

a. Sketch the plot of ΔG° vs T for this reaction in the temperature interval 1450–2100°K. Account for the changes in state noted above.

b. Assume that the standard state of O_2 is changed to 1 mm Hg and show how the curve in part a is modified.

8. At 1100°K and 1 atm, the metals A and B form an ideal liquid solution in the range from $X_B = 0$ to $X_B = 0.46$. In the range from $X_B = 0.52$ to $X_B = 1$, they form an ideal solid solution. In the range from $X_B = 0.46$ to $X_B = 0.52$, a mixture of solid and liquid solutions is stable. Sketch carefully the activity curve of B over the whole composition range, using the pure solid metal as the standard state. Calculate the activity of metastable pure liquid B.

9. **a.** A crucible containing 171.3 g of silver is placed in the reaction chamber of a Sieverts's apparatus. The temperature is adjusted to 1050°C. The chamber is evacuated and then filled with 50.9 cm³(STP) of argon which measures the *hot* (or *dead*) *volume* of the chambers. A mercury manometer indicates then a pressure of 650 mm. The chamber is then reevacuated and filled with oxygen. 354.2 cm³(STP) are needed to build a pressure of 650 mm Hg. Calculate the corresponding solubility of oxygen.

b. e_O^O has been measured and found to be approximately equal to 0.09 at 1050°C. Calculate at this temperature the solubility of oxygen under 1 atm air and under 4 atm O_2.

c. The same reaction chamber is used now with 150 g of iron at a temperature of 1600°C. The hot volume at 650 mm Hg is now equal to 47.5 cm³(STP). Knowing that the solubility of nitrogen under 1 atm is 0.045%, calculate the number of cubic centimeters of N_2(STP) necessary to build up a pressure of 650 mm Hg in the reaction chamber.

10. Determine the amount of V_2O_3(solid) which forms as a result of the addition of 2.5 kg of vanadium per tonne to a low carbon steel bath at 1600°C containing 0.06% oxygen. Also determine the final contents of oxygen and vanadium present in the bath as solute elements.
Data: The Gibbs free energy of formation of V_2O_3(s) at 1600°C is equal to −185,850 cal/mol. Vanadium melts at 2175°K with a heat of fusion equal to 5000 cal/mol. In liquid iron at 1600°C, $\gamma_V^\infty = 0.08$. Moreover,

$$\tfrac{1}{2}O_2(g) = \underline{O}(\%), \qquad \Delta G^\circ = -28{,}000 - 0.69T$$

The effects of interaction coefficients are neglected.

11. **a.** A sample of austenitic iron–carbon alloy is equilibrated at 1200°K in a furnace in which passes a flow of CO and CO_2 at a total pressure of 1 atm. After equilibration, the sample is quenched and analyzed for carbon. The experimental results of three runs are

Run no.	%CO_2	%C
1	20.5	0.12
2	8.4	0.35
3	5.2	0.55

Calculate the interaction coefficient e_C^C, the equilibrium constant K, and the ΔG° of the reaction $2CO = \underline{C}(\%) + CO_2$. Assume that the percentages of CO and CO_2 are not appreciably affected by their passage in the furnace.

b. Can you justify this assumption?

c. Calculate the activity coefficient of carbon γ_C^∞ at infinite dilution in austenite. (Pure graphite is the reference state.)

Data: At 1200°K, the Gibbs free energies of formation of CO and CO_2 are, respectively, −52,050 and −94,680 cal/mol.

12. A flow of gases passes through a furnace at 1823°K. At the inlet the partial pressures of the species present in the gas are $p_{H_2S} = 3.04 \times 10^{-3}$, $p_{H_2} = 0.513$, and $p_{Ar} = 0.507$. Calculate, assuming equilibrium and that the total pressure of the gas remains unchanged, the ratio of p_{H_2S} to p_{H_2} in the furnace, if

a. the formation of HS is neglected;

b. the formation of HS is not neglected.

The furnace contains a crucible of liquid iron. The following data are obtained after achieving equilibrium:

%S in Fe	$10^3 \times (p_{H_2S}/p_{H_2})$ in furnace
0.25	0.587
0.52	1.19
1.05	2.46
2.19	4.35
4.51	7.38

Deduce the interaction coefficient e_S^S and the $\Delta G°$ of the reaction $\frac{1}{2}S_2(g) = \underline{S}(\%)$. *Data*: $H_2(g) + \frac{1}{2}S_2(g) = H_2S(g)$, $\Delta G° = -21{,}530 + 11.73T$ cal; $\frac{1}{2}H_2(g) + \frac{1}{2}S_2(g) = HS(g)$, $\Delta G° = 18{,}560 - 3.67T$ cal.

References

1. E. A. Guggenheim, *Proc. Roy. Soc. London Ser. A* **148,** 304 (1935).
2. G. Borelius, *Ann. Physik* **24,** 489 (1935).
3. P. Franklin, *Methods of Advanced Calculus*. McGraw-Hill, New York, 1944.
4. C. Wagner, *Thermodynamics of Alloys*. Addison-Wesley, Reading, MA, 1962.
5. *The Chipman Conference* (J. F. Elliott and T. R. Meadowcroft, eds.). MIT Press, 1965, pp. xvii–xxi.
6. C. H. P. Lupis and J. F. Elliott, *Acta Met.* **14,** 529–538 (1966).
7. H. K. Hardy, *Acta Met.* **1,** 202–209 (1953).
8. C. H. P. Lupis and J. F. Elliott, *Acta Met.* **15,** 265–276 (1967).
9. L. S. Darken, *Trans. Met. Soc. AIME* **239,** 80–89 (1967).
10. R. O. Williams, *Trans. Met. Soc. AIME* **245,** 2565–2570 (1969).
11. C. W. Bale and A. D. Pelton, *Met. Trans.* **5,** 2323–2337 (1974).
12. C. H. P. Lupis and J. F. Elliott, *Trans. Met. Soc. AIME* **233,** 257–258 (1965).
13. C. H. P. Lupis and J. F. Elliott, *Trans. Met. Soc. AIME* **233,** 829–830 (1965).
14. J. F. Elliott, M. Gleiser, and V. Ramakrishna, *Thermochemistry for Steelmaking*. Addison-Wesley, Reading, MA, 1963, Vol. 2, p. 514.
15. J. Humbert and J. F. Elliott, *Trans. Met. Soc. AIME* **218,** 1076 (1960).

VIII Binary Phase Diagrams

1. General Features

In Section II.4.5 we established that the variance of a system in equilibrium may be calculated by the formula

$$\vartheta = m + 2 - \varphi \tag{1}$$

In a binary system ($m = 2$) there is a maximum of four coexisting phases ($\varphi = 4$). However, if the pressure is arbitrarily fixed, for example at 1 atm, there can be only three coexisting phases, this coexistence occurring at a single fixed temperature. Keeping the pressure arbitrarily fixed, the case of two phases in equilibrium corresponds to a monovariant system ($\vartheta = 1$), and that of a single phase to a bivariant system ($\vartheta = 2$).

In Chapter I we derived from the second law of thermodynamics that a system is in equilibrium at a given temperature and pressure when its Gibbs free energy is at a minimum. For a binary system containing two phases α and β, this criterion of equilibrium was expressed in Chapter II in terms of chemical potentials; the conditions are

$$\mu_1^\alpha(X_2^\alpha, T, P) = \mu_1^\beta(X_2^\beta, T, P) \tag{2}$$

$$\mu_2^\alpha(X_2^\alpha, T, P) = \mu_2^\beta(X_2^\beta, T, P) \tag{3}$$

At a given temperature and pressure, these two equations may be solved for the compositions X_2^α and X_2^β of the two phases α and β. If the pressure is kept constant and the temperature allowed to vary, a plot of T vs X_2 shows two lines (X_2^α as a function of T, X_2^β as a function of T) yielding the phases' composition boundaries. We note that if the pressure does not vary over a considerable range (e.g., up to a few kilobars), its effect on the thermodynamic properties of condensed systems is generally negligible. Consequently, for the sake of simplicity and convenience, in this chapter we shall consider it to be fixed at 1 atm. In cases where this assumption is not warranted, the necessary changes in the calculations can be readily introduced (e.g., reference states have then to be distinguished from standard states; see Chapter IV).

Solving the system of equations (2) and (3) may be done graphically by the common tangent construction described in Section II.4.4 (see Fig. 1). Moreover, the "lever rule" seen in Section III.3.1 applies here. Let X_2 represent the overall concentration of component 2 in the heterogeneous system, x the relative proportion (in moles) of phase α in the mixture, and G, G^α, and G^β the molar Gibbs free energies of the system and of the phases α and β, respectively. Then,[1]

$$X_2 = xX_2^\alpha + (1 - x)X_2^\beta \tag{4}$$

$$G = xG^\alpha + (1 - x)G^\beta \tag{5}$$

[1] In terms of number of moles, the results of equations (4) and (5) are immediate:

$$X_2 = \frac{n_2}{n} = \frac{n_2^\alpha}{n} + \frac{n_2^\beta}{n} = \frac{n_2^\alpha}{n^\alpha}\frac{n^\alpha}{n} + \frac{n_2^\beta}{n^\beta}\frac{n^\beta}{n} = X_2^\alpha x + X_2^\beta(1 - x)$$

and

$$nG = n^\alpha G^\alpha + n^\beta G^\beta \qquad \text{or} \qquad G = xG^\alpha + (1 - x)G^\beta$$

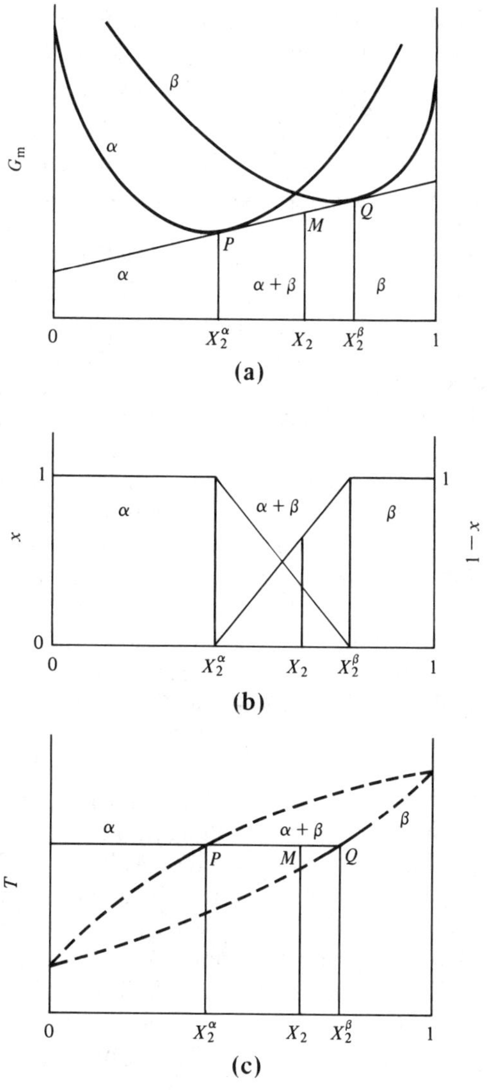

Figure 1. General features of phase diagrams. The "lever rule" determines that the amount x of phase α in the two-phase region α + β is equal to MQ/PQ.

With the notation of Fig. 1, equation (4) leads to

$$x = \frac{MQ}{PQ} = \frac{X_2^\beta - X_2}{X_2^\beta - X_2^\alpha}, \qquad 1 - x = \frac{PM}{PQ} = \frac{X_2 - X_2^\alpha}{X_2^\beta - X_2^\alpha} \tag{6}$$

Consequently, we may state the following:

1. For an overall concentration X_2 of component 2 in the system between 0 and X_2^α, only the phase α is present and the Gibbs free energy of the system is identical to that of the phase α at the composition X_2.
2. For values of X_2 varying between X_2^α and X_2^β, the system contains two phases α and β of fixed compositions X_2^α and X_2^β, and their relative proportions are given by the lever rule. The Gibbs free energy of the system is represented in Fig. 1a by the straight line PQ.

3. For values of X_2 between X_2^β and 1, only the phase β is present and the Gibbs free energy of the system is that of the phase β at the concentration X_2.

When the pressure is fixed and the temperature varies, in a diagram of temperature versus composition, the compositions X_2^α and X_2^β define domains where single phases exist and domains where two phases coexist (Fig. 1c). This type of diagram is referred to as a *phase diagram*.

If, in the binary system 1–2, three phases α, β, and γ can be observed, at a given pressure and temperature their free energy curves may have the relative positions described in Fig. 2a, where the free energy curve of β does not intersect the common tangent to the curves of α and γ. In that case, for any overall concentration X_2 between X_2^α and X_2^γ, the free energy of a mixture of α and γ phases (represented by the point M) is lower than the free energies of mixtures of α and β, or β and γ (represented by the points M' and M''). Thus, at equilibrium the phase β will be unstable. Note, however, that in certain instances (e.g., large activation energies for the phase γ) the phase β may be present as a metastable phase; the system is then in stable equilibrium with respect to small composition fluctuations, but unstable with respect to large composition fluctuations (see Chapter III).

In the case of Fig. 2b, where the free energy curve of β intersects the common tangent to the curves of α and γ, the domains of (most stable) equilibrium are defined by the two common tangents to the curves α and β, and β and γ, respectively: the two-phase regions α + β and β + γ are separated by a single phase region β.

The above constructions are valid for systems which present any number of phases (Fig. 3). Note that the Gibbs free energy of the system in its most stable equilibrium may be represented by a string, tightly stretched around portions of the free energy curves and anchored at the points A and B corresponding to the pure components 1 and 2 in their states of lowest free energies.

Since it is the relative positions of the free energy curves which determine phase diagram characteristics, it is clear that a simple translation of all curves parallel to the Gibbs free energy axis will leave these characteristics unchanged. But we have seen in Chapter I that the absolute values of the Gibbs free energy

Figure 2. Phase β as (**a**) metastable and (**b**) stable.

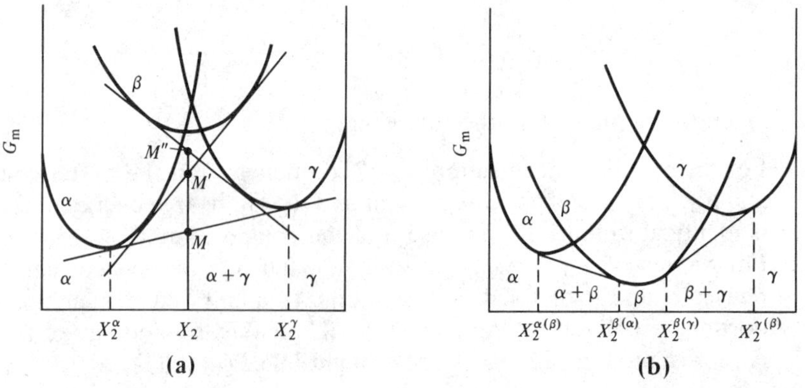

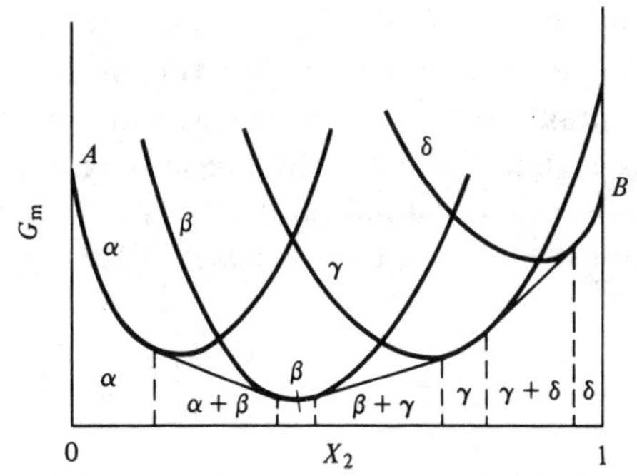

Figure 3. The Gibbs free energy of the system in its most stable equilibrium may be represented by a string tightly stretched around portions of the free energy curves and anchored at the points A and B.

of *each* element are arbitrary. Changing, for instance, the value of the chemical potential $\mu_1^{\circ\alpha}$ of component 1 when it is pure in the α structure by an amount $\Delta\mu_1^{\circ}$ should leave the phase diagram characteristics of the system 1–2 unchanged. We hasten to add that the change $\Delta\mu_1^{\circ}$ should be brought to bear on the free energy values of 1 in other structures as well, so that the difference $\mu_1^{\circ(\alpha)} - \mu_1^{\circ(\beta)}$

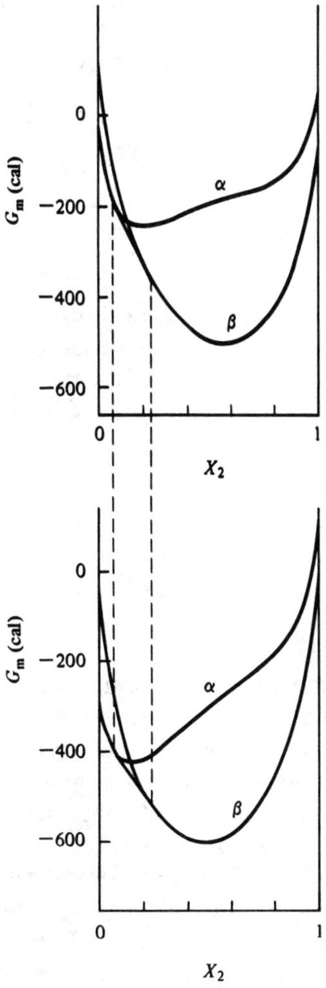

Figure 4. The composition of the points having a common tangent is unaffected by a change in the absolute value of $\mu_1^{\circ(\alpha)}$ (or $\mu_1^{\circ(\beta)}$).

measuring the relative stability of the two phases α and β remains unchanged. The appearance of the free energy curves diagram (see Fig. 4) is affected by the change $\Delta\mu_1^\circ$ (it does not result in a simple translation of the free energy curves), but the points with a common tangent (and therefore the phase diagram) remain unchanged because in the calculation of the equilibrium points [e.g., equation (2)] only the difference $\mu_1^\alpha - \mu_1^\beta$ intervenes and $(\mu_1^\alpha + \delta\mu_1^\circ) - (\mu_1^\beta + \delta\mu_1^\circ) = \mu_1^\alpha - \mu_1^\beta$.

2. Ideal and Nearly Ideal Systems

Let us consider a binary system 1–2 for which the liquid and solid solutions are both ideal. (What we shall describe for the solid–liquid equilibrium is valid for the equilibrium of any other two phases, liquid–gas or solid–solid). The relative positions of their free energy curves are described in a series of diagrams in Fig. 5. At high temperatures (e.g., T_a) the free energy of the liquid phase is lower than that of the solid at all compositions. On cooling, the free energy curves of both the liquid and solid phases move upward, but that of the liquid phase moves upward more rapidly because its entropy is generally higher than that of the solid phase.

Note that the distance between the intercepts of the free energy curves on the ordinate axes are

$$\Delta\mu_i^{\circ(s\to l)} = \mu_i^{\circ(l)} - \mu_i^{\circ(s)} = \Delta G_{f,i}^\circ = \Delta H_{f,i}^\circ - T\Delta S_{f,i}^\circ \tag{7}$$

where $i = 1$ or 2, and if $C_{p,i}^{(l)} = C_{p,i}^{(s)}$,

$$\Delta\mu_i^{\circ(s\to l)} \simeq \Delta S_{f,i}^\circ(T_{f,i} - T) \tag{8}$$

since at $T = T_{f,i}$, $\Delta G_{f,i}^\circ = 0$ and $\Delta H_{f,i}^\circ = T_{f,i}\Delta S_{f,i}^\circ$. Generally, the entropy of fusion ΔS_f° has very nearly the same value for components of the same nature (e.g., for many metals, it is close to 2.2 cal/°K) and only components of comparable nature can form ideal or nearly ideal solutions. Thus, if we suppose that it is component 2 which has the higher melting temperature ($T_{f,2} > T_{f,1}$),

$$\Delta\mu_2^{\circ(s\to l)} > \Delta\mu_1^{\circ(s\to l)} \tag{9}$$

(in algebraic values). At all temperatures, the two free energy lines have the same curvature (since they both represent ideal solutions) but are slightly tilted relative to each other because of unequal distances between their intercepts on the two vertical axes [equation (9)].

On cooling from T_a, the two curves first meet when the temperature reaches $T_{f,2}$ at the point representing the pure component 2. However, all other alloys are still liquid because of the tilt of the free energy curve of the liquid phase relative to that of the solid. Below $T_{f,2}$ the curves intersect (at only one point, because of similar curvatures and different tilts). The common tangent construction shows that for X_2 between 0 and X_2^l all alloys are liquid, for X_2 between X_2^s and 1 all alloys are solid, and for X_2 between X_2^l and X_2^s both phases coexist with liquid and solid solutions corresponding to the compositions X_2^l and X_2^s. On further cooling, the point of intersection as well as the compositions X_2^l, X_2^s move toward the left (solutions richer in component 1) until the temperature reaches $T_{f,1}$ where

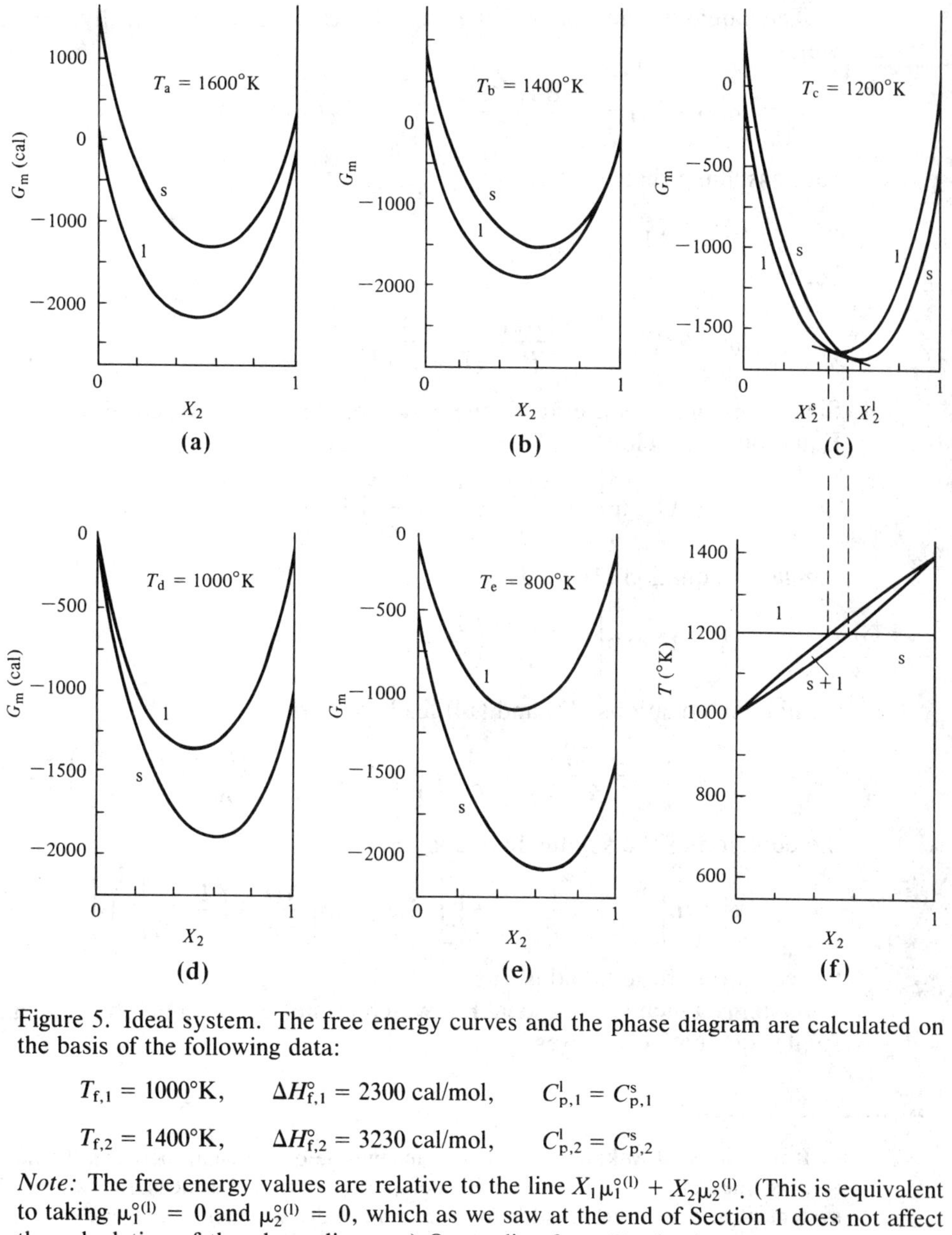

Figure 5. Ideal system. The free energy curves and the phase diagram are calculated on the basis of the following data:

$$T_{f,1} = 1000°K, \quad \Delta H^{\circ}_{f,1} = 2300 \text{ cal/mol}, \quad C^{l}_{p,1} = C^{s}_{p,1}$$

$$T_{f,2} = 1400°K, \quad \Delta H^{\circ}_{f,2} = 3230 \text{ cal/mol}, \quad C^{l}_{p,2} = C^{s}_{p,2}$$

Note: The free energy values are relative to the line $X_1\mu_1^{\circ(l)} + X_2\mu_2^{\circ(l)}$. (This is equivalent to taking $\mu_1^{\circ(l)} = 0$ and $\mu_2^{\circ(l)} = 0$, which as we saw at the end of Section 1 does not affect the calculation of the phase diagram.) On cooling from T_a, the free energy curve of the solid moves downward past the free energy curve of the liquid. (From b to c the scale of the ordinate axis has been changed by a factor of 2.)

the two curves meet at the freezing point of pure component 1 and all alloys are solid. At lower temperatures, only homogeneous solid solutions are stable.

These results can be assembled in a phase diagram (Fig. 5f). The three domains, liquid, liquid + solid, and solid, are separated by two boundaries known as the *liquidus* and *solidus*. The diagram resembles a *single lens*, which is characteristic of ideal, or nearly ideal, systems.

The liquidus and solidus lines can be calculated analytically. Equation (2) yields[2]

$$\ln(a_1^{(l)}/a_1^{(s)}) = -\frac{\Delta\mu_1^{\circ(s\to l)}}{RT} \tag{10}$$

and, assuming that

$$C_{p,1}^{(l)} = C_{p,1}^{(s)} \tag{11}$$

becomes

$$\ln(a_1^{(l)}/a_1^{(s)}) = -\frac{\Delta H_{f,1}^{\circ}}{R}\left(\frac{1}{T} - \frac{1}{T_{f,1}}\right) \tag{12}$$

Since the solutions are ideal, the activities may be replaced by mole fractions. Equation (12) yields

$$X_1^{l} = X_1^{s} \exp\left[-\frac{\Delta H_{f,1}^{\circ}}{R}\left(\frac{1}{T} - \frac{1}{T_{f,1}}\right)\right] \tag{13}$$

Similarly, equation (3) yields

$$X_2^{l} = X_2^{s} \exp\left[-\frac{\Delta H_{f,2}^{\circ}}{R}\left(\frac{1}{T} - \frac{1}{T_{f,2}}\right)\right] \tag{14}$$

Combining equations (13) and (14) leads to

$$X_1^{s} \exp\left[-\frac{\Delta H_{f,1}^{\circ}}{R}\left(\frac{1}{T} - \frac{1}{T_{f,1}}\right)\right] + X_2^{s} \exp\left[-\frac{\Delta H_{f,2}^{\circ}}{R}\left(\frac{1}{T} - \frac{1}{T_{f,2}}\right)\right] = 1 \tag{15}$$

the equation of the solidus line, and

$$X_1^{l} \exp\left[\frac{\Delta H_{f,1}^{\circ}}{R}\left(\frac{1}{T} - \frac{1}{T_{f,1}}\right)\right] + X_2^{l} \exp\left[\frac{\Delta H_{f,2}^{\circ}}{R}\left(\frac{1}{T} - \frac{1}{T_{f,2}}\right)\right] = 1 \tag{16}$$

the equation of the liquidus line.

Systems which are not ideal but which do not depart much from ideality have similar qualitative features.

It is of interest to know the order of magnitude of the width between the liquidus and solidus lines of an ideal system. By comparing it with the measured width of a given system, it may then be possible to ascertain rapidly whether or not this system deviates substantially from ideality.

Let us characterize this width by the distance $X_2^{s} - X_2^{l}$ at the temperature $T_0 = \frac{1}{2}(T_{f,1} + T_{f,2})$. If we assume that $\Delta S_{f,1}^{\circ} \simeq \Delta S_{f,2}^{\circ} = \Delta S_f^{\circ}$, it is then easily verified that

$$\frac{\Delta H_{f,2}^{\circ}}{R}\left(\frac{1}{T_0} - \frac{1}{T_{f,2}}\right) \simeq -\frac{\Delta H_{f,1}^{\circ}}{R}\left(\frac{1}{T_0} - \frac{1}{T_{f,1}}\right) \simeq \frac{\Delta S_f^{\circ}}{R}\left(\frac{T_{f,2} - T_{f,1}}{T_{f,2} + T_{f,1}}\right)$$

$$\simeq \frac{\Delta S_f^{\circ}}{2R}\frac{\Delta T_f}{T_0} \tag{17}$$

---

[2] We recall that we have assumed the pressure to be fixed at 1 atm. If it is not, reference states should be substituted to standard states in the equations of this chapter: e.g., $\Delta\mu_1^{*(s\to l)}$ instead of $\Delta\mu_1^{\circ(s\to l)}$ in equation (10).

where $\Delta T_f = T_{f,2} - T_{f,1}$. Consequently,

$$X_2^s - X_2^l \simeq \left[\exp\left(\frac{\Delta S_f^\circ}{2R}\frac{\Delta T_f}{T_0}\right) - 1\right] \Big/ \left[\exp\left(\frac{\Delta S_f^\circ}{2R}\frac{\Delta T_f}{T_0}\right) + 1\right] \tag{18}$$

But it may easily be shown that

$$\frac{e^x - 1}{e^x + 1} = \frac{x}{2}\left(1 - \frac{x^2}{12} + \cdots\right) \simeq \frac{x}{2} \tag{19}$$

and it is therefore justified to rewrite equation (18)

$$X_2^s - X_2^l \simeq \frac{\Delta S_f^\circ}{4R}\frac{\Delta T_f}{T_0} \tag{20}$$

The width of the two-phase domain is thus proportional to the entropy of fusion and to the difference of the melting temperatures.

The Cu–Ni system, for example, exhibits a single lens type of diagram. $\Delta S_{f,Cu}^\circ = 2.30$, $\Delta S_{f,Ni}^\circ = 2.44$, $T_{f,Cu} = 1356.5$°K, and $T_{f,Ni} = 1736$°K. Application of equation (20) leads to $X_2^s - X_2^l \simeq 0.07$. Values estimated from the phase diagrams in the literature are of the order of 0.14 (although Hultgren et al. [1] report that while the liquidus is well established, the solidus is probably too low); indeed, activity data show nonnegligible positive deviations in both the solid and liquid phases [1].

In the case of the Ge–Si system, $\Delta S_{f,Ge}^\circ = 6.28$, $\Delta S_{f,Si}^\circ = 7.18$, $T_{f,Ge} = 1210.4$°K, and $T_{f,Si} = 1685$°K. Application of equation (20) yields $X_2^s - X_2^l \simeq 0.28$, whereas the experimental value [2] is close to 0.32. Consequently, one can expect the Ge–Si system to exhibit nearly ideal behavior in both solid and liquid phases. This is also true of a system such as the NiO–MgO system. The entropies of fusion are 5.6 and 6.0 cal/°K and the temperatures of fusion are 2233°K and 3073°K, for NiO and MgO respectively. This yields a width equal to 0.23, in very good agreement with the experimental determinations [3].

In the preceding calculations, we have assumed for simplicity that the heat capacities of the pure solid and liquid i are equal. The corresponding expression of $\Delta\mu_i^{\circ(s\to l)}$ may easily be corrected if this assumption is unwarranted.

$$\begin{aligned}\Delta\mu_i^{\circ(s\to l)} &= (H_i^{\circ(l)} - H_i^{\circ(s)}) - T(S_i^{\circ(l)} - S_i^{\circ(s)})\\ &= \left(\Delta H_{f,i}^\circ + \int_{T_{f,i}}^{T} \Delta C_{p,i}^\circ\, dT\right) - T\left(\Delta S_{f,i}^\circ + \int_{T_{f,i}}^{T} \frac{\Delta C_{p,i}^\circ}{T}\, dT\right)\end{aligned} \tag{21}$$

where $\Delta C_{p,i}^\circ = C_p^{\circ(l)} - C_p^{\circ(s)}$. If the heat capacities are represented by polynomials of the form

$$C_p = a + bT + cT^{-2} \tag{22}$$

it may be easily demonstrated that

$$\begin{aligned}\frac{\Delta\mu_i^{\circ(s\to l)}}{RT} &= \frac{\Delta H_{f,i}^\circ}{R}\left(\frac{1}{T} - \frac{1}{T_{f,i}}\right) + \frac{\Delta a_i}{R}\left(1 - \frac{T_{f,i}}{T} + \ln\frac{T_{f,i}}{T}\right)\\ &\quad - \frac{\Delta b_i}{2RT}(T - T_{f,i})^2 - \frac{\Delta c_i}{2R}\left(\frac{1}{T} - \frac{1}{T_{f,i}}\right)^2\end{aligned} \tag{23}$$

3. Minima and Maxima

Let us now consider a nonideal system for which the solid and liquid free energy curves have different curvatures (because of different values of G^E) and have lost most of their tilt relative to one another because of reasonably close melting temperatures. In this case, the free energy curves of the liquid and solid phases in the freezing range may intersect at two points instead of one. Figure 6 shows that at these temperatures this gives rise to two composition ranges where both the solid and liquid phases coexist. On cooling, the free energy curve of the liquid moves upward relative to that of the solid and their intersections move closer to each other until they merge at a composition X_{2a}. At this point (and temperature) the free energy curves are tangent to each other and on the corresponding phase diagram yield a minimum for the solidus and liquidus lines. A liquid solution of

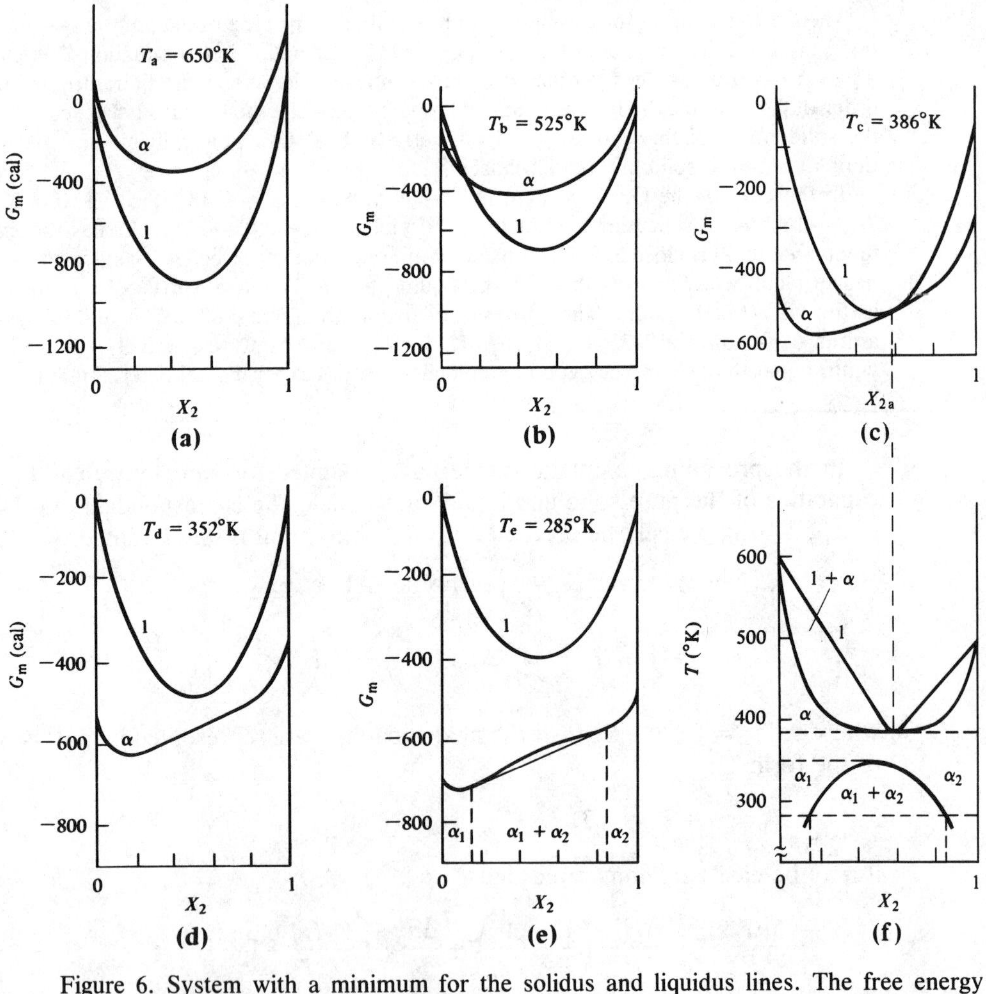

Figure 6. System with a minimum for the solidus and liquidus lines. The free energy curves and the phase diagram are calculated on the basis of the following data: $T_{f,1} = 600°K$, $\Delta H^\circ_{f,1} = 1300$ cal/mol, $T_{f,2} = 500°K$, $\Delta H^\circ_{f,2} = 1100$ cal/mol; $C^l_p = C^\alpha_p$ for both components 1 and 2; the liquid solution is ideal and the solid solution is regular with $\Omega^\alpha = 1400$ cal. (The note in the legend of Fig. 5 applies here as well.)

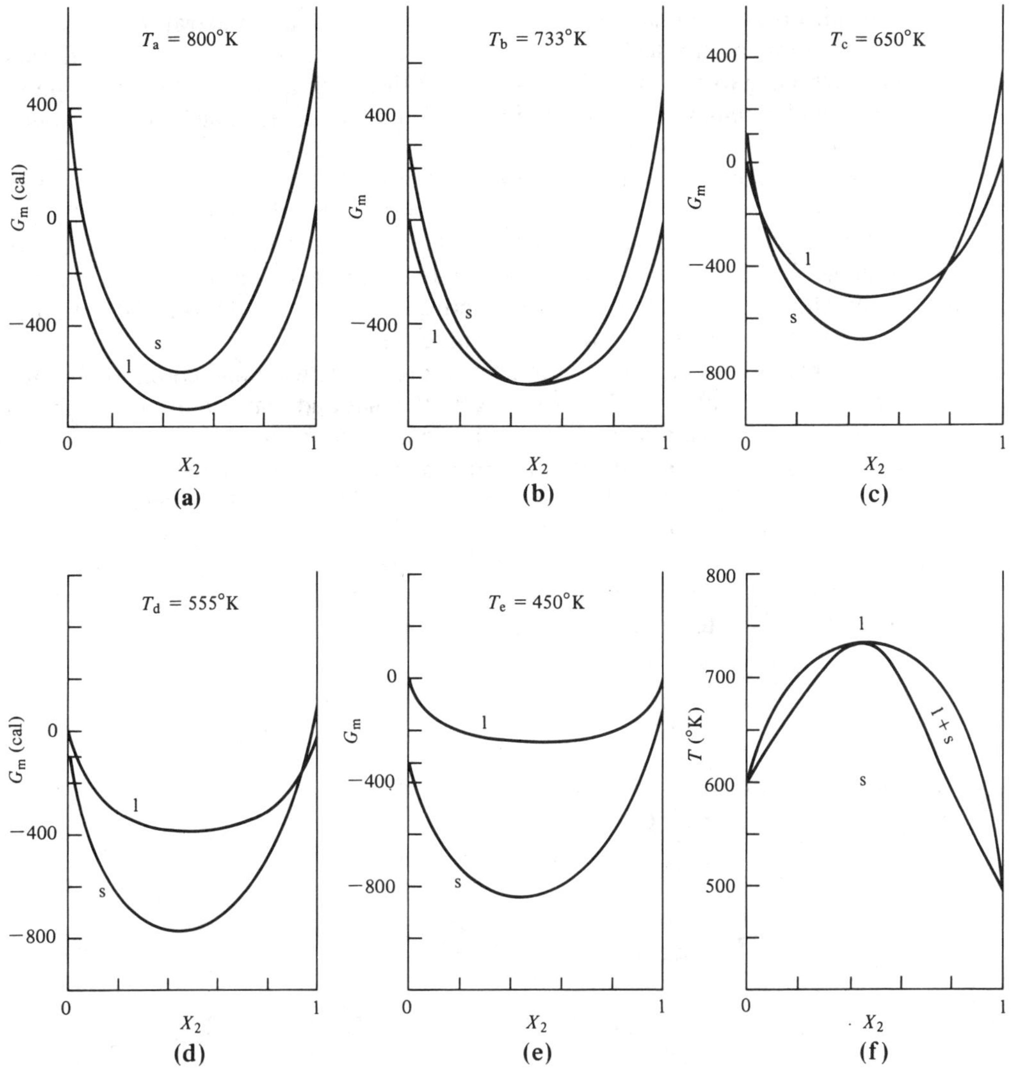

Figure 7. System with a maximum for the solidus and liquidus lines. The free energy curves and phase diagrams are calculated on the basis of the following data: $T_{f,1}$ = 600°K, $\Delta H^\circ_{f,1}$ = 1300 cal/mol, $T_{f,2}$ = 500°K, $\Delta H^\circ_{f,2}$ = 1000 cal/mol; $C_p^l = C_p^s$ for both components 1 and 2; the liquid solution is regular with Ω^l = 1500 cal and the solid solution is ideal.

composition X_{2a} solidifies *congruently*; i.e., the solid precipitating from the liquid has the same composition as the liquid.

As shown in Fig. 7, it is also possible to have a maximum instead of a minimum for the liquidus and solidus lines. A maximum is generally observed if the free energy curve of the liquid has a less pronounced curvature than the free energy curve of the solid. Conversely, a minimum is observed if the free energy curve of the solid has a less pronounced curvature than the liquid's. This occurs, for instance, if the solid phase shows positive deviations from ideality. We recall, then, that at the critical point the free energy curve is rather flat since its second and third order derivatives are equal to zero. Thus, at temperatures close to the critical temperature the free energy curve will remain relatively flat. It is therefore not surprising to find many systems exhibiting both a miscibility gap in the solid phase and a minimum in the solidus and liquidus lines. Figure 8 shows such a case for the Au–Ni system.

The existence of a minimum in the solidus and liquidus lines does not, however, ensure positive deviations from ideality. Cases where the free energy of the liquid has stronger negative deviations from ideality than the solid will also show such minima. Instead of displaying at lower temperatures a miscibility gap, the corresponding systems often display ordering or intermediate phase formation. The Au–Cu system provides such an example (see Fig. 9).

The preceding remarks do not pertain only to solid–liquid equilibria but to the equilibrium of any two phases α and β. Minima and maxima are often observed in liquid–gas equilibria and are called minimum boiling points and maximum boiling points; they are also referred to as *azeotropes*. For example, water–ethanol alcohol (H_2O–C_2H_5OH) displays a minimum boiling point while water–hydrochloric acid (H_2O–HCl) and water–nitric acid (H_2O–HNO_3) display maximum boiling points.

It may be noted that the position of the minimum or maximum of these phase boundaries is determined by the conditions

$$G^{(\alpha)} = G^{(\beta)} \tag{24}$$

Figure 8. The gold–nickel system [1].

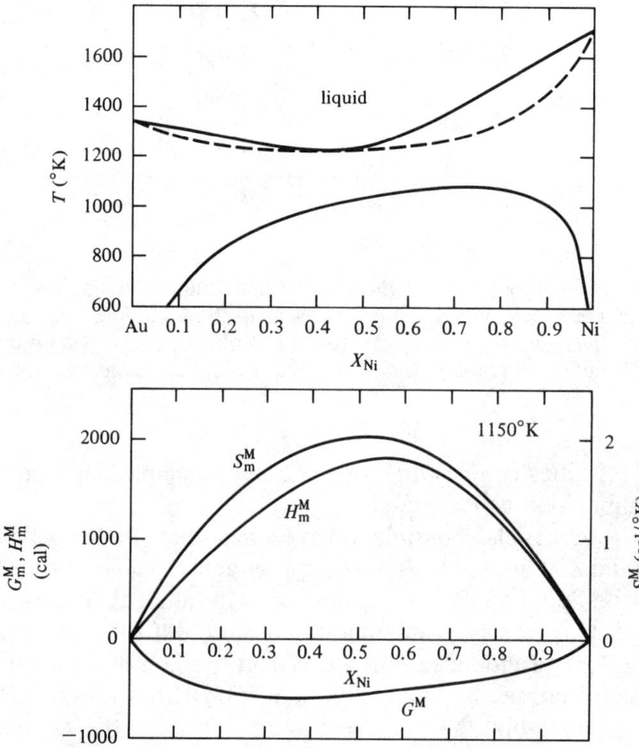

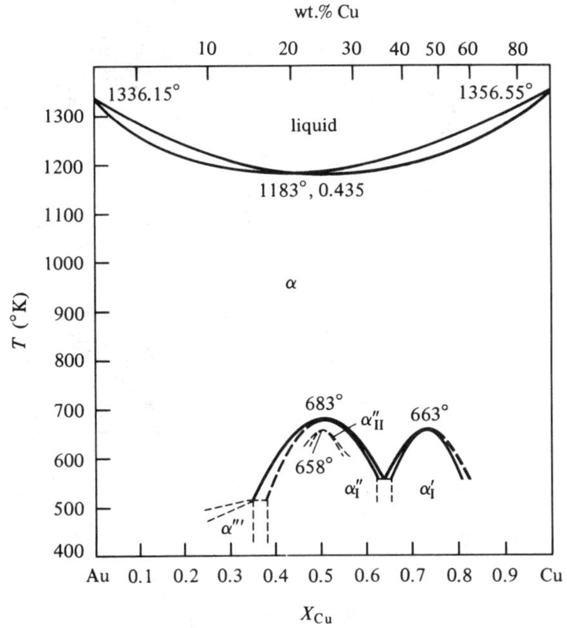

Figure 9. The gold–copper system [1].

and

$$\frac{\partial G^{(\alpha)}}{\partial X_2} = \frac{\partial G^{(\beta)}}{\partial X_2} \tag{25}$$

(For an application of these equations to the Au–Ni system, see Problem 3.)

4. Eutectic Points

So far we have examined cases where both the free energy curves of the solid and the liquid phases were convex ($\partial^2 G/\partial X_2^2 > 0$) for all compositions in the freezing range of temperatures. In Fig. 8, for example, the solid solution of the Au–Ni system has a convex free energy curve between the melting temperature of Ni (1725°K) and the minimum freezing temperature of 1224°K, although below the critical temperature (1093°K) the free energy curve is concave between the two spinodal points. We shall now examine the case in which the free energy curve of the liquid phase intersects a free energy curve of the solid phase which is no longer convex throughout the composition range.

This is illustrated in Fig. 10. At low temperatures such as T_e the alloys are homogeneous solids in the composition range $(0, X_2^{\alpha_1})$ and $(X_2^{\alpha_2}, 1)$; in the composition range $(X_2^{\alpha_1}, X_2^{\alpha_2})$, i.e., within the miscibility gap, we have a heterogeneous mixture of two phases of identical structures (e.g., fcc or bcc) but different compositions, $X_2^{\alpha_1}$ and $X_2^{\alpha_2}$. At higher temperatures T_c the miscibility gap becomes unstable with respect to the liquid phase, and instead of having a range of composition where the phases α_1 and α_2 coexist we have ranges of composition where the stable configurations are α_1 + liquid, liquid, and liquid + α_2. Between the temperatures T_c and T_e there is a temperature T_d at which the common tangent

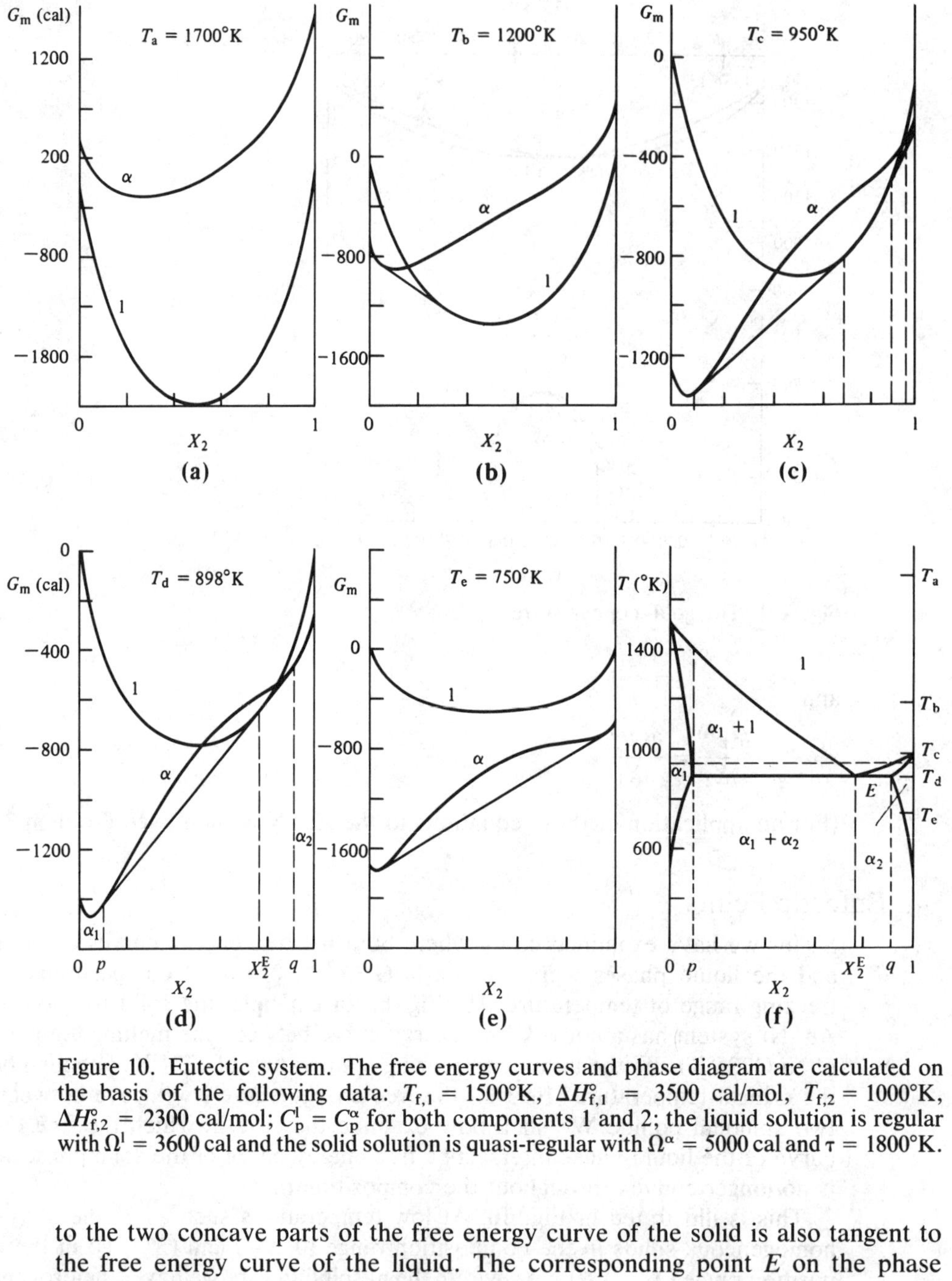

Figure 10. Eutectic system. The free energy curves and phase diagram are calculated on the basis of the following data: $T_{f,1} = 1500°K$, $\Delta H^{\circ}_{f,1} = 3500$ cal/mol, $T_{f,2} = 1000°K$, $\Delta H^{\circ}_{f,2} = 2300$ cal/mol; $C_p^l = C_p^\alpha$ for both components 1 and 2; the liquid solution is regular with $\Omega^l = 3600$ cal and the solid solution is quasi-regular with $\Omega^\alpha = 5000$ cal and $\tau = 1800°K$.

to the two concave parts of the free energy curve of the solid is also tangent to the free energy curve of the liquid. The corresponding point E on the phase diagram is called the *eutectic point*.

It may be noted that the *eutectic temperature* is the lowest temperature at which an alloy of the system 1–2 is still wholly or partially liquid. The composition X_2^E at which an alloy can still be wholly liquid at this temperature, is called the *eutectic composition*. At the eutectic point E, the following equilibrium prevails:

$$\text{liquid phase} = \text{solid phase } \alpha_1 + \text{solid phase } \alpha_2 \qquad (26)$$

where the phases α_1 and α_2 have the composition marked p and q in Fig. 10. The transformation corresponding to equation (26) is called the *eutectic reaction*.

Eutectic points may also be obtained in systems where the component metals freeze in different crystal structures α and β (see Fig. 11). At low temperatures T_e there is a composition range, determined by the common tangent to the free energy curves of the two solid solutions α and β, where both α and β coexist. Raising the temperature of the system, it is seen that there is a temperature at which the free energy curves of the three phases α, β, and liquid have a common tangent. This situation is very similar to that observed in Fig. (10) and is also said to correspond to a eutectic point. The eutectic reaction is then

$$\text{liquid phase} = \text{solid phase } \alpha + \text{solid phase } \beta \tag{27}$$

Figure 11. Eutectic system with three different structures. The free energy curves and phase diagram are calculated on the basis of the following data: $T_{f,1} = 1500$°K, $T_{f,2} = 2000$°K, $\Delta H^\circ_{f,1} = 3000$ cal/mol, $\Delta H^\circ_{f,2} = 4000$ cal/mol, $\Delta G_1^{\circ(\alpha\to\beta)} = 2000 + 0.5T$ cal/mol, $\Delta G_2^{\circ(\alpha\to\beta)} = -3000 - 0.5T$ cal/mol; $C_p^\alpha = C_p^\beta = C_p^l$ for both components 1 and 2; the solutions are assumed regular with $\Omega^l = \Omega^\alpha = 2000$ cal and $\Omega^\beta = 4000$ cal.

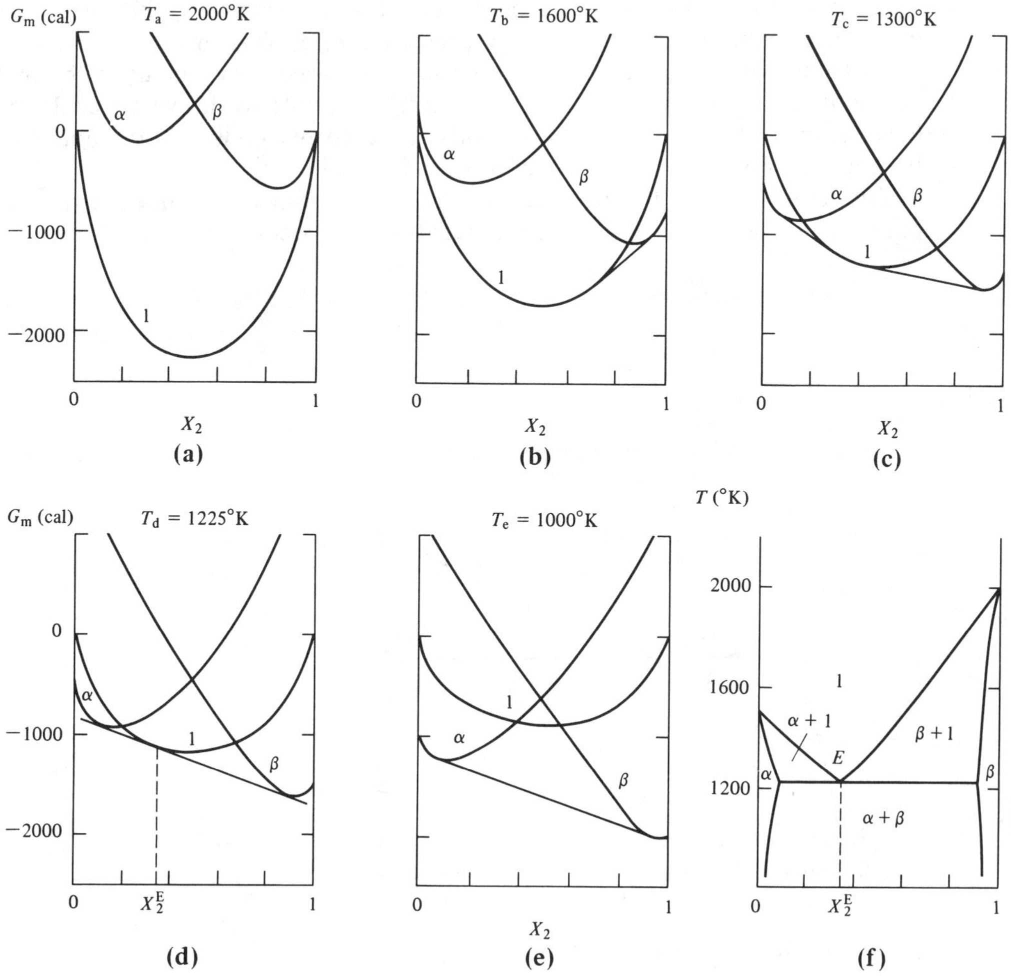

The equilibrium diagrams of several metallic systems are of the simple eutectic type shown in Figs. 10 or 11. Examples are Ag–Cu (Fig. 12), Al–Si, Cd–Zn, Pb–Sb, and Pb–Sn. It should be noted that it is often difficult to ascertain with confidence whether the two solid solutions observed in a binary belong to the case of a miscibility gap (Fig. 10) or are described by two different free energy curves (Fig. 11). For example, in the case of the Ag–Cu system, both freeze in the fcc structure, but they have rather different lattice parameters: 4.09 Å for Ag and 3.61 Å for Cu at 298°K. By very rapid cooling from the molten state it has been possible to obtain solid solutions with lattice spacings intermediate between those of Ag and Cu [4]. However, to ensure unequivocally that a single free energy curve (rather than two) corresponds to these solutions, one would have to study the thermodynamic properties of these solutions at temperatures above the critical temperature. If it exists, this critical temperature is likely to be much above the eutectic one, and the driving force for the precipitation of the liquid phase would be of such a magnitude that it would prevent the study of these metastable alloys.

There are many cases where the solubility of one of the components (say component 2) in the other (component 1) is negligible in the solid state. The structure α of solid component 1 cannot accommodate then any significant amount of atoms of component 2 and the free energy curve of phase α is now shrunk to a negligible width (i.e., the free energy of α is raised very rapidly with small additions of 2). In such cases, it is generally possible to derive an analytical expression of that branch of the liquidus curve corresponding to the boundary between the domains of stability of α + liquid and liquid.

For convenience, let us assume that $C^{l}_{p,1} = C^{s}_{p,1}$. Since component 1 remains pure, $a^{s}_{1} = 1$ at all temperatures, and equation (12) becomes

$$\ln X^{l}_{1} + \ln \gamma^{l}_{1} = -\frac{\Delta H^{\circ}_{f,1}}{R}\left(\frac{1}{T} - \frac{1}{T_{f,1}}\right) \tag{28}$$

Figure 12. The silver–copper system [1].

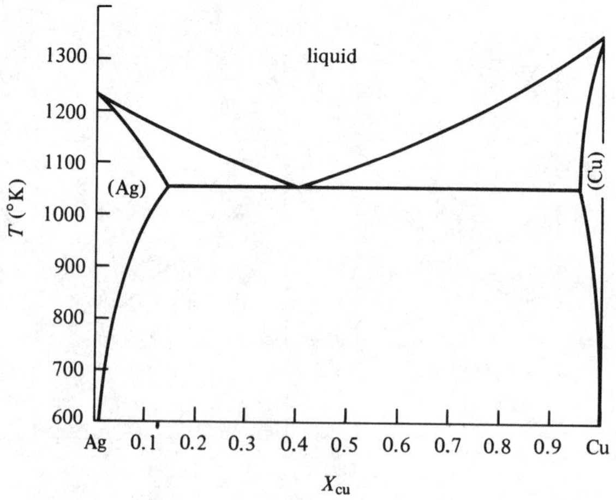

which is the equation of the liquidus branch. For example, if the liquid solution is regular,

$$\ln(1 - X_2^l) + \frac{\Omega^l}{RT}(X_2^l)^2 = -\frac{\Delta H_{f,1}^\circ}{R}\left(\frac{1}{T} - \frac{1}{T_{f,1}}\right) \tag{29}$$

If the solubility of 1 in the β structure of solid 2 is also negligible, the other branch of the liquidus curve may also be described analytically. Assuming again that $C_{p,2}^l = C_{p,2}^s$, the equation of this other branch is

$$\ln X_2^l + \frac{\Omega^l}{RT}(1 - X_2^l)^2 = -\frac{\Delta H_{f,2}^\circ}{R}\left(\frac{1}{T} - \frac{1}{T_{f,2}}\right) \tag{30}$$

The two branches of the liquidus intersect at the eutectic point (X_2^E, T^E). Consequently, the eutectic composition and temperature may be solved by observing that equations (29) and (30) must be satisfied simultaneously at this point. The Ag–Si and Na–Rb systems are such simple eutectic diagrams with no mutual miscibility in the solid state. (The assumption of a regular solution for the liquid phase, however, is not necessarily justified.)

It is possible for a solid phase γ to play the role of the liquid phase in a eutectic-type system. In that case, the point at which

$$\text{solid phase } \gamma = \text{solid phase } \alpha + \text{solid phase } \beta \tag{31}$$

is called *eutectoid* instead of eutectic. The solid phases α and β may have the same structure (in which case they correspond to α_1 and α_2 of a miscibility gap). A *monotectic* is associated with the case in which one of the solid phases in reaction (27) is replaced by another liquid phase:

$$\text{liquid phase } l_1 = \text{solid phase } \alpha + \text{liquid phase } l_2 \tag{32}$$

The Cu–Pb phase diagram [1] presents such a monotectic transformation. (See Problem 6.)

5. Peritectic Points

As seen in the preceding section, the characteristic of a eutectic point is a configuration in which the free energy curves of the liquid and of the two solid solutions have a common tangent and for which the equilibrium point of the liquid phase (point of tangence of the liquid's curve) is located between the equilibrium points of the two solid solutions. The case of a *peritectic point* is described in Figs. 13 and 14. Its characteristic is similar to that of a eutectic point in the fact that, here too, the free energy curves of the liquid and of the two solid solutions have a common tangent; however, in contrast, the equilibrium point of the liquid phase is no longer between those of the two solid solutions. We also recall that when the temperature is *raised* above the eutectic temperature, the free energy curve of the liquid "breaks through" the common tangent of the two solid solutions' curves; lowering the temperature below the eutectic temperature, the liquid is no longer stable. Figures 13 and 14 show that when the temperature is *lowered* below the peritectic temperature, it is one of the solid solution curves

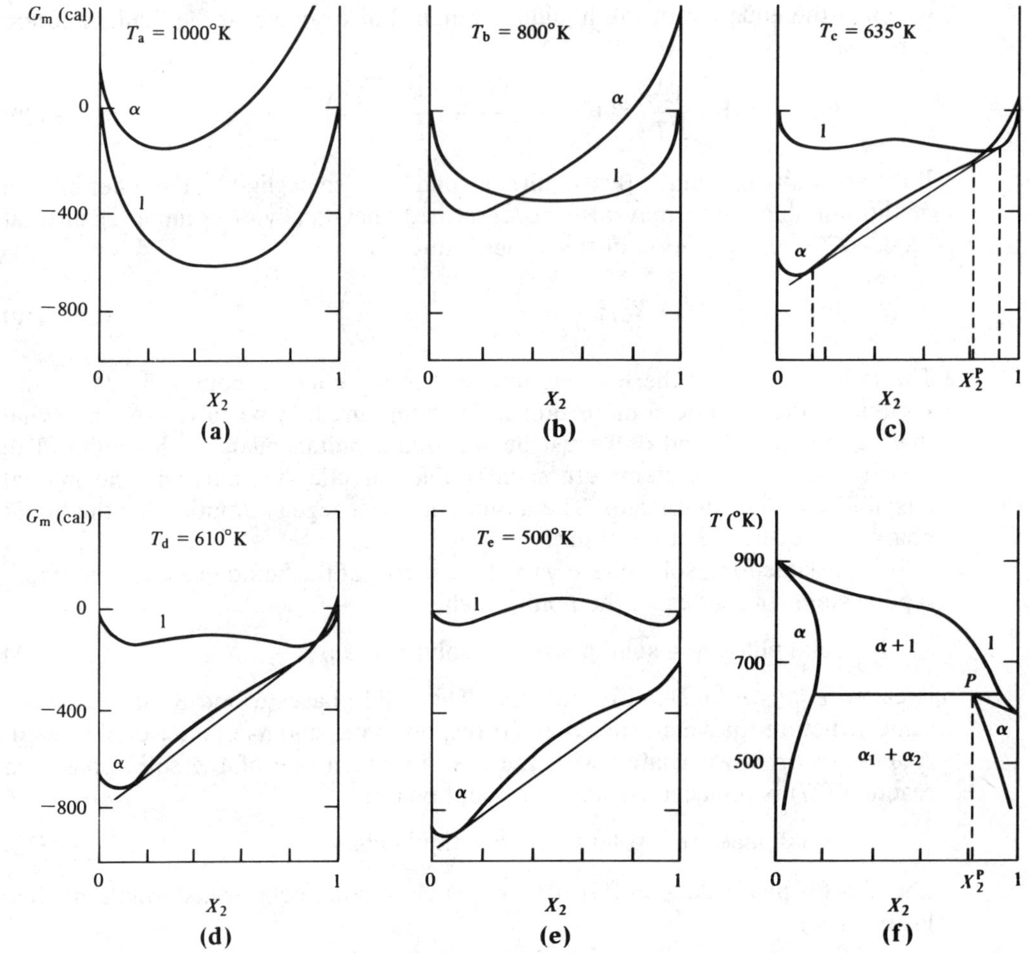

Figure 13. Peritectic system. The free energy curves and phase diagrams are calculated on the basis of the following data: $T_{f,1} = 900°K$, $\Delta H^\circ_{f,1} = 2000$ cal/mol, $T_{f,2} = 600°K$, $\Delta H^\circ_{f,2} = 1300$ cal/mol; $C^l_p = C^\alpha_p$ for both components 1 and 2; both liquid and solid solutions are regular with $\Omega = 3000$ cal.

which breaks through the common tangent to the curves of the other solid solution and of the liquid. Raising the temperature above the peritectic temperature yields only one two-phase domain.

The peritectic reaction may be represented either by

$$\text{liquid phase} + \text{solid phase } \alpha_1 = \text{solid phase } \alpha_2 \tag{33}$$

or by

$$\text{liquid phase} + \text{solid phase } \alpha = \text{solid phase } \beta \tag{34}$$

The first equation corresponds to the presence of a miscibility gap (see Fig. 13), the critical temperature of which is above the peritectic temperature. The second reaction corresponds to the case described in Fig. 14.

Peritectic type diagrams occur mainly when the melting points of the components differ widely. This may be seen by referring to Fig. 13 or 14. The Ag–Pt and Co–Cu systems (see Fig. 15) are examples of peritectic-type diagrams.

It is possible for a solid phase γ to play the role of the liquid phase in a peritectic-type system. In that case, the point at which

$$\text{solid phase } \gamma + \text{solid phase } \alpha_1 = \text{solid phase } \alpha_2 \tag{35}$$

or

$$\text{solid phase } \gamma + \text{solid phase } \alpha = \text{solid phase } \beta \tag{36}$$

is called *peritectoid* instead of peritectic.

Figure 14. Peritectic system with three different structures. The free energy curves and the phase diagram are calculated on the basis of the following data: $T_{f,1} = 1000°K$, $T_{f,2} = 600°K$, $\Delta H^\circ_{f,1} = 2500$ cal/mol, $\Delta H^\circ_{f,2} = 1300$ cal/mol, $\Delta G_1^{\circ(\beta\to l)} = 2000 - 2.5T$ cal/mol, $\Delta G_2^{\circ(\alpha\to l)} = 800 - 2.4T$, $C_p^\alpha = C_p^\beta = C_p^l$ for both components 1 and 2; the solutions are assumed regular with $\Omega^\alpha = 3200$ cal, $\Omega^\beta = 1820$ cal, and $\Omega^l = 3090$ cal.

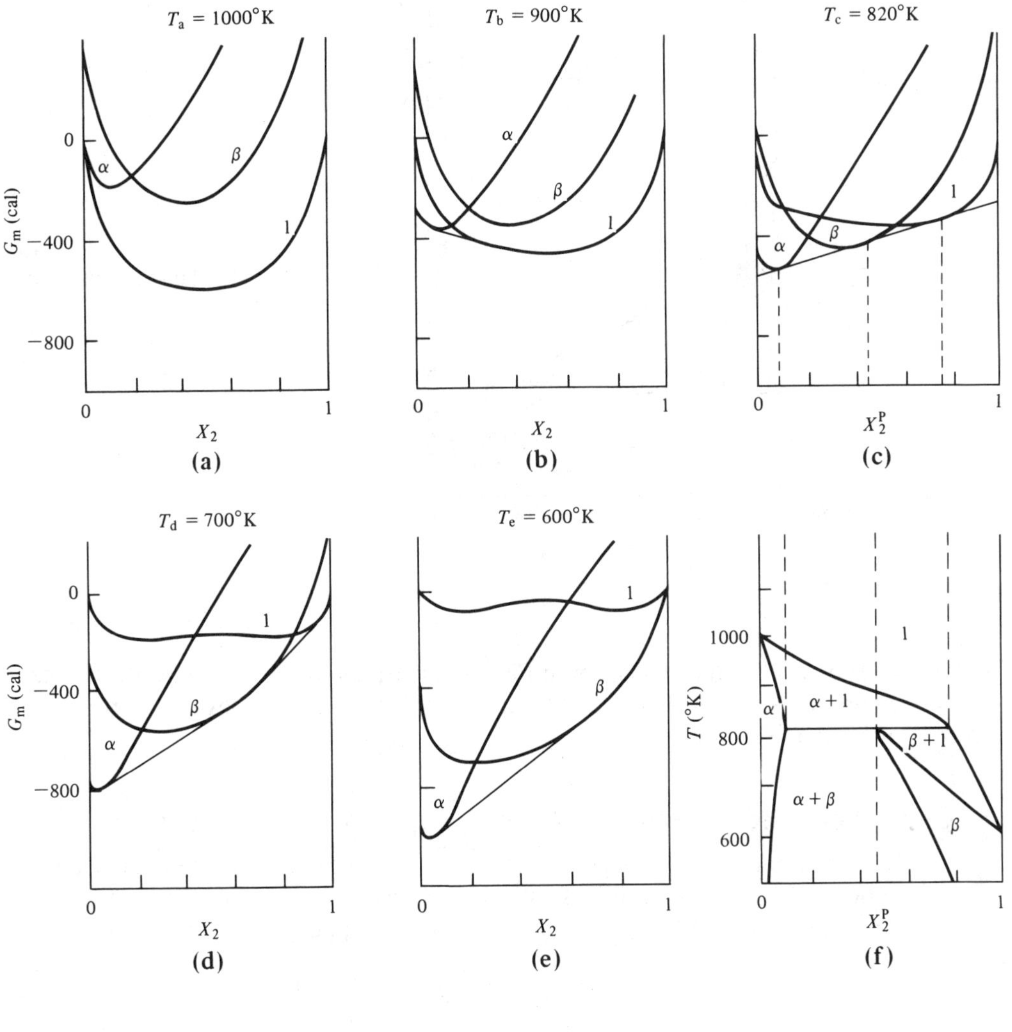

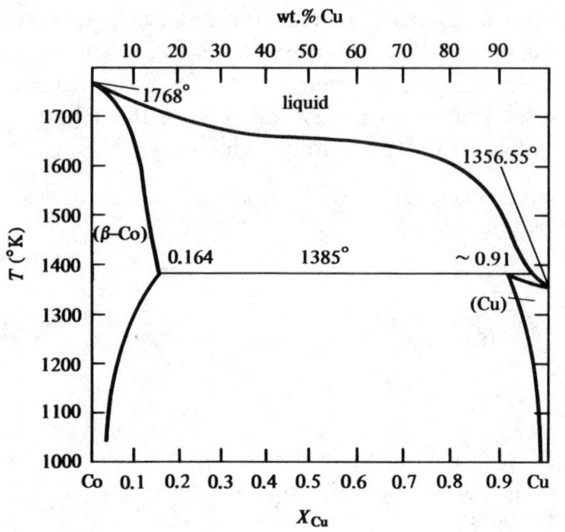

Figure 15. Phase diagram of the cobalt–copper system [1].

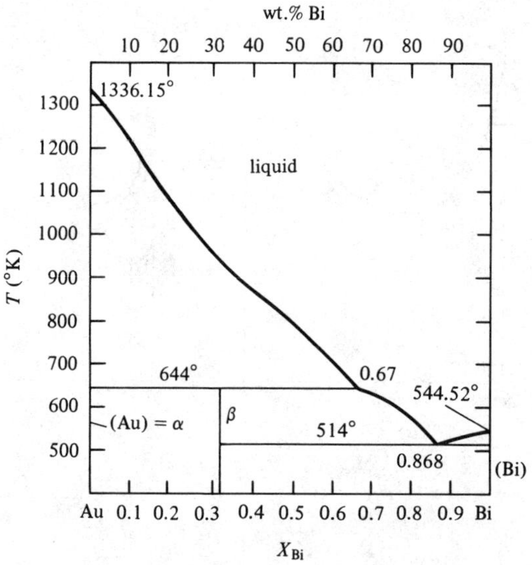

Figure 16. Phase diagram of the Au–Bi system [1].

The Au–Bi phase diagram in Fig. 16 illustrates the case of a peritectic reaction where the solid phases (α and β) are stoichiometric, i.e., have free energy curves of negligible width.

6. Correspondences Between Various Types of Phase Diagrams

Small changes in the shapes and relative positions of the free energy curves of various phases can have a drastic effect on the type of equilibrium diagram a system exhibits. Assuming that the solid and liquid phases are regular solutions

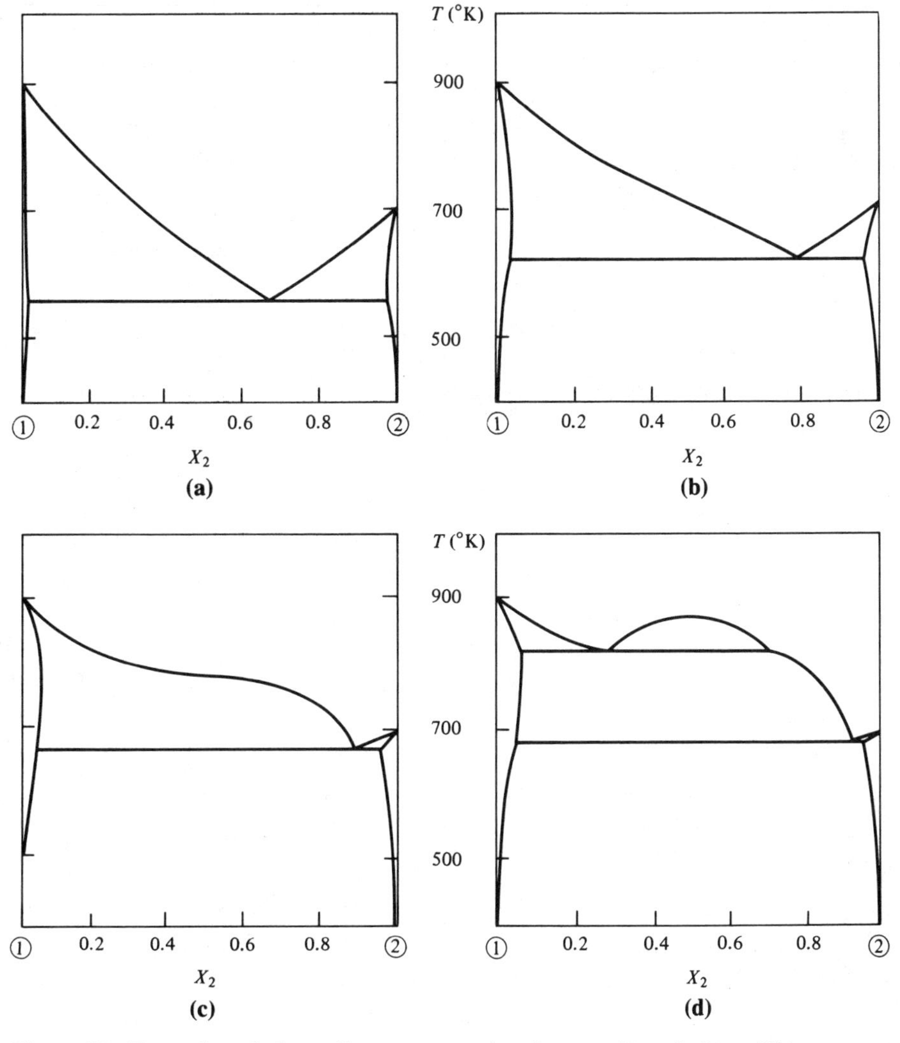

Figure 17. Examples of phase diagrams associated to regular solutions. The parameter Ω^l is equal to 4500 cal in all four cases, but that of the solid solution Ω^s increases from **(a)** 1000 cal to **(b)** 2000 cal, to **(c)** 3000 cal, and finally to **(d)** 3500 cal.

characterized by the parameters Ω^s and Ω^l, Figs. 17 and 18 illustrate the effects of changes in the values of these parameters. Similar changes in the nature of a phase diagram may be observed as an effect of pressure.

7. Complex Phase Diagrams

So far we have dealt only with relatively simple systems. These, however, show all the main features of binary diagrams, and systems with many more phases than the two or three considered previously can be analyzed in the same way.

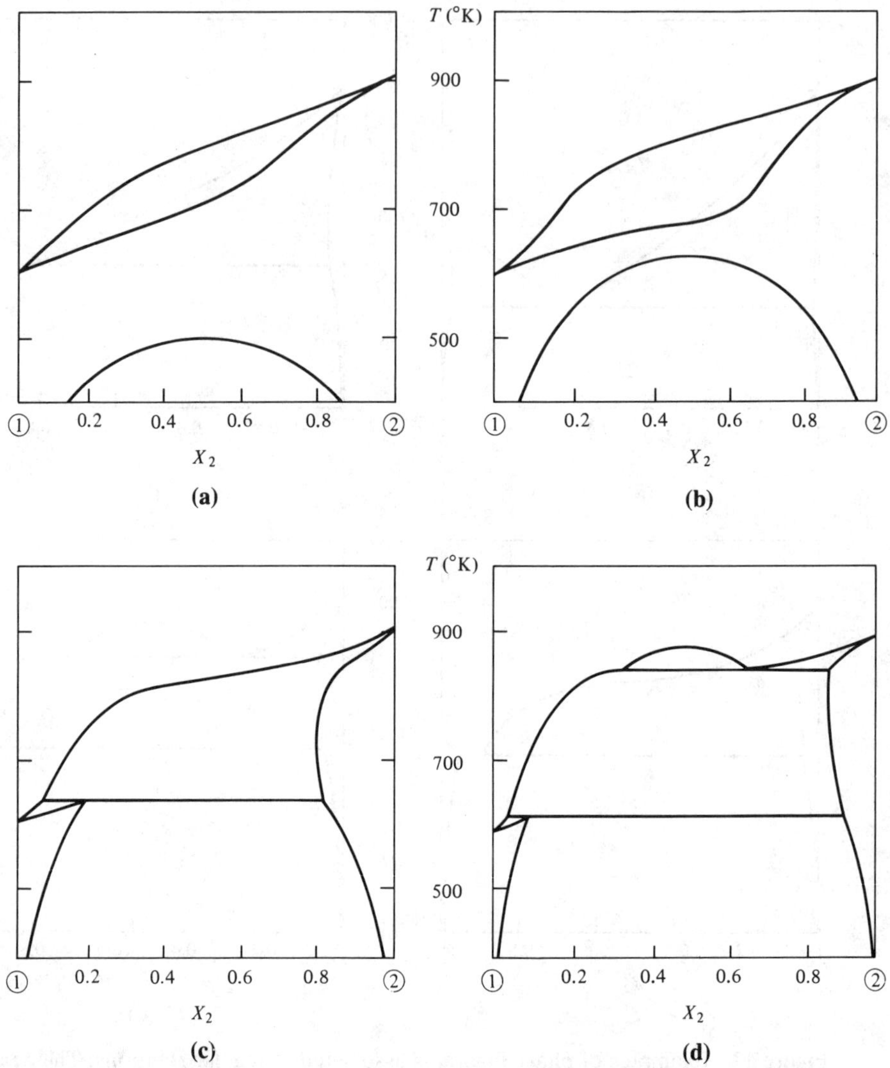

Figure 18. Examples of phase diagrams associated to regular solutions. The parameters Ω^l and Ω^s are kept equal in all cases, but increase in value from **(a)** 2000 cal to **(b)** 2500 cal, to **(c)** 3000 cal, and finally to **(d)** 3500 cal.

Figure 19 shows the example of the Fe–C system. It consists of a combination of a peritectic, a eutectic, and a eutectoid. The dashed lines in the figure correspond to metastable equilibria with cementite Fe_3C. (Cementite is often observed in iron alloys instead of graphite because it nucleates more easily than graphite.)

Another example is shown in Fig. 20, the Ga–Sb system. The maximum in the liquidus curve is equivalent to that studied in Section 3, the difference being that the phase β is now essentially stoichiometric (i.e., of negligible width).

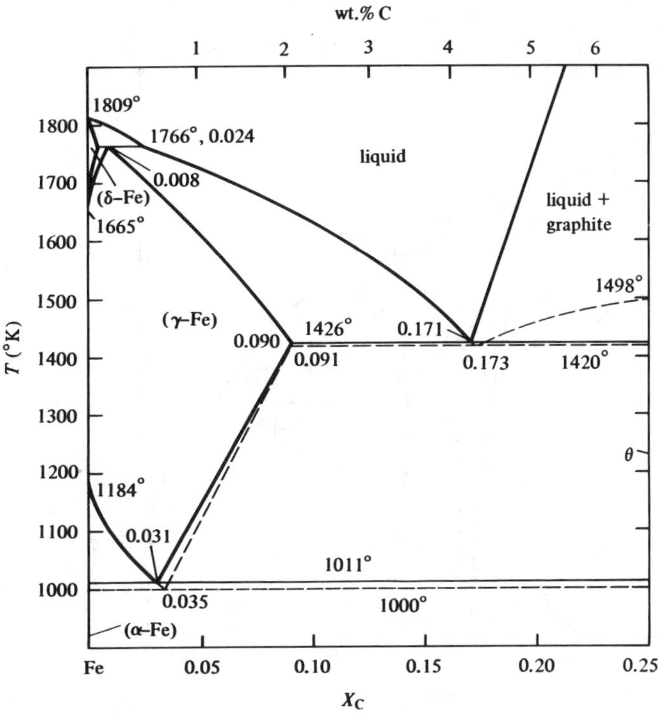

Figure 19. Phase diagram of the iron–carbon system [1]. The dashed lines correspond to equilibria with respect to cementite (θ) rather than graphite.

Figure 20. Phase diagram of the gallium–antimony system [1].

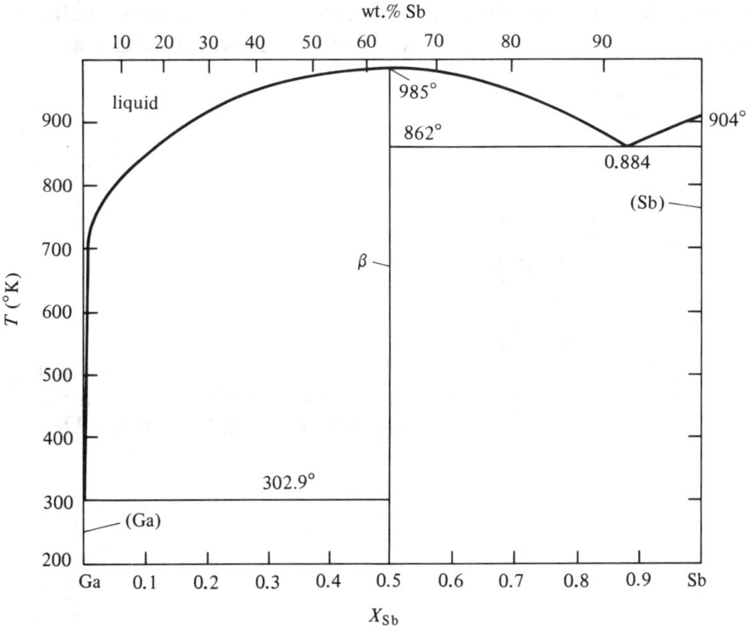

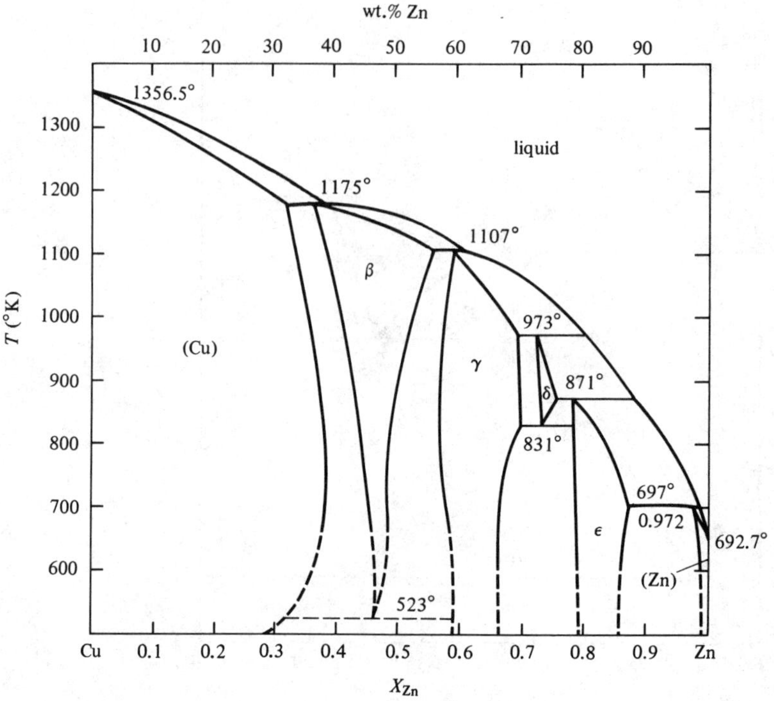

Figure 21. Phase diagram of the copper–zinc system [1].

The Cu–Zn system shown in Fig. 21 exhibits a relatively large number of phases. Because of the wide difference in the melting points of Cu and Zn (1083°C and 419°C), these phases appear in a succession of peritectic reactions. In addition, the δ phase which forms at about 70% Zn and 700°C becomes unstable at lower temperatures and decomposes by a eutectoid reaction into the γ and ε phases.

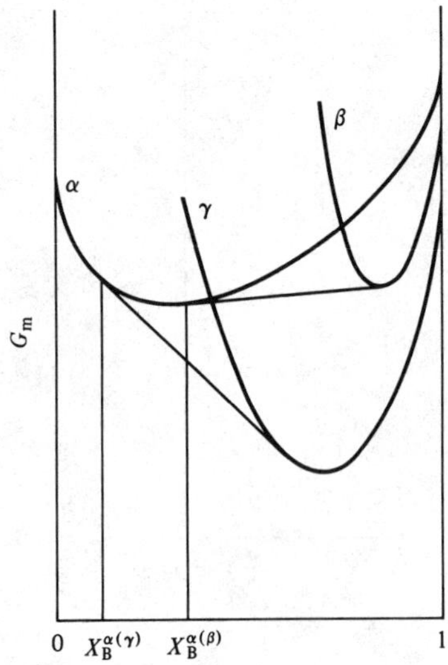

Figure 22. The solubility of B in α is larger in the case of the α–β equilibrium than in the case of the α–γ equilibrium when β is metastable with respect to γ.

A general observation should be made here. It concerns the solubility of a component B in a phase α in equilibrium with either a phase β or γ. If the phase β is metastable with respect to γ, the solubility of B is higher in case of the α–β equilibrium than for the α–γ equilibrium. Actually, when the stability of γ increases the solubility of B in α decreases. This may easily be understood through Fig. 22. One may verify, for example, that in the Fe–C diagram of Fig. 19 the solubility of carbon is higher in the γ and liquid phases when these phases are in equilibrium with cementite than when they are in equilibrium with graphite (the dashed lines are to the right of the corresponding solid lines).

8. Calculation of Phase Diagrams

8.1. Numerical Techniques for the Calculation of Phase Boundaries

The calculation of phase diagrams is reduced essentially to the calculation of phase boundaries in the equilibrium between two phases. At a given temperature T the equilibrium compositions X_2^α and X_2^β of two phases α and β are given by equations (2) and (3), which may be rewritten

$$\ln\frac{1 - X_2^\alpha}{1 - X_2^\beta} + \ln\gamma_1^\alpha - \ln\gamma_1^\beta - \frac{1}{RT}\Delta\mu_1^{\circ(\alpha\to\beta)} = 0 \tag{37a}$$

$$\ln\frac{X_2^\alpha}{X_2^\beta} + \ln\gamma_2^\alpha - \ln\gamma_2^\beta - \frac{1}{RT}\Delta\mu_2^{\circ(\alpha\to\beta)} = 0 \tag{37b}$$

where $\Delta\mu_i^{\circ(\alpha\to\beta)}$ is the difference in the values of the chemical potential of i in the two structures α and β when it is pure and at 1 atm pressure (i.e., in a Raoultian standard state).

The logarithms of the activity coefficients may usually be written as polynomials of the mole fraction X_2 (see Chapter VII), and the system of equations (37) is thus seen to involve both logarithmic expressions of the unknowns X_2^α and X_2^β as well as various powers of these unknowns. Except in a few simple cases such as those studied in Sections 2 and 3, an analytic solution of X_2^α and X_2^β as a function of T is not possible, and numerical solutions must be sought. These are now performed by computers.

A trial and error method was used by Rudman [5]. It consists in assigning arbitrary values to the couple of unknowns so that the whole range of possible compositions is covered, and then selecting the couple of values that best fit the system of equations (37). This procedure is time consuming, and an alternative method based on the Newton–Raphson iteration technique was offered by Kaufman and Bernstein [6]. Their method consists in selecting a couple of approximate equilibrium compositions $X_2^{\alpha'}$ and $X_2^{\beta'}$, and then calculating a more precise solution $X_2^{\alpha''}$, $X_2^{\beta''}$:

$$X_2^{\alpha''} = X_2^{\alpha'} + \delta X_2^\alpha \tag{38a}$$

$$X_2^{\beta''} = X_2^{\beta'} + \delta X_2^\beta \tag{38b}$$

where the mole fraction increments δX_2^α and δX_2^β are the solutions of a system of linear equations obtained by replacing the mole fractions $X_2^{\alpha''}$ and $X_2^{\beta''}$ in equations

(37) by their expressions from equations (38) and linearizing the results:

$$-(\psi_\alpha/X_1^{\alpha'})\,\delta X_2^\alpha + (\psi_\beta/X_1^{\beta'})\,\delta X_2^\beta + \ln(X_1^{\alpha'}/X_1^{\beta'})$$
$$+ \ln\gamma_1^\alpha - \ln\gamma_1^\beta - (1/RT)\Delta\mu_1^{\circ(\alpha\to\beta)} = 0 \quad (39a)$$
$$(\psi_\alpha/X_2^{\alpha'})\,\delta X_2^\alpha - (\psi_\beta/X_2^{\beta'})\,\delta X_2^\beta + \ln(X_2^{\alpha'}/X_2^{\beta'})$$
$$+ \ln\gamma_2^\alpha - \ln\gamma_2^\beta - (1/RT)\Delta\mu_2^{\circ(\alpha\to\beta)} = 0 \quad (39b)$$

The symbol ψ represents the stability function defined by Lupis and Gaye [7]:

$$\psi_\nu = \frac{X_1^\nu X_2^\nu}{RT}\frac{\partial^2 G^{M(\nu)}}{\partial(X_2^\nu)^2} = 1 + X_1^\nu\frac{\partial \ln\gamma_1^\nu}{\partial X_1^\nu} = 1 + X_2^\nu\frac{\partial \ln\gamma_2^\nu}{\partial X_2^\nu} \qquad (\nu = \alpha \text{ or } \beta) \quad (40)$$

The linear system of equations (39) can be easily solved for δX_2^α and δX_2^β and provides the values of $X_2^{\alpha''}$ and $X_2^{\beta''}$. These are then taken as initial values of $X_2^{\alpha'}$, $X_2^{\beta'}$ and the procedure is repeated until the mole fraction increments are small enough, i.e., until the equilibrium compositions are obtained to any desired degree of precision, provided that the iteration converges. Kaufman and Bernstein calculate the phase boundaries by steps corresponding to a set of temperatures differing by ΔT. At each step, the selected starting compositions are the equilibrium compositions determined at the preceding step. Whenever the iteration diverges, or converges too slowly, the interval of temperature ΔT is reduced by one-half. The number of iterations necessary to obtain an acceptable solution depends on the precision required and the departure of the initial points from the equilibrium points.

A different and more efficient method has been presented by Gaye and Lupis [8,9]. It substitutes for the iteration procedure a direct calculation of the equilibrium mole fraction increments between two steps.

The equilibrium concentrations at a temperature T being known, the equilibrium concentrations at the temperature $T + \Delta T$ are computed by a series expansion limited to the second order terms:

$$X_2^\nu(T+\Delta T) = X_2^\nu(T) + \left(\frac{dX_2^\nu}{dT}\right)_T \Delta T$$
$$+ \frac{1}{2}\left(\frac{d^2X_2^\nu}{dT^2}\right)_T (\Delta T)^2 \qquad (\nu = \alpha \text{ or } \beta) \quad (41)$$

$(dX_2^\nu/dT)_T$ and $(d^2X_2^\nu/dT^2)_T$ are the slope and curvature of the boundary corresponding to the ν phase at the temperature T (when considering the phase diagram as a plot of X_2 vs T). Their expressions are derived in Section 8.2.

Once the first and second derivatives of X_2^ν with respect to temperature have been obtained, equation (41) can be used. The temperature increment ΔT is fixed at each step and made small enough so that the approximation inherent in the application of equation (41) is acceptable. This is done by imposing the condition that the ratios of the second order terms to the first order terms in equation (41) are smaller than an arbitrary small number r:

$$\left|\frac{d^2X_2^\nu}{dT^2}\frac{(\Delta T)^2}{2}\right| < r\left|\frac{dX_2^\nu}{dT}\Delta T\right| \qquad (\nu = \alpha \text{ or } \beta) \quad (42)$$

Since this does not ensure that the sum of the higher order terms neglected in equation (41) is indeed negligible, as a safety against a possible divergence of the results from the exact equilibrium compositions, the Newton–Raphson iteration technique described above is applied at fixed temperature intervals. These tests can be performed at relatively few temperatures (e.g., every 5°), and because the starting compositions for each iteration are in general quite close to the equilibrium compositions, no appreciable increase in computer time results from the use of the iteration technique.

8.2. Slopes and Curvatures of Phase Boundaries

To calculate the derivatives dX_2/dT and d^2X_2/dT^2, we note that if we call f the function represented by the left member of equation (37a) or (37b)

$$\left(\frac{df}{dT}\right)_{\mathrm{eq}} = \frac{\partial f}{\partial X_2^\alpha}\frac{dX_2^\alpha}{dT} + \frac{\partial f}{\partial X_2^\beta}\frac{dX_2^\beta}{dT} + \frac{\partial f}{\partial T} = 0 \tag{43}$$

$(df/dT)_{\mathrm{eq}}$ is the derivative of f along the equilibrium phase boundaries and is identical to zero since f is identical to zero along these boundaries. $(\partial f/\partial X_2^\nu)$ are partial derivatives of f at constant T, $(\partial f/\partial T)$ is the partial derivative of f at constant composition, and dX_2^ν/dT is the slope of the ν boundary. Applying equation (43) to equations (37) thus leads to

$$-\frac{\psi_\alpha}{X_1^\alpha}\frac{dX_2^\alpha}{dT} + \frac{\psi_\beta}{X_1^\beta}\frac{dX_2^\beta}{dT} + \frac{\partial}{\partial T}(\ln\gamma_1^\alpha - \ln\gamma_1^\beta) + \frac{\Delta H_1^{\circ(\alpha\to\beta)}}{RT^2} = 0 \tag{44a}$$

$$\frac{\psi_\alpha}{X_2^\alpha}\frac{dX_2^\alpha}{dT} - \frac{\psi_\beta}{X_2^\beta}\frac{dX_2^\beta}{dT} + \frac{\partial}{\partial T}(\ln\gamma_2^\alpha - \ln\gamma_2^\beta) + \frac{\Delta H_2^{\circ(\alpha\to\beta)}}{RT^2} = 0 \tag{44b}$$

Solving for dX_2^α/dT and dX_2^β/dT yields

$$\frac{dX_2^\alpha}{dT} = \frac{X_1^\alpha X_2^\alpha}{(X_2^\alpha - X_2^\beta)\psi_\alpha}\left\{X_1^\beta\left[\frac{\partial}{\partial T}(\ln\gamma_1^\alpha - \ln\gamma_1^\beta) + \frac{\Delta H_1^{\circ(\alpha\to\beta)}}{RT^2}\right]\right.$$
$$\left. + X_2^\beta\left[\frac{\partial}{\partial T}(\ln\gamma_2^\alpha - \ln\gamma_2^\beta) + \frac{\Delta H_2^{\circ(\alpha\to\beta)}}{RT^2}\right]\right\} \tag{45}$$

and a parallel expression for dX_2^β/dT which may be immediately obtained from equation (45) by interchanging the indices α and β. The second derivatives are obtained in a similar way, by differentiation of equations (44) with respect to temperature. The resulting expressions of $d^2X_2^\alpha/dT^2$ is

$$\frac{d^2X_2^\alpha}{dT^2} = \frac{X_1^\alpha X_2^\alpha}{(X_2^\alpha - X_2^\beta)\psi_\alpha}\left\{X_1^\beta\left[\frac{\partial^2}{\partial T^2}(\ln\gamma_1^\alpha - \ln\gamma_1^\beta) + \frac{\partial}{\partial T}\left(\frac{\Delta H_1^{\circ(\alpha\to\beta)}}{RT^2}\right)\right]\right.$$
$$\left. + X_2^\beta\left[\frac{\partial^2}{\partial T^2}(\ln\gamma_2^\alpha - \ln\gamma_2^\beta) + \frac{\partial}{\partial T}\left(\frac{\Delta H_2^{\circ(\alpha\to\beta)}}{RT^2}\right)\right]\right\} - \frac{2}{\psi_\alpha}\frac{\partial\psi_\alpha}{\partial T}\frac{dX_2^\alpha}{dT}$$
$$- \left(\frac{1}{\psi_\alpha}\frac{\partial\psi_\alpha}{\partial X_2^\alpha} + \frac{(X_1^\alpha X_2^\beta/X_2^\alpha) + (X_2^\alpha X_1^\beta/X_1^\alpha)}{X_2^\alpha - X_2^\beta}\right)\left(\frac{dX_2^\alpha}{dT}\right)^2$$
$$+ \frac{X_1^\alpha X_2^\alpha\psi_\beta/X_1^\beta X_2^\beta\psi_\alpha}{X_2^\alpha - X_2^\beta}\left(\frac{dX_2^\beta}{dT}\right)^2 \tag{46}$$

An expression for $d^2X_2^\beta/dT^2$ is obtained from equation (46) by interchanging the indices α and β.

The case in which the β phase is a stoichiometric compound can be obtained from the preceding equations, but it is just as simple to deduce it directly. If, for example, the β phase corresponds to the compound A_xB_y, the following reaction may be considered:

$$xA(\text{in structure } \alpha) + yB(\text{in structure } \alpha) = A_xB_y(\text{phase } \beta) \tag{47}$$

This reaction may be characterized by the standard free energy of formation of the compound:

$$\Delta G^\circ = G^\circ_{A_xB_y} - x\mu_A^{\circ(\alpha)} - y\mu_B^{\circ(\alpha)} \tag{48a}$$

With the notation adopted in this chapter, this equation becomes

$$\Delta G^{\circ(\alpha\to\beta)} = G^\circ(\beta) - X_1^\beta\mu_1^{\circ(\alpha)} - X_2^\beta\mu_2^{\circ(\alpha)} \tag{48b}$$

where

$$G^\circ(\beta) = G(\beta) = X_1^\beta\mu_1^\beta + X_2^\beta\mu_2^\beta \tag{49}$$

since the compound is stoichiometric. Since X_2^β is now known (and fixed), only one equation needs to be studied to determine X_2^α. It is immediately deduced:

$$\Delta G^\circ/RT = X_1^\beta \ln a_1^\alpha + X_2^\beta \ln a_2^\alpha \tag{50}$$

Proceeding as in the general case, i.e., differentiating equation (50) with respect to temperature, yields

$$\frac{dX_2^\alpha}{dT} = \frac{X_1^\alpha X_2^\alpha}{(X_2^\alpha - X_2^\beta)\psi_\alpha}\left[\frac{\partial}{\partial T}(X_1^\beta \ln \gamma_1^\alpha + X_2^\beta \ln \gamma_2^\alpha) + \frac{\Delta H^{\circ(\alpha\to\beta)}}{RT^2}\right] \tag{51}$$

and

$$\frac{d^2X_2^\alpha}{dT^2} = \frac{X_1^\alpha X_2^\alpha}{(X_2^\alpha - X_2^\beta)\psi_\alpha}\left[\frac{\partial^2}{\partial T^2}(X_1^\beta \ln \gamma_1^\alpha + X_2^\beta \ln \gamma_2^\alpha) + \frac{\partial}{\partial T}\left(\frac{\Delta H^{\circ(\alpha\to\beta)}}{RT^2}\right)\right]$$
$$- \frac{2}{\psi_\alpha}\frac{\partial\psi_\alpha}{\partial T}\frac{dX_2^\alpha}{dT} - \left(\frac{1}{\psi_\alpha}\frac{\partial\psi_\alpha}{\partial X_2^\alpha} + \frac{(X_1^\alpha X_2^\beta/X_2^\alpha) + (X_1^\beta X_2^\alpha/X_1^\alpha)}{X_2^\alpha - X_2^\beta}\right)\left(\frac{dX_2^\alpha}{dT}\right)^2 \tag{52}$$

If the β phase is the pure component 2, the condition for equilibrium yielding X_2^α is

$$\ln X_2^\alpha + \ln \gamma_2^\alpha - (1/RT)\Delta\mu_2^{\circ(\alpha\to\beta)} = 0 \tag{53}$$

By the same technique of differentiation, or by taking X_2^β equal to 1 in equations (51) and (52), we obtain

$$\frac{dX_2^\alpha}{dT} = -\frac{X_2^\alpha}{\psi_\alpha}\left[\frac{\partial \ln \gamma_2^\alpha}{\partial T} + \frac{\Delta H_2^{\circ(\alpha\to\beta)}}{RT^2}\right] \tag{54}$$

$$\frac{d^2X_2^\alpha}{dT^2} = -\frac{X_2^\alpha}{\psi_\alpha}\left[\frac{\partial^2 \ln \gamma_2^\alpha}{\partial T^2} + \frac{\partial}{\partial T}\left(\frac{\Delta H_2^{\circ(\alpha\to\beta)}}{RT^2}\right)\right] - \frac{2}{\psi_\alpha}\frac{\partial\psi_\alpha}{\partial T}\frac{dX_2^\alpha}{dT}$$
$$- \left(\frac{1}{\psi_\alpha}\frac{\partial\psi_\alpha}{\partial X_2^\alpha} - \frac{1}{X_2^\alpha}\right)\left(\frac{dX_2^\alpha}{dT}\right)^2 \tag{55}$$

The slope dT/dX_2^ν and curvature $d^2T/(dX_2^\nu)^2$ of the $\nu/(\alpha + \beta)$ ($\nu = \alpha$ or β) boundary can be readily obtained from the expressions for dX_2^ν/dT and $d^2X_2^\nu/(dT)^2$ since

$$\frac{dT}{dX_2^\nu} = \frac{1}{dX_2^\nu/dT} \tag{56}$$

and

$$\frac{d^2T}{(dX_2^\nu)^2} = -\frac{d^2X_2^\nu/(dT)^2}{(dX_2^\nu/dT)^3} \tag{57}$$

A few remarks concerning the shapes of the phase boundaries can be drawn from equations (45) and (56). First, the slope dT/dX_2^ν is roughly proportional to the stability function ψ_ν. It becomes zero (the boundary becomes horizontal) when the stability function becomes equal to zero, that is, at a critical point of the phase ν. In general, nearly horizontal phase boundaries correspond to a phase ν of small stability. For example, a near horizontal liquidus exists in systems in which a metastable miscibility gap can be readily obtained with small undercooling. A steeper phase boundary generally corresponds to a phase ν of larger stability. For example, eutectic systems with steep liquidus curves characterize stable liquid phases.

Second, when the phase boundaries have a common point ($X_2^\alpha = X_2^\beta$), they are tangent at that point and the tangent is horizontal. It should be noted that this property does not apply to the limits $X_2 = 0$ or 1; in these cases the system of equations (37) is singular. The tangent is still horizontal, however, at the congruent transformation of a stoichiometric compound [cf. equation (51)].

8.3. Calculation of the Boundaries in the Vicinity of Some Invariant Points

In the method of calculations outlined in Section 8.1, the determination of a series of pairs of equilibrium compositions is based upon the knowledge of one pair of equilibrium points. These starting points may be a couple of equilibrium compositions determined either experimentally or from previous calculations. For example, in a eutectic diagram of three phases—liquid, α, and β—the starting points for the calculation of the α–β equilibrium can be taken as the equilibrium mole fractions of the α and β phases at the eutectic temperature; these would be obtained from the calculation of the liquid–α and liquid–β equilibria.

It may be noted that at the end points of the composition range ($X_2 = 0$ or 1) equations (45) and (46) cannot be used. At the critical point of a miscibility gap or at the congruent transformation point (e.g., melting point) of a compound, dX_2^ν/dT is infinite and the series expansion (41) cannot be used either. However, as it is desirable to start calculation of the boundaries at these points, the difficulty may be circumvented by calculating equilibrium points in the immediate vicinity of these singular points through different equations. At these new equilibrium compositions the regular numerical techniques can again be applied. The necessary equations will now be derived.

8.3.1. Dilute Solutions

When for the pure component 1, phases α and β coexist at the temperature T_1^t, the starting equilibrium points are selected on the tangents to the phase boundaries at infinite dilution of component 2, and for a temperature close to T_1^t. To calculate their slopes, we note that in equation (44a)

$$(\psi_\nu/X_1^\nu)_{X_2^\nu\to 0} = 1 \qquad (\nu = \alpha \text{ or } \beta) \tag{58}$$

$$\left(\frac{\partial \ln \gamma_1^\nu}{\partial T}\right)_{X_2^\nu\to 0} = 0 \qquad (\nu = \alpha \text{ or } \beta) \tag{59}$$

and

$$\frac{\Delta H_1^{\circ(\alpha\to\beta)}}{R(T_1^{\text{t}})^2} = \frac{\Delta S_1^{\circ(\alpha\to\beta)}}{RT_1^{\text{t}}} \tag{60}$$

Consequently, equation (44a) becomes

$$\left(\frac{dX_2^\beta}{dT}\right)_{X_2^\beta\to 0} - \left(\frac{dX_2^\alpha}{dT}\right)_{X_2^\alpha\to 0} + \frac{\Delta S_1^{\circ(\alpha\to\beta)}}{RT_1^{\text{t}}} = 0 \tag{61}$$

Because of L'Hospital's rule

$$\left(\frac{X_2^\beta}{X_2^\alpha}\right)_{X_2^\nu\to 0} = \left(\frac{dX_2^\beta}{dX_2^\alpha}\right)_{X_2^\nu\to 0} \tag{62}$$

equation (37b) yields

$$\left(\frac{dX_2^\beta}{dX_2^\alpha}\right)_{X_2^\nu\to 0} = \frac{\gamma_2^{\infty(\alpha)}}{\gamma_2^{\infty(\beta)}} \exp\left(-\frac{\Delta\mu_2^{\circ(\alpha\to\beta)}}{RT_1^{\text{t}}}\right) \tag{63}$$

where $\gamma_2^{\infty(\nu)}$ is the value of the activity coefficient of component 2 at infinite dilution. Combining equations (61) and (63) yields

$$\left(\frac{dX_2^\alpha}{dT}\right)_{X_2\to 0} = \frac{\gamma_2^{\infty(\beta)}(\Delta S_1^{\circ(\alpha\to\beta)}/RT_1^{\text{t}})}{\gamma_2^{\infty(\beta)} - \gamma_2^{\infty(\alpha)} \exp(-\Delta\mu_2^{\circ(\alpha\to\beta)}/RT_1^{\text{t}})} \tag{64a}$$

and

$$\left(\frac{dX_2^\beta}{dT}\right)_{X_2^\beta-0} = \frac{\gamma_2^{\infty(\alpha)}(\Delta S_1^{\circ(\alpha\to\beta)}/RT_1^{\text{t}}) \exp(-\Delta\mu_2^{\circ(\alpha\to\beta)}/RT_1^{\text{t}})}{\gamma_2^{\infty(\beta)} - \gamma_2^{\infty(\alpha)} \exp(-\Delta\mu_2^{\circ(\alpha\to\beta)}/RT_1^{\text{t}})} \tag{64b}$$

It should be observed that $\gamma_2^\nu \exp(\mu_2^{\circ(\nu)}/RT)$ is independent of the standard state selected for the activity coefficient of 2, since

$$\mu_2^\nu = (\mu_2^{\circ(\nu)} + RT \ln \gamma_2^\nu) + RT \ln X_2^\nu \tag{65}$$

Let us select the same standard state for 2 in both the α and β phases.

$$\mu_2^\alpha = \mu_2^\circ + RT \ln \theta_2^\alpha X_2^\alpha \tag{66a}$$

$$\mu_2^\beta = \mu_2^\circ + RT \ln \theta_2^\beta X_2^\beta \tag{66b}$$

In equations (63) and (64) $\Delta\mu_2^{\circ(\alpha\to\beta)}$ vanishes, resulting in much simpler expressions. For example, equation (63) becomes

$$\frac{(dT/dX_2^\alpha)_{X_2^\alpha\to 0}}{(dT/dX_2^\beta)_{X_2^\beta\to 0}} = \frac{\theta_2^{\infty(\alpha)}}{\theta_2^{\infty(\beta)}} \tag{67}$$

We deduce that the ratio of the slopes of the liquidus and solidus lines at $X_1 = 1$

is equal to the ratio of the activity coefficients of 2 at infinite dilution in the liquid and solid phases. Let us take, for example, the case of the iron–carbon phase diagram. Graphite is our standard state for the activity of carbon in both the liquid and bcc phases. From Fig. 19, we estimate that

$$\left(\frac{dT/dX_C^l}{dT/dX_C^\delta}\right)_{X_{Fe}=1} \simeq \frac{1}{8} \tag{68}$$

Therefore,

$$\left(\frac{\theta_C^{\infty(l)}}{\theta_C^{\infty(\delta)}}\right)_{1809°K} \simeq \frac{1}{8} \tag{69}$$

The reader may verify that this result is in excellent agreement with the expressions proposed by Gaye and Lupis [8]:

$$\ln \gamma_C^{\infty(\delta)} = -5.04 + (12{,}080/T) \tag{70}$$

$$\ln \gamma_C^{\infty(l)} = -2.2 + (3140/T) \tag{71}$$

The addition of a solute 2 changes the melting point of a solvent 1. At infinite dilution this change is measured by the slope of the liquidus, $(dT/dX_2^l)_{X_2\to 0}$. It is of interest to examine whether there is a limit to the reduction of the fusion temperature that no solute 2 can exceed. We note that equation (64b) may be rewritten

$$\left(\frac{dT}{dX_2^l}\right)_{X_2^l\to 0} = \left(\frac{\theta_2^{\infty(l)}}{\theta_2^{\infty(s)}} - 1\right)\frac{R(T_{f,1})^2}{\Delta H_{f,1}^\circ} \tag{72}$$

and, since activity coefficients are always positive, we deduce that

$$\lim\left(\frac{dT}{dX_2^l}\right)_{X_2^l\to 0} = -\frac{R(T_{f,1})^2}{\Delta H_{f,1}^\circ} = -\frac{RT_{f,1}}{\Delta S_{f,1}^\circ} \tag{73}$$

This limit depends only on the properties of the solvent. For example, we note that for iron

$$\lim\left(\frac{dT}{dX_2^l}\right)_{X_2^l\to 0} = -(1.987/1.824) \times 1809 = -1970°K \tag{74}$$

Thus, an addition of 1 at.% of any solute cannot depress the melting point of iron by more than about 20°C.

It is clear that equations (70)–(74) are valid for the reduction in transformation temperature of any phase, and not just the liquid.

8.3.2. Critical Point of a Miscibility Gap

At the critical point of a binary miscibility gap, the following equations are satisfied:

$$\frac{\partial^2 G}{\partial X_2^2} = \frac{\partial^3 G}{\partial X_2^3} = 0 \tag{75}$$

In terms of partial molar excess enthalpy and entropy, and by elimination of the critical temperature T_C, equations (75) yield the critical composition X_2^C as solution of

$$\frac{\partial H_2^E}{\partial X_2}\left(\frac{\partial^2 S_2^E}{\partial X_2^2}+\frac{R}{X_2^2}\right)=\frac{\partial^2 H_2^E}{\partial X_2^2}\left(\frac{\partial S_2^E}{\partial X_2}-\frac{R}{X_2}\right) \tag{76}$$

Once this equation has been solved for X_2^C, T_C is obtained by

$$T_C=\frac{\partial H_2^E/\partial X_2}{(\partial S_2^E/\partial X_2)-(R/X_2)} \tag{77}$$

Alternatively, we may calculate the coordinates of the critical point by noting that, analytically, equation (76) is equivalent to

$$\left(\frac{\partial T_C}{\partial X_2}\right)_C=0 \tag{78}$$

where T_C has the expression of equation (77). Consequently, as a convenient numerical technique, T_C can be considered the function of X_2 defined by equation (77); the composition X_2^C then corresponds to the extremum of this function.

Use of the direct method requires that the initial compositions X_2^α and X_2^β have different values. A pair of such compositions may be determined at a temperature close to the critical temperature (e.g., different from T_C by 1°K) by the approximate relationship

$$(X_2^\beta - X_2^C)^2 = (X_2^C - X_2^\alpha)^2$$

$$=\left\{6\left[R\left(\frac{1}{X_1}+\frac{1}{X_2}\right)-\frac{\partial^2 S^E}{\partial X_2^2}\right]\Big/\left[2RT\left(\frac{1}{X_1^3}+\frac{1}{X_2^3}\right)+\frac{\partial^4 G^E}{\partial X_2^4}\right]\right\}_{\substack{X_2=X_2^C\\ T=T_C}}(T_C-T) \tag{79}$$

This estimate is obtained by approximating the Gibbs free energy of the solution in the vicinity of the critical point by the leading terms of its Taylor series expansion with respect to temperature and composition. The values of X_2^α and X_2^β thus obtained become initial points for the Newton–Raphson iteration technique. They determine more accurate values of the equilibrium compositions at the same temperature. These are the starting points of the direct method, which then performs the bulk of the calculation of the miscibility gap's boundary.

8.3.3. Congruent Transformation of a Compound

At the congruent transformation point of a stoichiometric compound β (see Fig. 23), the first order derivative dT/dX_2^α is equal to zero. The second order derivative may be calculated by differentiating equation (50), or through equations (52) and (57). The result is

$$\left(\frac{d^2T}{d(X_2^\alpha)^2}\right)_{T=T^{(\alpha\to\beta)}}=\frac{\psi_\alpha R(T^{(\alpha\to\beta)})^2}{X_1^\beta X_2^\beta(\Delta H^\circ - H^{E(\alpha)})}$$

$$=\frac{T^{(\alpha\to\beta)}}{\Delta H^{(\alpha\to\beta)}}\left(\frac{\partial^2 G^{(\alpha)}}{\partial(X_2^\alpha)^2}\right)_{X_2^\alpha=X_2^\beta} \tag{80}$$

where ΔH° is the standard enthalpy of formation of the compound [reaction (47)],

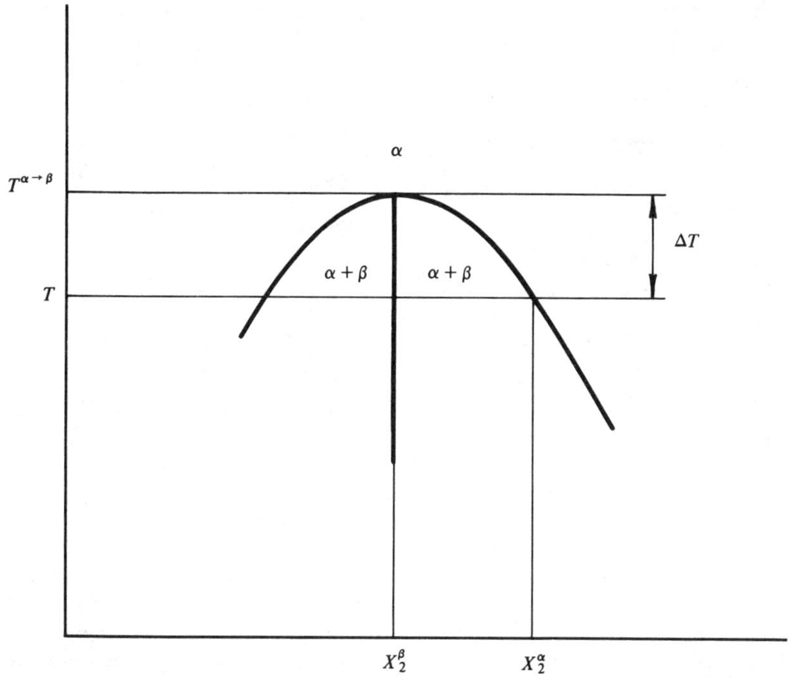

Figure 23. Phase boundaries in the vicinity of the congruent transformation point of a compound.

and $\Delta H^{(\alpha\to\beta)}$ is the enthalpy of transformation of α to β (from a mixture α of the same composition as β). In the vicinity of the congruent transformation point (i.e., neglecting terms of the third order in $X_2^\alpha - X_2^\beta$),

$$DT = T^{(\alpha\to\beta)} - T = -\frac{1}{2}\left(\frac{d^2T}{d(X_2^\alpha)^2}\right)_{T=T^{(\alpha\to\beta)}}(X_2^\alpha - X_2^\beta)^2 \tag{81}$$

or

$$(X_2^\alpha - X_2^\beta)^2 = -2\frac{DT}{T^{(\alpha\to\beta)}}\frac{\Delta H^{(\alpha\to\beta)}}{[\partial^2 G^{(\alpha)}/\partial(X_2^\alpha)^2]_{X_2^\alpha=X_2^\beta}} \tag{82a}$$

which may also be written

$$(X_2^\alpha - X_2^\beta)^2 = -\frac{2DT\Delta S^{(\alpha\to\beta)}}{[\partial^2 G^{(\alpha)}/\partial(X_2^\alpha)^2]_{X_2^\alpha=X_2^\beta}} \tag{82b}$$

As in the case of a miscibility gap near its critical point, the value of X_2^α obtained by equation (82) for an arbitrary DT becomes the starting point of the Newton–Raphson iteration technique, which in turn, determines the starting point of the direct method.

8.4. Example of Application of the Numerical Techniques

An example of a phase diagram calculation is shown in Fig. 24 for the Al–Zn system. This system exhibits a eutectic, a eutectoid, and a miscibility gap. From an analysis of various

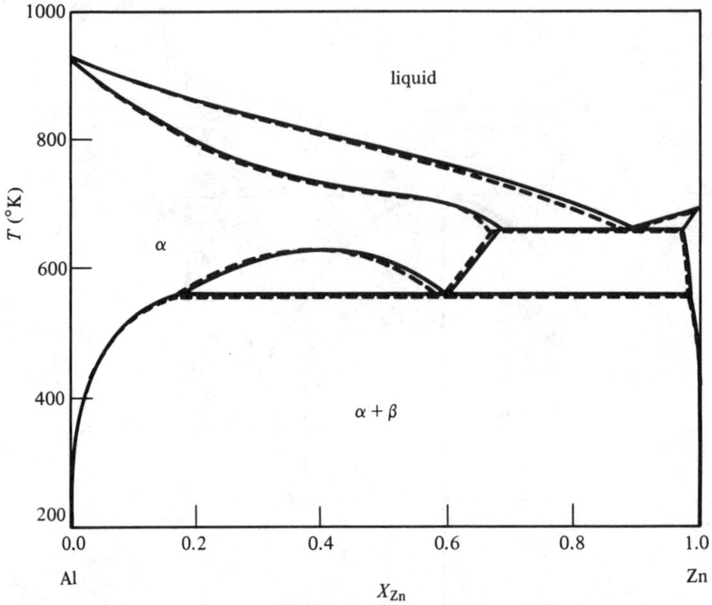

Figure 24. Phase diagram of the aluminum–zinc system. The dashed line is reported by Hultgren et al. [1] and the continuous line is calculated and plotted by computer [8].

data the following parameters were chosen for the liquid, α, and β phases [10]:

$$G^{E(l)} = X_{Al}X_{Zn}3250[1 - (T/1625)] \tag{83}$$

$$G^{E(\alpha)} = X_{Al}X_{Zn}(3150X_{Al} + 2300X_{Zn})[1 - (T/4000)] \tag{84}$$

$$G^{E(\beta)} = X_{Al}X_{Zn}5200[1 - (T/4000)] \tag{85}$$

In addition, the $\Delta\mu^{\circ(\alpha\to\beta)}$ of each component was estimated as

$$\Delta\mu_{Al}^{\circ(\alpha\to\beta)} = 120 \tag{86}$$

$$\Delta\mu_{Zn}^{\circ(\alpha\to\beta)} = -760 + 0.7T \tag{87}$$

Better estimates may now be available [11]; however, because of the small composition range where β is stable, the phase boundaries are not very sensitive to changes in the estimates of equations (86) and (87). The calculated diagram agrees with that proposed by Hansen and Anderko [12] on the basis of experimental determinations within the range of experimental accuracy. Total computer time for the calculations on Carnegie-Mellon University's Univac 1108 computer was of the order of 10 sec [9] (but it should be remembered in evaluating calculation times that different computers work at different speeds).

8.5. Calculation of the Thermodynamic Parameters of a Phase

The numerical calculations we have discussed so far concerned the compositions of two phases in mutual equilibrium assuming that their thermodynamic properties (e.g., G^M or H^M) are known. Often the converse problem is of equal importance: The phase boundaries are known, but the thermodynamic properties of the phases are not. These may be deduced as follows.

Consider the system of equations (37). If we were to assume that both phases α and β are regular, we would have only two parameters to determine, Ω^α and Ω^β. A pair of equilibrium compositions at a single temperature T would yield two linear equations in Ω^α and Ω^β [since $\ln \gamma_i = \Omega/RT(1 - X_i)^2$], and the values of these two parameters could be immediately deduced.

Equations (37) may also be solved at other temperatures. If they yield comparable values of Ω^α and Ω^β, the assumption of regular solutions for the phases α and β is then adequate. If the values of Ω^α and Ω^β are not comparable, the assumption is inadequate and more parameters are generally necessary to describe the thermodynamic properties of the phases [13].

One may try the assumption of quasi-regular solutions, subregular solutions, more general polynomial forms, or many other formalisms (see Section VII.2). Almost any formalism, with as many parameters as desired, can be used by solving equations (37) at an adequate number of selected temperatures. The significance of the results, however, should be questioned.

In general, one cannot know a priori the number of parameters necessary to describe the properties of a phase. The methods of statistical analysis are necessary here. For example, a minimum number of parameters are entered first, and their values are determined by a multiple regression analysis; the calculated results can then be compared with the initial data by the statistical measure of the level of significance of the fit. If this level is judged too low, an additional parameter is entered and the procedure repeated. Calculations are stopped when the fit is considered adequate.

Obviously, if the boundaries of a phase are known only within a small range of temperatures and compositions (e.g., because that is the only range where it is stable), it will not be possible, in general, to deduce the properties of that phase at substantially different temperatures and compositions with much confidence. Extrapolations must be recognized as educated guesses at best.

It must also be noted that quite often an investigator has at his disposal fragmentary information of both the phase boundaries and the properties of single phases (e.g., activity data) and that to this information is attached an experimental error. For example, in applying equations (37) X_2^α and X_2^β can be either unknowns or data points within a certain concentration interval. The problem is to deduce the parameters which best describe the information at hand and allow reasonable interpolations and extrapolations of the phase boundaries. In its general form the problem is quite complex, but a number of powerful techniques, such as the *simplex* method [14], may be adopted to solve it.

Problems

1. System A–B is assumed to behave ideally in both its liquid and solid phases. A melts at 1500°K with a heat of fusion equal to 3500 cal/mol; $C_{p,A}^{(l)} - C_{p,A}^{(s)}$ is approximately constant and equal to 1.1 cal/°K mol. B melts at 2300°K with an unknown heat of fusion; $C_{p,B}^{(l)}$ is assumed to be equal to $C_{p,B}^{(s)}$. In cooling a liquid solution of composition $X_B = 0.22$, the first solid crystals appear at a temperature equal to 1700°K. Calculate the heat of fusion of B consistent with these data and assumptions.

2. A and B form a eutectic. The melting points of A and B are 1400 and 1000°K and their heats of fusion are 3200 and 2500 cal/mol. At 1200°K the solubility of B in solid A is $X_B^s = 0.05$, and in liquid A it is $X_B^l = 0.40$. Estimate the activities of A and B

at the concentration $X_B^l = 0.70$ and the temperatures 1200 and 1600°K. Justify your estimates.

3. The Au–Ni phase diagram is shown in Fig. 8. The thermodynamic properties of the pure components are for Au, $T_f = 1336°K$, $\Delta H_f^\circ = 2995$ cal/mol, $\Delta C_p^{(s\to l)} = 1.34 - 1.24 \times 10^{-3}T$ cal/°K mol; and for Ni, $T_f = 1726°K$, $\Delta H_f^\circ = 4210$ cal/mol, $\Delta C_p^{(s\to l)} = 3.20 - 1.80 \times 10^{-3}T$ cal/°K mol. From activity data in the solid state and the coordinates of the critical point in the miscibility gap, the following estimate of the Gibbs free energy of the solid phase has been derived:

$$G_m^{E(s)} = X_{Au}X_{Ni}(5770X_{Au} + 9150X_{Ni} - 3400X_{Au}X_{Ni})[1 - (T/2660)] \quad \text{cal}$$

It is assumed that the excess Gibbs free energy of the liquid solution can be approximated by

$$G_m^{E(l)} = X_{Au}X_{Ni}(\alpha X_{Au} + \beta X_{Ni})$$

a. Sketch the free energy curves of the solid and liquid phases at 1000°K and 1300°K. Sketch also the activity of Ni across the composition range at the same temperatures.
b. Estimate the parameters α and β from the position of the minimum in the solidus and liquidus lines ($X_{Ni} = 0.42$, $T = 1223°K$).

4. A and B have negligible mutual solid solubilities in the solid state and their phase diagram shows a eutectic transformation. The liquid phase is assumed to be quasi-regular, and at 1 atm it is described by the equation

$$G_m^{E(l)} = X_A X_B 500[1 - (T/3000)] \quad \text{cal}$$

The pure components have the following characteristics: for A,

$$T_f = 1500°K, \quad \Delta H_f^\circ = 3600 \text{ cal/mol}, \quad \Delta C_p^{\circ(s\to l)} = 0$$

and for B,

$$T_f = 1200°K, \quad \Delta H_f^\circ = 2650 \text{ cal/mol}, \quad \Delta C_p^{\circ(s\to l)} = 0$$

a. Plot the phase diagram of the A–B system at 1 atm. Be quantitative, especially in your determination of the eutectic point.
b. Recalculate the phase diagram at a pressure of 10^4 atm. Assume that the molar volumes of solids A and B at their melting points are equal to 8.6 and 9.5 cm³/mol, respectively. In the liquid state $V_m^{(l)} = 9X_A + 10X_B + 0.5X_AX_B$ cm³. Neglect the effects of the coefficients of expansion and compressibility.

5. **a.** The Na–Rb phase diagram exhibits a eutectic transformation and negligible solid solubilities [1]. The eutectic point is at $T = 268.65°K$, $X_{Rb} = 0.821$. Assuming a subregular solution [equation (VII.41)], calculate the coefficients of that formalism.
b. At 320°K, it is determined that the composition of the liquidus is $X_{Rb} = 0.55$. Is this determination consistent with the formalism chosen above? Can you reconcile the calculations through the τ parameter, i.e., by writing $G_m^{E(l)} = X_1X_2(A_{21}X_1 + A_{12}X_2)[1 - (T/\tau)]$?

For Na

$$T_f = 371.0°K, \quad \Delta H_f^\circ = 621 \text{ cal/mol}, \quad \Delta C_p^{(s\to l)} \simeq 0$$

For Rb

$$T_f = 312.64°K, \quad \Delta H_f^\circ = 524 \text{ cal/mol}, \quad \Delta C_p^{(s\to l)} \simeq 0$$

6. The Cu–Pb phase diagram shows a monotectic transformation [1]. The miscibility gap in the liquid phase has a critical point at 1253°K and $X_{Pb} \simeq 0.35$. At the monotectic temperature 1227°K the boundaries of the miscibility gap are at $X_{Pb} = 0.147$ and $X_{Pb} \simeq 0.67$. There are practically no mutual solid solubilities. To represent the thermodynamic properties of the liquid phase, the following formalism is proposed [equation (VII.42)]: $G_m^{E(l)} = X_1X_2(A_{21}X_1 + A_{12}X_2 + A_{22}X_1X_2)$, with Cu ≡ 1 and Pb ≡ 2.
   a. From the asymmetry of the miscibility gap, deduce the parameter A_{22} [using equations (III.64) or (III.70)].
   b. Calculate the values of the other two parameters A_{21} and A_{12} from the coordinates of the critical point.
   c. Plot as precisely as you can the Gibbs free energy curves of the solid and liquid phases at 1227°K.

7. The Co–Cu phase diagram shows a peritectic transformation. The solid phase exhibits a miscibility gap the boundaries of which have the compositions $X_{Cu} = 0.164$ and $X_{Cu} \simeq 0.91$ at the peritectic temperature 1385°K [1]. Assume a subregular solution for the solid phase and calculate the parameters of that formalism [equation (VII.41); in equation (VII.44) and (45), $A_{22} = 0$].

   From this determination and the assumption of that formalism, deduce the composition and temperature of the metastable critical point. Are these results compatible with the phase diagram described by Hultgren et al. [1]?

8. Derive equation (80).

9. The phase diagram of the Ga–Sb system is shown in Fig. 20. The liquid phase may be assumed regular; its molar enthalpy of mixing is equal to −250 cal at $X_{Ga} = X_{Sb} = 0.5$. The heat of fusion of Sb is equal to 4690 cal/mol.
   a. Sketch a plot of the free energy curves vs X_{Sb} at $T = 900$°K.
   b. Draw, for the same temperature of 900°K, a plot of the activity of Sb vs X_{Sb}, using pure liquid Sb as the standard state. Be quantitative.
   c. Estimate the $\Delta G°$ at 985°K of the following reaction:

      Ga(l) + Sb(l) = GaSb(β)

      (i.e., the standard Gibbs free energy of formation of GaSb).

10. The behavior of carbon in liquid iron at 1560°C has been characterized by the following equations: $\ln \gamma_C = -0.37 + 7X_C + 11.7X_C^2$ and $H_C^E = 5.4 + 15.0X_C + 17.25X_C^2$ kcal, for which the standard state of carbon is pure graphite. Calculate the slope of the liquid–graphite boundary at 1560°C, and compare it with that shown in Fig. 19.

11. Assume that B has negligible solubility in solid A and calculate as concisely as you can the slope of the liquidus of the A–B system at $X_B = 0$. Compare your result to the slopes shown in six such phase diagrams found in the literature [e.g., 1]. As far as possible, choose components A with quite different temperatures and entropies of fusion.

References

1. R. Hultgren, P. D. Desai, D. T. Hawkins, M. Gleiser, and K. K. Kelley, *Selected Values of the Thermodynamic Properties of Binary Alloys*. Am. Soc. Metals, Metals Park, OH, 1973.
2. C. D. Thurmond, *J. Phys. Chem.* **57**, 827 (1953).

3. O. Kubaschewski, *Trans. Brit. Cer. Soc.* **60,** 67 (1961).
4. P. Duwez, private communication (1963) to W. B. Pearson, *Handbook of Lattice Spacings and Structures of Metals*. Pergamon, 1967, Vol. 2.
5. P. S. Rudman, *Thermodynamic Analysis and Synthesis of Phase Diagrams, I, Advances in Materials Research*. Interscience, New York, 1969, Vol. 4.
6. L. Kaufman and H. Bernstein, *Computer Calculations of Phase Diagrams*. Academic, New York, 1970.
7. C. H. P. Lupis and H. Gaye, *Proceedings of a Symposium Held at Brunel University and the National Physical Laboratory, July 1971* (O. Kubaschewski, ed.). Her Majesty's Stationery Office, London, 1972, pp. 469–482.
8. H. Gaye and C. H. P. Lupis, *Scr. Met.* **4,** 685–692 (1970).
9. H. Gaye and C. H. P. Lupis, *Met. Trans. A* **6,** 1049–1056 (1975).
10. H. Gaye, Ph.D. thesis, Carnegie-Mellon University, 1971.
11. L. Kaufman, *Proceedings of a Symposium Held at Brunel University and the National Physical Laboratory, July 1971* (O. Kubaschewski, ed.). Her Majesty's Stationery Office, London, 1972, pp. 373–402.
12. M. Hansen and K. Anderko, *Constitution of Binary Alloys*. McGraw-Hill, New York, 1958.
13. R. Hiskes and W. A. Tiller, *Mater. Sci. Eng.* **2,** 320–330 (1967–1968); **4,** 163–184 (1969).
14. D. J. Wilde, *Optimum Seeking Methods*. Prentice-Hall, Englewood Cliffs, NJ, 1964.

Selected Bibliography

Texts

P. Gordon, *Principles of Phase Diagrams in Materials Systems*. McGraw-Hill, New York, 1968.

A. Prince, *Alloy Phase Equilibria*. Elsevier, New York, 1966.

F. N. Rhines, *Phase Diagrams in Metallurgy*. McGraw-Hill, New York, 1956.

A. M. Alper, ed., *Phase Diagrams: Materials Science and Technology. Vol. 1—Theory, Principles and Techniques of Phase Diagrams*. Academic, New York, 1970.

W. Hume-Rothery, J. W. Christian, and W. B. Pearson, *Metallurgical Equilibria Diagrams*. Institute of Physics, London, 1952.

G. C. Carter, ed., *Applications of Phase Diagrams in Metallurgy and Ceramics*. Nat. Bur. of Standards Special Publ. 496, Washington, DC, 1978, 2 vols. Proceedings of a workshop held at the National Bureau of Standards, Gaithersburg, Maryland, January 10–12, 1977. An excellent source of information for phase diagram compilation activities, experimental methods, and computational techniques.

Film

Phase Diagrams from Free Energy Curves, a 16-mm color movie (of animated sequences) with sound, 22 min, written and produced by C. H. P. Lupis, 1975; distributed (for sale or rent) by Audio-Visual Services of the Pennsylvania State University, University Park, PA 16802. The material in the first sections of this chapter is presented in the film in essentially the same order. Students have usually found it a very helpful learning aid.

Compilations

R. Hultgren, P. D. Desai, D. T. Hawkins, M. Gleiser, K. K. Kelley, and D. D. Wagman, *Selected Values of the Thermodynamic Properties of Binary Alloys*. Am. Soc. Metals, Metals Park, OH, 1973. Provides both activity data and phase diagrams.

M. Hansen and K. Anderko, *Constitution of Binary Alloys,* 2nd ed. McGraw-Hill, New York, 1958.

R. P. Elliott, *Constitution of Binary Alloys, First Supplement.* McGraw-Hill, New York, 1965.

F. A. Shunk, *Constitution of Binary Alloys, Second Supplement.* McGraw-Hill, New York, 1969.

F. A. Shunk, *Constitution of Alloys,* a series of reports continuing the Hansen series. Part 1, May 1968; Part 2, May 1969; Part 3, May 1970; available from the Illinois Institute of Technology, Chicago.

W. G. Moffatt, ed., *Handbook of Binary Phase Diagrams.* Business Growth Services, General Electric Co., Schenectady, NY, 1976. Two volumes, loose-leaf compilation with periodic updating. Complements the Hansen–Elliott–Schunk series.

J. F. Elliott, M. Gleiser, and V. Ramakrishna, *Thermochemistry for Steelmaking.* Addison-Wesley, Reading, MA, 1963, Vol. 2.

E. M. Levin, C. R. Robbins, and H. F. McMurdie, *Phase Diagrams for Ceramists.* Am. Ceram. Soc., Columbus, OH, 1964; 1st suppl., 1969; 2nd suppl., 1975.

IX Analytic Expressions for the Thermodynamic Functions of Dilute Multicomponent Metallic Solutions

In Chapter VII we presented various analytic expressions to represent the composition dependence of various thermodynamic properties of binary solutions. The advantages of such expressions are readily apparent for the storage and retrieval of data, for interpolations and extrapolations (Chapter VII), and for the calculation of phase boundaries (Chapter VIII). Two-dimensional diagrams are inadequate to represent the composition dependence of the properties of multicomponent solutions, and analytic expressions become then indispensable. In this chapter, we shall focus on the formalism of interaction coefficients for dilute solutions. The concept underlying the formalism has already been seen in the case of binary mixtures (see Section VII.1); its convenience should be better appreciated in the case of many components. However, before attacking the subject of interaction coefficients, a generalization of Raoult's and Henry's laws to the case of multicomponent solutions is in order.

1. Raoult's and Henry's Laws for Multicomponent Solutions

For a solution of m components, we shall choose as independent variables the mole fractions $X_2, X_3, \ldots, X_m$ of the solutes 2, 3, . . . , m (reserving the number 1 for the solvent). As for a binary system, a solution is ideal in a given composition range if in that range the activity of one of its components is proportional to its mole fraction. We shall see that because of the Gibbs–Duhem equation, this entails that the activities of the other components are also proportional to their mole fractions. For the solvent 1 of an ideal solution, a multidimensional plot of its activity a_1 vs the mole fractions $X_2, X_3, \ldots, X_m$, yields the hyperplane of equation $a_1 = X_1$ (or $a_1 = 1 - X_2 - X_3 - \cdots - X_m$). For a real solution, Raoult's law is defined by stating that the hyperplane tangent to the activity surface a_1 at the point $X_1 = 1$ (or $X_2 = X_3 = \cdots X_m = 0$) is identical to the hyperplane of ideal mixing. This may be expressed by equations which are similar to equations (VI.47) and (VI.50) for binary mixtures:

$$\left(\frac{\partial \ln a_1}{\partial X_j}\right)_{X_1 \to 1} = -1 \qquad \text{for} \quad j = 2, 3, \ldots, m \tag{1}$$

or

$$(\partial \ln \gamma_1 / \partial X_j)_{X_1 \to 1} = 0 \qquad \text{for} \quad j = 2, 3, \ldots, m \tag{2}$$

Note that each of the partial derivatives in equation (1) or (2) is a binary property (of the binary system $1 - j$). Thus the generalization of Raoult's law to a multicomponent system introduces no new information.

Whereas Raoult's law is necessarily defined at the point $X_1 = 1$, the concept of Henry's law should apply to the behavior of a very dilute element in a solution of any arbitrary composition with respect to the other elements. In other words, for the element i Henry's law is defined at any given point α' where X_i and a_i are zero (see Fig. 1). The law may be expressed by the following equations:

$$(\partial a_i / \partial X_i)_{\text{at } \alpha'} = \text{finite value different from zero} \tag{3a}$$

and

$$(\partial a_i / \partial X_j)_{\text{at } \alpha'} = 0 \qquad \text{for} \quad j = 2, 3, \ldots, m, \quad j \neq i \tag{3b}$$

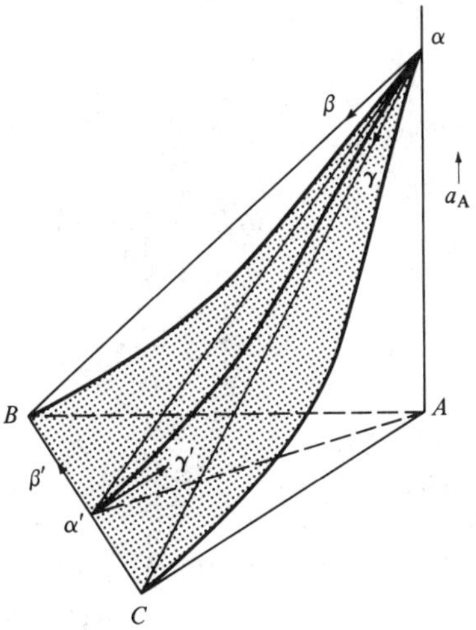

Figure 1. Illustration of Raoult's and Henry's laws for a ternary system. Raoult's law is defined at the point α by the tangent plane αβγ, which is identical to the plane of ideal mixing α*BC*. Henry's law is defined at any point α′ on *BC* by the tangent plane α′β′γ′, which contains *BC* and is neither horizontal nor vertical.

In contrast to the case of Raoult's law, the validity of Henry's law in the various binaries i–1, i–j, does not ensure its validity in the multicomponent system 1–i–···–j. A geometric interpretation of both Raoult's law and Henry's law for a ternary system is given in Fig. 1.

A necessary consequence of equations (3a) and (3b) is

$$\left(X_i \frac{\partial \ln \gamma_i}{\partial X_j}\right)_{\text{at } \alpha'} = 0 \qquad \text{for} \quad j = 2, 3, \ldots, i, \ldots, m \tag{4}$$

Writing the Gibbs–Duhem equation

$$X_1\, d \ln \gamma_1 + \sum_{i=2}^{m} X_i\, d \ln \gamma_i = 0 \tag{5}$$

it is clear that if the solutes 2, 3, . . . , m obey Henry's law at the point $X_1 = 1$, then Raoult's law is obeyed by the solvent 1. The special case of *Henry's first order law* adds the restriction that each derivative $(\partial \ln \gamma_i/\partial X_j)_{\text{at } \alpha'}$ is finite.

2. Dilute Ternary Solutions

For reasons of clarity, before attacking the case of multicomponent solutions, we shall restrict our presentation of interaction coefficients to ternary solutions.

2.1. Approximation of a Series by a Polynomial

As in the case of binary systems (Section VII.1), polynomials are the most widely used analytic functions to represent integral or partial molar excess properties. Again, for dilute solutions, the theoretical justification of these polynomials may be considered to be the assumed existence of a Taylor series. Expanded at the point of infinite dilution, the series may be written in terms of the mole fractions of the solutes X_2 and X_3:

$$Y = (Y)_{X_1\to 1} + \left(\frac{\partial Y}{\partial X_2}\right)_{X_1\to 1} X_2 + \left(\frac{\partial Y}{\partial X_3}\right)_{X_1\to 1} X_3$$
$$+ \frac{1}{2!}\left[\left(\frac{\partial^2 Y}{\partial X_2^2}\right)_{X_1\to 1} X_2^2 + 2\left(\frac{\partial^2 Y}{\partial X_2\,\partial X_3}\right)_{X_1\to 1} X_2X_3 + \left(\frac{\partial^2 Y}{\partial X_3^2}\right)_{X_1\to 1} X_3^2\right]$$
$$+ \cdots + \frac{1}{n!}\left[\left(\frac{\partial Y}{\partial X_2}\right)_{X_1\to 1} X_2 + \left(\frac{\partial Y}{\partial X_3}\right)_{X_1\to 1} X_3\right]^{(n)} + \cdots \tag{6}$$

where the symbolic power notation means

$$\left[\left(\frac{\partial Y}{\partial X_2}\right)_{X_1\to 1} X_2 + \left(\frac{\partial Y}{\partial X_3}\right)_{X_1\to 1} X_3\right]^{(n)} = \left(\frac{\partial^n Y}{\partial X_2^n}\right)_{X_1\to 1} X_2^n + \cdots$$
$$+ \frac{n!}{(n-i)!i!}\left(\frac{\partial^n Y}{\partial X_2^{n-i}\,\partial X_3^i}\right)_{X_1\to 1} X_2^{n-i}X_3^i + \cdots + \left(\frac{\partial^n Y}{\partial X_3^n}\right)_{X_1\to 1} X_3^n \tag{7}$$

The series may also be written with a remainder:

$$Y = (Y)_{X_1\to 1} + \sum_{k=1}^{n} \frac{1}{k!}\left[\left(\frac{\partial Y}{\partial X_2}\right)_{X_1\to 1} X_2 + \left(\frac{\partial Y}{\partial X_3}\right)_{X_1\to 1} X_3\right]^{(k)} + R_n \tag{8}$$

where

$$R_n = \frac{1}{(n+1)!}\left[\left(\frac{\partial Y}{\partial X_2}\right)_{\xi X_2, \xi X_3} X_2 + \left(\frac{\partial Y}{\partial X_3}\right)_{\xi X_2, \xi X_3} X_3\right]^{(n+1)} \tag{9}$$

and ξ is a number between 0 and 1.

If the estimated experimental error on Y is ΔY, the remainder R_n in equation (8) may be neglected for a value of n such that

$$R_n \leq \Delta Y \tag{10}$$

For a fixed value of n (for instance, for an approximation by a polynomial of degree $n = 2$), R_n decreases when the dilution of the solution increases. R_n also depends on the magnitudes of the derivatives of Y, and these are a reflection of the strength of the interactions of the elements in the solution. *Thus, for comparable experimental errors, polynomials of higher order are warranted by solutions of lower dilution and higher departures from ideality.*

2.2. *Application to* ln γ_2; *Free Energy Interaction Coefficients*

It should be noted at the start that the assumption of an analytic form for any thermodynamic property establishes definite analytic forms for other properties through the use of relevant thermodynamic relationships. For instance, adopting an analytic form for the activity coefficient of one solute yields, through the Gibbs–Duhem equation, analytic forms for the activity coefficients of the solvent and the other solutes. In addition, the relationship between integral and partial molar quantities yields the Gibbs free energy of the solution, and relevant derivatives with respect to temperature and pressure yield other properties such as entropy, enthalpy, and volume. Directing the analytic study to the behavior of a solute rather than to the behavior of the solvent, or of the solution, is justified by the fact that generally it is the solute which is under direct experimental observation. Indeed, a large fraction of our most accurate data concerns the study of metalloids (C, O, S, N, . . .) as solutes. This is partly because the control of their concentrations is industrially important, and partly because their activities are relatively easy to measure by equilibrating the alloys with suitable gas phases (H_2–CH_4, CO–CO_2, H_2–H_2O, H_2–H_2S, N_2, . . .).

Identifying the function Y in Section 2.1 with $\ln \gamma_2$ implies the existence of a Taylor series for $\ln \gamma_2$, that is, the existence of all the partial derivatives of $\ln \gamma_2$ with respect to X_2 and X_3. The previous section has shown that the existence of $\ln \gamma_2^\infty$ (which is the derivative of order zero) is equivalent to the applicability of Henry's zeroth order law. Similarly, it has been noted that for metallic solutions, Henry's first order law (establishing the existence of the derivative of order 1) is generally obeyed. The assumption of a Taylor series for $\ln \gamma_2$ can therefore be viewed as equivalent to the assumption of obedience to Henry's laws of all orders (zero, first, and nth), that is, to a generalized Henry's law at the composition $X_1 = 1$.

The infinite series representation of $\ln \gamma_2$ takes the form

$$G_2^E/RT = \ln \gamma_2 = \sum_{n_2, n_3 = 0}^{\infty} J_{n_2, n_3}^{(2)} X_2^{n_2} X_3^{n_3} \tag{11}$$

or

$$\ln \gamma_2 = J_{0,0}^{(2)} + J_{1,0}^{(2)} X_2 + J_{0,1}^{(2)} X_3 + J_{2,0}^{(2)} X_2^2 + J_{1,1}^{(2)} X_2 X_3 + J_{0,2}^{(2)} X_3^2 + \cdots \tag{12}$$

The coefficient $J_{0,0}^{(2)}$ is the value of $\ln \gamma_2$ at infinite dilution of all solutes: $\ln \gamma_2^\infty$. In calculating the derivative,

$$J_{1,0}^{(2)} = (\partial \ln \gamma_2 / \partial X_2)_{X_1 \to 1} \tag{13}$$

X_3 is being kept equal to zero, and thus $J_{1,0}^{(2)}$ is a binary coefficient. It is recognized as the self-interaction coefficient $\epsilon_2^{(2)}$ (see Section VII.1). The coefficient

$$J_{0,1}^{(2)} = (\partial \ln \gamma_2 / \partial X_3)_{X_1 \to 1} \tag{14}$$

is a ternary quantity and a measure of how the presence of 3 affects the behavior of 2. It is designated by $\epsilon_2^{(3)}$ (read "epsilon 3 on 2").

$$\epsilon_2^{(3)} = (\partial \ln \gamma_2 / \partial X_3)_{X_1 \to 1} \tag{15}$$

By far the most important of the interaction coefficients, it was introduced by Wagner [1] and extensively used by Chipman and co-workers [2]. It should be noted that neither the subscript nor the superscript of ϵ can refer to the solvent 1. The latter is implied as the medium in which the interaction 2–3 is operating. Of course, this interaction will be different in different solvents.

The coefficients of the second order terms are $J_{2,0}^{(2)}$, $J_{1,1}^{(2)}$, and $J_{0,2}^{(2)}$ (for each, the sum of the subscripts is equal to 2). Introduced by Lupis and Elliott [3,4], they have been designated by the symbol ρ. $J_{2,0}^{(2)}$ is the binary interaction coefficient $\rho_2^{(2)}$, and $J_{1,1}^{(2)}$ and $J_{0,2}^{(2)}$ are ternary interaction coefficients designated respectively by $\rho_2^{(2,3)}$ and $\rho_2^{(3)}$. Thus, the series expression of $\ln \gamma_2$ may be written

$$\ln \gamma_2 = \ln \gamma_2^\infty + \epsilon_2^{(2)} X_2 + \epsilon_2^{(3)} X_3 + \rho_2^{(2)} X_2^2 + \rho_2^{(2,3)} X_2 X_3 + \rho_2^{(3)} X_3^2 + \cdots \tag{16}$$

In a similar way, for solute 3,

$$G_3^E/RT = \ln \gamma_3 = \sum_{n_2,n_3=0}^{\infty} J_{n_2,n_3}^{(3)} X_2^{n_2} X_3^{n_3} \quad (17)$$

or

$$\ln \gamma_3 = \ln \gamma_3^\infty + \epsilon_3^{(2)} X_2 + \epsilon_3^{(3)} X_3 + \rho_3^{(2)} X_2^2 + \rho_3^{(2,3)} X_2 X_3 + \rho_3^{(3)} X_3^2 + \cdots \quad (18)$$

Note that in a coefficient J, the subscripts refer to the elements which are "added" and the superscript refers to the solute the behavior of which is being analyzed. By contrast, in the ϵs and ρs, the subscripts and superscripts play the inverse role. The first convention is more practical than the second when dealing with high order interaction coefficients (and, we shall see, in the case of several solutes) but less practical for the usual case of coefficients of order 1 and 2 (ϵ and ρ).

$\ln \gamma_1$ and G^E may also be expanded into Taylor series:

$$G_1^E/RT = \ln \gamma_1 = \sum_{n_2,n_3=0}^{\infty} J_{n_2,n_3}^{(1)} X_2^{n_2} X_3^{n_3} \quad (19)$$

$$G_m^E/RT = \sum_{n_2,n_3=0}^{\infty} \phi_{n_2,n_3} X_2^{n_2} X_3^{n_3} \quad (20)$$

Their coefficients are not called interaction coefficients, although simply related to them through the Gibbs–Duhem equation. For instance, it will be shown that equation (20) may be rewritten

$$G_m^E = RT(X_2 \ln \gamma_2^\infty + X_3 \ln \gamma_3^\infty + \tfrac{1}{2}\epsilon_2^{(2)} X_2^2 + \epsilon_2^{(3)} X_2 X_3 + \tfrac{1}{2}\epsilon_3^{(3)} X_3^2 + \cdots) \quad (21)$$

2.3. *Enthalpy and Entropy Interaction Coefficients*

As for the free energy functions, enthalpy and entropy functions can also be expanded in Taylor series. For example,

$$H_2^E = \sum_{n_2,n_3}^{\infty} L_{n_2,n_3}^{(2)} X_2^{n_2} X_3^{n_3} = H_2^{E\infty} + \eta_2^{(2)} X_2 + \eta_2^{(3)} X_3 + \cdots \quad (22)$$

and

$$S_2^E = \sum_{n_2,n_3}^{\infty} K_{n_2,n_3}^{(2)} X_2^{n_2} X_3^{n_3} = S_2^{E\infty} + \sigma_2^{(2)} X_2 + \sigma_2^{(3)} X_3 + \cdots \quad (23)$$

Among these enthalpy and entropy interaction coefficients, the first order ones, η and σ, are the most useful. We stress the definitions of $\eta_2^{(3)}$ and $\sigma_2^{(3)}$:

$$\eta_2^{(3)} = (\partial H_2^E/\partial X_3)_{X_1 \to 1} \quad (24)$$

$$\sigma_2^{(3)} = (\partial S_2^E/\partial X_3)_{X_1 \to 1} \quad (25)$$

There are relations between free energy, enthalpy, and entropy interaction coefficients. Since

$$\ln \gamma_2 = G_2^E/RT = (H_2^E/RT) - (S_2^E/R) \quad (26)$$

then

$$J^{(2)}_{n_2,n_3} = (L^{(2)}_{n_2,n_3}/RT) - (K^{(2)}_{n_2,n_3}/R) \tag{27}$$

and in particular,

$$\epsilon_2^{(3)} = (\eta_2^{(3)}/RT) - (\sigma_2^{(3)}/R) \tag{28}$$

η is generally obtained either through the temperature dependence of ϵ, or by calorimetric measurements; σ is generally obtained as the difference between ϵ and η/RT [equation (28)], and as such, often bears a large uncertainty.

2.4. Qualitative Atomistic Interpretation of the Interaction Coefficients

Coefficients such as $\ln \gamma_2^\infty$, $\epsilon_2^{(3)}$, and $\rho_2^{(3)}$ may be given a simple atomistic interpretation through the *central atoms model* [5] of a metallic solution. A quantitative evaluation of these coefficients by the model will be developed in Chapters XV and XVI. In this section, the presentation will be limited to a very qualitative interpretation of the results; it should be helpful in giving a more intuitive understanding of the significance of these coefficients, which so far have been presented as an abstract mathematical convenience.

The model assumes that the forces between atoms are short range and neglects the effect of non-nearest neighbors. This assumption of short range forces appears realistic for metallic solutions and whether one, two, or three nearest neighbor shells around an atom have to be considered does not change the overall qualitative features of the model. The properties of the solution are obtained by summing the contributions of each atom in the solution.

Let us designate by Z the coordination number, that is, the number of atoms in the nearest neighbor shell. It is possible to show that $\ln \gamma_B^\infty$ compares essentially the energy of an atom B surrounded by Z atoms A to the energy of an atom B surrounded by Z atoms B (i.e., B when it is pure). Choosing now a reference energy state for any central atom corresponding to a nearest-neighbor shell of all A atoms, the model also shows that ϵ_B^B analyzes the contribution of a configuration which has a central atom B surrounded by an atom B and $Z - 1$ atoms A, and that ϵ_B^C analyzes the configuration of an atom B surrounded by one atom C and $Z - 1$ atoms A. Physically, this last configuration implies another configuration in which the surrounding atoms of the central atom C contain at least one B atom (see Fig. 2). Consequently, this suggests a relation between ϵ_B^C and ϵ_C^B; indeed, we shall see that

$$\epsilon_C^B = \epsilon_B^C \tag{29}$$

$\rho_B^{(B,C)}$ describes the contribution of an atom B surrounded by an atom B, an atom C, and $Z - 2$ atoms A. Some of these configurations (see Fig. 2) entail the presence of an atom C surrounded by 2 atoms B and $Z - 2$ atoms A, a situation described by $\rho_C^{(B)}$. Again, we shall establish that as a consequence of the Gibbs–Duhem equation there is a relation between $\rho_B^{(B,C)}$ and $\rho_C^{(B)}$:

$$\rho_B^{(B,C)} + \epsilon_C^B = 2\rho_C^{(B)} + \epsilon_B^B \tag{30}$$

The probability of a configuration of i atoms B, j atoms C, and $Z - i - j$ atoms A around an atom A, B, or C, is a function of the concentrations X_B and X_C. For an ideal solution, this probability is equal to $[Z!/(Z - i - j)!i!j!]X_A^{(Z-i-j)}X_B^iX_C^j$; for nonideal solutions, this expression generally represents a correct order of magnitude. Thus, for very dilute solutions the probability of nearest-neighbor shells with several solute atoms B or C is

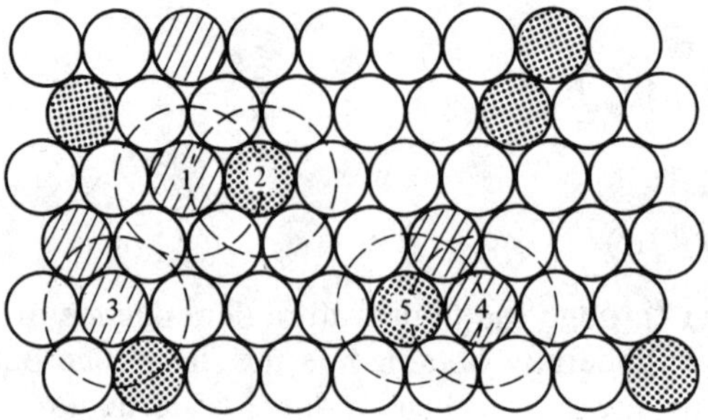

Figure 2. Illustration of various atomic configurations for the central atoms model. Configuration 1 corresponds to ϵ_B^C, 2 to ϵ_C^B, 3 and 4 to $\rho_B^{B,C}$, and 5 to ρ_C^B. Open circles, A; hatched circles, B; dotted circles, C.

negligible and the corresponding interaction coefficients which describe those configurations should have a negligible weight in the thermodynamic expression of the properties of the solution.

Consider, for example, ϵ_B^C. The probability that an atom B has a C atom as a neighbor is proportional to X_C. Correspondingly, in a property of B such as $\ln \gamma_B$, ϵ_B^C is weighted by X_C; i.e., it is the product $X_C \epsilon_B^C$ which appears in the expansion of $\ln \gamma_B$. However, if we consider a function such as G^E, we are no longer focusing on any kind of atom in particular, but on a sum of all these atoms. Therefore, the probability of the configuration of 1 C together with $Z - 1$ A must be multiplied by the probability of finding a B atom in its center. The latter is proportional to X_B. Consequently, ϵ_B^C should be expected to appear first in the development of G^E as a term in $X_B X_C$ [see equation (21)]. This intuitive reasoning is easily extended to other coefficients.

Note that in very dilute solutions the configurations which have B and C atoms as nearest neighbors have a very small probability. Practically, then, an atom B never interacts with an atom C and only configurations of a binary type, all As with one B, or one C, have to be considered. The solution may then be studied as a sum of two binaries, according to the formalism of a very dilute solution:

$$G^E \simeq (RTX_B \ln \gamma_B^\infty) + (RTX_C \ln \gamma_C^\infty) \tag{31}$$

If the concentrations of B and C are increased (e.g., X_B and $X_C \simeq 0.05$), then the probabilities of having two solute atoms appear in the same configuration may become sizable and, correspondingly, the terms in ϵ_B^B, ϵ_C^C, and ϵ_B^C may have to be taken into account—but not necessarily those in ρ. This argument may be extended to solutions with a large number of components and it will be seen that for many systems of metallurgical interest, the alloys can be considered as sums of binaries and ternaries, occasionally quaternaries. It is this very feature of the formalism which greatly simplifies the study of dilute solutions and explains its success.

It is important to recall our assumption of short-range atomic interactions. For, if the forces around a central atom are long range, many shells would have to be considered, and possibly the whole solution itself. In that case, each solute atom is always interacting with all the other atoms and the same approach is no longer possible. This occurs for electrolytic solutions, where the forces are ionic. (Ionic potentials decrease with the distance r as $1/r$, whereas metallic potentials decrease roughly as $1/r^6$.) Correspondingly, it is found that the excess properties of an electrolytic solution do not accept Taylor expansion series of the type described above. In particular, Henry's first order law is not obeyed.

In the formalism of interaction coefficients, the ϵ coefficients are by far the most useful and it may be worthwhile to dwell a little longer on the significance of their values. If the couple of atoms B and B, or B and C, have a much stronger attraction for each other than for atoms A, the configurations in which they are present have a relatively low energy and will appear more often than would have been calculated on the basis of a random distribution of atoms. ϵ_B^B and ϵ_B^C are then found to be negative. If the solute atoms strongly repel each other, the opposite is true: Configurations including two solute atoms have a relatively high energy and are less probable than if calculated on the basis of a random distribution; ϵ_B^B and ϵ_B^C are then positive.

On the basis of the same model, enthalpy and entropy interaction coefficients of low order can also be rather simply described. An enthalpy interaction coefficient such as η_B^C analyzes the change in the depth of the "potential well" of the central atom B when surrounded by 1 atom C and $Z - 1$ atoms A on the one hand, and when surrounded by Z atoms A on the other; an entropy interaction coefficient such as σ_B^C analyzes the change in the curvature of the potential well of the central atom B for the same change in the composition of the nearest-neighbor shell.

It may be added that since the interatomic distance between the atoms is practically insensitive to the composition of the solution, a change in the depth of the potential well can be expected to be reflected by a change in its curvature; e.g., if the potential is deeper, it will appear more needlelike (it will have a higher curvature). Consequently, enthalpy and entropy interaction coefficients will tend to be roughly proportional:

$$\eta_B^C = \tau\sigma_B^C \tag{32}$$

τ, the coefficient of proportionality, has the dimension of a temperature and should be independent of the nature of the solutes B and C. Such a correlation was empirically observed by Chipman and Corrigan [6]. Noting that equations (28) and (32) lead to

$$\epsilon_B^C = \frac{\eta_B^C}{R}\left(\frac{1}{T} - \frac{1}{\tau}\right) \tag{33}$$

Lupis and Elliott [7] also regarded τ as the temperature at which the system would become ideal if its properties could be linearly extrapolated from the experimental range of temperatures (see Fig. 3). Indeed, in a large majority of cases, the observed values of τ lie a few hundred degrees higher than the temperatures at which the systems were studied [7].

2.5. Reciprocal Relations Between Interaction Coefficients

In our atomistic interpretation of interaction coefficients, we have been led to expect reciprocal relations between these coefficients. The most important one concerns the equality of $\epsilon_2^{(3)}$ and $\epsilon_3^{(2)}$; at infinite dilution of both solutes, the effect of 3 on the behavior of 2 is identical to the effect of 2 on the behavior of 3. We shall now give two demonstrations of this relation, in order to familiarize the reader with the type of calculations they involve.

2.5.1. First Demonstration

The first demonstration is based on a Maxwell relationship:

$$\left(\frac{\partial\mu_2}{\partial n_3}\right)_{n_1,n_2} = \left(\frac{\partial\mu_3}{\partial n_2}\right)_{n_1,n_3} \tag{34}$$

which may be rewritten

$$\left(\frac{\partial \ln a_2}{\partial n_3}\right)_{n_1,n_2} = \left(\frac{\partial \ln a_3}{\partial n_2}\right)_{n_1,n_3} \tag{35}$$

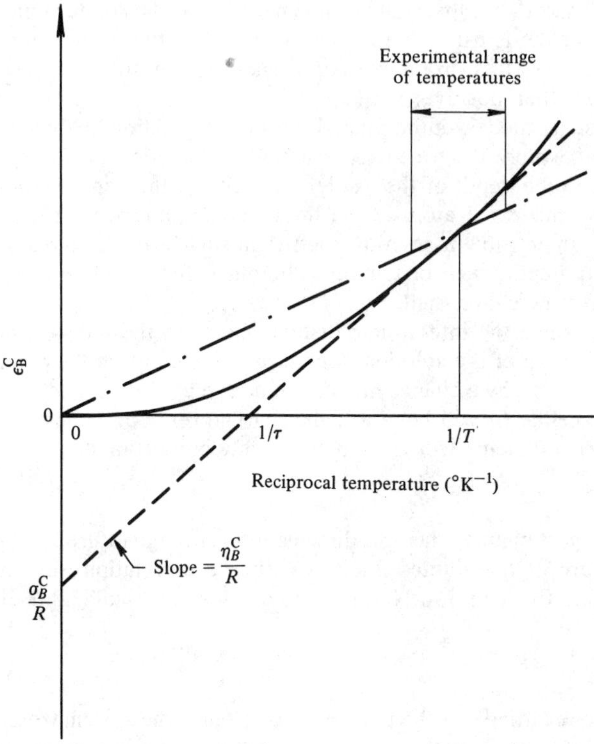

Figure 3. Temperature dependence of the free energy interaction coefficient ϵ_B^C, the τ parameter [7]. ———, proposed curve; – – –, the Chipman and Corrigan line; —·—, regular solution line.

To change coordinates, we note that

$$\left(\frac{\partial \ln a_2}{\partial n_3}\right)_{n_1,n_2} = \left(\frac{\partial \ln a_2}{\partial X_2}\right)\left(\frac{\partial X_2}{\partial n_3}\right)_{n_1,n_2} + \left(\frac{\partial \ln a_2}{\partial X_3}\right)\left(\frac{\partial X_3}{\partial n_3}\right)_{n_1,n_2} \tag{36}$$

and since

$$X_2 = \frac{n_2}{n_1 + n_2 + n_3} = \frac{n_2}{n}, \qquad X_3 = \frac{n_3}{n_1 + n_2 + n_3} = \frac{n_3}{n} \tag{37}$$

then

$$\left(\frac{\partial X_2}{\partial n_3}\right)_{n_1,n_2} = -\frac{X_2}{n}, \qquad \left(\frac{\partial X_3}{\partial n_3}\right)_{n_1,n_2} = \frac{1 - X_3}{n} \tag{38}$$

Consequently,

$$\left(\frac{\partial \ln a_2}{\partial n_3}\right)_{n_1,n_2} = \frac{1}{n}\left[-X_2\left(\frac{\partial \ln a_2}{\partial X_2}\right) + (1 - X_3)\frac{\partial \ln a_2}{\partial X_3}\right]$$

$$= \frac{1}{n}\left[-X_2\left(\frac{1}{X_2} + \frac{\partial \ln \gamma_2}{\partial X_2}\right) + (1 - X_3)\frac{\partial \ln \gamma_2}{\partial X_3}\right]$$

or

$$\left(\frac{\partial \ln a_2}{\partial n_3}\right)_{n_1,n_2} = \frac{1}{n}\left[\left(-1 - X_2\frac{\partial \ln \gamma_2}{\partial X_2}\right) + (1 - X_3)\frac{\partial \ln \gamma_2}{\partial X_3}\right] \tag{39}$$

Similarly,

$$\left(\frac{\partial \ln a_3}{\partial n_2}\right)_{n_1,n_3} = \frac{1}{n}\left[\left(-1 - X_3\frac{\partial \ln \gamma_3}{\partial X_3}\right) + (1 - X_2)\frac{\partial \ln \gamma_3}{\partial X_2}\right] \tag{40}$$

and application of equation (35) leads to

$$-X_2\frac{\partial \ln \gamma_2}{\partial X_2} + (1 - X_3)\frac{\partial \ln \gamma_2}{\partial X_3} = -X_3\frac{\partial \ln \gamma_3}{\partial X_3} + (1 - X_2)\frac{\partial \ln \gamma_3}{\partial X_2} \tag{41}$$

At the limit where $X_2 \to 0$ and $X_3 \to 0$ (i.e., $X_1 \to 1$)

$$\left(\frac{\partial \ln \gamma_2}{\partial X_3}\right)_{X_1\to 1} = \left(\frac{\partial \ln \gamma_3}{\partial X_2}\right)_{X_1\to 1}$$

or

$$\epsilon_2^{(3)} = \epsilon_3^{(2)} \tag{42}$$

We wish to emphasize the fact that the partial derivative $(\partial/\partial n_2)_{n_1,n_3}$ is very different from $(1/n)(\partial/\partial X_2)_{n,X_3}$. This is the source of frequent mistakes; the two derivatives have very different meanings.

2.5.2. Second Demonstration

We shall now write the series developments of $\ln \gamma_1$, $\ln \gamma_2$, and $\ln \gamma_3$ and apply to them the Gibbs–Duhem equation in order to obtain the reciprocal relationships between the coefficients of these series:

$$\ln \gamma_1 = J_{1,0}^{(1)}X_2 + J_{0,1}^{(1)}X_3 + J_{2,0}^{(1)}X_2^2 + J_{1,1}^{(1)}X_2X_3 + J_{0,2}^{(1)}X_3^2 + \cdots \tag{43}$$

$$\ln \gamma_2 = \ln \gamma_2^\infty + \epsilon_2^{(2)}X_2 + \epsilon_2^{(3)}X_3 + \cdots \tag{44}$$

$$\ln \gamma_3 = \ln \gamma_3^\infty + \epsilon_3^{(2)}X_2 + \epsilon_3^{(3)}X_3 + \cdots \tag{45}$$

The standard state of the solvent 1 being Raoultian, the term $J_{0,0}^{(1)}$ has been omitted in the development of $\ln \gamma_1$ since it is equal to zero: $(\gamma_1)_{X_1\to 1} = 1$.

The Gibbs–Duhem equation may be written in the form

$$X_1\, d \ln \gamma_1 + X_2\, d \ln \gamma_2 + X_3\, d \ln \gamma_3 = 0 \tag{46}$$

and, consequently, in the form

$$(1 - X_2 - X_3)\frac{\partial \ln \gamma_1}{\partial X_2} + X_2\frac{\partial \ln \gamma_2}{\partial X_2} + X_3\frac{\partial \ln \gamma_3}{\partial X_2} = 0 \tag{47}$$

Differentiating equations (43)–(45) with respect to X_2 yields the $\partial \ln \gamma_i/\partial X_2$ terms, which we substitute in equation (47):

$$(1 - X_2 - X_3)\,(J_{1,0}^{(1)} + 2\,J_{1,0}^{(1)}X_2 + J_{1,1}^{(1)}X_3 + \cdots)$$
$$+ X_2(\epsilon_2^{(2)} + \cdots) + X_3(\epsilon_3^{(2)} + \cdots) = 0 \tag{48}$$

This equation is of the form

$$A + BX_2 + CX_3 + DX_2^2 + \cdots = 0 \tag{49}$$

It is also an identity, which means that it must remain valid whatever the values of X_2 and X_3. This is possible only if all the coefficients $A, B, C, D, \ldots$ are equal to zero. Consequently, we collect separately all the terms independent of X_2 and X_3, all the terms in X_2, all those in X_3, in X_2^2, etc., and write that their respective sums are equal to zero.

There is only one term independent of X_2 and X_3, and that is $J_{1,0}^{(1)}$. Thus,

$$J_{1,0}^{(1)} = 0 \tag{50}$$

This result means that Raoult's law is obeyed for the binary system 1–2 (see Section 1). It could have been expected from the start: the very writing of equation (44) assumes obedience to Henry's law, which is a sufficient condition for obedience to Raoult's law.

Collecting the terms in X_2 yields

$$2J_{2,0}^{(1)} + \epsilon_2^{(2)} = 0$$

or

$$J_{2,0}^{(1)} = -\tfrac{1}{2}\epsilon_2^{(2)} \tag{51}$$

$J_{2,0}^{(1)}$ is the binary coefficient $(\partial \ln \gamma_2/\partial X_2)_{X_1 \to 1}$, and relation (51) was already obtained in the study of binary solutions.

Collecting the terms in X_3 yields

$$J_{1,1}^{(1)} + \epsilon_3^{(2)} = 0$$

or

$$\epsilon_3^{(2)} = -J_{1,1}^{(1)} = -\left(\frac{\partial^2 \ln \gamma_1}{\partial X_2\, \partial X_3}\right)_{X_1 \to 1} \tag{52}$$

This is a new result. Because of the symmetry with respect to the subscripts 2 and 3 it is also clear that

$$\epsilon_2^{(3)} = -J_{1,1}^{(1)} \tag{53}$$

and thus

$$\epsilon_2^{(3)} = \epsilon_3^{(2)} \tag{54}$$

Obviously, equation (53) may also be obtained by writing the Gibbs–Duhem equation in the form

$$(1 - X_2 - X_3)\frac{\partial \ln \gamma_1}{\partial X_3} + X_2\frac{\partial \ln \gamma_2}{\partial X_3} + X_3\frac{\partial \ln \gamma_3}{\partial X_3} = 0 \tag{55}$$

and proceeding as for equation (52).

Note that the development of $\ln \gamma_1$ is now

$$\ln \gamma_1 = -\tfrac{1}{2}\epsilon_2^{(2)}X_2^2 - \epsilon_2^{(3)}X_2X_3 - \tfrac{1}{2}\epsilon_3^{(3)}X_3^2 + \cdots \tag{56}$$

It is easy to find the development of G^E in terms of interaction coefficients, since

$$G^E/RT = (1 - X_2 - X_3)\ln \gamma_1 + X_2 \ln \gamma_2 + X_3 \ln \gamma_3 \tag{57}$$

and the developments of ln γ_1, ln γ_2, and ln γ_3 are known. Applying equations (44), (45), and (56) to equation (57) leads to

$$G^E/RT = X_2 \ln \gamma_2^\infty + X_3 \ln \gamma_3^\infty + \tfrac{1}{2}\epsilon_2^{(2)}X_2^2 + \epsilon_2^{(3)}X_2X_3 + \tfrac{1}{2}\epsilon_3^{(3)}X_3^2 + \cdots \quad (58)$$

(Note the factor $\frac{1}{2}$ in front of $\epsilon_2^{(2)}$ and $\epsilon_3^{(3)}$. There is often a tendency to overlook the contribution of ln γ_1 to G^E and to multiply by 2 the terms in ϵ.)

The previous demonstration can easily be generalized to higher order interaction coefficients. It is found [4] that

$$J_{n_2,n_3}^{(2)} = J_{n_2,n_3}^{(1)} - \frac{n_2+1}{n_2+n_3} J_{n_2+1,n_3}^{(1)} \quad (59)$$

and

$$J_{n_2,n_3}^{(3)} = J_{n_2,n_3}^{(1)} - \frac{n_3+1}{n_2+n_3} J_{n_2,n_3+1}^{(1)} \quad (60)$$

With $n_2 = n_3 = 1$, equation (59) becomes

$$\rho_2^{(2,3)} = -\epsilon_2^{(3)} - J_{2,1}^{(1)} \quad (61)$$

and with $n_2 = 2$, $n_3 = 0$, equation (60) becomes

$$\rho_3^{(2)} = -\tfrac{1}{2}\epsilon_2^{(2)} - \tfrac{1}{2}J_{2,1}^{(1)} \quad (62)$$

Thus,

$$\rho_2^{(2,3)} + \epsilon_2^{(3)} = 2\rho_3^{(2)} + \epsilon_2^{(2)} = -J_{2,1}^{(1)} \quad (63)$$

Similarly,

$$\rho_3^{(2,3)} + \epsilon_2^{(3)} = 2\rho_2^{(3)} + \epsilon_3^{(3)} = -J_{1,2}^{(1)} \quad (64)$$

The general relationship between the coefficients of G^E and ln γ_1 may also be noted:

$$\phi_{n_2,n_3} = -J_{n_2,n_3}^{(1)}/(n_2 + n_3 - 1) \quad (65)$$

The geometrical significance of ϵ_i^j and ρ_i^j is illustrated in Fig. 4, which shows the effect of j (2 or 3) on the activity coefficient of i (3 or 2) at low concentrations of i. The two slopes at the origin $\epsilon_2^{(3)}$ and $\epsilon_3^{(2)}$ are identical but the curvatures $\rho_2^{(3)}$ and $\rho_3^{(2)}$ are not.

2.6. *Examples of Application*

We now apply our formalism to the case of nitrogen in liquid iron at 1600°C. In pure liquid iron, the nitrogen solubility under 1 atm is equal to 0.045% or, in mole fraction, 1.8×10^{-3}. We have seen in Section VII.5.2 that in this case nitrogen follows Sieverts's law (because, at least up to 1 atm of nitrogen, the terms in $\epsilon_N^N X_N$ and $\rho_N^N X_N^2$ are negligible, and $\varphi_N \simeq 1$).

Let us add carbon to the bath to produce the concentration X_C. The problem is to find the new solubility of nitrogen.

To calculate the activity coefficient of nitrogen φ_N (based on a Henrian standard state) we use equation (18):

$$\ln \varphi_N = \epsilon_N^N X_N + \epsilon_N^C X_C + \rho_N^N X_N^2 + \rho_N^{C,N} X_C X_N + \rho_N^C X_C^2 \quad (66)$$

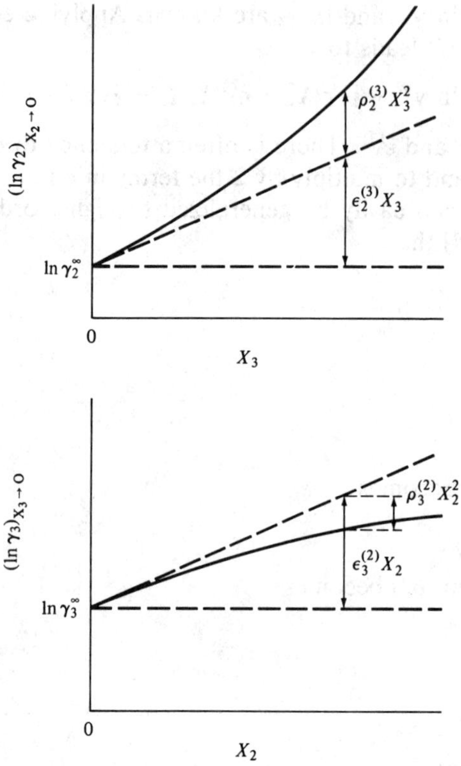

Figure 4. Geometrical significance of some interaction coefficients. The slopes at the origin are equal ($\epsilon_2^{(3)} = \epsilon_3^{(2)}$), but the curvatures are not [$\rho_2^{(3)} \neq \rho_3^{(2)}$; see equations (63) and (64)].

All the terms containing X_N may be neglected because X_N, the new solubility of nitrogen, is likely to remain small. Thus equation (66) becomes

$$\ln \varphi_N = \epsilon_N^C X_C + \rho_N^C X_C^2 \tag{67}$$

The values of ϵ_N^C and ρ_N^C are, respectively, equal to 5.86 and 11.7 [8]. For a carbon content of 0.02 (about 0.43% in weight)

$$\ln \varphi_N = 5.86 \times 0.02 + 11.7 \times (0.02)^2$$

$$= 0.117 + 0.005 = 0.122$$

and

$$\varphi_N = 1.13$$

At equilibrium, under 1 atm nitrogen, the presence of carbon does not affect the activity of nitrogen (since the chemical potential of nitrogen remains equal to that of the gas phase), and consequently

$$(X_N \varphi_N)_{\text{ternary}} = (X_N \varphi_N)_{\text{binary}}$$

or (68)

$$(X_N)_{\text{ternary}} = \frac{1.8 \times 10^{-3} \times 1}{1.13} = 1.6 \times 10^{-3} \quad \text{or} \quad 0.040\%$$

If the carbon content is 0.20 (about 5.1% in weight),

$$\ln \varphi_N = 5.86 \times 0.20 + 11.7 \times (0.20)^2$$
$$= 1.172 + 0.468 = 1.64$$

and

$$\varphi_N = 5.15$$

The new solubility of nitrogen is then

$$X_N = \frac{1.8 \times 10^{-3}}{5.15} = 0.35 \times 10^{-3} \quad \text{or} \quad 0.009\%$$

Thus, the solubility of nitrogen in pig iron is only about one-fifth its value in low alloy steels.

It should be noted that at low concentration of carbon the contribution of the second order interaction coefficient ρ_N^C is negligible. In Fig. 5 this corresponds to the fact that in this range of concentration the curve $\ln \varphi_N$ vs X_C may be approximated by its tangent, the slope of which is equal to ϵ_C^N.

Obviously, whenever the actual experimental curve $\ln \varphi_2$ vs X_3 is available over the region of interest, it is better to measure $\ln \varphi_2$ directly on this curve than to use the previous formalism, which after all is only its parametric representation. Figure 5 illustrates several of these curves for nitrogen in liquid iron alloys at 1600°C. The elements which raise its activity coefficient, such as Al, As, C, Co, Cu, Ni, P, S, Sb, Si, and Sn, decrease its solubility; those which lower its activity coefficient, such as Cb, Cr, Mn, Mo, Ta, Ti, V, and W, increase it [9, 10].

The effect of these elements is dependent on the temperature and may be analyzed through the enthalpy interaction coefficients. For instance, at 1500°C in the case of carbon, the first order enthalpy interaction coefficient η_N^C is equal to 35.27 kcal and the second order enthalpy interaction coefficients λ_N^C is equal to 260 kcal [8]. Consequently, equation (28) may be applied in the form

$$\epsilon_T = \epsilon_{1873^\circ K} + \frac{\eta}{R}\left(\frac{1}{T} - \frac{1}{1873}\right)$$

or

$$(\epsilon_N^C)_{1500^\circ C} = 5.86 + \frac{35270}{1.987}\left(\frac{1}{1773} - \frac{1}{1873}\right) = 6.40 \tag{69}$$

Similarly,

$$(\rho_N^C)_{1500^\circ C} = 11.7 + \frac{260000}{1.987}\left(\frac{1}{1773} - \frac{1}{1873}\right) = 15.6$$

Thus, at $X_C = 0.20$,

$$\ln \varphi_N = 6.4 \times 0.2 + 15.6 \times (0.2)^2 = 1.90$$

or

$$\varphi_N = 6.7$$

Knowing that the solubility of nitrogen in pure liquid iron at 1500°C is equal to

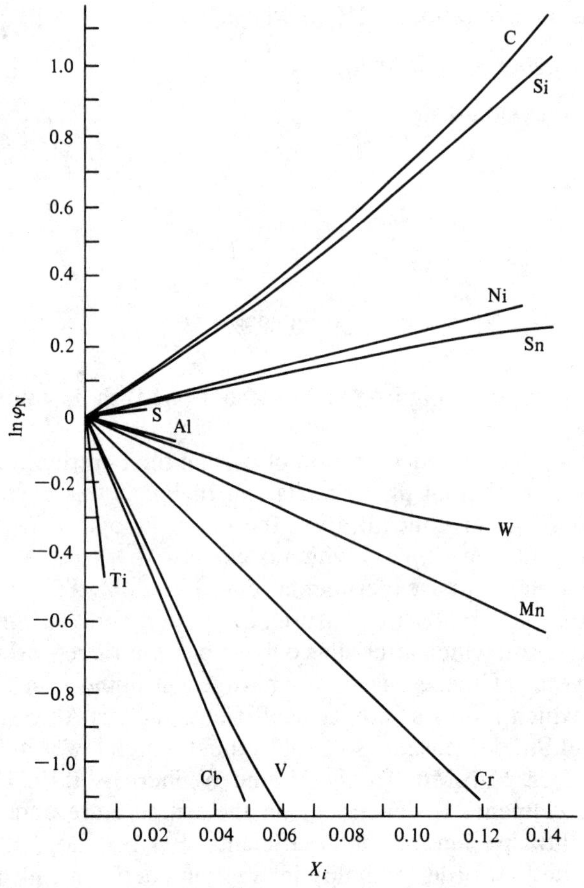

Figure 5. Effect of solute elements i on the activity coefficient of nitrogen in liquid iron at 1600°C.

1.76×10^{-3}, or 0.044%, at a carbon concentration $X_C = 0.20$, it is $(1.76 \times 10^{-3})/6.7 = 0.26 \times 10^{-3}$, or 0.0065%.

We now turn to another example. Liquid silver and iron are practically immiscible. A crucible containing both elements is placed in a furnace at 1550°C and under neutral atmosphere (argon). Two layers are formed: one of silver at the bottom, the other of iron at the top. We add a small quantity of silicon to the charge; it distributes itself between the two layers in such a way that at equilibrium

$$\mu_{Si}^{(Ag)} = \mu_{Si}^{(Fe)} \tag{70}$$

or, with a Raoultian standard state for silicon in both layers,

$$a_{Si}^{(Ag)} = a_{Si}^{(Fe)} \tag{71}$$

Thus,

$$X_{Si}^{(Ag)}/X_{Si}^{(Fe)} = \gamma_{Si}^{(Fe)}/\gamma_{Si}^{(Ag)} \tag{72}$$

If we now also add a small quantity of carbon, the latter, nearly insoluble in silver, goes almost entirely into the iron layer. However, because of the presence of carbon in iron the activity of silicon is changed and a redistribution occurs.

ϵ_C^{Si} in Fe is known to be positive, and as

$$\ln \gamma_{Si} = \ln \gamma_{Si}^{\infty} + \epsilon_{Si}^{Si} X_{Si} + \epsilon_{Si}^{C} X_C + \cdots \tag{73}$$

the effect of a carbon addition is to raise the activity coefficient and thus the activity and chemical potential of silicon. Consequently, there will be a net transfer of silicon to the silver layer in which the silicon has a lower chemical potential. Equilibrium will be reached when the two chemical potentials will be equal again.

From an atomistic viewpoint, both silicon and carbon may be thought to behave as positively charged ions in liquid iron. They repel each other, and thus on addition of carbon some of the silicon leaves the iron for the relatively more accommodating silver neighbors.

It may be observed that if the thermodynamics of the Ag–Si system are known, then a sampling of the two layers yields $X_{Si}^{(Ag)}$ and $X_{Si}^{(Fe)}$ and, through equation (72), $\gamma_{Si}^{(Fe)}$. Repeating the experiment at different compositions allows the determination of $\ln \gamma_{Si}^{\infty}$, ϵ_{Si}^{Si}, and ϵ_{Si}^{C} in Fe. The method has often been used, in particular by Chipman and co-workers [11]. Note that only one layer needs to be sampled and analyzed, a mass balance equation yielding the composition of the other. However, whenever possible a direct determination of both compositions—providing the added check of the mass balance—is far more satisfactory.

By this method as well as others, the following values for the interaction coefficients of the Fe–Si–C system at 1600°C have been determined [12]: $\epsilon_C^C = 11$, $\epsilon_{Si}^{Si} = 13$, $\epsilon_{Si}^{C} = \epsilon_C^{Si} = 5.6$, $\rho_C^C = -5.6$, and $\rho_C^{C,Si} = 23$. To emphasize the significance of these values, let us assume that we are interested in the behavior of carbon in the solution and how it is affected by the presence of silicon. A good measure of the effect is ϵ_C^{Si}. It describes the rate at which silicon additions raise the activity coefficient of carbon at high dilution of both solutes. Thus, at $X_C = 0.02$ an addition of Si to give $X_{Si} = 0.05$ raises $\ln \varphi_C$ by nearly 0.28 (5.6×0.05), that is, multiplies φ_C by 1.32. However, at $X_C = 0.18$, which is near saturation of carbon (0.205 in pure liquid iron at 1600°C), the same addition has a much larger effect, since then [see equation (18) or (75)]

$$\left(\frac{\partial \ln \varphi_C}{\partial X_{Si}}\right)_{\substack{X_C = 0.18 \\ X_{Si} \simeq 0}} = \epsilon_C^{Si} + \rho_C^{C,Si} X_C = 5.6 + 23 \times 0.18 = 9.7 \tag{74}$$

The same addition of silicon now raises $\ln \varphi_C$ by 0.49 (= 9.7×0.05) or multiplies φ_C by 1.63.

To calculate more exactly the activity coefficient of carbon, one would of course use the equation

$$\ln \varphi_C = \epsilon_C^C X_C + \epsilon_C^{Si} X_{Si} + \rho_C^C X_C^2 + \rho_C^{C,Si} X_C X_{Si} + \rho_C^{Si} X_{Si}^2 \tag{75}$$

Appendix 5 lists ϵ interaction coefficients for liquid iron at 1600°C. It may be seen that the coefficients have high values when strong interactions between the two solutes 2 and 3 can be expected, for example, on account of the existence of a compound quite stable under different conditions; e.g., ϵ_O^{Al} is strongly negative and alumina (Al_2O_3) has a large negative free energy of formation.

The coefficients ρ have been introduced only recently [3,4] and are obviously less important than the ϵ; thus, so far, they have been measured for only a few systems. However, this number is rapidly increasing. Appendix 5 also lists some of their values. Higher order coefficients are generally not warranted by our present experimental accuracy.

3. Dilute Multicomponent Solutions

3.1. Second Order Free Energy Interaction Coefficients

In the case of a dilute multicomponent solution 1–2–3–···–m, where 1 is the solvent and 2, 3, . . . , m are the solutes, the behavior of a solute species such as i is dependent on the concentrations of all solutes, and this dependence may be studied, as in the case of a ternary solution, through the concept of interaction coefficients.

Expanding into a Taylor series the logarithm of the activity coefficient of the solute i with respect to the mole fractions of the solutes yields

$$\ln \gamma_i = \sum_{n_2} \cdots \sum_{n_j} \cdots \sum_{n_m} J^{(i)}_{n_2,\ldots,n_j,\ldots,n_m} X_2^{n_2} \cdots X_j^{n_j} \cdots X_m^{n_m} \tag{76}$$

or

$$\ln \gamma_i = \ln \gamma_i^\infty + \sum_{j=2}^{m} \left(\frac{\partial \ln \gamma_i}{\partial X_j}\right)_{X_1\to 1} X_j + \sum_{j=2}^{m} \frac{1}{2}\left(\frac{\partial^2 \ln \gamma_i}{\partial X_j^2}\right)_{X_1\to 1} X_j^2$$
$$+ \sum_{j=2}^{m-1} \sum_{k>j}^{m} \left(\frac{\partial^2 \ln \gamma_i}{\partial X_j\, \partial X_k}\right)_{X_1\to 1} X_j X_k + \cdots \tag{77}$$

If the solution is sufficiently dilute that all the second order terms may be neglected, equation (77) becomes

$$\ln \gamma_i = \ln \gamma_i^\infty + \epsilon_i^{(2)} X_2 + \epsilon_i^{(3)} X_3 + \cdots + \epsilon_i^{(m)} X_m \tag{78}$$

Even if m, the total number of solutes, is large, the activity coefficient of i is obtained as a summation of terms involving only ternary systems. The convenience of this result has made the use of this formalism very widespread.

Often, however, the concentrations of the solutes warrant the retention also of the second order terms. Consequently, we now introduce the second order free energy interaction coefficient $\rho_i^{(j,k)}$:

$$\rho_i^{(j,k)} = J^{(i)}_{n_2=0,\ldots,n_j=1,\ldots,n_k=1,\ldots,n_m=0} = (\partial^2 \ln \gamma_i/\partial X_j\, \partial X_k)_{X_1\to 1} \qquad (j \neq k) \tag{79}$$

The coefficients $\rho_i^{(j,k)}$ represent properties of quaternary systems 1–i–j–k, whereas the coefficients $\epsilon_i^{(j)}$ and $\rho_i^{(j)}$ represent properties of ternary systems 1–i–j (or binary systems for $i = j$).

Equation (77) may be rewritten

$$\ln \gamma_i = \ln \gamma_i^\infty + \sum_{j=2}^{m} \epsilon_i^{(j)} X_j + \sum_{j=2}^{m} \rho_i^{(j)} X_j^2 + \sum_{j=2}^{m-1} \sum_{k>j}^{m} \rho_i^{(j,k)} X_j X_k + \cdots \tag{80}$$

As previously emphasized, to replace the Taylor infinite series representation of $\ln \gamma_i$ by a polynomial of degree n, the remainder R_n should be smaller than the experimental error on $\ln \gamma_i$. However, it must be noted that in general this experimental error does not change appreciably with the number m of solutes, whereas the remainder R_n increases as m increases. Thus, the situation often arises where the concentration levels of the solutes warrant the use of only linear terms (in ϵ) for binary or ternary systems, but necessitate the use of quadratic terms (in ρ) for multicomponent systems. For example, it is clear that at a concentration of $X_2 = 0.1$ a binary solution may be considered to be dilute, but that a multicomponent solution can not be considered as such when $X_2 = X_3 = X_4 = \cdots = X_m = 0.1$.

For illustration purposes, let us calculate the solubility of nitrogen in an iron alloy of composition: $X_{Cr} = 0.15$, $X_{Ni} = 0.10$, $X_{Si} = 0.04$. Appendix 5 lists some values of the coefficients $\rho_N^{i,j}$ for the interactions of i and j on nitrogen in liquid iron at 1600°C. Equation (80) becomes

$$\ln \varphi_N = \epsilon_N^{Cr} X_{Cr} + \epsilon_N^{Ni} X_{Ni} + \epsilon_N^{Si} X_{Si} + \rho_N^{Cr} X_{Cr}^2 + \rho_N^{Ni} X_{Ni}^2 + \rho_N^{Si} X_{Si}^2 + \rho_N^{Cr,Ni} X_{Cr} X_{Ni} + \rho_N^{Cr,Si} X_{Cr} X_{Si} + \rho_N^{Ni,Si} X_{Ni} X_{Si} \quad (81)$$

yielding

$$\begin{aligned}\ln \varphi_N &= [-10 \times 0.15 + 1.5 \times 0.10 + 5.9 \times 0.04] \\ &\quad + [5.0 \times (0.15)^2 + 2 \times (0.10)^2 + 2.8 \times (0.04)^2] \\ &\quad + [-0.5 \times 0.10 \times 0.15 - 21 \times 0.15 \times 0.04 - 3.4 \times 0.10 \times 0.04] \\ &= -1.124\end{aligned}$$

or

$$\varphi_N = 0.325$$

The solubility of nitrogen in the alloy is thus 3.08 (or 1/0.325) times its value in pure iron (i.e., $X_N = 5.5 \times 10^{-3}$ instead of 1.8×10^{-3}).

Enthalpy and entropy interaction coefficients may be defined for multicomponent solutions in a manner similar to that in which they were defined for ternary solutions. The definitions and symbols relative to all commonly used free energy, enthalpy, and entropy interaction coefficients are summarized in Table 1.

3.2. Reciprocal Relationships

If the thermodynamic behavior of one species is known across the composition range of a multicomponent system 1–2–⋯–m, the behavior of all the other species may then be deduced through the integration of the Gibbs–Duhem equation. To understand this better, it may be noted that the single Gibbs–Duhem equation

$$\sum_{i=2}^{m} X_i \, d \ln \gamma_i = 0 \quad (82)$$

in fact generates $m - 1$ independent equations

$$\sum_{i=2}^{m} X_i \frac{\partial \ln \gamma_i}{\partial X_j} = 0 \qquad \text{for } j = 2, 3, \ldots, m \quad (83)$$

Consequences of this result are the reciprocal relationships among interaction coefficients. It may be demonstrated that [4]

$$J^{(i)}_{n_2,\ldots,n_i,\ldots,n_m} = J^{(1)}_{n_2,\ldots,n_i,\ldots,n_m} - \left[(n_i + 1) \Big/ \sum_{j=2}^{m} n_j\right] J^{(1)}_{n_2,\ldots,n_i+1,\ldots,n_m} \quad (84)$$

Table 1. Definitions of Interaction Coefficients[a]

Order	General designation	Free energy[b] X	Free energy[b] wt%	Entropy X	Entropy wt%	Enthalpy X	Enthalpy wt%	System for obtaining property
Zero	$J^{(i)}_{0,0,\ldots,n_j=0,\ldots,0}$	$\ln \gamma_i^{\infty}$ $\ln \varphi_i = 0$[c]	$\log f_i = 0$	$S_i^{E\infty}$	$\mathcal{S}_i^{E\infty} = 0$	$H_i^{E\infty}$	$\mathcal{H}_i^{E\infty} = 0$	1–i binary
First	$J^{(i)}_{0,0,\ldots,n_i=1,\ldots,0}$	$\epsilon_i^{(i)}$	$e_i^{(i)}$	$\sigma_i^{(i)}$	$s_i^{(i)}$	$\eta_i^{(i)}$	$h_i^{(i)}$	1–i binary
	$J^{(i)}_{0,0,\ldots,n_j=1,\ldots,0}$	$\epsilon_i^{(j)}$	$e_i^{(j)}$	$\sigma_i^{(j)}$	$s_i^{(j)}$	$\eta_i^{(j)}$	$h_i^{(j)}$	1–i–j ternary
Second	$J^{(i)}_{0,0,\ldots,n_j=2,\ldots,0}$	$\rho_i^{(i)}$	$r_i^{(i)}$	$\pi_i^{(i)}$	$p_i^{(i)}$	$\lambda_i^{(i)}$	$l_i^{(i)}$	1–i binary
	$J^{(i)}_{0,0,\ldots,n_j=2,\ldots,0}$	$\rho_i^{(j)}$	$r_i^{(j)}$	$\pi_i^{(j)}$	$p_i^{(j)}$	$\lambda_i^{(j)}$	$l_i^{(j)}$	1–i–j ternary
	$J^{(i)}_{0,0,\ldots,n_j=1,n_k=1,\ldots,0}$	$\rho_i^{(j,k)}$	$r_i^{(j,k)}$	$\pi_i^{(j,k)}$	$p_i^{(j,k)}$	$\lambda_i^{(j,k)}$	$l_i^{(j,k)}$	1–i–j–k quaternary
Higher	$J^{(i)}_{0,0,\ldots,n_i=3,\ldots,0}$	—	—	—	—	—	—	1–i binary
	$J^{(i)}_{0,0,\ldots,n_j=3,\ldots,0}$	—	—	—	—	—	—	1–i–j ternary

General relationship:

$$J^{(i)}_{n_2,\ldots,n_j,\ldots,n_m} = \frac{L^{(i)}_{n_2,\ldots,n_j,\ldots,n_m}}{RT} - \frac{K^{(i)}_{n_2,\ldots,n_j,\ldots,n_m}}{R}$$

Representative equations:

$$\ln \gamma_i = \ln \gamma_i^{\infty} + \sum_{j=2}^{m} \epsilon_i^{(j)} X_j + \sum_{j=2}^{m} \rho_i^{(j)} X_j^2 + \sum_{j=2}^{m-1} \sum_{k>j}^{m} \rho_i^{(j,k)} X_j X_k + O(X^3)$$

$$\log f_i = \sum_{j=2}^{m} e_i^{(j)}(\% j) + \sum_{j=2}^{m} r_i^{(j)}(\% j)^2 + \sum_{j=2}^{m-1} \sum_{k>j}^{m} r_i^{(j,k)}(\% j)(\% k) + O(\%^3)$$

[a] At $X_1 \to 1$, or wt% $1 \to 100$. [b] $G_i^E/RT = \ln \gamma_i$. [c] $\ln \varphi_i = \ln(\gamma_i/\gamma_i^{\infty})$.

which, for the appropriate values of the n_js, leads to

$$\rho_i^{(j,k)} + \epsilon_k^{(j)} = \rho_j^{(k,i)} + \epsilon_i^{(k)} = \rho_k^{(i,j)} + \epsilon_j^{(i)} \tag{85}$$

There are equivalent relationships among enthalpy and entropy interaction coefficients. A summary of these is given in Table 2.

It may be noted that in the Taylor series expansion of G^E/RT

$$G^E/RT = \sum_{n_2=0}^{\infty} \cdots \sum_{n_j=0}^{\infty} \cdots \sum_{n_m=0}^{\infty} \phi_{n_2,\ldots,n_j,\ldots,n_m} X_2^{n_2} \cdots X_j^{n_j} \cdots X_m^{n_m} \tag{86}$$

Table 2. Reciprocal Relationships Among Interaction Coefficients

Free energy (G_i^E/RT)	Entropy	Enthalpy
$\epsilon_i^{(j)} = \epsilon_j^{(i)}$	$\sigma_i^{(j)} = \sigma_j^{(i)}$	$\eta_i^{(j)} = \eta_j^{(i)}$
$\rho_i^{(i,j)} + \epsilon_i^{(j)} = 2\rho_j^{(i)} + \epsilon_i^{(i)}$	$\pi_i^{(i,j)} + \sigma_i^{(j)} = 2\pi_j^{(i)} + \sigma_i^{(i)}$	$\lambda_i^{(i,j)} + \eta_i^{(j)} = 2\lambda_j^{(i)} + \eta_i^{(i)}$
$\rho_i^{(j,k)} + \epsilon_j^{(k)} = \rho_j^{(i,k)} + \epsilon_i^{(k)}$	$\pi_i^{(j,k)} + \sigma_j^{(k)} = \pi_j^{(i,k)} + \sigma_i^{(k)}$	$\lambda_i^{(j,k)} + \eta_j^{(k)} = \lambda_j^{(i,k)} + \eta_i^{(k)}$
$= \rho_k^{(i,j)} + \epsilon_i^{(j)}$	$= \pi_k^{(i,j)} + \sigma_i^{(j)}$	$= \lambda_k^{(i,j)} + \eta_i^{(j)}$

Generally,

$$J^{(i)}_{n_2,\ldots,n_i,\ldots,n_m} = J^{(1)}_{n_2,\ldots,n_i,\ldots,n_m} - \frac{n_i + 1}{\sum_{j=2}^{m} n_j} J^{(1)}_{n_2,\ldots,n_i+1,\ldots,n_m}$$

the property

$$G^{E}/RT = \sum_{i=1}^{m} X_i \ln \gamma_i \tag{87}$$

leads directly to the relation [4]

$$\phi_{n_2,\ldots,n_i,\ldots,n_m} = -J^{(1)}_{n_2,\ldots,n_i,\ldots,n_m} \Big/ \left[\left(\sum_{j=2}^{m} n_j\right) - 1\right] \tag{88}$$

4. Interaction Coefficients on a Weight Percent Basis

It is often practical to use weight percents instead of mole fractions as composition coordinates. In Chapter VII, we defined the activity coefficient f_i, the excess enthalpy and entropy functions $\mathcal{H}_i^{E}$ and $\mathcal{S}_i^{E}$, and the interaction coefficients contained in their Taylor series expansions in the case of binary solutions. For multicomponent solutions, these series include the concentrations of the other solutes:

$$\log f_i = \sum_{j=2}^{m} e_i^{(j)}\%j + \sum_{j=2}^{m} r_i^{(j)}(\%j)^2 + \sum_{j=2}^{m-1}\sum_{k>j}^{m} r_i^{(j,k)}(\%j)(\%k) + \cdots \tag{89}$$

$$\mathcal{H}_i^{E} = \sum_{j=2}^{m} h_i^{(j)}\%j + \sum_{j=2}^{m} l_i^{(j)}(\%j)^2 + \sum_{j=2}^{m-1}\sum_{k>j}^{m} l_i^{(j,k)}(\%j)(\%k) + \cdots \tag{90}$$

$$\mathcal{S}_i^{E} = \sum_{j=2}^{m} s_i^{(j)}\%j + \sum_{j=2}^{m} p_i^{(j)}(\%j)^2 + \sum_{j=2}^{m-1}\sum_{k>j}^{m} p_i^{(j,k)}(\%j)(\%k) + \cdots \tag{91}$$

The definition of these interaction coefficients may be found in Table 1. Note the parallel between the functions and coefficients defined on the two scales: $\ln \phi \rightarrow \log f$, $\epsilon \rightarrow e$, $\sigma \rightarrow s$, $\rho \rightarrow r$, etc.

Relations between the free energy, enthalpy, and entropy interaction coefficients are immediately deduced from the formula relating $\log f_i$, $\mathcal{H}_i^{E}$, and $\mathcal{S}_i^{E}$:

$$2.3RT \log f_i = \mathcal{H}_i^{E} - T\mathcal{S}_i^{E} \tag{92}$$

For example,

$$e_i^{(j)} = \frac{h_i^{(j)}}{2.3RT} - \frac{s_i^{(j)}}{2.3R} \tag{93}$$

The conversion relationship between $\epsilon_i^{(j)}$ and $e_i^{(j)}$ may be obtained exactly as in the case of the binary coefficients $\epsilon_i^{(i)}$ and $e_i^{(i)}$ (Sections V.4.1). The result is

$$\epsilon_i^{(j)} = 230\,\frac{M_j}{M_1}\,e_i^{(j)} + \frac{M_1 - M_j}{M_1} \tag{94}$$

Other conversion relationships are similarly obtained. They are listed in Table 3.

The reciprocal relationship $\epsilon_i^{(j)} = \epsilon_j^{(i)}$ may be transposed to the coefficients $e_i^{(j)}$ and $e_j^{(i)}$ through equation (94):

$$e_j^{(i)} = \frac{M_j}{M_i}\,e_i^{(j)} + 0.434 \times 10^{-2}\,\frac{M_i - M_j}{M_i} \tag{95}$$

[The last term in equations (94) and (95) has commonly been omitted in the literature. This error, which was brought to the attention of thermodynamicists

Table 3. Conversion Relationships for Interaction Coefficients Based on Mole Fractions and Weight Percents

	Free energy	Enthalpy	Entropy
Function	$2.3 \log f_i = \ln \gamma_i + \ln \frac{X_i}{(\%i)} + \ln \frac{100M_i}{\gamma_i^\infty M_1}$	$\mathcal{H}_i^E = H_i^E - H_i^{E\infty}$	$\mathcal{S}_i^E = S_i^E - 2.3R \log \frac{100M_iX_i}{M_1\%i} - S_i^{E\infty}$
Interaction coefficients:			
		first order	
	$\epsilon_i^{(j)} = 230 \frac{M_j}{M_1} e_i^{(j)} + \frac{M_1 - M_j}{M_1}$	$\eta_i^{(j)} = 100 \frac{M_j}{M_1} h_i^{(j)}$	$\sigma_i^{(j)} = 100 \frac{M_j}{M_1} s_i^{(j)} - R \frac{M_1 - M_j}{M_1}$
		second order	
	$\rho_i^{(j)} = \frac{2.3 \times 10^2}{M_1^2} [10^2 M_j^2 r_i^{(j)} + M_j(M_1 - M_j)e_i^{(j)}] + \frac{1}{2}\left(\frac{M_1 - M_j}{M_1}\right)^2$	$\lambda_i^{(j)} = \frac{10^2}{M_1^2} [10^2 M_j^2 l_i^{(j)} + M_j(M_1 - M_j)h_i^{(j)}]$	$\pi_i^{(j)} = \frac{10^2}{M_1^2} [10^2 M_j^2 p_i^{(j)} + M_j(M_1 - M_j)s_i^{(j)}] - \frac{1}{2}R\left(\frac{M_1 - M_j}{M_1}\right)^2$
	$\rho_i^{(j,k)} = \frac{2.3 \times 10^2}{M_1^2} [10^2 M_j M_k r_i^{(j,k)} + M_j(M_1 - M_k)e_i^{(j)} + M_k(M_1 - M_j)e_i^{(k)}] + \frac{(M_1 - M_j)(M_1 - M_k)}{M_1^2}$	$\lambda_i^{(j,k)} = \frac{10^2}{M_1^2} [10^2 M_j M_k l_i^{(j,k)} + M_j(M_1 - M_k)h_i^{(j)} + M_k(M_1 - M_j)h_i^{(k)}]$	$\pi_i^{(j,k)} = \frac{10^2}{M_1^2} [10^2 M_j M_k p_i^{(j,k)} + M_j(M_1 - M_k)s_i^{(j)} + M_k(M_1 - M_j)s_i^{(k)}] - \frac{R(M_1 - M_j)(M_1 - M_k)}{M_1^2}$

by Lupis and Elliott [13], may be especially serious in the use of equation (95).] Similarly, the relationships $\eta_i^{(j)} = \eta_j^{(i)}$ and $\sigma_i^{(j)} = \sigma_j^{(i)}$ yield

$$h_j^{(i)} = \frac{M_j}{M_i} h_i^{(j)} \tag{96}$$

$$s_j^{(i)} = \frac{M_j}{M_i} s_i^{(j)} + \frac{R}{100} \frac{M_j - M_i}{M_i} \tag{97}$$

5. Application to Deoxidation Reactions

In Section 2.6 we gave several examples of the use of interaction coefficients. We now add another example pertaining to the calculation of deoxidation reactions.

Deoxidation is an important step in the refining of steel and considerable efforts have been made to obtain the data required for its control. A deoxidation reaction may be written

$$x\underline{\mathrm{M}}(\%) + y\underline{\mathrm{O}}(\%) = \mathrm{M}_x\mathrm{O}_y \tag{98}$$

where M is the deoxidizing element added to the solution. If K is the equilibrium constant of reaction (98), we have

$$K = \frac{a_{\mathrm{M}_x\mathrm{O}_y}}{a_{\mathrm{M}}^x a_{\mathrm{O}}^y} = \frac{a_{\mathrm{M}_x\mathrm{O}_y}}{[\%\ \mathrm{M}]^x f_{\mathrm{M}}^x [\%\ \mathrm{O}]^y f_{\mathrm{O}}^y} \tag{99}$$

or

$$\log a_{\mathrm{M}_x\mathrm{O}_y} - x(\log \%\mathrm{M} + \log f_{\mathrm{M}}) - y(\log \%\mathrm{O} + \log f_{\mathrm{O}}) = \log K \tag{100}$$

Since $\log f_{\mathrm{M}}$ and $\log f_{\mathrm{O}}$ are functions of %M and %O, equation (100) yields %O as a function of %M.

Plots of %O in solution (and in equilibrium with the oxide $\mathrm{M}_x\mathrm{O}_y$) as a function of %M generally show a minimum (see Fig. 6), that is, an optimum quantity of deoxidizer: above a certain concentration of M the level of oxygen rises ("overkill"). The existence and the position of this minimum is easily found by further examination of equation (100).

If the oxide $\mathrm{M}_x\mathrm{O}_y$ is essentially stoichiometric (and therefore independent of %O or %M), differentiation of equation (100) with respect to %M yields

$$x\left(\frac{0.434}{\%\mathrm{M}} + \frac{\partial \log f_{\mathrm{M}}}{\partial \%\mathrm{M}}\right) + y\left(\frac{0.434}{\%\mathrm{O}}\frac{\partial \%\mathrm{O}}{\partial \%\mathrm{M}} + \frac{\partial \log f_{\mathrm{O}}}{\partial \%\mathrm{M}}\right) = 0 \tag{101}$$

At the minimum

$$(\partial \%\mathrm{O}/\partial \%\mathrm{M})_{\min} = 0 \tag{102}$$

Consequently,

$$(\%\mathrm{M})_{\min} = -\frac{0.434x}{y(\partial \log f_{\mathrm{O}}/\partial \%\mathrm{M}) + x(\partial \log f_{\mathrm{M}}/\partial \%\mathrm{M})} \tag{103}$$

If we adopt a formalism limited to the first order interaction coefficients, equation

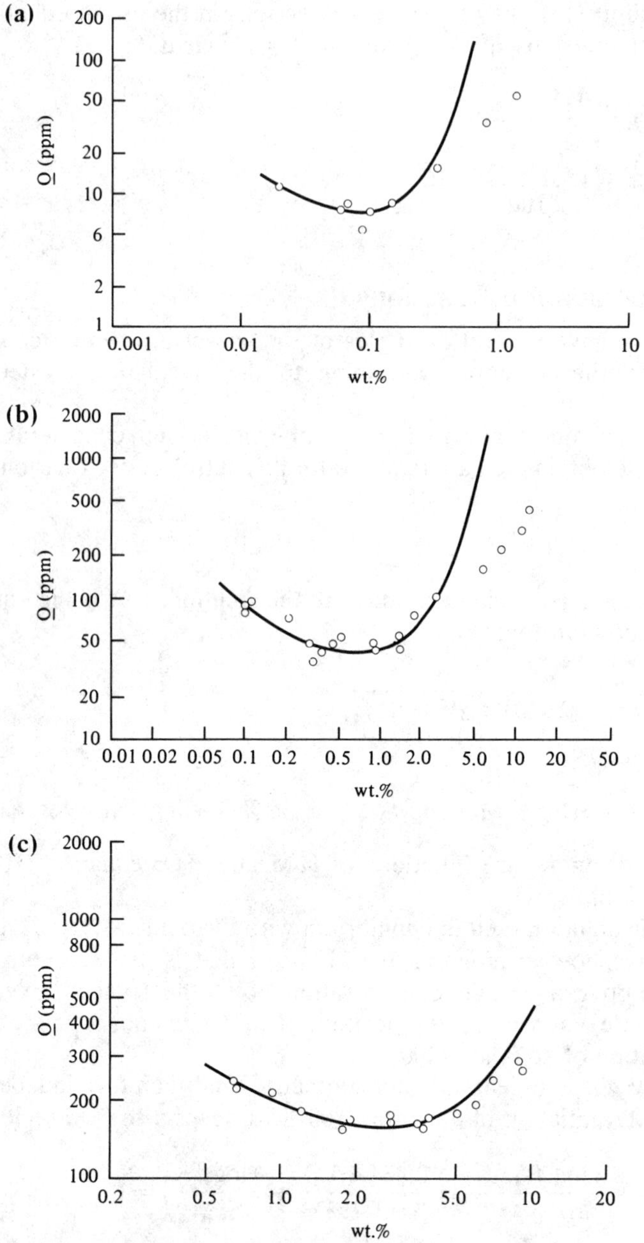

Figure 6. Solubility of oxygen vs **(a)** aluminum, **(b)** titanium, and **(c)** vanadium contents. The curves are calculated by Smith and Kirkaldy [15] on the basis of a first order interaction formalism with **(a)** $\epsilon_O^{Al} = -360$, **(b)** $\epsilon_O^{Ti} = -87$, and **(c)** $\epsilon_O^{V} = -27.7$. The data are from Fruehan [16] **(a)**, **(b)** and Kontopoulos [17] **(c)**.

(103) becomes

$$(\%M)_{min} = \frac{-0.434x}{ye_O^M + xe_M^M} \tag{104}$$

A deoxidizer has great affinity for oxygen and this results in an interaction coefficient e_O^M which is large and *negative* (in absolute values, it is generally much larger than e_M^M). Consequently, we verify that $(\%M)_{min}$ is indeed a positive quantity.

The same analysis could have been made in terms of mole fractions. The result would have been

$$(X_M)_{min} = \frac{-x}{y\epsilon_O^M + x\epsilon_M^M} \tag{105}$$

Depending on the range of concentration of M considered, the second order interaction coefficients r_O^M, or ρ_O^M, may have to be introduced in the equations above. Figure 6 shows the effects of Al, V, and Ti on the concentration of oxygen in solution [14]. The solid curves are calculated by Smith and Kirkaldy [15] on the basis of a formalism limited to first order interaction coefficients. In the case of Al and Ti it is clear that disagreement with the data in the high concentration range is due (at least partly) to the second order interaction coefficients. (r_O^{Al} and r_O^{Ti} are evaluated respectively as 1.7 and 0.031; on a mole fraction basis, ρ_O^{Al} and ρ_O^{Ti} are calculated to be 8600 and 500 [10].) In some cases, it is even possible that the size of the higher order coefficients (order 2 and above) suppresses the minimum.

Problems

1. **a.** Develop the Taylor series of $\ln \gamma_1$ with respect to X_2 and X_3, the mole fractions of the solutes 2 and 3, up to the third order. Express the coefficients of the series in terms of the interaction coefficients ϵ and ρ.
   b. Extend this expression to the case of a quaternary system.

2. A steel contains 0.6% C, 1% Mn, 0.3% Si, 0.8% Cr (by weight). Calculate the value of the activity of carbon at 1000°C, when the standard state of carbon is graphite, and also when it is Henrian at a concentration of 1%. At 1000°C, $\gamma_C^\infty = 7.9$, $\epsilon_C^C = 8.7$, $\epsilon_C^{Mn} = -4.3$, $\epsilon_C^{Si} = 11.2$, $\epsilon_C^{Cr} = -12$.

3. The activity of sulfur in liquid iron alloys at 1600°C has been determined by studying the equilibrium between sulfur in the melt and gaseous atmospheres containing H_2 and H_2S. For the reaction

   $$\underline{S}(\%) + H_2 = H_2S$$

   it has been found that $\log K = -2.577$ and $e_S^S = -0.030$. Estimate the interaction coefficients e_S^{Al} and e_S^C from the following data:

$(p_{H_2S}/p_{H_2}) \times 10^3$	%S	%Al	%C
1.03	0.351	1.03	—
1.00	0.302	1.91	—
0.95	0.256	2.78	—
1.19	0.300	2.16	0.56
1.42	0.311	2.21	1.04
1.74	0.289	2.08	1.95

   Show all calculations and graphs.

4. **a.** In an investigation of the interaction of carbon on the activity coefficient of silicon, liquid Fe–Si–C alloys of different compositions were equilibrated at 1420°C through the medium of a bath of liquid silver. Silver is practically insoluble in the

iron alloys, and iron and carbon are also insoluble in liquid silver. Derive the following formula used by the investigators:

$$\left(\frac{\partial \ln \gamma_{Si}}{\partial X_C}\right)_{X_{Si}} = -\left[1 + X_{Si}\left(\frac{\partial \ln \gamma_{Si}}{\partial X_{Si}}\right)_{X_C=0}\right]\left(\frac{\partial \ln X_{Si}}{\partial X_C}\right)_{a_{Si}}$$

b. In a run involving three different alloys, after sufficient time for equilibration, the system is quenched. An analysis of the alloys yields the following results:

	alloy 1	alloy 2	alloy 3
X_C	0.0023	0.0214	0.0412
X_{Si}	0.1519	0.1407	0.1298

It is known that in liquid iron at 1420°C $\epsilon_{Si}^{Si} \simeq 13$. Estimate ϵ_{Si}^{C}. State carefully your assumptions.

5. On the basis of a statistical model, the following coordinate has been proposed:

$$z_i = n_i \bigg/ \left(n_1 + \sum_{j=2}^{m} n_j v_j\right) \qquad (j = 2, 3, \ldots, m)$$

where n_1 and n_j are, respectively, the number of moles of solvent 1 and solute j, and $v_j = +1$ for a substitutional element j and -1 for an interstitial element j. The corresponding activity coefficient is

$$\psi_i = a_i / z_i$$

A first order interaction may be defined as

$$\theta_i^{(j)} = \left(\frac{\partial \ln \psi_i}{\partial z_j}\right)_{z_i, z_j \to 0}$$

Find the relationship between $\theta_i^{(j)}$ and $\theta_j^{(i)}$.

6. Experiments on the behavior of carbon and silicon in liquid iron at 1600°C have yielded the following information. At high dilutions of both carbon and silicon, ϵ_C^{Si} has been measured and found to be equal to 5.6. The effect of silicon on the solubility of graphite has also been determined; the slope of $\ln X_C$ vs X_{Si} at $X_{Si} \to 0$ and graphite saturation measures -3.55. From these results, deduce estimates of $\rho_C^{C,Si}$ and ρ_{Si}^{C}. The solubility of graphite in liquid iron is $X_C = 0.205$; other data on the Fe–C binary may be found in Appendix 5.

7. The solubility of nitrogen in pure liquid iron under 1 atm nitrogen is 0.045% at 1600°C and 0.046% at 1800°C. Find the solubility of nitrogen in an iron alloy containing 10% Cr, 4% V, 2% Si, and 0.5% C, at 1600°C and 1800°C.

How would you plot the temperature dependence of the nitrogen solubility in such an alloy, and what is the analytic value of the slope of the line defined by your plot at 1600°C?

8. Y is an element with a detrimental effect on the physical properties of the solvent A and it is desired to reduce its concentration to the level of 0.1 at.% when A is liquid at the temperature T. B and C are known to form the compounds BY and CY, which are mutually immiscible. At T

$$\underline{B} + \underline{Y} = BY(s), \qquad K_B = 10^5$$

$$\underline{C} + \underline{Y} = CY(s), \qquad K_C = 10^6$$

Assume the initial concentration of Y is 1 at.%. The relative prices of B and C are 1 and α. Find the optimum economic solution to achieve the level of 0.1 at.% for Y (e.g., the relative quantities of B and C which are to be added to the bath) in the cases below. Discuss your results in terms of α in each of the following cases.

a. Assume the solution is ideal.

b. Element D is available. Its price if $\alpha/5$ and $\epsilon_Y^D = 10$. The other interaction coefficients are assumed negligible.

c. The presence of D improves the quality of the alloy yielded by the bath and increases its price by $\alpha/5.2$ (nearly matching the cost of D). Assume this price increase is proportional to the content of D and find the most desirable composition.

d. Element D is not available, but a thermodynamic study reveals that $\epsilon_Y^B = -20$, $\epsilon_Y^C = -30$, $\epsilon_Y^Y \simeq 0$, and $\epsilon_C^B \simeq 0$.

Note that Henrian standard states are adopted for these reactions:

$$\varphi_i = a_i/X_i \qquad \text{and} \qquad (\varphi_i)_{X_i \to 0} = 1$$

(1 at.% corresponds to $X = 0.01$.)

Figure 7. Solubility of oxygen in liquid silver alloys [18].

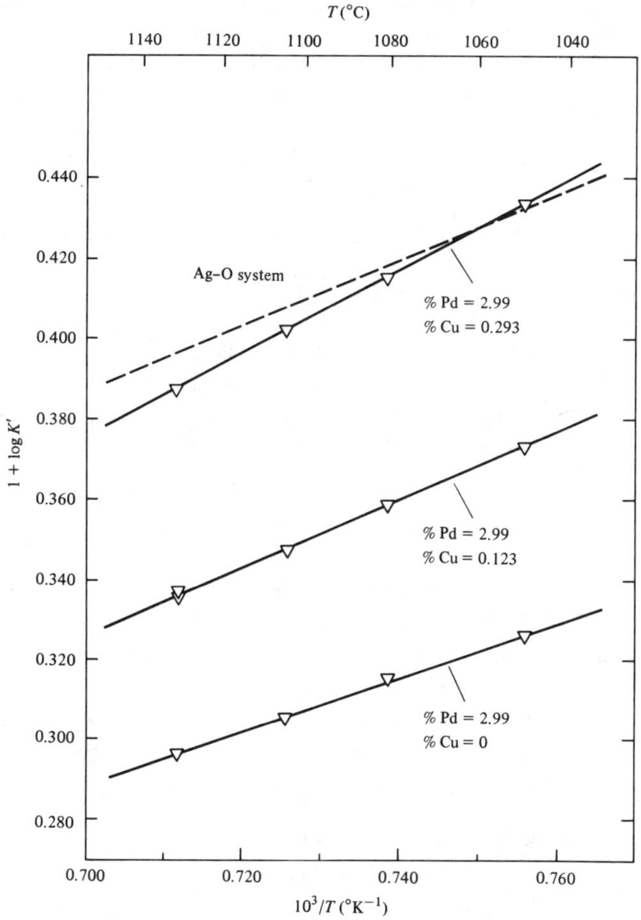

9. Figure 7 summarizes some data on the solubility of oxygen in silver alloys. By definition

$$K' = \%O/(p_{O_2})^{1/2}$$

Assume that e_O^O is negligible.

a. Estimate the coefficients e_O^{Pd} and e_O^{Cu} at 1100°C.

b. Estimate the coefficients h_O^{Pd} and h_O^{Cu}.

c. Express analytically the solubility of oxygen (at $p_{O_2} = 1$ atm) as a function of T, % Pd, and % Cu.

10. Derive the conversion relationship between $\rho_i^{(j)}$ and $r_i^{(j)}$ shown in Table 3.

11. Determine the quantity of SiO_2(s) which forms as a result of the addition of 6 kg of silicon per tonne to a steel bath at 1600°C containing 0.025% oxygen. Also determine the final contents of oxygen and silicon present in the bath as solute elements. The effects of the contents of the other solutes in the bath are assumed negligible, $a_{SiO_2} = 1$, and

$$\underline{Si}(\%) + 2\underline{O}(\%) = SiO_2(s), \qquad \Delta G° = -139{,}070 + 53.0T$$

The relevant interaction coefficients are listed in Appendix 5.

12. Mori and Moro-Oka [19] measured the effect of various elements i at 1600°C on the solubility of nitrogen in a liquid stainless steel alloy analyzing 18% Cr, 8% Ni by the parameter $b_N^i = (\partial \log f_N/\partial \%i)_{\%i\to 0,\ \%Cr=18,\ \%Ni=8}$. For niobium and tantalum, they found $b_N^{Nb} = -0.043$ and $b_M^{Ta} = -0.019$. Verify that these results are in good agreement with the first and second order free energy coefficients listed in Appendix 5 and associated with Nb and Ta.

References

1. C. Wagner, *Thermodynamics of Alloys*. Addison-Wesley, Reading, MA, 1952, pp. 47–53.
2. J. Chipman, *J. Iron Steel Inst.* **180,** 97 (1955).
3. C. H. P. Lupis and J. F. Elliott, *J. Iron Steel Inst.* **203,** 739 (July 1965).
4. C. H. P. Lupis and J. F. Elliott, *Acta Met.* **14,** 529–538 (1966).
5. C. H. P. Lupis and J. F. Elliott, *Acta Met.* **15,** 265–276 (1967).
6. J. Chipman and D. A. Corrigan, *Application of Fundamental Thermodynamics to Metallurgical Processes* (G. R. Fitterer, ed.). Gordon and Breach, New York, 1967, pp. 23–37.
7. C. H. P. Lupis and J. F. Elliott, *Trans. Met. Soc. AIME* **236,** 130 (1966).
8. D. W. Gomersall, A. McLean, and R. G. Ward, *Trans. Met. Soc. AIME* **242,** 1309–1315 (1968).
9. E.-H. Foo and C. H. P. Lupis, *Acta Met.* **21,** 1409–1430 (1973).
10. G. K. Sigworth and J. F. Elliott, *Met. Sci.* **8,** 298–310 (1974).
11. D. Schroeder and J. Chipman, *Trans. Met. Soc. AIME* **230,** 1492–1494 (1964).
12. C. H. P. Lupis, *Acta Met.* **16,** 1365–1375 (1968).
13. C. H. P. Lupis and J. F. Elliott, *Trans. Met. Soc. AIME* **233,** 257–258 (1965).
14. S. E. Feldman and J. S. Kirkaldy, *Can. Met. Quart.* **13,** 625–630 (1974).
15. P. N. Smith and J. S. Kirkaldy, *A Samarin Memorial Symposium*. Acad. of Sci., Moscow, 1972.
16. R. J. Fruehan, *Met. Trans.* **1,** 3403–3410 (1970).
17. A. Kontopoulos, Ph.D. thesis, McMaster University, 1971.
18. C. H. P. Lupis, Sc.D. thesis, MIT, 1965.
19. T. Mori and A. Moro-Oka, Kyoto University Publ., Kyoto, Japan, 1966.

X Multicomponent Solutions and Phase Diagrams

In Chapter VIII we systematically analyzed the characteristics of binary phase diagrams. A similar presentation of the characteristics of ternary and multicomponent phase diagrams would be extremely long and lies outside the scope of this book. However, an introduction to the subject should be helpful and is therefore presented in Sections 1 and 2. For further study the reader is referred to Refs. [1–4], which provide an excellent coverage of the subject. In Section 3 we examine various possible analytic representations of the Gibbs free energy of a multicomponent phase, and in Section 4 we present means of calculating a multicomponent phase diagram.

1. General Features of Ternary Phase Diagrams

In Section II.5.3 we established that the variance ϑ of a system of m components and φ phases is calculated by the formula

$$\vartheta = m + 2 - \varphi$$

In a ternary system $m = 3$ and consequently a maximum of five coexistent phases is possible. However, for condensed systems (metallic ones in particular) if the pressure does not vary over a considerable range (e.g., up to a few kilobars), its effect is generally negligible. We shall therefore consider that the pressure is fixed at 1 atm. The variance of the system is then decreased by one unit and becomes

$$\vartheta = m + 1 - \varphi$$

The maximum number of phases is now 4. A two-phase equilibrium $\alpha \rightleftarrows \beta$ is bivariant. If the temperature is selected arbitrarily, the two-phase equilibrium becomes monovariant and only three phases can coexist. Thus, in a three-dimensional phase diagram of T vs X_2 and X_3 ($X_1 = 1 - X_2 - X_3$) the compositions X_2^α, X_3^α and X_2^β, X_3^β are represented by two surfaces, which become two lines in an isothermal section. To examine this further, let us first present traditional methods of graphical representation.

1.1. Graphical Representation

Since the composition of a ternary alloy is determined by only two coordinates, such as X_2 and X_3, it is possible to represent it in a system of cartesian coordinates, as illustrated in Fig. 1a. The composition of an alloy has to lie within the triangle ABC. The line BC corresponds to the binary system 2–3 (and has the equation $X_1 = 1 - X_2 - X_3 = 0$), the line AB to the binary 1–2, and the line AC to the binary 1–3. This method of representation is useful when the three components do not play symmetrical roles (for example when 1 is the solvent and 2 and 3 remain dilute in the composition range of interest). It is often used in chemical engineering systems.

When the three components play analogous roles, it is preferable to use the equilateral triangle shown in Fig. 1b. In such a triangle, the sum of the perpendiculars from any point M inside the triangle to its sides is a constant equal to the height of the triangle. It is upon this property that the representation is based. The distances Ma, Mb, Mc to the sides BC, AC, and AB are measures, respectively, of X_1, X_2, and X_3. If the perpendicular AP from the vertex A to the base BC is divided into equal parts and a series of lines are drawn through the divisions

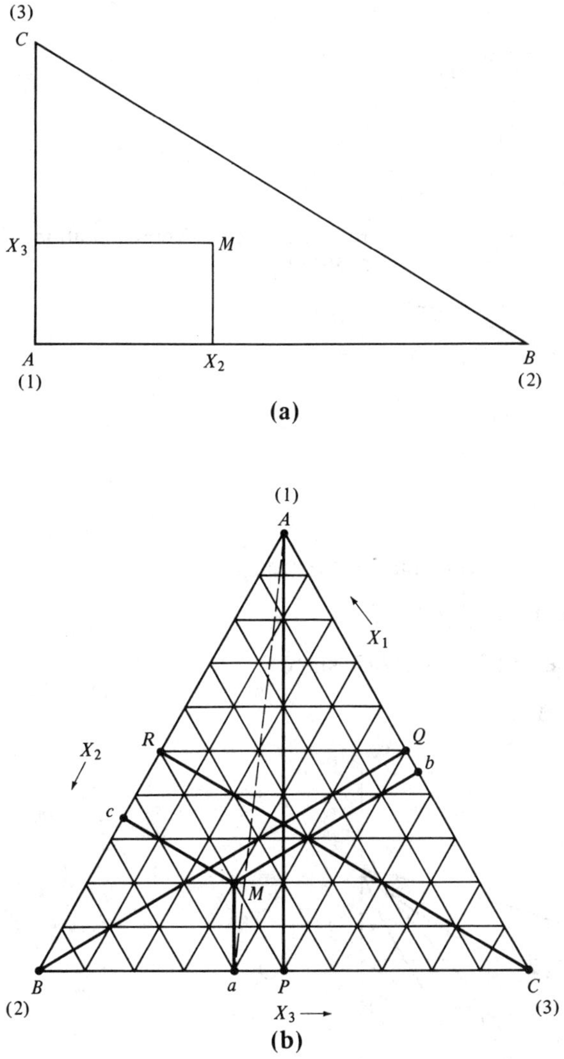

Figure 1. Concentration triangle for the graphical representation of ternary phase diagrams.

at right angles to AP, these lines represent compositions at a constant distance from BC and therefore at constant X_1. If similar lines are drawn through the perpendiculars BQ and CR, a triangular mesh is obtained which allows a quick reading of any composition.

We note that the vertices A, B, C represent the pure components 1, 2, 3 and that the edges AB, BC, CA represent the binaries 1–2, 2–3, and 3–1. Alloys on a line drawn from a vertex (identifying a component) have a constant ratio of the other two components: e.g., along Aa, all the alloys have the same X_2/X_3.

It is useful to represent any composition X_1, X_2, X_3 by a vector $\mathbf{X}$. If we choose an orthonormal basis to represent these vectors (axes perpendicular to each other and with the same scale of measurement) then the compositions of the pure components 1, 2, 3 are represented in Fig. 2 by the vectors $\mathbf{0A}$ (1,0,0), $\mathbf{0B}$ (0,1,0), and $\mathbf{0C}$ (0,0,1). The concentration triangle is then the equilateral triangle ABC.

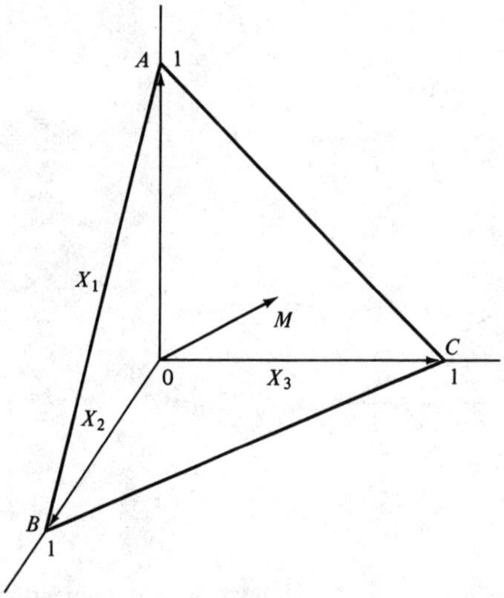

Figure 2. Vector representation of the composition.

Any composition $\mathbf{0M}$ (X_1, X_2, X_3) may be written

$$X_1\mathbf{0A} + X_2\mathbf{0B} + X_3\mathbf{0C} = \mathbf{0M} \tag{1}$$

and the point M must be in triangle ABC, since

$$X_1 + X_2 + X_3 = 1 \tag{2a}$$

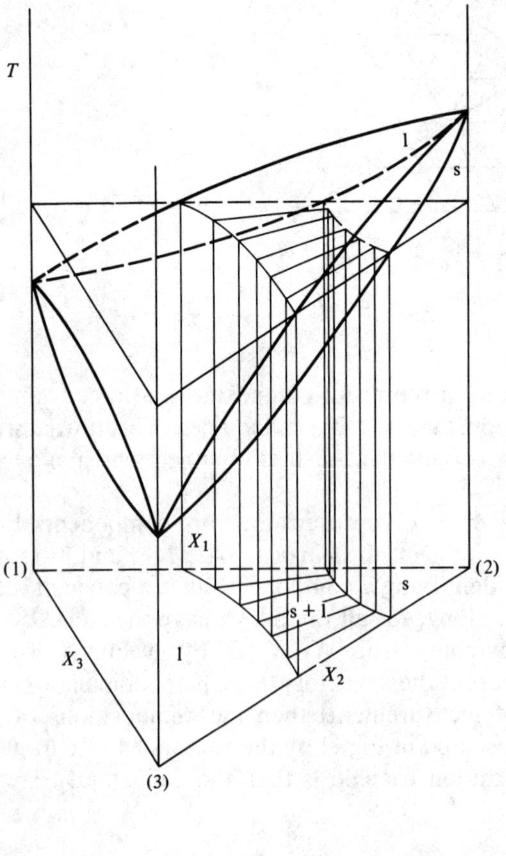

Figure 3. Example of a simple ternary diagram: the elements 1–3 mix almost ideally.

and

$$0 \leq X_i \leq 1 \qquad (i = 1, 2, 3) \tag{2b}$$

Equation (2a) is indeed the analytic representation of the plane ABC, and the inequalities (2b) restrict the point M to the interior of the triangle ABC in the plane of that triangle.

If different scales are used for the three axes $0A$, $0B$, $0C$, then the triangle ABC is no longer equilateral. It is also possible to use as basis vectors $\mathbf{0A}$, $\mathbf{0B}$, $\mathbf{0C}$, which are not perpendicular to each other. Although the principles we shall expose remain the same, it is more convenient to use an orthonormal basis with the associated equilateral triangle (sometimes referred to as the *Gibbs concentration triangle*), and we adopt it for the remainder of this chapter.

To represent the effect of temperature, an axis is added perpendicular to the composition triangle. The sides of the prism thus obtained correspond to the three binary phase diagrams.

1.2. Examples of Ternary Phase Diagrams

For illustrative purposes, let us consider a ternary phase diagram where components 1–3 mix almost ideally (see Fig. 3). The binaries are therefore shaped like single lenses (see Chapter VIII) and the ternary diagram consists of two simple surfaces, the liquidus and the solidus. In the two-phase region, a liquid

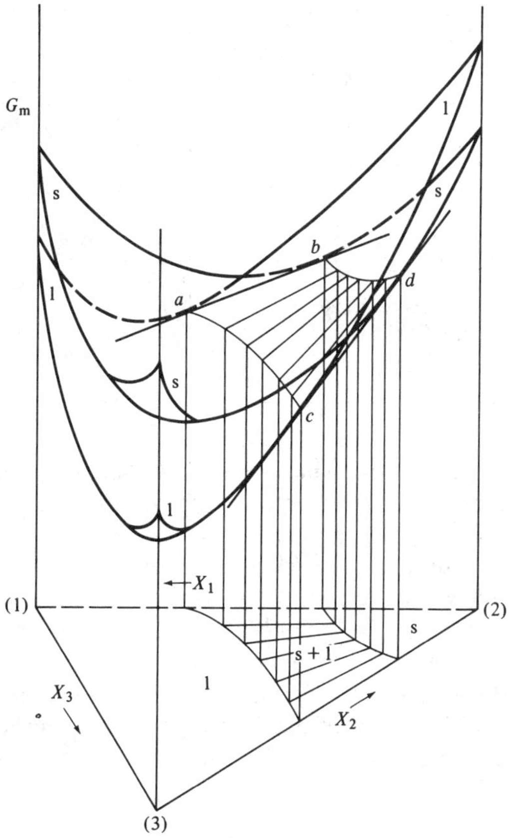

Figure 4. Free energy surfaces corresponding to the isothermal section shown in Fig. 3.

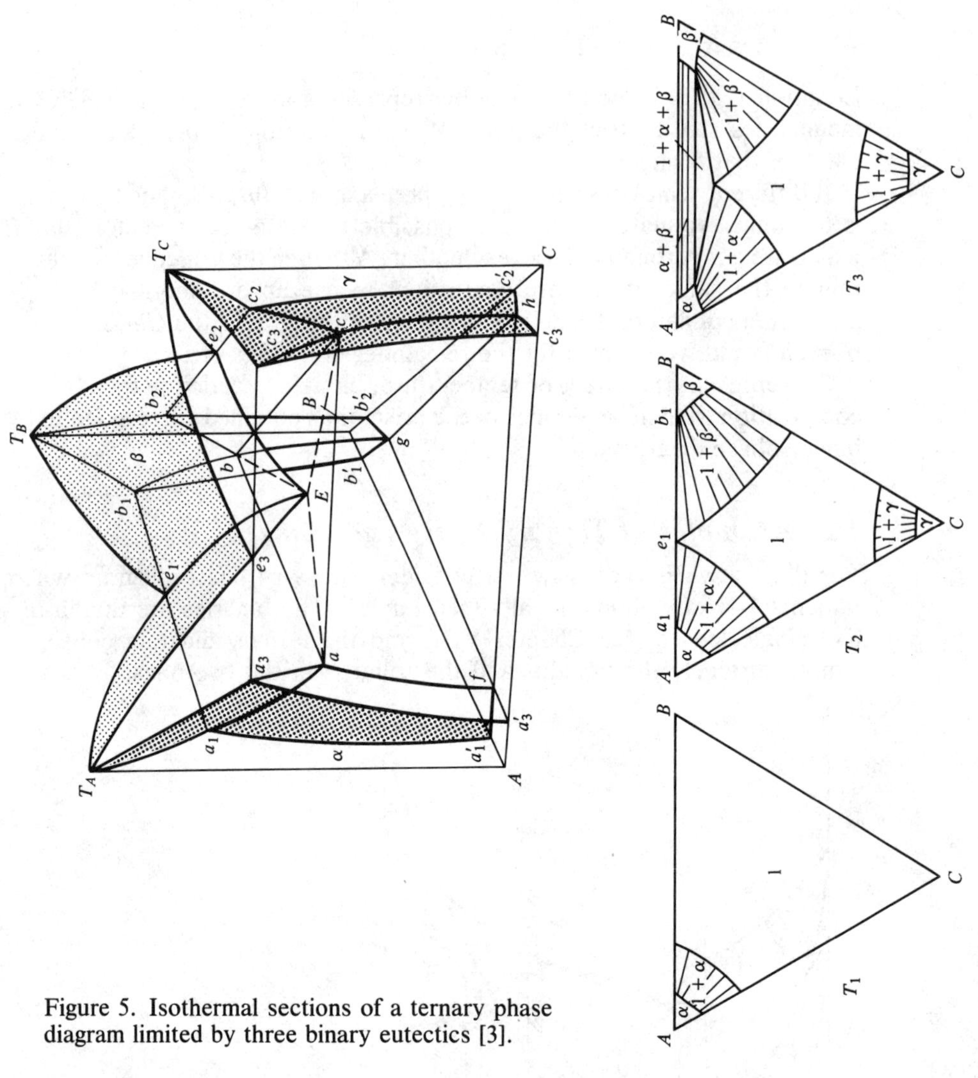

Figure 5. Isothermal sections of a ternary phase diagram limited by three binary eutectics [3].

alloy is in equilibrium with a solid alloy. The line joining the two points in the phase diagram must be horizontal, since at equilibrium the two alloys must be at the same temperature (see Chapter II), and is called a *tie line*.[1] Figure 3 shows an isothermal section of the phase diagram with a series of tie lines.

For binary systems, we have seen how the phase diagram can be derived from the construction of a common tangent to free energy curves. The same construction can be extended to ternary (and multicomponent) systems. Figure 4 illustrates the free energy surfaces of the system considered in Fig. 3 at the temperature of the isothermal section. There are a series of planes which are tangent to the free energy surfaces of both the liquid and solid phases (the planes "roll" from *ab* to

[1] It is also occasionally referred to as a *conode*, a term originating in the translation of the German word "Konode."

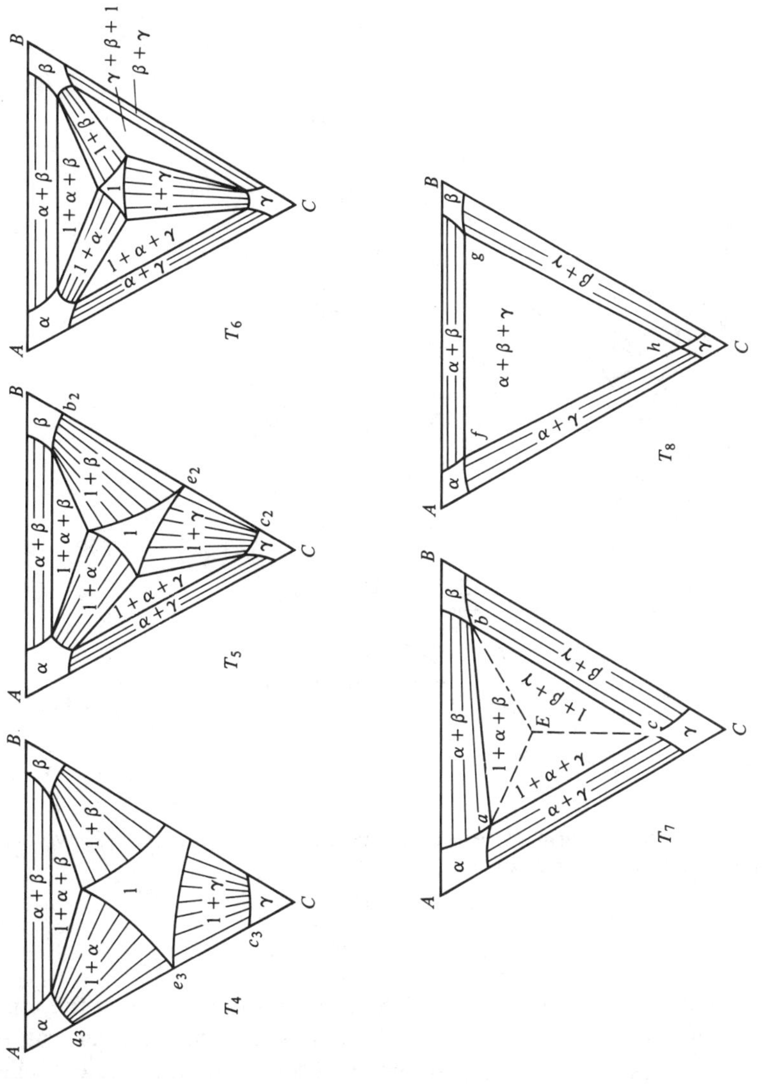

cd). In these planes, the lines which join their contact points on the two surfaces are the tie lines; when projected on the concentration triangle, they are identical to those shown in Fig. 3.

Another example of a ternary phase diagram is shown in Fig. 5. The three binary systems are characterized by eutectic transformations. A series of isothermal sections at decreasing temperatures gives a reasonable description of the three-dimensional surface. We note the appearance of three-phase domains (e.g., at T_3, T_5). They are always bounded by tie lines and share an edge with two-phase domains. With a single-phase domain, three-phase domains can only share a point in the isothermal section (or a curved line in the three-dimensional diagram). The point E (at T_7) corresponds to a four-phase equilibrium. We have shown through the phase rule that, since the pressure is arbitrarily fixed, this equilibrium is indeed invariant (there is no degree of freedom left).

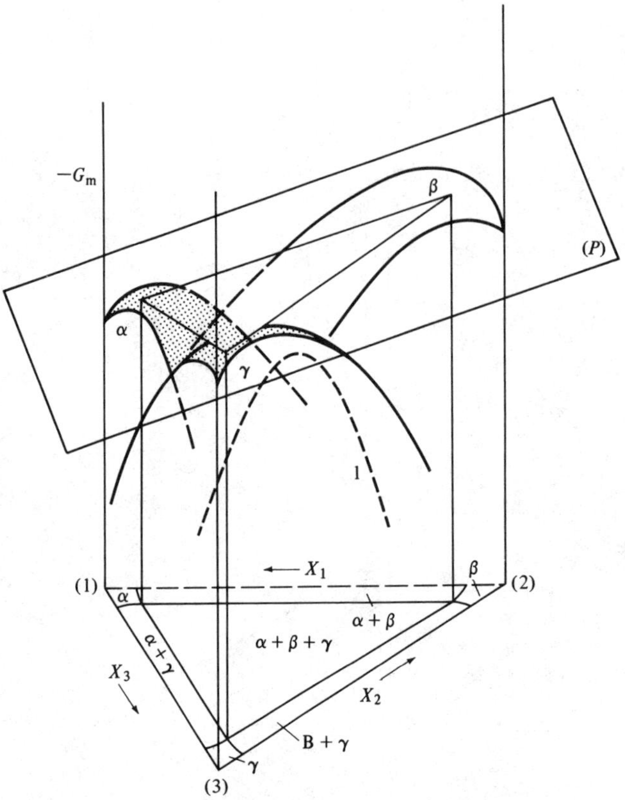

Figure 6. Free energy surfaces below the eutectic temperature T_E. (They correspond to the isothermal section $T = T_8$ of Fig. 5.)

Figure 6 illustrates the position of the free energy surfaces at the temperature $T < T_E$. This time $-G_m$ instead of $+G_m$ is plotted along the vertical axis. This makes the process of visualizing stable equilibria somewhat easier. Imagine that we wrap on this system of surfaces a flexible membrane (such as these plastic stretching ones which have become popular in many kitchens!). The points resting on this membrane have a maximum $-G_m$ (or a minimum G_m) for their composition and represent stable equilibria. If they rest on the free energy surfaces, they correspond to a single phase system. If they rest between surfaces, they correspond to equilibria between two or three phases.

Also in Fig. 6 let us note, at $T < T_E$, the plane P which sits on the three free energy surfaces of the α, β, and γ phases and is tangent to these surfaces. This case corresponds to T_8 in Fig. 5. As the temperature varies, the free energy surfaces move relative to each other. At $T = T_E$ the free energy surface of the fourth phase (liquid) has been raised relative to the three others and all four surfaces are now tangent to a common plane. The four phases are then in equilibrium (case of T_7 in Fig. 5). At $T > T_E$ a common tangent plane to all four surfaces does not exist, and other planes tangent to three surfaces, α, β, l, or α, γ, l, or β, γ, l, must then be considered (e.g., at T_6 and T_5).

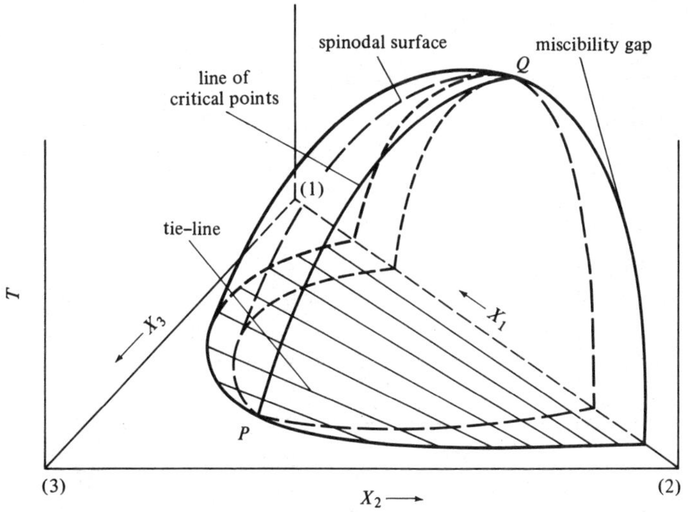

Figure 7. Miscibility gap, spinodal surface, and line of critical temperatures in a ternary system.

A Gibbs free energy surface is not always uniformly concave. When it is not, it generates a miscibility gap (as in the case of binary systems). Figure 7 illustrates the phase diagram of a ternary miscibility gap. The critical line *PQ* corresponds to the point at which the tie lines in isothermal sections become of vanishingly small length. It also belongs to the spinodal surface. In Chapter XI we shall deal more extensively with the subject of miscibility gaps for ternary and multicomponent systems.

In Figs. 3–7, we have seen various examples of equilibria between two, three, or four phases. The compositions of these phases may now be read; to find their amounts, we must in turn consider the extension of the lever rule of binary systems to ternary ones.

1.3. Lever Rule

Consider at the temperature *T* the two-phase equilibrium $\alpha \rightleftarrows \beta$ (or, in the case of a miscibility gap, $\alpha_1 \rightleftarrows \alpha_2$). If X_1, X_2, X_3 are the overall composition coordinates of the system (point *M* in Fig. 8), we wish to determine the amounts and compositions of the two individual phases α and β which constitute that system *M*. With the vector notation adopted in Section 1.1, we must have

$$\mathbf{0M} = n^{\alpha}\mathbf{0P} + n^{\beta}\mathbf{0Q} \tag{3}$$

where **0*P*** and **0*Q*** are the compositions of the two phases α and β. Equation (3) shows that *M* must be on the line *PQ* (since $n^{\alpha} + n^{\beta} = 1$) and is the center of gravity of *PQ*, where *P* bears the weight n^{α} and *Q* the weight n^{β}. Consequently, the tie line *PQ* passing by *M* yields the two compositions X_i^{α} at *P* and X_i^{β} at *Q*,

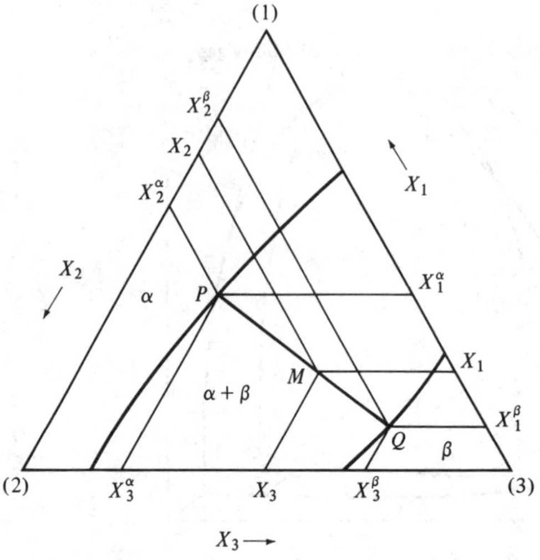

Figure 8. Illustration of the lever rule for a two-phase equilibrium:

$$n^\alpha = \frac{MQ}{PQ} = \frac{X_i^\beta - X_i}{X_i^\beta - X_i^\alpha}$$

$$n^\beta = \frac{PM}{PQ} = \frac{X_i - X_i^\alpha}{X_i^\beta - X_i^\alpha}$$

and the numbers of moles of α and β:

$$n^\alpha = \frac{MQ}{PQ} = \frac{X_i^\beta - X_i}{X_i^\beta - X_i^\alpha}, \quad n^\beta = \frac{PM}{PQ} = \frac{X_i - X_i^\alpha}{X_i^\beta - X_i^\alpha} \quad (i = 1, 2 \text{ or } 3) \tag{4}$$

If M is a mixture of three phases α, β, γ represented respectively by the points P, Q, R, in Fig. 9, then

$$\mathbf{0M} = n^\alpha \mathbf{0P} + n^\beta \mathbf{0Q} + n^\gamma \mathbf{0R} \tag{5}$$

Equation (5) shows that if we attach the weights n^α to P, n^β to Q, and n^γ to R, then M is the center of gravity of the tie triangle PQR. As a consequence, the areas of the triangles MQR, MPR, and MPQ are proportional to n^α, n^β, n^γ, or

$$n^\alpha = \frac{\Delta MQR}{\Delta PQR}, \quad n^\beta = \frac{\Delta MPR}{\Delta PQR}, \quad n^\gamma = \frac{\Delta MPQ}{\Delta PQR} \tag{6}$$

(where the triangle is a notation for area).

It is also convenient to obtain graphically the values of n^α, n^β, n^γ in a slightly different way. Let

$$\mathbf{0S} = \frac{n^\beta}{n^\beta + n^\gamma} \mathbf{0Q} + \frac{n^\gamma}{n^\beta + n^\gamma} \mathbf{0R} \tag{7}$$

then

$$\mathbf{0M} = n^\alpha \mathbf{0P} + (n^\beta + n^\gamma) \mathbf{0S} \tag{8}$$

The point S is at the intersection of PM and QR (see Fig. 9).[2] Equation (8) shows that M may be considered as the center of gravity of P and S, i.e., we may use the common lever rule and regard M as a mixture of the alloys P and S. S itself

[2] S, Q, R are colinear because of equation (7), and M, P, S are colinear because of equation (8).

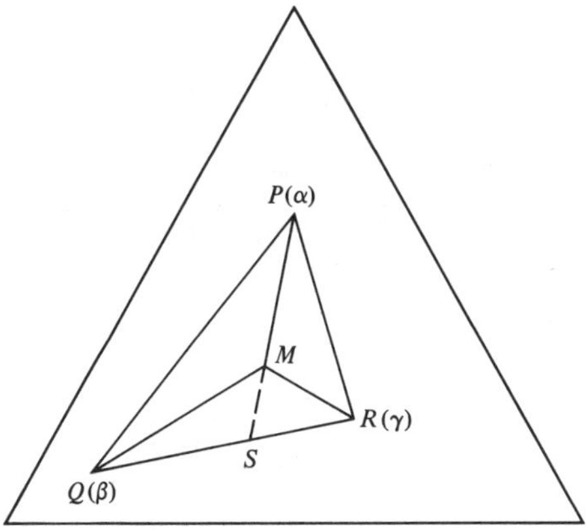

Figure 9. Illustration of the lever rule for a ternary alloy consisting of the phases α, β, and γ:

$$n^{\alpha} = \frac{MS}{PS} = \frac{\Delta MQR}{\Delta PQR}$$

is the center of gravity of the lever QR. Thus,

$$n^{\alpha} = MS/PS \tag{9a}$$

and since

$$n^{\beta} + n^{\gamma} = PM/PS \tag{9b}$$

we also have

$$n^{\beta} = \frac{RS}{QR}\frac{PM}{PS}, \qquad n^{\gamma} = \frac{QS}{QR}\frac{MS}{PS} \tag{9c}$$

In the case of a four-phase equilibrium, the results of the tie triangle are readily extended. The point M in Fig. 9 could just as well represent the composition of a liquid phase in equilibrium with the α, β, and γ phases represented by P, Q, and R.

1.4. Four-Phase Equilibria

Let us briefly consider the cases shown in Fig. 10. In Fig. 10a the liquid phase is in equilibrium with α, β, and γ:

$$l = \alpha + \beta + \gamma \tag{10}$$

In Fig. 10b $L(l)$ lies outside the triangle but $R(\gamma)$ lies inside the triangle LPQ, and we have the equilibrium

$$\gamma = l + \alpha + \beta \tag{11}$$

In Fig. 10c there is no phase lying inside the triangle formed by the compositions of the other three. In this case, the equilibrium is of the type

$$l + \beta = \alpha + \gamma \tag{12}$$

By analogy with the terminology of binary phase diagrams, equation (10) is said to represent a eutectic reaction, and equation (12) is said to represent a peritectic reaction. Equation (11) can be considered a mixture of eutectic and peritectic reactions, and is sometimes referred to as a *quasi-peritectic* reaction

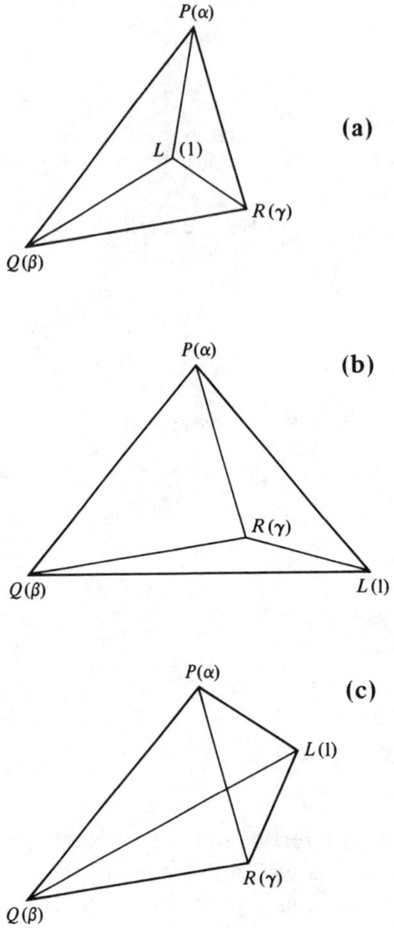

Figure 10. Three four-phase equilibria: **(a)** $l = \alpha + \beta + \gamma$, **(b)** $\gamma = \alpha + \beta + l$, **(c)** $l + \beta = \alpha + \gamma$.

[3]. In Rhines's text [2], equilibria (10), (12), and (11) are listed as class I, II, and III reactions, respectively.

An example of a ternary phase diagram with a eutectic reaction such as (10) has been shown in Fig. 5. Other examples (including peritectic and quasi-peritectic equilibria) may be found in the texts of Rhines [2] and Prince [3] with a profusion of well-drawn figures.

2. Notes on the Graphical Representation of Multicomponent Phase Diagrams

The complexity of a system increases very rapidly with the number of its components. We have seen that the phase diagram of a ternary system requires, at constant pressure, a three-dimensional model. A quaternary system requires a four-dimensional one, and a system of m components an m-dimensional model.

To represent the composition of a quaternary system, a technique very similar to that of the Gibbs equilateral triangle is generally used. The representation is now based upon an equilateral tetrahedron: all edges have the same length and all faces are equilateral triangles. If we consider a point M within the tetrahedron (see Fig. 11), the distances of M to the four faces of the tetrahedron add to a constant independent of the position of M. Thus, the perpendiculars Ma, Mb, Mc, Md to the four faces BCD, ACD, ABD, ABC are used to measure the concentrations of the components 1, 2, 3, 4. The corners (vertices) of the tetrahedron represent the pure components of the system, the edges its binary subsystems and the faces its ternary ones.

This method of representation may be generalized to any number m of components. An equilateral polyhedron of $m-1$ dimensions is used. Again its corners represent the pure components, each edge passing by two corners represents the binary system corresponding to these corners, and so on. The concentrations X_i ($i = 1, 2, \ldots, m$) of any representative point M of a system are obtained by measuring the distances of M to the $(m-2)$-dimensional faces of the polyhedron.

The lever rule may be generalized just as well. For example, if M is a mixture of four phases α, β, γ, δ represented by the points P, Q, R, and S (Fig. 12), M is the center of gravity of the polyhedron $PQRS$ when the weight n^α (number of moles of the phase α) is attached to P, n^β to Q, n^γ to R, and n^δ to S. n^α may be measured as the ratio of the volumes $MQRS$ to $PQRS$, and similarly for n^β, n^γ, n^δ. Other methods of representation as well as rules governing the phase diagram of an m-component system and its phase equilibria may be found in the classical text of Palatnik and Landau [4].

Let us note that two-dimensional sections of a multidimensional phase diagram

Figure 11. Concentration equilateral tetrahedron for the graphical representation of a quaternary system. The perpendiculars Ma, Mb, Mc, Md to the four faces BCD, ACD, ABD, ABC measure the concentrations of components 1–4.

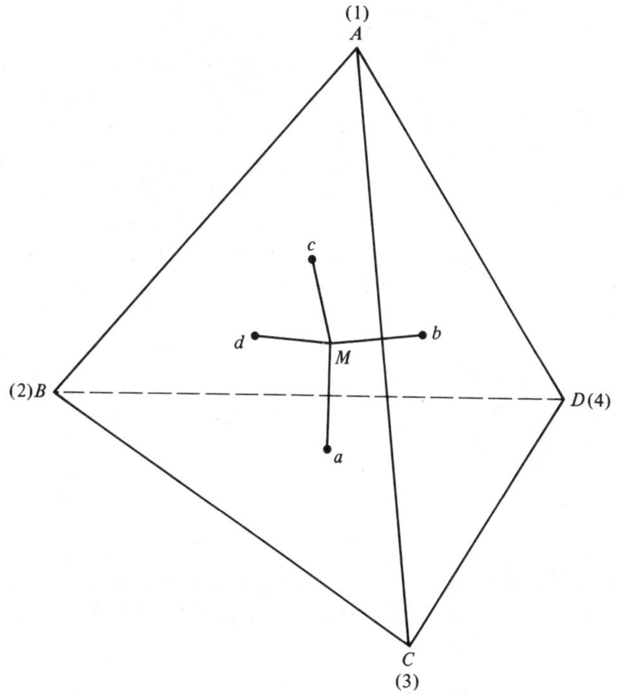

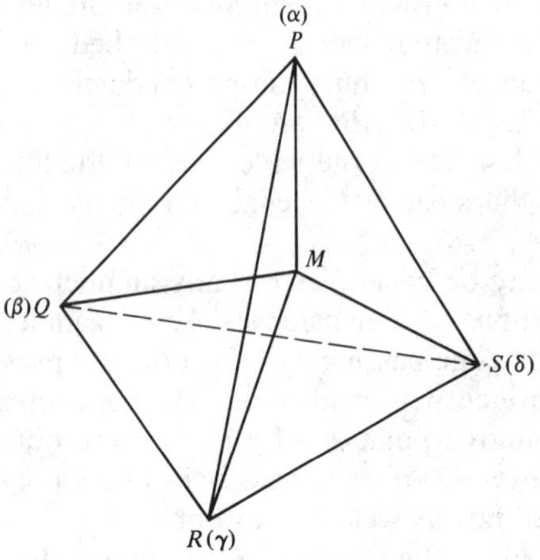

Figure 12. Illustration of the lever rule in a quaternary system: n^α is equal to the ratio of the volumes of the polyhedra *MQRS* and *PQRS*, n^β to the ratio of volumes *MPRS* and *PQRS*, and so forth.

are extremely useful. Generally, these sections are made either parallel to the temperature axis (so that a pseudobinary system is obtained) or perpendicular to it (isothermal section). Furthermore, to reduce the number of dimensions, the sections are also made at constant ratio of two concentration variables (e.g., X_i/X_j) or at fixed concentrations of certain components. Occasionally, only the sum of the concentrations of certain components is taken into account: this occurs most often when these components play very analogous roles (e.g., basic oxides such as MgO and CaO in silicate systems [5]).

3. Representation and Calculation of Gibbs Free Energies

In Chapter IX we saw that the formalism of interaction coefficients provided a convenient analytic representation of the properties of a dilute multicomponent phase. In this section we shall seek other analytic representations which are useful over the entire range of concentrations. For simplicity, we shall restrict these analytic representations to polynomials and adopt the mole fraction as our composition coordinate. These choices are warranted by statistical thermodynamics models of substitutional solutions; for interstitial solutions, other choices may occasionally be more convenient and these are best dictated by appropriate models.

If the behavior of one component has been experimentally determined over a large composition range, then the behavior of all the other components may be deduced through the Gibbs–Duhem equation, and the integral excess free energy per mole of solution is obtained:[3]

$$G^{\mathrm{E}} = RT \sum_{i=1}^{m} X_i \ln \gamma_i \qquad (13)$$

However, if no individual component is the subject of a particular determination, then, for symmetry reasons, it is more satisfactory to adopt as the basis an analytic

[3] For brevity, we shall omit the subscript *m* in subsequent equations.

representation of G^E and then to deduce the activity coefficients of the components. We shall examine both approaches.

3.1. *Analytic Representation of the Integral Gibbs Free Energy*

At the risk of redundancy, we recall that one may write for any particular phase

$$G = \sum_{i=1}^{m} X_i\mu_i^* + RT \sum_{i=1}^{m} X_i \ln X_i + G^E \tag{14}$$

where μ_i^* is the Gibbs free energy of the pure component i in the same structure as that of the phase under consideration, and that, indeed, the problem of an analytical representation of G is that of finding one for G^E.

The degree of the polynomial representing G^E, and therefore the number of parameters introduced in this representation, depends on the amount of data we have and on the accuracy of these data. For example, if only n data points are available, it would be totally unwarranged to use a representation with a number of parameters larger than n. Also, given the experimental uncertainty attached to each point, the useful number of parameters is generally much smaller than n. The determination of the optimum number of parameters is a classical problem of statistics; various regression analyses exist as well as numerous computer programs to perform them. Often, the judgment of an experienced investigator can readily provide the answer.

The simplest expression of G^E corresponds to ideal mixing: $G^E = 0$. This case seldom arises, but it is very useful for rough approximations. The next simplest expression is that of a regular solution:

$$G^E = \sum_{i=1}^{m-1} \sum_{j>i}^{m} X_i X_j \Omega_{ij} \tag{15}$$

For example, for a ternary system it would be

$$G^E = X_1X_2\Omega_{12} + X_1X_3\Omega_{13} + X_2X_3\Omega_{23} \tag{16}$$

Each term $X_iX_j\Omega_{ij}$ represents the contribution of the binary system i–j. Thus, in this model the multicomponent system is obtained as a sum of all the binary contributions.

In the regular solution model, the excess entropy is equal to zero. The quasi-regular solution model retains some of the simplicity of the regular model but no longer assumes that S^E is equal to zero [6]. It yields the expression

$$G^E = \sum_{i,j} X_iX_j\Omega_{ij}[1 - (T/\tau_{ij})] \tag{17}$$

Often, because of insufficient data on the entropy, it is further assumed that all the τ_{ij} are identical to a single value τ.

For ternary systems, it is useful to consider the properties of the solution as the sum of binary contributions and specifically ternary ones. We shall write

$$G^E = G^E_{1-2} + G^E_{1-3} + G^E_{2-3} + G^E_{1-2-3} \tag{18}$$

where G^E_{i-j} represents an analytical expression of G^E for the binary system i–j [e.g., equations (VII.42) or (VII.43)], and G^E_{1-2-3} the specifically ternary contributions. For example, G^E_{1-2} could be equal to

$$G^E_{1-2} = X_1X_2(A_{210}X_1 + A_{120}X_2 + A_{220}X_1X_2) \tag{19}$$

G^E_{1-3} and G^E_{2-3} could be obtained from equation (19) by permutation of the indices. G^E_{1-2-3} could be written

$$G^E_{1-2-3} = X_1X_2X_3A_{111} \tag{20a}$$

or if additional parameters are needed,

$$G^E_{1-2-3} = X_1X_2X_3(A_{211}X_1 + A_{121}X_2 + A_{112}X_3) \tag{20b}$$

Obviously, this procedure can be extended to any number of parameters. However, especially in the case of metallic solutions, rarely more than three ternary parameters are presently warranted. We also note that the coefficients A may be considered temperature dependent and that it is often convenient to write them as $A[1 - (T/\tau)]$.

The case of a quaternary system is similar. We may write

$$G^E = G^E_{1-2} + G^E_{1-3} + G^E_{1-4} + G^E_{2-3} + G^E_{2-4} + G^E_{3-4} + G^E_{1-2-3} + G^E_{1-2-4} + G^E_{1-3-4} + G^E_{2-3-4} + G^E_{1-2-3-4} \tag{21}$$

where

$$G^E_{1-2-3-4} = X_1X_2X_3X_4A_{1111} \tag{22a}$$

or for additional parameters

$$G^E_{1-2-3-4} = X_1X_2X_3X_4(A_{2111}X_1 + A_{1211}X_1 + A_{1121}X_3 + A_{1112}X_4) \tag{22b}$$

and so forth (any number of parameters can be added this way).

It should be observed that the number p of coefficients increases very rapidly with the degree N of the polynomial chosen to represent the properties of the solution. Specifically, it may be shown [7] that

$$p = \sum_{q=2}^{N} C^q_m C^{q-1}_{N-1} = C^N_{m+N-1} - m \tag{23}$$

where q is the order of the coefficients and C^i_j is the combinatorial factor $j!/(j-i)!i!$. For example, in the case of a ternary solution if $N = 3$ (polynomial of order 3), the number of terms of order 2 is equal to 6 and there is only one specifically ternary parameter (A_{111}) for a total of 7 parameters. If the order of the polynomial is increased to 4, 12 coefficients are then necessary. For a quaternary system with a polynomial of order 3, 16 coefficients are needed, but with a polynomial of order 4, 31 coefficients are needed: 18 of order 2 (and thus binary parameters), 12 of order 3 (ternary ones), and only 1 of order 4 (specifically quaternary).

Note that in the representation of a multicomponent system, such as that described in equation (21) for a quaternary, it is not necessary that all the polynomials representing the binaries i–j or the ternaries i–j–k be of the same order. Certain subsystems may warrant many more terms than others and the approach above has the flexibility to accommodate this.

Although very convenient, this approach has one significant drawback. To understand it, consider equation (18) and let us assume that the binary solution 1–2 has been determined to be quasi-regular. We do not obtain an equivalent formula on replacing the binary contribution G^E_{1-2} in equation (18) by

$$G^E_{1-2} = X_1X_2(\alpha X_1 + \beta X_2) \tag{24}$$

or by

$$G^E_{1-2} = X_1X_2[\alpha + (\beta - \alpha)X_2] \tag{25}$$

since X_1 is not equal to $1 - X_2$ in the ternary system. Agreeing to a symmetric formalism [e.g., equation (24) rather than (25)] may remove some of this arbitrariness, but the real issue is different. These analytic techniques are very arbitrary in the first place, and their selection is entirely at the discretion of the investigator. What is incumbent upon him is to specify without ambiguity that selection.

Once an expression of the excess Gibbs free energy has been selected, the expressions of the activity coefficients can easily be derived through the following equation (see Section II.3.2.2):

$$G^E_i = RT \ln \gamma_i = G^E + \sum_{j=2}^{m} (\delta_{ij} - X_j)\frac{\partial G^E}{\partial X_j} \tag{26}$$

where δ_{ij} is Kronecker's symbol ($\delta_{ij} = 0$ for $i \neq j$ and $\delta_{ij} = 1$ for $i = j$).

For example, the reader may verify that if one selects a polynomial of order 4 for the G^E of a ternary solution [i.e., equations (18), (19), and (20b)], then the expression for $\ln \gamma_1$ becomes

$$\ln \gamma_1 = (\ln \gamma_1)_{1-2} + (\ln \gamma_1)_{1-3} + (\ln \gamma_1)_{1-2-3} \tag{27}$$

where

$$RT(\ln \gamma_1)_{1-2} = 2X_1(1 - X_1)X_2A_{210} + (1 - 2X_1)X_2^2A_{120} + X_1(2 - 3X_1)X_2^2A_{220} \tag{28a}$$

$$RT(\ln \gamma_1)_{1-3} = 2X_1(1 - X_1)X_3A_{201} + (1 - 2X_1)X_3^2A_{102} + X_1(2 - 3X_1)X_3^2A_{202} \tag{28b}$$

and

$$RT(\ln \gamma_1)_{1-2-3} = X_2X_3[-(2X_2A_{021} + 2X_3A_{012} + 3X_2X_3A_{022}) + X_1(2 - 3X_1)A_{211} + (1 - 3X_1)(X_2A_{121} + X_3A_{112})] \tag{28c}$$

Expressions for $\ln \gamma_2$ and $\ln \gamma_3$ can be readily obtained by permutation of the indices (e.g., for $\ln \gamma_2$, $X_1 \to X_2$, $X_2 \to X_3$, $X_3 \to X_1$, and $A_{ijk} \to A_{kij}$). For a more complete description of these relations, the reader is referred to Gaye and Lupis [7].

To illustrate the equations above, the ternary system Bi–Pb–Zn at 793°K is considered. Its properties may be represented by the equations [7]

$$G^{E}/RT = (G^{E}/RT)_{\text{Bi-Pb}} + (G^{E}/RT)_{\text{Bi-Zn}} + (G^{E}/RT)_{\text{Pb-Zn}} + (G^{E}/RT)_{\text{Bi-Pb-Zn}} \tag{29}$$

where

$$(G^{E}/RT)_{\text{Bi-Pb}} = X_{\text{Bi}}X_{\text{Pb}}[(-0.69 \pm 0.01)X_{\text{Bi}} + (-0.635 \pm 0.01)X_{\text{Pb}} + (-0.35 \pm 0.02)X_{\text{Bi}}X_{\text{Pb}}] \tag{30}$$

$$(G^{E}/RT)_{\text{Bi-Zn}} = X_{\text{Bi}}X_{\text{Zn}}[(1.5 \pm 0.15)X_{\text{Bi}} + (3.3 \pm 0.15)X_{\text{Zn}} + (-3.2 \pm 0.4)X_{\text{Bi}}X_{\text{Zn}}] \tag{31}$$

$$(G^{E}/RT)_{\text{Pb-Zn}} = X_{\text{Pb}}X_{\text{Zn}}[(2.6 \pm 0.1)X_{\text{Pb}} + (4.5 \pm 0.1)X_{\text{Zn}} + (-1.6 \pm 0.2)X_{\text{Pb}}X_{\text{Zn}}] \tag{32}$$

and

$$(G^{E}/RT)_{\text{Bi-Pb-Zn}} = X_{\text{Bi}}X_{\text{Pb}}X_{\text{Zn}}[(2.5 \pm 0.25)X_{\text{Bi}} + (2.3 \pm 0.2)X_{\text{Pb}} + (-0.2 \pm 0.2)X_{\text{Zn}}] \tag{33}$$

The isoactivity curves resulting from this representation and calculated through equations (27) and (28) are in excellent agreement with the data and calculations of Valenti et al. [8] (see Fig. 13).

Figure 13. Isoactivity curves of **(a)** Zn, **(b)** Bi, and **(c)** Pb in the ternary system Zn–Bi–Pb at 793°K. The dotted lines are reported by Valenti et al. [8].

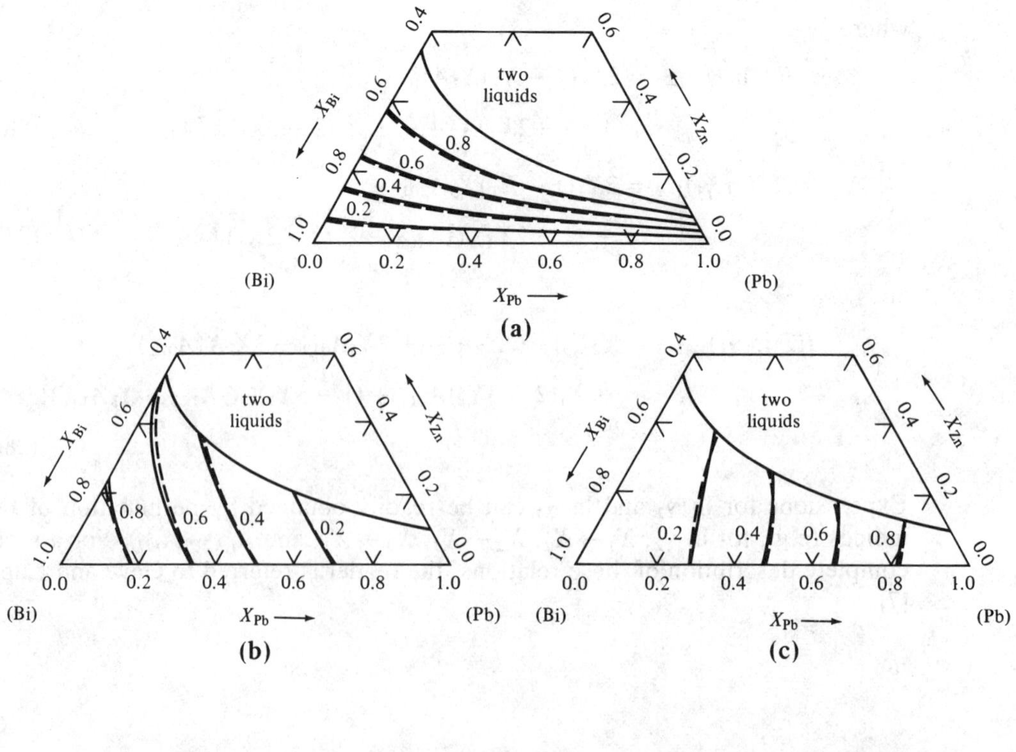

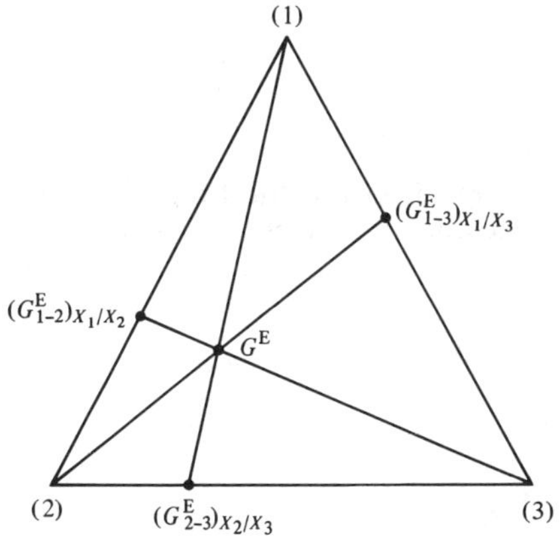

Figure 14. Geometrical representation of terms appearing in Kohler's equation (34).

It should be observed that in the case where binary data exist but no ternary ones do, the Gibbs free energy of the ternary system may be estimated from the limiting binary systems through equation (18) with G^E_{1-2-3} taken to be zero [e.g., $A_{111} = 0$ in equation (20)]. This method of estimation is, of course, no more than an interpolation technique.

There are many other formulas which take into account the limiting binary systems in different ways. One of these formulas is due to Kohler [9]. It may be written

$$G^E = (X_1 + X_2)^2(G^E_{1-2})_{X_1/X_2} + (X_1 + X_3)^2(G^E_{1-3})_{X_1/X_3} + (X_2 + X_3)^2(G^E_{2-3})_{X_2/X_3} \tag{34}$$

where $(G^E_{i-j})_{X_i/X_j}$ is the value of G^E in the binary i–j calculated at the composition X_i/X_j. Figure 14 shows more explicitly at which compositions the limiting binaries are calculated.

Another formula, valid for any number of components, has been proposed by Colinet [10]:

$$G^E = \tfrac{1}{2} \sum_{\substack{i=1\\ i\neq j}}^{m} \sum_{j=1}^{m} \frac{X_i}{1 - X_j} (G^E_{i-j})_{X_j} \tag{35}$$

The compositions at which the excess free energies of the binaries are calculated are shown in Fig. 15 in the case of a ternary system.

It may be observed that the equations of both Kohler and Colinet lead to regular ternary solutions when the binary subsystems are assumed to be regular. To verify this more easily, we note that in equation (34)

$$(G^E_{i-j})_{X_i/X_j} = \frac{X_i}{X_i + X_j} \frac{X_j}{X_i + X_j} \Omega_{ij} \tag{36}$$

and in equation (35)

$$(G^E_{i-j})_{X_j} = X_j(1 - X_j)\Omega_{ij} \tag{37}$$

A good review of many such analytical expressions may be found in an article by Ansara [11].

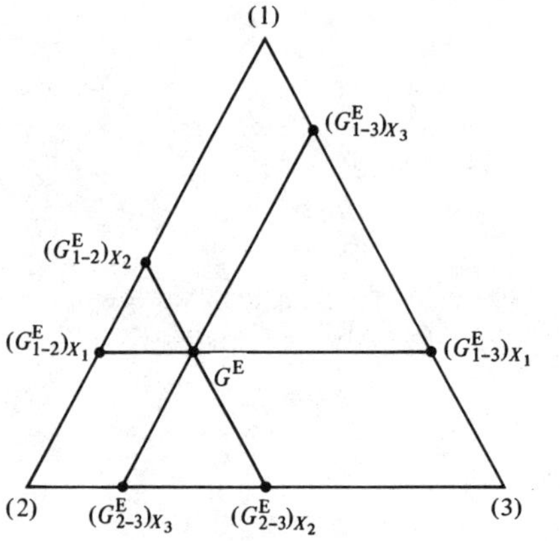

Figure 15. Geometrical representation of terms appearing in Colinet's equation (35).

3.2. Analytic Representation of Activities

In the introduction to Section 3 we remarked that if the thermodynamic activity of one component has been determined over a given composition range, then the activities of all the other components may be deduced in that range through integration of the Gibbs–Duhem equation,

$$\sum_{i=1}^{m} X_i \, d \ln \gamma_i = 0 \tag{38}$$

In Chapter VII a graphical integration of this equation was presented in the case of a binary solution. Graphical integrations are also possible in the case of multicomponent systems. However, as we shall see in Section 3.3 they are very lengthy and tedious to perform, even in the case of a ternary solution. In contrast, an analytical integration is much easier to accomplish.

Let us assume that it is the activity of component 1 that has been measured in a solution of m components and that by methods of statistical analysis we have determined the coefficients $J^{(1)}_{n_2,\ldots,n_i,\ldots,n_m}$ of the polynomial

$$\ln \gamma_1 = \sum_{n_2} \cdots \sum_{n_j} \cdots \sum_{n_m} J^{(1)}_{n_2,\ldots,n_j,\ldots,n_m} X_2^{n_2} \cdots X_j^{n_j} \cdots X_m^{n_m} \tag{39}$$

and its order $N = \sum_{j=2}^{m} n_j$. By the very definition of N it follows that

$$J^{(1)}_{n_2,\ldots,n_j,\ldots,n_m} = 0 \quad \text{for} \quad \sum_{j=2}^{m} n_j > N \tag{40}$$

In our study of interaction coefficients (Chapter IX, Sections 2.5 and 3.2), we saw that the coefficients $J^{(i)}$ and ϕ of the polynomials representing $\ln \gamma_i$ and G^E/RT in terms of the variables $X_2, \ldots, X_j, \ldots, X_m$ are related to those of $\ln \gamma_1$ by the equations

$$J^{(i)}_{n_2,\ldots,n_i,\ldots,n_m} = J^{(1)}_{n_2,\ldots,n_i,\ldots,n_m} - \frac{n_i + 1}{\sum n_j} J^{(1)}_{n_2,\ldots,n_i+1,\ldots,n_m} \tag{41}$$

and

$$\phi_{n_2,\ldots,n_i,\ldots,n_m} = \frac{-1}{(\sum n_j) - 1} J^{(1)}_{n_2,\ldots,n_i,\ldots,n_m} \tag{42}$$

These can be exploited to yield the analytical representation of ln γ_i and G^E/RT. In particular, we note that because of equation (40)

$$J^{(i)}_{n_2,\ldots,n_j,\ldots,n_m} = J^{(1)}_{n_2,\ldots,n_j,\ldots,n_m} \qquad \text{for} \quad \sum_{j=2}^{m} n_j = N \tag{43}$$

and also that the polynomials expressing ln γ_i and G^E/RT have the same order N. Moreover, if the representation is to hold over the entire composition diagram and we choose Raoultian reference states for all the components, then

$$(\ln \gamma_i)_{X_i=1} = 0 \qquad (i = 2, 3, \ldots, m) \tag{44}$$

and consequently,

$$\sum_{p=0}^{N} J^{(i)}_{0,\ldots,n_i=p,\ldots,0} = 0 \qquad (i = 2, 3, \ldots, m) \tag{45}$$

Because $J^{(1)}_{0,\ldots,n_i=0,\ldots,0}$ and $J^{(1)}_{0,\ldots,n_i=1,\ldots,0}$ are both equal to zero (by Raoult's law), equation (45) may be transformed by equation (41) into

$$J^{(i)}_{0,\ldots,0,\ldots,0} = \phi_{0,\ldots,n_i=1,\ldots,0} = \sum_{p=2}^{N} \frac{1}{p-1} J^{(1)}_{0,\ldots,n_i=p,\ldots,0} \tag{46}$$

Thus, given the polynomial representing ln γ_1, equations (41), (42), and (46) readily yield the polynomials representing G^E/RT and ln γ_i. All these calculations can be easily performed by computers (and such a program is provided by Gaye [7]).

3.3. Graphical Integration of the Gibbs–Duhem Equation

Occasionally, a graphical integration may be warranted. There are several techniques for performing it and we may cite here those of Darken [12], Wagner [13], Schuhmann [14], Gokcen [15], and Gaye and Lupis [7]. We shall briefly present the last one as an example of the calculations that these techniques involve.

We again assume that the activity of component 1 has been determined and that we wish to calculate the excess free energy and the activity coefficients of the other components at a composition $X_1, X_2, \ldots, X_m$ characterized by the point P in Fig. 16. Let O be the point representing the pure component 1 and Q the point on the line OP corresponding to $X_1 = 0$. The position of any point M

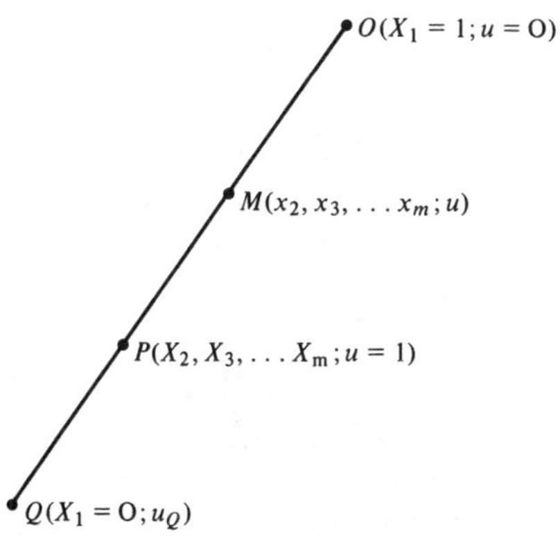

Figure 16. Integration path for the Gibbs–Duhem equation; $u = x_i/X_i$ for $i = 2, 3, \ldots, m$.

on the line OP may be fixed by the parameter u:

$$OM = uOP \tag{47a}$$

or, if the mole fractions of the alloy M are designated x_i

$$x_i = uX_i \quad \text{for} \quad i = 2, 3, \ldots, m \tag{47b}$$

Equation (26) may be rewritten

$$\ln \gamma_1 = (G^{E}/RT) - \sum_{j=2}^{m} x_j \frac{\partial(G^{E}/RT)}{\partial x_j} \tag{48}$$

and we note now that along the line OP, G^{E} is a function of u only:

$$G^{E}(x_2, \ldots, x_m) = G^{E}(uX_2, \ldots, uX_m) = g(u) \tag{49}$$

Since

$$\frac{dg}{du} = \sum_{j=2}^{m} \frac{\partial G^{E}}{\partial x_j} \frac{dx_j}{du} = \frac{1}{u} \sum_{j=2}^{m} x_j \frac{\partial G^{E}}{\partial x_j} \tag{50}$$

equation (48) becomes

$$\ln \gamma_1 = \frac{g}{RT} - \frac{u}{RT} \frac{dg}{du} \tag{51}$$

or

$$\frac{d}{du}\left(\frac{g}{uRT}\right) = -\frac{\ln \gamma_1}{u^2} \tag{52}$$

Integrating it between O and P yields

$$\left(\frac{g}{uRT}\right)_{u=1} - \left(\frac{g}{uRT}\right)_{u=0} = -\int_0^1 \frac{\ln \gamma_1}{u^2}\, du \tag{53}$$

Moreover, since

$$g/RT = x_1 \ln \gamma_1 + \sum_{i=2}^{m} x_i \ln \gamma_i = x_1 \ln \gamma_1 + u \sum_{i=2}^{m} X_i \ln \gamma_i \tag{54}$$

for $u = 0$

$$(g/uRT)_{u=0} = \sum_{i=2}^{m} X_i \ln \gamma_i^{\infty} \tag{55}$$

because $\ln \gamma_1$ is of the order of u^2 (Raoult's law). Consequently, equation (53) becomes

$$\frac{G^{E}}{RT} = \sum_{i=2}^{m} X_i \ln \gamma_i^{\infty} - \int_0^1 \frac{\ln \gamma_1}{u^2}\, du \tag{56a}$$

When the experimental data correspond to a solution dilute in component 1, the integration may be performed more conveniently between the points P and Q, the excess free energy being usually known at the point Q (since a subsystem of $m - 1$ components is integrated before the system of m components). The equivalent to equation (56a) is then

$$\frac{G^{\mathrm{E}}}{RT} = \sum_{i=2}^{m} X_i(\ln \gamma_i)_Q + \int_1^{u_Q} \frac{\ln \gamma_1}{u^2}\, du \tag{56b}$$

The calculation of $\ln \gamma_i$ is based on

$$\ln \gamma_i = \ln \gamma_1 + \frac{\partial(G^{\mathrm{E}}/RT)}{\partial X_i} \tag{57}$$

which is an immediate consequence of equations (26). After some manipulations it may be shown to yield [7]

$$\ln \gamma_i = \ln \gamma_i^{\infty} + \ln \gamma_1 - \int_0^1 \frac{1}{u} \frac{\partial \ln \gamma_1}{\partial X_i}\, du \tag{58a}$$

or

$$\ln \gamma_i = (\ln \gamma_i)_Q + \ln \gamma_1 + \int_1^{u_Q} \frac{1}{u} \frac{\partial \ln \gamma_1}{\partial X_i}\, du \tag{58b}$$

Plots of $\ln \gamma_1$ vs X_i along sets of lines where X_i varies and all the X_js remain constant provide values of the derivatives $\partial \ln \gamma_1/\partial X_i$. Plots of the integrands in equations (56) and (58), along pseudobinary lines (joining given points of the concentration diagram to the corner 0 of the pure component 1) and vs the corresponding parameter u, yield the values of the integrals and, consequently, of G^{E} and $\ln \gamma_i$.

It is estimated that in the case of a ternary system such a graphical integration, over the whole composition diagram and at points distant by 0.1 mol fraction, would require about 215 curves (20 showing the variations of $\ln \gamma_1$ vs X_2 and X_3 and yielding $\partial \ln \gamma_1/\partial X_2$ and $\partial \ln \gamma_1/\partial X_3$, and 195 showing the variations of the integrands). The other methods referenced in this section require similar (and often additional) work. Whenever possible, an analytical integration is generally far more expedient.

4. Calculation of Multicomponent Phase Diagrams

As already illustrated in Section 1.2, Fig. 4, knowledge of the Gibbs free energies of the phases of a given multicomponent system allows the determination of its phase diagram. Theoretically, this is fairly simple and is just a generalization of the principles studied for binary systems. However, whereas drawing a common tangent to two curves is relatively easy, constructing a common tangent plane to two surfaces of a ternary system is far more difficult, and the difficulty increases with the number of components. Because analytic solutions generally do not exist, numerical techniques based on the calculating power of a computer had to be developed.

In Section VIII.8 methods were presented for the calculation of the compositions of two phases in equilibrium in a binary system. In a ternary system, if three phases are in equilibrium at a given temperature, their compositions are fixed and can be simply obtained as intersections of two tie lines (determined from the calculation of two-phase equilibria). In a quaternary or higher order system, the domains corresponding to the equilibrium coexistence of more than two phases may still be deduced from the equilibrium between two phases (e.g., as the intersection of two surfaces). Thus, a method to calculate the composition of two phases in equilibrium is sufficient to calculate a ternary phase diagram in its entirety, and is necessary for the calculation of higher order phase diagrams.

The method that we presented in some detail for binary systems consists in solving two equations expressing the equality of the chemical potentials of each component in the two phases. A method based on the same principle has been used by Kaufman et al. [16,17], Hurle and Pike [18], and Ansara et al. [19]. These methods become somewhat cumbersome with a large number of components and a direct minimization of the system's Gibbs free energy has appeared preferable. Spencer et al. [20] used this principle to adapt a simplex technique developed by Nelder and Mead [21], and Gaye and Lupis [22] also used this principle to develop a stepwise procedure that will be briefly presented below.

4.1. Formulation of the Conditions for Equilibrium Between Two Phases by Direct Minimization of the Gibbs Free Energy

Consider an isothermal section of the boundaries of a two-phase field in a multicomponent phase diagram, as schematized in Fig. 17. A point P in that field describes a system of composition described by the mole fractions X_i° ($i = 1, 2, \ldots, m$), and consisting of n_α mol of phase α and n_β mol of phase β. The compositions of these phases are represented by the points Q_α and Q_β and described by the mole fractions X_i^α and X_i^β. The calculation of the phase equilibrium consists in determining, for any point P of the two-phase field, the $2m + 2$ unknowns n_α, n_β, X_i^α, and X_i^β (for $i = 1, 2, \ldots, m$) which define the tie line $Q_\alpha Q_\beta$.

A mass balance provides $m + 2$ relations, or constraints, among the $2m + 2$ unknowns:

$$X_1^\nu = 1 - \sum_{i=2}^{m} X_i^\nu \qquad \text{for} \quad \nu = \alpha, \beta \tag{59a}$$

$$n_\alpha + n_\beta = 1 \tag{59b}$$

$$n_\alpha X_i^\alpha + n_\beta X_i^\beta = X_i^\circ \qquad \text{for} \quad i = 2, 3, \ldots, m \tag{59c}$$

The m remaining relations necessary to determine the tie line $Q_\alpha Q_\beta$ have to be obtained by stating that the system is in equilibrium. Consequently, the values of the mole fractions X_i^α, X_i^β and the amounts of phases n_α, n_β must be such that they minimize the total Gibbs free energy of the system G_s while obeying the

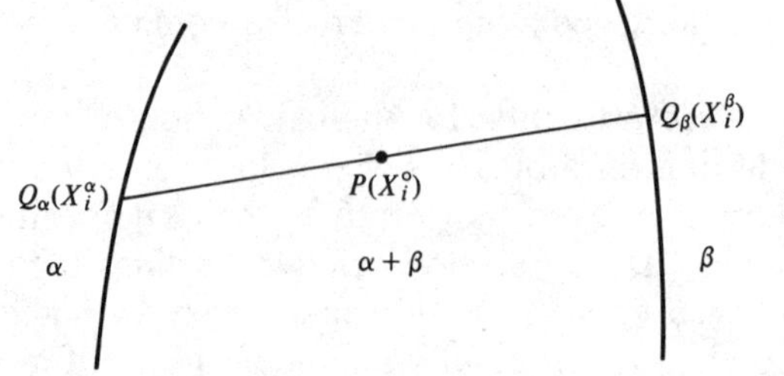

Figure 17. Section of a multicomponent phase diagram.

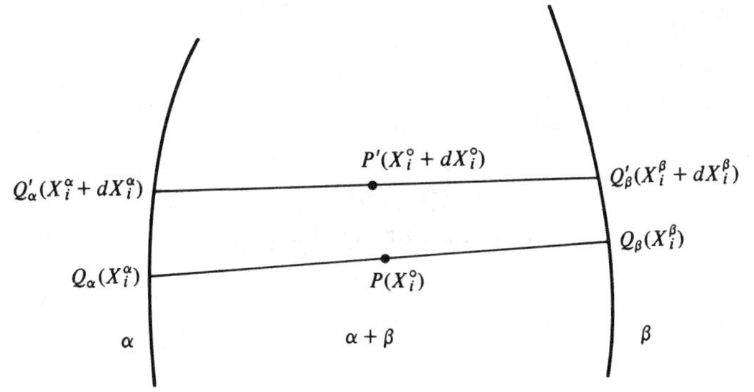

Figure 18. Representation of two neighboring equilibrium states in a multicomponent system.

constraints (59). Note that

$$G_s = n_\alpha G^\alpha + n_\beta G^\beta \tag{60}$$

where G^α and G^β represent the molar free energies of the phases α and β. At a given temperature T, G^ν (for $\nu = \alpha$ or β) can be expressed in terms of the variables X_i ($i = 1, 2, \ldots, m$) by means of the equations presented in Section 3.1. The objective function to be minimized, G_s, therefore contains both logarithmic and polynomial terms, and the constraints (59c) are nonlinear. Rather time consuming search techniques or gradient methods are necessary to solve the problem formulated in this manner. One such technique is presented by Gaye and Lupis [22], although for the bulk of the calculations these authors use the simpler stepwise procedure developed in Section 4.2.

4.2. Stepwise Calculation of an Isothermal Section

Let us consider the two neighboring equilibrium states represented in Fig. 18 by the points P and P'. The free energy difference DG_s associated with these states, i.e., when going from the known equilibrium state X_i^α, X_i^β of overall composition X_i° to the unknown equilibrium state of overall composition $X_i^\circ + dX_i^\circ$, can be approximated by a second order Taylor series in terms of the variables dn_α, dX_i^α, and dX_i^β (for $i = 2, \ldots, m$). Noting that $dn_\beta = -dn_\alpha$,

$$\begin{aligned} DG_s = {} & (G^\alpha - G^\beta)\, dn_\alpha + \sum_{i=2}^{m} (n_\alpha G_i'^\alpha\, dX_i^\alpha + n_\beta G_i'^\beta\, dX_i^\beta) \\ & + \sum_{i=2}^{m} (G_i'^\alpha\, dX_i^\alpha - G_i'^\beta\, dX_i^\beta)\, dn_\alpha \\ & + \tfrac{1}{2} \sum_{i,j=2}^{m} (n_\alpha G_{ij}''^\alpha\, dX_i^\alpha\, dX_j^\alpha + n_\beta G_{ij}''^\beta\, dX_i^\beta\, dX_j^\beta) \end{aligned} \tag{61}$$

where $G_i'^\nu$ and $G_{ij}''^\nu$ represent the first and second order derivatives of the molar free energy in the phase ν with respect to mole fractions, evaluated at the point Q_ν. They are equal to

$$G_i'^\nu = \mu_i^{*(\nu)} - \mu_1^{*(\nu)} + RT \ln \frac{X_i^\nu}{X_1^\nu} + \frac{\partial G^{E\nu}}{\partial X_i^\nu} \qquad (\nu = \alpha, \beta;\ i = 2, \ldots, m) \tag{62}$$

and

$$G''^{\nu}_{ij} = RT\left(\frac{\delta_{ij}}{X_i^\nu} + \frac{1}{X_1^\nu}\right) + \frac{\partial^2 G^{E\nu}}{\partial X_i^\nu \, \partial X_j^\nu} \qquad (\nu = \alpha, \beta;\, i = 2, \ldots, m) \tag{63}$$

where δ_{ij} is Kronecker's symbol.

In terms of the variables dn_α, dX_i^α, and dX_i^β the constraints expressing the mass balances (59) can be approximated by their differential forms, obtained by differentiation of equation (59c):

$$dX_i^\alpha = \frac{1}{n_\alpha + dn_\alpha}[(X_i^\beta - X_i^\alpha)\, dn_\alpha - n_\beta\, dX_i^\beta + dX_i^\circ + dn_\alpha\, dX_i^\beta] \qquad (i = 2, \ldots, m) \tag{64}$$

Both constraints (59a) and (59b) are implicitly satisfied in equations (61) and (64). The new $2m - 1$ variables dn_α, dX_i^α, and dX_i^β $(i = 2, \ldots, m)$, defining the unknown tie line $Q'_\alpha Q'_\beta$, must now minimize DG_s in equation (61) and obey the constraints (64). Replacing in equation (61) the $m - 1$ variables dX_i^α by their expressions in terms of the m variables dn_α and dX_i^β (for $i = 2, \ldots, m$) given in equation (64), the solution is obtained by writing that the derivatives of DG_s with respect to the latter m variables are zero. This results in a system of m linear equations with respect to the m variables dn_α and dX_i^β $(i = 2, \ldots, m)$.

These equations may be expressed in matrix and vector notation:

$$(\mathrm{A})\mathbf{Z} = \boldsymbol{B1} + \boldsymbol{B2} \tag{65}$$

where the elements of the matrix (A) and vectors $\boldsymbol{Z}$, $\boldsymbol{B1}$, and $\boldsymbol{B2}$ are

$$A_{11} = \sum_{i,j=2}^{m} (X_i^\beta - X_i^\alpha)(X_j^\beta - X_j^\alpha) G''^{\alpha}_{ij} \tag{66a}$$

$$A_{1i} = A_{i1} = n_\alpha(G'^{\alpha}_i - G'^{\beta}_i) - n_\beta \sum_{j=2}^{m} (X_j^\beta - X_j^\alpha) G''^{\alpha}_{ij} \qquad (i = 2, \ldots, m) \tag{66b}$$

$$A_{ij} = A_{ji} = n_\beta(n_\alpha G''^{\beta}_{ij} + n_\beta G''^{\alpha}_{ij}) \qquad (i, j = 2, \ldots, m) \tag{66c}$$

$$Z_1 = dn_\alpha \tag{67a}$$

$$Z_i = dX_i^\beta \qquad \text{for} \quad i = 2, \ldots, m \tag{67b}$$

$$B1_1 = n_\alpha[(G^\beta - G^\alpha) - \sum_{i=2}^{m} (X_i^\beta - X_i^\alpha) G'^{\alpha}_i] \tag{68a}$$

$$B1_i = n_\alpha n_\beta (G'^{\alpha}_i - G'^{\beta}_i) \qquad \text{for} \quad i = 2, \ldots, m \tag{68b}$$

$$B2_1 = -\sum_{i,j=2}^{m} (X_i^\beta - X_i^\alpha) G''^{\alpha}_{ij}\, dX_j^\circ \tag{69a}$$

$$B2_i = n_\beta \sum_{j=2}^{m} G''^{\alpha}_{ij}\, dX_j^\circ \qquad \text{for} \quad i = 2, \ldots, m \tag{69b}$$

The calculation of the phase boundaries has thus been reduced to the solution of a system of m linear equations yielding the m quantities dn_α and dX_i^β (for $i = 2, \ldots, m$). Once these have been obtained the remaining variables dX_i^α (for $i = 2, \ldots, m$) can be calculated using the constraint equations.

It may be noted that the right hand side of equation (65) has been separated into two vectors $\boldsymbol{B1}$ and $\boldsymbol{B2}$. The vector $\boldsymbol{B1}$ measures the deviation of the calculated points Q_α and Q_β from the true equilibrium points (because of errors inherent in the quadratic approximation of the Gibbs free energy surface), whereas the vector $\boldsymbol{B2}$ represents the effect of changing the overall composition (its elements are proportional to the length of the vector $\boldsymbol{dX}^\circ$). This is convenient because calculating the equilibrium compositions corresponding to P' yields a check on the equilibrium compositions corresponding to P. If the difference is found to be larger than the precision required, the equation $\mathbf{Z} = (A)^{-1}\mathbf{B1}$ may be used in an iteration, the overall composition remaining unchanged. Should this iteration diverge or converge too slowly, a search method (mentioned above) is used as a last resort.

Other details of the program of calculations can be found in the original reference [22]. It is interesting, however, to dwell further on the problem of the starting points.

In the stepwise method described here, the determination of a series of pairs of equilibrium compositions is based on the knowledge of one pair of equilibrium points. These starting points may be a couple of equilibrium compositions determined either experimentally or from previous calculations. In general, it would be natural to select a pair of predetermined equilibrium points in the $m-1$ component system $1\text{–}2\text{–}\cdots\text{–}(m-1)$. However, these cannot be directly used in the stepwise calculations since, when one of the mole fractions is zero, the matrix (A) of equation (65) is singular. Nevertheless, starting points for the stepwise method can be found on the limiting tangents, for $X_m \to 0$, to the section of the boundary investigated.

4.3. Slopes of the Phase Boundaries at Infinite Dilution of Component m

Let us consider the couple of equilibrium points Q_α and Q_β at $X_m = 0$ (see Fig. 19) and a point Q''_α in the plane tangent to the α–$(\alpha+\beta)$ boundary at Q_α. The tie line going through Q''_α intersects the plane tangent to the β–$(\alpha+\beta)$ boundary in Q''_β, and the points Q''_α and Q''_β constitute an approximation of the equilibrium points Q'_α and Q'_β.

The coordinates of the points Q''_α and Q''_β are

$$X_i^\nu = [X_i^\nu]_{Q_\nu} + dX_i^\nu \qquad \text{for} \quad i = 1, \ldots, m \tag{70}$$

Let us assume for convenience that all the increments dX_i^α are zero except dX_1^α, dX_m^α, and dX_k^α (where k represents an arbitrary component). Given dX_m^α, the equilibrium conditions

$$\mu_i^\alpha = \mu_i^\beta \qquad \text{for} \quad i = 1, 2, \ldots, m \tag{71a}$$

rewritten in terms of the activity coefficients γ_i,

$$\mu_i^{*\alpha} + RT \ln X_i^\alpha \gamma_i^\alpha = \mu_i^{*\beta} + RT \ln X_i^\beta \gamma_i^\beta \tag{71b}$$

provide (through differentiation) m relations allowing the calculation of the increments dX_k^α and dX_j^β (for $j = 2, \ldots, m$), which determine the points Q''_α and Q''_β:

$$\left(\frac{\partial \ln \gamma_1^\alpha}{\partial X_k^\alpha} - \frac{1}{X_1^\alpha}\right)_{Q_\alpha} dX_k^\alpha + \left(\frac{\partial \ln \gamma_1^\alpha}{\partial X_m^\alpha} - \frac{1}{X_1^\alpha}\right)_{Q_\alpha} dX_m^\alpha - \sum_{j=2}^{m} \left(\frac{\partial \ln \gamma_1^\beta}{\partial X_j^\beta} - \frac{1}{X_1^\beta}\right)_{Q_\beta} dX_j^\beta = 0 \tag{72a}$$

$$\left(\frac{\partial \ln \gamma_i^\alpha}{\partial X_k^\alpha} + \frac{\delta_{ik}}{X_k^\alpha}\right)_{Q_\alpha} dX_k^\alpha + \left(\frac{\partial \ln \gamma_i^\alpha}{\partial X_m^\alpha}\right)_{Q_\alpha} dX_m^\alpha - \sum_{j=2}^{m} \left(\frac{\partial \ln \gamma_i^\beta}{\partial X_j^\beta} + \frac{\delta_{ij}}{X_i^\beta}\right)_{Q_\beta} dX_j^\beta = 0 \qquad \text{for} \quad i = 2, \ldots, m-1 \tag{72b}$$

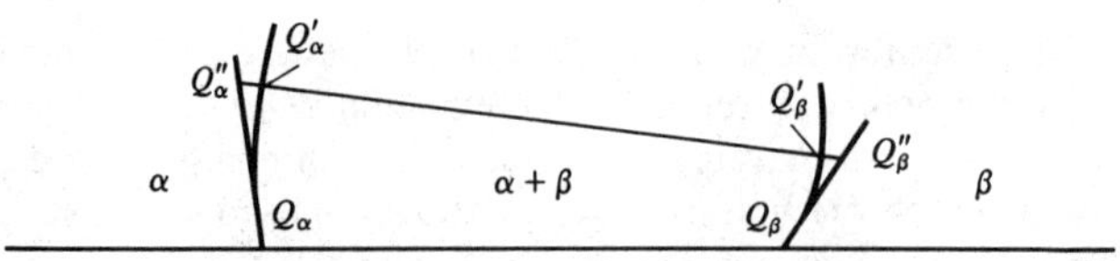

Figure 19. Schematic representation of the phase boundaries for small values of X_m.

and

$$dX_m^\beta = dX_m^\alpha \frac{(\gamma_m^\alpha)_{Q_\alpha}}{(\gamma_m^\beta)_{Q_\beta}} \exp \frac{\mu_m^{*\alpha} - \mu_m^{*\beta}}{RT} \tag{73}$$

[Equation (73) was obtained by use of L'Hospital's rule.]

The variable dX_m^β is directly obtained by equation (73). The variable dX_k^α can be calculated by addition of equations (72a) and (72b) weighted by the appropriate factors X_i^β (for $i = 1, \ldots, m-1$) to take advantage of the Gibbs–Duhem relation:

$$\sum_{i=1}^{m-1} \left(X_i^\beta \frac{\partial \ln \gamma_i^\beta}{\partial X_j^\beta} \right)_{Q_\beta} = 0 \quad \text{for} \quad j = 2, \ldots, m \tag{74}$$

The result is

$$dX_k^\alpha = \frac{dX_m^\beta + dX_m^\alpha \left[\sum_{i=1}^{m-1} X_i^\beta (\partial \ln \gamma_i^\alpha / \partial X_m^\alpha)_{Q_\alpha} - (X_1^\beta / X_1^\alpha) \right]}{(X_1^\beta / X_1^\alpha) - (X_k^\beta / X_k^\alpha) - \sum_{i=1}^{m-1} X_i^\beta (\partial \ln \gamma_i^\alpha / \partial X_k^\alpha)_{Q_\alpha}} \tag{75}$$

The remaining $m-2$ variables can be obtained by solution of the system of $m-2$ linear equations obtained from equation (72b):

$$\sum_{j=2}^{m-1} \left(\frac{\partial \ln \gamma_i^\beta}{\partial X_j^\beta} + \frac{\delta_{ij}}{X_i^\beta} \right)_{Q_\beta} dX_j^\beta = \left(\frac{\partial \ln \gamma_i^\alpha}{\partial X_k^\alpha} + \frac{\delta_{ij}}{X_k^\alpha} \right)_{Q_\alpha} dX_k^\alpha + \left(\frac{\partial \ln \gamma_i^\alpha}{\partial X_m^\alpha} \right)_{Q_\alpha} dX_m^\alpha - \left(\frac{\partial \ln \gamma_i^\beta}{\partial X_m^\beta} \right)_{Q_\beta} dX_m^\beta \quad \text{for} \quad i = 2, \ldots, m-1 \tag{76}$$

In the case of ternary systems, equations (73), (75), and (76) can be transformed into

$$\frac{dX_3^\beta}{dX_3^\alpha} = \left(\frac{\gamma_3^\alpha}{\gamma_3^\beta} \right)_{X_3 \to 0} \exp \left(\frac{\mu_3^{*\alpha} - \mu_3^{*\beta}}{RT} \right) \tag{77}$$

$$\frac{\psi^\beta}{X_2^\beta} dX_2^\beta = \frac{X_1^\beta}{X_2^\beta - X_2^\alpha} dX_3^\alpha - \left(\frac{X_1^\alpha}{X_2^\beta - X_2^\alpha} + \frac{\partial \ln \gamma_2^\beta}{\partial X_3^\beta} \right) dX_3^\beta \tag{78}$$

and

$$\frac{\psi^\alpha}{X_2^\alpha} dX_2^\alpha = \left(\frac{X_1^\beta}{X_2^\beta - X_2^\alpha} - \frac{\partial \ln \gamma_2^\alpha}{\partial X_3^\alpha} \right) dX_3^\alpha - \frac{X_1^\alpha}{X_2^\beta - X_2^\alpha} dX_3^\beta \tag{79}$$

where ψ^α and ψ^β represent the stability functions of the binary system 1–2 in the phases α and β [see equation (VIII.40)]. These equations are easy to exploit and provide a convenient way to predict the effect of small additions of component 3 on the binary phase diagram 1–2.

An interesting case is that of an α phase concentrated in component 1 and a β phase concentrated in component 2, which occurs quite commonly for solid–solid equilibria, and occasionally for liquid miscibility gaps. Then $(\gamma_3^\alpha)_{X_3\to 0}$ and $(\gamma_3^\beta)_{X_3\to 0}$ can be approximated by $(\gamma_3^{\infty\alpha})_{1-3}$ and $(\gamma_3^{\infty\beta})_{2-3}$, respectively, representing the values of the activity coefficient of 3 at infinite dilution in the binary 1–3 in the phase α and in the binary 2–3 in the phase β; in addition the stability functions are nearly unity and $\partial \ln \gamma_2^\alpha / \partial X_3^\alpha$ may be approximated by the ternary interaction coefficient $\epsilon_2^{(3)}$ in the phase α. Using these limiting values yields

$$\frac{dX_2^\alpha}{dX_3^\alpha} = -X_2^\alpha \left[\epsilon_2^{(3)\alpha} + \frac{(\gamma_3^{\infty\alpha})_{1-3}}{(\gamma_3^{\infty\beta})_{2-3}} \exp\left(\frac{\mu_3^{*\alpha} - \mu_3^{*\beta}}{RT} \right) \right] \tag{80}$$

Equation (80) shows that the effect of small additions of component 3 on the solubility of component 2 in 1 is in general small, since dX_2^α/dX_3^α is proportional to X_2^α. The solubility will be substantially increased when $\epsilon_2^{(3)}$ has a large negative value.

Another application of equations (77)–(79) occurs in the calculation of the effect of a small addition of component 3 on the eutectic (or peritectic) temperature of the binary system 1–2. This calculation is presented in detail in the next chapter. Before attacking it, let us conclude this section on the calculation of phase diagrams by stressing certain points.

4.4. *Conclusions*

The phase diagram is a large reservoir of basic information which generally is very poorly tapped because the calculations necessary to retrieve this information are much too complex to be carried out by traditional means. The use of computer techniques to calculate phase boundaries from free energy data is essential to full exploitation of the thermodynamic data.

a. Computer techniques constitute a necessary tool for checking the consistency of experimental data on activities and phase diagrams. Until recently, compilations of thermodynamic properties and phase diagrams were conducted independently of each other (and often regrettably so).
b. Because the determination of phase boundaries is experimentally easier to perform than the measurement of activities, many more data can be used to test theoretical models of solutions.
c. These techniques allow a more rational way of devising sets of critical experiments for the accurate determination of phase diagrams. It must be realized that the complete experimental investigation of even a simple ternary system may cost several man-years. Hence, a combination of computation and critical experiments is crucial in the determination of ternary and higher order phase diagrams.
d. Interpolations or extrapolations of data are much easier. This is especially helpful in ranges where direct measurements are difficult or impossible (e.g., for metastable equilibria, at high temperatures where many side reactions become unavoidable, and at low temperatures where reactions may be so slow as to take several months or years). In addition, these extrapolations can be performed according to various assumptions and the sensitivity of the results to these assumptions can be readily assessed.

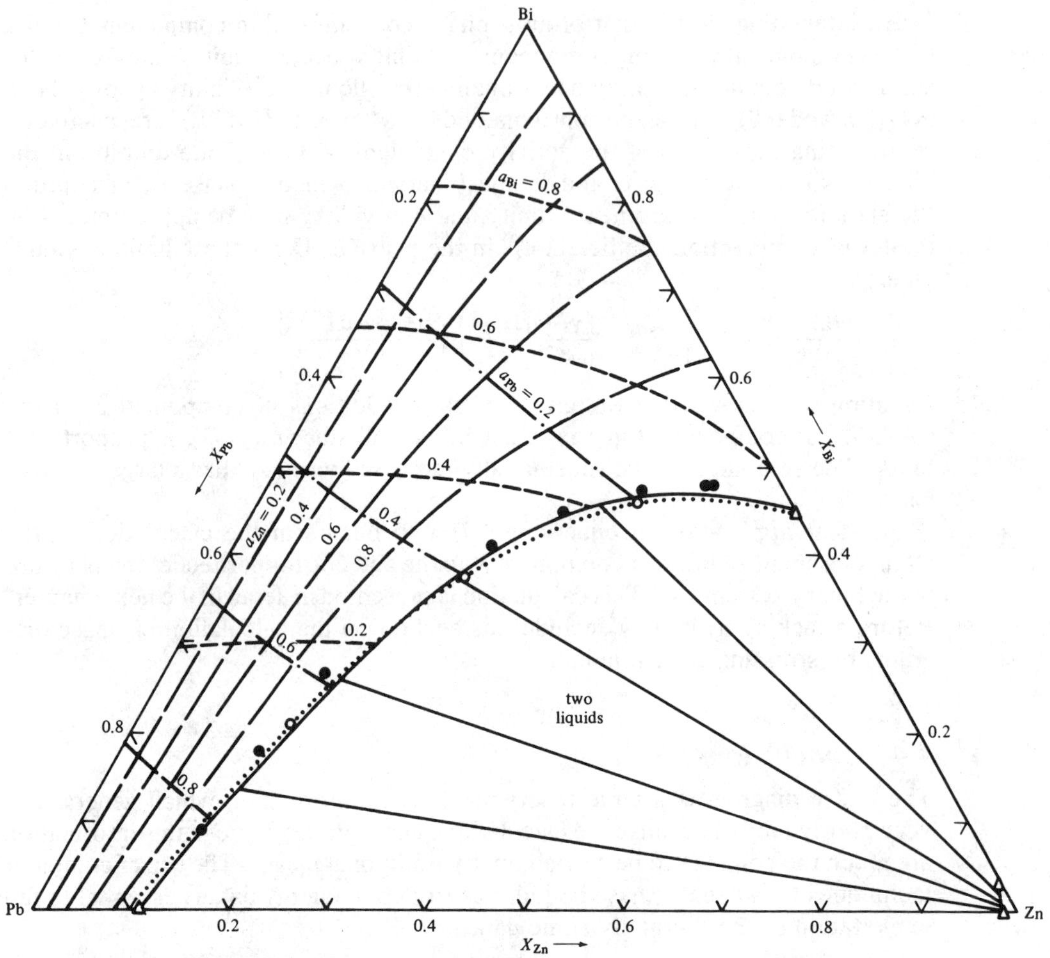

Figure 20. Isoactivity curves and isothermal section of the lead–zinc–bismuth system at 520°C. The experimental points on the boundaries of the miscibility gap were reported by Hansen et al. [23] (triangles), Seith et al. [24] (black circles), and Valenti et al. [8] (white circles). The dotted line is also reported by Valenti et al. [8] and the continuous line is calculated by computer.

e. These computer techniques facilitate greatly the storage and retrieval of data. This is especially important for systems with large number of components.

Let us observe that widespread use of computers for the calculation of phase diagrams is possible only if the calculations can be made with sufficient efficiency to bear a reasonable cost. It appears that such efficiency is now reasonably well achieved. A concrete example may illustrate this; the miscibility gap of the Bi–Pb–Zn system at 793°K (see Fig. 20) was calculated with only 7 sec of computer time on Carnegie-Mellon University's UNIVAC 1108 [22]. The formalism used was that of equations (29)–(33), but the time needed to perform the calculations is practically independent of the complexity of the expressions describing the excess free energies of the phases.

Problems

1. For $G^E_{1-2-3} = A_{111}X_1X_2X_3$, calculate the expression of $\ln \gamma_2$ with the formalism of equations (18), (19), and (27).

2. **a.** Consider the ternary regular solution: $G^E_m = X_1X_2\Omega_{12} + X_1X_3\Omega_{13} + X_2X_3\Omega_{23}$. Calculate the expression for $\ln \gamma_1$.
   b. Apply equations (19) and (28) with $A_{210} = A_{120} = \Omega_{12}$, $A_{201} = A_{102} = \Omega_{13}$, $A_{021} = A_{012} = \Omega_{23}$, and all other As equal to 0, in order to recalculate the expression of $\ln \gamma_1$. Why are the results different? Explain why the results become identical in the case where $\Omega_{12} + \Omega_{13} + \Omega_{23} = 0$.

3. From the results of equations (29)–(33) on the representation of the excess Gibbs free energy of the Bi–Pb–Zn liquid solution, calculate $\ln \gamma_{Pb}$ and a_{Pb} at 793°K and $X_{Pb} = 0.3$, $X_{Zn} = 0.1$.

4. An analysis of the thermodynamic properties of the three binary liquid solutions Ag–Cu, Ag–Au, and Au–Cu at 1050°K yields $(\gamma^\infty_{Ag})_{\text{in Cu}} = 5.48$, $(\gamma^\infty_{Cu})_{\text{in Ag}} = 6.76$, $(\gamma^\infty_{Au})_{\text{in Ag}} = 0.21$, $(\gamma^\infty_{Au})_{\text{in Cu}} = 0.076$. Assume that all three binary solutions are subregular and derive an analytical expression for the interpolation of the values of $(\gamma_{Au})_{X_{Au}\to 0}$, at infinite dilution of Au in the liquid Ag–Cu solution. Apply it to the value of $(\gamma_{Au})_{X_{Au}\to 0, X_{Cu}=0.4}$.

5. At 1273°K, the excess free energies of the Ag–Pb, Ag–Cu, and Cu–Pb liquid solutions are characterized by the following values:

Ag–Pb binary:	$\gamma^\infty_{Ag} = 2.03$,	$\gamma^\infty_{Pb} = 0.92$
Ag–Cu binary:	$\gamma^\infty_{Ag} = 3.97$,	$\gamma^\infty_{Cu} = 4.28$
Cu–Pb binary:	$\gamma^\infty_{Cu} = 6.94$,	$\gamma^\infty_{Pb} = 8.37$

   Assume a subregular solution and calculate by the method of equation (18) (with G^E_{1-2-3} assumed equal to 0) the value of G^E_m at $X_1 = X_3 = \frac{1}{4}$, $X_2 = \frac{1}{2}$. Also calculate the value of G^E_m at the same temperature and composition by the approximations of Kohler and Colinet.

6. Consider the effect of a component 3 on the solubility of carbon in liquid iron in equilibrium with graphite. Assume that 3 has no solubility in graphite. Determine an analytic expression for the slope of that solubility dX_C/dX_3 for $X_3 \to 0$ by applying equation (79). Rederive this result *directly*, i.e., bypassing equation (79). How would this equation be transformed if the concentration of carbon is sufficiently low to justify a formalism of first and second order interaction coefficients?

7. Redo Problem 6 substituting cementite (Fe_3C) for graphite.

References

1. G. Masing, *Introduction to the Theory of Three-Component Systems* (B. A. Rogers, transl.). Dover, New York, 1960.
2. F. N. Rhines, *Phase Diagrams in Metallurgy*. McGraw-Hill, New York, 1956.
3. A. Prince, *Alloy Phase Equilibria*. Elsevier, New York, 1966.
4. L. S. Palatnik and A. I. Landau, *Phase Equilibria in Multicomponent Systems*. Holt, Rinehart and Winston, New York, 1964.

5. J. F. Elliott, M. Gleiser, and V. Ramakrishna, *Thermochemistry for Steelmaking*. Addison-Wesley, Reading, MA, 1963, Vol. 2 (e.g., p. 606).
6. C. H. P. Lupis and J. F. Elliott, *Acta Met.* **15,** 265–276 (1967).
7. H. Gaye, Ph.D. thesis, Carnegie-Mellon University, Pittsburgh, PA, 1971.
8. V. Valenti, L. Oleari, and M. Fiorani, *Gazz. Chim. Ital.* **86,** 920–941 (1956).
9. F. Kohler, *Monatsh. Chemie* **91,** 738 (1960).
10. C. Colinet, D.E.S., Fac. des Sci., Université de Grenoble, France, 1967.
11. I. Ansara, *Metallurgical Symposium, 1971* (O. Kubaschewski, ed.). Her Majesty's Stationery Office, London, 1972, pp. 403–421.
12. L. S. Darken, *J. Am. Chem. Soc.* **72,** 2909 (1950).
13. C. Wagner, *Thermodynamics of Alloys*. Addison-Wesley, Reading, MA, 1952, pp. 19–22.
14. R. Schuhmann, Jr., *Acta Met.* **3,** 219 (1955).
15. N. A. Gokcen, *J. Phys. Chem.* **64,** 401 (1960).
16. L. Kaufman and H. Bernstein, *Computer Calculations of Phase Diagrams*. Academic, New York, 1970.
17. L. Kaufman and H. Nesor, *Met. Trans.* **5,** 1617–1629 (1974).
18. D. T. J. Hurle and E. R. Pike, *J. Mater. Sci. Eng.* **1,** 399–402 (1966).
19. I. Ansara, P. Desré, and E. Bonnier, *J. Chim. Phys.* **66,** 297 (1969).
20. J. F. Counsel, E. B. Lees, and P. J. Spencer, *Met. Sci.* **5,** 210 (1971).
21. J. A. Nelder and R. Mead, *Comp. J.* **7,** 308 (1965).
22. H. Gaye and C. H. P. Lupis, *Met. Trans.* **6A,** 1057–1064 (1975).
23. M. Hansen and K. Anderko, *Constitution of Binary Alloys*. McGraw-Hill, New York, 1958.
24. W. Seith, H. Johnson, and J. Wagner, *Z. Metallk.* **46,** 773–779 (1955).

Selected Bibliography

Texts

L. S. Palatnik and A. I. Landau, *Phase Equilibria in Multicomponent Systems*. Holt, Rinehart and Winston, New York, 1964.

A. Prince, *Alloy Phase Equilibria*. Elsevier, New York, 1966.

F. N. Rhines, *Phase Diagrams in Metallurgy*. McGraw-Hill, New York, 1956.

G. Masing, *Ternary Systems, Introduction to the Theory of Three-Component Systems* (B. A. Rogers, transl.). Dover, New York, 1960.

Films

Reading Ternary Phase Diagrams, 8 min.

Ternary Diagrams Derived from Binaries, 6 min.

Isothermal Sections with Simple Ternary Eutectic, 8 min.

Isothermal Sections with Solid Solution, 8 min.

All four films available from Audio-Visual Services, The Pennsylvania State University, University Park, PA 16802, for sale or rental, 16 mm or super 8 mm, in color, with sound, animated computer graphics; very helpful in understanding temperature-composition relationships, tie lines, lever rule, phase rule, etc.

Compilations

Applications of Phase Diagrams in Metallurgy and Ceramics (G. C. Carter, ed.). Nat. Bur. of Standards Spec. Publ. 496, Washington, DC, 1978, 2 Vols. Proceedings of a workshop held at the National Bureau of Standards, Gaithersburg, Maryland, 10–12

January 1977. An excellent source of information; see, in Vol. 2, the articles on phase diagram compilation activities, and especially the article by G. C. Carter, pp. 36–89.

W. Guertler, M. Guertler, and E. Anastasiadias, *A Compendium of Constitutional Ternary Diagrams of Metallic Systems* (in German, 1958); available from the Nat. Tech. Inf. Serv., Washington, DC, as document TT 69-55069.

Metals Handbook, Metallography, Structures and Phase Diagrams, 8th ed. Am. Soc. Metals, Metals Park, OH, 1973, Vol. 8. Approximately 400 binary and 50 ternary phase diagrams.

E. M. Levin, C. R. Robbins, and H. F. McMurdie, *Phase Diagrams for Ceramists.* American Ceramic Society, Columbus, OH, 1964; first supplement, 1969; second supplement, 1975.

A. Muan and E. F. Osborn, *Phase Equilibria among Oxides in Steelmaking.* Addison-Wesley, Reading, MA, 1965.

XI Stability of Multicomponent Solutions and Effects of a Third Component on Some Invariant Points of Binary Systems

In Section III.3 we derived a criterion for the stability of a binary solution with respect to infinitesimal composition fluctuations (i.e., with respect to spinodal decomposition). In Section 1 of this chapter we shall generalize this study to the case of a multicomponent solution by extending the method presented in Section III.3.2 [1]. The stability conditions that we shall find lead in Section 2 to the definition of the stability function ψ.

Unstable solutions yield miscibility gaps. In binary systems, these are characterized by critical points, in ternary systems by critical lines. The slope of such a line at infinite dilution of one solute, that is, the effect of small additions of a third component on the critical point of a binary system, is of particular interest. It will be calculated and discussed in Section 3. Finally, in Section 4 we shall examine the effect of small additions of a third component on other binary invariant points, such as eutectic and peritectic points.

1. Stability Conditions for a Multicomponent Solution

1.1. Derivation

Consider the decomposition of a homogeneous solution of n_1 mol of component 1, n_2 mol of component 2, . . . , n_m mol of component m, into the phases α, β, . . . , ν, . . . , θ, which—at least initially—can be arbitrarily close in composition to that of the original phase. Each phase ν contains $\lambda_\nu(n_1 + dn_1)$ mol of component 1, $\lambda_\nu(n_2 + dn_2)$ mol of component 2, . . . , $\lambda_\nu(n_m + dn_m)$ mol of component m. Elementary mass balances yield

$$\sum_{\nu=\alpha}^{\theta} \lambda_\nu \, dn_i^\nu = 0 \qquad \text{for} \quad i = 1, 2, \ldots, m \tag{1}$$

and

$$\sum_{\nu=\alpha}^{\theta} \lambda_\nu = 1 \tag{2}$$

The Gibbs free energy of the phase ν is

$$G^\nu = G[\lambda_\nu(n_1 + dn_1^\nu), \ldots, \lambda_\nu(n_i + dn_i^\nu), \ldots, \lambda_\nu(n_m + dn_m^\nu)] \tag{3}$$

and since it is an extensive property, that is, a homogeneous function of degree 1,

$$G^\nu = \lambda_\nu G(n_1 + dn_1^\nu, \ldots, n_i + dn_i^\nu, \ldots, n_m + dn_m^\nu) \tag{4}$$

Expanding G into a Taylor series, we obtain

$$G^\nu = \lambda_\nu G(n_1, \ldots, n_i, \ldots, n_m) + \sum_{i=1}^{m} \mu_i \, dn_i^\nu + \sum_{i,j=1}^{m} \tfrac{1}{2} G_{ij} \, dn_i^\nu \, dn_j^\nu + \cdots \tag{5}$$

where G_{ij} denotes the second derivative of G with respect to n_i and n_j. We note that $G_{ij} = G_{ji}$.

Spinodal decomposition into phases α, . . . , ν, . . . , θ does not occur if it is accompanied by an increase in the Gibbs free energy of the total system. Therefore, with respect to infinitesimal fluctuations in composition and at constant temperature and pressure, the criterion of stability is

$$DG = \sum_{\nu=\alpha}^{\theta} \lambda_\nu G(n_1 + dn_1^\nu, \ldots, n_i^\nu + dn_i, \ldots, n_m + dn_m^\nu)$$

$$- G(n_1, \ldots, n_i, \ldots, n_m) \geq 0$$

$$= \sum_{\nu=\alpha}^{\theta} \lambda_\nu \left[G(n_1, \ldots, n_i, \ldots, n_m) + \sum_{i=1}^{m} \mu_i \, dn_i^\nu + \tfrac{1}{2} \sum_{i,j=1}^{m} G_{ij} \, dn_i^\nu \, dn_j^\nu + \cdots \right]$$

$$- G(n_i, \ldots, n_i, \ldots, n_m) \geq 0 \tag{6a}$$

This condition is readily simplified by the application of equations (1) and (2):

$$DG = \sum_{\nu=\alpha}^{\theta} \lambda_\nu \sum_{i,j=1}^{m} \tfrac{1}{2} G_{ij} \, dn_i^\nu \, dn_j^\nu + \text{higher order terms} \geq 0 \tag{6b}$$

The consideration of higher order terms is only important when the first summation is equal to zero; in that case the first nonvanishing term yields the sign of DG. Dismissing this particular case, the necessary and sufficient condition for stable equilibrium is

$$\delta^2 G = \sum_{\nu=\alpha}^{\theta} \lambda_\nu \sum_{i,j=1}^{m} \tfrac{1}{2} G_{ij} \, dn_i^\nu \, dn_j^\nu \geq 0 \tag{7}$$

All the λ_ν, dn_i^ν, dn_j^ν are arbitrary within the restrictions of equations (1) and (2) and the inequality $\lambda_\nu \geq 0$. The cumbersome condition (7) may then be replaced by the equivalent but simpler conditions that the quadratic form associated to the symmetric matrix G_{ij} be always positive or null:

$$\sum_{i,j=1}^{m} G_{ij} x_i x_j \geq 0 \tag{8a}$$

or

$$(x_1 \quad x_2 \quad \cdots \quad x_m) \begin{pmatrix} G_{11} & G_{12} & \cdots & G_{1m} \\ G_{21} & G_{22} & \cdots & G_{2m} \\ \vdots & \vdots & & \vdots \\ G_{m1} & G_{m2} & \cdots & G_{mm} \end{pmatrix} \begin{pmatrix} x_1 \\ x_2 \\ \vdots \\ x_m \end{pmatrix} \geq 0 \tag{8b}$$

regardless of the values of the dummy variables x_2. Condition (8) is obviously sufficient, for then all the terms in $\delta^2 G$ are positive or null. To demonstrate that it is necessary, we show that if one of the quadratic terms $\sum G_{ij} \, dn_i^\alpha \, dn_j^\alpha$ is negative, a possible decomposition path may be found for the system. For instance, in accordance with the restrictions of equation (1), we may choose

$$dn_i^\beta = -(\lambda_\alpha/\lambda_\beta) \, dn_i^\alpha \qquad \text{for} \quad i = 1, 2, \ldots, m \tag{9a}$$

and

$$dn_i^\nu = 0 \qquad \text{for all} \quad \nu \neq \alpha \text{ or } \beta \quad \text{and} \quad i = 1, 2, \ldots, m \tag{9b}$$

Consequently,

$$\delta^2 G = \lambda_\alpha \sum_{i,j=1}^{m} \tfrac{1}{2} G_{ij} \, dn_i^\alpha \, dn_j^\alpha + \lambda_\beta \sum_{i,j=1}^{m} \tfrac{1}{2} G_{ij} \, dn_i^\alpha \, dn_j^\alpha \, (\lambda_\alpha/\lambda_\beta)^2 \tag{10a}$$

or

$$\delta^2 G = \lambda_\alpha [1 + (\lambda_\alpha/\lambda_\beta)] \sum_{i,j=1}^{m} \tfrac{1}{2} G_{ij}\, dn_i^\alpha\, dn_j^\alpha \tag{10b}$$

Since $\lambda_\alpha[1 + (\lambda_\alpha/\lambda_\beta)]$ is positive, $\delta^2 G$ is negative, a result contrary to our criterion of stability.

There are several equivalent sets of necessary and sufficient conditions to express that the quadratic form associated with the G_{ij} matrix is invariably positive or null. One of them is to write that the eigenvalues of the matrix are all positive or null. Another is that the coefficients of the characteristic equation (of which the eigenvalues are the roots) are alternatively positive and negative. However, both sets of conditions lead to analytic expressions which from a practical viewpoint are quite cumbersome. We present below another set of mathematical conditions that results in simpler thermodynamic expressions.

Hancock demonstrated [2] that necessary and sufficient conditions are that each of the following diagonal determinants be always positive or null:

$$G_{11} \geq 0 \tag{11-1}$$

$$\begin{vmatrix} G_{11} & G_{12} \\ G_{21} & G_{22} \end{vmatrix} \geq 0 \tag{11-2}$$

$$\vdots$$

$$\begin{vmatrix} G_{11} & G_{12} & \cdots & G_{1m} \\ G_{21} & G_{22} & \cdots & G_{2m} \\ \vdots & \vdots & & \vdots \\ G_{m1} & G_{m2} & \cdots & G_{mm} \end{vmatrix} \geq 0 \tag{11-m}$$

The last condition is trivial; the determinant associated with the G_{ij} matrix is always null because of the Gibbs–Duhem equation:

$$n_1 G_{1i} + n_2 G_{2i} + \cdots + n_m G_{mi} = 0 \qquad (i = 1, 2, \ldots, m) \tag{12}$$

The number of conditions is thus reduced to $m - 1$. The order in which the components are chosen in the set of conditions (11) is immaterial. When a solution is dilute, it is generally convenient to express these conditions in terms of the solutes rather than in terms of the solvent 1. Conditions (11) may be then rewritten

$$\begin{vmatrix} \dfrac{\partial\mu_2}{\partial n_2} & \dfrac{\partial\mu_2}{\partial n_3} & \cdots & \dfrac{\partial\mu_2}{\partial n_i} \\ \dfrac{\partial\mu_3}{\partial n_2} & \dfrac{\partial\mu_3}{\partial n_3} & \cdots & \dfrac{\partial\mu_3}{\partial n_i} \\ \vdots & \vdots & & \vdots \\ \dfrac{\partial\mu_i}{\partial n_2} & \dfrac{\partial\mu_i}{\partial n_3} & \cdots & \dfrac{\partial\mu_i}{\partial n_i} \end{vmatrix} \geq 0 \qquad (i = 2, 3, \ldots, m) \tag{13-i}$$

In terms of mole fractions X_i, activities a_i, and activity coefficients $\gamma_i = a_i/X_i$, we note that

$$\frac{\partial\mu_k}{\partial n_l} = \sum_{p=2}^{m} \frac{\partial\mu_k}{\partial X_p}\frac{\partial X_p}{\partial n_l} = RT \sum_{p=2}^{m} \frac{\partial \ln a_k}{\partial X_p}\frac{\partial X_p}{\partial n_l}$$

$$= \frac{RT}{n} \sum_{p=2}^{m} \left(\frac{\delta_{kp}}{X_p} + \frac{\partial \ln \gamma_k}{\partial X_p} \right) (\delta_{lp} - X_p) \tag{14}$$

where δ_{lp} is Kronecker's symbol ($\delta_{lp} = 1$ when $l = p$, $\delta_{lp} = 0$ when $l \neq p$). By relatively simple algebraic manipulations [1], it is possible to show that conditions (13) may be transformed to yield for $i = 2, 3, \ldots, m$

$$\begin{vmatrix} \frac{1}{X_2} + \frac{\partial \ln \gamma_2}{\partial X_2} & \frac{\partial \ln \gamma_2}{\partial X_3} & \cdots & \frac{\partial \ln \gamma_2}{\partial X_i} & \frac{\partial \ln \gamma_2}{\partial X_{i+1}} & \cdots & \frac{\partial \ln \gamma_2}{\partial X_m} \\ \frac{\partial \ln \gamma_3}{\partial X_2} & \frac{1}{X_3} + \frac{\partial \ln \gamma_3}{\partial X_3} & \cdots & \frac{\partial \ln \gamma_3}{\partial X_i} & \frac{\partial \ln \gamma_3}{\partial X_{i+1}} & \cdots & \frac{\partial \ln \gamma_3}{\partial X_m} \\ \vdots & \vdots & & \vdots & \vdots & & \vdots \\ \frac{\partial \ln \gamma_i}{\partial X_2} & \frac{\partial \ln \gamma_i}{\partial X_3} & \cdots & \frac{1}{X_i} + \frac{\partial \ln \gamma_i}{\partial X_i} & \frac{\partial \ln \gamma_i}{\partial X_{i+1}} & \cdots & \frac{\partial \ln \gamma_i}{\partial X_m} \\ X_{i+1} & X_{i+1} & \cdots & X_{i+1} & X_1 + X_{i+1} & \cdots & X_{i+1} \\ \vdots & \vdots & & \vdots & \vdots & & \vdots \\ X_m & X_m & \cdots & X_m & X_m & \cdots & X_1 + X_m \end{vmatrix} \geq 0 \qquad (15\text{-}i)$$

More specifically, condition (15-m) is

$$\begin{vmatrix} \frac{1}{X_2} + \frac{\partial \ln \gamma_2}{\partial X_2} & \frac{\partial \ln \gamma_2}{\partial X_3} & \cdots & \frac{\partial \ln \gamma_2}{\partial X_m} \\ \frac{\partial \ln \gamma_3}{\partial X_2} & \frac{1}{X_3} + \frac{\partial \ln \gamma_3}{\partial X_3} & \cdots & \frac{\partial \ln \gamma_3}{\partial X_m} \\ \vdots & \vdots & & \vdots \\ \frac{\partial \ln \gamma_m}{\partial X_2} & \frac{\partial \ln \gamma_m}{\partial X_3} & \cdots & \frac{1}{X_m} + \frac{\partial \ln \gamma_m}{\partial X_m} \end{vmatrix} \geq 0 \qquad (15\text{-}m)$$

Table 1 lists conditions (15) for binary, ternary, and quaternary systems.

1.2. Existence of a Most Restrictive Condition

Since there is only one condition for the stability of a binary system, but $m - 1$ conditions for an m-component system, the probability of violating a stability condition, and thus forming a miscibility gap, increases with the number of components. This result is also intuitive, since the atoms of the solution have now many more ways of redistributing themselves and introducing complexities in the form of the free energy hypersurface. It is interesting to examine whether there is a stability condition which is most likely to be violated, that is, whether there is a most restrictive thermodynamic condition [3]. The existence of such a condition would greatly simplify the application of the stability criteria, since only one condition could then be considered instead of $m - 1$.

For a ternary system 1–2–3 the two stability conditions are

$$D^{(1)} = G_{22} \geq 0 \qquad (16\text{-}1)$$

$$D^{(2)} = G_{22}G_{33} - G_{23}^2 \geq 0 \qquad (16\text{-}2)$$

In a composition diagram, these two conditions define two domains of instability. Starting at a point where the solution is stable (for instance at a point where the

Table 1. Stability Conditions (15-i) for Binary, Ternary, and Quaternary Systems

Binary system	
$i = 2$	$\dfrac{1}{X_2} + \dfrac{\partial \ln \gamma_2}{dX_2} \geq 0$
Ternary system	
$i = 2$	$(1 - X_2)\left(\dfrac{1}{X_2} + \dfrac{\partial \ln \gamma_2}{\partial X_2}\right) - X_3 \dfrac{\partial \ln \gamma_2}{\partial X_3} \geq 0$
$i = 3$	$\left(\dfrac{1}{X_2} + \dfrac{\partial \ln \gamma_2}{\partial X_2}\right)\left(\dfrac{1}{X_3} + \dfrac{\partial \ln \gamma_3}{\partial X_3}\right) - \dfrac{\partial \ln \gamma_2}{\partial X_3} \dfrac{\partial \ln \gamma_3}{\partial X_2} \geq 0$
Quaternary system	
$i = 2$	$(1 - X_2)\left(\dfrac{1}{X_2} + \dfrac{\partial \ln \gamma_2}{\partial X_2}\right) - X_3 \dfrac{\partial \ln \gamma_2}{\partial X_3} - X_4 \dfrac{\partial \ln \gamma_2}{\partial X_4} \geq 0$
$i = 3$	$(1 - X_2 - X_3)\left[\left(\dfrac{1}{X_2} + \dfrac{\partial \ln \gamma_2}{\partial X_2}\right)\left(\dfrac{1}{X_3} + \dfrac{\partial \ln \gamma_3}{\partial X_3}\right) - \dfrac{\partial \ln \gamma_2}{\partial X_3} \dfrac{\partial \ln \gamma_3}{\partial X_2}\right]$
	$- X_4 \left[\dfrac{\partial \ln \gamma_2}{\partial X_4}\left(\dfrac{1}{X_3} + \dfrac{\partial \ln \gamma_3}{\partial X_3} - \dfrac{\partial \ln \gamma_3}{\partial X_2}\right) + \dfrac{\partial \ln \gamma_3}{\partial X_4}\left(\dfrac{1}{X_2} + \dfrac{\partial \ln \gamma_2}{\partial X_2} - \dfrac{\partial \ln \gamma_2}{\partial X_3}\right)\right] \geq 0$
$i = 4$	$\left(\dfrac{1}{X_2} + \dfrac{\partial \ln \gamma_2}{\partial X_2}\right)\left(\dfrac{1}{X_3} + \dfrac{\partial \ln \gamma_3}{\partial X_3}\right)\left(\dfrac{1}{X_4} + \dfrac{\partial \ln \gamma_4}{\partial X_4}\right) + \dfrac{\partial \ln \gamma_2}{\partial X_3} \dfrac{\partial \ln \gamma_3}{\partial X_4} \dfrac{\partial \ln \gamma_4}{\partial X_2} + \dfrac{\partial \ln \gamma_2}{\partial X_4} \dfrac{\partial \ln \gamma_3}{\partial X_2} \dfrac{\partial \ln \gamma_4}{\partial X_3}$
	$- \left(\dfrac{1}{X_2} + \dfrac{\partial \ln \gamma_2}{\partial X_2}\right) \dfrac{\partial \ln \gamma_3}{\partial X_4} \dfrac{\partial \ln \gamma_4}{\partial X_3} - \left(\dfrac{1}{X_3} + \dfrac{\partial \ln \gamma_3}{\partial X_3}\right) \dfrac{\partial \ln \gamma_2}{\partial X_4} \dfrac{\partial \ln \gamma_4}{\partial X_2}$
	$- \left(\dfrac{1}{X_4} + \dfrac{\partial \ln \gamma_4}{\partial X_4}\right) \dfrac{\partial \ln \gamma_2}{\partial X_3} \dfrac{\partial \ln \gamma_3}{\partial X_2} \geq 0$

solution is dilute), we gradually change the concentrations until condition (16-1) or (16-2) is violated. The boundary of the domain $D^{(2)}$ must be reached first [4]. Indeed, if we assume that it is the boundary of the domain $D^{(1)}$ which is reached first, at this point $G_{22} = 0$ and the second condition is necessarily violated ($D^{(2)} = -G_{23}^2 \geq 0$); this would contradict our original assumption.

If the boundaries of the two domains have a common point, they must also have a common tangent. For if the two lines were to cross each other, as illustrated in Fig. 1a, any point M on the line QP would be such that $D^{(1)} = 0$ and $D^{(2)} > 0$. As remarked, these results are incompatible. Thus, the two lines must be tangent at their common point Q, as illustrated in the example of Fig. 1b.

The reasoning associated with Fig. 1a assumes that Q is not a singular point for either boundary line. If it is, the lines are not necessarily tangent. The singularity may be of two types. In the first case, the lines meet but do not cross and are not tangent to each other (e.g., points Q' and Q'' in Fig. 1b). In the second, the tangent at Q for $D^{(1)}$ or $D^{(2)}$ is not single valued. Other types of singularity are unlikely because of the analytic forms that $D^{(1)}$ and $D^{(2)}$ usually have. A brief discussion on the occurrence of these singularities is given by Gaye and Lupis [3].

The result found for ternary systems may be generalized to multicomponent systems. Let us designate by $D^{(r)}$ the diagonal determinant of order r associated with matrix G_{ij}. The demonstration for an m-component system is based on the mathematical theorem stating that if $D^{(r)} = 0$, then the product $D^{(r-1)}D^{(r+1)} \leq$

0 [5]. Again, starting from a point where the solution is stable, the composition is gradually changed until the first boundary $D^{(r)} = 0$ is met. It is impossible to have $1 < r < m - 1$ and $D^{(r-1)}D^{(r+1)} < 0$, for then either $D^{(r-1)}$ or $D^{(r+1)}$ would be negative and $D^{(r)} = 0$ would not be the first boundary reached. Also, r cannot be equal to 1, for if $D^{(1)} = 0$, then $D^{(2)}$ is negative, as seen in equation (16-2). Thus, r has to be equal to $m - 1$ and the first boundary reached is that of $D^{(m-1)}$. Consequently, the inequality

$$D^{(m-1)} = \begin{vmatrix} G_{22} & G_{23} & \cdots & G_{2m} \\ G_{32} & G_{33} & \cdots & G_{3m} \\ \vdots & \vdots & & \vdots \\ G_{m2} & G_{m3} & \cdots & G_{mm} \end{vmatrix} \geq 0 \tag{17}$$

is the most restrictive thermodynamic condition.

To complete this demonstration, we must consider the possibility that at the point where the first boundary is met, not only $D^{(r)} = 0$ but also $D^{(r-1)}$ or $D^{(r+1)} = 0$. It may be shown that at this point, if two consecutive diagonal determinants $D^{(j-1)}$ and $D^{(j)}$ are equal to zero, then $D^{(j+1)}$ is also equal to zero [6]. Hence, $D^{(j+2)} = \cdots = D^{(m-1)} = 0$, and this point belongs to the boundary of the $D^{(m-1)}$ domain. It may also be noted that at this common point the boundaries must be tangent. Otherwise, it would be possible to pass from a stable region to an unstable one by crossing first a boundary other than $D^{(m-1)} = 0$; as shown above, this is impossible. However, as in the case of a ternary system, an exception to that common tangent rule occurs when the common point is singular (most frequently located at a boundary of the composition diagram).

The condition $D^{(m-1)} \geq 0$ is necessary but not sufficient for the stability of the phase under consideration. Indeed, the domain bound by $D^{(m-1)} = 0$ may contain "holes," i.e., it may be constituted by regions which are not simply connected. In this case, the other conditions $D^{(r)}$ should also be calculated. However, it is sufficient to calculate their sign at only one point of the hole in order to determine if the phase is stable or unstable at any other point of the hole. Consider the example of Fig. 2. The boundaries $D^{(1)}$ and $D^{(2)}$ correspond to a regular ternary

Figure 1. Slopes of the boundary lines $D^{(1)} = 0$ and $D^{(2)} = 0$ at a common point. **(a)** The curves are not tangent to each other at Q (impossible), and **(b)** a possible case.

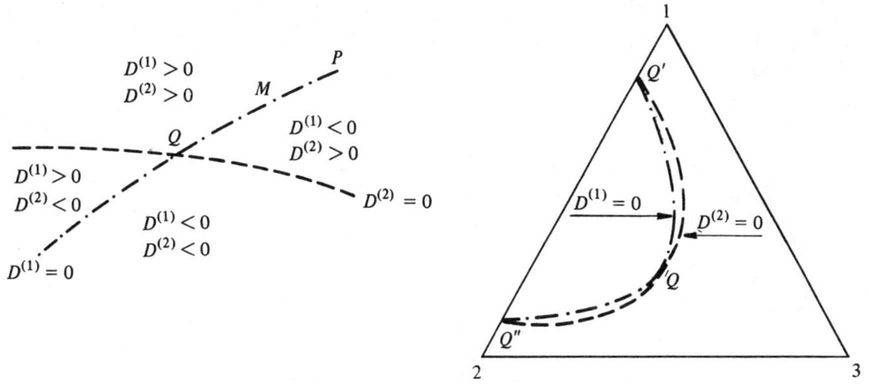

(a) (b)

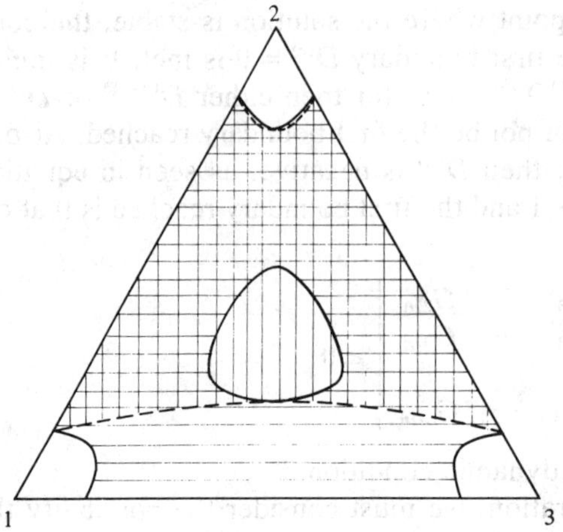

Figure 2. Illustration of a case where one of the stability domains is not simply connected. The case is that of a regular solution where $\Omega_{12} = \Omega_{13} = \Omega_{23}$ at the temperature $T = \Omega_{12}/4R$. The continuous line represents the equation $D^{(2)} = 0$ and the dashed line $D^{(1)} = 0$. The domain hatched horizontally is that of instability with respect to $D^{(2)}$ and the domain hatched vertically is that of instability with respect to $D^{(1)}$.

solution where $\Omega_{12} = \Omega_{13} = \Omega_{23}$ at the temperature $T_C = \Omega_{12}/4R$. In the hole, $D^{(2)} > 0$ and the phase is unstable ($D^{(1)} < 0$). We note that it is not possible to reach any point of that hole from any composition where the phase is stable (near the corners of the triangle) without crossing first the boundary $D^{(2)} = 0$. This result is in agreement with our conclusion that $D^{(m-1)}$ is the most restrictive stability condition.

The important result to be retained is then the following. Given a stable solution of m components, if we change its composition so that it decomposes into two or more phases, we need only consider the $D^{(m-1)}$ condition in order to calculate the concentrations at which the solution becomes unstable.

The following section describes a stability function ψ [7] whose definition is closely related to the $D^{(m-1)}$ condition. Some applications of this condition can be found in examples of the use of the ψ function.

2. Stability Function ψ

2.1. Definition

In Chapter III we introduced the stability function ψ of a binary solution. We shall now generalize this function to include the case of a multicomponent solution.

We define the stability function ψ of an m-component phase by

$$\psi = \left(\frac{n}{RT}\right)^{m-1} \frac{X_1 X_2 \cdots X_i \cdots X_m}{X_1^2} \begin{vmatrix} G_{22} & G_{23} & \cdots & G_{2m} \\ G_{32} & G_{33} & \cdots & G_{3m} \\ \vdots & \vdots & & \vdots \\ G_{m2} & G_{m3} & \cdots & G_{mm} \end{vmatrix} \tag{18a}$$

or

$$\psi = \left(\frac{n}{RT}\right)^{m-1} \frac{X_1 X_2 \cdots X_i \cdots X_m}{X_1^2} D^{(m-1)} \tag{18b}$$

In the right member of this equation, component 1 appears to play a special role. It is possible, however, to demonstrate that this is not so and that the value of ψ remains unchanged regardless of which mole fractions are selected as independent variables [7]. In other words, if the right member of equation (18) were identified as ψ_1 we would have

$$\psi_1 = \psi_2 = \cdots = \psi_i = \cdots = \psi_m = \psi \tag{19}$$

The function ψ may be given other equivalent forms [6]:

$$\psi = X_1 X_2 \cdots X_m \begin{vmatrix} \dfrac{\partial^2(G^M/RT)}{\partial X_2^2} & \cdots & \dfrac{\partial^2(G^M/RT)}{\partial X_2\,\partial X_i} & \cdots & \dfrac{\partial^2(G^M/RT)}{\partial X_2\,\partial X_m} \\ \vdots & & \vdots & & \vdots \\ \dfrac{\partial^2(G^M/RT)}{\partial X_i\,\partial X_2} & \cdots & \dfrac{\partial^2(G^M/RT)}{\partial X_i^2} & \cdots & \dfrac{\partial^2(G^M/RT)}{\partial X_i\,\partial X_m} \\ \vdots & & \vdots & & \vdots \\ \dfrac{\partial^2(G^M/RT)}{\partial X_m\,\partial X_2} & \cdots & \dfrac{\partial^2(G^M/RT)}{\partial X_m\,\partial X_i} & \cdots & \dfrac{\partial^2(G^M/RT)}{\partial X_m^2} \end{vmatrix} \tag{20}$$

$$\psi = \begin{vmatrix} \dfrac{\partial \ln a_2}{\partial \ln X_2} & \dfrac{\partial \ln a_2}{\partial \ln X_3} & \cdots & \dfrac{\partial \ln a_2}{\partial \ln X_m} \\ \dfrac{\partial \ln a_3}{\partial \ln X_2} & \dfrac{\partial \ln a_3}{\partial \ln X_3} & \cdots & \dfrac{\partial \ln a_3}{\partial \ln X_m} \\ \vdots & \vdots & & \vdots \\ \dfrac{\partial \ln a_m}{\partial \ln X_2} & \dfrac{\partial \ln a_m}{\partial \ln X_3} & \cdots & \dfrac{\partial \ln a_m}{\partial \ln X_m} \end{vmatrix} \tag{21}$$

and

$$\psi = \begin{vmatrix} 1 + X_2 \dfrac{\partial \ln \gamma_2}{\partial X_2} & X_3 \dfrac{\partial \ln \gamma_2}{\partial X_3} & \cdots & X_m \dfrac{\partial \ln \gamma_2}{\partial X_m} \\ X_2 \dfrac{\partial \ln \gamma_3}{\partial X_2} & 1 + X_3 \dfrac{\partial \ln \gamma_3}{\partial X_3} & \cdots & X_m \dfrac{\partial \ln \gamma_2}{\partial X_m} \\ \vdots & \vdots & & \vdots \\ X_2 \dfrac{\partial \ln \gamma_m}{\partial X_2} & X_3 \dfrac{\partial \ln \gamma_m}{\partial X_3} & \cdots & 1 + X_m \dfrac{\partial \ln \gamma_m}{\partial X_m} \end{vmatrix} \tag{22}$$

In equation (20), ψ may be seen to be proportional to the total curvature (inverse of the product of the principal radii of curvature) of the hypersurface describing the free energy of mixing as a function of the mole fractions $X_2, X_3, \ldots, X_m$.

We also observe that ψ is dimensionless, positive for a stable solution, and equal to 1 for an ideal solution. In addition, there are no analytic discontinuities in the values of ψ when the concentrations of one of the solutes vanishes (or upon addition of a new solute).

2.2. *Applications*

2.2.1. Binary Solutions

In a binary solution, ψ reduces to

$$\psi = X_1 X_2 \frac{d^2(G^M/RT)}{dX_2^2} = 1 + X_1 \frac{d \ln \gamma_1}{dX_1} = 1 + X_2 \frac{d \ln \gamma_2}{dX_2} \tag{23}$$

Figure 3 illustrates composition ranges where ψ is positive or negative. We note that to determine the limits of a miscibility gap the curve G^M must be known near both compositions a and b. However, often only one portion of the curve is known, e.g., for $0 < X_2 < e$. The existence of the spinodal point c, at $\psi = 0$, affirms the existence of a miscibility gap and locates the point a in the range $0 < X_2 < c$. Thus, the sign of ψ may be used as a local criterion for determining the presence of a miscibility gap and for obtaining a qualitative estimate of its position. This conclusion can be readily extended to multicomponent systems; indeed, it is in these cases that it should prove most valuable.

The stability condition $\psi \geq 0$ may also be used to provide upper and lower bounds to interaction parameters [8]. For example, if the formalism adopted in the range $0 < X_2 < X_2'$ is limited to the first order interaction coefficient

$$\ln \gamma_2 = \ln \gamma_2^\infty + \epsilon_2^{(2)} X_2 \tag{24}$$

then, necessarily,

$$\psi = 1 + X_2 \epsilon_2^{(2)} \geq 0 \qquad \text{for} \quad 0 < X_2 < X_2' \tag{25}$$

and

$$\epsilon_2^{(2)} \geq -1/X_2' \tag{26}$$

If the inequality (26) is not respected, then one of the following statements can be asserted:

a. The solution is unstable in the range $(0, X_2')$.
b. The formalism chosen is inadequate; e.g., a higher order polynomial is necessary in the range $(0, X_2')$.

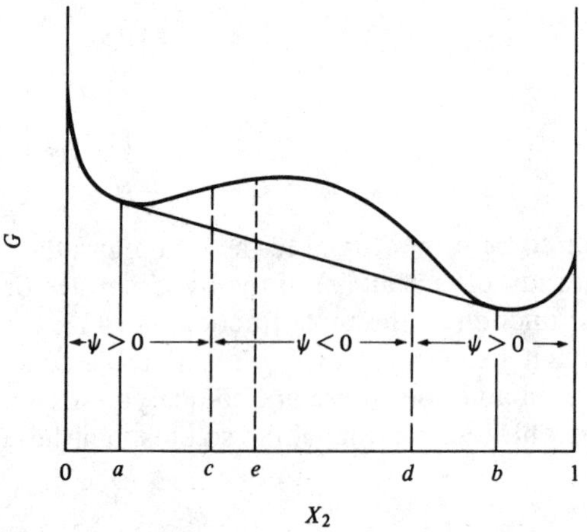

Figure 3. Illustration of the ranges of composition for positive ψ for a system exhibiting a miscibility gap.

c. The measured value of $\epsilon_2^{(2)}$ is erroneous; it should be larger than the lower bound $-1/X'_2$.

Similar conditions can be found for second order interaction parameters [8].

Occasionally, it is convenient to calculate ψ in terms of a different composition coordinate. For example, Chipman's lattice ratio z_i is defined as

$$z_i = n_i \Big/ \sum_{j=1}^{m} v_j n_j \tag{27}$$

where $v_j = +1$ for substitutional solutes and -1 for interstitial ones [9]. For a binary solution, ψ would become

$$\psi = (1 + 2z_2)\left(1 + z_2 \frac{d \ln \phi_2}{dz_2}\right) \tag{28}$$

where ϕ_2 is the activity coefficient of 2 ($\phi_2 = a_2/z_2$).

As an example of an application, let us consider the activity of sulfur in iron at 1500°C. Ban-ya and Chipman [9] showed that in a plot of $\ln \phi_S$ vs z_S a straight line of slope

$$\theta_S^S = (d \ln \phi_S/dz_S)_{z_S \to 0} = -5.3 \tag{29}$$

represents quite well the data which extend to $z_S = 0.16$. The condition $\psi \geq 0$ readily establishes a spinodal at $z_S = 0.18$ ($X_S = 0.14$), and therefore a miscibility gap at $z_S < 0.18$. However, on the one hand, the accepted phase diagram [10,11] shows no miscibility gap at this temperature, and on the other hand, it is unlikely that one should find a strong deviation from the straight line $\ln \phi_2$ vs z_S in the range between 0.16 and 0.18. Consequently, a reexamination of the data appears warranted.

Not only negative values of ψ are of interest. In Chapter III we saw that strong positive values of ψ can shed some light on the structure of a solution and the possible existence of clusters.

2.2.2. Ternary Solutions

In ternary solutions, equation (22) becomes

$$\psi = \left(1 + X_2 \frac{\partial \ln \gamma_2}{\partial X_2}\right)\left(1 + X_3 \frac{\partial \ln \gamma_3}{\partial X_3}\right) - X_2 X_3 \frac{\partial \ln \gamma_2}{\partial X_3} \frac{\partial \ln \gamma_3}{\partial X_2} \tag{30}$$

We note that for $X_3 \to 0$ this is reduced to expression (23) in the binary case.

As for binary systems, the sign of ψ can yield upper and lower bounds to the values of interaction parameters. For example, assuming that the formalism adopted for $\ln \gamma_i$ is limited to first order interaction coefficients, the following conditions must be respected for a solution to be stable:

$$\psi = 1 + X_2 \epsilon_2^{(2)} + X_3 \epsilon_3^{(3)} + X_2 X_3 [\epsilon_2^{(2)} \epsilon_3^{(3)} - (\epsilon_2^{(3)})^2] \geq 0 \tag{31}$$

Let us consider the case of the Fe–Ti–S system. At 1600°C, $\epsilon_{Ti}^{Ti} = 2.7$ [12], $\epsilon_S^S = -3.3$ [9] and $\epsilon_S^{Ti} = -14$ [13]. Applying inequality (31), we see that for $X_{Ti} = 0.10$ the maximum permissible concentration of sulfur is 0.035 (2.1%). In addition, we should note that these compositions correspond to the spinodal line and that the miscibility gap calculated on that basis should extend to lower concentrations of sulfur and titanium.

The existence of such a miscibility gap would be in doubt if the values of the coefficients in inequality (32) were themselves in doubt. However, the determination by Ban-ya and Chipman of ϵ_S^{Ti} is based on very careful experimentation [13], and its reliability should be excellent. Moreover, small adjustments in the values of ϵ_S^S and ϵ_{Ti}^{Ti} and the introduction of second order interaction coefficients seem to have minor influences in the calculations. Consequently, the existence of a miscibility gap appears quite probable. We note that this miscibility gap could be metastable if another phase, such as titanium sulfide, were more stable in the composition range of interest.

The effect of tin on the activity of sulfur provides another example of application of condition (31). Ban-ya and Chipman determined at 1550°C a value of ϵ_S^{Sn} equal to 50.9 [13] but stated that this value is too high in comparison to data of other systems. Condition (31) provides a more definite reason for rejecting it; for example, with $\epsilon_{Sn}^{Sn} = -0.31$ [12], we calculate that their alloy assaying 3.44% Sn and 1.17% S would be associated with a negative value of ψ.

Again, in performing such calculations ψ may be expressed in coordinates other than mole fractions. In these examples the use of Chipman's lattice ratio z_i [equation (27)] is convenient. Inequality (31) would become

$$1 + z_2\theta_2^{(2)} + z_3\theta^{(3)} + z_2 z_3[\theta_2^{(2)}\theta_3^{(3)} - (\theta_2^{(3)})^2] \geq 0 \tag{32}$$

The spinodal line $\psi = 0$ is useful for a qualitative estimate of the position and shape of a miscibility gap. Figure 4 illustrates it in the case of the Pb–Zn–Ag system. The spinodal line is calculated at various temperatures on the basis of binary data and the quasi-chemical model of ternary solutions [3]. It is in excellent agreement with the miscibility gap determined experimentally by Seith and Helmhold [14]. We stress that in order to calculate the miscibility gap, the free energy surface must be known fron one end of the composition domain (zinc corner) to the other (lead corner). Although this could be done for the Pb–Zn–Ag system, it is often not the case: activity data may be available only in a local concentration area, e.g., in the zinc-rich corner *or* the lead-rich corner. Calculation of the spinodal line then yields information that could not otherwise be obtained.

For ternary systems, as for binaries, the search for high positive values and peaks of the function ψ would be helpful in the study of short range order in solutions. However, depending on the type of solutions investigated, the function

Figure 4. Illustration of the method for predicting miscibility gaps. The dashed lines are the predicted spinodal lines, equation (30) [3], and the solid lines are the isothermal section of the miscibility gap deduced from the experimental points (white circles) of Seith and Helmhold [14].

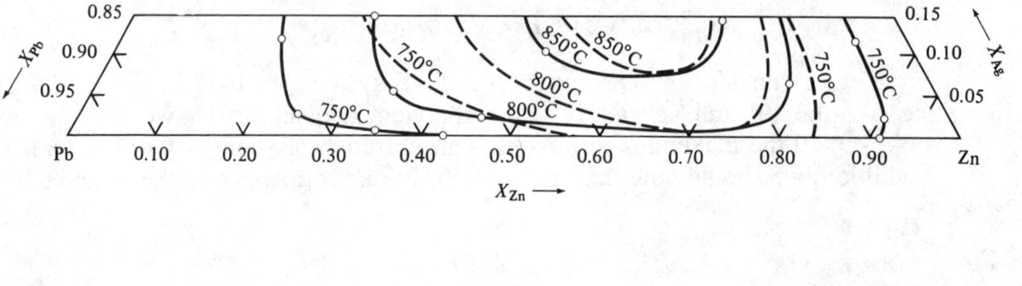

ψ alone may not provide the information sought. In ternary systems, ψ measures, at a point of the Gibbs free energy surface, the product of its two principal curvatures.[1] If the surface is nearly isotropic, i.e., if the three components play similar roles, then the use of ψ is probably sufficient. It may not be if the curvature in one direction is much more pronounced than in another direction. For instance, in an ionic structure A–B–O, where A and B are two metal cations, the curvature is likely to be very high in a direction measuring a change in the oxygen concentration from the value pertaining to the stoichiometry of the oxide. On the contrary, in a direction measuring a change in the cations ratio, the curvature may be very low. The product of these two curvatures, or ψ, could then assume a large range of values and its use alone would not be sufficient.

The previous considerations on the use of the ψ function for binary and ternary systems can be readily extended to systems of any number of components.

3. Critical Lines and Surfaces

3.1. Analytic Derivation

In the preceding section we have developed stability criteria which, when violated, ensure the presence of miscibility gaps. In the present section we shall calculate the corresponding critical points. Their study allows a more precise determination of the miscibility gaps and an estimate of the orientations of their tie lines.

At a fixed pressure, there is only one critical point in a metastable binary solution (see Chapter III), but in a ternary metastable solution there is a line of critical points (see Fig. X.7), and in an m-component system there is an $(m-2)$-dimensional surface of critical points. In order to find their equations, let us consider in Fig. 5a line AB intersecting the spinodal surface of a multidimensional phase diagram at a critical point C. A point P on the portion AC of the line within the spinodal surface (and therefore within the miscibility gap surface) represents an unstable solution with respect to infinitesimal fluctuations. The tie line passing through P provides the compositions of the two equilibrium phases θ' and θ''. When P is at C the two phases become indistinguishable, and beyond C on the portion CB of the line, the solution is stable and only one phase θ is present.

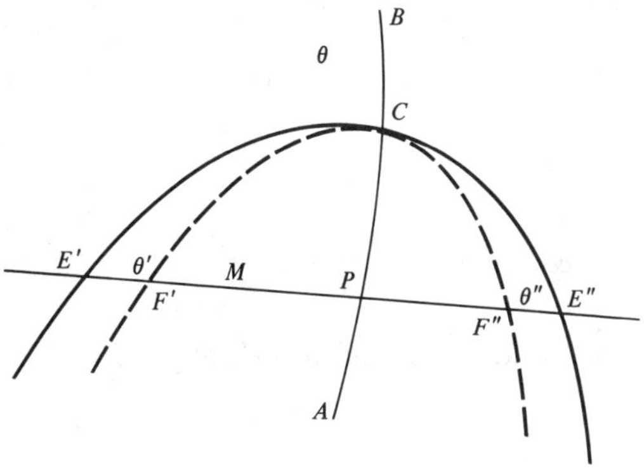

Figure 5. Representation of a multidimensional miscibility gap.

[1] If λ' and λ'' are the two eigenvalues of the 2×2 matrix whose general element is $(1/RT)(\partial^2 G^M/\partial X_i\, \partial X_j)$ $(i, j = 2, 3)$, ψ is equal to $X_1 X_2 X_3 \lambda' \lambda''$.

The solution P contains n_i° moles of component i. The number of moles n_i of component i in a solution whose representative point M is on the tie line passing by P is such that

$$n_i = n_i^\circ + \lambda\alpha_i \qquad \text{for} \quad i = 1, 2, \ldots, m \tag{33}$$

where α_i is the direction cosine of the tie line relative to the axis measuring the concentration of i and λ is a parameter defining the position of M on the tie line.

Along the tie line the Gibbs free energy of the solution is a function of λ only:

$$G(n_1^\circ + \lambda\alpha_1, \ldots, n_i^\circ + \lambda\alpha_i, \ldots, n_m^\circ + \lambda\alpha_m) = g(\lambda) \tag{34}$$

and its variation is illustrated by the curve a in Fig. 6. It shows two equilibrium points E' and E'' with a common tangent and two inflexion points F' and F''. When P moves to C, the tie line becomes tangent to the spinodal surface and to the miscibility gap surface, and the points E', E'', F', and F'' have merged into C (curve b in Fig. 6). When P is beyond C, on the portion CB of the line AB, for any line passing by P the free energy presents a positive curvature (curve c in Fig. 6).

It is readily recognized that the critical point C on the curve b is a point at which

$$\frac{d^2g}{d\lambda^2} = 0 \tag{35}$$

$$\frac{d^3g}{d\lambda^3} = 0 \tag{36}$$

and

$$\frac{d^4g}{d\lambda^4} \geq 0 \tag{37}$$

It may also be noted that

$$\frac{dg}{d\lambda} = \sum_{i=1}^{m} \frac{\partial G}{\partial n_i}\frac{\partial n_i}{\partial \lambda} = \sum_{i=1}^{m} \alpha_i \frac{\partial G}{\partial n_i} \tag{38}$$

$$\frac{d^2g}{d\lambda^2} = \sum_{j=1}^{m} \frac{\partial}{\partial n_j}\left(\sum_{i=1}^{m} \alpha_i \frac{\partial G}{\partial n_i}\right)\frac{\partial n_j}{\partial \lambda} = \sum_{i,j=1}^{m} \frac{\partial^2 G}{\partial n_i\, \partial n_j}\alpha_i\alpha_j = \sum_{i,j=1}^{m} G_{ij}\alpha_i\alpha_j \tag{39}$$

and similarly,

$$\frac{d^3g}{d\lambda^3} = \sum_{i,j,k=1}^{m} G_{ijk}\alpha_i\alpha_j\alpha_k \tag{40}$$

$$\frac{d^4g}{d\lambda^4} = \sum_{i,j,k,l=1}^{m} G_{ijkl}\alpha_i\alpha_j\alpha_k\alpha_l \tag{41}$$

where G_{ij}, G_{ijk}, G_{ijkl} are the second, third, and fourth order derivatives of G. Thus, at the critical point

$$\sum_{i,j=1}^{m} G_{ij}\alpha_i\alpha_j = 0 \tag{42}$$

$$\sum_{i,j,k=1}^{m} G_{ijk}\alpha_i\alpha_j\alpha_k = 0 \tag{43}$$

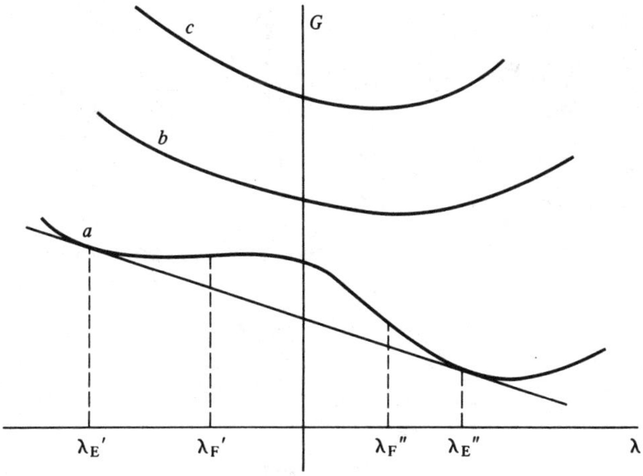

Figure 6. Variation of the free energy as a function of the parameter λ.

and

$$\sum_{i,j,k,l=1}^{m} G_{ijkl}\alpha_i\alpha_j\alpha_k\alpha_l \geq 0 \tag{44}$$

This property must be satisfied independently of the path AB, and the critical point may be defined by any set of values α_i satisfying equations (42) and (43) and inequality (44). Practically, inequality (44) is very rarely checked, the nature of the system ensuring its verification.

In Section 1.1 we saw that condition (42) may be rewritten

$$\psi = \begin{vmatrix} 1 + X_2 \dfrac{\partial \ln \gamma_2}{\partial X_2} & \cdots & X_i \dfrac{\partial \ln \gamma_2}{\partial X_i} & \cdots & X_m \dfrac{\partial \ln \gamma_2}{\partial X_m} \\ \vdots & & \vdots & & \vdots \\ X_2 \dfrac{\partial \ln \gamma_i}{\partial X_2} & \cdots & 1 + X_i \dfrac{\partial \ln \gamma_i}{\partial X_i} & \cdots & X_m \dfrac{\partial \ln \gamma_i}{\partial X_m} \\ \vdots & & \vdots & & \vdots \\ X_2 \dfrac{\partial \ln \gamma_m}{\partial X_2} & \cdots & X_i \dfrac{\partial \ln \gamma_m}{\partial X_i} & \cdots & 1 + X_m \dfrac{\partial \ln \gamma_m}{\partial X_m} \end{vmatrix} = 0 \tag{45}$$

Condition (43) may be transformed in various different ways. It is convenient to write it [15]

$$\phi = \begin{vmatrix} 1 + X_2 \dfrac{\partial \ln \gamma_2}{\partial X_2} & \cdots & X_i \dfrac{\partial \ln \gamma_2}{\partial X_i} & \cdots & X_m \dfrac{\partial \ln \gamma_2}{\partial X_m} \\ \vdots & & \vdots & & \vdots \\ X_2 \dfrac{\partial \ln \gamma_{m-1}}{\partial X_2} & \cdots & X_i \dfrac{\partial \ln \gamma_{m-1}}{\partial X_i} & \cdots & X_m \dfrac{\partial \ln \gamma_{m-1}}{\partial X_m} \\ X_2 \dfrac{\partial \psi}{\partial X_2} & \cdots & X_i \dfrac{\partial \psi}{\partial X_i} & \cdots & X_m \dfrac{\partial \psi}{\partial X_m} \end{vmatrix} = 0 \tag{46}$$

Equations (45) and (46) define the locus of critical points.

In a ternary system, these equations become

$$\psi = \left(1 + X_2 \frac{\partial \ln \gamma_2}{\partial X_2}\right)\left(1 + X_3 \frac{\partial \ln \gamma_3}{\partial X_3}\right) - X_2 X_3 \frac{\partial \ln \gamma_2}{\partial X_3} \frac{\partial \ln \gamma_3}{\partial X_2} = 0 \tag{47}$$

$$\phi = \left(1 + X_2 \frac{\partial \ln \gamma_2}{\partial X_2}\right) \frac{\partial \psi}{\partial X_3} - X_2 \frac{\partial \ln \gamma_2}{\partial X_3} \frac{\partial \psi}{\partial X_2} = 0 \tag{48}$$

Equation (48) shows that at a given temperature the critical point of the isothermal spinodal line may be identified as the point on this line ($\psi = 0$) which has a tangent of slope

$$\frac{dX_3}{dX_2} = -\frac{\partial\psi/\partial X_2}{\partial\psi/\partial X_3} = -\frac{1 + X_2(\partial \ln \gamma_2/\partial X_2)}{X_2\,(\partial \ln \gamma_2/\partial X_3)} = -\frac{X_3(\partial \ln \gamma_3/\partial X_2)}{1 + X_3(\partial \ln \gamma_3/\partial X_3)} \tag{49}$$

3.2. *Effects on a Binary Critical Point of Small Additions of a Third Component*

The problem of the influence of a third component on the mutual miscibility of two phases of the same structure (e.g., two liquids) may be studied quite conveniently by determining the effect of an added substance 3 on the critical temperature of a binary mixture 1–2. Consequently, let us determine the slope of the critical line (dT/dX_3) at the binary critical point Q.

The slope of the critical line at any point on the line may be obtained by differentiating equations (45) and (46):

$$\frac{\partial\psi}{\partial X_2} dX_2 + \frac{\partial\psi}{\partial X_3} dX_3 + \frac{\partial\psi}{\partial T} dT = 0 \tag{50}$$

$$\frac{\partial\phi}{\partial X_2} dX_2 + \frac{\partial\phi}{\partial X_3} dX_3 + \frac{\partial\phi}{\partial T} dT = 0 \tag{51}$$

and solving for dX_2/dX_3 and dT/dX_3. However, at the binary point Q the third order derivative of G with respect to X_2 is equal to zero (see Section III.4) and

$$\left(\frac{\partial\psi}{\partial X_2}\right)_Q = 0 \tag{52}$$

Consequently, equation (50) yields

$$\left(\frac{dT}{dX_3}\right)_Q = -\frac{(\partial\psi/\partial X_3)_Q}{(\partial\psi/\partial T)_Q} \tag{53}$$

If the Gibbs free energy of the solution under consideration is regular,

$$G^E = X_1 X_2 \Omega_{12} + X_1 X_3 \Omega_{13} + X_2 X_3 \Omega_{23} \tag{54}$$

then it is straightforward to show that equation (53) becomes

$$\left(\frac{dT}{dX_3}\right)_Q = \frac{(\Omega_{13} - \Omega_{23})^2 - \Omega_{12}^2}{2R\Omega_{12}} \tag{55}$$

A discussion of the results of this equation may be found in Defay and Prigogine [4]. We note here that if component 3 has similar behaviors in 1 and 2 (e.g., about equal solubility in 1 and 2), then

$$\Omega_{13} \simeq \Omega_{23} \tag{56}$$

and

$$\left(\frac{dT}{dX_3}\right)_Q \simeq -\frac{\Omega_{12}}{2R} = -T_Q \tag{57}$$

(see Section III.4.1). Consequently, component 3 would lower the solution's critical temperature and thus increase the mutual solubility of 1 and 2.

If component 3 has very little solubility in one of the components, e.g., 2, then

$$\Omega_{23} \gg \Omega_{13} \quad \text{and} \quad \Omega_{23} \gg \Omega_{12} \tag{58}$$

Consequently,

$$\left(\frac{dT}{dX_3}\right)_Q = \frac{\Omega_{23}^2}{2R\Omega_{12}} > 0 \tag{59}$$

Component 3 would, therefore, increase the critical temperature and decrease the mutual solubility of 1 and 2. This is, for instance, the effect of carbon on liquid iron–copper solutions. Carbon has a substantial solubility in liquid iron but a very small one in copper. However, the ternary liquid solution Fe–Cu–C is not regular, and for a more quantitative estimate we turn again to equation (53). Recalling the expression of ψ for a ternary solution (equation 30), we readily establish that

$$\left(\frac{\partial\psi}{\partial X_3}\right)_Q = X_2^Q\left(\frac{\partial^2 \ln \gamma_2}{\partial X_2\,\partial X_3} - \frac{\partial \ln \gamma_2}{\partial X_3}\frac{\partial \ln \gamma_3}{\partial X_2}\right)_Q \tag{60}$$

and

$$\left(\frac{\partial\psi}{\partial T}\right)_Q = X_2^Q\left(\frac{\partial^2 \ln \gamma_2}{\partial X_2\,\partial T}\right)_Q = -\frac{X_2^Q}{RT_Q^2}\left(\frac{\partial H_2^E}{\partial X_2}\right)_Q \tag{61}$$

Thus,

$$\left(\frac{\partial T}{\partial X_3}\right)_Q = RT_Q^2\left(\frac{\partial^2 \ln \gamma_2}{\partial X_2\,\partial X_3} - \frac{\partial \ln \gamma_2}{\partial X_3}\frac{\partial \ln \gamma_3}{\partial X_2}\right)_Q \Bigg/ \left(\frac{\partial H_2^E}{\partial X_2}\right)_Q \tag{62}$$

As an example of the use of this expression, let us then consider the effect of carbon on the iron–copper system. This system exhibits a metastable miscibility gap in the liquid phase and its critical point was experimentally determined by Nakagawa [16] through magnetic susceptibility measurements on the supercooled liquid. It lies at a temperature of 1420°C and a mole fraction of Cu equal to 0.5. The activity coefficient of Cu may be expressed [16]

$$\ln \gamma_{Cu} = \frac{3690}{T} X_{Fe}^2 (1.660 - 2.118X_{Fe} + 1.608X_{Fe}^2) \tag{63}$$

and yields

$$\left(\frac{\partial H_{Cu}^E}{\partial X_{Cu}}\right)_Q = -6350 \quad \text{cal} \tag{64}$$

To evaluate the numerator of equation (63) it is expedient to adopt a formalism limited to second order interaction coefficients, even though the solution is not dilute in copper. This leads to

$$\frac{\partial^2 \ln \gamma_{Cu}}{\partial X_{Cu}\,\partial X_C} = \rho_{Cu}^{Cu,C} \tag{65}$$

$$\left(\frac{\partial \ln \gamma_{Cu}}{\partial X_C}\right)_Q = \epsilon_{Cu}^{C} + \rho_{Cu}^{Cu,C} X_{Cu} = \epsilon_{Cu}^{C} + \tfrac{1}{2}\rho_{Cu}^{Cu,C} \tag{66}$$

$$\left(\frac{\partial \ln \gamma_C}{\partial X_{Cu}}\right)_Q = \epsilon_C^{Cu} + 2\rho_C^{Cu} X_{Cu} = \epsilon_{Cu}^{C} + \rho_C^{Cu} \tag{67}$$

Activity data in the ternary Fe–Cu–C system were determined at low concentration of copper by Koros and Chipman [17]. Their results lead to a value of 4.2 for the interaction coefficient ϵ_C^{Cu} at 1550°C. This value may also be used to estimate the second order coefficients through the central atoms model of interstitial solutions (see Chapter XVI); this yields $\rho_C^{Cu} = 0.7$ and $\rho_{Cu}^{Cu,C} = -4.5$. Consequently,

$$\left(\frac{\partial^2 \ln \gamma_{Cu}}{\partial X_{Cu}\,\partial X_C} - \frac{\partial \ln \gamma_{Cu}}{\partial X_C}\frac{\partial \ln \gamma_C}{\partial X_{Cu}}\right)_{\substack{T = 1550°C \\ X_{Cu}=0.5,\ X_C \to 0}} \simeq -14.1 \tag{68}$$

Assuming that $S^E \simeq 0$, at 1420°C this value becomes

$$\left(\frac{\partial^2 \ln \gamma_{Cu}}{\partial X_{Cu}\,\partial X_C} - \frac{\partial \ln \gamma_{Cu}}{\partial X_C}\frac{\partial \ln \gamma_C}{\partial X_{Cu}}\right)_Q \simeq -16 \tag{69}$$

Thus, the effect of carbon on the critical temperature of the iron–copper solution may now be evaluated:

$$\left(\frac{\partial T}{\partial X_C}\right)_Q \simeq \frac{1.987 \times (1693)^2 \times 16}{6350} \simeq 15{,}000°K \tag{70a}$$

or

$$\left(\frac{\partial T}{\partial \%C}\right)_Q \simeq 700°K \tag{70b}$$

At 1450°C (slightly above the liquidus temperature of the iron–copper phase diagram), Maddocks and Clausen [18] and Iwase et al. [19] found that the presence of carbon led to an immiscibility gap. Maddocks and Clausen estimated the necessary amount to be slightly above 0.05% C while Iwase et al. determined it to be of the order of 0.2–0.3% C. Equation (70b) yields

$$\%C = \Delta T/700 = (1450 - 1420)/700 \simeq 0.04 \tag{71}$$

in good agreement with these measurements.

Another example of application of equation (62) may be given in the case of the Zn–Pb–Ag system. On the basis of the activity coefficients of Pb and Ag [3], the reader may verify that silver raises the critical temperature of the Pb–Zn miscibility gap by the amount

$$\left(\frac{dT}{dX_{Ag}}\right)_Q \simeq 770°K \tag{72}$$

The distance in terms of X_{Ag} separating the critical points of the two immiscibility boundaries at 800 and 850°C is then of the order of

$$\Delta X_{Ag} \simeq (850 - 800)/770 = 0.065 \tag{73}$$

In Fig. 4 it is seen to be approximately equal to 0.07; this is in excellent agreement with the calculated value in equation (73).

4. Effect of Small Additions of a Third Component on the Eutectic and Peritectic Temperatures of Binary Systems

In the preceding section we have analyzed the effects of small additions of a third component on a binary critical point. Similarly, we shall now study the effect of third component additions on binary eutectic and peritectic points. This study is of special interest in the case of eutectics. The eutectic composition of a binary system corresponds to the lowest temperature at which the liquid phase is still stable; consequently, it is useful in numerous applications to examine whether minor additions of alloying elements depress or raise this temperature. For example, in the development of brazing alloys, the depression of a eutectic temperature is generally beneficial. To refractory bricks based on the CaO–MgO system (whose phase diagram exhibits a simple eutectic), oxides which lower the eutectic temperature are detrimental.

The cases of peritectics and eutectoids can be examined in the same perspective. In the Fe–C system, the peritectic and eutectoid temperatures correspond respectively to the highest and lowest temperatures at which the γ phase (austenite) is stable. The effect of alloying elements on these temperatures is important for the heat treatment of steels.

4.1. Analytic Derivation

Let us consider the binary eutectic system 1–2 in Fig. 7 and the effect of a small addition of a component 3 at the temperature $T_E + dT$. The liquid of composition M is in equilibrium with both solid phases α and β. As the temperature varies,

Figure 7. Graphical illustration of the effect of small additions of a third component on the eutectic temperature of a binary system. The ternary section is drawn at the temperature $T_E + dT$.

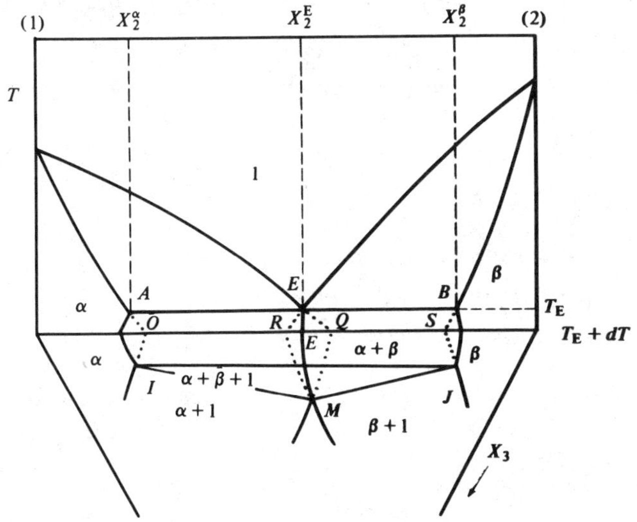

the point M describes a line EM, and it is the orientation of this line at the eutectic point E that we wish to calculate.

To find the composition of the point M we examine the metastable equilibria α + liquid, described by the lines OI–QM, and β + liquid, described by the lines SJ–RM. The point M is at the intersection of the lines QM and RM, which, for infinitesimal increments, may be approximated by their tangents at the points Q and R.

In terms of mole fractions, the equations of the lines QM and RM are

$$X_3 = \frac{dX_3^{l(\alpha)}}{dX_2^{l(\alpha)}}(X_2 - X_2^{l(\alpha)}) \tag{74}$$

$$X_3 = \frac{dX_3^{l(\beta)}}{dX_2^{l(\beta)}}(X_2 - X_2^{l(\beta)}) \tag{75}$$

where the upperscript $l(\nu)$ identifies the ν phase ($\nu = \alpha$ or β) in equilibrium with the liquid phase. The coordinates of the point M are obtained by solving equations (74) and (75):

$$X_2^M = \left(X_2^{l(\alpha)}\frac{dX_3^{l(\alpha)}}{dX_2^{l(\alpha)}} - X_2^{l(\beta)}\frac{dX_3^{l(\beta)}}{dX_2^{l(\beta)}}\right) \Big/ \left(\frac{dX_3^{l(\alpha)}}{dX_2^{l(\alpha)}} - \frac{dX_3^{l(\beta)}}{dX_2^{l(\beta)}}\right) \tag{76}$$

$$X_3^M = \left[\frac{dX_3^{l(\alpha)}}{dX_2^{l(\alpha)}}\frac{dX_3^{l(\beta)}}{dX_2^{l(\beta)}} \Big/ \left(\frac{dX_3^{l(\alpha)}}{dX_2^{l(\alpha)}} - \frac{dX_3^{l(\beta)}}{dX_2^{l(\beta)}}\right)\right](X_2^{l(\alpha)} - X_2^{l(\beta)}) \tag{77}$$

To calculate $X_2^{l(\alpha)} - X_2^{l(\beta)}$, or RQ, we note that

$$X_2^{l(\alpha)} - X_2^{E} = \frac{dX_2^{l(\alpha)}}{dT}dT \tag{78}$$

$$X_2^{l(\beta)} - X_2^{E} = \frac{dX_2^{l(\beta)}}{dT}dT \tag{79}$$

so that

$$X_2^{l(\alpha)} - X_2^{l(\beta)} = \left(\frac{dX_2^{l(\alpha)}}{dT} - \frac{dX_2^{l(\beta)}}{dT}\right)dT \tag{80}$$

Moreover, since X_3^M is dX_3 (increment of 3 from E to M), equation (77) becomes

$$dX_3 = \left[\frac{dX_3^{l(\alpha)}}{dX_2^{l(\alpha)}}\frac{dX_3^{l(\beta)}}{dX_2^{l(\beta)}} \Big/ \left(\frac{dX_3^{l(\alpha)}}{dX_2^{l(\alpha)}} - \frac{dX_3^{l(\beta)}}{dX_2^{l(\beta)}}\right)\right]\left(\frac{dX_2^{l(\alpha)}}{dT} - \frac{dX_2^{l(\beta)}}{dT}\right)dT \tag{81}$$

or

$$\left(\frac{dT}{dX_3}\right)_E = \left(\frac{dX_2^{l(\alpha)}}{dX_3^{l(\alpha)}} - \frac{dX_2^{l(\beta)}}{dX_3^{l(\beta)}}\right) \Big/ \left(\frac{dX_2^{l(\beta)}}{dT} - \frac{dX_2^{l(\alpha)}}{dT}\right) \tag{82}$$

Similarly, the slope of the projected line EM is

$$\left(\frac{dX_2}{dX_3}\right)_E = \left(\frac{dX_2^{l(\alpha)}}{dX_3^{l(\alpha)}}\frac{dX_2^{l(\beta)}}{dT} - \frac{dX_2^{l(\beta)}}{dX_3^{l(\beta)}}\frac{dX_2^{l(\alpha)}}{dT}\right) \Big/ \left(\frac{dX_2^{l(\beta)}}{dT} - \frac{dX_2^{l(\alpha)}}{dT}\right) \tag{83}$$

The denominator of equations (82) and (83) can be easily read off the phase diagram (see Fig. 8):

$$\frac{dX_2^{l(\beta)}}{dT} - \frac{dX_2^{l(\alpha)}}{dT} = \left|\frac{RQ}{dT}\right| \tag{84}$$

To calculate the numerators, it is convenient to recall the results of Chapter X, Section 4.3, yielding the slopes of ternary two-phase boundaries at infinite dilution of one of the components [equations (77)–(79)]. For the α–liquid equilibrium

$$\frac{dX_3^{\alpha(l)}}{dX_3^{l(\alpha)}} = \frac{\gamma_3^{\infty(l,E)}}{\gamma_3^{\infty(\alpha,A)}} \exp \frac{\mu_3^{*(l)} - \mu_3^{*(\alpha)}}{RT} \tag{85}$$

$$\frac{\psi^l}{X_2^l} \frac{dX_2^{l(\alpha)}}{dX_3^{l(\alpha)}} = \frac{X_1^l}{X_2^l - X_2^\alpha} \frac{dX_3^{\alpha(l)}}{dX_3^{l(\alpha)}} - \left(\frac{X_1^\alpha}{X_2^l - X_2^\alpha} + \frac{\partial \ln \gamma_2^l}{\partial X_3^l} \right) \tag{86}$$

In these equations $dX_i^{\alpha(l)}$ and $dX_i^{l(\alpha)}$ are respective increments in the mole fractions of i in the α and liquid phases in equilibrium with respect to each other. γ_2^l is the activity coefficient of 2 in the liquid phase, and $\gamma_3^{\infty(l,E)}$, $\gamma_3^{\infty(\alpha,A)}$ are respectively the activity coefficients of 3 at infinite dilution in the liquid phase of eutectic composition E and in the α phase of composition A (see Fig. 7). $\mu_3^{*(\nu)}$ is the reference value of the chemical potential of component 3 in the ν phase. Combining equations (85) and (86) and noting that X_1^l and X_2^l are the mole fractions of 1 and 2 at the eutectic point, X_1^E and X_2^E, we find

$$\frac{\psi^l}{X_2^E} \frac{dX_2^{l(\alpha)}}{dX_3^{l(\alpha)}} = \frac{X_1^E}{X_2^E - X_2^\alpha} \frac{\gamma_3^{\infty(l,E)}}{\gamma_3^{\infty(\alpha,A)}} \exp \frac{\mu_3^{*(l)} - \mu_3^{*(\alpha)}}{RT} - \frac{X_1^\alpha}{X_2^E - X_2^\alpha} - \frac{\partial \ln \gamma_2^l}{\partial X_3^l} \tag{87}$$

The β liquid equilibrium yields a similar equation:

$$\frac{\psi^l}{X_2^E} \frac{dX_2^{l(\beta)}}{dX_3^{l(\beta)}} = \frac{X_1^E}{X_2^E - X_2^\beta} \frac{\gamma_3^{\infty(l,E)}}{\gamma_3^{\infty(\beta,B)}} \exp \frac{\mu_3^{*(l)} - \mu_3^{*(\beta)}}{RT} - \frac{X_1^\beta}{X_2^E - X_2^\beta} - \frac{\partial \ln \gamma_2^l}{\partial X_3^l} \tag{88}$$

Subtracting equation (87) from (88) and combining the difference with equations (82) and (84), we obtain

$$\begin{aligned} \left(\frac{dT}{dX_3} \right)_E = \frac{1}{|RQ/dT|} \frac{X_1^E X_2^E}{\psi^l} \Bigg[& \frac{1}{X_2^\beta - X_2^E} \left(\frac{\gamma_3^{\infty(l,E)}}{\gamma_3^{\infty(\beta,B)}} \exp \frac{\mu_3^{*(l)} - \mu_3^{*(\beta)}}{RT_E} - 1 \right) \\ & + \frac{1}{X_2^E - X_2^\alpha} \left(\frac{\gamma_3^{\infty(l,E)}}{\gamma_3^{\infty(\alpha,A)}} \exp \frac{\mu_3^{*(l)} - \mu_3^{*(\alpha)}}{RT_E} - 1 \right) \Bigg] \end{aligned} \tag{89}$$

A similar equation for dX_2/dX_3 can be readily derived from equations (83), (84), (87), and (88). Its expression, however, is somewhat cumbersome and need not be written here.

4.2. *Alternative Forms and Consequences*

Several observations concerning equation (89) should be made. First, the term RQ/dT may be expressed analytically. Through equation (45) of Section VIII.8.1, it is straightforward to demonstrate that

$$\frac{dX_2^{l(\beta)}}{dT} - \frac{dX_2^{l(\alpha)}}{dT} = \left| \frac{RQ}{dT} \right| = \frac{X_1^E X_2^E}{\psi^l} \frac{(X_2^\beta - X_2^\alpha)}{(X_2^\beta - X_2^E)(X_2^E - X_2^\alpha)} \frac{\Delta H^E}{RT_E^2} \tag{90}$$

where

$$\Delta H^E = H^{(l,E)} - \left(\frac{X_2^E - X_2^\alpha}{X_2^\beta - X_2^\alpha} H^{(\beta,B)} + \frac{X_2^\beta - X_2^E}{X_2^\beta - X_2^\alpha} H^{(\alpha,A)} \right) \tag{91a}$$

or

$$\Delta H^{\mathrm{E}} = H^{(\mathrm{l},E)} - \left(\frac{AE}{AB} H^{(\beta,B)} + \frac{EB}{AB} H^{(\alpha,A)}\right) \tag{91b}$$

$H^{(\mathrm{l},E)}$, $H^{(\beta,B)}$, and $H^{(\alpha,A)}$ identify, respectively, the molar enthalpies of the liquid, β, and α phases at the points E, B, and A. From the lever rule, AE/AB and EB/AB are seen as the amounts of the phases β and α associated with the eutectic reaction

$$\alpha + \beta = \text{liquid} \tag{92}$$

Thus, ΔH^{E} is the enthalpy of that reaction per mole of the binary 1–2. Equation (89) may then be rewritten

$$\left(\frac{dT}{dX_3}\right)_{\mathrm{E}} = \frac{RT_{\mathrm{E}}^2}{\Delta H^{\mathrm{E}}}\left[\frac{X_2^E - X_2^\alpha}{X_2^\beta - X_2^\alpha}\left(\frac{\gamma_3^{\infty(\mathrm{l},E)}}{\gamma_3^{\infty(\beta,B)}} \exp\frac{\mu_3^{*(\mathrm{l})} - \mu_3^{*(\beta)}}{RT_{\mathrm{E}}} - 1\right)\right.$$
$$\left. + \frac{X_2^\beta - X_2^E}{X_2^\beta - X_2^\alpha}\left(\frac{\gamma_3^{\infty(\mathrm{l},E)}}{\gamma_3^{\infty(\alpha,A)}} \exp\frac{\mu_3^{*(\mathrm{l})} - \mu_3^{*(\alpha)}}{RT_{\mathrm{E}}} - 1\right)\right] \tag{93a}$$

or

$$\left(\frac{dT}{dX_3}\right)_{\mathrm{E}} = -\frac{RT_{\mathrm{E}}^2}{\Delta H^{\mathrm{E}}}\left[1 - \frac{X_2^E - X_2^\alpha}{X_2^\beta - X_2^\alpha}\left(\frac{\gamma_3^{\infty(\mathrm{l},E)}}{\gamma_3^{\infty(\beta,B)}} \exp\frac{\mu_3^{*(\mathrm{l})} - \mu_3^{*(\beta)}}{RT_{\mathrm{E}}}\right)\right.$$
$$\left. - \frac{X_2^\beta - X_2^E}{X_2^\beta - X_2^\alpha}\left(\frac{\gamma_3^{\alpha(\mathrm{l},E)}}{\gamma_3^{\infty(\alpha,A)}} \exp\frac{\mu_3^{*(\mathrm{l})} - \mu_3^{*(\alpha)}}{RT_{\mathrm{E}}}\right)\right] \tag{93b}$$

Second, because activity coefficients and exponentials are always positive, it is readily apparent that lowering the eutectic temperature of a given binary system 1–2 by any ternary addition cannot exceed a certain limit. This limit E_{12} is

$$E_{12} = \left(\frac{dT}{dX_3}\right)_{\mathrm{E,lim}} = \frac{-1}{|RQ/dT|}\,\frac{X_1^E X_2^E}{\psi^{\mathrm{l}}}\,\frac{(X_2^\beta - X_2^\alpha)}{(X_2^\beta - X_2^E)\,(X_2^E - X_2^\alpha)}$$
$$= -RT_{\mathrm{E}}^2/\Delta H^{\mathrm{E}} \tag{94}$$

Third, the exponential factors in equations (89) and (93) may be easily calculated [e.g., equations (VIII.21)–(VIII.23)]. We recall here that (assuming $P = 1$ atm)

$$\exp\frac{\mu_3^{*(\mathrm{l})} - \mu_3^{*(\alpha)}}{RT_{\mathrm{E}}} \simeq \exp\frac{\Delta H_{\mathrm{f},3}^\circ - T_{\mathrm{E}}\,\Delta S_{\mathrm{f},3}^\circ}{RT_{\mathrm{E}}} = \exp\frac{\Delta S_{\mathrm{f},3}^\circ}{R}\left(\frac{T_{\mathrm{f},3}}{T_{\mathrm{E}}} - 1\right) \tag{95}$$

where $T_{\mathrm{f},3}$ is the temperature of fusion of component 3 in the α phase (which may correspond to a metastable equilibrium) and $\Delta H_{\mathrm{f},3}^\circ$, $\Delta S_{\mathrm{f},3}^\circ$ are the corresponding enthalpy and entropy changes. Equations (93) and (95) show that solutes with temperatures of fusion very different from the eutectic temperature are more likely to have strong effects on the eutectic temperature of the binary system 1–2.

Fourth, if the same reference state is chosen for the activity of 3 in all three phases α, β, l, the exponential factors in equations (93) are eliminated. Equation (93b) may be rewritten

$$\left(\frac{dT}{dX_3}\right)_{\mathrm{E}} = E_{12}\left(1 - \frac{X_2^E - X_2^\alpha}{X_2^\beta - X_2^\alpha}\,\frac{\theta_3^{\infty(\mathrm{l},E)}}{\theta_3^{\infty(\beta,B)}} - \frac{X_2^\beta - X_2^E}{X_2^\beta - X_2^\alpha}\,\frac{\theta_3^{\infty(\mathrm{l},E)}}{\theta_3^{\infty(\alpha,A)}}\right) \tag{96}$$

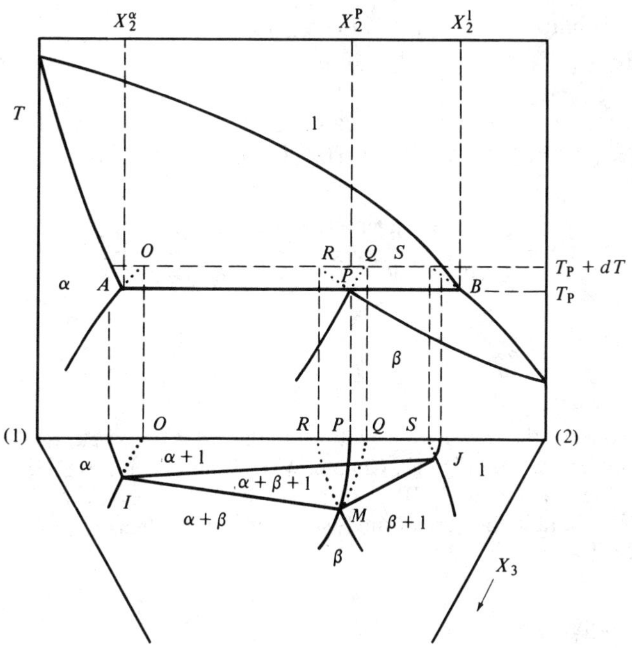

Figure 8. Graphical illustration of the effect of small additions of a third component on the peritectic point of a binary system. The ternary section is drawn at the temperature $T_P + dT$.

where we now designate the activity coefficients of 3 by θ_3 in order to recall the convention that they correspond to the same reference state.

Fifth, in cases where β corresponds to the pure component 2 in a structure where components 1 and 3 have neglibible solubilities, equations (89) and (96) become

$$\left(\frac{dT}{dX_3}\right)_E = E_{12}\left(1 - \frac{X_1^E}{X_1^\alpha}\frac{\theta_3^{\infty(l,E)}}{\theta_3^{\infty(\alpha,A)}}\right) \tag{97}$$

When components 1–3 have negligible solubilities in the solid state, the expression for the effect of 3 on the eutectic temperature of the binary 1–2 reduces to

$$\left(\frac{dT}{dX_3}\right)_E = E_{12} = \frac{-1}{|RQ/dT|}\frac{1}{\psi^l} = -\frac{RT_E^2}{\Delta H^E} \tag{98}$$

Finally, the effect of small additions of a third component on the peritectic temperature of a binary system may readily be determined from the preceding equations. With the phases identified in Fig. 8, dT/dX_3 at the peritectic point P is obtained by interchanging the upperscript β and l and noting that RQ/dT has a sign opposite to that in the eutectic case. The rise in peritectic temperature of the system 1–2 by any element 3 cannot exceed a limit P_{12} equal to

$$P_{12} = \left(\frac{dT}{dX_3}\right)_{P,\lim} = \frac{1}{|RQ/dT|}\frac{X_1^P X_2^P}{\psi^\beta}\frac{(X_2^l - X_2^\alpha)}{(X_2^l - X_2^P)(X_2^P - X_2^\alpha)} \tag{99a}$$

or

$$P_{12} = -RT_P^2/\Delta H^P \tag{99b}$$

where ΔH^P is the enthalpy of the peritectic reaction

$$\alpha + l = \beta \tag{100}$$

per mole of the binary 1–2, and may be calculated by an equation analogous to equation (91). (ΔH^P is negative whereas ΔH^E is positive.) dT/dX_3 may then be expressed

$$\left(\frac{dT}{dX_3}\right)_P = P_{12}\left(1 - \frac{X_2^P - X_2^\alpha}{X_2^l - X_2^\alpha}\frac{\theta_3^{\infty(\beta,P)}}{\theta_3^{\infty(l,B)}} - \frac{X_2^l - X_2^P}{X_2^l - X_2^\alpha}\frac{\theta_3^{\infty(\beta,P)}}{\theta_3^{\infty(\alpha,A)}}\right) \tag{101}$$

4.3. Examples

We shall now apply some of the preceding equations to the Ag–Cu system, which presents a simple eutectic equilibrium (see Fig. 12, Chapter VIII), and to the Fe–C peritectic and eutectoid equilibria (Fig. 19, Chapter VIII). In the case of the Ag–Cu system we shall examine the effect of Sn, and in the case of the Fe–C system the effects of Si. The pressure will be assumed to be 1 atm.[2]

4.3.1. Ag–Cu–Sn System

The α and β phases in the Ag–Cu system have the same fcc structure and are part of a miscibility gap. At the eutectic temperature, 1052°K, $X_{Cu}^\alpha = 0.141$, $X_{Cu}^E = 0.399$, and $X_{Cu}^\beta = 0.951$. The slopes of the phase boundaries at the eutectic point yield

$$|RQ/dT| = 0.5 \times 10^{-2} \quad °K^{-1} \tag{102}$$

Moreover, a straightforward extrapolation of the data compiled by Hultgren et al. [20] shows that

$$(\psi^l)_{X_{Cu}=0.399} = \left(1 + X_{Cu}^l \frac{\partial \ln \gamma_{Cu}^l}{\partial X_{Cu}^l}\right)_{X_{Cu}=0.399} = 0.24 \tag{103}$$

Consequently,

$$\left(\frac{dT}{dX_3}\right)_{E,\lim} = \frac{-1}{0.5 \times 10^{-2}} \times \frac{0.399 \times 0.601}{0.24} \times \frac{(0.951 - 0.141)}{(0.951 - 0.399)(0.399 - 0.141)} \tag{104a}$$

or

$$E_{12} \simeq -1140°K \tag{104b}$$

Thus, the maximum lowering of the eutectic temperature caused by the addition of 1 at.% of any ternary element is of the order of −11°C.

From the data compiled by Hultgren et al. [20] on the Ag–Sn and Cu–Sn systems, Lupis [21] estimated the following values for the activity coefficient of Sn in the ternary system:[3]

$$\gamma_{Sn}^{\infty(\alpha,A)} = 1.28, \qquad \gamma_{Sn}^{\infty(\beta,B)} = 0.17, \qquad \gamma_{Sn}^{\infty(l,E)} = 0.16 \tag{105}$$

In order to calculate the exponential factor in equation (93), we note that $T_{f,Sn} = 505.06°K$,

---

[2] Actually, pressure has little effect on the phase diagrams of condensed systems until it reaches at least a few hundred atmospheres.

[3] These estimates are based on an analytic interpolation method. Lupis also examined [21] the effects of Zn and Au on the Ag–Cu eutectic.

$\Delta H^{\circ}_{f,Sn} = 1680$ cal/mol [22], and $\Delta C^{\circ}_{p,Zn} = C^{\circ(l)}_{p,Zn} - C^{\circ(s)}_{p,Zn} = 2.88 - 0.063T$ cal/mol °K [23]. As $\Delta C^{\circ}_{p} \neq 0$, equation (95) is not strictly applicable. More accurately, equation (23) of Chapter VIII should be used. It yields

$$\exp \frac{\mu_3^{\circ(l)} - \mu_3^{\circ(s)}}{RT_E} = 0.482 \tag{106}$$

Equation (93b) thus becomes

$$\left(\frac{dT}{dX_{Sn}}\right)_E = -1140 \left[1 - 0.319 \left(\frac{0.16}{0.17} \times 0.482\right) - 0.681 \left(\frac{0.16}{1.28} \times 0.482\right)\right]$$

$$= -928°\text{K} \tag{107a}$$

or, on a weight percent basis,

$$\left(\frac{dT}{d\%\text{Sn}}\right)_E = -7°\text{C} \tag{107b}$$

This result may be compared with the diagram shown in Fig. 9 and given by Chang et al. [24]. The agreement is excellent.

4.3.2. Fe–C–Si System

In the Fe–C phase diagram, the peritectic equilibrium occurring at 1769°K is characterized by the following compositions: $X_C^{\delta} = 0.0046$, $X_C^{\gamma} = X_C^{P} = 0.0083$, and $X_C^{l} = 0.0233$ [10]. To evaluate the stability function ψ^{γ} we note that for dilute solutions ψ^{γ} may be approximated by

$$\psi^{\gamma} = 1 + X_C \epsilon_C^C \tag{108a}$$

where ϵ_C^C is the self-interaction coefficient of carbon. At 1769°K it is roughly equal to 7 [25]. Consequently,

$$\psi^{\gamma} = 1 + 0.0083 \times 7 = 1.06 \tag{108b}$$

A reading of the Fe–C phase diagram [10,20] yields a value of dT/RQ of the order of 2600°K. Thus, equation (99a) predicts that no element 3 can raise the peritectic temperature by more than

$$P_{12} = \left(\frac{dT}{dX_3}\right)_{P,\text{lim}} = 2600 \times \frac{0.0083 \times 0.9917}{1.06} \times \frac{0.0187}{0.0190 \times 0.0037} = 6800°\text{K} \tag{109}$$

(or 68°C/at.%). P_{12} may also be calculated on the basis of the enthalpy of the peritectic

Figure 9. Liquidus surface of the Ag–Cu–Sn system [24]. The temperatures quoted are in degrees Celsius.

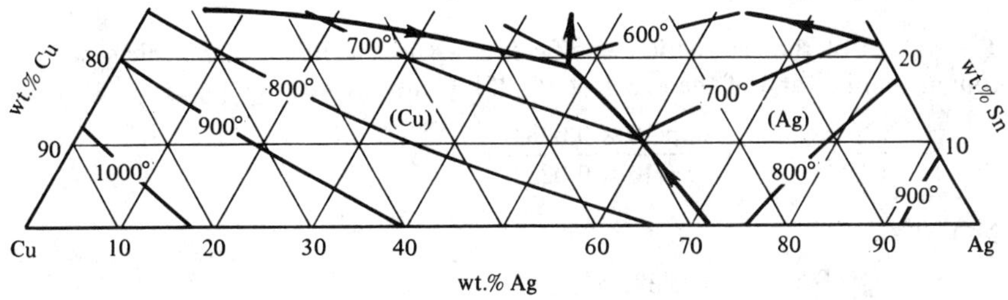

reaction [equation (99b)]. A rearrangement of the terms of equation (91b) leads to

$$\Delta H^P = X^{\delta}_{\mathrm{Fe}} \frac{PB}{AB} (H^{\circ\gamma}_{\mathrm{Fe}} - H^{\circ\delta}_{\mathrm{Fe}}) + X^{\mathrm{l}}_{\mathrm{Fe}} \frac{AP}{AB} (H^{\circ\gamma}_{\mathrm{Fe}} - H^{\circ\mathrm{l}}_{\mathrm{Fe}})$$

$$+ H^{\mathrm{M}(\gamma,P)} - \frac{AP}{AB} H^{\mathrm{M}(\mathrm{l},B)} - \frac{PB}{AB} H^{\mathrm{M}(\delta,A)} \qquad (110)$$

where H^{M} designates the enthalpy of mixing (or excess enthalpy) with respect to graphite and

$$\frac{AP}{AB} = \frac{X^P_{\mathrm{C}} - X^{\delta}_{\mathrm{C}}}{X^{\mathrm{l}}_{\mathrm{C}} - X^{\delta}_{\mathrm{C}}} = 0.198, \qquad \frac{PB}{AB} = \frac{X^{\mathrm{l}}_{\mathrm{C}} - X^P_{\mathrm{C}}}{X^{\mathrm{l}}_{\mathrm{C}} - X^{\delta}_{\mathrm{C}}} = 0.802 \qquad (111)$$

With the data of Orr and Chipman for the standard enthalpies of iron [26] and those of Hultgren et al. [20] for the excess enthalpies, equation [110] yields

$$\Delta H^{\mathrm{P}} = 0.9954 \times 0.802 \times (-276) + 0.9767 \times 0.198 \times (-3548)$$

$$+ 88 - 25 - 92 = -935 \quad \text{cal} \qquad (112)$$

Consequently, equation (99b) yields

$$P_{12} = \frac{1.987 \times (1769)^2}{935} = 6650°\mathrm{K} \qquad (113)$$

In view of the accuracy of the data, the value is in excellent agreement with that found in equation (109).

To illustrate equation (101) on the effect of a specific solute on the peritectic temperature, let us now consider silicon as an alloying element. Adopting the same reference state for Si in the δ and γ phases, a necessary condition for equilibrium between the two phases is

$$X^{\delta}_{\mathrm{Si}} \theta^{\delta}_{\mathrm{Si}} = X^{\gamma}_{\mathrm{Si}} \theta^{\gamma}_{\mathrm{Si}} \qquad (114)$$

In the dilute range, because of L'Hospital's rule, this condition yields

$$\left(\frac{dX^{\delta}_{\mathrm{Si}}}{dX^{\gamma}_{\mathrm{Si}}}\right)_{X_{\mathrm{Si}}\to 0} = \left(\frac{dT/dX^{\gamma}_{\mathrm{Si}}}{dT/dX^{\delta}_{\mathrm{Si}}}\right)_{X_{\mathrm{Si}}\to 0} = \left(\frac{\theta^{\infty(\gamma)}_{\mathrm{Si}}}{\theta^{\infty(\delta)}_{\mathrm{Si}}}\right)_{X_{\mathrm{Fe}}\to 1} \qquad (115)$$

A reading of the slopes of the boundaries in the Fe–Si phase diagram [27,28] yields at 1673°K

$$(\theta^{\infty(\gamma)}_{\mathrm{Si}}/\theta^{\infty(\delta)}_{\mathrm{Si}})_{X_{\mathrm{Fe}}\to 1} \simeq \tfrac{4}{3} \qquad (116)$$

and at 1809°K

$$(\theta^{\infty(\delta)}_{\mathrm{Si}}/\theta^{\infty(\mathrm{l})}_{\mathrm{Si}})_{X_{\mathrm{Fe}}\to 1} \simeq \tfrac{4}{3} \qquad (117)$$

Assuming that these ratios do not vary strongly with the temperature, we deduce that at 1769°K

$$(\theta^{\infty(\gamma)}_{\mathrm{Si}}/\theta^{\infty(\mathrm{l})}_{\mathrm{Si}})_{X_{\mathrm{Fe}}\to 1} \simeq \tfrac{16}{9} \qquad (118)$$

The values of the interaction coefficients $\epsilon^{\mathrm{Si}}_{\mathrm{C}}$ in the liquid and γ phases at 1769°K are roughly similar and of the order of 10 [25]. Consequently,

$$\frac{\theta^{\infty(\gamma,P)}_{\mathrm{Si}}}{\theta^{\infty(\mathrm{l},B)}_{\mathrm{Si}}} = \frac{16}{9} \times \frac{\exp(10 \times 0.0083)}{\exp(10 \times 0.0233)} \simeq 1.53 \qquad (119)$$

Similarly,

$$\theta^{\infty(\gamma,P)}_{\mathrm{Si}}/\theta^{\infty(\delta,A)}_{\mathrm{Si}} \simeq 1.38 \qquad (120)$$

Equation (101) then yields

$$\left(\frac{dT}{dX_{Si}}\right)_P = 6700(1 - 0.198 \times 1.53 - 0.802 \times 1.38) = -2740°K \tag{121a}$$

or

$$\left(\frac{dT}{d\%Si}\right)_P = -54°C \tag{121b}$$

Large errors may originate from the reading of slopes in somewhat inaccurate phase diagrams, especially when the boundaries exhibit pronounced curvatures. Consequently, we stress that the result above should be considered as more qualitative than quantitative. Even as such, it is quite useful and in agreement with the observation in the literature that silicon depresses the peritectic temperature of the Fe–C system [29].

Similar observations on the effects of various solutes on the α–γ–graphite eutectoid equilibrium can be readily explained. The eutectoid temperature is equal to 1011°K and the compositions of the α and γ phases are $X_C^\alpha = 0.0009$ and $X_C^E = 0.031$. The value of $|RQ/dT|$ may be read from the phase diagram and is roughly evaluated at 4.5×10^{-4} $°K^{-1}$. ψ^γ is found to be equal to 1.33 on the basis of the thermodynamic expressions proposed by Foo and Lupis [30]. Consequently, equation (94a) yields

$$E_{12} = \left(\frac{dT}{dX_3}\right)_{E,\lim} = -1720°K \tag{122}$$

We note that the calculation of the enthalpy of the eutectoid reaction would lead to the expression

$$\Delta H^E = X_{Fe}^E\,(H_{Fe}^{\circ\gamma} - H_{Fe}^{\circ\alpha}) + H^{M(\gamma,E)} - \frac{X_{Fe}^E}{X_{Fe}^\alpha} H^{M(\alpha,\,A)} \tag{123}$$

which, through the data compiled by Orr and Chipman [26] and Hultgren et al. [20], becomes

$$\Delta H^E = 0.969 \times 916 + 327 - 22 = 1193 \quad \text{cal} \tag{124}$$

Thus, equation (94) yields

$$E_{12} = -\frac{1.987 \times (1011)^2}{1193} = -\ 1702°K \tag{125}$$

The value is in excellent agreement with that found in equation (122) and is based on more accurate data.

Equation (97) becomes

$$\left(\frac{dT}{dX_3}\right)_E = -1700\left(1 - 0.97\,\frac{\theta_3^{\infty(\gamma,E)}}{\theta_3^{\infty(\alpha,A)}}\right) \tag{126}$$

Elements which stabilize the ferrite field are generally such that

$$\left(\frac{dT/dX_3^\gamma}{dT/dX_3^\alpha}\right)_{X_{Fe}\to 1} = \left(\frac{\theta_3^{\infty(\gamma)}}{\theta_3^{\infty(\alpha)}}\right)_{X_{Fe}\to 1} > 1 \tag{127}$$

For example, in the case of silicon a reading of the slopes of the α–γ boundaries at 1184°K in the Fe–Si phase diagram [27] yields

$$\left(\frac{\theta_{Si}^{\infty(\gamma)}}{\theta_{Si}^{\infty(\alpha)}}\right)_{X_{Fe}\to 1} \simeq 1.4 \tag{128}$$

We shall assume that this value is essentially the same at 1011°K. Taking into account the interaction coefficients ϵ_C^{Si}, which are here rather high ($\simeq$13), equation (126) becomes

$$\left(\frac{dT}{dX_{Si}}\right)_E = -1700\{1 - 0.97 \times 1.4 \exp[(0.031 - 0.0009) \times 13]\} = 1714°K \quad (129a)$$

or, on a weight percent basis

$$\left(\frac{dT}{d\%Si}\right)_E = 34°C \quad \text{or} \quad 61°F \quad (129b)$$

This result is in good agreement with the reported effect of silicon [29].

In the calculations pertaining to the Fe–C–3 eutectoid equilibria, we note that because of the high dilution of carbon the effect of the interaction coefficients may be neglected unless these coefficients have unusually high values (as in the case of silicon). Moreover, since the value of X_{Fe}^E/X_{Fe}^α (0.97) in equation (97) or (126) is so close to 1, we deduce from equation (127) that additions of elements which stabilize the bcc phase (α) raise the Fe–C eutectoid temperature, while those which stabilize the fcc phase (γ) depress that temperature. Deviations from this simple result would be due to large interaction coefficients.

In the case of the α–γ–Fe_3C eutectoid equilibrium, the results are virtually the same. The limit E'_{12} is also roughly equal to -1700°K and equation (97) becomes

$$\left(\frac{dT}{dX_3}\right)_{E'} = -1700\left(1 - 0.863\,\frac{\theta_3^{\infty(\gamma,E')}}{\theta_3^{\infty(\alpha,A)}}\right) \quad (130)$$

Again, ferrite stabilizers are likely to raise the eutectoid temperature (1000°K) while austenite stabilizers are likely to depress it. These general predictions are in good agreement with the data in the literature [29].

4.4. *Conclusions*

A simple thermodynamic equation has been derived which expresses the effect of small additions of a third component on the eutectic or peritectic temperature of a binary system. The equations necessary to calculate the effect of such additions on the eutectic or peritectic composition have also been derived. However, they are somewhat more cumbersome and need not be exploited here.

We demonstrated that there is an upper limit on lowering the eutectic temperature, or raising the peritectic temperature, that no ternary element can exceed, a limit that depends only on the properties of the binary system. For the eutectic, it is equal to

$$\left(\frac{dT}{dX_3}\right)_{E,\lim} = -\frac{RT_E^2}{\Delta H^E} \quad (131a)$$

and for the peritectic it is

$$\left(\frac{dT}{dX_3}\right)_{P,\lim} = -\frac{RT_P^2}{\Delta H^P} \quad (131b)$$

where ΔH^E and ΔH^P are the enthalpies of the eutectic and peritectic reactions per mole of the binary 1–2. This result is entirely analogous to the more familiar limit placed on the depression of the melting point of a pure component 1 by any solute 2:

$$\left(\frac{dT}{dX_2}\right)_{\text{f,lim}} = -\frac{RT_{\text{f},1}^2}{\Delta H_{\text{f},1}} \tag{132}$$

where $\Delta H_{\text{f},1}$ is the heat of fusion of component 1 (per mole) and $T_{\text{f},1}$ is its temperature of fusion (see Section VIII.8.3.1).

Where data on the heats of reaction are unavailable, an alternative form of equation (131) can be applied. It is based on the slopes of the boundaries in the phase diagram at the eutectic or peritectic point and the value of the stability function.

The determination of the effect of a given solute on the eutectic or peritectic temperature hinges on the values of the activity coefficients of that solute at infinite dilution in the phases in mutual equilibrium. In the absence of ternary data, various approximations can be made to estimate the activity coefficients. Some of these approximations have been illustrated through a study of the effect of tin on the silver–copper eutectic temperature and of silicon on the eutectoid and peritectic temperatures of the iron–carbon system. The results are fairly sensitive to the estimates of the activity coefficients, but on a semiquantitative basis fairly reliable predictions can generally be obtained.

Problems

1. Demonstrate the passage from equation (53) to equation (55) in the case of a ternary regular solution.

2. The critical point of the isothermal section of a miscibility gap has a tangent dX_3/dX_2 measured in equation (49). Find a substitute expression for dX_3/dX_2 in terms of the second order derivatives of G_m^E with respect to X_2 and X_3.

3. At 1123°K the excess Gibbs free energy of an fcc solid solution of Cu, Ni, and Fe may be represented by [31]

$$G_m^E = 2164X_1^2X_2 + 3658X_1X_2^2 - 3127X_2^2X_3 - 1001X_2X_3^2$$
$$+ 6760X_1^2X_3 + 9589X_1X_3^2 + 1281X_1X_2X_3 \quad \text{cal}$$

where Cu = 1, Ni = 2, Fe = 3. From the ternary diagram representing the miscibility gap associated to that solution [31], it may be read that the critical point is at $X_2 = 0.35$, $X_3 = 0.17$, and that at this point the boundary of the miscibility gap has a slope dX_3/dX_2 equal to 1 ± 0.15.

a. Calculate at the critical point the partial derivatives $\partial^2G_m^E/\partial X_2^2$, $\partial^2G_m^E/\partial X_3^2$, and $\partial^2G_m^E/\partial X_2\,\partial X_3$.

b. Verify that at the critical point the value of ψ is indeed 0 (or close to it).

c. Verify through the equation derived in Problem 2 that a calculated value of the slope at the critical point is, within the accuracy of the reading, 1 ± 0.15.

4. **a.** In a ternary solution 1–2–3, assume that the activity of component 1 is known in the entire composition range and that the activity of 2 is also known in the binary solution 2–3. Let $s = X_3/(X_1 + X_3)$ and $t = X_3/(X_2 + X_3)$. Demonstrate that

$$\ln \gamma_2(s, t) - \ln \gamma_2(s = 1, t) = \int_1^s \left(\frac{t}{s}\right)^2 \left(\frac{\partial \ln \gamma_1}{\partial t}\right)_s ds \qquad \text{for constant } t$$

Explain how this equation allows the determination of ln γ_2 over the entire composition range and explain how you would proceed.

b. If the integration limit were to be taken as $s = 0$ instead of $s = 1$, show that the integral approaches a finite value for this limit. Calculate that value in terms of the interaction coefficients ϵ_i^j.

5. a. Demonstrate that the effect of small additions of a solute 3 on the eutectic composition X_2^E of a binary system 1–2 may be represented by

$$\left(\frac{dX_2}{dX_3}\right)_E = -\frac{X_2^E}{\psi^l}\left(\frac{\partial \ln \gamma_2^l}{\partial X_3^l}\right) - E_{12}\left[\frac{EB}{AB}\left(\frac{\theta_3^{\infty(l,E)}}{\theta_3^{\infty(\alpha,A)}} - \frac{X_1^\alpha}{X_1^E}\right)\frac{dX_2^{l(\beta)}}{dT} + \frac{AE}{AB}\left(\frac{\theta_3^{\infty(l,E)}}{\theta_3^{\infty(\beta,B)}} - \frac{X_1^\beta}{X_1^E}\right)\frac{dX_2^{l(\alpha)}}{dT}\right]$$

b. Show that in the case in which 1 and 3 have negligible solubilities in β this equation reduces to

$$\left(\frac{dX_2}{dX_3}\right)_E = -\frac{X_2^E}{\psi^l}\left(\frac{\partial \ln \gamma_3^l}{\partial X_3^l}\right) + \left(\frac{dT}{dX_3}\right)_E \frac{dX_2^{l(\beta)}}{dT}$$

c. Apply this to the case of the Fe–C(graphite) eutectoid. Select a solute (e.g., silicon) and verify that the calculations and the published information are in agreement within experimental error. State all the assumptions entering your estimates. (Note that if the phase boundaries are known, the preceding equation may provide an estimate of the coefficient $\epsilon_2^{(3)}$.)

References

1. G. Bernard, R. Hocine, and C. H. P. Lupis, *Trans. Met. Soc. AIME* **239,** 1600–1604 (1967).
2. H. Hancock, *Theory of Maxima and Minima.* Ginn, Boston, MA, 1917, p. 91.
3. H. Gaye and C. H. P. Lupis, *Trans. Met. Soc. AIME* **245,** 2543–2546 (1969).
4. I. Prigogine and R. Defay, *Chemical Thermodynamics* (D. H. Everett, transl.). Longman, New York, 1954, p. 250.
5. T. Muir, *A Treatise in the Theory of Determinants* (revised by W. H. Metzler). Banta, 1930, p. 370.
6. H. Gaye and C. H. P. Lupis, unpublished.
7. C. H. P. Lupis and H. Gaye, *Metallurgical Chemistry, Proceedings of a Symposium held at Brunel University and the National Physical Laboratory, July 1971* (O. Kubaschewski, ed.). Her Majesty's Stationery Office, London, 1972, pp. 469–482.
8. C. H. P. Lupis, *Acta Met.* **16,** 1365–1375 (1968).
9. S. Ban-ya and J. Chipman, *Trans. Met. Soc. AIME* **242,** 940–946 (1968).
10. J. F. Elliott, M. Gleiser, and V. Ramakrishna, *Thermochemistry for Steelmaking.* Addison-Wesley, Reading, MA, 1960, Vol. 2.
11. M. Hansen, *Constitution of Binary Alloys,* 2nd ed. McGraw-Hill, New York, 1958, pp. 704–708.
12. G. K. Sigworth and J. F. Elliott, *Met. Sci.* **8,** 298–310 (1974).
13. S. Ban-ya and J. Chipman, *Trans. Met. Soc. AIME* **245,** 133–145 (1969).
14. W. Seith and G. Helmhold, *Z. Metallk.* **42,** 138 (1951).
15. H. Gaye, Ph.D. thesis, Carnegie-Mellon University, Pittsburgh, PA, 1971, Appendix B.
16. Y. Nakagawa, *Acta Met.* **6,** 704–711 (1958).

17. P. J. Koros and J. Chipman, *J. Met.* **8,** 1102–1104 (1956).
18. W. R. Maddocks and G. E. Clausen, *Iron Steel Inst. London Spec. Rep.* No. 14, 97 (1936).
19. K. Iwase, M. Okamoto, and T. Amemiya, *Sci. Rep. Tohoku Univ.* **26,** 618 (1937).
20. R. Hultgren, P. D. Desai, D. T. Hawkins, M. Gleiser, and K. K. Kelley, *Selected Values of the Thermodynamic Properties of Binary Alloys.* Am. Soc. Metals, Metals Park, OH, 1973.
21. C. H. P. Lupis, *Met. Trans.* **9B,** 231–239 (1978).
22. R. Hultgren, P. D. Desai, D. T. Hawkins, M. Gleiser, and K. K. Kelley, *Selected Values of the Thermodynamic Properties of the Elements.* Am. Soc. Metals, Metals Park, OH, 1973.
23. K. K. Kelley, *High-Temperature Heat Content, Heat Capacity and Entropy Data for the Elements and Inorganic Compounds,* Bulletin 584, Bur. of Mines. U.S. GPO, Washington, DC, 1960.
24. Y. A. Chang, D. Goldberg, and J. Neumann, *J. Phys. Chem. Ref. Data* **6,** 621–673 (1977).
25. E-Hsin Foo and C. H. P. Lupis, *Acta Met.* **21,** 1409–1430 (1973).
26. R. L. Orr and J. Chipman, *Trans. Met. Soc. AIME* **239,** 630–633 (1967).
27. W. A. Fischer, K. Lorenz, H. Fabritius, A. Hoffmann, and G. Kalwa, *Arch. Eisenhüttenw.* **37,** 79–86 (1966).
28. W. Köster and T. Godecke, *Z. Metallk.* **59,** 602–605 (1968).
29. E. C. Bain and H. W. Paxton, *Alloying Elements in Steel,* 2nd ed. American Soc. Metals, Metals Park, OH, 1961, p. 112.
30. E-Hsin Foo and C. H. P. Lupis, Proc. Int. Conf. Sci. Technol. Iron and Steel, suppl. to *Trans. Iron Steel Inst. Japan* **11,** 404–408 (1971).
31. J. F. Counsell, E. B. Lees, and P. J. Spencer, *Metallurgical Chemistry Symposium 1971.* (O. Kubaschewski, ed.). Her Majesty's Stationery Office, London, 1972, p. 451.

XII Thermodynamic Functions Associated with Compounds

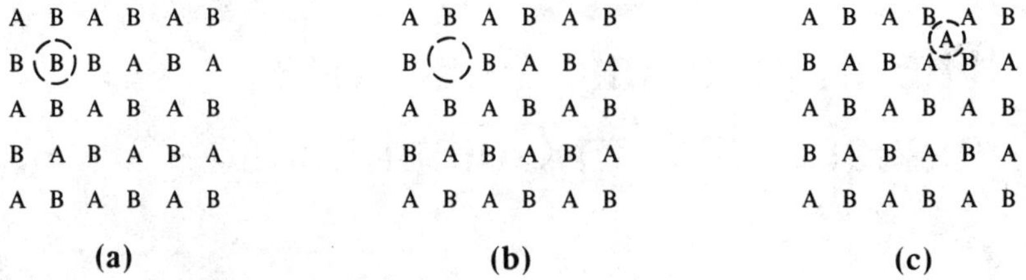

Figure 1. Examples of points defects in a stoichiometric compound. **(a)** substitutional; **(b)** vacancy; **(c)** interstitial.

1. Stoichiometric and Nonstoichiometric Compounds

Two elements A and B may mix to form an ordered structure where the atoms A and B occupy well-defined lattice sites. A completely ordered structure exists only if the ratio of the numbers of atoms A and B is equal to the ratio of relatively small integers ν_A and ν_B. The mixture is then designated by the formula $A_{\nu_A}B_{\nu_B}$. For example, if $\nu_A = 2\nu_B$ it could be designated A_2B.

Deviations from such formulas correspond to compositions which are not compatible with complete order and occur through local misarrangements or *defects*.[1] The study of defects in solids is an important branch of materials science but lies outside the scope of this text. Nevertheless, for purposes of illustration, we shall note only that there are three main types of defects—substitutional, vacancy, and interstitial (Fig. 1)—although many other types of defects occur. Swalin [1] gives a good introduction to the subject.

In a plot of the molar Gibbs free energy of the ordered phase vs composition (Fig. 2), the narrowness of that curve, or the steepness of its branches, indicates the difficulty with which defects are introduced in that structure. At a given temperature and pressure, this phase has a range of stability which is defined relative to the stability of other phases. If this range has negligible width, then

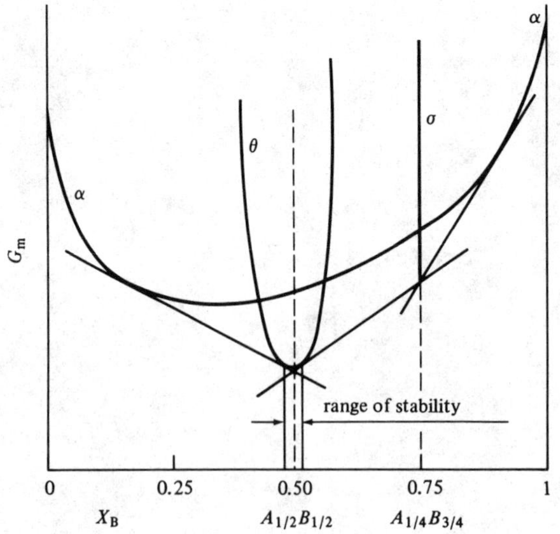

Figure 2. Free energy curves of compounds. The compound $A_{1/2} \cdot B_{1/2}$ (or AB) is nonstoichiometric whereas the compound $A_{1/4}B_{3/4}$ (or AB_3) is stoichiometric.

[1] Defects may also occur even when the composition of the mixture corresponds to that of complete order.

its structure, for all practical purposes, exists at a single composition and is referred to as a *stoichiometric compound*. If the range of stability is not negligible, the phase corresponds to a *nonstoichiometric compound*.

Let us consider the iron–oxygen phase diagram (Fig. 3). At 1000°C, hematite (Fe_2O_3) is stoichiometric, magnetite (Fe_3O_4) is nonstoichiometric, and wüstite (FeO) is also nonstoichiometric. In the case of wüstite, we note that the com-

Figure 3. Iron–oxygen phase diagram. (*After Elliott et al.* [2].)

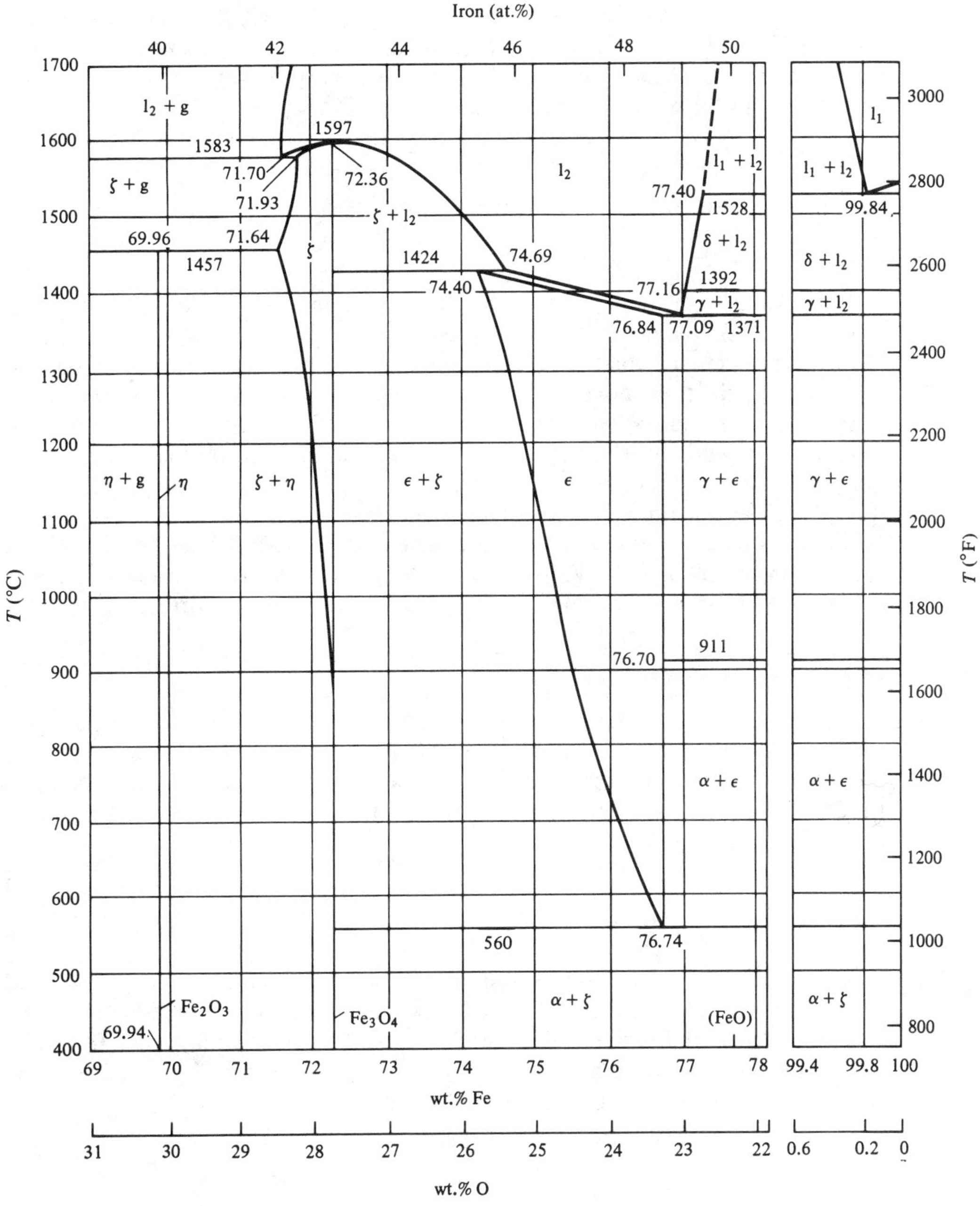

position of the completely ordered structure ($X_{Fe} = X_O = 0.5$ for FeO) does not fall within the range of stability of the wüstite phase. We also observe that magnetite becomes stoichiometric at lower temperatures. At what temperature its range of stability becomes negligible is an arbitrary matter.

In the thermodynamic analysis of a system, it is often convenient to describe the system in terms of certain of its compounds without reference to their constituents. For example, the Fe–O phase diagram (or a major part of it) may be adequately and conveniently described as the FeO–Fe_2O_3 system, i.e., as mixtures of FeO and Fe_2O_3 instead of Fe and O. If these compounds are to be considered as components having their own individuality, they must have their proper thermodynamic functions, such as chemical potential and activity. The definition and application of these functions necessitate some attention and are the subject of the succeeding sections.

2. Chemical Potential of a Compound

2.1. Binary Systems

Taking again the example of the iron–oxygen system, let us consider the compound FeO and examine the possible definition of its chemical potential, μ_{FeO}. A possible confusion that should be immediately dispelled originates from the necessary distinction between FeO and the wüstite phase. Whereas FeO denotes a specific composition, we must specify a *range* of composition for the wüstite phase. Thus, while we shall establish that it is legitimate to consider the chemical potentials of components such as FeO or $Fe_{0.95}O$, the chemical potential of a phase such as wüstite is a meaningless concept.

So far, we have studied the thermodynamics of a binary system A–B in terms of its elements A and B. We shall now attempt to describe the same system in terms of other species (or compounds), which can be real or virtual.

2.1.1. Fundamental Equations

Let us select the entities $U \equiv A_{\nu_A^U}A_{\nu_B^U}$ and $V \equiv A_{\nu_A^V}B_{\nu_B^V}$. For convenience, we shall write these formulas in such a way that the sums of the indices ν_A and ν_B are equal to 1; ν_B^U and ν_B^V represent, then, the compositions of U and V in mole fraction of B. The numbers of moles of A and B and of U and V are related by simple mass balances:

$$n_A = \nu_A^U n_U + \nu_A^V n_V \tag{1a}$$

$$n_B = \nu_B^U n_U + \nu_B^V n_V \tag{1b}$$

The Gibbs free energy of the system may be analyzed either in terms of n_A and n_B or in terms of n_U and n_V. Formally, the thermodynamic functions of U and V have similar definitions to those of A and B and one may pass from one description of the system to the other by a simple change of variables. The definitions of the chemical potentials of U and V are

$$\mu_U = \left(\frac{\partial G}{\partial n_U}\right)_{P,T,n_V} \tag{2a}$$

$$\mu_V = \left(\frac{\partial G}{\partial n_V}\right)_{P,T,n_U} \tag{2b}$$

Recalling that

$$(dG)_{P,T} = \mu_A\, dn_A + \mu_B\, dn_B \tag{3}$$

we deduce the relation

$$\left(\frac{\partial G}{\partial n_U}\right)_{P,T,n_V} = \mu_A \left(\frac{\partial n_A}{\partial n_U}\right)_{P,T,n_V} + \mu_B \left(\frac{\partial n_B}{\partial n_U}\right)_{P,T,n_V} \tag{4}$$

which through equation (1) becomes

$$\mu_U = \nu_A^U \mu_A + \nu_B^U \mu_B \tag{5a}$$

Similarly, we would have

$$\mu_V = \nu_A^V \mu_A + \nu_B^V \mu_B \tag{5b}$$

In a plot of the Gibbs free energy vs composition (Fig. 4), a tangent to the free energy curve of a phase may be determined by its two intercepts a and b on the A and B axes at $X_A = 1$ and $X_B = 1$. It may also be described by the intercepts u and v on the $A_{\nu_A^U}B_{\nu_B^U}$ and $A_{\nu_A^V}B_{\nu_B^V}$ axes corresponding to the compositions $X_B = \nu_B^U$ and $X_B = \nu_B^V$. We have demonstrated in Section II.3 that the intercepts a and b represent the partial molar properties μ_A and μ_B. Equations (5) show that u and v represent μ_U and μ_V.

In the particular case where we would select the species V as corresponding to B, the binary solution would be described as a solution of atoms B and (real or virtual) "molecules" or clusters $U \equiv A_{\nu_A^U}B_{\nu_B^U}$. Then, n_V would be equal not to n_B but to the number of free atoms B which are not part of the species U. Indeed,

Figure 4. The graphical method of intercepts determines the chemical potentials of A and B as well as the chemical potentials of compounds such as $U \equiv A_{\nu_A^U}B_{\nu_B^U}$ and $V \equiv A_{\nu_A^V}B_{\nu_B^V}$. (*Note*: $\nu_A + \nu_B = 1$ for U and V.)

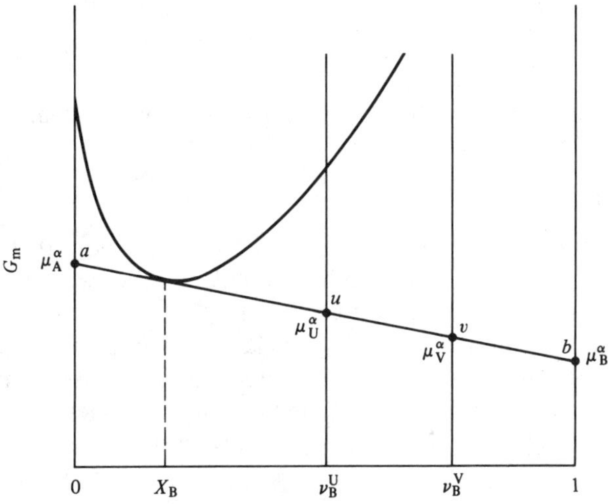

the mass balances in equation (1) yield

$$n_A = \nu_A^U n_U \tag{6a}$$

$$n_B = \nu_B^U n_U + n_V \tag{6b}$$

However, equation (5b) or the graphical method of intercepts shows that μ_B would equal to μ_V.

At constant pressure and temperature, the Gibbs free energy is a homogeneous function of n_A and n_B (see Section II.1). Because the change of coordinates defined by equations (1) is linear and without a constant term, it is clear that G is also a homogeneous function of n_U and n_V. Consequently,

$$G = n_U\mu_U + n_V\mu_V \tag{7}$$

and

$$S\,dT - V\,dP + n_U\,d\mu_U + n_V\,d\mu_V = 0 \tag{8}$$

Equation (5a) may be interpreted as representing the equilibrium of the chemical reaction

$$\nu_A^U A + \nu_B^U B = A_{\nu_A^U}B_{\nu_B^U} \quad \text{(or U)} \tag{9}$$

between the species A, B, and U of the same phase. However, this would be misleading since equations (5) are merely the consequences of a formalism and not the consequences of any physical transformation.

2.1.2. Equilibrium Between Phases

At a given P and T let us now assume equilibrium between two phases α and σ of the binary system A–B. The conditions for equilibrium (see Chapter II) may be written

$$\mu_A^\alpha = \mu_A^\sigma \tag{10a}$$

$$\mu_B^\alpha = \mu_B^\sigma \tag{10b}$$

or

$$\mu_U^\alpha = \mu_U^\sigma \tag{11a}$$

$$\mu_V^\alpha = \mu_V^\sigma \tag{11b}$$

If the phase σ is characterized by the compound U, it may be advantageous to replace equation (11a) by

$$\nu_A^U\mu_A^\alpha + \nu_B^U\mu_B^\alpha = \mu_U^\sigma \tag{12}$$

which is obtained by combining equations (11a) and (5a). We note that equation (12) is not sufficient to determine the equilibrium between the two phases α and σ. It must be used with another condition such as (10a), (10b), or (11b).

If the compound σ is stoichiometric, condition (12) becomes sufficient to assure equilibrium between the phases α and σ. This is understood by observing that in the attainment of equilibrium the changes δn_A^α and δn_B^α are no longer both arbitrary (see Section II.4) but are related by the stoichiometric composition of the σ phase:

$$\delta n_A^\alpha/\nu_A^U = \delta n_B^\alpha/\nu_B^U = -\delta n_U^\sigma \tag{13}$$

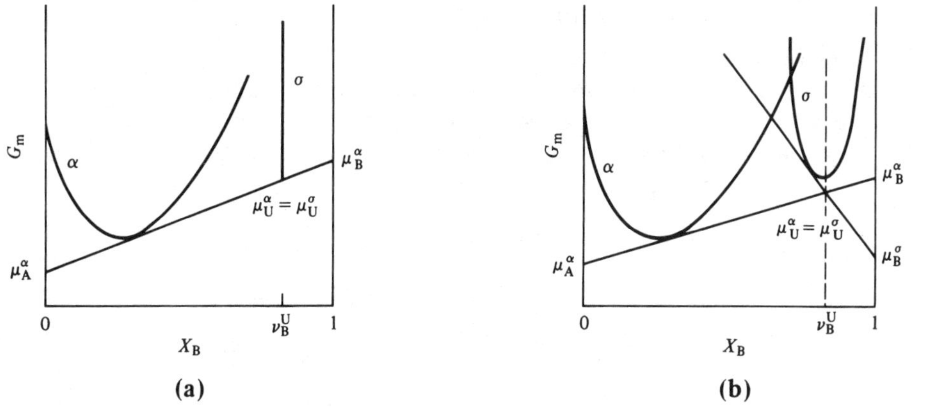

Figure 5. **(a)** A case in which only one condition is sufficient to express the equilibrium of the stoichiometric phase σ with the phase α:

$$\mu_U^\sigma = \mu_U^\alpha \qquad \text{or} \qquad \mu_U^\sigma = \nu_A^U \mu_A^\alpha + \nu_B^U \mu_B^\alpha$$

(b) A case in which the phase σ is not stoichiometric. The condition of case **(a)** is no longer sufficient.

One degree of freedom having been lost, only one condition, instead of two, is necessary. The same result may be reached through the graphical construction of the common tangent (see Fig. 5). We note that in this case μ_U^σ is equal to the molar Gibbs free energy of the σ phase.

2.1.3. Remarks

At the beginning of Section 2.1.1 we chose to write the formula for compounds A_xB_y with the restriction $x + y = 1$. This restriction is convenient for the graphical illustration of some of the equations derived above. However, the case $x + y = 1$ is not fundamentally different, since

$$A_{\lambda x}B_{\lambda y} = \lambda A_xB_y \tag{14}$$

and

$$\mu_{A_{\lambda x}B_{\lambda y}} = \lambda \mu_{A_xB_y} \tag{15}$$

For instance, $\mu_{AB_3} = 4\mu_{A_{1/4}B_{3/4}}$.

In Fig. 2 the equilibria between the σ and θ phases and between the σ and α phases correspond to two distinct common tangents. Each is associated with different values of the chemical potentials of A and B. Consequently, we see that across the composition range of the binary system A–B, the chemical potential of A or B may vary quite abruptly at the composition of a stoichiometric compound (e.g., σ in Fig. 2) or in the range of stability of a nonstoichiometric compound (e.g., θ).

For example, we may consider the case of magnetite, Fe_3O_4, which is stoichiometric at 650°C. When it coexists with the wüstite phase, its oxygen potential p_{O_2} is of the order of 10^{-22}; when it coexists with hematite, Fe_2O_3, its oxygen potential is equal to 10^{-12}, a substantially different value [3].

2.2. *Multicomponent Systems*

The results we derived for binary systems is readily extended to systems of any number of components in a straightforward manner. Let A, B, C, . . . be the elements present in a system and U, V, W, . . . the real or virtual species (or compounds) selected to describe the system. These species are mixtures of atoms A, B, C, . . . corresponding to the equations

$$\nu_A^U A + \nu_B^U B + \nu_C^U C + \cdots = A_{\nu_A^U} B_{\nu_B^U} C_{\nu_C^U} \equiv U \tag{16a}$$

$$\nu_A^V A + \nu_B^V B + \nu_C^V C + \cdots = A_{\nu_A^V} B_{\nu_B^V} C_{\nu_C^V} \equiv V \tag{16b}$$

$$\nu_A^W A + \nu_B^W B + \nu_C^W C + \cdots = A_{\nu_A^W} B_{\nu_B^W} C_{\nu_C^W} \equiv W \tag{16c}$$

$$\vdots \qquad\qquad \vdots$$

We may rewrite these equations in matrix notation:

$$\begin{pmatrix} U \\ V \\ W \\ \vdots \end{pmatrix} = \begin{pmatrix} \nu_A^U & \nu_B^U & \nu_C^U & \cdots \\ \nu_A^V & \nu_B^V & \nu_C^V & \cdots \\ \nu_A^W & \nu_B^W & \nu_C^W & \cdots \\ \vdots & \vdots & \vdots & \end{pmatrix} \begin{pmatrix} A \\ B \\ C \\ \vdots \end{pmatrix} \tag{17}$$

For convenience in the graphical interpretation of the results, the sum of the subscripts of any compound formula will be taken as equal to 1; e.g., $\nu_A^U + \nu_B^U + \nu_C^U + \cdots = 1$. The number of selected species must be identical to the number of elements, and a species may be a pure component ($U \equiv A$ if $\nu_A^U = 1$, $\nu_B^U = 0$, $\nu_C^U = 0, \ldots$).

The chemical potentials of the compounds U, V, W may be defined by the matrix equation

$$\begin{pmatrix} \mu_U \\ \mu_V \\ \mu_W \\ \vdots \end{pmatrix} = \begin{pmatrix} \nu_A^U & \nu_B^U & \nu_C^U & \cdots \\ \nu_A^V & \nu_B^V & \nu_C^V & \cdots \\ \nu_A^W & \nu_B^W & \nu_C^W & \cdots \\ \vdots & \vdots & \vdots & \end{pmatrix} \begin{pmatrix} \mu_A \\ \mu_B \\ \mu_C \\ \vdots \end{pmatrix} \tag{18a}$$

or

$$\boldsymbol{\mu}^\dagger = (\nu)\boldsymbol{\mu} \tag{18b}$$

Figure 6 gives a geometrical interpretation of μ_U in a ternary system. Consider the plane intercepting the axes A, B, C at μ_A, μ_B, μ_C. The chemical potential μ_U is seen as the intercept of that plane with the vertical axis U at the composition $X_A = \nu_A^U$, $X_B = \nu_B^U$, $X_C = \nu_C^U$.

The coordinates associated with μ_A, μ_B, μ_C, . . . are the numbers of moles n_A, n_B, n_C, . . . Similarly, we associate with μ_U, μ_V, μ_W, . . . the numbers of moles n_U, n_V, n_W, . . . defined by the mass balances

$$\begin{pmatrix} n_A \\ n_B \\ n_C \\ \vdots \end{pmatrix} = \begin{pmatrix} \nu_A^U & \nu_A^V & \nu_A^W & \cdots \\ \nu_B^U & \nu_B^V & \nu_B^W & \cdots \\ \nu_C^U & \nu_C^V & \nu_C^W & \cdots \\ \vdots & \vdots & \vdots & \end{pmatrix} \begin{pmatrix} U \\ V \\ W \\ \vdots \end{pmatrix} \tag{19a}$$

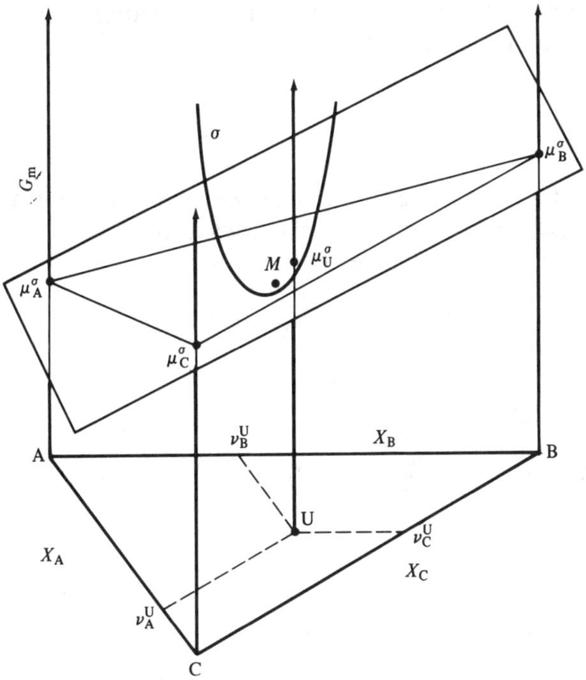

Figure 6. Graphical method of intercepts for a ternary system. The plane tangent to the Gibbs free energy surface of the σ phase at the point M intercepts the axis corresponding to the compound $U \equiv A_{\nu_A^U}B_{\nu_B^U}C_{\nu_C^U}$ at μ_U^σ.

or

$$\boldsymbol{n} = (\nu)^t \boldsymbol{n}\dagger \tag{19b}$$

where $(\nu)^t$ is the transposed matrix of (ν).

The Gibbs free energy of the system may be described in terms of the components A, B, C, . . . or U, V, W, ... Since

$$G = \boldsymbol{n} \cdot \boldsymbol{\mu} \tag{20}$$

equation (19b) yields

$$G = (\nu)^t \boldsymbol{n}\dagger \cdot \boldsymbol{\mu} \tag{21a}$$

Through a classical result of linear algebra, we may rewrite this equation

$$G = \boldsymbol{n}\dagger \cdot (\nu)\, \boldsymbol{\mu} \tag{21b}$$

Equation (18b) then yields

$$G = \boldsymbol{n}\dagger \cdot \boldsymbol{\mu}\dagger \tag{22}$$

We could similarly establish that at constant temperature and pressure

$$dG = \boldsymbol{\mu} \cdot d\boldsymbol{n} = \boldsymbol{\mu}\dagger \cdot d\boldsymbol{n}\dagger \tag{23}$$

and

$$\boldsymbol{n} \cdot d\boldsymbol{\mu} = \boldsymbol{n}\dagger \cdot d\boldsymbol{\mu}\dagger = 0 \tag{24}$$

Equation (23) could also have been used for an alternative definition of the chemical potential of a compound such as U:

$$\mu_U = \left(\frac{\partial G}{\partial n_U}\right)_{P,T,n_V} \tag{25}$$

Equation (24) represents the Gibbs–Duhem equation with either set of components.

If at a given temperature and pressure several phases $\alpha, \beta, \ldots, \sigma, \ldots$ coexist, the conditions for equilibrium (see Chapter II) are

$$\mu_A^\alpha = \mu_A^\beta = \cdots = \mu_A^\sigma = \cdots \tag{26a}$$

$$\mu_B^\alpha = \mu_B^\beta = \cdots = \mu_B^\sigma = \cdots \tag{26b}$$

$$\mu_C^\alpha = \mu_C^\beta = \cdots = \mu_C^\sigma = \cdots \tag{26c}$$

They may also be written in terms of the species U, V, W. . . :

$$\mu_U^\alpha = \mu_U^\beta = \cdots = \mu_U^\sigma = \cdots \tag{27a}$$

$$\mu_V^\alpha = \mu_V^\beta = \cdots = \mu_V^\sigma = \cdots \tag{27b}$$

$$\mu_W^\alpha = \mu_W^\beta = \cdots = \mu_W^\sigma = \cdots \tag{27c}$$

A set of equilibrium conditions may also be written by interchanging some of equations (26) for some of equations (27), taking into account equations (18). For example, if U is a compound of small or negligible range of stoichiometry, we may associate it with the phase σ and write

$$\nu_A^U \mu_A^\alpha + \nu_B^U \mu_B^\alpha + \nu_C^U \mu_C^\beta = \mu_U^\sigma \tag{28}$$

as a combination of

$$\mu_U^\sigma = \nu_A^U \mu_A^\sigma + \nu_B^U \mu_B^\sigma + \nu_C^U \mu_C^\sigma \tag{29}$$

and

$$\mu_A^\alpha = \mu_A^\sigma, \qquad \mu_B^\alpha = \mu_B^\sigma, \qquad \mu_C^\beta = \mu_C^\sigma \tag{30}$$

Equation (29) is a case of equations (18), and equations (30) are part of the set of equations (26). We recognize that equation (28) expresses the equilibrium of the chemical reaction

$$\nu_A^U A + \nu_B^U B + \nu_C^U C = A_{\nu_A^U} B_{\nu_B^U} C_{\nu_C^U} \quad \text{(or U)} \tag{31}$$

where A and B are in the α phase and C is in the β phase.

One may analyze the equilibrium of a system in terms of the formalism of chemical reactions, or in terms of free energy curves (or surfaces) and equalities of chemical potentials (e.g., through the common tangent construction). This is a matter of convenience. In the case of several components and phases, the approach through chemical reactions is often simpler.

The equations in this section were mainly developed for solid and liquid phases. They can also be applied to gaseous phases but are then of much less interest.

3. Activity of a Compound

To define the activity of a compound A_xB_y we may use the general definition of the activity function presented in Chapter IV:

$$\mu_{A_xB_y} = \mu^*_{A_xB_y} + RT \ln a_{A_xB_y} \tag{32}$$

where $\mu^*_{A_xB_y}$ is a function of the temperature and pressure representing the chemical potential of A_xB_y in an arbitrary reference state. To define more completely the activity function, the choice of this reference state must be specified.

3.1. Reference States

In the case of a stoichiometric compound, composition is no longer a variable. There is only one natural choice for the reference state, that corresponding to $\mu^*_{A_xB_y}$ equal to the molar Gibbs free energy of the compound A_xB_y (see Fig. 7a). Whenever the compound A_xB_y is stable, we then see that its chemical potential is identical to $\mu^*_{A_xB_y}$ and, consequently, that its activity is equal to 1.

When A_xB_y is nonstoichiometric, there are several options. These are illustrated in Fig. 7b by the points N_1, N_2, N_3 (for $x + y = 1$). Choosing N_1 yields an activity of 1 to $a_{A_xB_y}$ at the stoichiometric composition, that is when $X_B = y$. However, when the phase σ associated to A_xB_y is in equilibrium with the phase to its left (α) or to its right (β), $a_{A_xB_y}$ is different from 1. Choosing N_2 for reference state, the activity of A_xB_y is no longer equal to 1 at the stoichiometric composition; it is equal to 1 when the phases σ and α are in equilibrium, that is for an overall concentration X_B between X'_B and y. The case of N_3 for reference state is similar.

The choice of the reference or standard state is generally dictated by convenience. If, in an investigation, equilibrium of the phase σ with the phase α is likely to occur more often than with the phase β, then the standard state corresponding

Figure 7. Reference states for the chemical potential of A_xB_y in the σ phase. **(a)** $\mu^{*(\sigma)}_{A_xB_y}$ is equal to the molar Gibbs free energy of the stoichiometric compound. **(b)** The points N_1, N_2, and N_3 illustrate three different selections of the reference state.

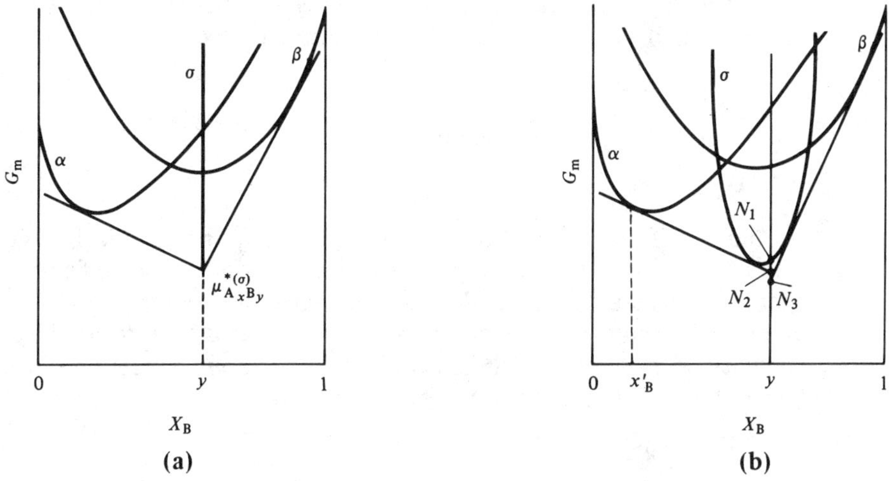

to N_2 should be selected. However, whichever selection is made should be clearly stated. Failure to do so may result in serious inconsistencies and errors. For example, let us consider the following reactions:

$$\mathrm{Fe + CO_2 = FeO + CO} \qquad (\Delta G^\circ_{33}) \tag{33}$$

$$\mathrm{2FeO + CO_2 = Fe_2O_3 + CO} \qquad (\Delta G^\circ_{34}) \tag{34}$$

$$\mathrm{2Fe + 3CO_2 = Fe_2O_3 + 3CO} \qquad (\Delta G^\circ_{35}) \tag{35}$$

The third reaction may be calculated as a combination of the first two, i.e.,

$$\Delta G^\circ_{35} = 2\Delta G^\circ_{33} + \Delta G^\circ_{34} \tag{36}$$

However, this is correct only if FeO has the same standard state in the first two reactions. Reaction (33), involving the equilibrium of the iron phase with the wüstite phase, is likely to be associated with a standard state of FeO represented above by the point N_2. Similarly, reaction (34) is likely to be associated with the standard state for FeO represented by N_3. With these selections of standard states, equation (36) is incorrect since the terms in μ°_{FeO} do not cancel.

To avoid such errors, it is often convenient to adopt the notation introduced by Worrell and Chipman [4] for the standard states of nonstoichiometric compounds. $\overline{\mathrm{A}}_x\mathrm{B}_y$ and $\mathrm{A}_x\overline{\mathrm{B}}_y$ are associated, respectively, with standard states at A-rich and at B-rich boundaries of the homogeneous field (i.e., at N_2 and at N_3). In addition, although the standard state associated with the stoichiometric composition of $\mathrm{A}_x\mathrm{B}_y$ (i.e., at N_1) is less commonly chosen, we shall represent it by the symbol $\overline{\mathrm{A}_x\mathrm{B}_y}$.

3.2. Composition Dependence

Figure 8b shows the variation in the activity of $\mathrm{A}_x\mathrm{B}_y$, corresponding to the case of Fig. 7b, across the entire composition range. The shape of the curve—passing through a maximum and equal to zero at *both* ends of the diagram—is very unlike the shapes of the curves for elements A and B in Fig. 8a. However, the behavior of $a_{\mathrm{A}_x\mathrm{B}_y}$ may be easily understood by applying the method of intercepts to the plot of Fig. 7b. It may also be understood through the equation

$$a_{\mathrm{A}_x\mathrm{B}_y} = K a_{\mathrm{A}}^x a_{\mathrm{B}}^y \tag{37}$$

which is associated with the relation

$$\mu_{\mathrm{A}_x\mathrm{B}_y} = x\mu_{\mathrm{A}} + y\mu_{\mathrm{B}} \tag{38}$$

In Chapter III we deduced that in a stable system the chemical potential and activity of a component i are monotone increasing with the mole fraction of that component. This may seem at odds with the maximum exhibited by $a_{\mathrm{A}_x\mathrm{B}_y}$ in Fig. 8b. In order to resolve this apparent contradiction, let us first recall the condition for stability with respect to infinitesimal composition fluctuations in a binary system 1–2 at constant temperature and pressure:

$$G_{22} = (\partial\mu_2/\partial n_2)_{n_1} \tag{39a}$$

In the case where the binary system A–B is described by the species U ≡ A and

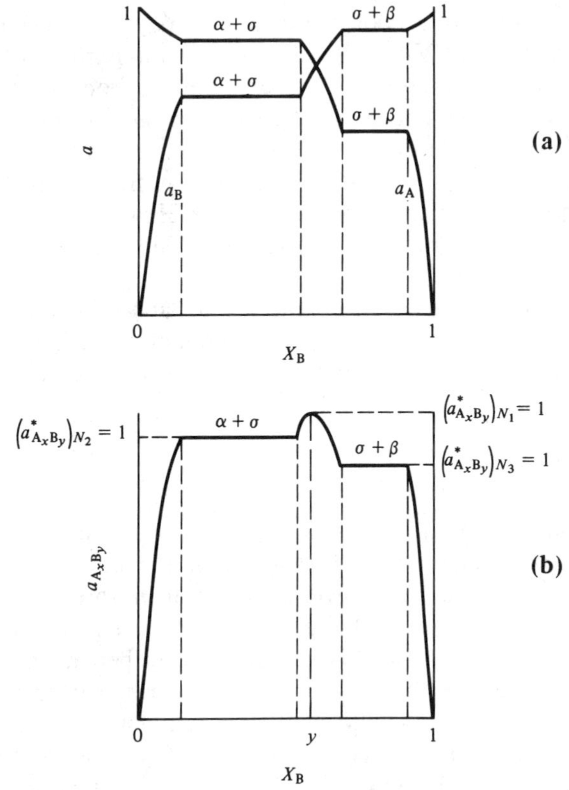

Figure 8. Activities of A, B, and A_xB_y as functions of composition. The curves correspond to the case of Fig. 7b. The absolute values of $a_{A_xB_y}$ depend on the reference state selected (N_1, N_2, or N_3, identified in Fig. 7b).

$V \equiv A_xB_y$, this condition becomes

$$(\partial \mu_V / \partial n_V)_{n_U} \geq 0 \tag{39b}$$

It may be rewritten

$$\left(\frac{\partial \mu_V}{\partial n_V}\right)_{n_U} = \frac{d\mu_V}{dX_V}\left(\frac{\partial X_V}{\partial n_V}\right)_{n_U} = \frac{d\mu_V}{dX_V}\frac{X_U}{n} \geq 0 \tag{40}$$

where $n = n_U + n_V = n_A + n_B$ for $x + y = 1$. In addition,

$$n_A = n_U + xn_V \tag{41a}$$

$$n_B = yn_V \tag{41b}$$

for which we deduce

$$X_V = X_B/y \tag{42a}$$

$$X_U = 1 - (X_B/y) \tag{42b}$$

Condition (40) then becomes

$$(y - X_B)\frac{d \ln a_{A_xB_y}}{dX_B} \geq 0 \tag{43}$$

We verify that $a_{A_xB_y}$ does increase for $X_B < y$ and decrease for $X_B > y$.

A difficulty with the composition coordinates X_U or X_V should be noted. In the case selected with $U \equiv A$ and $V \equiv A_xB_y$, X_V becomes *negative* in the composition range $X_B > y$ [equation (42b)]. This is explained by observing that at $X_B = y$, all the atoms A are in the species A_xB_y and there are no longer "free" atoms A (or species U). For $X_B > y$, there may be "free" atoms B, but if we wish to analyze the binary system in terms of $U \equiv A$ and $V \equiv A_xB_y$, n_U must be considered as an algebraic quantity which becomes negative. However, in this range, it generally becomes more appropriate to describe the system in terms of other species.

These considerations on the activity, reference state, and composition coordinate of a compound can be easily extended to systems of order higher than 2. We stress, however, that even more caution is then necessary to specify without ambiguity the reference state of a compound.

4. Applications

Let us consider the ternary system Fe–Si–O at 1600°C and 1 atm. If the concentration of silicon and oxygen are small, the stable phase is metallic: liquid iron with silicon and oxygen as solutes. If, however, the concentrations of silicon, oxygen, and iron are comparable, the stable phase is a liquid silicate slag which is ionic in character. The metallic phase is generally described in terms of its elements, Fe, Si, and O, whereas the ionic phase is described in terms of the compounds SiO_2, FeO, and Fe_2O_3. The choice of these compounds is based upon the fact that in slags, silicon is present as the cation Si^{+4}, oxygen as the anion O^{2-} and iron as the ferrous and ferric cations Fe^{+2}, Fe^{+3}. Since the solution must be electrically neutral, the selection of the species SiO_2, FeO, and Fe_2O_3 ensures this neutrality and simplifies the description of the slag phase.

As another example, let us consider the Fe–Si–Mn–O system along the orthosilicate joint SiO_4Fe_2–SiO_4Mn_2, that is, the pseudobinary system defined by the restrictions

$$n_{Si} = \tfrac{1}{4}n_O = \tfrac{1}{2}(n_{Fe} + n_{Mn}) \tag{44}$$

A natural choice of components to describe this slag system consists of SiO_4Fe_2 and SiO_4Mn_2 (also written $SiO_2 \cdot 2FeO$ and $SiO_2 \cdot 2MnO$). However, it turns out that a better choice of species is $Si_{1/2}O_2Fe$ and $Si_{1/2}O_2Mn$. It is better because at 1150°C the activities of these components are then, within experimental error, linear functions of $n_{Si_{1/2}O_2Fe}/(n_{Si_{1/2}O_2Fe} + n_{Si_{1/2}O_2Mn})$ or $n_{Fe}/(n_{Fe} + n_{Mn})$ [5]. This behavior may be readily interpreted as due to a random mixture of Fe^{+2} and Mn^{+2} cations on a sublattice of octahedral sites in the silicate network. A straight line for the activity of SiO_4Fe_2 would be interpreted by a random interchange of *pairs* of cations Fe^{+2} with *pairs* of cations Mn^{+2}. This is not the case, and the data for the activity of SiO_4Fe_2 are fitted by a parabola, in agreement with equation (15), which yields

$$\mu_{SiO_4Fe_2} = 2\mu_{Si_{1/2}O_2Fe} \tag{45}$$

$$a_{SiO_4Fe_2} = a^2_{Si_{1/2}O_2Fe} \tag{46}$$

It may be recalled here that we already encountered a somewhat similar situation in our discussion of nitrogen solubility and adherence to Henry's law (Section VI.4.3).

In general, then, a certain choice of species to describe a solution may result in a considerable simplification of the formalism and in an easier structural interpretation. The transformation of variables outlined in Section 2.2 is linear and well adapted to computer manipulations. It could be useful in the study of some complex systems and may be applied as follows.

Assume that for a certain solution the experimental data are expressed in terms of the species U′, V′, W′, . . . with which we associate the vectors $\boldsymbol{n}'$ and $\boldsymbol{\mu}'$ to describe the number of moles and chemical potential of each species (see Section 2.2). Theoretical considerations suggest that the choice of the species U″, V″, W″, . . . may be more relevant. In order to relate the vectors $\boldsymbol{n}''$, $\boldsymbol{\mu}''$ associated with U″, V″, W″, . . . to $\boldsymbol{n}'$ and $\boldsymbol{\mu}'$, we note that equations (18) and (19) yield

$$\boldsymbol{\mu}' = (\nu')\boldsymbol{\mu}, \qquad \boldsymbol{\mu}'' = (\nu'')\boldsymbol{\mu} \tag{47}$$

and

$$\boldsymbol{n} = (\nu'^{t})\boldsymbol{n}' = (\nu''^{t})\boldsymbol{n}'' \tag{48}$$

where (ν') and (ν'') are the matrices describing the formulas of the species U′, V′, W′, . . . and U″, V″, W″, . . . in terms of the elements A, B, C, . . . (see Section 2.2). From equations (47) and (48), we readily obtain

$$\boldsymbol{\mu}'' = (\nu'')(\nu')^{-1}\boldsymbol{\mu}' \tag{49}$$

$$\boldsymbol{n}'' = (\nu''^{t})^{-1}(\nu'^{t})\boldsymbol{n}' \tag{50}$$

The relevance of the new choice of species may be analyzed in terms of the activities or activity coefficients. For example, a "better" choice may be characterized by the introduction of a smaller number of parameters to represent the experimental data at the same level of statistical significance (e.g., *F* level).

5. Summary

A thermodynamic system may be analyzed in terms of its elements A, B, C . . . or in terms of certain compounds $A_xB_yC_z$. . . . The latter description is natural for a gaseous phase, when these compounds can be identified as gaseous molecular species, or for condensed systems when these compounds are stoichiometric. This description is still convenient in the case of nonstoichiometric compounds, but the distinction between a phase and a nonstoichiometric compound may lead to some confusion in the thermodynamic formalism associated with the system. It is also convenient for certain solutions such as silicate glasses, and it may result not only in a simplification of the mathematical expressions describing the system but also in easier structural interpretation of the data. The thermodynamic functions associated with these compounds can be defined in a straightforward manner and without ambiguity, but it is important in the case of activities to clearly specify the reference or standard states adopted. The graphical method of intercepts offers a very convenient way of illustrating the formalism associated with these compounds and the relation of this formalism to that associated with the elements of the system.

Problems

1. The maximum amount the temperature of fusion of a species A can be lowered by small additions of any species B has been expressed in equation (VIII.73): $(dT/dX_B)_{X_B \to 0} = -R(T_{f,A})^2/\Delta H^\circ_{f,A}$. Is this result still valid, or is it in need of modification, if the same physical system is now represented in terms of the species A_2 and B?

2. In a given solution A–B, both solutes A and B are assumed to obey Henry's (zeroth order) law. The solution is now described by the species $U \equiv A$ and $V \equiv A_xB_y$. Do the solutes U and V also obey Henry's law? Discuss this in terms of the values of x and y.

References

1. R. A. Swalin, *Thermodynamics of Solids*, 2nd ed., Wiley, New York, 1972, Chs. 14 and 15.
2. J. F. Elliott, M. Gleiser, and V. Ramakrishna, *Thermochemistry for Steelmaking*, Addison-Wesley, Reading, MA, 1963, Vol. 2, p. 406.
3. A. Muan and E. F. Osborn, *Phase Equilibria Among Oxides in Steelmaking*, Addison-Wesley, Reading, MA, 1965, p. 28.
4. W. L. Worrell and J. Chipman, *J. Phys. Chem.* **68,** 860 (1964).
5. K. Schwerdtfeger, A. Muan, and L. S. Darken, *Trans. Met. Soc. AIME* **236,** 201–211 (1966).

XIII Surfaces and Surface Tensions

In deriving the conditions for equilibrium which apply to a heterogeneous system (Chapter II), we neglected the effects of the interfaces separating the various phases present. More specifically, in considering the heterogeneous system, we assumed that the variations of its energy which are dependent on the interfaces are very small and negligible in comparison with the variations which depend on the bulk properties of the phases. We shall now examine the influence and properties of these interfaces.

If we consider, for example, a metal–gas interface, the metal atoms at that interface are in a higher energy state than in the bulk of the metal. Consequently, the metal tries to assume a shape which minimizes its number of surface atoms; e.g., if no other force is present, it will be spherical. The force which minimizes this surface area is called *surface tension*. It will be defined more precisely below. In the case of an alloy, the composition of the interface is generally quite different from that of the bulk. If a given solute is more abundant at the interface, it may be said that this solute is preferentially *adsorbed* or *surface active*. Phenomena dealing primarily with surface tensions will be covered in this chapter, while phenomena dealing primarily with adsorption will be covered in Chapter XIV. Clearly, however, both types of phenomena are related, and the distinction introduced here is merely one of convenience.

In this chapter, as in the next, our thermodynamic treatment of interfaces follows essentially the method of Gibbs [1]. Also, we continue to assume that the only work performed by a system is done against pressures. Gravitational, electric, or magnetic forces are not considered. More importantly, we note that since the stress tensor within a crystalline solid is not hydrostatic, the assumption of an isotropic pressure limits the validity of this treatment in its applications to solids. Nevertheless, a rigorous treatment of the thermodynamics of solids is very complex and outside the scope of this text.

1. Fundamental Equations

So far in considering heterogeneous systems we have assumed, implicitly or explicitly, that two phases in contact are homogeneous up to a dividing geometrical surface. Obviously, this is not the case. The atoms in the vicinity of the interfaces are in a different environment than in the bulk; their densities and energies are different. The physical interface is three dimensional. However, it is at most only a few atoms thick, of the order of 10^{-9} m (or a few angstroms).

1.1. Temperature and Chemical Potentials at the Interface

Let us now imagine a geometric surface Σ lying in the region of heterogeneity between two phases α and β. *The surface Σ is chosen such that its points are similarly situated with respect to the conditions of adjacent elements.* With this stipulation, the directions of the normals to Σ are well determined, although the exact position of Σ is still arbitrary: other surfaces parallel to Σ and still within the physical interface also satisfy our definition.

A closed surface θ is generated by a normal to Σ moving along a closed curve within Σ. The surface θ cuts the surface Σ and includes a part of the homogeneous bulk phases α and β on each side (see Fig. 1). Two other surfaces parallel to Σ,

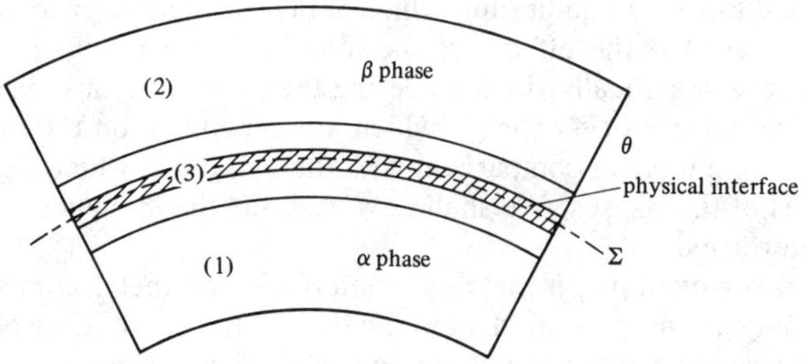

Figure 1. Physical interface and Gibb's geometric dividing surface.

very close to each other but enclosing the physical interface, divide the system within θ into three parts. Parts 1 and 2 contain only homogeneous portions of the phases α and β, while part 3 contains the surface Σ and the heterogeneity of the interphase region.

We recall (see Chapter II) that the criterion of equilibrium of a system may be written

$$(\delta E)_{S,V,n_i} \geq 0 \tag{1}$$

Applying this condition to the system 3 and keeping its boundaries fixed, we have

$$(\delta E^{(3)})_{S^{(3)},n_i^{(3)}} \geq 0 \tag{2}$$

As seen in Chapter II, the equality sign corresponds to a reversible equilibrium. Consequently, if we consider reversible variations in system 3, the variation in its energy must be of the form

$$\delta E^{(3)} = A_0\,\delta S^{(3)} + \sum_{i=1}^{m} A_i\,\delta n_i^{(3)} \tag{3}$$

since $\delta E^{(3)}$ must vanish with $\delta S^{(3)}$ and all the $\delta n_i^{(3)}$. The coefficients A_i are partial derivatives of $E^{(3)}$ with respect to $S^{(3)}$ and the $n_i^{(3)}$. It is clear that A_0 must be $T^{(3)}$, the temperature of the surface layer. *Similarly, although chemical potentials were defined only in the case of homogeneous systems, by a natural extension of the term, A_i may be called the chemical potential of i at the interface.* Consequently, equation (3) becomes

$$\delta E^{(3)} = T^{(3)}\,\delta S^{(3)} + \sum_{i=1}^{m} \mu_i^{(3)}\,\delta n_i^{(3)} \tag{4}$$

Let us now apply the criterion of equilibrium (1) to the system enclosed by the surface θ:

$$(\delta E^{(1)} + \delta E^{(2)} + \delta E^{(3)})_{S,V,n_i} \geq 0 \tag{5}$$

For variations in which all the surfaces are fixed,

$$\delta E^{(1)} = T^{(1)}\,\delta S^{(1)} + \sum_{i=1}^{m} \mu_i^{(1)}\,\delta n_i^{(1)} \tag{6a}$$

$$\delta E^{(2)} = T^{(2)}\,\delta S^{(2)} + \sum_{i=1}^{m} \mu_i^{(2)}\,\delta n_i^{(2)} \tag{6b}$$

Since for the total system S and n_i are fixed

$$\delta S = \delta S^{(1)} + \delta S^{(2)} + \delta S^{(3)} = 0 \tag{7}$$

$$\delta n_i = \delta n_i^{(1)} + \delta n_i^{(2)} + \delta n_i^{(3)} = 0 \qquad (i = 1, \ldots, m) \tag{8}$$

Proceeding exactly as in Section II.4.2, we can readily see that the criterion of equilibrium (5) leads to the equilibrium conditions

$$T^{(1)} = T^{(2)} = T^{(3)} \tag{9}$$

$$\mu_i^{(1)} = \mu_i^{(2)} = \mu_i^{(3)} \qquad (i = 1, \ldots, m) \tag{10}$$

when the systems (1)–(3) allow heat and mass transfers but have no moving boundaries.

In Chapter II we disregarded the effects of interfaces and established the equilibrium conditions of equal temperatures and equal chemical potentials for each species in the coexisting phases. We have now demonstrated that the same conditions (9) and (10) must be obeyed when the heterogeneous interfaces are taken into account.

1.2. Model System

We shall now replace the system enclosed by the surface θ by a model system for which we assume that the phase α and β are homogeneous right up to the dividing surface Σ (see Fig. 2). Quantities relating to the part of the system containing α will be identified by a prime and those pertaining to the part of the system containing β by a double prime.

Let c_i denote the concentration of a component i in moles per unit volume. In the subsystem of volume V', the concentration is assumed uniform and

$$n_i' = c_i' V' \tag{11a}$$

Similarly,

$$n_i'' = c_i'' V'' \tag{11b}$$

For the model system to be stoichiometrically equivalent to the real system, it is necessary to assume that the dividing surface contains a certain number of moles n_i^σ such that the total number of moles of i in the real system n_i is equal

Figure 2. (**a**) Real and (**b**) model systems. In the model system, the phases α and β are assumed homogeneous up to the dividing surface Σ.

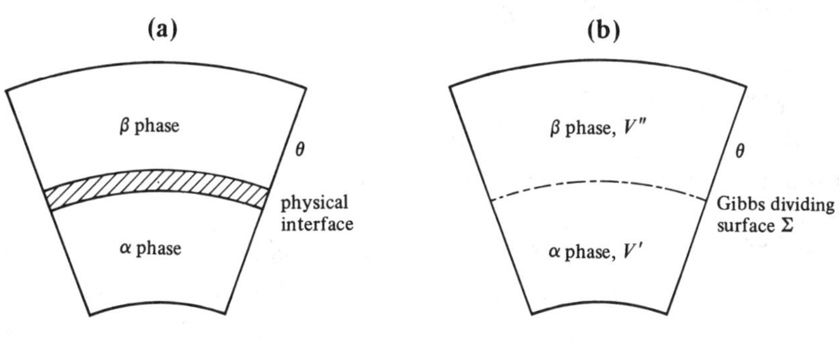

to

$$n_i = n_i' + n_i'' + n_i^\sigma \tag{12}$$

The surface excess number of moles, or the number of moles adsorbed at the surface, is thus defined by

$$n_i^\sigma = n_i - n_i' - n_i'' \tag{13}$$

where n_i' and n_i'' are defined by equations (11). n_i^σ divided by the area s of the surface Σ enclosed by θ yields the superficial density of i, or the *adsorption* of i:

$$\Gamma_i = n_i^\sigma / s \tag{14}$$

Γ_i may be positive or negative.

Other surface excess quantities may be similarly defined. For example,

$$E^\sigma = E - E' - E'' \tag{15}$$

$$S^\sigma = S - S' - S'' \tag{16}$$

We note, however, that

$$V^\sigma = V - V' - V'' = 0 \tag{17}$$

We shall see that the surface excess quantities Γ_i, E^σ, S^σ are dependent upon the exact location of the surface Σ. This dependence will be examined in Section 1 of Chapter XIV.

1.3. Surface Tension

Let us consider reversible variations in the energy of the system enclosed by the surface θ for which all the boundaries and physical interface remain fixed. We have seen [as in equation (4)] that δE may be written

$$\delta E = T\,\delta S + \sum_{i=1}^{m} \mu_i\,\delta n_i \tag{18}$$

Equation (15), related to the model system, yields

$$\delta E^\sigma = \delta E - \delta E' - \delta E'' \tag{19}$$

For homogeneous phases of fixed boundaries

$$\delta E' = T\,\delta S' + \sum_{i=1}^{m} \mu_i\,\delta n_i' \tag{20a}$$

$$\delta E'' = T\,\delta S'' + \sum_{i=1}^{m} \mu_i\,\delta n_i'' \tag{20b}$$

Substituting equations (18) and (20) in equation (19), we obtain

$$\delta E^\sigma = T(\delta S - \delta S' - \delta S'') + \sum_{i=1}^{m} \mu_i(\delta n_i - \delta n_i' - \delta n_i'') \tag{21}$$

or, through equations (16) and (17),

$$\delta E^{\sigma} = T\,\delta S^{\sigma} + \sum_{i=1}^{m} \mu_i\,\delta n_i^{\sigma} \tag{22}$$

The excess quantities E^{σ}, S^{σ}, n_i^{σ} are determined both by the state of the physical system considered and by the imaginary surfaces by which they are defined. For the variations associated with equation (22), we imagined that all these surfaces remain fixed. It is obvious, however, that the form of the surfaces which lie in the region of homogeneity on either side of the interface cannot affect these values. The complete expression of δE^{σ} for all reversible variations therefore depends only on the position and form of the surface Σ.

Let us first assume that the form of Σ remains unchanged but that it is translated along its normal. The quantities E^{σ}, S^{σ}, and n_i^{σ} depending on the exact location of Σ will vary because of the way they were defined. However, they will still satisfy equation (22) since that equation is valid for any given position of Σ. Equation (22) also remains unaffected if both the surface Σ and the physical system enclosed by θ are changed in position *while their relative positions remain the same*. Consequently, variations in the position of Σ, either by translation or rotation, which keep its form unchanged do not affect the validity of equation (22).

Variations in the form of Σ, if the portion of Σ considered is sufficiently small, may be analyzed by changes in its area s and two principal curvatures c_1, c_2. Equation (22) must then be rewritten

$$\delta E^{\sigma} = T\,\delta S^{\sigma} + \sum_{i=1}^{m} \mu_i\,\delta n_i^{\sigma} + \sigma\,\delta s + C_1\,\delta c_1 + C_2\,\delta c_2 \tag{23}$$

where σ, C_1, and C_2 are the partial derivatives of E with respect to s, c_1, and c_2, and are determined by the initial state of the system and the position and form of Σ. Equation (23) now determines δE^{σ} for all possible variations in the system enclosed by θ.

Gibbs [1] rewrites equation (23)

$$\delta E^{\sigma} = T\,\delta S^{\sigma} + \sum_{i=1}^{m} \mu_i\,\delta n_i^{\sigma} + \sigma\,\delta s + \tfrac{1}{2}(C_1 + C_2)\,\delta(c_1 + c_2) + \tfrac{1}{2}(C_1 - C_2)\,\delta(c_1 - c_2) \tag{24}$$

and proceeds to show that it is possible to choose the exact position of the geometrical surface Σ in the physical interface in such a way that the term $\frac{1}{2}(C_1 + C_2)\,\delta(c_1 + c_2)$ vanishes. Moreover, Gibbs shows that because the physical interface is generally very thin with respect to the bulk of the adjoining phases, the term $\frac{1}{2}(C_1 - C_2)\,\delta(c_1 - c_2)$ is negligible. Consequently, equation (24) becomes

$$\delta E^{\sigma} = T\,\delta S^{\sigma} + \sum_{i=1}^{m} \mu_i\,\delta n_i^{\sigma} + \sigma\,\delta s \tag{25}$$

Let us now reconsider reversible variations in the energy E of the system enclosed by θ:

$$dE = dE' + dE'' + dE^{\sigma} \tag{26}$$

For homogeneous phases (the boundaries of which need no longer be kept fixed) we have

$$dE' = T\,dS' + \sum_{i=1}^{m} \mu_i\,dn_i' - P'\,dV' \tag{27a}$$

$$dE'' = T\,dS'' + \sum_{i=1}^{m} \mu_i\,dn_i'' - P''\,dV'' \tag{27b}$$

Incorporating equations (25) and (27) in equation (26), we obtain

$$dE = T(dS' + dS'' + dS^{\sigma}) + \sum_{i=1}^{m} \mu_i(dn_i' + dn_i'' + dn_i^{\sigma})$$
$$- P'\,dV' - P''\,dV'' + \sigma\,ds \tag{28}$$

or

$$dE = T\,dS + \sum_{i=1}^{m} \mu_i\,dn_i - P'\,dV' - P''\,dV'' + \sigma\,ds \tag{29}$$

This equation yields the definition of the surface tension σ:

$$\sigma = (\partial E/\partial s)_{S,n_i,V',V''} \tag{30}$$

We note that the volumes V' and V'' are defined by the position of the geometrical surface Σ which must be located at the position which makes the term $C_1 + C_2$ vanish.

1.4. Equilibrium Conditions for the Pressures

In Section 1.1, we showed that the temperatures and the chemical potentials of each species must be the same in both phases α and β. To investigate the equilibrium conditions related to the pressures, let us imagine that the dividing surface moves a uniform distance δl along its normal; the entropy S, the volume V, and the number of moles n_i are kept constant. We note that

$$\delta V' = s\,\delta l = -\delta V'' \tag{31}$$

and

$$\delta s = (c_1 + c_2)s\,\delta l \tag{32}$$

Consequently, equation (29) yields

$$\delta E = (P'' - P')s\,\delta l + \sigma(c_1 + c_2)s\,\delta l$$
$$= s\,\delta l\,[\sigma(c_1 + c_2) - (P' - P'')] \tag{33}$$

The criterion of equilibrium [equation (1)] thus leads to the equilibrium condition

$$P' - P'' = \sigma(c_1 + c_2) \tag{34a}$$

or, in terms of the radii of curvatures,

$$P' - P'' = \sigma\left(\frac{1}{r_1} + \frac{1}{r_2}\right) \tag{34b}$$

This replaces the equation of the equality of pressures which was derived in Chapter II for the case where the effects of interfaces were neglected. It is attributed to Laplace [2] and bears his name. We note that the curvatures, or radii of curvature, are positive when their centers lie on the side associated to P'.

For a sphere equation (34) becomes

$$P' - P'' = 2\sigma/r \tag{35}$$

and for a planar surface

$$P' = P'' \tag{36}$$

This last case is of special interest. Indeed, we observe that equation (29) becomes

$$dE = T\,dS + \sum_{i=1}^{m} \mu_i\,dn_i - P\,dV + \sigma\,ds \tag{37}$$

and that

$$\sigma = (\partial E/\partial s)_{S,n_i,V} \tag{38}$$

Only the total volume need be kept constant. The position of the geometrical surface Σ no longer affects the definition of σ. We also verify that the determination of Σ by the condition $C_1 + C_2 = 0$ is no longer possible since both curvatures c_1 and c_2 are equal to zero, and in equation (23) the terms $C_1\,\delta c_1$, $C_2\,\delta c_2$ vanish.

2. Mechanical Equivalence of the Model System

In Section 1 we introduced the surface tension by examining all possible variations of a surface energy due to changes in the position and form of the interface between two phases α and β. This surface energy was defined as an excess quantity necessary to make our model system equivalent to the real system. We may now examine whether the model system can be made mechanically equivalent to the real system. These conditions for equivalence yield another presentation of the surface tension which may provide the reader with a better physical understanding of this quantity.

2.1. General Procedure and Definition of the Surface Tension

In 1805 Young [3] observed that from a mechanical standpoint the system of two phases α and β in Fig. 1 behaves as if it consisted of two homogeneous phases separated by a membrane uniformly stretched and of infinitesimal thickness. To define the tension at a point M of the membrane let us imagine a curve PQ passing by M (see Fig. 3) and dividing the surface into two regions. Across an element δu of the curve at M, one region exerts a force $\sigma\,\delta u$ tangential to the surface. σ is called the *surface tension* or the *interfacial tension* at the point M. The surface is in a state of uniform tension (uniformly stretched) if

a. σ is perpendicular to the dividing line and has the same value irrespective of the direction of the line, and
b. σ has the same value at all points of the surface.

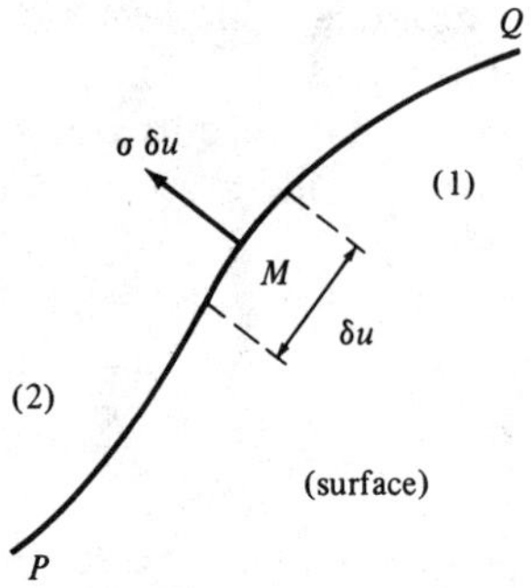

Figure 3. Surface tension force. If region (1) is assumed to be removed, an element δu of the boundary PQ of region (2) at M is subjected to a force $\sigma\,\delta u$ perpendicular to the boundary and in the tangent plane to the surface at M.

The mechanical equivalence of the real and ideal systems will be established at a point M of the interface if at this point we write two equations: one on the equality of forces, the other on the equality of moments. These two equations determine both the magnitude of the surface tension and the position of the dividing surface.

It is instructive to detail the calculations in the case of a cylindrical surface of constant curvature. It will become clear that the same approach can be extended to other types of surfaces.

2.2. Case of a Cylindrical Surface of Constant Curvature

Figure 4 shows a section perpendicular to the axis of a cylindrical interface. The dividing surface Σ (i.e., the membrane) is cut by the plane of the section along a curve of radius r and center 0. Along the z axis, the forces exerted on the plane by the system above that plane consist of uniform pressures P' from a to c and P'' from d to b; from c to d, i.e., in the region of the physical interface, the distribution of forces is nonuniform and at least some parts of this region are subjected to tensions. We shall measure the forces positively for tensions and negatively for pressures. In order to establish the mechanical equivalence of the real and model systems enclosed by the boundaries $PQRS$, we shall have to write that the sums of their forces are equal and that the sums of their moments are also equal. We note that the forces arising from hydrostatic pressures between a and c and between b and d are the same in both systems and therefore play no role in determining the conditions of equivalence. It is thus sufficient to consider the region between c and d.

2.2.1. Conditions Pertaining to Forces and Moments

Let us identify a point M in the physical interface layer by its polar coordinates ρ and φ. The force exerted on a small area $\rho\,d\rho\,d\varphi$ at M is $F = t(\rho)\rho\,d\rho\,d\varphi$. The moment with respect to the center 0 is the vectorial product $\mathbf{0M} \times \boldsymbol{F}$. For symmetry reasons, it is obvious that only projections along the x axis need be considered. The projection on this axis of the moment associated with t is $t(\rho)\rho^2\,d\rho\cos\varphi\,d\varphi$. Its integration from $-\theta/2$ to $+\theta/2$ yields $2\sin(\theta/2)t(\rho)\rho^2\,d\rho$.

The equation pertaining to the equality of forces is

$$\theta\int_{\rho_c}^{\rho_\sigma}(-P')\rho\,d\rho + \theta\int_{\rho_\sigma}^{\rho_d}(-P'')\rho\,d\rho + \sigma\rho_\sigma\theta = \theta\int_{\rho_c}^{\rho_d} t\rho\,d\rho \tag{39}$$

or

$$\sigma\rho_\sigma = \int_{\rho_c}^{\rho_\sigma} (t + P')\rho\, d\rho + \int_{\rho_\sigma}^{\rho_d} (t + P'')\rho\, d\rho \tag{40}$$

Similarly, the equality of moments yields (after canceling the term $2 \sin \theta/2$):

$$\sigma\rho_\sigma^2 = \int_{\rho_c}^{\rho_\sigma} (t + P')\rho^2\, d\rho + \int_{\rho_\sigma}^{\rho_d} (t + P'')\rho^2\, d\rho \tag{41}$$

Multiplying equation (40) by ρ_σ and substracting it from equation (41), we obtain

$$\int_{\rho_c}^{\rho_\sigma} (t + P')(\rho - \rho_\sigma)\rho\, d\rho + \int_{\rho_\sigma}^{\rho_d} (t + P'')(\rho - \rho_c)\rho\, d\rho = 0 \tag{42}$$

It is convenient to change origin and measure positions from the point c along the z axis. We note that

$$\rho - \rho_\sigma = z - z_\sigma \tag{43a}$$

or

$$\rho = \rho_\sigma + z - z_\sigma = r[1 + c(z - z_\sigma)] \tag{43b}$$

where r is equal to ρ_σ the radius of curvature of Σ and c is equal to $1/r$, the curvature of Σ. Equations (40) and (42) become

$$\sigma = \int_0^{z_\sigma} (t + P')[1 + c(z - z_\sigma)]\, dz + \int_{z_\sigma}^{z_d} (t + P'')[1 + c(z - z_\sigma)]\, dz \tag{44}$$

Figure 4. Cylindrical interface of constant curvature.

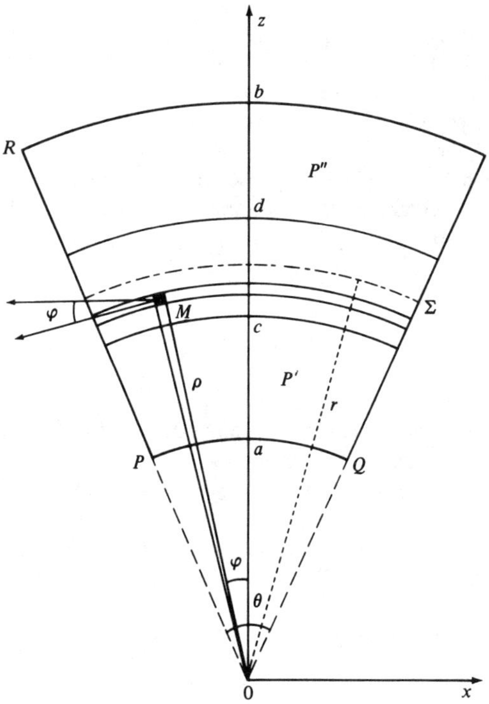

and

$$\int_0^{z_\sigma} (t + P')[1 + c(z - z_\sigma)](z - z_\sigma)\, dz + \int_{z_\sigma}^{z_d} (t + P'')[1 + c(z - z_\sigma)](z - z_\sigma)\, dz = 0 \tag{45}$$

These equations determine the magnitude of the surface tension σ and the position of the dividing surface z_σ. The particular dividing surface satisfying equation (45) is often referred to as the *surface of tension*.

We observe that the case of a planar interface would correspond to $c = 0$. Equations (44) and (45) would then become

$$\sigma = \int_0^{z_\sigma} (t + P')\, dz + \int_{z_\sigma}^{z_d} (t + P')\, dz \tag{46}$$

$$\int_0^{z_\sigma} (t + P')(z - z_\sigma)\, dz + \int_{z_\sigma}^{z_d} (t + P'')(z - z_\sigma)\, dz = 0 \tag{47}$$

2.2.2. Consequences of the Assumption on the Thinness of the Interface Relative to the Radius of Curvature

Let us reconsider equations (44) and (45). By inspection (e.g., for constant t) it is clear that although both σ and z_σ are uniquely determined they both depend on the curvature c.

We shall now make an important assumption. In general (that is, in most cases of practical interest), the thickness of the interface is very small compared to the radius of curvature ($z_d \ll r$). We shall assume that in these cases equation (45) (related to the equality of moment for a curved interface) may be replaced by equation (47) (related to the equality of moment for a plane interface).

Let us justify this assumption. Equation (45) may be rewritten

$$\int_0^{z_\sigma} (P' + t)(z_\sigma - z)\, dz - \frac{1}{r}\int_0^{z_\sigma} (P' + t)(z_\sigma - z)^2\, dz = \int_{z_\sigma}^{z_d} (P'' + t)(z - z_\sigma)\, dz + \frac{1}{r}\int_{z_\sigma}^{z_d} (P'' + t)(z - z_\sigma)^2\, dz \tag{48}$$

We may now show that in each member of this equation the second term is negligible compared to the first when $z_d \ll r$. Because of the theorem of the mean,[1] we have

$$\frac{1}{r}\int_0^{z_\sigma} (P' + t)(z_\sigma - z)^2\, dz \le \frac{1}{r}\int_0^{z_\sigma} |P' + t|(z_\sigma - z)^2\, dz \le \frac{z_d}{r}\int_0^{z_\sigma} |P' + t|(z_\sigma - z)\, dz \tag{49}$$

The integral $I_1 = \int_0^{z_\sigma} |P' + t|(z_\sigma - z)\, dz$ is unlikely to be much larger than the integral $I_2 = \int_0^{z_\sigma} (P' + t)(z_\sigma - z)\, dz$ because the physical interface is not likely to have substantial portions of it subjected to a pressure exceeding P' ($t' < - P'$). Consequently, the product $I_1(z_d/r)$ for $z_d/r \ll 1$ may be neglected in comparison to I_2. Similarly, the second integral in the right-hand member of equation (48) may be neglected in comparison to the first. Equation (48) (or 45) then becomes equivalent to equation (47). It may also be noted that

[1] The theorem shows that $\int_a^b f(x)g(x)\, dx = g(\xi) \int_a^b f(x)\, dx$ where $a<\xi<b$, when $f(x)$ remains of constant sign in the interval (a,b) [4].

if equations (45) and (47) are equivalent, equations (44) and (46) are also equivalent. Indeed, multiplying equation (47) by c and substracting it from equation (44) yields equation (46).

Consequently, we see that when the thickness of the physical interface is much smaller than its radius of curvature, the surface tension σ is independent of that radius of curvature. σ is not, however, independent of the position of the dividing surface z_σ.

Consider now the cylindrical system defined by the base $PQRS$ in Fig. 4 and of height h. We shall demonstrate that at constant S and n_i the change in the energy of the system (i.e., its work) associated with the movement of its boundaries is of the form $-P'\,\delta V' - P''\,\delta V'' + \sigma\,\delta s$, and contains no term for the curvature if the position of Σ is determined by equation (47).

The volume V', which extends from PQ to Σ, may be written as the sum of two terms—the volume v' extending from PQ to the boundary c of the physical interface and the volume extending from that boundary to Σ:

$$V' = v' + \int_0^{z_\sigma} r[1 + c(z - z_\sigma)]\theta h\, dz \tag{50}$$

or

$$V' = v' + s\int_0^{z_\sigma} [1 + c(z - z_\sigma)]\, dz \tag{51}$$

where s is the area of the Gibbs dividing surface Σ enclosed in the system. Similarly,

$$V'' = v'' + s\int_{z_\sigma}^{z_d} [1 + c(z - z_\sigma)]\, dz \tag{52}$$

Let the boundaries of the system move. The dividing surface may also change its area and curvature. Equations (51) and (52) yield

$$\delta V' = \delta v' + \delta s\int_0^{z_\sigma} [1 + c(z - z_\sigma)]\, dz + s\,\delta c\int_0^{z_\sigma} (z - z_\sigma)\, dz \tag{53}$$

$$\delta V'' = \delta v'' + \delta s\int_0^{z_\sigma} [1 + c(z - z_\sigma)]\, dz + s\,\delta c\int_0^{z_\sigma} (z - z_\sigma)\, dz \tag{54}$$

In the real system, the consequent energy variation is

$$(\delta E)_{S,n_i} = -P'\,\delta v' - P''\,\delta v'' + \delta s\int_0^{z_d} t[1 + c(z - z_\sigma)]\, dz + s\,\delta c\int_0^{z_d} t(z - z_\sigma)\, dz \tag{55}$$

We shall now verify that this energy variation is equal to that in the model system written in the form

$$(\delta E)_{S,n_i} = -P'\,\delta V' - P''\,\delta V'' + \sigma\,\delta s \tag{56}$$

Using equations (53) and (54), equation (56) becomes

$$\begin{aligned}(\delta E)_{S,n_i} = {} & -P'\,\delta v' - P''\,\delta v'' + \sigma\,\delta s \\ & + \delta s\left\{\int_0^{z_\sigma} (-P')[1 + c(z - z_\sigma)]\, dz + \int_{z_\sigma}^{z_d} (-P'')[1 + c(z - z_\sigma)]\, dz\right\} \\ & + s\,\delta c\left[\int_0^{z_\sigma} (-P')(z - z_\sigma)\, dz + \int_{z_\sigma}^{z_d} (-P'')(z - z_\sigma)\, dz\right]\end{aligned} \tag{57}$$

Equating the second members of equations (55) and (57), we obtain

$$\sigma\,\delta s = \delta s\left\{\int_0^{z\sigma}(P' + t)[1 + c(z - z_\sigma)]\,dz + \int_{z\sigma}^{zd}(P'' + t)[1 + c(z - z_\sigma)]\,dz\right\}$$

$$+ s\,\delta c\left[\int_0^{z\sigma}(P' + t)(z - z_\sigma)\,dz + \int_{z\sigma}^{zd}(P'' + t)(z - z_\sigma)\,dz\right] \qquad (58)$$

We verify through equation (44) that the sum in the first brackets is equal to σ and through equation (47) that the sum in the second brackets is equal to zero. Consequently, the variation in the energy of the system $(\delta E)_{S,n_i}$ may indeed be written in the form (56), but only when the position of the dividing surface coincides with that of the surface of tension, i.e., when it is fixed by equation (47). We recall that this equation translates the equality of the sums of moments in the real and model systems when the thickness of the physical interface is much smaller than its radius of curvature. For any other position of the dividing surface, the curvature term in the expression of δE does not vanish. For example, if z_σ is *not* much smaller than r and is fixed by equation (45), the sum in the brackets multiplying $s\,\delta c$ in equation (58) is not equal to zero.

2.2.3. Equilibrium Condition

We have verified that the total change in the energy of the system may be written

$$dE = T\,dS + \sum_{i=1}^{m}\mu_i\,dn_i - P'\,dV' - P''\,dV'' + \sigma\,ds \qquad (59)$$

The criterion of equilibrium is

$$(\delta E)_{S,V'+V'',n_i} \geq 0 \qquad (60)$$

Let us consider that the interface moves a distance δl along its normal. Then,

$$\delta V' = s\,\delta l, \qquad \delta V'' = -s\,\delta l, \qquad \delta s = cs\,\delta l \qquad (61)$$

and

$$(\delta E)_{S,V'+V'',n_i} = -P's\,\delta l + P''s\,\delta l + \sigma cs\,\delta l \qquad (62)$$

Consequently, at equilibrium

$$s\,\delta l(-P' + P'' + \sigma c) \geq 0 \qquad (63)$$

Since δl may be either positive or negative, we must have

$$P' - P'' = \sigma(1/r) \qquad (64)$$

This condition is equivalent to that yielded by equation (34b) for $r_1 = r$ and $r_2 = \infty$.

2.2.4. Dependence of the Surface Tension on the Position of the Dividing Surface

We have shown that the condition expressing the equality of the sums of forces in the real and model systems defines the surface tension:

$$\sigma = \int_0^{z\sigma}(P' + t)[1 + c(z - z_\sigma)]\,dz + \int_{z\sigma}^{zd}(P'' + t)[1 + c(z - z_\sigma)]\,dz \qquad (65)$$

If the equality of the sums of moments is not written to determine z_σ, σ becomes a function of z_σ. We note that

$$\frac{d\sigma}{dz_\sigma} = (P' - P'') - c\left[\int_0^{z_\sigma} (P' + t)\, dz + \int_{z_\sigma}^{zd} (P'' + t)\, dz\right]$$

$$- c^2\left[\int_0^{z_\sigma} (P' + t)(z - z_\sigma)\, dz + \int_{z_\sigma}^{zd} (P'' + t)(z - z_\sigma)\, dz\right] \quad (66)$$

If the dividing surface is selected to correspond to the surface of tension, that is, if z_σ is selected to correspond to the value determined by equation (47) (equality of moments in the case where $z_\sigma \ll r$), then the term in the second brackets is equal to zero. Moreover, we have demonstrated that the term in the first bracket becomes then equal to σ. Consequently, at the surface of tension, equation (66) becomes

$$d\sigma/dz_\sigma = (P' - P'') - c\sigma \quad (67)$$

At equilibrium, and for the surface of tension, equation (64) is obeyed. Therefore, for this value of z_σ,

$$d\sigma/dz_\sigma = 0 \quad (68)$$

It is straightforward to establish that for the same value of z_σ

$$d^2\sigma/dz_\sigma^2 = \sigma c^2 \quad (69)$$

Consequently, σ takes its *minimum* value at the surface of tension. It is possible to determine (theoretically) the surface of tension by this property.

---

3. Gibbs Adsorption Equation

Let us reconsider in Fig. 1 the system enclosed in the surface θ and assume its equilibrium. It is obvious that if we were to double the size of the system by doubling the area of its α–β interface, then the number of moles of the system, its energy, and its entropy would also be doubled. Similarly, in the associated model system (Fig. 2) the volumes V' and V'' would be doubled as well. Consequently, we see that the energy of the system is a homogeneous function of degree 1 with respect to the variables S, V', V'', n_i, and s and we may write

$$E = S\left(\frac{\partial E}{\partial S}\right)_{V',V'',n_i,s} + V'\left(\frac{\partial E}{\partial V'}\right)_{S,V'',n_i,s} + V''\left(\frac{\partial E}{\partial V''}\right)_{S,V',n_i,s}$$

$$+ \sum_{i=1}^{m} n_i\left(\frac{\partial E}{\partial n_i}\right)_{S,V',V'',n_j,s} + s\left(\frac{\partial E}{\partial s}\right)_{S,V',V'',n_i} \quad (70)$$

Considering equation (29), the various partial derivatives in equation (70) are readily identified and equation (70) becomes

$$E = TS - P'V' - P''V'' + \sum_{i=1}^{m} n_i\mu_i + \sigma s \quad (71)$$

The properties of the homogeneous bulk phases lead to

$$E' = TS' - P'V' + \sum_{i=1}^{m} n_i' \mu_i \tag{72a}$$

$$E'' = TS'' - P''V'' + \sum_{i=1}^{m} n_i'' \mu_i \tag{72b}$$

Substracting equations (72) from (71), we obtain

$$E^\sigma = TS^\sigma + \sum_{i=1}^{m} n_i^\sigma \mu_i + \sigma s \tag{73}$$

Recalling equation (25),

$$dE^\sigma = T\, dS^\sigma + \sum \mu_i\, dn_i^\sigma + \sigma\, ds \tag{74}$$

We can readily establish that differentiation of equation (73) leads to

$$S^\sigma\, dT + \sum_{i=1}^{m} n_i^\sigma\, d\mu_i + s\, d\sigma = 0 \tag{75}$$

or

$$d\sigma = -\frac{S^\sigma}{s}\, dT - \sum_{i=1}^{m} \Gamma_i\, d\mu_i \tag{76}$$

This equation is known as the *Gibbs adsorption equation*. We shall later see how it can be exploited to deduce the composition of the interface from the isothermal composition dependence of the surface tension.

4. Surface Tension and the Thermodynamic Potential Ω

4.1. Thermodynamic Equations

The thermodynamic potential function Ω is defined as the difference

$$\Omega = A - \sum_{i=1}^{m} n_i \mu_i \tag{77}$$

where A is the Helmholtz free energy. For bulk homogeneous phases, $\sum n_i \mu_i$ is equal to the Gibbs free energy. Consequently, for these bulk homogeneous phases

$$\Omega' = -P'V' \tag{78a}$$

$$\Omega'' = -P''V'' \tag{78b}$$

The surface excess potential Ω^σ is then equal to

$$\Omega^\sigma = \Omega - \Omega' - \Omega'' = A + P'V' + P''V'' - \sum_{i=1}^{m} n_i \mu_i \tag{79}$$

or, using equation (71),

$$\Omega^\sigma = \sigma s \tag{80}$$

This relation may be considered a two-dimensional equivalent of (78). We also note that since

$$\Omega^\sigma = A^\sigma - \sum_{i=1}^{m} n_i^\sigma \mu_i \tag{81}$$

we obtain

$$\sigma = \frac{A^\sigma}{s} - \sum_{i=1}^{m} \Gamma_i \mu_i \tag{82}$$

We shall see that for planar surfaces it is occasionally possible to choose the position of the dividing surface so as to make null the last term of equation (82) (e.g., in the case of a pure component). In such cases, the surface tension becomes equal to a surface free energy per unit area. However, we emphasize that it is in general distinct from that quantity.

Equation (80) shows that the surface tension may be defined as the surface excess of the potential Ω per unit area. This definition is of special interest in the case of solids [5]. Let us elaborate briefly on that point.

4.2. Surfaces of Solids

The interfaces considered so far were assumed to behave mechanically as a membrane stretched uniformly and isotropically by a force which is the same at all points and for all directions in the surface, and which acts normally to any section taken through the surface (see Section 2.1). The origin of that force has been explained in Section 2. The tensor of pressures which, in the bulk of a phase such as a liquid or gas, reduces to an isotropic pressure evolves as the surface is approached, its components parallel to the surface diminishing progressively in value and possibly changing sign. The difference between its resultant and the force which would have been exerted if the tensor had remained uniform is the surface tension.

The surface of a solid is also a region where the stress tensor differs from that in the bulk. Generally, these stress tensors are not isotropic and, consequently, the surface tension is not necessarily isotropic either. (We note, however, that since it appears as a difference between two nonisotropic tensors, it *could* be isotropic.) For solids then, the concept of surface tension as a mechanical force becomes difficult. Experimentally, its direct measurement is only possible when the atoms of the solid can achieve a certain mobility, e.g., in metals at temperatures very close to their melting point. In these cases, the surface tension is thought to behave isotropically [5].

Since the concept of the surface tension as mechanical force is no longer adequate for solids, one may wish to define it through the work necessary to increase a surface area without doing any volume work. This is equivalent to using the definition

$$\sigma = (\partial E/\partial s)_{S,V',V'',n_i,\mathrm{rev}} \tag{83}$$

or

$$\sigma = (\partial A/\partial s)_{T,V',V'',n_i,\mathrm{rev}} \tag{84}$$

However, equation (84) (or 83) is not simply applicable to a solid whose state of internal stress varies from point to point. T, V', V'', and n_i are not the only variables which have to remain constant in the derivation. These problems are avoided by defining the surface tension through integral quantities, i.e., through the potential Ω [equation (80) or (82)].

Let us observe that, in this definition of σ, σ does depend on the exact location of the geometric dividing surface except in the case of planar surfaces. By proceeding as in Section 2.2.4, it is readily established that when the position of the geometric interface is moved along its normal, σ passes through a minimum. The general condition for mechanical equilibrium,

$$d\sigma/dz_\sigma = (P' - P'') - \sigma(c_1 + c_2) \tag{85}$$

then becomes Laplace's equation,

$$P' - P'' = \sigma(c_1 + c_2) \tag{86}$$

The location of the geometric interface is generally chosen to correspond to that minimum.

5. Variance of a Two-Phase System and Effects of the Interface's Curvature

5.1. *Variance of a Two-Phase System*

Let us consider a system of two phases α and β separated by an interface. We have seen that for all practical purposes (i.e., for all thermodynamic considerations) this system may be replaced by a model system for which the interface is reduced to a geometric surface and the phases are homogeneous up to that geometric surface. The state of each phase may be characterized by the independent variables: temperature, pressure, and composition ($X_2, X_3, \ldots, X_m$). To define the state of both phases, we therefore need the values of $2(m + 1)$ variables.

If the two phases are separated by a planar interface, the conditions for equilibrium are

$$\mu_i^\alpha = \mu_i^\beta \qquad (i = 1, 2, \ldots, m) \tag{87}$$

$$T^\alpha = T^\beta \tag{88}$$

$$P^\alpha = P^\beta \tag{89}$$

These are $m + 2$ equations and we see that the variance of this system at equilibrium is

$$\vartheta = 2(m + 1) - (m + 2) = m \tag{90}$$

The same result is obtained by direct application of the phase rule derived in Chapter II ($\vartheta = m + 2 - \varphi$ where $\varphi = 2$).

If the two phases are now separated by a curved interface, we have established that equilibrium conditions (87) and (88) still prevail but equation (89) is replaced by Laplace's equation:

$$P^\alpha - P^\beta = \sigma(c_1 + c_2) \tag{91}$$

Consequently, the mean curvature, $c = \frac{1}{2}(c_1 + c_2)$ has to be identified as a new independent variable. [Alternatively, if we still wish to analyze the two phases of the system by $2m + 2$ variables, equation (89) can no longer be counted as establishing a relation among two variables.] The variance of the two-phase system becomes

$$\vartheta = m + 1 \tag{92}$$

We shall now examine consequences of the interface's curvature, starting with the case of a two-phase system of one component.

5.2. *Effect of Curvature on the Vapor Pressure of a Pure Species*

Consider the equilibrium of a pure liquid with its vapor. If the interface separating the phases is flat, at any given temperature the vapor pressure is determined. However, if the interface is curved, as in the case of droplets, at any given temperature the vapor pressure of the liquid droplets depends on their radius r. Let us derive this dependence.

If two phases α and β are in equilibrium,

$$\mu_1^\alpha = \mu_1^\beta \tag{93}$$

and for any reversible change

$$\delta\mu_1^\alpha = \delta\mu_1^\beta \tag{94}$$

At constant temperature, this condition becomes

$$V_1^\alpha\, \delta P^\alpha = V_1^\beta\, \delta P^\beta \tag{95}$$

However,

$$P^\alpha - P^\beta = 2\sigma/r \tag{96}$$

where r is the mean radius of curvature $[r = 1/c = \frac{1}{2}(1/c_1 + 1/c_2)]$. For changes between equilibrium states

$$\delta P^\alpha - \delta P^\beta = \delta(2\sigma/r) \tag{97}$$

Combining equations (95) and (97), we obtain

$$\frac{V_1^\beta - V_1^\alpha}{V_1^\alpha}\,\delta P^\beta = \delta(2\sigma/r) \tag{98}$$

Equation (98) may be applied to the equilibrium of any two phases, but we now return to the particular case of liquid droplets in equilibrium with their vapor. We identify α as the liquid phase and β as the gaseous phase. We may neglect the molar volume of the liquid in front of the molar volume of the gas and assume that the latter is ideal. Equation (98) then becomes

$$\frac{RT}{V_1^l}\frac{\delta P^g}{P^g} = \delta\left(\frac{2\sigma}{r}\right) \tag{99}$$

Assuming further that the pressure dependence of the molar volume of the liquid species 1 is negligible, equation (99) may be integrated between the equilibrium associated with an interface of mean curvature $1/r$ and the equilibrium associated with a flat interface ($r = \infty$). The result is

$$\ln \frac{P^g}{(P^g)_{r=\infty}} = \frac{V_1^l}{RT}\left(\frac{2\sigma}{r}\right) \tag{100}$$

This equation is often referred to as Thomson's equation [6].

We observe that the vapor pressure of a liquid drop increases when its size decreases. Consequently, large drops will grow at the expense of smaller drops.

At 1300°K, the surface tension of pure liquid silver is equal to 0.900 N/m (or 900 dyn/cm) and its molar volume is equal to 11.68 cm^3/mol [7]. Equation (100) then yields

$$\ln \frac{(P^g)_r}{(P^g)_{r=\infty}} = \frac{11.68 \times 10^{-6} \times 2 \times 0.9}{1.987 \times 4.185 \times 1300 \times r} = \frac{1.95 \times 10^{-9}}{r} \tag{101}$$

It shows that droplets of radii $r = 10^{-8}$ m (100 Å) would have a vapor pressure 1.21 times higher than that of a flat interface (1.7×10^{-5} atm instead of 1.4×10^{-5} atm). For $r = 10^{-7}$ m (0.1 μm) the vapor pressure increase is only of the order of 2%.

5.3. Effect of Curvature on the Boiling Point of a Pure Species

Let us now consider that the phase β is subjected to a constant pressure P^β. Then, for reversible changes, equation (94) becomes

$$-S_1^\alpha\,\delta T + V_1^\alpha\,\delta P^\alpha = -S_1^\beta\,\delta T \tag{102}$$

$$V_1^\alpha\,\delta P^\alpha = (S_1^\alpha - S_1^\beta)\,\delta T = (H_1^\alpha - H_1^\beta)\,\delta T/T \tag{103}$$

We apply this equation to the equilibrium of a bubble of gas (α) in a liquid (β). Assuming that the gas behaves ideally, equation (103) becomes

$$\frac{\delta T}{T^2} = \frac{R}{\Delta H_{vap}}\frac{\delta P^g}{P^g} \tag{104}$$

But Laplace's equation yields

$$P^g - P^l = 2\sigma/r \tag{105}$$

and

$$\delta P^g = \delta(2\sigma/r) \tag{106}$$

Equation (104) may then be rewritten

$$\frac{\delta T}{T^2} = \frac{R}{\Delta H_{vap}}\frac{\delta(2\sigma/r)}{P^l + (2\sigma/r)} \tag{107}$$

Neglecting the temperature and curvature dependence of the heat of vaporization, the integration of equation (107) yields

$$\left(\frac{1}{T}\right)_{r=\infty} - \left(\frac{1}{T}\right)_r = \frac{R}{\Delta H_{vap}}\ln\left(1 + \frac{2\sigma/r}{P^l}\right) \tag{108}$$

Let us apply this equation to the boiling of water at $P^l = 1$ atm. $(T)_{r=\infty} = 373$°K, $\Delta H_{vap} = 40.4$ kJ/mol, and $\sigma = 55.46 - 0.215(T - 373)$ dyn/cm (or N/m $\times 10^{-3}$).

For $r = 10^{-6}$ m (1 μm), T is found to be equal to 394°K, an increase of 21°C over that of the flat interface. If the water is free of dust particles and the walls of the containing vessel are very smooth, much larger superheating is possible.

The effect of curvature is much more pronounced for a gas bubble than for a liquid droplet. This should not be surprising. The effect of curvature corresponds to a pressure effect, which is much larger for gases than for liquids (since the molar volume of a gas is much larger than that of a liquid).

5.4. Effect of Curvature on the Solubility of a Pure Species

Let us assume that the phase α consists only of species i and is in equilibrium with a phase β of components 1, 2, . . . , i, . . . , m. (That could be the case, for example, of graphite in equilibrium with a liquid iron alloy.) Considering reversible changes at constant T, P^β, and X_j^β (for $j \neq 1$ and i), we have

$$\delta\mu_i^\beta = \delta\mu_i^\alpha = V_i^\alpha\, \delta P^\alpha = V_i^\alpha\, \delta(2\sigma/r) \tag{109}$$

and assuming that V_i^α is practically independent of the pressure,

$$(\mu_i^\beta)_r - (\mu_i^\beta)_{r=\infty} = V_i^\alpha(2\sigma/r) \tag{110}$$

$(\mu_i^\beta)_r$ is the chemical potential of i in the β phase in equilibrium with a particle or droplet of α of radius r. Equation (110) may be rewritten

$$\ln \frac{(a_i^\beta)_r}{(a_i^\beta)_{r=\infty}} = \frac{V_i^\alpha}{RT}\left(\frac{2\sigma}{r}\right) \tag{111}$$

If we assume that the activity coefficient of i in the β phase is not a strong function of X_i equation (111) yields

$$\ln \frac{(X_i^\beta)_r}{(X_i^\beta)_{r=\infty}} \simeq \frac{V_i^\alpha}{RT}\frac{2\sigma}{r} \tag{112}$$

We see that the solubility of i in the β phase increases with the curvature of the α phase. This effect is relatively small, however. In the case of graphite in a liquid iron alloy, it may be verified that the increase in solubility is detectable only for radii smaller than about 0.1 μm.

5.5. Effect of Curvature on the Chemical Potential of a Solute

Let us now consider the more general case of two multicomponent phases α and β in equilibrium with respect to each other. For reversible variations in which the temperature and the composition of the α phase are fixed, the state of the α phase is dependent upon a single parameter, e.g., P^α. We may then write for a solute i of the phase α

$$\delta\mu_i^\alpha = \overline{V}_i^\alpha\, \delta P^\alpha \tag{113}$$

δP^α may be related to the change in σ/r by

$$\delta P^\alpha - \delta P^\beta = \delta(2\sigma/r) \tag{114}$$

However, in order to eliminate δP^β a second equation is required. It may be obtained as follows.

For reversible changes

$$\delta\mu_j^\alpha = \delta\mu_j^\beta \tag{115}$$

or

$$\overline{V}_j^\alpha\, \delta P^\alpha = \sum_{k=2}^{m} \left(\frac{\partial \mu_j^\beta}{\partial X_k^\beta}\right)_{P^\beta, T, X_k^\beta} \delta X_k^\beta + \overline{V}_j^\beta\, \delta P^\beta \tag{116}$$

Multiplying both sides of this equation by X_j^β and summing for all j ($j = 1, \ldots, m$), we obtain

$$\left(\sum_{j=1}^{m} X_j^\beta \overline{V}_j^\alpha\right) \delta P^\alpha = \sum_{k=2}^{m} \left(\sum_{j=1}^{m} X_j^\beta \frac{\partial \mu_j^\beta}{\partial X_k^\beta}\right) \delta X_k^\beta + V_m^\beta\, \delta P^\beta \tag{117}$$

Because of the Gibbs–Duhem equation the sum in parentheses on the right-hand side of equation (117) is equal to zero. Consequently,

$$\left(\sum_{j=1}^{m} X_j^\beta \overline{V}_j^\alpha\right) \delta P^\alpha = V_m^\beta\, \delta P^\beta \tag{118}$$

Combining equations (113), (114), and (118) we obtain

$$\delta\mu_i^\alpha = \left[\overline{V}_i^\alpha \Big/ \left(1 - \sum_{j=1}^{m} X_j^\beta \frac{\overline{V}_j^\alpha}{V_m^\beta}\right)\right] \delta\left(\frac{2\sigma}{r}\right) \tag{119}$$

This equation may be readily integrated if we assume that the various partial and molar volumes remain practically constant. If the phase β is a gas and α is a liquid or solid, then $\overline{V}_j^\alpha \ll V_m^\beta$, and equation (119) becomes

$$(\mu_i^\alpha)_r - (\mu_i^\alpha)_{r=\infty} \simeq \overline{V}_i^\alpha \frac{2\sigma}{r} \tag{120}$$

5.6. Remarks

We have noted that the curvature effect calculated above generally becomes important only when the radii of curvature are quite small, say 10^{-7} m (0.1 μm) or smaller. We may also observe that if the radii are of the order of 10^{-9} m (10 Å), the number of molecules in the phase considered is too small for our thermodynamic formalism to apply.

Although the effect of curvature on a thermodynamic property may be small, its role may still be very important. Consider the case of droplets, or small solid inclusions, of a pure species in a large second phase. If initially all inclusions are of the same size, they can be in equilibrium but their equilibrium is unstable. Indeed, if an inclusion shrinks (by transferring an infinitesimal quantity of mass to the large second phase), the chemical potential of the pure species in the inclusion will increase, which in turn will lead to further shrinkage.

The same instability may occur if the inclusions considered are not pure (or not stoichiometric). However, in such cases, the resulting changes in composition may decrease the chemical potential of a given component in the inclusion and oppose the effect of curvature. For further reading on the subject, Defay and Prigogine [5] is recommended.

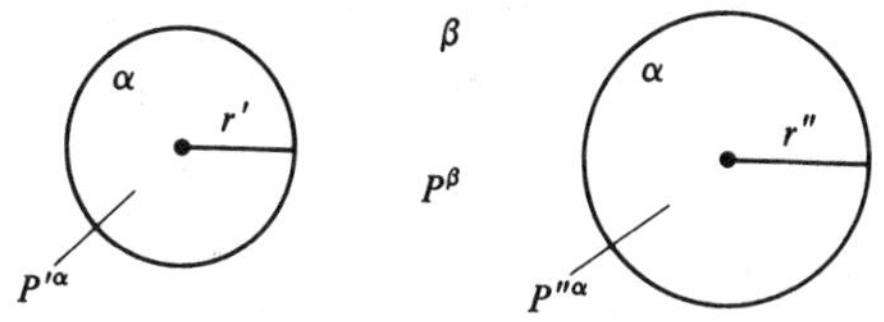

Figure 5. Ostwald ripening. The two inclusions are at the same temperature and have identical structures and compositions, but they are subjected to different pressures because of their different curvatures. They are not in mutual equilibrium, and the larger inclusion will grow at the expense of the smaller one.

In our derivation of Thomson's equation (120), $(\mu_i^\alpha)_r$ and $(\mu_i^\alpha)_{r=\infty}$ represent the chemical potential of a component i in a particle of radius r or ∞ at equilibrium with the phase β. Let us now consider two inclusions of identical compositions, temperatures, and structures, but different radii of curvature (r' and r'') and in contact with the same β phase (Fig. 5). The two inclusions cannot be in equilibrium; because of Laplace's equation, they are subjected to different pressures P'^α and P''^α:

$$P'^\alpha - P^\beta = 2\sigma/r', \qquad P''^\alpha - P^\beta = 2\sigma/r'' \tag{121}$$

Consequently,

$$P'^\alpha - P''^\alpha = 2\sigma\left(\frac{1}{r'} - \frac{1}{r''}\right) \tag{122}$$

Moreover, since

$$(\mu_i^\alpha)_{P'^\alpha} - (\mu_i^\alpha)_{P''^\alpha} = \int_{P''^\alpha}^{P'^\alpha} \overline{V}_i^\alpha\, dP \simeq \overline{V}_i^\alpha(P'^\alpha - P''^\alpha) \tag{123}$$

we obtain

$$(\mu_i^\alpha)_{r'} - (\mu_i^\alpha)_{r''} = \overline{V}_i^\alpha 2\sigma\left(\frac{1}{r'} - \frac{1}{r''}\right) \tag{124}$$

Equation (124) shows that the chemical potential of i in the smaller inclusion is higher than in the larger inclusion and that consequently the larger inclusion would grow at the expense of the smaller one. This phenomenon is known as *Ostwald ripening*. However, in cases where the resulting mass transfers change the compositions of the inclusions, these composition changes may oppose the effect of curvature.

For $r'' = \infty$ (flat interface), equation (124) becomes identical to equation (120). Nevertheless, although the formal results are the same, their significances are not, since equation (124) does not describe the effect of curvature on the *equilibrium* state of a phase.

In our derivations of the equations analyzing the effect of curvature, σ and r appeared as a single parameter σ/r, and we did not have to consider the dependence of σ on r. The effect of curvature on σ may be studied much as its effect on other thermodynamic properties. The discussion of the results, however, would be much lenghier and instead we refer the reader to a comprehensive article by Koenig [8]. It appears that this effect is generally small: for water and $r = 10^{-7}$

m, the difference is of the order of 0.4%, and for $r = 10^{-8}$ m, it is of the order of 3%. Consequently, the effect of curvature on σ is often neglected, and justifiably so.

6. Equilibrium Shape of a Crystal

6.1. Geometric Description of a Crystal

We shall now derive some preliminary equations related to the geometry of a crystal that will be useful in our thermodynamic treatment of the equilibrium shape of a solid. From some point 0 in the interior of a crystal let us draw lines normal to all possible orientations of the crystal faces (see Fig. 6). On each line we select a point $Q^{(\nu)}$ at a distance $h^{(\nu)} = 0Q^{(\nu)}$ and construct a plane perpendicular to the line $0Q^{(\nu)}$. Each plane so constructed defines a face of area $A^{(\nu)}$ limited by its intersections with other planes. Let N be the number of faces of the closed polyhedron we obtain. The polyhedron is entirely defined by the heights $h^{(1)}, \ldots, h^{(\nu)}, \ldots, h^{(N)}$ of its faces and the directions of the lines along which they are measured.

For fixed directions of the lines (determined by the lattice of the crystal), the volume V' of the crystal may be considered to be a function of the variables $h^{(\nu)}$. It is obvious that if we were to multiply these heights by a factor λ we would obtain

$$V'(\lambda h^{(1)}, \ldots, \lambda h^{(N)}) = \lambda^3 V'(h^{(1)}, \ldots, h^{(N)}) \tag{125}$$

Consequently, V' is a homogeneous function of degree 3 with respect to the variables $h^{(\nu)}$ and we may write (see Section II.2)

$$V' = \frac{1}{3} \sum_{\nu=1}^{N} \frac{\partial V'}{\partial h^{(\nu)}} h^{(\nu)} \tag{126}$$

If a straight line is drawn from 0 to each corner of the polyhedron, the crystal will be divided into N pyramids of height $h^{(\nu)}$, base $A^{(\nu)}$, and volume $\frac{1}{3}A^{(\nu)}h^{(\nu)}$. Equation (126) clearly shows that the area $A^{(\nu)}$ is equal to

$$A^{(\nu)} = \partial V'/\partial h^{(\nu)} \tag{127}$$

We note that

$$\frac{\partial A^{(\nu)}}{\partial h^{(j)}} = \frac{\partial A^{(j)}}{\partial h^{(\nu)}} = \frac{\partial^2 V'}{\partial h^{(j)}\, \partial h^{(\nu)}} \tag{128}$$

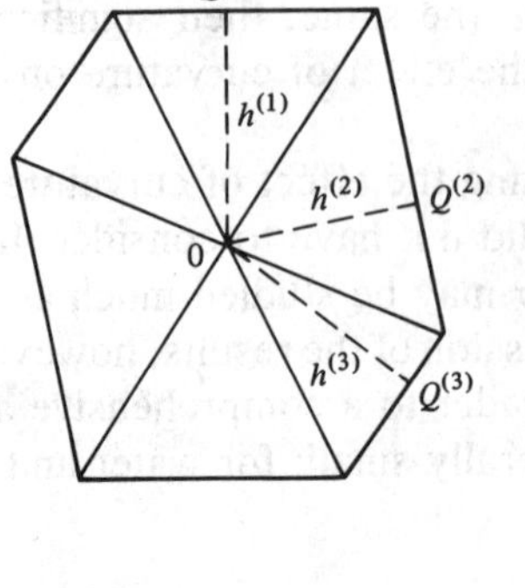

Figure 6. Geometric parameters describing a crystal.

It may also be observed that each area $A^{(\nu)}$ is a homogeneous function of degree 2:

$$A^{(\nu)}(\lambda h^{(1)}, \ldots, \lambda h^{(N)}) = \lambda^2 A^{(\nu)}(h^{(1)}, \ldots, h^{(N)}) \tag{129}$$

Consequently,

$$A^{(\nu)} = \frac{1}{2} \sum_{j=1}^{N} \frac{\partial A^{(\nu)}}{\partial h^{(j)}} h^{(j)} = \frac{1}{2} \sum_{j=1}^{N} \frac{\partial A^{(j)}}{\partial h^{(\nu)}} h^{(j)} \tag{130}$$

6.2. *Wulff's Relationships*

The faces of the polyhedron having different crystallographic orientations also have different atomic densities and, in general, different interfacial tensions. Let us now consider a crystal at equilibrium with another phase. For all reversible changes, the variation in the energy of this system is

$$\delta E = T\,\delta S - P'\,\delta V' - P''\,\delta V'' + \sum_{i=1}^{m} \mu_i\,\delta n_i + \sum_{j=1}^{N} \sigma^{(j)}\,\delta A^{(j)} \tag{131}$$

The corresponding variation in the Helmholtz free energy is

$$\delta A = -S\,\delta T - P'\,\delta V' - P''\,\delta V'' + \sum_{i=1}^{N} \mu_i\,\delta n_i + \sum_{j=1}^{N} \sigma^{(j)}\,\delta A^{(j)} \tag{132}$$

Imagine a reversible transformation at constant temperature, constant volumes V' and V'', and constant n_i. In other words, only the shape of the crystal is allowed to vary while maintaining its volume constant. Since the Helmholtz free energy must be minimum at equilibrium, we must have

$$\delta\phi = \sum_{j=1}^{N} \sigma^{(j)}\,\delta A^{(j)} = 0 \tag{133}$$

where the term $\delta\phi$ is introduced for convenience, and

$$\delta V' = 0 \tag{134}$$

These two equations express the problem of minimizing the function $\phi = \sum_{j=1}^{N} \sigma^{(j)} A^{(j)}$ at constant value of the function V'. This problem may therefore be treated by the classical method of Lagrange multipliers [9]. Adopting the variables $h^{(\nu)}$ to describe the transformation, we may write

$$\frac{\partial\phi}{\partial h^{(\nu)}} - \eta \frac{\partial V'}{\partial h^{(\nu)}} = 0 \qquad \text{for} \quad \nu = 1, 2, \ldots, N \tag{135}$$

where η is a Lagrange multiplier. Using equations (127) and (133), we obtain

$$\sum_{j=1}^{N} \left(\sigma^{(j)} \frac{\partial A^{(j)}}{\partial h^{(\nu)}} \right) - \eta\, A^{(\nu)} = 0 \tag{136}$$

Through equation (130), it becomes

$$\sum_{j=1}^{N} \sigma^{(j)} \frac{\partial A^{(j)}}{\partial h^{(\nu)}} - \frac{\eta}{2} \sum_{j=1}^{N} \frac{\partial A^{(j)}}{\partial h^{(\nu)}} h^{(j)} = 0 \tag{137a}$$

or

$$\sum_{j=1}^{N} \frac{\partial A^{(j)}}{\partial h^{(\nu)}} \left(\sigma^{(j)} - \frac{\eta}{2} h^{(j)} \right) = 0 \qquad \text{for} \quad \nu = 1, 2, \ldots, N \tag{137b}$$

This set of conditions represents N homogeneous equations of N variables $X^{(j)} = \sigma^{(j)} - (\eta/2)h^{(j)}$ ($j = 1, \ldots, N$). One possible solution is $x^{(1)} = 0$, $x^{(2)} = 0$, $\ldots$, $x^{(N)} = 0$; i.e.,

$$\frac{\sigma^{(1)}}{h^{(1)}} = \frac{\sigma^{(2)}}{h^{(2)}} = \cdots = \frac{\sigma^{(N)}}{h^{(N)}} = \frac{\eta}{2} \tag{138}$$

These relationships bear the name of Wulff, who established them in 1901 [10]. A full discussion of these relationships (e.g., whether they are necessary as well as sufficient, whether they correspond to an absolute minimum of the free energy) is outside the scope of this text. Instead, the reader is referred to articles by Herring [11,12] and Mullins [13].

6.3. Wulff Plots

The dependences of σ on the orientation of a crystal's surface and the consequent equilibrium shape of the crystal are conveniently summarized in a *Wulff plot*. In this type of diagram σ is plotted as a function of orientation. For a liquid or amorphous material such as glass the Wulff plot is obviously a sphere. However, for a crystal it will not be spherical and will adopt the symmetry of the crystal. Figures 7 and 8 show examples of σ plots (full lines). The consequent equilibrium shapes of the crystals are shown as broken lines.

A general method of constructing the equilibrium shape of a crystal from a σ plot is illustrated in Fig. 7 for a two-dimensional case, and it is readily extended to any three-dimensional case. If Mt is the perpendicular to a radius $0M$ of the σ plot, the equilibrium shape is obtained as the inner envelope of the Mt line as M describes the σ plot. Because of stability considerations, certain orientations

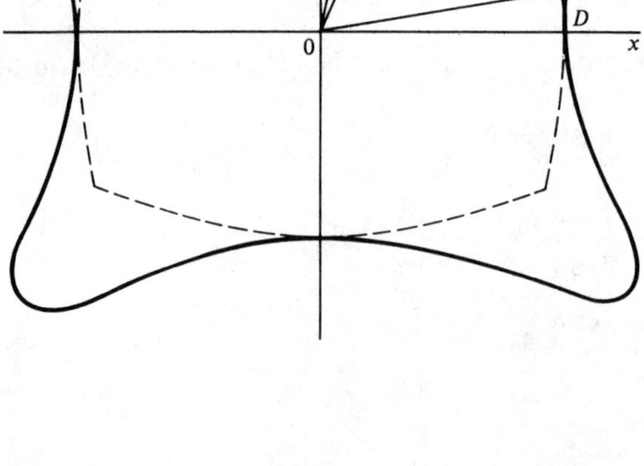

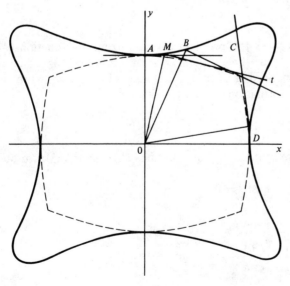

Figure 7. Wulff plot. The solid line shows the value of σ as a function of the orientation and the broken line represents the consequent equilibrium shape of the crystal. The orientations corresponding to the arc BC are unstable.

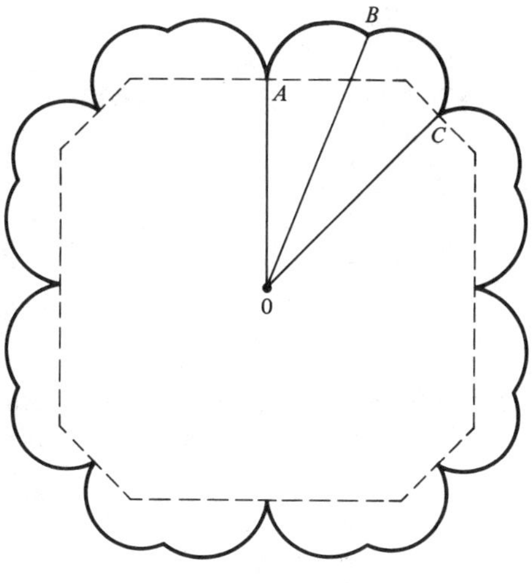

Figure 8. Wulff plot with cusps. The solid line shows the value of σ as a function of orientation while the broken line represents the consequent equilibrium shape. The surface normal to $0A$ is stable but that normal to $0B$ is unstable and may develop into a hill and valley structure.

of the $0M$ radii have no corresponding surface orientations in the equilibrium shape. For example, in the case of Fig. 7 and in the quadrangle of positive values for x and y, only the directions between $0A$ and $0B$ and between $0C$ and $0D$ correspond to stable surface orientations of the crystal shape.

A number of theorems relating the shape of crystals to characteristics of the σ plot have been derived by Herring [12]. We shall confine our presentation here to just a few of these results.

If $0A$ is a radius vector of the σ plot in whose neighborhood the derivatives of σ are continuous (i.e., no cusp is present; see Fig. 7), a necessary and sufficient condition for the direction $0A$ to be among the normals of the equilibrium shape is that the σ plot nowhere passes inside the sphere drawn through the origin and tangent to the σ plot at A. A continuously curved surface with normals in the neighborhood of A is then stable.

The equilibrium shape of a crystal will include a flat surface of finite extent and direction normal to a vector $0A$ of the σ plot if the σ plot has a pointed cusp at A and there is a sphere passing through A and the origin 0 which lies entirely inside the σ plot. Such a case is shown in Fig. 8.

Herring [12] has pointed out that the equilibrium shape is of direct practical interest only for very small crystals. For large crystals, a significant alteration in shape requires the transfer of a large number of atoms and the reduction in free energy may not be sufficient to achieve it. Metastable equilibrium states are more likely to be reached. It is therefore of interest to examine whether the free energy of a plane surface may be lowered through faceting, that is, by rearranging the atoms into hills and valleys of a size large with respect to atomic dimensions but still small macroscopically. Herring established that if a given macroscopic surface of a crystal does not coincide in orientation with some position of the equilibrium shape (e.g., surface of direction normal to the vector $0B$ in Fig. 8), there exists a hill-and-valley structure with a lower energy than that of the given flat surface. If the orientation of the given surface is included among those of the equilibrium shape, no faceting can be more stable. For further information on the shapes of crystals, the references by Herring and Mullins [11–14] are recommended.

7. The Equation of Laplace for a Crystal

Let us consider a crystal of volume V' and surface area s'. With the notation adopted in Section 6 we have

$$V' = \tfrac{1}{3} \sum_{\nu=1}^{N} A^{(\nu)} h^{(\nu)} \tag{139}$$

$$s' = \sum_{\nu=1}^{N} A^{(\nu)} \tag{140}$$

We assume that the crystal is in equilibrium with another phase of volume V''. At constant temperature, total volume, and number of moles n_i, the criterion of equilibrium may be written

$$(\delta A)_{T,V,n_i} \geq 0 \tag{141}$$

or

$$-(P' - P'')\,\delta V' + \sum_{\nu=1}^{N} \sigma^{(\nu)}\,\delta A^{(\nu)} \geq 0 \tag{142}$$

for which we used

$$\delta V' = -\delta V'' \tag{143}$$

We now imagine a reversible variation in the state of the system in which all the heights $h^{(j)}$ of the crystal are kept constant except for $h^{(i)}$. We may then write

$$\delta V' = \frac{\partial V'}{\partial h^{(i)}}\,\delta h^{(i)} = A^{(i)}\,\delta h^{(i)} \tag{144}$$

and

$$\delta A^{(\nu)} = \frac{\partial A^{(\nu)}}{\partial h^{(i)}}\,\delta h^{(i)} = \frac{\partial A^{(i)}}{\partial h^{(\nu)}}\,\delta h^{(i)} \tag{145}$$

because of equation (128). The criterion of equilibrium (142) becomes

$$-(P' - P'')A^{(i)}\,\delta h^{(i)} + \sum_{\nu=1}^{N} \sigma^{(\nu)} \frac{\partial A^{(i)}}{\partial h^{(\nu)}}\,\delta h^{(i)} \geq 0 \tag{146}$$

or

$$\delta h^{(i)}\left[-(P' - P'')A^{(i)} + \sum_{\nu=1}^{N} \frac{\sigma^{(\nu)}}{h^{(\nu)}} \frac{\partial A^{(i)}}{\partial h^{(\nu)}} h^{(\nu)}\right] \geq 0 \tag{147}$$

Because of Wulff's relationships

$$\frac{\sigma^{(\nu)}}{h^{(\nu)}} = \frac{\eta}{2} \quad \text{for} \quad \nu = 1, \ldots, N \tag{148}$$

and because of equation (130)

$$\frac{1}{2} \sum_{\nu=1}^{N} \frac{\partial A^{(i)}}{\partial h^{(\nu)}} h^{(\nu)} = A^{(i)} \tag{149}$$

equation (147) becomes

$$A^{(i)}\,\delta h^{(i)}[-(P' - P'') + \eta] \geq 0 \tag{150}$$

Since $\delta h^{(i)}$ can be either positive or negative, we must therefore have

$$P' - P'' = \eta = 2\frac{\sigma^{(1)}}{h^{(1)}} = \cdots = 2\frac{\sigma^{(\nu)}}{h^{(\nu)}} = \cdots = 2\frac{\sigma^{(N)}}{h^{(N)}} \tag{151}$$

Comparing this equation with the equation of Laplace (34), we see that $\sigma^{(\nu)}/h^{(\nu)}$ plays the same role as the product of σ by the mean radius of curvature (c or $1/r$). In particular, our equations in Section 5 on the effects of curvature remain valid after substitution of σ/r by $\sigma^{(\nu)}/h^{(\nu)}$.

8. Equilibrium at a Line of Contact of Three Phases

8.1. Condition for Equilibrium

Figure 9 shows three phases α, β, γ in mutual equilibrium. The three surfaces along which they meet have a common line of contact perpendicular to the plane of the drawing. Along each boundary there is a surface tension which tends to reduce the area of that boundary. We shall assume that the surface forces are the only ones present and that they are independent of the orientation of the boundary to which they apply. The three forces are in mechanical equilibrium if

$$\sigma_{\alpha\beta}\boldsymbol{t}_{\alpha\beta} + \sigma_{\alpha\gamma}\boldsymbol{t}_{\alpha\gamma} + \sigma_{\beta\gamma}\boldsymbol{t}_{\beta\gamma} = 0 \tag{152}$$

where $\boldsymbol{t}_{\alpha\beta}$ is a unit vector in the α–β boundary normal to the line of intersection; $\boldsymbol{t}_{\alpha\gamma}$ and $\boldsymbol{t}_{\beta\gamma}$ have similar definitions. The familiar force triangle allows us to express condition (152) in the form

$$\sigma_{\alpha\beta}/\sin\theta_\gamma = \sigma_{\alpha\gamma}/\sin\theta_\beta = \sigma_{\beta\gamma}/\sin\theta_\alpha \tag{153}$$

where θ_α, θ_β, θ_γ are the angles determined by the three forces, as shown in Fig. 9.

We note that if the three interfacial tensions are equal, the angles θ are all

Figure 9. Equilibrium at a line of contact between three phases α, β, γ. The line of contact is normal to the plane of the drawing. The interfacial tensions are assumed to be the only forces present.

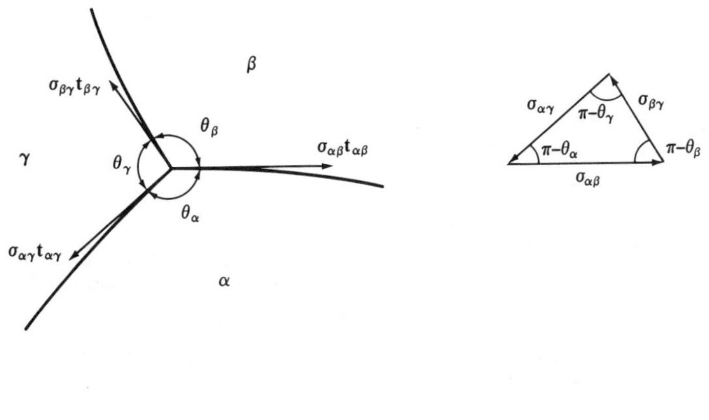

equal to 120°. For example, if a mixture of water and soap is agitated in a closed bottle, the boundaries of the resulting bubbles generally meet at 120°.

8.2. Contact Angle

If a liquid drop rests on a solid substrate in a gaseous atmosphere, by convention the angle of contact θ shown in Fig. 10 (and measured "through" the liquid) characterizes the wetting of the substrate by the drop. A projection on the horizontal axis of the forces acting on the line of contact at the point 0 eliminates the contribution of the gravity force and yields the following condition for equilibrium:

$$\sigma_{sg} = \sigma_{ls} + \sigma_{lg} \cos \theta \tag{154}$$

This equation was first derived by Young [3], but often it is also referred to as the Young–Dupré equation [15]. The ratio

$$k = (\sigma_{sg} - \sigma_{sl})/\sigma_{lg} \tag{155}$$

is sometimes called the *wetting coefficient*. If $k \leq -1$, the solid is not wetted. If $k \geq 1$, the solid is completely wetted. If $-1 < k < 1$, k is equal to $\cos \theta$ and the solid is only partially wetted.

The *sessile drop technique* is based upon the configuration shown in Fig. 10. An analysis of the drop's profile, originally due to Bashforth and Adams [16], shows that the curvature at the top of the drop yields a measure of the liquid–gas interfacial tension. The technique then consists in determining various geometric characteristics of the drop's profile [17] and in deducing from these both σ_{lg} and the contact angle θ. At high temperatures and controlled atmospheres, the liquid drop and its substrate are in a furnace and the profile is obtained on photographs taken through a telemicroscope (see Fig. 11).

The Young–Dupré equation (154) is deceptively simple. There are both conceptual and experimental difficulties associated with it. On the theoretical front, a rigorous thermodynamic derivation of this equation has been provided by Johnson [18]. Experimentally, the angle θ is very difficult to obtain with adequate reproducibility and there is a large hysteresis associated with the roughness of the substrate and whether the angle is advancing or receding. Wenzel [19] attempted to explain it by developing an expression between the macroscopic roughness of the solid's surface r, the observed contact angle ϕ, and the true contact angle θ:

$$\cos \phi = r \cos \theta \tag{156}$$

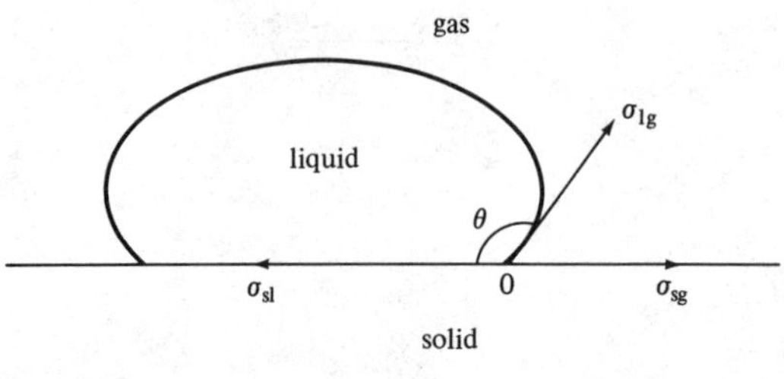

Figure 10. Contact angle θ of a liquid drop resting on a solid substrate.

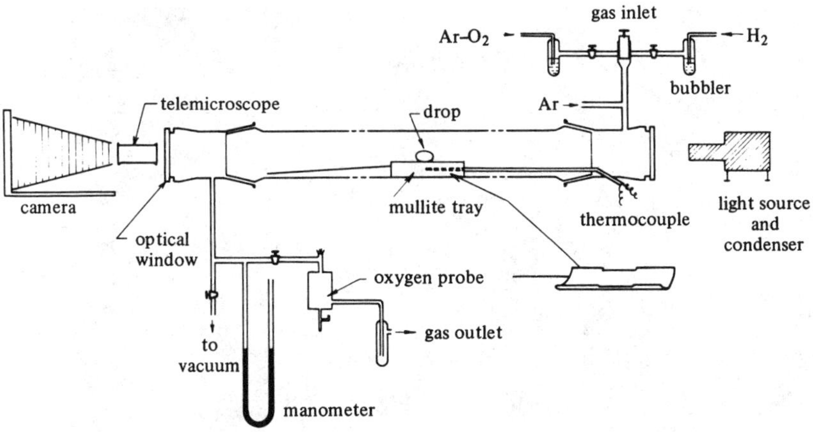

Figure 11. Sketch of a sessile drop apparatus [17].

(r is defined as the ratio of the true wetted area of the solid to the apparent area.) A more general analysis of this hysteresis effect has been given by Gallois [20].

8.3. Phase Distribution in a Polycrystalline Solid

Interfacial tensions play a determining role in the distribution of phases in a solid (see, for example, Smith [21]). Let us consider the case where small amounts of a second phase β are present in an otherwise homogeneous polycrystalline alloy of phase α. The phase β generally nucleates and grows at a grain boundary (see Fig. 12) because a decrease in the total interfacial area is thus achieved. In the configuration shown in Fig. 13, a *dihedral angle* φ is defined. Its value is obtained from the equilibrium of the interfacial tensions at 0:

$$\sigma_{\alpha\alpha} = 2\sigma_{\alpha\beta} \cos(\varphi/2) \tag{157}$$

Depending on the relative magnitudes of $\sigma_{\alpha\alpha}$ and $\sigma_{\alpha\beta}$, the angle φ can take any value from 0° to 180°. Some consequent shapes of the β inclusions are illustrated in Fig. 14.

We note that if $\sigma_{\alpha\alpha}/2\sigma_{\alpha\beta} \geq 1$, the dihedral angle is equal to 0. This is of great practical importance in the case where β is a phase of low melting point. Then, the second phase β wets completely the grains of phase α and causes the solid to disintegrate along its grain boundaries. This phenomenon is called *hot shortness*.

8.4. Torque Component in Grain Boundaries

A grain boundary is said to be *coherent* if one can trace a continuity of lattice structure from one grain to the next across the boundary (see Fig. 15). If the angle of misorientation between the two lattice structures is very small, there are generally a small number of defects (dislocation, vacancies) associated with the grain boundary. When the angle increases, the number of defects increases. Eventually, the misorientation becomes such that it is no longer possible to trace any con-

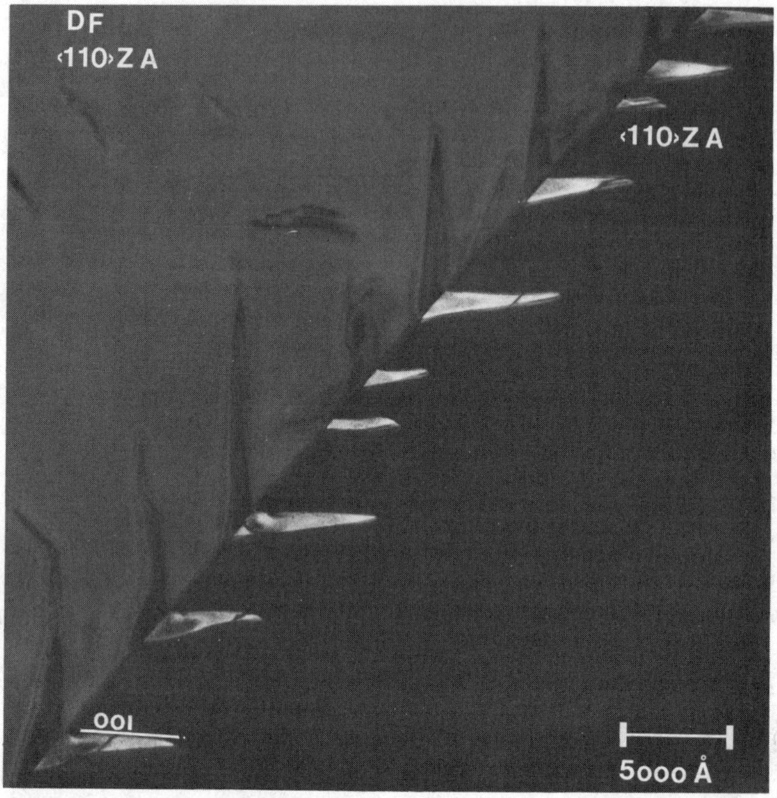

Figure 12. Precipitation of a second phase at the grain boundary of a Ni–18% V alloy heated for 12 hr at 800°C [22].

tinuity between the lattices of the two adjacent grains and the boundary is said to be *incoherent*.

Generally, the interfacial tension of a large-angle grain boundary (e.g., with an angle of misorientation larger than 20° or 25°) is practically independent of that angle. On the other hand, the interfacial tension of a small-angle grain boundary is a sensitive function of misorientation.

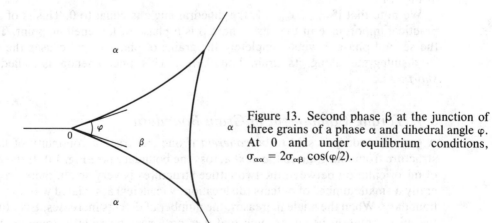

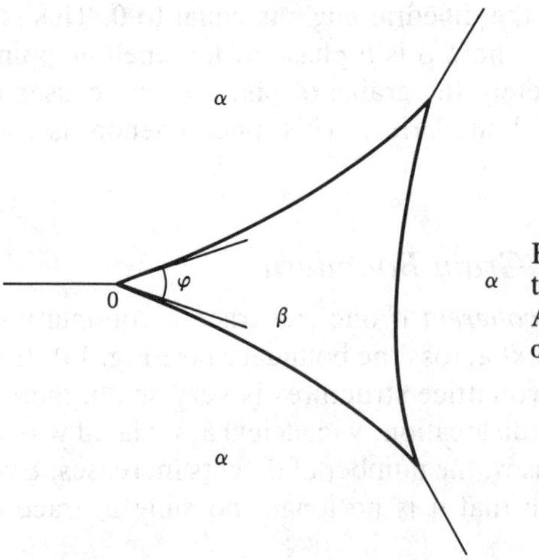

Figure 13. Second phase β at the junction of three grains of a phase α and dihedral angle φ. At 0 and under equilibrium conditions, $\sigma_{\alpha\alpha} = 2\sigma_{\alpha\beta}\cos(\varphi/2)$.

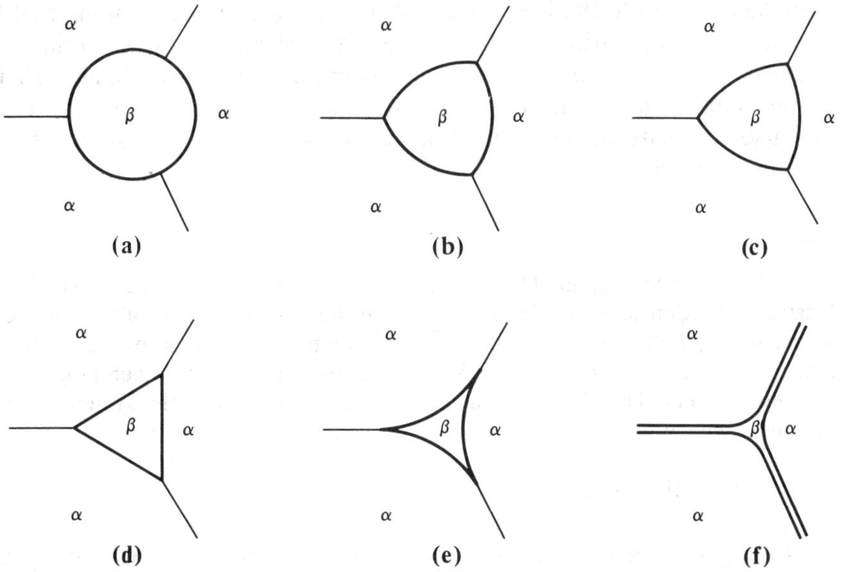

Figure 14. Shape of an inclusion as a function of the dihedral angle φ: (**a**) 180°, (**b**) 135°, (**c**) 90°, (**d**) 60°, (**e**) 30°, (**f**) 0°.

Herring [24] has established that at the junction of three grain boundaries condition (152) is not adequate if the interfacial tensions of the grain boundaries are dependent upon orientation. Instead, equation (152) must be replaced by

$$\sum_{i=1}^{3} \sigma_i \boldsymbol{t}_i + \sum_{i=1}^{3} \frac{\partial \sigma_i}{\partial \alpha_i} \boldsymbol{t}_i \times \boldsymbol{n} = 0 \qquad (158)$$

where σ_i is the interfacial tension of the grain boundary i, $\boldsymbol{t}_i$ is a unit vector in that

Figure 15. Coherent grain boundary in gold [23]. (**a**) A high magnification of a symmetric [110] boundary inclination and (**b**) a line-drawing rendition of the lattice fringes appearing in (**a**). In (**b**) the termination of fringes are indicated by solid dots and the symmetric [110] inclination is denoted by the dashed line. The spacing between the (200) fringes is 2.04 Å.

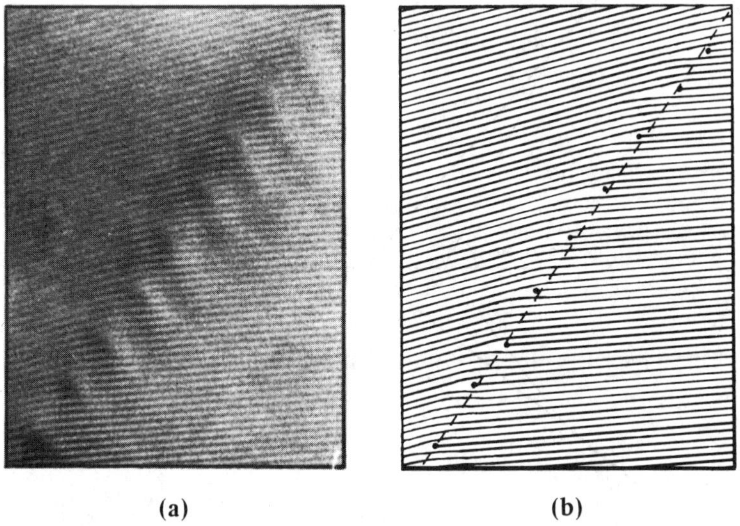

boundary normal to the line of interaction, $\boldsymbol{n}$ is a unit vector along that line, and α_i is an angle measuring the crystallographic orientation of the boundary i. (The vectorial product $\boldsymbol{t}_i \times \boldsymbol{n}$ is a unit vector perpendicular to both $\boldsymbol{t}_i$ and $\boldsymbol{n}$.) The term $\partial\sigma_i/\partial\alpha_i$ implies that each interface strives not only to contract with a tension σ_i but also to rotate to an orientation of lower interfacial tension with a torque measured by $\partial\sigma_i/\partial\alpha_i$.

To demonstrate equation (158), we shall follow a procedure basically similar to that of Herring [24]. Consider, in Fig. 16, an infinitesimal displacement of the intersection line from 0 to Q, parallel to itself and in the plane of interface 1. In this displacement, interface 2 acquires an angle at B, interface 3 at C; BQ and CQ are both much larger than $0Q$ but still infinitesimal. The change in the Helmholtz free energy of the system accompanying the displacement is

$$\delta A = \delta\left(\sum_i \sigma_i A_i\right) \tag{159}$$

if all the grains are constituted by the same phase. Per unit length along the line of intersection (normal to the drawing), equation (159) becomes

$$\delta A = \sigma_1 0Q + \sigma_2(BQ - B0) + \sigma_3(CQ - C0) + \delta\sigma_2\, BQ + \delta\sigma_3\, CQ \tag{160}$$

For the triangle $BQ0$ we may write

$$\frac{Q0}{\sin\delta\alpha_2} = \frac{B0}{\sin(\alpha_2 + \delta\alpha_2)} = \frac{BQ}{\sin\alpha_2} = \frac{B0 - BQ}{\sin(\alpha_2 + \delta\alpha_2) - \sin\alpha_2} \tag{161}$$

Moreover, to the first order variation

$$\sin\delta\alpha_2 = \delta\alpha_2 \tag{162}$$

and

$$\sin(\alpha_2 + \delta\alpha_2) - \sin\alpha_2 = \cos\alpha_2\,\delta\alpha_2 \tag{163}$$

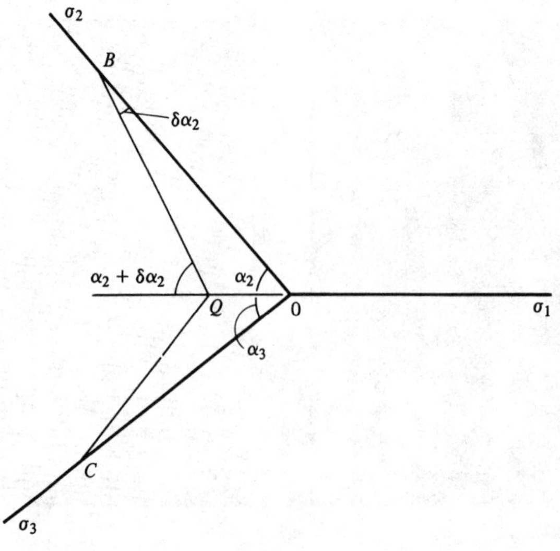

Figure 16. Infinitesimal displacement of three interfaces intersecting along a line. The line of intersection is normal to the plane of the drawing.

Consequently, equation (161) yields

$$B0 - BQ = Q0 \cos \alpha_2 \tag{164a}$$

Similarly,

$$C0 - CQ = Q0 \cos \alpha_3 \tag{164b}$$

Equation (160) becomes

$$\delta A = (\sigma_1 - \sigma_2 \cos \alpha_2 - \sigma_3 \cos \alpha_3)Q0 + \frac{\partial \sigma_2}{\partial \alpha_2} \delta\alpha_2\, BQ + \frac{\partial \sigma_3}{\partial \alpha_3} \delta\alpha_3\, CQ \tag{165}$$

But equations (161) and (164a) show that

$$BQ\, \delta\alpha_2 = Q0 \sin \alpha_2 \tag{166a}$$

Similarly,

$$CQ\, \delta\alpha_3 = Q0 \sin \alpha_3 \tag{166b}$$

Thus,

$$\delta A = \left(\sigma_1 - \sigma_2 \cos \alpha_2 - \sigma_3 \cos \alpha_3 + \sin \alpha_2 \frac{\partial \sigma_2}{\partial \sigma_2} + \sin \alpha_3 \frac{\partial \sigma}{\partial \alpha_3}\right) Q0 \tag{167}$$

At equilibrium $\delta A = 0$, and consequently

$$(\sigma_1 - \sigma_2 \cos \alpha_2 - \sigma_3 \cos \alpha_3) + \left(\sin \alpha_2 \frac{\partial \sigma_2}{\partial \alpha_2} + \sin \alpha_3 \frac{\partial \sigma_3}{\partial \alpha_3}\right) = 0 \tag{168}$$

Equation (168) is the projection of the vector equation (158) on the plane of interface 1. Similar calculations for virtual displacements in the planes of the other interfaces establish the other projections of the vector equation (158). The terms in the second parenthesis of equation (158) (or 168) are important in the case of grain boundaries of small misorientation and, if neglected, can lead to serious errors [24,25].

A description of the experimental techniques measuring interfacial tensions is beyond the scope of this text. A good survey of these techniques may be found in a text by Murr [26]. Here we shall comment only briefly on the *zero-creep technique*, a widespread technique of measuring the surface tension of solids.[2]

The zero-creep technique is based upon the fact that thin wires or foils, when heated to temperatures close to their melting points, will contract if the surface tension is larger than the static stress associated with a load (such as the weight of the wire or foil) and will elongate (creep) if it is smaller. By loading the specimens at several points [27], one obtains a number of data points and the dependence of strain rate versus load. The load corresponding to the zero-creep point then allows the calculation of the surface free energy. The presence of grain boundaries complicates the analysis somewhat in that a second unknown, the interfacial tension of a grain boundary, enters the calculations. However, a relationship between the two interfacial tensions (grain boundary and metal–gas) may be obtained by studying the grooving of the grain boundary emerging at the surface of the specimen and measuring the associated dihedral angle (e.g., see

[2] For reasons of convenience, a solid–gas interface is often referred to as the "surface" of the solid, or as its "free surface."

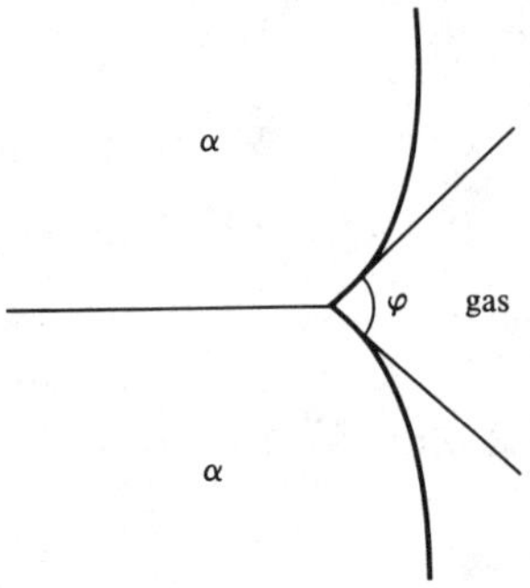

Figure 17. Grain boundary intersecting a free surface and normal to it. At equilibrium, condition (152) yields $\sigma_{\alpha\alpha} = 2\sigma_{\alpha g}\cos(\varphi/2)$.

Fig. 17). Consequently, a combination of the zero-creep technique and dihedral angle measurements yields both the metal–gas and grain boundaries interfacial tensions.

9. Representative Values of Interfacial Tensions

Table 1 lists surface tension values of liquid metals at their fusion temperatures [28]. These values were obtained by a variety of methods and under reducing or neutral conditions (e.g., under hydrogen, argon, and vacuum). For any given metal, there is often considerable scatter between the values obtained by different investigators, the most frequent and important source of errors being the presence of impurities. It is usually difficult to obtain an accuracy much better than 2%.

Table 1. Surface Tension Values (dyn/cm) of Liquid Metals at their Melting Points

Li 398 *454*	Be 1390 *1560*											B	C	N	O	F	Ne
Na 191 *371*	Mg 559 *922*											Al 914 *933*	Si 865 *1685*	P	S	Cl	Ar
K 115 *336*	Ca 361 *1112*	Sc	Ti 1650 *1943*	V 1950 *2175*	Cr 1700 *2130*	Mn 1090 *1517*	Fe 1872 *1809*	Co 1873 *1768*	Ni 1778 *1726*	Cu 1360 *1357*	Zn 782 *693*	Ga 718 *303*	Ge 621 *1210*	As	Se 106 *494*	Br	Kr
Rb 85 *313*	Sr 303 *1041*	Y	Zr 1480 *2125*	Nb 1900 *2740*	Mo 2250 *2890*	Tc	Ru 2250 *2523*	Rh 2000 *2233*	Pd 1500 *1825*	Ag 903 *1234*	Cd 570 *594*	In 556 *430*	Sn 544 *505*	Sb 367 *904*	Te 180 *723*	I	Xe
Cs 70 *302*	Ba 277 *1002*	La 720 *1193*	Hf 1630 *2500*	Ta 2150 *3287*	W 2500 *3680*	Re 2700 *3453*	Os 2500 *3300*	Ir 2250 *2716*	Pt 1800 *2042*	Au 1140 *1336*	Hg 498 *234*	Tl 464 *577*	Pb 468 *601*	Bi 378 *545*	Po	At	Rn
Fr	Ra	Ac															
			Ce 740 *1071*	Pr	Nd 689 *1289*	Pm	Sm	Eu	Gd 810 *1585*	Tb	Dy	Ho	Er	Tm	Yb	Lu	
			Th 978 *2028*	Pa	U 1550 *1405*	Np	Pu 550 *913*	Am	Cm	Sk	Cf	Es	Fm	Md	No	Lw	

Surface tension values from a review article by Allan [28]. The numbers in italics represent fusion temperatures (°K); they were taken from the compilation of Hultgren et al [31].

Aqueous solutions have surface tension values generally below those of liquid metals. Water itself has a surface tension equal to 77 dyn/cm at 0°C. Organic liquids have surface tensions of the order of 20–40 dyn/cm [29]. The surface tensions of liquid metallic oxides and silicate glasses have orders of magnitude comparable to those of liquid metals. Table 2 illustrates the range of values encountered for the surface tensions of some halides, oxides, and sulfides [30].

Solid metals at their melting points have surface tension values about 20% higher than those of the corresponding liquids. For a large-angle grain boundary, the interfacial tension is about one-third that of the metal's free surface [32].

We also note that interfacial tension values are temperature dependent, and usually decrease with temperature. For example, $d\sigma/dT$ for many cubic metals is of the order of -0.45 dyn/cm °K.

For a more complete survey of interfacial tension values, correlations with physical properties and other methods of estimation, the reader is referred to the text by Murr [26]. However, in order to provide the reader with some physical insight into the atomic origin of surface energies, we shall briefly present the results of a simple model for liquid metals. It is convenient to introduce the *molar surface energy* of the liquid σ_M:

$$\sigma_M = a(\sigma - T\,d\sigma/dT) \tag{169}$$

where a is the average area occupied by an atom at the surface of the metal.

Table 2. Surface Tensions of Some Liquid Halides, Oxides, and Sulfides

Substance	T (°C)	σ (dyn/cm)
NaCl	1000	98
NaBr	900	91
NaI	700	84
KCl	900	90
KBr	800	85
KI	800	69
RbCl	828	89
RbBr	831	81
RbI	772	72
CsF	826	96
CsCl	830	78
CsBr	808	72
CsI	821	63
$NaNO_3$	350	115
KNO_3	350	109
$PbCl_2$	500	137
FeO (in equilibrium with Fe)	1420	585
Al_2O_3	2050	690
SiO_2	1800	307
GeO_2	1150	259
P_2O_5	100	60
B_2O_3	1000	82
Cu_2S	1200	400
NiS	1200	577
CoS	1200	488
PbS	1200	200
Sb_2S_3	1200	94

Data from Richardson [30].

Assuming that the energy of the liquid may be calculated as a sum of pairwise interaction energies, the following relationship is then derived:

$$\sigma_M = (m/Z)\Delta H_{vap} \tag{170}$$

where m is the number of broken bonds when an atom is brought from the bulk to the surface, Z is the number of nearest neighbors to an atom in the bulk, and ΔH_{vap} is the enthalpy of vaporization of the liquid (at 0°K). If the liquid is assumed to have an fcc configuration, m/Z is equal to 0.25, and if it is assumed to have a bcc configuration, m/Z is equal to 0.36; in both cases the surface plane is chosen to be the closest packed plane of the lattice.

A plot of σ_M vs ΔH_{vap}, from Allen [33], is shown in Fig. 18. The correlation between the two quantities is well verified, but for most liquid metals the ratio is closer to an average value of 0.15.

Several factors may explain this quantitative discrepancy. A likely one is that surface atoms, being subject to a one-sided attraction, lie closer to their underlying neighbors than they do in bulk. Consequently, surface atoms must be bound to their sublaying atoms by greater binding energies (e.g., greater by about 13% in Oriani's estimate [34]). More complex factors are related to the electronic gradient at the surface [35].

The correlation of ΔH_{vap} with σ has been modified by Shapksi [36] and Grosse [37]. They proposed a correlation of the surface tension σ of liquid metals at their melting points with ΔH_{vap} divided by $(V_m)^{2/3}$, where V_m is the molar volume of the liquid (extrapolated to 0°K):

$$(\sigma)_{T_f} = c[\Delta H_{vap}/(V_m)^{2/3}]_{0°K} \tag{171}$$

Figure 18. Experimental test of equation (170). (*From Allen* [*33*].)

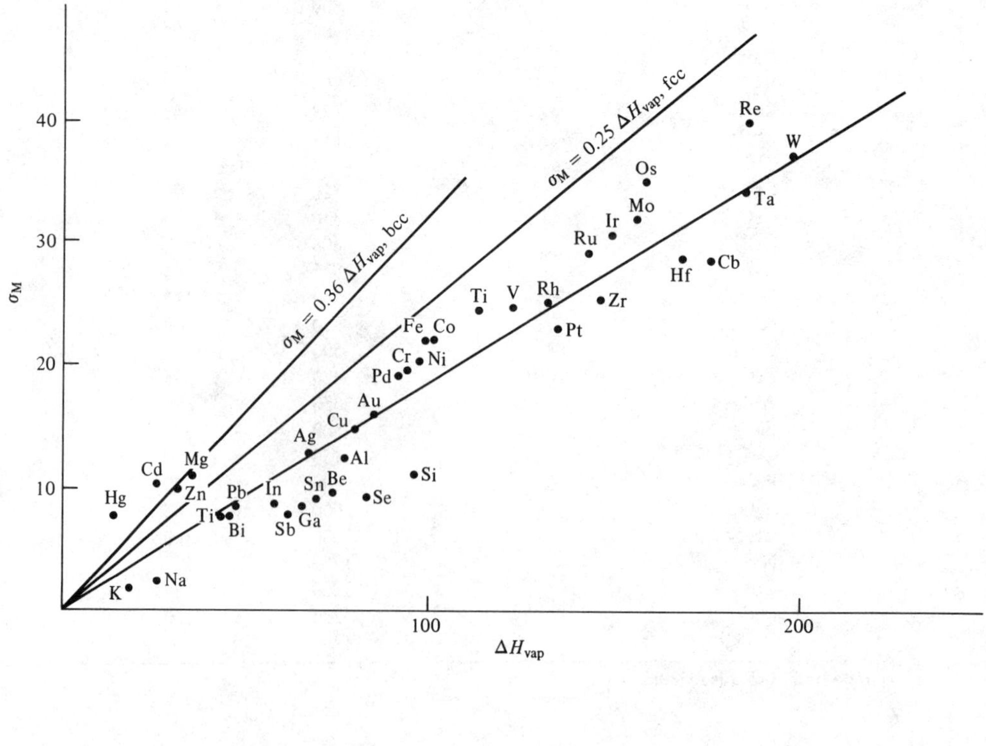

This correlation has been justified by Miedema [38] on the basis of discontinuities in the electron density at the boundary of a Wigner–Seitz cell.

10. Summary

In Chapter II we derived conditions for the equilibrium of a heterogeneous system without taking into account the interfaces between its phases. In this chapter we established that the equality of temperatures and the equality of the chemical potentials of any component are not affected by consideration of the interface. However, if the interface is curved, the pressures on each side of that interface are no longer equal.

A physical interface is a region of heterogeneity of finite thickness. However, thermodynamically, a system of two phases separated by such an interface may be replaced by a model system in which the two phases in contact are homogeneous right up to a dividing geometrical surface Σ. For the model system to be equivalent to the real system, the position of Σ is not arbitrary and a surface tension force σ acting in the surface Σ has to be introduced. At equilibrium, the pressures on both sides of Σ are related by the equation of Laplace:

$$P' - P'' = \sigma\left(\frac{1}{r_1} + \frac{1}{r_2}\right) \tag{172}$$

where r_1 and r_2 are the two principal radii of curvature. We note that the pressure is higher on the side of Σ where the centers of curvature are lying. Also, this equation is not rigorously valid for a solid, since the stress tensor within a crystalline solid is not isotropic and does not reduce to a pressure term.

Straightforward balances in the model system introduce surface excess quantities, e.g.,

$$n_i^\sigma = n_i - n_i' - n_i'' \tag{173}$$

The potential Ω, defined as

$$\Omega = A - \sum_i n_i \mu_i \tag{174}$$

has a corresponding surface excess potential Ω^σ which is simply related to the surface tension

$$\Omega^\sigma = \sigma s \tag{175}$$

where s is the area of the interface considered. We have seen that this equation provides a more convenient definition of the surface tension of a solid than the definition based on the Helmholtz free energy,

$$\sigma = \left(\frac{\partial A}{\partial s}\right)_{T, V', V'', n_i} \tag{176}$$

In calculating the variance of a two-phase system, the mean curvature of the dividing surface yields an additional degree of freedom. The chemical potentials of the components depend on that curvature. In particular, this dependence gives rise to "Ostwald ripening."

The surface tension of a solid depends on the lattice orientation of its surface.

Wulff's relationships,

$$\frac{\sigma^{(1)}}{h^{(1)}} = \frac{\sigma^{(2)}}{h^{(2)}} = \cdots = \frac{\sigma^{(N)}}{h^{(N)}} \tag{177}$$

where the $h^{(i)}$ are defined in Section 6.1, determine the equilibrium shapes of small crystals. Surface tensions also affect the morphologies of inclusions and grain boundaries. We have analyzed the equilibrium conditions for these forces and the consequences of these conditions in some simple cases. For a detailed investigation of this interesting but complex subject, it is again recommended that the reader consult the references cited in Section 8. In addition, we note that a promising new treatment of that problem has been given by Hoffman and Cahn [39].

Problems

1. On blowing air through a water–soap mixture, bubbles are formed and carried away. The air–liquid interfacial tension is equal to 40 dyn/cm. Calculate in mm Hg the difference in the pressures inside a bubble of 1 mm radius and in the air surrounding the bubble.

2. A liquid aluminum alloy with a density of 2.350 g/cm³ is held in a furnace at 700°C under neutral atmosphere. A capillary tube is introduced in the metallic bath and the liquid alloy is observed to rise in the capillary. The liquid completely wets the inner walls of the tube and the bottom of the meniscus (approximately spherical) is measured to be at a height of 18.6 cm with respect to the surface of the liquid alloy. The inside diameter of the capillary tube has been measured to be equal to 0.80 mm. Calculate the surface tension of the liquid alloy.

3. In a room at 26°C liquid mercury has been spilled. Calculate the vapor pressure of droplets of 0.1 mm radius, knowing that bulk liquid mercury at that temperature has a vapor pressure equal to 0.002 mm Hg and a density of 13.5 g/cm³.

4. A two-dimensional Wulff plot for a solid alloy shows a simple square. Draw the corresponding two-dimensional equilibrium shape of a small crystal.

5. **a.** The work of adhesion W of a phase α to a phase β has been defined as the work per unit surface that must be exerted to separate the two phases, thereby creating two new interfaces α–gas and β–gas. Assume that α and β are two pure components A and B with no mutual miscibility, and that the gas is adsorbed neither by A nor by B. Calculate W in terms of interfacial tensions.
b. If α is a pure liquid A, and β is a pure solid B, is it possible to measure W by the sessile drop technique?
c. Should the results be modified if α and β are not pure components, but solutions?

6. Girifalco and Good [40] have estimated the value of the interfacial tension σ_{AB} of two immiscible liquids A and B by the equation $\sigma_{AB} = \sigma_A + \sigma_B - 2\phi(\sigma_A\sigma_B)^{1/2}$, where ϕ is a function of the molar volumes of the two liquids and, empirically, is found to vary between 0.5 and 1.15. Assume that the equation is still valid when one of the phase is solid, and explain why water wets most clean metals.

7. **a.** A liquid B is placed on a liquid A. Assuming that the two liquids are immiscible, derive a criterion in terms of the surface tensions σ_A, σ_B, and σ_{AB} for the liquid B to spread on A.

b. At 20°C, the surface tension of benzene is equal to 28.9 dyn/cm, that of water is equal to 72.8 dyn/cm, and the interfacial tension of benzene on water is equal to 35.0 dyn/cm. Initially, there is little mass transfer between the two phases, and the surface tensions have the values quoted above. With time, however, the surface tension of water in contact with benzene decreases to a value of 62.2 dyn/cm, while the other two surface tensions remain essentially unaffected. Describe what happens to the spreading of the liquid film of benzene.

8. A nickel specimen is held under helium at 1200°C. The dihedral angle φ of a grain boundary perpendicular to the free surface of the specimen is measured to be equal to 157.5°. Knowing that the surface free energy of nickel is 2280 erg/cm^2 at 1060°C and decreases with T at the rate of -0.55 erg/(cm^2 °K), calculate the grain boundary's interfacial tension at 1200°C.

9. The water–ice interfacial tension has been evaluated at 22 dyn/cm. In very clean water, supercooling has been observed to reach 40°C before crystallization occurs. Estimate the corresponding critical radius for nucleation of a spherical particle of ice. The enthalpy of fusion of ice is 1436 cal/mol. (The equations of Section III.6.4 may be used.)

10. **a.** The rate of nucleation for crystallization from a melt has been expressed by the equation [41] $I = n(kT/h)\exp(-Q_D/kT)\exp(-\Delta G^*/kT)$ where n is the number of molecules of liquid per cubic centimeter, k is Boltzmann's constant, h is Planck's constant, Q_D is the activation energy for self-diffusion, and ΔG^* is the activation energy for nucleation [or $G(r_c) - G(0)$ in Section III.6.4]. Show that when the undercooling $\Delta T = T_f - T$ increases, the rate of nucleation passes by a maximum. Neglect the temperature dependence of σ and of the molar volume.

b. For gold, the solid–liquid interfacial tension has been estimated as 0.132 N/m. Gold melts at 1336°K with a heat of fusion equal to 2955 cal/mol. The density of solid gold is 19 g/cm^3. The activation energy for self-diffusion may be assumed to be of the order of 10 kcal/mol. The maximum undercooling has been observed to be about 260°K. Check whether the result derived above agrees semiquantitatively with these data.

References

1. J. W. Gibbs, *Collected Works*. Yale University Press, New Haven, CT, 1948, Vol. 1.
2. P. S. Laplace, *Mécanique Céleste*. Paris, 1806, Vol. 10 suppl.
3. T. Young, *Phil. Trans. Roy. Soc. London* **95,** 65 (1805); reprinted with additions in *Works of Dr. Young* (Peacock, ed.). London, 1855, Vol. 1, p. 418.
4. W. Fulks, *Advanced Calculus,* 2nd ed. Wiley, New York, 1969, p. 126.
5. R. Defay and I. Prigogine, with A. Bellemans and D. H. Everett, *Surface Tension and Adsorption*. Wiley, New York, 1966.
6. W. Thomson (Lord Kelvin), *Phil. Mag.* **42,** 448 (1871).
7. G. Bernard and C. H. P. Lupis, *Met. Trans.* **2,** 555 (1971).
8. F. O. Koenig, *J. Chem. Phys.* **18,** 449 (1950).
9. T. L. Hill, *An Introduction to Statistical Thermodynamics*. Addison-Wesley, Reading, MA, 1960, p. 481.
10. G. Wulff, *Z. Krist.* **34,** 449 (1901).
11. C. Herring, *Structure and Properties of Solid Surfaces* (R. Gomer and C. S. Smith, ed.). University of Chicago Press, Chicago, 1952.
12. C. Herring, *Phys. Rev.* **82,** 87 (1951).

13. W. W. Mullins, *Metal Surfaces: Structure, Energetics and Kinetics*. Am. Soc. Metals, Metals Park, OH, 1963, Ch. 2.
14. P. G. Shewmon and W. M. Robertson, *Metal Surfaces: Structure, Energetics and Kinetics*. Am. Soc. Metals, Metals Park, OH, 1963, Ch. 3.
15. A. Dupré, *Théorie Mécanique de la Chaleur*. Paris, 1869, p. 207.
16. T. Bashforth and J. C. Adams, *An Attempt to Test the Theory of Capillary Action*. Cambridge University Press, Cambridge, England, 1883.
17. G. Bernard and C. H. P. Lupis, *Met. Trans.* **2,** 555–559 (1971).
18. R. E. Johnson, *J. Phys. Chem.* **63,** 1655 (1959).
19. R. N. Wenzel, *Ind. Eng. Chem.* **28,** 988 (1936).
20. B. Gallois, Ph.D. thesis, Carnegie-Mellon University, 1979.
21. C. S. Smith, *Trans. Am. Inst. Min. Mat. Eng.* **175,** 15 (1948).
22. D. Laughlin and R. J. Rioja, private communication, Carnegie-Mellon University, Pittsburgh, 1979.
23. F. Cosandey, Y. Komem, C. L. Bauer, and C. B. Carter, *Scripta Met.* **12,** 577–582 (1978).
24. C. Herring, *Physics of Powder Metallurgy* (W. E. Kingston, ed.). McGraw-Hill, New York, 1951.
25. M. McLean and B. Gale, *Phil. Mag.* **20,** 1033 (1969).
26. L. E. Murr, *Interfacial Phenomena in Metals and Alloys*. Addison-Wesley, Reading, MA, 1975.
27. E. D. Hondros, *Proc. Roy Soc. London Ser. A* **286,** 479 (1965).
28. B. C. Allan, *Liquid Metals, Chemistry and Physics* (S. Z. Beer, ed.). Dekker, New York, 1972, Ch. 4.
29. *Handbook of Chemistry and Physics,* 60th ed., CRC Press, Boca Raton, FL, 1979, p. F-46.
30. F. D. Richardson, *Physical Chemistry of Melts in Metallurgy*. Academic, London, 1974, Vol. 2, p. 140.
31. R. Hultgren, P. D. Desai, D. T. Hawkins, M. Gleiser, K. K. Kelley, and D. D. Wagman, *Selected Values of the Thermodynamic Properties of the Elements*. Am. Soc. Metals, Metals Park, OH, 1973.
32. R. A. Swalin, *Thermodynamics of Solids,* 2nd ed. Wiley, New York, 1972, p. 244.
33. B. C. Allen, *Trans. Met. Soc. AIME* **227,** 1175 (1963).
34. R. A. Oriani, *J. Chem. Phys.* **18,** 575 (1950).
35. V. K. Semechenko, *Surface Phenomena in Metals and Alloys*. Addison-Wesley, Reading, MA, 1962.
36. A. S. Shapski, *J. Chem. Phys.* **16,** 386–389 (1948).
37. A. A. V. Grosse, *J. Inorg. Nucl. Chem.* **6,** 1349 (1964).
38. A. R. Miedema, *Z. Metallk.* **69,** 183 (1978).
39. D. W. Hoffman and J. W. Cahn, *Surface Sci.* **31,** 368 (1972).
40. L. A. Girifalco and R. J. Good, *J. Phys. Chem.* **61,** 904 (1959).
41. A. W. Adamson, *Physical Chemistry of Surfaces*. Wiley, New York, 1976, p. 377.

Selected Bibliography

J. W. Gibbs, *Collected Works*. Yale University Press, New Haven, CT, 1948, Vol. 1.

R. Defay and I. Prigogine, with A. Bellemans and D. H. Everett, *Surface Tension and Adsorption*. Wiley, New York, 1966.

L. E. Murr, *Interfacial Phenomena in Metals and Alloys*. Addison-Wesley, Reading, MA, 1975.

N. K. Adams, *The Physics and Chemistry of Surfaces,* 3rd ed. Oxford University Press, London, 1941.

A. W. Adamson, *Physical Chemistry of Surfaces,* 3rd ed. Wiley, New York, 1976.

R. Aveyard and D. A. Haydon, *An Introduction to the Principles of Surface Chemistry.* Cambridge University Press, Cambridge, England, 1973.

J. J. Bikerman, *Surface Chemistry, Theory and Applications,* Academic, New York, 1958.

Chemistry and Physics of Interfaces, Vol. 1 based on a Symposium on Interfaces, 15–16 June 1964; Vol. 2 based on the 5th Annual State of the Art Symposium on Interfaces, 11–12 June 1968. Am. Chem. Soc., Washington, DC, 1965, 1971.

XIV Adsorption

In examining the properties of interfaces, we devoted most of the preceding chapter to the concept of surface tension and we considered it primarily as a mechanical force. In this chapter we shall focus our analysis on the composition of an interface and how it differs from the bulk compositions of the adjoining phases. Obviously, however, surface tensions and surface compositions are not unrelated: the Gibbs adsorption equation (Section XIII.3) relates them.

To describe the composition of an interface without ambiguity is not a straightforward task. Several thermodynamic functions are useful, but they must be defined with care.

1. Surface Excess Quantities and the Position of the Dividing Surface

In Section XIII.1 we showed that a system of two phases α and β separated by a three-dimensional interface may be replaced by a model system consisting of the phases α and β assumed homogeneous right up to a two-dimensional dividing surface Σ (see Fig. 1). We also showed that for the model system to be mechanically equivalent to the real system the surface Σ must be in a nonarbitrary position, often referred to as *the surface of tension*.

A surface excess quantity, such as the adsorption of component i, is defined by an equation of the type

$$\Gamma_i = \frac{n_i^\sigma}{s} = \frac{1}{s}(n_i - n_i' - n_i'') = \frac{1}{s}(n_i - c_i'V' - c_i''V'') \quad (1)$$

Figure 1. **(a)** Real and **(b)** model systems. In the model system, the phases α and β are assumed to have constant concentrations up to the dividing surface Σ. The number of moles adsorbed n_i^σ is equal to the area in black (lower left figure).

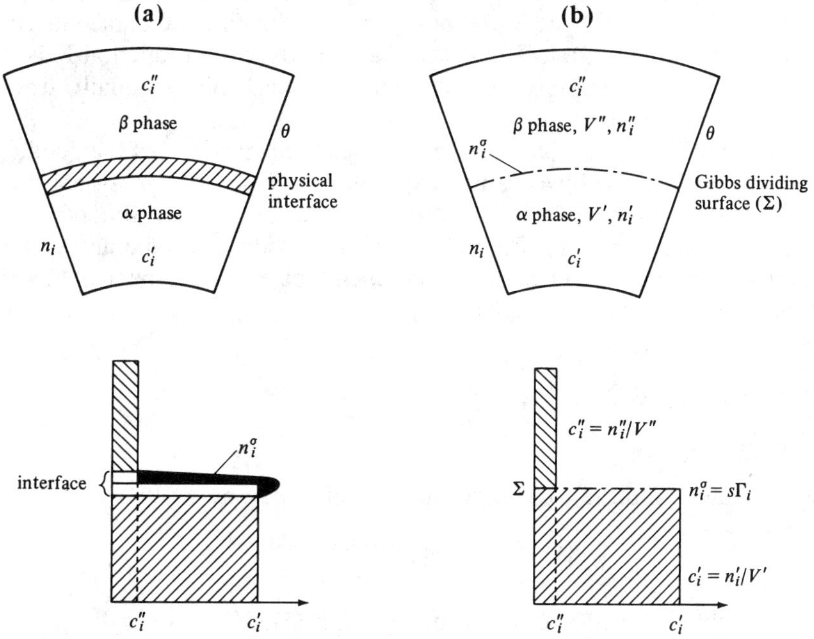

The terms appearing in this equation are identified in Fig. 1 (and in Section XIII.1.2). It is obvious that any change in the position of the dividing surface Σ would affect the value of Γ_i defined by equation (1). However, if Σ is defined as the surface of tension, there is then theoretically no ambiguity on the value of Γ_i or any other surface excess quantity.

In the particular case of a planar interface, we have noted that the mechanical equivalence of the real system and the model system is no longer dependent upon the exact position of Σ. Thus, the position of Σ may be chosen arbitrarily without affecting the consistency of the model system. It is worthwhile to examine whether this arbitrariness has a significant effect on the values of surface excess quantities such as Γ_i.

Equation (1) may be rewritten

$$\Gamma_i = \frac{1}{s}\left[n_i - c_i'V - (c_i'' - c_i')V''\right] \tag{2}$$

If the position of the dividing surface is moved along its normal by an increment δz, such that V' increases by the amount $s\,\delta z$, the value of Γ_i changes by the amount

$$\delta\Gamma_i = (c_i'' - c_i')\,\delta z \tag{3}$$

Let us consider, for example, the interface of a liquid silver–copper alloy at 1108°C in equilibrium with a gas phase. A concentration of silver in the alloy equal to $X_{Ag} = 0.3$ corresponds to $c'_{Ag} = 0.032$ mol/cm^3. The concentration of silver in the gas phase c''_{Ag} is negligible. A shift of 10 Å in the position of the dividing surface creates a change:

$$\begin{aligned}\delta\Gamma_{Ag} &= -0.032 \times 10^{-7} \quad \text{mol/cm}^2 \\ &= -3.2 \times 10^{-5} \quad \text{mol/m}^2\end{aligned} \tag{4}$$

This change may be compared with an experimental value of Γ_{Ag} equal to 2.3×10^{-5} mol/m^2 [1] and corresponding to a conventional choice of the dividing surface (see Section 2). We thus see that the adsorption function Γ_i is extremely sensitive to the position of the dividing surface. This is equally true of other surface excess properties.

Clearly, it is not possible to define physically the position of the dividing surface with a precision which would not affect significantly the values of surface excess properties. Consequently, one must seek thermodynamic functions which are invariant with respect to the position of the dividing surface and can still satisfactorily describe surface phenomena. Such functions have been defined by Gibbs [2] and will now be presented.

2. Relative Adsorptions

2.1. *Definition*

Like Γ_i in equation (2), the adsorption Γ_1 of 1 may be written

$$\Gamma_1 = \frac{1}{s}\left[n_1 - c_1'V - (c_1'' - c_1')V''\right] \tag{5}$$

In the right-hand sides of equations (2) and (5), V'' is the only quantity which depends on the position of the dividing surface Σ. Eliminating V'' between these two equations, we obtain

$$\Gamma_i - \Gamma_1 \frac{c_i' - c_i''}{c_1' - c_1''} = \frac{1}{s}\left[(n_i - c_i'V) - (n_1 - c_1'V)\frac{c_i' - c_i''}{c_1' - c_1''}\right] \tag{6}$$

Although both Γ_i and Γ_1 depend on the position of Σ, we see that their combination in the left-hand side of equation (6) is independent of that position. This combination is called the *relative adsorption of i with respect to 1* and is denoted $\Gamma_i^{(1)}$:

$$\Gamma_i^{(1)} = \Gamma_i - \Gamma_1 \frac{c_i' - c_i''}{c_1' - c_1''} \tag{7}$$

Since $\Gamma_i^{(1)}$ is independent of the position of Σ, we may choose this position to correspond to a value of Γ_1 equal to zero (see Fig. 2). Thus, the Gibbs relative adsorption $\Gamma_i^{(1)}$ is also the adsorption of i at the surface where the adsorption of 1 is zero.

2.2. Simplified Form

In the case of a solid–gas or a liquid–gas interface, the concentrations of the various components in the gas phase are usually negligible compared to those in the solid or liquid. Consequently, equation (7) may be rewritten

$$\Gamma_i^{(1)} = \Gamma_i - \Gamma_1(c_i'/c_1') \tag{8}$$

or

$$\Gamma_i^{(1)} = \Gamma_i - \Gamma_1(X_i'/X_1') \tag{9}$$

In order to gain a better intuitive understanding of the relative adsorptions, let us place the dividing surface Σ immediately below the region of heterogeneity (on the side of the solid or liquid; see Fig. 3). We would then have

$$\Gamma_i^* = \frac{n_i^\sigma}{s} \simeq \frac{n_i - n_i'}{s} \tag{10}$$

Figure 2. Position of the dividing surface Σ which gives zero adsorption of component 1 for two concentration profiles. n_1^σ is the algebraic sum of the hatched area (counted negative) and the black area (counted positive).

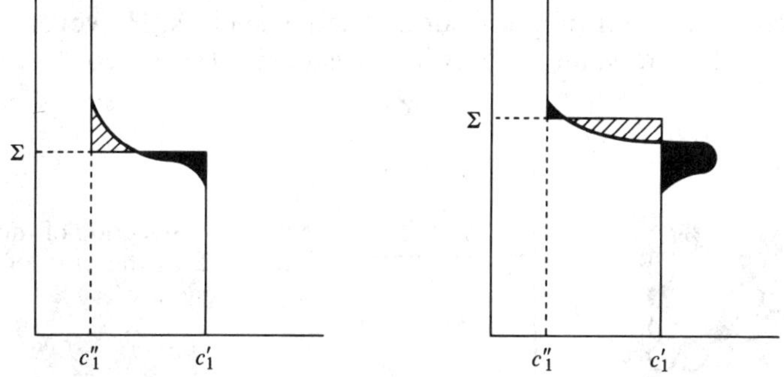

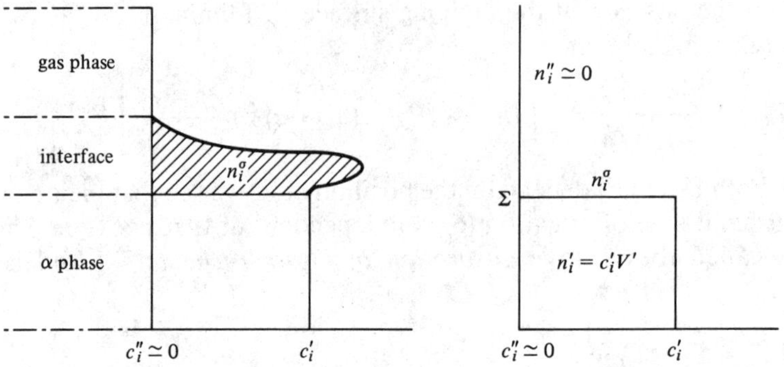

Figure 3. Convenient selection of the dividing surface Σ in the case of a solid–gas or liquid–gas interface.

and n_i^σ would simply be the number of moles of i in the region of heterogeneity (surface layer). The relative adsorption of i with respect to 1 would become

$$\Gamma_i^{(1)} = \Gamma_i^* - \Gamma_1^*(X_i'/X_1') \tag{11}$$

It would be equal to zero when

$$\frac{\Gamma_i^*}{\Gamma_1^*} = \frac{X_i'}{X_1'} \tag{12}$$

that is, when the ratio of the concentrations of i and 1 in the surface layer is identical to that in the bulk of the solid or liquid. If $\Gamma_i^{(1)}$ is positive, the surface layer is relatively richer than the bulk in component i. Conversely, if $\Gamma_i^{(1)}$ is negative, the surface layer is relatively poorer than the bulk in component i.

We note that if we represent our system by a model in which the region of heterogeneity is confined to one monolayer (see Fig. 4), Γ_i^* would be equal to

$$\Gamma_i^* = \frac{n^{(m)}}{s} X_i^{(m)} \tag{13}$$

where $n^{(m)}/s$ is the number of moles in the monolayer per unit area, and where $X_i^{(m)}$ is the mole fraction of i in the monolayer. $\Gamma_i^{(1)}$ would become

$$\Gamma_i^{(1)} = \frac{n^{(m)}}{s}\left(X_i^{(m)} - X_1^{(m)}\frac{X_i'}{X_1'}\right) \tag{14}$$

Because of its simplicity, the monolayer model has been widely adopted to interpret many surface phenomena. Physically, it also appears to represent fairly well the adsorption behavior of many surface-active species. However, its indiscriminate use may lead to some serious inconsistencies [3].

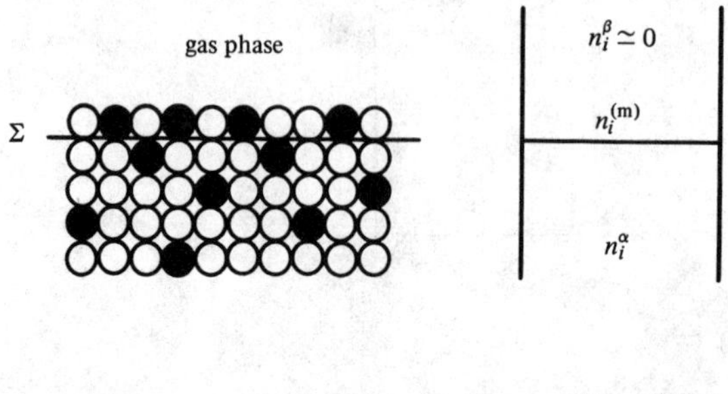

Figure 4. Selection of the dividing surface Σ in the case of a monolayer model.

3. Relative Functions and the Gibbs Adsorption Equation

As with the relative adsorption of i with respect to 1, it is possible to define other relative functions which are independent of the position of the dividing surface. For example, the relative entropy with respect to component 1 is defined as

$$s^{(1)} = s^{\sigma} - \Gamma_1 \frac{s' - s''}{c_1' - c_1''} \tag{15}$$

where s' and s'' are the entropy densities in the phases α and β (e.g., $s' = S'/V'$) and s^{σ} is the surface excess entropy per unit area ($s^{\sigma} = S^{\sigma}/s$).[1]

In Section 1 we emphasized that it is only in the case of planar surfaces that the position of the dividing surface is arbitrary. In addition, under most experimental conditions under which adsorptions and other surface properties are being measured, the curvature of the interface being investigated is sufficiently small that a planar interface may be safely assumed. Consequently, we shall restrict our presentation in the rest of this chapter to the case of planar surfaces exclusively.

For each homogeneous phase on either side of an interface, the Gibbs–Duhem equation may be written

$$s'\,dT - dP + \sum_{i=1}^{m} c_i'\,d\mu_i = 0 \tag{16a}$$

$$s''\,dT - dP + \sum_{i=1}^{m} c_i''\,d\mu_i = 0 \tag{16b}$$

Consequently,

$$(s' - s'')\,dT + \sum_{i=1}^{m} (c_i' - c_i'')\,d\mu_i = 0 \tag{17}$$

The Gibbs adsorption equation (Section XIII.3) may also be written

$$s^{\sigma}\,dT + \sum_{i=1}^{m} \Gamma_i\,d\mu_i + d\sigma = 0 \tag{18}$$

Eliminating $d\mu_1$ between equations (17) and (18), we obtain

$$d\sigma = -\left(s^{\sigma} - \Gamma_1 \frac{s' - s''}{c_1' - c_1''}\right) dT - \sum_{i=2}^{m} \left(\Gamma_i - \Gamma_1 \frac{c_i' - c_i''}{c_1' - c_1''}\right) d\mu_i \tag{19}$$

or

$$d\sigma = -s^{(1)}\,dT - \sum_{i=2}^{m} \Gamma_i^{(1)}\,d\mu_i \tag{20}$$

We note that for a binary system

$$\Gamma_2^{(1)} = -\left(\frac{\partial\sigma}{\partial\mu_2}\right)_T = -\frac{1}{RT}\left(\frac{\partial\sigma}{\partial \ln a_2}\right)_T \tag{21}$$

[1] s without superscript denotes an area.

Thus, $\Gamma_2^{(1)}$ may be obtained from the dependence of the surface tension on the activity of the solute 2 at any given temperature. This equation is the basis for most adsorption measurements associated with liquid solutions.

4. Reduced Adsorptions

Component 1 plays a special role in the relative adsorptions we have defined. More symmetric measures of adsorption are often more attractive. We recall that in order to find a thermodynamic function which not only measures the adsorption of i but is also independent of the position of the dividing surface, we eliminated the quantity V'' between the expressions of Γ_i and Γ_1 in equations (2) and (5). It is also possible to eliminate V'' between equation (2) and the following equation, which expresses the adsorption of all the components at the interface:

$$\Gamma = \frac{n^\sigma}{s} = \frac{1}{s}(n - c'V' - c''V'') = \frac{1}{s}[n - c'V - (c'' - c')V''] \tag{22}$$

The result is

$$\frac{n_i^\sigma}{s} - \frac{n^\sigma}{s}\left(\frac{c_i'' - c_i'}{c'' - c'}\right) = \frac{1}{s}\left[(n_i - c_i'V) - (n - c'V)\left(\frac{c_i'' - c_i'}{c'' - c'}\right)\right] \tag{23}$$

The *reduced adsorption of i*,

$$\overline{\Gamma}_i = \Gamma_i - \Gamma\left(\frac{c_i'' - c_i'}{c'' - c'}\right) \tag{24}$$

is thus a function which is independent of the position of the dividing surface and possesses the required symmetry.

We note that

$$\sum_{i=1}^{m} \overline{\Gamma}_i = 0 \tag{25}$$

It is also straightforward to verify that

$$\Gamma_i^{(1)} = \overline{\Gamma}_i - \overline{\Gamma}_1\left(\frac{c_i' - c_i''}{c_1' - c_1''}\right) \tag{26}$$

The reduced entropy $\overline{S}$ may be similarly defined:

$$\overline{S} = s^\sigma - \Gamma\frac{s' - s''}{c' - c''} \tag{27}$$

It is easily shown that the Gibbs adsorption equation then becomes

$$d\sigma = -\overline{S}\,dT - \sum_{i=1}^{m} \overline{\Gamma}_i\,d\mu_i \tag{28}$$

Reduced functions are of special interest in the case of a solid–gas or liquid–gas interface. Because densities in a gas phase are generally much smaller than in the solid or liquid state (except near the critical point), $\overline{\Gamma}_i$ becomes

$$\overline{\Gamma}_i = \Gamma_i - X_i'\Gamma \tag{29}$$

where X_i' is the mole fraction of i in the condensed phase. With the choice of the dividing surface at the boundary between the surface layer and the solution, $\overline{\Gamma}_i$ takes a simple and convenient form:

$$\overline{\Gamma}_i = \frac{n^{(s)}}{s}(X_i^{(s)} - X_i') \tag{30}$$

where $n^{(s)}/s$ is the total number of moles in the surface layer per unit area. We see that a positive sign for $\overline{\Gamma}_i$ means that concentration of i in the surface layer is higher than in the bulk. The segregation of i is no longer calculated relative to a given component and this is a decided advantage.

5. Alternative Thermodynamic Treatment of a Planar Interface

Cahn [4] has proposed a somewhat different procedure for examining the thermodynamic properties of a planar interface. His treatment has the advantages of simplicity and convenience. However, it does not apply to curved surfaces.

Let us consider in Fig. 5 a system θ of two phases α and β separated by a planar interface. The system is divided in three parts. Parts (1) and (2) contain only homogeneous portions of the phases α and β while part (3) contains all of the heterogeneous region. The boundaries of part (3) lie within the homogeneous regions of the α and β phases but are close to the surface layer.

In Section XIII.3, we saw that the energy of the system θ is a homogeneous function of degree 1 with respect to the variables S, V, n_i, and s. Consequently,

$$E = TS - PV + \sum_{i=1}^{m} \mu_i n_i + \sigma s \tag{31}$$

or

$$\sigma = \frac{1}{s}\left[(E - TS + PV) - \left(\sum_{i=1}^{m} n_i \mu_i\right)\right] \tag{32}$$

$E - TS + PV$ is the Gibbs free energy of the system θ, while $\sum_{i=1}^{m} n_i \mu_i$ would be its Gibbs free energy if there were no interface. Thus, σ may be considered as an excess Gibbs free energy per unit area. Moreover, since

$$(E^{\alpha} - TS^{\alpha} - PV^{\alpha}) - \sum_{i=1}^{m} n_i^{\alpha}\mu_i = 0 \tag{33}$$

$$(E^{\beta} - TS^{\beta} - PV^{\beta}) - \sum_{i=1}^{m} n_i^{\beta}\mu_i = 0 \tag{34}$$

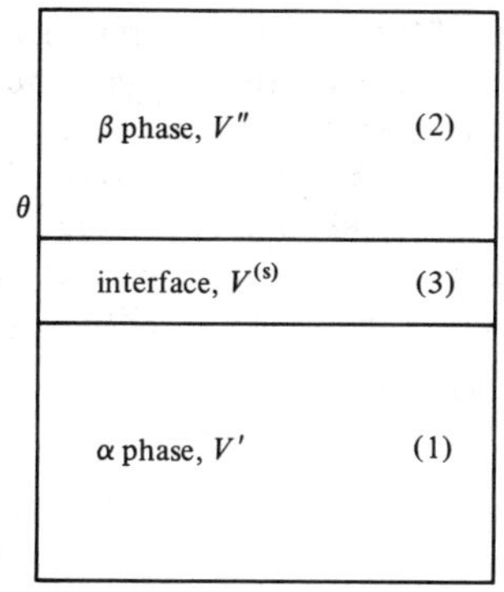

Figure 5. Illustration for Cahn's treatment of planar interfaces. Parts (1) and (2) contain homogeneous portions of the phases α and β, while part (3) contains all the interface's heterogeneity.

we see that the homogeneous parts of the phases α and β have no influence on the value of σ. Therefore, equation (32) applies to the surface layer defined with a sufficient thickness to penetrate into the homogeneous regions of the phases α and β:

$$\sigma = \frac{1}{s}\left[(E^{(s)} - TS^{(s)} + PV^{(s)}) - \sum_{i=1}^{m} n_i^{(s)}\mu_i\right] \tag{35}$$

The energy $E^{(s)}$ of the three-dimensional surface layer being a homogeneous function of degree 1 with respect to the variables $S^{(s)}$, $V^{(s)}$, $n_i^{(s)}$, and s, we may also write (see Section II.2)

$$S^{(s)}\,dT - V^{(s)}\,dP + \sum_{i=1}^{m} n_i^{(s)}\,d\mu_i + s\,d\sigma = 0 \tag{36}$$

or

$$\frac{S^{(s)}}{s}\,dT - \frac{V^{(s)}}{s}\,dP + \sum_{i=1}^{m} \frac{n_i^{(s)}}{s}\,d\mu_i + d\sigma = 0 \tag{37}$$

For each homogeneous phase, the Gibbs–Duhem equation yields

$$S^{\alpha}\,dT - V^{\alpha}\,dP + \sum_{i=1}^{m} n_i^{\alpha}\,d\mu_i = 0 \tag{38}$$

$$S^{\beta}\,dT - V^{\beta}\,dP + \sum_{i=1}^{m} n_i^{\beta}\,d\mu_i = 0 \tag{39}$$

Equations (37)–(39) form a system of linear equations with respect to $d\sigma$ and the $m + 2$ variables dT, dP, and $d\mu_i$. However, as seen by the phase rule only m of the $m + 2$ variables are independent. Any two of these, X and Y, may be eliminated. Through Cramer's rule [5] it may be demonstrated [4] that we obtain

$$d\sigma = -[S/X, Y]\,dT + [V/X, Y]\,dP - \sum_{i=1}^{m} [n_i/X, Y]\,d\mu_i \tag{40}$$

where the quantities $[Z/X, Y]$ (with $Z = S$, V, or n_i) are equal to

$$[Z/X, Y] = \begin{vmatrix} Z^{(s)}/s & X^{(s)}/s & Y^{(s)}/s \\ Z^{\alpha} & X^{\alpha} & Y^{\alpha} \\ Z^{\beta} & X^{\beta} & Y^{\beta} \end{vmatrix} \Bigg/ \begin{vmatrix} X^{\alpha} & Y^{\alpha} \\ X^{\beta} & Y^{\beta} \end{vmatrix} \tag{41}$$

The terms $[Z/X, Y]$ have some important properties.

5.1. Invariance with Respect to the Boundaries of the Surface Layer

Let us demonstrate that these terms are independent of the positions of the surface layer's boundaries (as long as these boundaries remain in the homogeneous regions of the adjoining phases). This invariance is a consequence of the properties of determinants.

Assume that the boundary of the surface layer with the phase α is moved along its normal by a small increment. The entropy S^{α}, volume V^{α}, and number of moles n_i^{α} increase by λS^{α}, λV^{α}, λn_i^{α}, while the corresponding quantities in the surface layer ($S^{(s)}$, $V^{(s)}$, $n_i^{(s)}$) decrease by the same amounts. The new value of a term such as $[Z/X, Y]$ is then

$$[Z/X, Y]^* = \begin{vmatrix} (Z^{(s)} - \lambda Z^{\alpha})/s & (X^{(s)} - \lambda X^{\alpha})/s & (Y^{(s)} - \lambda Y^{\alpha})/s \\ Z^{\alpha} + \lambda Z^{\alpha} & X^{\alpha} + \lambda X^{\alpha} & Y^{\alpha} + \lambda Y^{\alpha} \\ Z^{\beta} & X^{\beta} & Y^{\beta} \end{vmatrix} \times \begin{vmatrix} X^{\alpha} + \lambda X^{\alpha} & Y^{\alpha} + \lambda Y^{\alpha} \\ X^{\beta} & Y^{\beta} \end{vmatrix}^{-1} \tag{42}$$

or

$$[Z/X, Y]^* = (1 + \lambda) \begin{vmatrix} \frac{Z^{(s)}}{s} - \frac{\lambda}{s} Z^\alpha & \frac{X^{(s)}}{s} - \frac{\lambda}{s} X^\alpha & \frac{Y^{(s)}}{s} - \frac{\lambda}{s} Y^\alpha \\ Z^\alpha & X^\alpha & Y^\alpha \\ Z^\beta & X^\beta & Y^\beta \end{vmatrix}$$

$$\times \left[(1 + \lambda) \begin{vmatrix} X^\alpha & Y^\alpha \\ X^\beta & Y^\beta \end{vmatrix} \right]^{-1} \tag{43}$$

After canceling the term $1 + \lambda$ and noting that the value of a determinant remains unchanged when a multiple of one row is added to another, we see that the quantity $[Z/X, Y]^*$ is identical to $[Z/X, Y]$. Consequently, all the terms $[Z/X, Y]$ are invariant with respect to the positions of the surface layer's boundaries.

5.2. Choice of the Two Dependent Variables X and Y

Let us select X and Y as V and n_1. Equation (40) becomes

$$d\sigma = -[S/V, n_1]\, dT - \sum_{i=2}^{m} [n_i/V, n_1]\, d\mu_i \tag{44}$$

since a determinant with two identical columns is equal to zero. The term $[n_i/V, n_1]$ is equal to

$$[n_i/V, n_1] = \frac{n_i^{(s)}}{s} - \frac{n_1^{(s)}}{s} \left(\frac{c_i^\beta - c_i^\alpha}{c_1^\beta - c_1^\alpha} \right) - \frac{V^{(s)}}{s} \left(\frac{c_i^\alpha c_1^\beta - c_1^\alpha c_i^\beta}{c_1^\beta - c_1^\alpha} \right) \tag{45}$$

If we were to model our physical system by a system in which the two boundaries of the surface layer are infinitely close (i.e., Gibbs's model), the volume of the surface layer $V^{(s)}$ would become zero and equation (45) would be reduced to

$$[n_i/V, n_1] = \Gamma_i^{(1)} \tag{46}$$

Similarly, $[S/V, n_1]$ would be equal to the relative entropy. Thus, we verify once again that these relative functions are independent of the position of the dividing surface.

We note, however, that in the treatment above it is not necessary to model the surface with an interface of infinitesimal thickness. Indeed, equations (44) and (45) have a clearer physical meaning without this assumption.

There are several other possible selections of X and Y, and, as noted by Cahn [4], this flexibility has some definite advantages. Consider, for example, two binary condensed phases α and β. At equilibrium and at a given temperature and pressure their compositions are fixed and can be obtained from the phase diagram. It is of interest to investigate the effect of temperature or pressure on the interfacial tension of this system. This is immediately found through equation (40):

$$d\sigma = -[S/n_1, n_2]\, dT + [V/n_1, n_2]\, dP \tag{47}$$

Reduced functions may also be found through this approach. Equation (37) may be rewritten

$$\frac{S^{(s)}}{s} dT - \frac{V^{(s)}}{s} dP + \frac{n^{(s)}}{s} d\mu_1 + \sum_{i=2}^{m} \frac{n_i^{(s)}}{s} d(\mu_i - \mu_1) + d\sigma = 0 \tag{48}$$

and equations (38) and (39) become

$$S^\alpha\, dT - V^\alpha\, dP + n^\alpha\, d\mu_1 + \sum_{i=2}^{m} n_i^\alpha\, d(\mu_i - \mu_1) = 0 \tag{49}$$

$$S^\beta\, dT - V^\beta\, dP + n^\beta\, d\mu_1 + \sum_{i=2}^{m} n_i^\beta\, d(\mu_i - \mu_1) = 0 \tag{50}$$

Eliminating from this system of linear equations the two dependent variables dP and $d\mu_1$ leads to

$$d\sigma = -[S/V, n]\, dT - \sum_{i=2}^{m} [n_i/V, n]\, d(\mu_i - \mu_1) \tag{51}$$

which for $V^{(s)} = 0$ may be verified to reduce to

$$d\sigma = -\overline{S}\, dT - \sum_{i=2}^{m} \overline{\Gamma}_i\, d(\mu_i - \mu_1) \tag{52}$$

or

$$d\sigma = -\overline{S}\, dT - \sum_{i=2}^{m} \overline{\Gamma}_i\, d\mu_i + \sum_{i=2}^{m} \overline{\Gamma}_i\, d\mu_1 \tag{53}$$

Moreover, since

$$\sum_{i=2}^{m} \overline{\Gamma}_i = -\overline{\Gamma}_1 \tag{54}$$

we see that equation (53) is identical to equation (28), that is, to

$$d\sigma = -\overline{S}\, dT - \sum_{i=1}^{m} \overline{\Gamma}_i\, d\mu_i \tag{55}$$

---

6. Perfect Solution Model of an Interface

6.1. Definition

For the case of bulk materials, an ideal solution has been defined by stating that the activities of its components are proportional to their mole fractions (see Section VI.2). In Chapter XV we shall see that this corresponds to a statistical model in which the atoms are distributed at random on lattice sites and there is no energy change associated with a rearrangement of the atoms. Generally, the behaviors of real solutions are approximated by such a model when the components' atoms are similar in size and in chemical characteristics.

The perfect solution model of an interface is somewhat analogous to the ideal solution model. It assumes that the two phases adjoining the interface are ideal and that the region of heterogeneity is restricted to a monolayer. In this monolayer, as in the bulk phases, the atoms are assumed to have similar sizes and to be distributed at random. The composition of the monolayer is, however, different from the composition of either phase. The development of the model [3,6] leads to the equations

$$\sigma - \sigma_i = \frac{RT}{\alpha} \ln \frac{X_i^{\mathrm{m}}}{X_i} \qquad \text{(for each solute } i\text{)} \tag{56}$$

where σ_i is the interfacial tension in the case of the pure component i, α is the area occupied by one mole in the monolayer and X_i^{m} is the mole fraction of i in the monolayer.

Equation (56) may also be derived without recourse to statistical thermodynamics [3]. Let us assume that the heterogeneous region has shrunk to a monolayer. Equation (35) becomes

$$\sigma s = G^{(m)} - \sum_{i=1}^{m} n_i^{(m)} \mu_i \tag{57}$$

s and $G^{(m)}$ are homogeneous functions of degree 1 with respect to the $n_i^{(m)}$ at constant T and P. They are also functions of σ, but we may neglect this dependence (in the same sense that the dependence of bulk properties on the pressure are usually negligible for liquids and solids). Equation (57) may then be rewritten

$$\sigma \sum_{i=1}^{m} n_i^{(m)} \left(\frac{\partial s}{\partial n_i^{(m)}}\right)_{P,T,n_j^{(m)}} = \sum_{i=1}^{m} n_i^{(m)} \left(\frac{\partial G^{(m)}}{\partial n_i^{(m)}}\right)_{P,T,n_j^{(m)}} - \sum_{i=1}^{m} n_i^{(m)} \mu_i \tag{58}$$

The partial derivative $\partial s/\partial n_i^{(m)}$ may be identified as the area α_i occupied by 1 mol of component i in the monolayer. The partial derivative $\partial G^{(m)}/\partial n_i^{m}$ may be considered to be a monolayer chemical potential $\mu_i^{(m)}$.[2] Equation (58) then yields

$$\sigma\alpha_i = \mu_i^{(m)} - \mu_i \tag{59}$$

If the components of the system have atoms (or molecules) or similar sizes and chemical characteristics, the molar areas occupied by different components are the same:

$$\alpha_i = \alpha \tag{60}$$

and the solutes behave ideally in the bulk:

$$\mu_i = \mu_i^*(T, P) + RT \ln X_i \tag{61}$$

By analogy, it is also assumed that they behave ideally in the monolayer and therefore that one can write

$$\mu_i^{(m)} = \mu_i^{(m)*}(T, P) + RT \ln X_i^{(m)} \tag{62}$$

Incorporating equations (60)–(62) into (59), we obtain

$$\sigma\alpha = \mu_i^{(m)*}(T, P) - \mu_i^*(T, P) + RT \ln(X_i^{(m)}/X_i) \tag{63}$$

When the system contains only the pure component i equation (63) yields

$$\sigma_i\alpha = \mu_i^{(m)*} - \mu_i^* \tag{64}$$

Consequently, equation (63) becomes

$$\sigma - \sigma_i = \frac{RT}{\alpha} \ln(X_i^{(m)}/X_i) \tag{65}$$

6.2. Consequences

Another form of equation (65) is

$$X_i^{(m)} = X_i e^{(\sigma - \sigma_i)\alpha/RT} \tag{66}$$

[2] This monolayer chemical potential is different from that defined in Section XIII.1 for the region of heterogeneity and according to Gibbs's method. Their values differ by $\sigma\alpha_i$.

Eliminating the mole fractions $X_i^{(m)}$ by

$$\sum_{i=1}^{m} X_i^{(m)} = 1 \tag{67}$$

we obtain the dependence of the surface tension on the composition of the bulk solution:

$$e^{-\sigma\alpha/RT} = \sum_{i=1}^{m} X_i e^{-\sigma_i\alpha/RT} \tag{68}$$

Combining equations (66) and (68), we also obtain the concentration of i in the monolayer as a function of the composition of the bulk solution:

$$X_i^{(m)} = X_i e^{-\sigma_i\alpha/RT} \Big/ \sum_{j=1}^{m} X_j e^{-\sigma_j\alpha/RT} \tag{69}$$

Let us consider a liquid solution in equilibrium with a gas phase. Neglecting the concentrations in the gas phase, the reduced adsorption $\overline{\Gamma}_i$ is equal to

$$\overline{\Gamma}_i = \Gamma_i - X_i\Gamma = \frac{1}{\alpha}(X_i^{(m)} - X_i) = \frac{X_i}{\alpha}(e^{(\sigma-\sigma_i)\alpha/RT} - 1) \tag{70}$$

Consequently, we see that the monolayer is enriched in the components which have a lower interfacial tension than that of the solution.

The perfect solution model may be applied to the Ag–Au system. The crystallographic radii of silver and gold are nearly identical (1.444 and 1.441 Å, respectively [7]) and the electronic constitutions of these atoms are very similar. Their phase diagram presents no compound (it is *lens type*) and the excess free

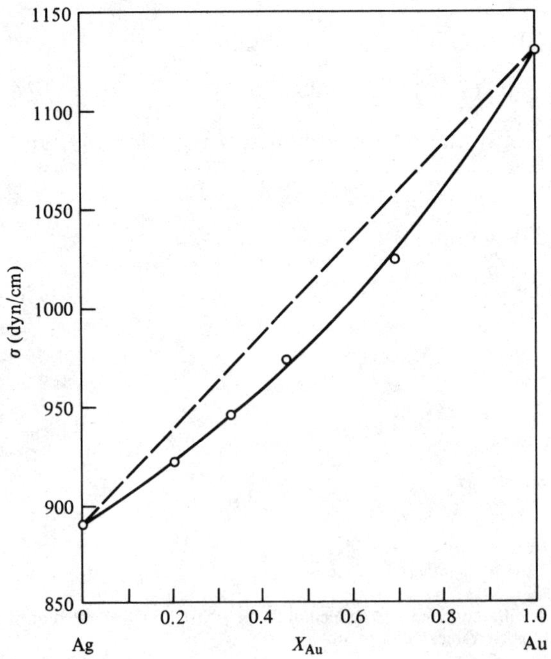

Figure 6. Surface tension of silver–gold alloys at 1108°C [9]. The solid line is predicted by the perfect solution model.

energy of their liquid solution is small (about −600 cal at 1350°K [8]). The molar volume V_m of silver being equal to 11.6 cm^3 at 1108°C, we may deduce the area α by the formula [6]

$$\alpha = \frac{1}{0.918}(N_0)^{1/3}(V_m)^{2/3} \tag{71}$$

where N_0 is Avogadro's number (6.023×10^{23}). The constant 0.918 is introduced on the assumption that the atoms occupy a close-packed configuration in the monolayer. At 1108°C, the parameter α/RT becomes equal to 4.1 m/N (or 4.1×10^{-3} cm/dyn). With this value, the perfect solution model's equation

$$e^{-\sigma\alpha/RT} = X_1 e^{-\sigma_1\alpha/RT} + X_2 e^{-\sigma_2\alpha/RT} \tag{72}$$

shows, in Fig. 6, a very satisfactory agreement with the data of Bernard and Lupis [9]. We note that according to the model, the surface is richer in silver than the bulk.

Within the accuracy of surface tension data (often 1–2%), the model is also fairly satisfactory in the cases of the Cu–Ni system [10], the Au–Bi system [11] and the alloys of Ge with Bi, In, Pb, and Sn [12].

7. Mixtures of Two Metals

By and large, the majority of alloy systems do not fit a model as simple as the perfect solution model. By analogy with the models of bulk solutions, a regular solution model would seem to be the next logical approximation. Indeed, such a model has been developed by Guggenheim [13]. However, the regular solution model for surfaces has met with considerably less success than its counterpart for bulk solutions. Two assumptions appear critical: the identification of the interface with a monolayer and the use of interaction energies with identical values in the bulk and in the monolayer. To analyze these assumptions, let us consider the possible crystalline structure of a metallic surface.

If an ideal metal crystal is cut through a crystallographic plane and separated in two, two identical surfaces are created. The surfaces would be considered ideal if all the atoms were to remain in the positions they initially occupied. However, this is unlikely. For example, it has been calculated [14] that for the (111) plane of fcc copper there is a 5.5% expansion, perpendicular to the surface, between the first and second layers and 1% expansion between the second and third. These figures become 13% and 3% for the (100) face, 20% and 5% for the (110) face. Although these results have not been confirmed experimentally and indeed depend on several assumptions (such as the form of the interaction potential between two atoms), they still represent a likely order of magnitude.

We also note that the energetic state of a surface atom is bound to be very different from the energetic state of a bulk atom. In an fcc metal, the number of nearest neighbors is equal to 12. In (111), (100), and (110) surfaces, this number is reduced to 9, 8, and 6 respectively. As a result, surface atoms have very high affinity for any foreign atoms or molecules because of their tendency to reactivate missing bonds and reestablish the symmetry of the energetic field to which they are subject in the bulk of the crystal. Clearly, this reactivity is quite dependent on the crystallographic orientation of the surface.

In the case of alloys, we observe that surface atoms of different components need not occupy the same plane. In ionic melts and crystals, it has often been postulated [15,16] that ions of higher polarizability protrude more out of the surface, setting up an ionic double layer at the interface. Depending on the nature of the metallic systems studied, a similar phenomenon may well occur, although probably less pronounced.

A regular solution model assumes that the atoms of different components are distributed at random and therefore are not too different in size and chemical characteristics. For this reason, as well as some elaborated above, a monolayer model appears improbable: the concentration gradient is likely to occur over several atomic layers. Defay and Prigogine have also shown that the monolayer assumption leads to some thermodynamic inconsistencies [3].

In the bulk, strong deviations from ideality have often been associated with the formation of more or less stable clusters (see Lupis et al. [17] and Section III.5). This is even more likely to occur at a surface. Since the bonding electrons of surface atoms are interacting with a smaller number of atoms, it appears indeed that surface atoms would have more flexibility in associating themselves with other surface atoms and in creating stabler clusters. A model based on such entities has been developed by Laty et al. [18] and has given encouraging results when applied to the surface tension data of Al–Cu, Fe–Si, and Ni–Si alloys. Progress in the development of the model should come with the deletion of the assumption of equal atomic interactions in the bulk and at the surface.

The assumption of molecular aggregates of fixed stoichiometry has also been exploited to explain maxima and minima in many surface tension isotherms. Kaufman [19] has observed that these extrema often correspond to the compositions of intermetallic compounds in the bulk solid phases. It may readily be seen that such an extremum must occur at a composition of the solution which is identical to that of the molecular aggregate being assumed. Recalling that

$$\left(\frac{\partial \sigma}{\partial X_2}\right)_T = -\Gamma_2^{(1)}\left(\frac{\partial \mu_2}{\partial X_2}\right)_T \tag{73}$$

we see that $\partial\sigma/\partial X_2$ is equal to zero when $\Gamma_2^{(1)}$ is equal to zero, that is, when

$$\frac{\Gamma_2}{\Gamma_1} = \frac{X_2^{(m)}}{X_1^{(m)}} = \frac{X_2}{X_1} \tag{74}$$

(For completeness, it may also be observed that $\partial\sigma/\partial X_2$ is also equal to zero at spinodal points, that is, when $\partial\mu_2/\partial X_2$ is equal to zero.)

A general satisfactory model for interfaces is still to be developed. On the one hand, too simple a model does not describe adequately adsorption behavior; on the other hand, too realistic a model would introduce too many unknown quantities and therefore too many parameters for the model to be useful. Another major handicap is the dearth of accurate data. Although many systems have been studied, serious discrepancies and large scatter plague the results. In large part, these experimental errors are due to the contamination of the surface by surface-active species. Their behavior deserves special description.

8. Surface-Active Species

8.1. Dilute Solutions

A solute is said to be *surface active* if even in very small quantity it has a large effect on the interfacial tension. For metals, elements of the group VIB, namely O, S, Se and Te, are generally among the strongest surface-active elements known.

Typical curves of $\Gamma_2^{(1)}$ vs X_2 and σ vs X_2 or $\ln a_2$ are shown in Figs. 7 and 8. The solutions being generally very dilute in component 2, the difference between $\Gamma_2^{(1)}$ and $\overline{\Gamma}_2$ is negligible. The *surface activity* of 2 is defined as the slope of σ vs X_2 at infinite dilution:

$$J_2 = -\left(\frac{\partial\sigma}{\partial X_2}\right)_{X_1\to 1} \tag{75}$$

Using the Gibbs adsorption equation and assuming Henry's law (Section VI.4) we obtain

$$\frac{J_2}{RT} = -\frac{1}{RT}\left(\frac{\partial\sigma}{\partial X_2}\right)_{X_1\to 1} = \Gamma_2^{(1)}\left(\frac{\partial \ln a_2}{\partial X_2}\right)_{X_1\to 1}$$
$$= \left(\frac{\Gamma_2^{(1)}}{X_2}\right)_{X_1\to 1} = \left(\frac{\overline{\Gamma}_2}{X_2}\right)_{X_1\to 1} \tag{76}$$

Experimentally, J_2 is found to be finite. Thus the slope of $\Gamma_2^{(1)}$ (or $\overline{\Gamma}_2$) vs X_2 is also finite. This result is sometimes referred to as *Henry's law for surfaces*.

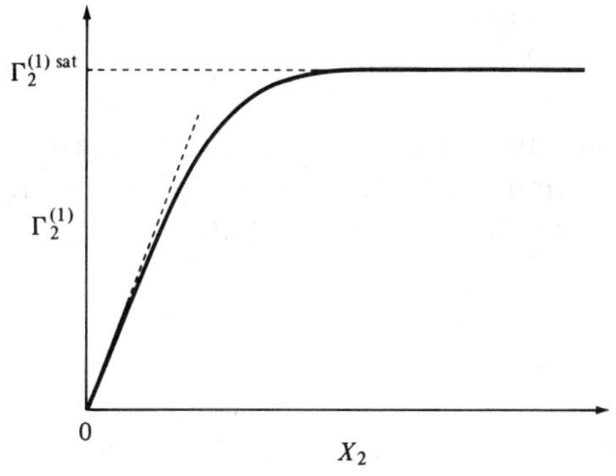

Figure 7. Typical composition dependence of the relative adsorption of a surface-active species.

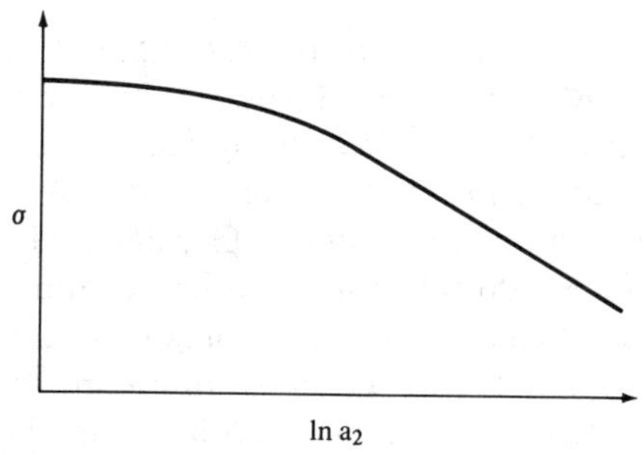

Figure 8. Typical dependence of the surface tension on the activity of a surface-active species.

The value of J_2 is high for surface-active species, often of the order of 10^3 N/m. For example, at 1550°C, the surface-activity of sulfur in liquid iron is of the order of 1.8×10^6 dyn/cm (1800 N/m); 20 ppm of sulfur (by weight) then reduces the surface tension of liquid iron by about 60 dyn/cm, or about 3% of its value. Extremely high purity systems are therefore necessary for the measurement of surface properties.

In analyzing adsorption behavior, the *adsorption coefficient* ν_2 may be introduced. It is defined as

$$\nu_2 = \frac{\bar{\Gamma}_2}{X_2} \frac{1}{k} \tag{77}$$

where k is a normalization constant designed to make ν_2 dimensionless. In dilute solutions, it may be chosen to be the number of atoms per unit area of the pure solvent 1 as calculated for a monolayer model. Equation (71) would yield

$$k = \Gamma_1^\circ = 0.918\,(N_0)^{-1/3}(V_{m,1})^{-2/3} \tag{78}$$

We note that in a monolayer model, ν_2 becomes equal to

$$\nu_2 = \frac{1}{\Gamma_1^\circ} \frac{n^m}{s} \left(\frac{X_2^{(m)}}{X_2} - 1 \right) \tag{79}$$

or

$$\nu_2 \simeq \frac{X_2^{(m)}}{X_2} - 1 \tag{80}$$

if the number of atoms at the surface is assumed not to vary with composition.

At infinite dilution, equation (76) yields

$$J_2 = (RT\Gamma_1^\circ)\nu_2^\infty \tag{81}$$

Selected values of ν_2^∞ are listed in Tables 1 and 2. We see that these values can be very high and that the concentration of a surface-active species at an interface is often several thousand times higher than that in the bulk.

8.2. Saturation Stage

A surface-active species is characterized not only by a high value of the surface activity J_2 but also by a saturation stage. This is manifested in Fig. 7 by a practically constant value of $\Gamma_2^{(1)}$ after a certain concentration is reached, and in Fig. 8 by a constant slope of σ vs $\ln a_2$.

Generally, the saturation stage corresponds to the formation of a two-dimensional compound. The natures and structures of such compounds have been investigated extensively in the case of solid surfaces by techniques such as low energy electron diffraction (LEED), field ion microscopy, and Auger spectroscopy [14]. The study of oxygen adsorption on nickel by Germer et al. and MacRae [22,23] yields results which are fairly typical of many other systems.

During the initial stages of adsorption, the oxygen atoms show a marked tendency to be localized at sites where they are in contact with the maximum number of surface atoms of the metal substrate. Occupancy on a site inhibits wholly or partly occupancy of an adjacent site. Saturation occurs through the formation of

Table 1. Adsorption Parameters for Elements of the Group VIB [20]

Solvent	$(RT\Gamma_0^1)_{T\,°K}$ (N/m)	Solute	T (°K)	ν_2^∞	$\Gamma_2^{(1)sat} \times 10^6$ (mol/m²)	$A_2^{(1)sat}$ (Å²/atom)	$X_2^c \times 10^6$
Fe	$(0.412)_{1823}$	O	1823	2,500	16.1	10.3	240
		S	1823	4,500	11.0	15.1	90
		Se	1823	27,000	11.2	14.7	10
		Te	1823	>27,000	11.1	15.0	<10
Cu	$(0.310)_{1373}$	O	1373	10,000	5.7	29	20
		S	1393	14,000	~11.4	~14.5	30
		Se	1423	14,000	14	11.9	40
		Te	1423	7,800	12	13.8	60
Ag	$(0.219)_{1253}$	O	1253	1,750	4.8	34	130
Pb	$(0.118)_{973}$	O	1023	650	4.9	34	520
		S	1023	520	—	—	—
		Se	1023	350	—	—	—
		Te	973	130	0.7	240	370

$A_2^{(1)sat}$ (Å²/atom) × 6025 = $1/\Gamma_2^{(1)sat}$ (m²/mol).
$1/\Gamma_1^0$ is 6.11 Å²/atom for Fe and Cu, 7.9 Å²/atom for Ag, and 11.4 Å²/atom for Pb.
$J_2 = -(d\sigma/dX_2)_{X_2\to 0} = (RT\Gamma_1^0)\,\nu_2^\infty$
$X_2^c = (\Gamma_2^{(1)sat}/\Gamma_1^0)(1/\nu_2^\infty) = (A_1^0/A_2^{(1)sat})(1/\nu_2^\infty)$.

a two-dimensional compound with a superlattice structure. For the (100) plane of nickel, this superlattice has the stoichiometry NiO. Each oxygen atom occupies twice the area of a metal atom. After the formation of this two-dimensional compound, no further adsorption of oxygen is observed.

On the (111) plane of nickel, saturation occurs in two stages. First a superlattice of the Ni_3O stoichiometry is formed. It appears quite stable and is not modified by heating the metal to a temperature close to its melting point. However, upon exposure to higher potentials of oxygen, the surface layer undergoes another rearrangement until the stoichiometry Ni_2O is reached.

Under the influence of the atoms of the surface-active species, the metal atoms of the substrate may undergo substantial displacement. Because the strength of their mutual interaction may exceed that of the metal–metal interactions, the atoms seek to adopt a new configuration which satisfies more fully the requirements of their relative sizes, stoichiometry and coordination numbers, directional bonding, etc. The surface is then said to be *reconstructed*.

Table 2. Adsorption Coefficients at Infinite Dilution ν_2^∞ for Liquid Iron at 1550°C [20][a]

	III	IVB	VB	VIB
	B: 6	C: 0.7	N: 200–800	O: 2500–5000
	Al: 9	Si: 2.9	P: 3	S: 4500
Cr: 2.8 Ni: 0.1 Cu: 11		Ge: 27	As: 850	Se: 27,000
		Sn: 700	Sb: 2400	Te: >27,000

[a] Many of the values are based on the pioneering experimental work of Kozakevitch and Urbain [21].

In the case of liquid metals, there is ample experimental evidence that a short range is retained. Similar adsorption phenomena with formation of superlattices of stoichiometry $A_{n-1}B$ may therefore be assumed. Clearly, however, such structures would not cover the entire surface as a coherent compound. Small ordered patches, separated by mobile boundaries, are more likely. Direct experimental observation of the surface structure and chemistry is, however, much more difficult than in the case of solid surfaces. The value of $\Gamma_2^{(1)}$ at saturation, $\Gamma_2^{(1)\mathrm{sat}}$, is expected to be primarily a function of the sizes of the metal cations and solute anions. In order to estimate $\Gamma_2^{(1)\mathrm{sat}}$ the area requirements of corresponding three-dimensional compounds may be determined. Although the limitations of this estimation technique are evident, it is nonetheless useful. As shown in Table 3, it indeed yields fairly satisfactory results, especially when one considers the uncertainty in the experimental data.

8.3. Models of Adsorption

Whether for prediction, interpretation, or description, models of statistical thermodynamics for surfaces are very useful. We shall briefly describe some of these models in the case of a surface-active species at a metal–gas interface.

A common assumption is the assimilation of the interface to one monolayer. We have seen that this assumption is unrealistic in the case of mixtures of two metals. However, in the case of a surface-active species, such as oxygen or sulfur, on a metal (a case where the characters of the solvent and solute atoms are widely different), the assumption seems much more acceptable. Indeed, Auger spectroscopy often shows that the segregation of the solute is limited to one or two atomic layers.

Table 3. Area Requirements at Saturation [20, 21, 24, 25, 26]

System 1–2	T (°K)	$A_2^{(1)\mathrm{sat}}$ (Å²/atom) experimental	calculated[a]	calculated[b]	Ionic radius r^-	r^+	Ionic charge Z^-	Z^+
Fe–O	1823	10.3	6.0	8 (111)	1.33	0.83	2	2
Fe–S	1823	15.1	10.5	12 (010)	1.74	0.83	2	2
Fe–Se	1823	14.7	13	13 (010)	1.91	0.83	2	2
Fe–Te	1823	15.0	14.6		2.11	0.83	2	2
Cu–O	1273	29	6.0	18 (100) 30 (111)	1.32	0.96	2	1
Cu–S	1393	14.5	10.5	14.5(100) 16.8(111)	1.74	0.96	2	1
Cu–Se	1423	11.9	13		1.91	0.96	2	1
Cu–Te	1423	13.8	14.6		2.11	0.96	2	1
Ag–O	1253	34	6.0	23 (100) 38 (111)	1.32	1.13	2	1

[a] On the basis of a monolayer of close-packed solute anions.
[b] On the basis of the solid compound.

8.3.1. Langmuir's Isotherm

Langmuir's model is based on the assumption that the surface has a number of well-defined sites which may be occupied by only one atom (or molecule) of the surface-active species. The adsorbed atoms have no mutual interaction. Formally, the resulting equation may be derived by considering

$$\underline{2} + V^s = 2^s \tag{82}$$

for which the dissolved element 2 reacts with a vacancy at the surface in order to form an occupied site. Applying the mass action law to the reaction (82), we obtain

$$K = \Gamma_2 / a_2(\Gamma_2^{sat} - \Gamma_2) \tag{83}$$

where a_2 is the activity of 2 in the solution (with a Henrian reference state), Γ_2 is the concentration of 2 at the surface, and $\Gamma_2^{sat} - \Gamma_2$ is the concentration of vacancies.

Neglecting the difference between $\Gamma_2^{(1)}$ and Γ_2, the Gibbs adsorption equation may be combined with equation (83) and integrated. This yields

$$\sigma = \sigma_1 - RT\Gamma_2^{(1)sat} \ln(1 + K a_2) \tag{84}$$

where σ_1 is the surface tension of the pure component 1. The form of equation (84) was found empirically by Szyszkowski in 1908 and bears his name.

As seen in Sections 7.1 and 7.2, the adsorption behavior of a surface-active species ($\Gamma_2^{(1)}$ vs X_2; Fig. 9) is characterized by a high slope at the origin and an early stage of saturation. Let us denote by X_2^c the abscissa of the point at which the horizontal asymptote $\Gamma_2^{(1)} = \Gamma_2^{(1)sat}$ intercepts the tangent at the origin. Since for dilute solutions $a_2 = X_2$, we deduce from equations (83) and (76)

$$K = \frac{1}{X_2^c} = \frac{J_2}{RT\Gamma_2^{(1)sat}} \tag{85}$$

In Langmuir's model, at a_2 or $X_2 = X_2^c$, $\Gamma_2^{(1)} = \frac{1}{2}\Gamma_2^{(1)sat}$. This disagrees with many experimental observations which yield a much steeper rise of the adsorption curve (i.e., at $X_2 = X_2^c$, $\Gamma_2^{(1)} > \Gamma_2^{(1)sat}$). However, equation (84) is in reasonable agreement with surface tension data [27]. This shows, therefore, that surface tension curves are not very sensitive to values of the adsorption function.

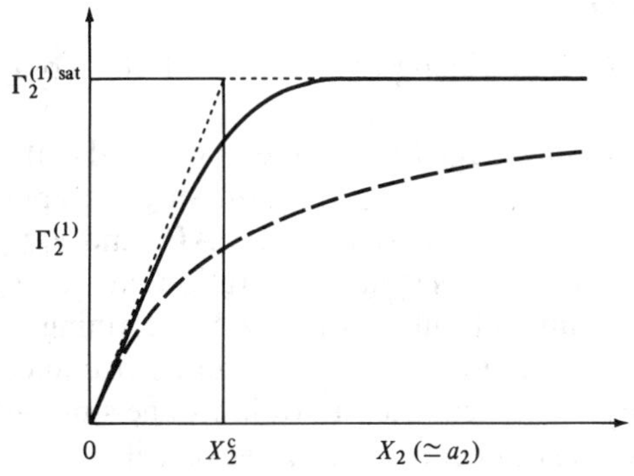

Figure 9. Composition dependence of the relative adsorption. The solid line corresponds schematically to the assumed behavior of surface-active species, the dashed line to Langmuir's isotherm [equation (83)].

Szyszkowski's equation has a convenient analytical form. Nonetheless, it is worthwhile to search for other equations which represent more accurately adsorption behavior.

8.3.2. Fowler and Guggenheim Model

In the model of Fowler and Guggenheim [28], as in Langmuir's model, the number of lattice sites at the surface is independent of composition and the adsorbed atoms are distributed at random. However, these adsorbed atoms now interact with each other.

Let us designate by θ the fraction of surface sites occupied by the solute:

$$\theta = \Gamma_2/\Gamma_1^\circ \tag{86}$$

In terms of θ, the derivation of the model yields the following equation:

$$Ka_2 = \frac{\theta}{1-\theta}\exp\left[-2\left(\frac{Z\omega}{kT}\right)\theta\right] \tag{87}$$

Z is the number of nearest neighbors within the surface layer. ω is equal to the regular solution parameter $u_{12} - \frac{1}{2}(u_{11} + u_{22})$, where u_{ij} is the bonding energy between two surface atoms i and j. K is a constant with respect to composition, but is temperature dependent. It may be interpreted through the model in terms of bonding energies and atomic partition functions and is identified here as equal to $1/X_2^c$ or $J_2/RT\Gamma_2^{(1)\text{sat}}$.

We note that for $\omega = 0$ equation (87) yields Langmuir's isotherm. $\omega > 0$ corresponds to a net attraction between the adsorbed solute atoms, and $\omega < 0$ to a repulsion.

An interesting feature of the model is that it predicts a transformation for the monolayer equivalent to that of a miscibility gap. The critical point is $\theta_c = \frac{1}{2}$, $T_c = Z\omega/2k$. Below the critical temperature, the monolayer phase is unstable and separates into two other stable phases of different compositions. Such transformations have now been found [29,30] and confirm, at least qualitatively, the model's prediction.

The Gibbs adsorption equation may be combined with equation (87) and integrated in terms of θ. The result is

$$\frac{\sigma_1 - \sigma}{RT\Gamma_1^\circ} = -\ln(1-\theta) - \left(\frac{Z\omega}{kT}\right)\theta^2 \tag{88}$$

Equations (87) and (88) offer a parametric representation of the composition dependence of σ.

In its application to the case of surface-active species on metals, the model has one severe limitation. We have seen that a saturation stage is approached which corresponds to stoichiometries such as A_3B, A_2B, AB, and $\theta_{max} = \frac{1}{4}, \frac{1}{3}$, and $\frac{1}{2}$, respectively. The model above predicts spatial saturation only at $\theta_{max} = 1$.

It may appear possible to circumvent this difficulty by defining θ as $\Gamma_2/\Gamma_2^{(\text{sat})}$. This would signify that only a fraction of the sites can be occupied by the solute atoms. Such an assumption is valid near the saturation stage since adsorbed atoms may exclude adjacent sites from occupancy. However, it is not valid at low concentrations; all sites are available then, not only $\frac{1}{4}$, $\frac{1}{3}$, or $\frac{1}{2}$ of them.

8.3.3. Other Models

In Section 6 we outlined some of the difficulties which hamper the development of a surface model in the case of mixtures of two metals. Some of the same difficulties prevent the development of a satisfactory model in the case of a surface-active species on a metal, namely, the difficulty of characterizing sufficiently well the crystalline features of the surface, the unknown natures and energies of the atomic bonds, and the dearth of accurate data.

Two types of adsorption behavior should be distinguished (Fig. 10). In the first (Fig. 10a), there is no reconstruction of the surface: the solute atoms of the surface-active species merely sit on the top of the metal surface without displacing any underlying atom. The chemisorption of CO on Pd [31,32] and the segregation of C on Ni [30,33,34] appear to be such examples. In the second type of adsorption behavior (Fig. 10b), sometimes called *corrosion chemisorption*, the surface monolayer is composed of both solvent and solute atoms and, because of their very large affinity, genuine two-dimensional compounds are formed at saturation. This type of behavior has been well documented for oxygen and sulfur on many different metals [35–38].

Bernard and Lupis [39] developed a simple monolayer model for the adsorption of a solute B where saturation is achieved with the formation of a compound $A_{n-1}B$ (Fig. 11). Low and high concentration approximations are derived separately because of mathematical difficulties. It is convenient to introduce the dimensionless quantities

$$p = a_2(J_2/RT\Gamma_1^\circ) \tag{89}$$

$$\varphi = (\sigma_1 - \sigma)/RT\Gamma_1^\circ \tag{90}$$

At high concentrations, the model yields

$$\ln p = \ln \frac{n\theta}{1 - n\theta} + 2n\theta(g/kT) \tag{91}$$

$$\varphi = -\frac{1}{n}\ln(1 - n\theta) + n\theta^2(g/kT) \tag{92}$$

θ is the fraction of B atoms (B or solute 2) in the monolayer or Γ_2/Γ_1° and n is the stoichiometric coefficient of the compound $A_{n-1}B$, or $\Gamma_1^\circ/\Gamma_2^{sat}$. g measures a net interatomic force between the adsorbed solute atoms B. It is negative for a net attractive force, positive for a repulsion. At low concentrations, the expressions

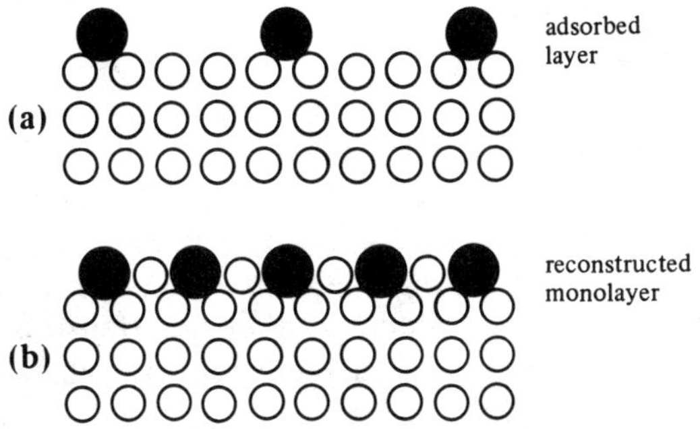

Figure 10. Examples of adsorption behaviors. **(a)** The atoms adsorbed merely sit on the top of the metal surface. **(b)** Both solute and solvent atoms are incorporated in the top layer.

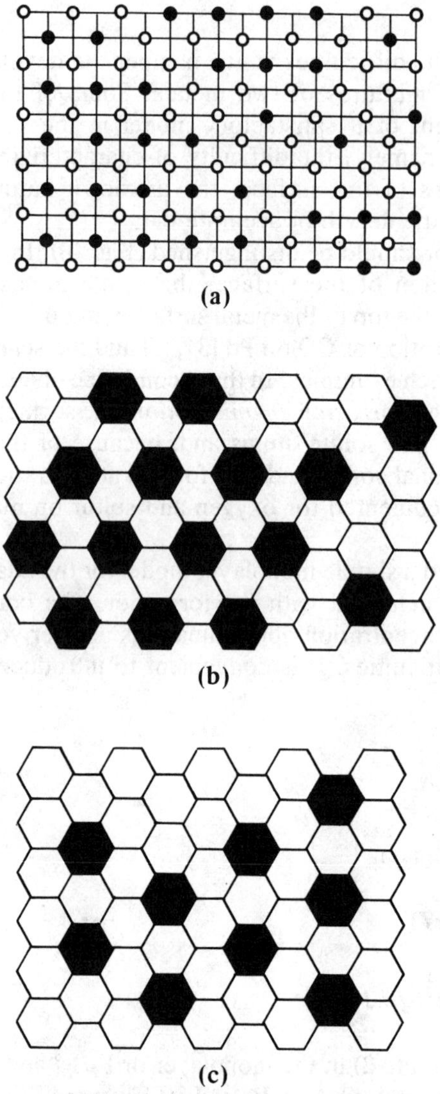

Figure 11. Adsorption with formation of a two-dimensional compound $A_{n-1}B$: **(a)** AB, **(b)** A_2B, and **(c)** A_3B. Filled areas denote B (solute), while open areas represent A (solvent).

for p and φ depend on the value of n. For $n = 2$

$$\ln p = \ln \theta + 3 \ln(1 - \theta) - 4 \ln(1 - 2\theta) + 2\theta(g/kT) \tag{93}$$

$$\varphi = 3 \ln(1 - \theta) - 2 \ln(1 - 2\theta) + \theta^2(g/kT) \tag{94}$$

and for $n = 4$

$$\ln p = \ln \theta + 11 \ln(1 - \theta) - 12 \ln(1 - 4\theta) + 2\theta(g/kT) \tag{95}$$

$$\varphi = 11 \ln(1 - \theta) - 3 \ln(1 - 4\theta) + \theta^2(g/kT) \tag{96}$$

A simple graphic interpolation between the high and low concentration approximations completes the description of the system. A first order transition similar to that of the Fowler and Guggenheim model is predicted for values of g/kT smaller than about -2.

Application of the model to various systems such as oxygen on silver and iron is in satisfactory agreement with the data. The parameter g is found to be negative, which means that net attractive forces between the adsorbed atoms of the solute must be assumed.

On the basis of a formalism developed by Hillert and Staffanson [40] for three-dimensional interstitial alloys, analytic expressions similar to those presented above [equations (91)–(96)] have been proposed by Guttman [41] in an excellent review article. The soundness of the theoretical basis for these expressions may be debated, but they have the advantage of spanning the entire concentration range.

The assumptions of most present models are oversimplified by necessity. However, before a general and realistic model can be developed, many more accurate data on the effects of surface-active species must be gathered.

8.4. Remarks

In our description of the adsorption of surface-active species, we have confined ourselves to metal–gas interfaces, that is, to "free" surfaces. Actually, there is abundant experimental evidence that very similar phenomena occur at grain boundaries (at least, high-angle grain boundaries). Elements which are surface active at free surfaces are also surface active at grain boundaries and the similarity of their adsorption isotherms is often striking [42]. This parallelism has been exploited by Guttmann [43], for example, in order to explain results on the temper embrittlement of iron alloys.

We applied the term "surface active" to elements which segregate strongly at the interface and decrease substantially its surface tension, even when present in minute quantities. It is clear that this definition is somewhat subjective: often,

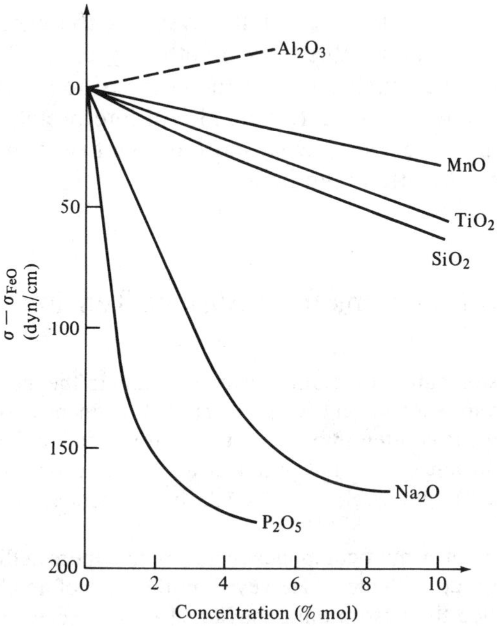

Figure 12. Effect of alloying additions on the surface tension of liquid FeO at 1420°C [46].

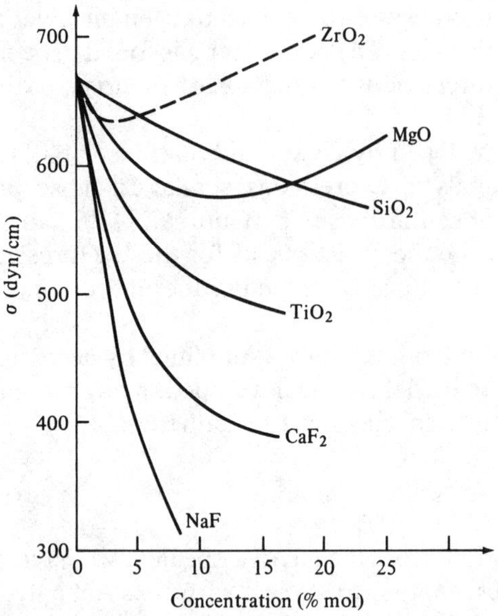

Figure 13. Effect of alloying additions on the surface tension of the liquid slag $CaO-Al_2O_3$ at temperatures between 1610 and 1680°C [47].

there is no obvious demarcation between elements which have strong and mild effects. From the examples which have been given (Table 2) it may be inferred that metallic solutes have a moderate effect on metallic surfaces; elements with strong effects are nonmetals. This is not always the case: the effect of Pb on the surface of liquid Cu is rather strong, since at 1100°C J is of the order of 70,000 dyn/cm [44,45] (and ν^{∞} of the order of 200). This should not be surprising since elements with low surface tensions can be expected to have large effects when associated with elements of high surface tensions. This may even be seen through equation (72) of the perfect solution model.

Although we focused our presentation on metallic systems, the general thermodynamic approach above is also applicable to many other systems. Oxides are an interesting class of systems and illustrations of the composition dependence of their surface tensions are shown in Figs. 12 and 13. An interpretation of the behavior of such systems is, however, somewhat more difficult than in the case of metallic systems and will not be attempted here.

9. Derivation of the Adsorption Functions from Surface Tension Data in Ternary Systems

Most adsorption and surface tension data for metallic systems pertain to the measurement of binary systems. However, from a technical viewpoint multicomponent alloys are of paramount importance. Moreover, it is often impossible to avoid the contamination of a surface with many solutes. For instance, it is not practically feasible to remove oxygen and sulfur from liquid iron at the 10 ppm level, even though their effects at this level are not negligible.

Measuring the surface tension of a multicomponent alloy is not more difficult than measuring the surface tension of a pure element. However, in the case of an alloy many more data are necessary to deduce the adsorption behavior of each species and the reduction of the data is more complex. We shall now briefly outline various methods by which adsorption functions may be obtained from surface tension data in the case of ternary systems.

9.1. Direct Method

At constant temperature, the Gibbs adsorption equation may be written

$$d\sigma = -\Gamma_2^{(1)}\, d\mu_2 - \Gamma_3^{(1)}\, d\mu_3 \tag{97}$$

Consequently,

$$\Gamma_2^{(1)} = -\left(\frac{\partial\sigma}{\partial\mu_2}\right)_{\mu_3} = -\frac{1}{RT}\left(\frac{\partial\sigma}{\partial \ln a_2}\right)_{a_3} \tag{98}$$

$$\Gamma_3^{(1)} = -\left(\frac{\partial\sigma}{\partial\mu_3}\right)_{\mu_2} = -\frac{1}{RT}\left(\frac{\partial\sigma}{\partial \ln a_3}\right)_{a_2} \tag{99}$$

In addition, we note that

$$\frac{\Gamma_2^{(1)}}{\Gamma_3^{(1)}} = -\left(\frac{\partial\mu_3}{\partial\mu_2}\right)_{\sigma} = -\left(\frac{\partial \ln a_3}{\partial \ln a_2}\right)_{\sigma} \tag{100}$$

The *direct method* ("direct" because it is a straightforward application of the Gibbs adsorption equation) is a graphical technique based upon these equations [48]. It assumes that the surface tension data are described by lines of constant value and that isoactivity curves are available. Figures 14 and 15 show such an example in the case of Ag–Au–Cu liquid alloys at 1381°K.

It is convenient to draw the sets of curves on tracing paper and juxtapose them. Drawing all sets of curves on the same figure with different colors (in order to distinguish them more easily) is also adequate. A curve of constant a_3 is selected; along this line the values of σ as a function of a_2 may be recorded and plotted. At any given point, the slope of this

Figure 14. Surface tension of Ag–Au–Cu liquid alloys at 1350°K [48].

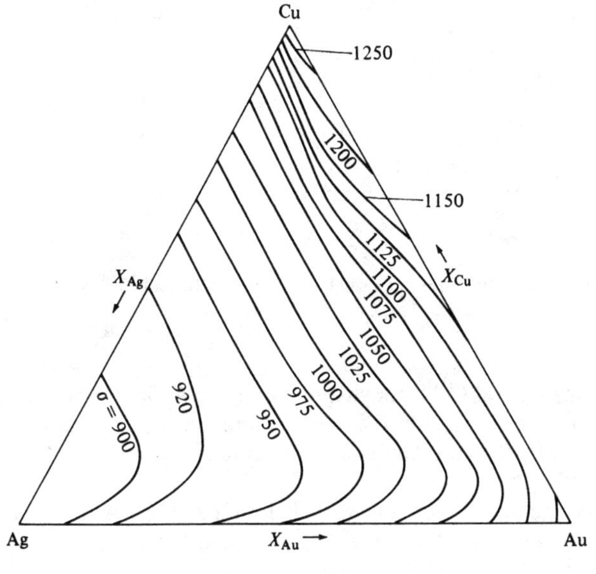

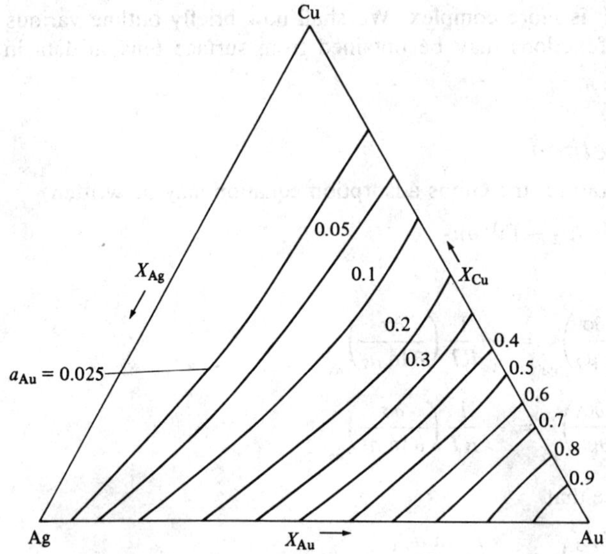

(a)

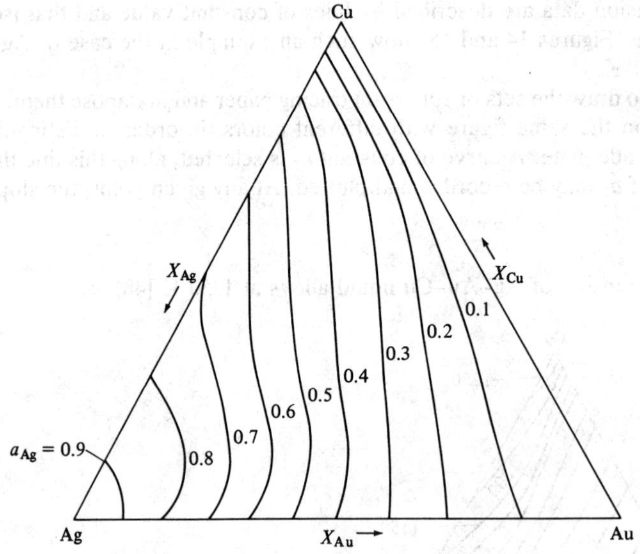

(b)

Figure 15. Isoactivity curves of the Ag–Au–Cu liquid solution at 1350°K [48].

curve yields the relative adsorption $\Gamma_2^{(1)}$ through equation (98). A similar procedure yields $\Gamma_3^{(1)}$. Equation (100) is convenient to verify the results along lines of constant σ.

Figure 16 shows the results of that technique in the case of the Ag–Au–Cu system as lines of constant $\Gamma_{Ag}^{(Cu)}$ and $\Gamma_{Au}^{(Cu)}$.

We recall that positive values of $\Gamma_{Ag}^{(Cu)}$ mean that the ratio of silver to copper at the surface is higher than in the bulk. $\Gamma_{Ag}^{(Cu)}$ is equal to zero when the surface adsorptions of

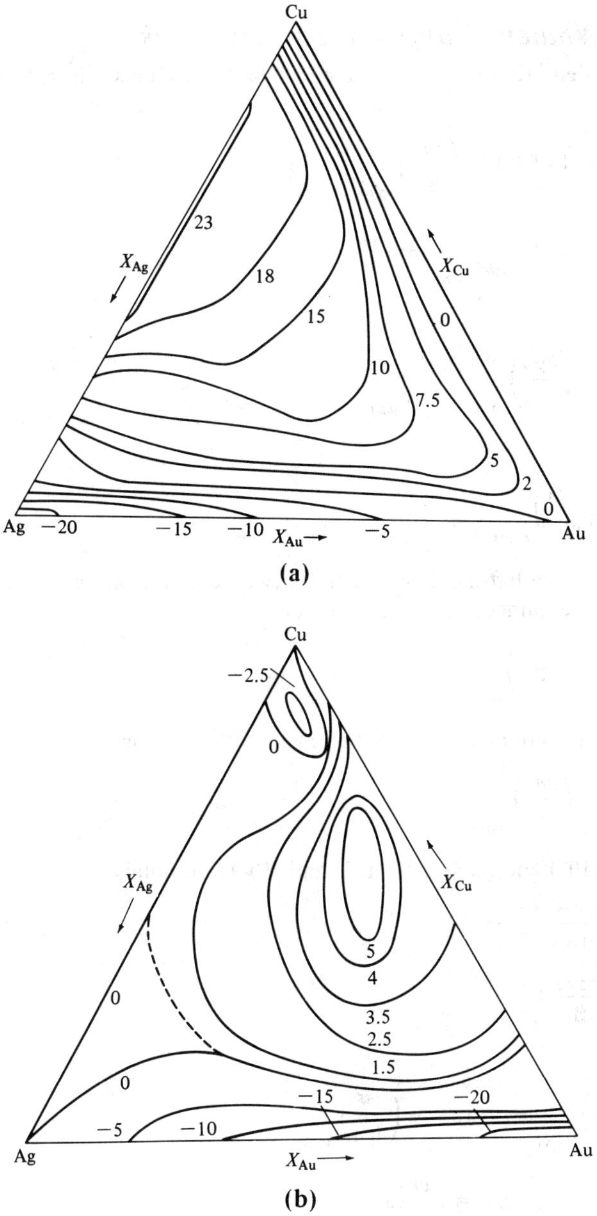

Figure 16. Curves of constant relative adsorption of **(a)** Ag and **(b)** Au with respect to Cu [$\Gamma^{(Cu)}$ in (mol/m^2) $\times$ 10^6] in the Ag–Au–Cu liquid solution at 1350°K [48].

Ag and Cu are in the same ratio as their mole fractions in the bulk;

$$\frac{\Gamma_{Ag}}{X_{Ag}} = \frac{\Gamma_{Cu}}{X_{Cu}} \tag{101}$$

At the point where the lines $\Gamma^{(Cu)}_{Ag} = 0$ and $\Gamma^{(Cu)}_{Au} = 0$ intersect, σ passes through an absolute minimum; the surface composition is then equal to that in the bulk:

$$\frac{\Gamma_{Ag}}{X_{Ag}} = \frac{\Gamma_{Au}}{X_{Au}} = \frac{\Gamma_{Cu}}{X_{Cu}} \tag{102}$$

9.2. *Method of Whalen, Kaufman, and Humenik*

In the method of Whalen, Kaufman, and Humenik [49] the Gibbs adsorption equation yields

$$-\left(\frac{\partial \sigma}{\partial \mu_2}\right)_{n_3/n_1} = \Gamma_2^{(1)} + \Gamma_3^{(1)} \left(\frac{\partial \mu_3}{\partial \mu_2}\right)_{n_3/n_1} \tag{103}$$

$$-\left(\frac{\partial \sigma}{\partial \mu_3}\right)_{n_2/n_1} = \Gamma_2^{(1)} \left(\frac{\partial \mu_2}{\partial \mu_3}\right)_{n_2/n_1} + \Gamma_3^{(1)} \tag{104}$$

We now note that

$$\left(\frac{\partial \mu_2}{\partial n_3}\right)_{n_1,n_2} = -\left(\frac{\partial \mu_2}{\partial n_2}\right)_{n_1,n_3} \left(\frac{\partial n_2}{\partial n_3}\right)_{\mu_2,n_1} \tag{105}$$

and

$$\left(\frac{\partial \mu_3}{\partial n_2}\right)_{n_1,n_3} = \left(\frac{\partial \mu_3}{\partial \mu_2}\right)_{n_1,n_3} \left(\frac{\partial \mu_2}{\partial n_2}\right)_{n_1,n_3} \tag{106}$$

Since the derivatives on the left-hand side of these two equations are equal (as both are equal to $\partial^2 G/\partial n_2\, \partial n_3$), we immediately deduce that

$$\left(\frac{\partial \mu_3}{\partial \mu_2}\right)_{n_1,n_3} = -\left(\frac{\partial n_2}{\partial n_3}\right)_{\mu_2,n_1} \tag{107}$$

(The derivation of this equation is due to Schuhmann [50].) Similarly,

$$\left(\frac{\partial \mu_2}{\partial \mu_3}\right)_{n_1,n_2} = -\left(\frac{\partial n_3}{\partial n_2}\right)_{\mu_3,n_1} \tag{108}$$

Combining equations (107) and (108) with (103) and (104), we obtain

$$\Gamma_2^{(1)} = \frac{\alpha_2 + \beta_{23}\alpha_3}{1 - \beta_{23}\beta_{32}} \tag{109}$$

$$\Gamma_3^{(1)} = \frac{\alpha_3 + \beta_{32}\alpha_2}{1 - \beta_{23}\beta_{32}} \tag{110}$$

where

$$\alpha_2 = -\left(\frac{\partial \sigma}{\partial \mu_2}\right)_{X_3/X_1}, \qquad \alpha_3 = -\left(\frac{\partial \sigma}{\partial \mu_3}\right)_{X_2/X_1} \tag{111}$$

$$\beta_{23} = \left(\frac{\partial n_2}{\partial n_3}\right)_{\mu_2,n_1}, \qquad \beta_{32} = \left(\frac{\partial n_3}{\partial n_2}\right)_{\mu_3,n_1} \tag{112}$$

Let us determine these various quantities at a point M (Fig. 17) of the composition triangle.

The line passing by the apex B of the triangle corresponds to a constant value of the ratio X_3/X_1. Plotting σ as a function of μ_2 along this line then yields α_2. α_3 is similarly deduced along a line passing by M and C.

β_{23} can be evaluated by finding the intercept N of the slope of the isoactivity curve (constant μ_2) with the BC edge of the triangle. β_{23} is then measured as NC/BN (or X_2/X_3). This construction is due to Schuhmann [50]. Its justification may be demonstrated along the following arguments. Consider the isoactivity curve as a function n_2 of n_3 at constant μ_2 and n_1. Any point on the slope of this curve at M has the same value of $(\partial n_2/\partial n_3)_{\mu_2,n_1}$. As $X_1 \to 0$, $\partial n_2/\partial n_3 \to n_2/n_3$ or X_2/X_3, and consequently the intercept N allows a convenient measure of β_{23}. Alternatively, it is possible to make a plot of n_2/n_1 vs n_3/n_1

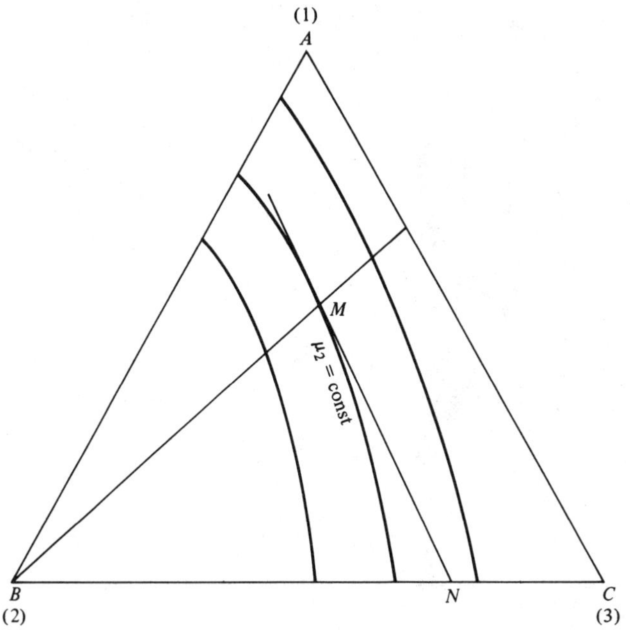

Figure 17. Determination of β_{23} (= NC/BN).

along the isoactivity curve and to take its slope. After a similar determination of β_{32}, equations (109) and (110) may be applied to deduce the values of $\Gamma_2^{(1)}$ and $\Gamma_3^{(1)}$.

Whalen et al. [49] have applied their technique to several Fe–C–X alloys. However, in an attempt to simplify the analysis, they assume that the term α_C is negligible. This may be shown to be incorrect and to lead to serious errors in the determination of $\Gamma_C^{(Fe)}$. For example, in the case of the Fe–C–Si system, they conclude that the addition of silicon leads to a desorption of carbon whereas proper analysis of the data leads to the opposite result. On balance, it seems that the direct method is faster and less cumbersome.

9.3. Graphic-Analytic Method

The two previous methods assumed the ready availability of isoactivity curves. Often, the thermodynamic properties of the bulk are known through analytic expressions and deducing from these expressions isoactivity curves is a lengthy task. In such cases, a further technique [48] is advantageous, based on the following expressions for the reduced adsorptions:

$$\frac{\bar{\Gamma}_2}{X_2} = -\frac{(1 - X_2)[1 + X_3(\partial \ln \gamma_3/\partial X_3)]}{\psi RT}\left(\frac{\partial\sigma}{\partial X_2}\right)_{X_3/X_1} + \frac{X_3(1 - X_3)(\partial \ln \gamma_2/\partial X_3)}{\psi RT}\left(\frac{\partial\sigma}{\partial X_3}\right)_{X_2/X_1} \tag{113}$$

$$\frac{\bar{\Gamma}_3}{X_3} = \frac{X_2(1 - X_2)(\partial \ln \gamma_3/\partial X_2)}{\psi RT}\left(\frac{\partial\sigma}{\partial X_2}\right)_{X_3/X_1} - \frac{(1 - X_3)[1 + X_2(\partial \ln \gamma_2/\partial X_2)]}{\psi RT}\left(\frac{\partial\sigma}{\partial X_3}\right)_{X_2/X_1} \tag{114}$$

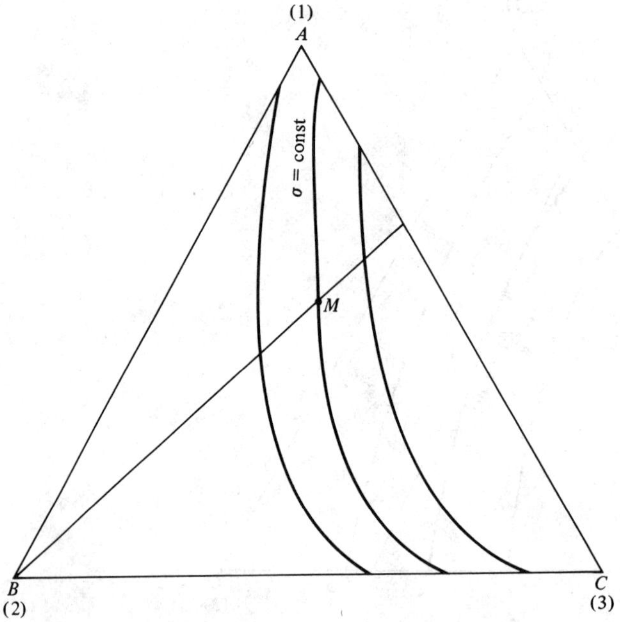

Figure 18. Determination of $(\partial\sigma/\partial X_2)_{X_3/X_1}$ at M. Along the line BM of constant X_3/X_1 the values of σ are recorded as a function of X_2. In the consequent plot, the slope of the curve at the composition X_2 corresponding to M yields $(\partial\sigma/\partial X_2)_{X_3/X_1}$.

for which we recall that the stability function ψ (Section XI.2) has been expressed as

$$\psi = \left(1 + X_2 \frac{\partial \ln \gamma_2}{\partial X_2}\right)\left(1 + X_3 \frac{\partial \ln \gamma_3}{\partial X_3}\right) - X_2 X_3 \frac{\partial \ln \gamma_2}{\partial X_3}\frac{\partial \ln \gamma_3}{\partial X_2} \tag{115}$$

Obtaining equations (113) and (114) from the Gibbs adsorption equation is straightforward, but the manipulations are somewhat too lengthy to warrant their reproduction here.

At a point M of the triangle with curves of constant σ (Fig. 18), a line BM is drawn and along that pseudobinary a plot of σ vs $\underline{X_2}$ may be deduced. The slope at M yields $(\partial\sigma/\partial X_2)_{X_3/X_1}$. The remaining steps to obtain $\overline{\Gamma}_2$ and $\overline{\Gamma}_3$ are evident.

The results of this technique for the Ag–Au–Cu system are shown in Fig. 19. Expressing adsorption behavior in terms of reduced adsorption rather than relative adsorption offers here a simpler interpretation of the data.

In the case of an ideal solution, equations (113) and (114) yield particularly simple expressions of $\overline{\Gamma}_2$ and $\overline{\Gamma}_3$. For example,

$$\overline{\Gamma}_2 = -\frac{X_2(1 - X_2)}{RT}\left(\frac{\partial\sigma}{\partial X_2}\right)_{X_3/X_1} \tag{116}$$

The surface is richer in component 2 than the bulk if the slope of σ vs X_2 along BM (constant X_3/X_1) is negative.

9.4. Analytic Method

The derivation of adsorption functions from surface tension data in a ternary system is somewhat reminiscent of a Gibbs–Duhem integration (Section X.3.3). The graphical work is time consuming, and a method exclusively analytic and supported by computer calculations may often be fastest and most reliable. The method may be based on the following

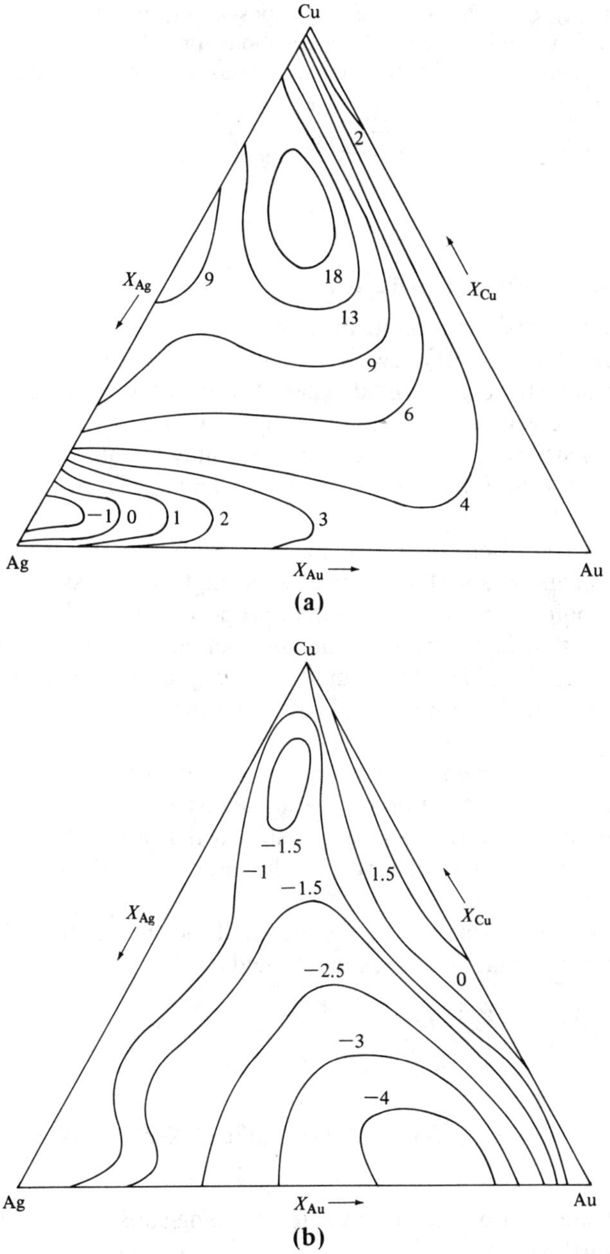

Figure 19. Curves of constant reduced adsorption of (**a**) Au and (**b**) Ag in the Ag–Au–Cu liquid solution at 1350°K [$\overline{\Gamma}$ in (mol/m^2) $\times$ 10^6]. (*From Gallois and Lupis* [*48*].)

equations:

$$\frac{\Gamma_2^{(1)}}{X_2} = -\frac{1 + X_3(\partial \ln \gamma_3/\partial X_3)}{\psi RT}\left(\frac{\partial \sigma}{\partial X_2}\right)_{X_3} + \frac{X_3(\partial \ln \gamma_3/\partial X_2)}{\psi RT}\left(\frac{\partial \sigma}{\partial X_3}\right)_{X_2} \tag{117}$$

$$\frac{\Gamma_3^{(1)}}{X_3} = \frac{X_2(\partial \ln \gamma_2/\partial X_3)}{\psi RT}\left(\frac{\partial \sigma}{\partial X_2}\right)_{X_3} - \frac{1 + X_2(\partial \ln \gamma_2/\partial X_2)}{\psi RT}\left(\frac{\partial \sigma}{\partial X_3}\right)_{X_2} \tag{118}$$

These are obtained simply by differentiating the Gibbs adsorption equation (97) with respect to X_2 and X_3 and by solving the resulting equations for $\Gamma_2^{(1)}$ and $\Gamma_3^{(1)}$.

In the case of an ideal solution, we note the simple form taken by the relative adsorption:

$$\Gamma_2^{(1)} = -\frac{X_2}{RT}\left(\frac{\partial\sigma}{\partial X_2}\right)_{X_3} = -\frac{1}{RT}\left(\frac{\partial\sigma}{\partial \ln X_2}\right)_{X_3} \tag{119}$$

---

10. Adsorption in Multicomponent Solutions

Although the interfacial tensions and adsorption characteristics of many binary systems have been determined, very few systematic investigations of ternary and higher order systems have been performed. These show that the effects of various solutes are often not merely additive. For example, although two solutes may separately be surface inactive in a given solvent, they may become surface active in the presence of each other. In liquid iron, C, Cr and C, Si are two pairs of such solutes [49].

The explanation of such behavior must lie in the different nature of atomic interactions in the bulk and at a surface. The case of the Fe–C–Si system appears to illustrate well this point. The thermodynamic properties of the liquid solution lead to the conclusion that in the bulk C and Si repel each other. On the other hand, surface properties (i.e., the strong mutual segregation of these solutes at the surface; see the end of Section 9.2) indicate a strong affinity and a bonding probably not unlike that of SiC.

Relatively little work has been done to model the adsorption characteristics of ternary and multicomponent solutions, presumably because of the dearth of accurate data. Nonetheless, Guttman [43] has shown that a substantial amount of data on temper embrittlement can be readily interpreted through a simple regular solution model.

Assuming different interactions among the atoms of the surface monolayer and of the bulk, the ternary regular solution model yields

$$X_2^{(m)} = k_2X_2/(X_1 + k_2X_2 + k_3X_3) \tag{120}$$

with

$$k_2 = \left(\frac{X_2^{(m)}}{X_2}\right)_{X_1\to 1} \exp(-\epsilon_2^{(2)}X_2 - \epsilon_2^{(3)}X_3)\exp(2\alpha_{12}^{(m)}X_2^{(m)} - \Omega_{23}^{(m)}X_3^{(m)}) \tag{121}$$

Expressions for $X_3^{(m)}$ and k_3 are obtained by interchanging subscripts 2 and 3. The parameter $\Omega_{23}^{(m)}$ is equal to

$$\Omega_{23}^{(m)} = \alpha_{23}^{(m)} - \alpha_{12}^{(m)} - \alpha_{13}^{(m)} \tag{122}$$

and

$$\alpha_{ij}^{(m)} = \frac{Z^{(m)}}{RT}\left[u_{ij}^{(m)} - \tfrac{1}{2}(u_{ii}^{(m)} + u_{jj}^{(m)})\right] \tag{123}$$

where $Z^{(m)}$ is the coordination number of an atom in the surface monolayer and $u_{ij}^{(m)}$ is the bonding energy between two atoms i and j in that monolayer.

10.1. Adsorption Interaction Coefficient $\xi_i^{(j)}$

It is often of considerable technological interest to examine whether the addition of one component j hinders or promotes the adsorption of another component i. For very dilute solutions, this effect may possibly be studied through the adsorption interaction coefficient $\xi_i^{(j)}$

$$\xi_i^{(j)} = \left(\frac{\partial \nu_i}{\partial X_j}\right)_{X_1 \to 1} \tag{124}$$

where the adsorption coefficient ν_i has been defined in equation (77).

We may assume that ν_i can be developed into a Taylor series with respect to the mole fractions of the solutes $X_2, X_3, \ldots, X_m$ and write

$$\nu_i = \nu_i^\infty + \sum_{j=2}^{m} \xi_i^{(j)} X_j + \cdots \tag{125}$$

We stress that in the case of surface-active species this equation would be valid only in the very dilute range, e.g., at $X_j < X_j^c$ (Section 8). Second order adsorption interaction coefficients do not seem presently warranted.

The coefficient $\xi_i^{(j)}$ is somewhat similar to the familiar free energy interaction coefficient for bulk solutions:

$$\epsilon_i^{(j)} = \left(\frac{\partial \ln \gamma_i}{\partial X_j}\right)_{X_1 \to 1} \tag{126}$$

We recall that we established (Section IX.2.5) a reciprocal relation between $\epsilon_i^{(j)}$ and $\epsilon_j^{(i)}$ (the two coefficients are equal). A reciprocal relation between $\xi_i^{(j)}$ and $\xi_j^{(i)}$ may also be derived. It originates from the Gibbs adsorption equation and the relation

$$\left(\frac{\partial^2 \sigma}{\partial X_i\, \partial X_j}\right)_{X_1 \to 1} = \left(\frac{\partial^2 \sigma}{\partial X_j\, \partial X_i}\right)_{X_1 \to 1} \tag{127}$$

The derivation is straightforward and yields

$$\xi_i^{(j)} = \xi_j^{(i)} + (\nu_i^\infty - \nu_j^\infty)(\epsilon_i^{(j)} + 1) \tag{128}$$

The equation could be valuable in checking the reliability of adsorption measurements (by Auger spectroscopy, for example).

In the concentration range where equation (125) is valid the interfacial tension σ may also be expanded into a Taylor series:

$$\sigma = \sigma_1 + \sum_{i=2}^{m} \left(\frac{\partial \sigma}{\partial X_i}\right)_{X_1 \to 1} X_i + \sum_{i=2}^{m} \frac{1}{2}\left(\frac{\partial^2 \sigma}{\partial X_i^2}\right)_{X_1 \to 1} X_i^2 + \sum_{i=2}^{m-1} \sum_{j>i}^{m} \left(\frac{\partial^2 \sigma}{\partial X_i\, \partial X_j}\right)_{X_1 \to 1} X_i X_j + \cdots \tag{129}$$

which, in terms of the interaction coefficients of adsorption may be rewritten

$$\sigma = \sigma_1 - \Gamma_1^\circ RT \left\{ \sum_{i=2}^{m} \nu_i^\infty X_i + \sum_{i=2}^{m-1} \sum_{j>i}^{m} [\xi_i^{(j)} + \nu_j^\infty(\epsilon_i^{(j)} + 1)] X_i X_j + \cdots \right\} \tag{130}$$

10.2 Examples

To illustrate the use of equation (130), let us assume that an investigator seeks to determine the surface activity of a component 2: $-(\partial\sigma/\partial X_2)_{X_1\to 1}$. Often, the contamination by a solute 3 is practically impossible to prevent, and what is being measured is the quantity

$$-\left(\frac{\partial\sigma}{\partial X_2}\right)_{X_3,X_2\to 0} \simeq -\left(\frac{\partial\sigma}{\partial X_2}\right)_{X_1\to 1} - \left(\frac{\partial^2\sigma}{\partial X_2\,\partial X_3}\right)_{X_1\to 1} X_3$$

$$\simeq \Gamma_1^\circ RT\{\nu_2^\infty + [\xi_2^{(3)} + \nu_3^\infty(\epsilon_2^{(3)} + 1)]X_3\} \qquad (131)$$

Consider the potential effect of 50 ppm of oxygen on the possible determination of the surface activity of silicon (or of ν_{Si}^∞) in liquid iron at 1550°C. The magnitude of the term $\nu_{\mathrm{O}}^\infty(\epsilon_{\mathrm{Si}}^{\mathrm{O}} + 1)X_{\mathrm{O}}$ alone is $2500 \times (-15 + 1) \times 1.7 \times 10^{-4}$, or about -6 ($\xi_{\mathrm{Si}}^{\mathrm{O}}$ is unknown); by comparison ν_{Si}^∞ is reported as 2.9. This example underscores the difficulty of measuring surface tensions accurately. Contamination problems are indeed the cause of many discrepancies in the literature.

Table 4 lists some estimates of the adsorption interaction coefficients $\xi_i^{(j)}$.

The regular solution model offers an expression of $\xi_2^{(3)}$, which may easily be derived from equations (120) and (121):

$$\xi_2^{(3)} = -(\nu_2^\infty + 1)[\nu_3^\infty(\Omega_{23}^{(m)} + 1) + (\Omega_{23}^{(m)} - \epsilon_2^{(3)})] \qquad (132)$$

Let us apply it to the case of the Fe–C–Si liquid solution at 1450°C. $\epsilon_{\mathrm{C}}^{\mathrm{Si}}$ is equal to 10.6 [51], and we may assume that the values of ν_{Si}^∞ and ν_{C}^∞ at 1550°C do not change appreciably at 1450°C; these values are 2.9 and 0.7, respectively. Consequently, equation (132) yields

$$\xi_{\mathrm{C}}^{\mathrm{Si}} = 38.6 - 6.6\Omega_{\mathrm{Si,C}}^{(m)} \qquad (133)$$

$\Omega_{\mathrm{Si,C}}^{(m)}$ is a relative measure of the bonding between Si and C at the surface [see equations (122) and (123)]. As noted earlier, the bond between Si and C is expected to be strong, resembling that of the SiC compound. Thus, $\Omega_{\mathrm{Si,C}}^{(m)}$ is expected to be negative and $\xi_{\mathrm{C}}^{\mathrm{Si}}$ positive. The data of Whalen et al. [49] lead indeed to an estimate of +120.

Table 4. Adsorption Interaction Coefficients [20]

Solvent	T (°K)	Solute 2	Solute 3	$\xi_2^{(3)}$	$\xi_3^{(2)}$
Fe	1623	Cr	C	50	65
	1700	Co	C	~200	~200
	1673	Ni	C	60	63
	1623	Mn	C	>0	
	1823	S	C	30,000	small
	1723	Si	C	120	90
	1823	W	C	>0	
	1823	S	Mn	−30,000	small
Ag	1381	O	Au	4,500	200

11. Heats of Adsorption and Effect of the Temperature on the Surface Tension

11.1. Standard State and Standard Heats of Adsorption

Let us consider the transfer of one mole of a solute 2 from the bulk of a phase α to its interface with another phase. This transfer may be represented in terms of a chemical reaction

$$\underline{2}(\alpha) = \underline{\underline{2}}(\text{ads}) \tag{134}$$

In Section XIII.1 we defined the chemical potential of a component at an interface and showed that at equilibrium it is equal to the chemical potential of that component in the bulk. Consequently, at equilibrium

$$\Delta G_{\alpha\to\sigma} = \mu_2^{\sigma} - \mu_2^{\alpha} = 0 \tag{135}$$

To define $\Delta G^{\circ}_{\alpha\to\sigma}$, the standard states of the solute 2 in the bulk and at the interface must be specified. For the bulk, it is convenient to select a Henrian reference state based on a mole fraction composition coordinate (Section VII.3.4):

$$\mu_2^{\circ(\alpha)} = (\mu_2^{\alpha} - RT \ln X_2^{\alpha})_{X_2^{\alpha}\to 0} \tag{136}$$

To define a similar standard state for the solute 2 at the interface, we note that $\Gamma_2^{(1)}/X_2$ tends towards a finite limit where $X_2 \to 0$. Consequently, we define $\mu_2^{\circ(\sigma)}$ by

$$\mu_2^{\circ(\sigma)} = (\mu_2^{\sigma} - RT \ln \Gamma_2^{(1)})_{\Gamma_2^{(1)}\to 0} \tag{137}$$

and $\Delta G^{\circ}_{\alpha\to\sigma}$ by the relation

$$\Delta G^{\circ}_{\alpha\to\sigma} = \mu_2^{\circ(\sigma)} - \mu_2^{\circ(\alpha)} \tag{138}$$

We see that at equilibrium

$$\Delta G^{\circ}_{\alpha\to\sigma} = -RT \ln K \tag{139}$$

where

$$K = (\Gamma_2^{(1)}/X_2^{\alpha})_{X_2^{\alpha}\to 0} \tag{140}$$

The associated standard enthalpy of reaction, i.e., the standard heat of adsorption from the bulk, is then defined by

$$\Delta H^{\circ}_{\alpha\to\sigma} = \frac{d(\Delta G^{\circ}_{\alpha\to\sigma}/T)}{d(1/T)} = -R\,\frac{d \ln K}{d(1/T)} \tag{141}$$

In the case of a gas–solid or gas–liquid interface, the transfer of one mole from the gas phase to the surface is often of more interest than the transfer of one mole from the bulk. The associated chemical reaction

$$2(\text{g}) = \underline{\underline{2}} \tag{142}$$

can be treated in an identical way. We note that it may also be considered to be the sum of the absorption reaction

$$2(\text{g}) = \underline{2}(\alpha) \tag{143}$$

and the adsorption reaction

$$\underline{2}(\alpha) = \underline{\underline{2}} \tag{144}$$

and that consequently

$$\Delta H^\circ_{g\rightarrow\sigma} = \Delta H^\circ_{g\rightarrow\alpha} + \Delta H^\circ_{\alpha\rightarrow\sigma} \tag{145}$$

The case where component 2 is present in the gas phase as a diatomic molecule offers no difficulty. For example, the reaction to be considered for the adsorption of oxygen from the gas phase to the surface of liquid silver is

$$\tfrac{1}{2}O_2(g) = \underline{\underline{O}} \tag{146}$$

The associated standard heat of adsorption has been found to be of the order of −34 kcal [24].

11.2. Isosteric Heat of Adsorption

Experimentally, it is found that for most gas–metal systems the heat of adsorption is a function of the coverage of the surface by the adsorbed species and that this heat generally decreases when the coverage increases. Several possible causes can be cited.

For example, if the surface of a solid is comprised of terraces, steps, and kinks (Fig. 20), a kink site would have a higher reactivity than a step or terrace site since an adsorbed atom would interact with more substrate atoms. A variation in the heterogeneity of the surface may also be due the reconstruction of the surface which often accompanies the adsorption process.

As can be expected, different crystallographic planes present different heats of adsorption. Table 5 shows the variation in the heat of adsorption of diatomic sulfur on silver and gold with the crystallographic orientation of the surface [14] (for a coverage $\theta = \frac{1}{2}$). It is observed that the heat of adsorption becomes more negative as the roughness of the surface increases (e.g., the metal atoms become less closely packed). The difference in the cases of the (111) and (110) planes is of the order of 10 kcal/mol, and thus far from negligible.

In addition to the heterogeneity of the surface, mutual interactions between adsorbed species are also responsible for changes in the heat of adsorption with coverage. Figure 21 is an example from Tracy et al. [31] which shows the decrease in the bond energy of CO on the (100) face of palladium. The decrease is interpreted as a repulsion between the dipoles of the CO molecules. For low coverages, this decrease appears to be minimized by the mobility of the adsorbed molecules.

Figure 20. Illustration of a terrace, step and kink at the surface of a solid.

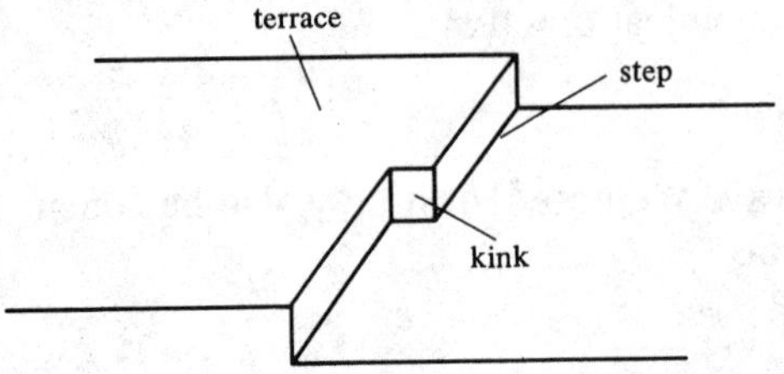

Table 5. Isosteric Heat of Adsorption of Sulfur on Silver and Gold [14]

Face	$(q_{st})_{\theta=\frac{1}{4}}$ [kcal/mol S_2 (gas)]	
	silver	gold
(111)	−54.4	−84
(100)	−58.6	−86
(110)	−66.2	−94

With increasing coverage, an average critical distance between the molecules is reached and the heat of adsorption decreases abruptly. Higher coverages correspond to the formation and compression of a two-dimensional structure identified by low-energy electron diffraction and denoted C(4×2)45°.

Having established that the heat of adsorption varies with coverage, it is important to specify the value of θ, or of the adsorption function $\Gamma_2^{(1)}$, at which this heat is being measured.

Let us consider a binary system for which the solute 2 is adsorbed at a gas–solid or gas–liquid interface:

$$2(g) = \underline{\underline{2}} \qquad (147)$$

Figure 21. Change in the bonding energy of CO adsorbed on the (100) face of palladium. (*After Tracy and Palmberg* [*31*].) *Note*: 1 eV = 23.062 kcal/mol.

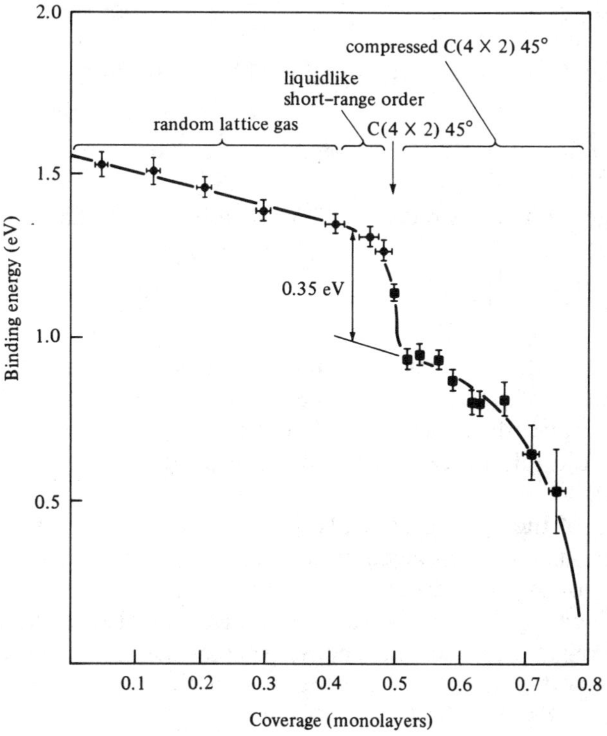

At equilibrium

$$\mu_2^g = \mu_2^\sigma \tag{148}$$

and

$$\mu_2^g - \mu_2^{\circ(g)} = \mu_2^\sigma - \mu_2^{\circ(g)} \tag{149}$$

or

$$R \ln f_2 = (\mu_2^\sigma - \mu_2^{\circ(g)})/T \tag{150}$$

For a planar interface, the system depends on two independent variables. We may select them as T and $\Gamma_2^{(1)}$. Differentiating equation (150) with respect to $1/T$ at constant $\Gamma_2^{(1)}$, we obtain

$$R\left(\frac{\partial \ln f_2}{\partial(1/T)}\right)_{\Gamma_2^{(1)}} = \left(\frac{\partial(\mu_2^\sigma/T)}{\partial(1/T)}\right)_{\Gamma_2^{(1)}} - \frac{d(\mu_2^{\circ(g)}/T)}{d(1/T)} \tag{151}$$

or

$$R\left(\frac{\partial \ln f_2}{\partial(1/T)}\right)_{\Gamma_2^{(1)}} = H_2^\sigma - H_2^{\circ(g)} = q_{st} \tag{152}$$

where H_2^σ represents the first partial derivative on the right hand side of equation (152). The difference $H_2^\sigma - H_2^{\circ(g)}$ is called the *isosteric heat of adsorption* and often given the symbol q_{st}. It is a function of $\Gamma_2^{(1)}$ (or θ). When $\Gamma_2^{(1)} \to 0$, q_{st} becomes equal to the standard heat of adsorption from the gas phase defined previously:

$$(q_{st})_{\Gamma_2^{(1)} \to 0} = \Delta H_{g\to\sigma}^\circ \tag{153}$$

For a diatomic gas such as oxygen, equation (152) becomes

$$\frac{R}{2}\left(\frac{\partial \ln f_{O_2}}{\partial(1/T)}\right)_{\Gamma_O^{(1)}} = H_O^\sigma - \tfrac{1}{2}H_{O_2}^{\circ(g)} = q_{st} \tag{154}$$

and, assuming oxygen in the gas phase behaves ideally, we obtain

$$\left(\frac{\partial \ln p_{O_2}^{1/2}}{\partial T}\right)_{\Gamma_O^{(1)}} = -\frac{q_{st}}{RT^2} \tag{155}$$

Plotting adsorption isotherms at various temperatures, equation (152) or (155) allows the determination of q_{st}. It is also possible to measure the heat of adsorption at various coverages directly in a calorimeter. The heat thus measured is not identical to q_{st} but the difference is generally negligible (see, e.g., Defay and Prigogine [3, p. 51]).

If the equilibrium of the system depends on more than two independent variables, other restrictions must be placed on the derivatives in equation (152). For example, in the case of oxygen adsorption by liquid silver, argon may be present in the gas phase; since argon is not adsorbed, the additional restriction may be taken to be that of constant pressure. If another species than oxygen is adsorbed, the derivatives may be taken at fixed coverages of both species. For nonplanar surfaces, the curvatures should be kept constant.

11.3. Physical and Chemical Adsorptions

The adsorption of gases on solids may generally be classified into two types of behavior: physical adsorption and chemical adsorption.

Physical adsorption may be characterized by a low isosteric heat, less than, roughly, 10 kcal/mol (in absolute value). The atomic forces binding adsorbed atoms to the surface are usually van der Waals forces. The electronic state of the atoms of the solid surface is not expected to undergo substantial change, and as a result the surface is not reconstructed. Physical adsorption is not very specific as to the nature of the solid involved and is easily reversible. This is characteristic of the adsorption of rare gases on solids and of molecular (i.e., nondissociative) adsorption of diatomic gases. For these gases, Henry's law for surfaces has often been verified, i.e.,

$$(\Gamma_i/p_i)_{p_i\to 0} = \text{nonzero finite value} \tag{156}$$

By contrast, chemical adsorption (or chemisorption) is characterized by large isosteric heats: of the order of 20–150 kcal/mol. The atoms of the gas and of the solid undergo substantial changes (as to the transfer or concentration of electrons) characteristic of ionic and covalent bonding. The surface of the solid is often reconstructed.

Chemisorption presents many similarities with a chemical reaction. One of the most notable is its specificity as to the nature of the atoms involved. For homopolar diatomic gases such as O_2, N_2, and H_2, chemical adsorption is dissociative. The associated heats of adsorption are generally higher than the heats of formation of the corresponding compounds.

At room temperature, chemisorption is generally not reversible; e.g., the adsorbed species cannot be removed by vacuum. However, chemisorption may become reversible at high temperatures.

11.4. Effect of Temperature on the Surface Activity

In equations (140) and (141) we established that

$$\Delta H^\circ_{\alpha\to\sigma} = -R\,\frac{d\ln(\Gamma_2^{(1)}/X_2^\alpha)_{X_2^\alpha\to 0}}{d(1/T)} \tag{157}$$

The limit of $\Gamma_2^{(1)}/X_2^\alpha$ has already been related to the surface activity through the Gibbs adsorption equation. We found [equation (76)] that

$$\left(\frac{\Gamma_2^{(1)}}{X_2^\alpha}\right)_{X_2^\alpha\to 0} = \frac{J_2}{RT} \tag{158}$$

Consequently, equation (157) yields

$$\frac{d\ln J_2}{dT} = \frac{\Delta H^\circ_{\alpha\to\sigma} + RT}{RT^2} \tag{159}$$

$\Delta H^\circ_{\alpha\to\sigma}$ is large and negative for surface-active species. Therefore, the surface activity J_2 of such species decreases when the temperature increases. For example, let us assume a standard enthalpy of adsorption equal to -50 kcal/mol and a temperature of 1000°K. Equation (159) shows that $d\ln J_2/dT$ is then equal

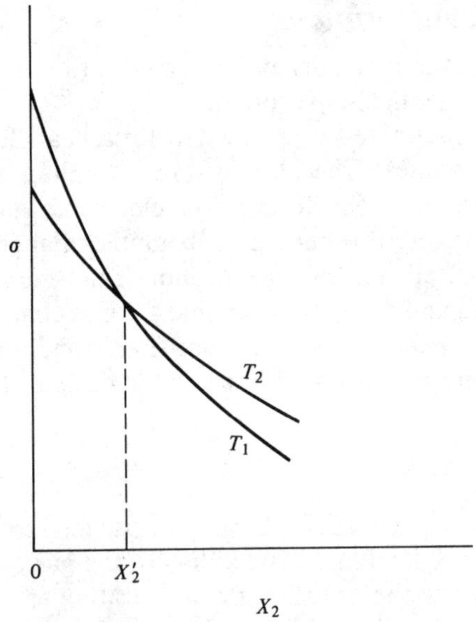

Figure 22. Dependence of the surface tension on temperature and composition ($T_2 > T_1$). The derivative of σ with respect to T is negative for $X_2 < X_2'$ and positive for $X_2 > X_2'$.

to $-0.024\ °K^{-1}$; a temperature increment of 30°C decreases the value of J_2 by a factor of 2.

In Section XIII.9 we noted that the surface tension of a pure metal generally decreases with temperature. The presence of a surface-active species decreases it further, but less rapidly at higher temperatures (Fig. 22). Above a certain concentration, the surface tension of the metal is then found to increase with temperature. This concentration is often of the order of parts per million. Consequently, even very low contamination levels can lead to errors in the sign and order of magnitude of $d\sigma/dT$ for "pure" metals. This is the source of many serious discrepancies in experimental results reported in the literature.

12. Summary

Most chemical reactions involve the transfer of mass across an interface and the composition of that interface often has a determining role in controlling the kinetics of these reactions. The composition of grain boundaries, of metal–ceramic joints and of other composites' interfaces may also have a strong effect on the mechanical properties of these systems. For these reasons, as well as many others, the study of adsorption phenomena has great scientific and technical importance. The recent development of experimental techniques, such as Auger spectroscopy, has spurred the interest of many researchers. Considerable progress is being achieved and more is expected in the near future.

The composition of an interface is generally expressed through the formalism of adsorption functions Γ_i based on Gibbs's concept of surface excess quantities. Because these excess quantities depend on the position of the dividing surface, relative and reduced adsorption functions are introduced. In the case of metal–gas interfaces, they take the simple form

$$\Gamma_i^{(1)} = \Gamma_i - (X_i/X_1)\Gamma_1 \tag{160}$$

$$\overline{\Gamma}_i = \Gamma_i - X_i \sum_{j=1}^{m} \Gamma_j = \Gamma_i - X_i\Gamma \tag{161}$$

The reduced function $\overline{\Gamma}_i$ has the advantage of symmetry with respect to the components of the system, although it yields a somewhat more cumbersome expression of the Gibbs adsorption equation. Cahn's treatment eliminates the arbitrariness of a dividing two-dimensional surface but is applicable only to planar surfaces.

The assumption that an interface's heterogeneity is confined to a single monolayer is extremely convenient and offers a simple interpretation of the formalism by which surfaces are studied and of many experimental results. It is, however, an unrealistic assumption,[3] especially for mixtures of like atoms. It appears more acceptable in the case of nonmetals such as oxygen and sulfur adsorbed on metals.

The perfect solution model of an interface is to surfaces what the ideal solution model is to bulk solutions—mostly a convenient reference. Its representative equation for the surface tensions of binary alloys is

$$e^{-\sigma\alpha/kT} = X_1 e^{-\sigma_1\alpha/kT} + X_2 e^{-\sigma_2\alpha/kT} \tag{162}$$

Surface tension and adsorption data for most alloys show large deviations from the model which are difficult to predict in any quantitative way. This is a reflection of the fact that interatomic surface forces are quite different from their corresponding forces in the bulk and are presently poorly known.

Surface-active species are characterized by a high ratio of their concentrations at the surface and in the bulk. At infinite dilution, this ratio has generally a nonzero finite limit:

$$\left(\frac{\Gamma_2^{(1)}}{X_2}\right)_{X_2\to 0} = \left(\frac{\overline{\Gamma}_2}{X_2}\right)_{X_2\to 0} = \text{nonzero finite value} \tag{163}$$

This limit is related to the slope at the origin of the curve σ vs X_2, i.e., to the surface activity J_2, through the Gibbs adsorption equation. The relation is

$$\left(\frac{\Gamma_2^{(1)}}{X_2}\right)_{X_2\to 0} = \left(\frac{\overline{\Gamma}_2}{X_2}\right)_{X_2\to 0} = \frac{J_2}{RT} \tag{164}$$

With increasing concentrations, a saturation stage is reached which may usually be likened to the formation of a two-dimensional compound.

Adsorption behaviors in multicomponent systems are poorly understood and there is a real dearth of accurate and systematic data. Contamination is a serious experimental problem, for impurities at the level of parts per million can have very significant effects. The adsorption interaction coefficient $\xi_i^{(j)}$ is convenient to reduce experimental data and to analyze discrepancies. A conversion rela-

[3] It may be added that, at least for regular solutions, the monolayer model has been shown to be incompatible with the Gibbs equation [3].

tionship between $\xi_i^{(j)}$ and $\xi_j^{(i)}$ also offers the possibility of a check on adsorption data. The proposed definition of $\xi_i^{(j)}$ is based on reduced adsorption functions but is readily adaptable to relative adsorption functions.

Adsorption data may be obtained from surface tension measurements through the Gibbs adsorption equation. For ternary and higher order systems, the derivation may be done by several graphical techniques. These are somewhat akin to the techniques of integrating the Gibbs–Duhem equation for bulk multicomponent solutions. These techniques are somewhat lengthy and tedious; an analytic solution (computer assisted) is often preferable.

Physical adsorption is generally associated with van der Waals forces, while chemical adsorption is associated with stronger bonding of a more chemical nature. Physisorption and chemisorption are perhaps best distinguished by the magnitude of the heats of adsorption. Because these heats are sensitive to surface composition, they are often defined at constant adsorption and bear the name of isosteric heats. At infinite dilution of a solute, the isosteric heat is related to the temperature dependence of the surface activity of that solute by equation (159).

Problems

1. Given a binary system of two phases α and β, calculate the effect of pressure on their interfacial tension at constant temperature.

2. **a.** A surface-active species 2 adsorbs on 1 by building a close-packed monolayer. If the atomic radius of 2 is 1.5 Å, what would be the value of Γ_2 at saturation coverage (in mol/m^2)?
   b. Assuming that at $T = 1000°K$ and $X_2 = 10^{-3}$ saturation has already been attained and $\sigma = 1200$ dyn/cm, estimate the value of σ at $X_2 = 10^{-2}$.

3. Assume that a solute 2 behaves ideally in the bulk of a solid 1, and that its adsorption on the surface of 1 obeys the equations (87) and (88) of the Fowler and Guggenheim model. Calculate, in terms of the parameters of the model (K and $\alpha = Z\omega/kT$), the quantities $(d\sigma/dX_2)_{X_2\to 0}$, $(d^2\sigma/dX_2^2)_{X_2\to 0}$, ν_2^∞, and $\xi_2^{(2)} = (d\nu_2/dX_2)_{X_2\to 0}$.

4. **a.** At 1100°C, the surface tensions of pure liquid copper and lead are 1.279 N/m and 0.377 N/m, respectively. The area occupied by a Cu atom at the surface is estimated at 6.11 Å^2. Assume that the surface properties of the liquid Cu–Pb alloys at 1100°C are those of a perfect solution and calculate the surface activity of Pb and the adsorption coefficient ν_{Pb}^∞ (with $k = \Gamma_{Pb}^\circ$).
   b. It has been found experimentally that ν_{Pb}^∞ at 1100°C is equal to 135. Discuss your previous result.

5. Assuming that adsorption is restricted to a monolayer, the difference $\pi = \sigma_{solvent} - \sigma_{solution} = \sigma_1 - \sigma$ may be interpreted as a surface pressure. For example, if a line barrier at a liquid–gas interface were to separate an area of pure solvent 1 from an area where the solute 2 is adsorbed, that barrier would be subjected to a force arising from the bombardment of that line by the adsorbed atoms 2. This is the two-dimensional equivalent of the pressure of a gas, which arises from the bombardment of its molecules against the walls of the container of that gas.

   The two-dimensional equivalent to the ideal gas law is $\pi A = RT$, where A is the area occupied by one mole of solute 2, or $1/\Gamma_2$. With a covolume correction to the ideal gas law, the equation of state is $\pi(A - A_{sat}) = RT$. Calculate the relation between the activity of 2 in the bulk and its concentration at the surface $\theta_2 = \Gamma_2/\Gamma_2^{sat}$, for both equations of state. (*Note*: these equations of state apply to dilute solutions.)

6. On the basis of a statistical model for surface-active species, Bernard and Lupis [39] derived the following equation of state:

$$[\sigma_1 - \sigma - (gA_{sat}/A^2)](A - A_{sat}) = RT$$

which is analogous to the van der Waals equation of state, and in which g accounts for interactions between adsorbed atoms of the solute 2. Calculate the dependence of $\ln a_2$ and of $\sigma_1 - \sigma$ on $\theta = \Gamma_2/\Gamma_2^{sat} = A_{sat}/A$. Deduce the dependence of the isosteric heat of adsorption on θ from the fact that, for oxygen adsorbed on liquid silver [39] or copper [26], g/kT has been found approximately constant.

7. The surface tension of pure liquid silver has been determined to be equal to 0.905 N/m at 1253°K and to 0.890 N/m at 1381°K. The effect of oxygen on the surface tension of liquid silver has also been measured and it has been calculated that ν_O^∞ equals 1750 at 1253°K and 550 at 1381°K.

a. Calculate the isosteric heat of adsorption of oxygen on liquid silver for very dilute concentrations of oxygen. Γ_{Ag}° equals 0.21×10^{-4} mol/m² and its temperature dependence may be neglected.

b. Estimate at what concentration of oxygen the surface tension of the liquid silver–oxygen solution shows an increase with an increase in temperature.

8. Derive the reciprocal relationship between $\xi_2^{(3)}$ and $\xi_3^{(2)}$ [equation (128)] from the Gibbs adsorption equation.

Figure 23. Effect of oxygen on the surface tension of liquid silver–gold alloys at 1108°C.

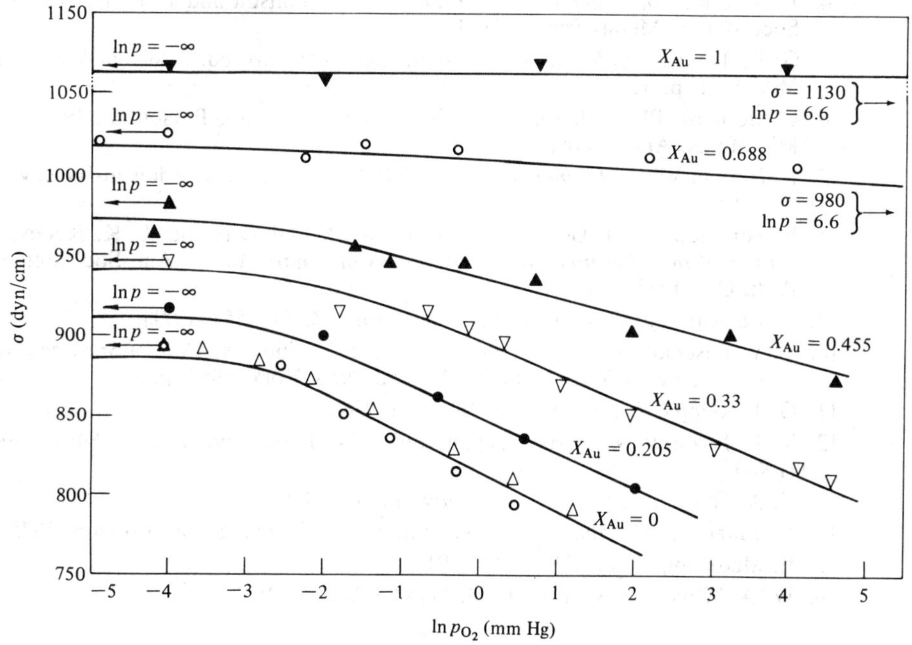

9. **a.** The following adsorption coefficients are defined:

$$\nu_2^* = \frac{\Gamma_2^{(1)}}{X_2}\frac{1}{k} \qquad \text{and} \qquad \xi_2^*(3) = \left(\frac{\partial \nu_2^*}{\partial X_3}\right)_{X_1\to 1}$$

Find the relationship between $\nu_2^{*\infty}$, $\xi_2^{*(3)}$ and ν_2^∞, $\xi_2^{(3)}$ and the conversion relationship between $\xi_2^{*(3)}$ and $\xi_3^{*(2)}$. Assume the same value of k for ν_2 and ν_2^*.

b. What would be the advantages and disadvantages of using the interaction adsorption coefficient $\theta_2^{(3)} = (\partial \ln \nu_2/\partial X_3)_{X_1\to 1}$ instead of $\xi_2^{(3)}$?

10. In analyzing the surface tension of Fe–C–S liquid solutions at 1450°C, Kozakevitch et al. [52] noted that, within the accuracy of their measurements, a single curve, σ vs a_S, represented well all the data, regardless of the concentration of carbon. Express the consequences of this fact to evaluate ξ_S^C.

11. Figure 23 shows some data of Bernard and Lupis [6,53] on the surface tensions of liquid Ag–Au–O alloys. From these data, it has been calculated that $\xi_O^{Au} = 4500$. Oxygen has no solubility in Au, and when dilute in liquid silver, follows the relationship $X_O = 1.83 \times 10^{-2}\,(p_{O_2})^{1/2}$ at 1381°K (p_{O_2} is measured in atmospheres); in addition $(\epsilon_O^{Au})_{\text{in Ag}} = 6.8$ [54]. Interpret the value of ξ_O^{Au} in terms of the regular solution model. Sketch the curves X_O^m, the concentration of oxygen at the surface assuming a monolayer model, vs X_O, the fraction of oxygen in the bulk, at $X_{Au} = 0$ and $X_{Au} = 0.1$.

References

1. P. Sebo, B. Gallois, and C. H. P. Lupis, *Met. Trans.* **8B,** 691–693 (1977).
2. J. W. Gibbs, *Collected Works*. Yale University Press, New Haven, 1948, Vol. 1.
3. R. Defay and I. Prigogine, with A. Bellemans and D. H. Everett, *Surface Tension and Adsorption*. Wiley, New York, 1966.
4. J. W. Cahn, *Interfacial Segregation* (W. C. Johnson and J. M. Blakely, ed.). Am. Soc. Metals, Metals Park, OH, 1979.
5. G. B. Thomas, *Calculus and Analytic Geometry*, 3rd ed. Addison-Wesley, Reading, MA, 1961, p. 420.
6. G. Bernard, Ph.D. thesis, Carnegie-Mellon University, Pittsburgh, 1970, University Microfilms, Ann Arbor, MI.
7. L. H. Van Vlack, *Elements of Materials Science*, 2nd ed. Addison-Wesley, Reading, MA, 1964.
8. R. Hultgren, P. D. Desai, D. T. Hawkins, M. Gleiser, and K. K. Kelley, *Selected Values of the Thermodynamic Properties of Binary Alloys*. Am. Soc. Metals, Metals Park, OH, 1973.
9. G. Bernard and C. H. P. Lupis, *Met. Trans.* **2,** 555–559 (1971).
10. V. V. Fesenko, V. N. Eremenko, and M. I. Vasiliu, *The Role of Surface Phenomena in Metallurgy*. N.Y. Consultants Bureau, New York, 1963, p. 31.
11. O. J. Kleppa, *J. Phys. Chem.* **60,** 446 (1956).
12. V. B. Lazarev, V. S. Arakalcyan, and A. V. Pershikov, *Russ. J. Phys. Chem.* **42,** 1 (1968).
13. E. A. Guggenheim, *Trans. Faraday Soc.* **41,** 50 (1945).
14. J. Oudar, *Physics and Chemistry of Surfaces*. Blakie & Son, London, 1975.
15. E. Madelung, *Phys. Z.* **20,** 494 (1919).
16. W. A. Weyl, *Trans. N.Y. Acad. Sci.* **12,** 245 (1950).

17. C. H. P. Lupis, H. Gaye, and G. Bernard, *Scripta Met.* **4,** 497–502 (1970).
18. P. Laty, J. C. Joud, and P. Desré, *Surface Sci.* **60,** 109 (1976); L. Goumiri, Ph.D. thesis, University of Grenoble, France, 1980.
19. S. M. Kaufman, *Acta Met.* **15,** 1089 (1967).
20. C. H. P. Lupis and B. Gallois, *Proceedings of an International Conference on the Physical Chemistry of Iron and Steelmaking, October 1978*. Versailles, France.
21. P. Kozakevitch and G. Urbain, *Mem. Sci. Rev. Met.* **58,** 401, 517, and 931 (1961).
22. L. H. Germer, R. M. Stern, and A. U. Mac Rae, *Metals Surfaces: Structure, Energetics and Kinetics*. Am. Soc. Met. 1963, p. 287.
23. A. U. Mac Rae, *Surface Sci.* **1,** 319 (1964).
24. G. Bernard and C. H. P. Lupis, *Met. Trans.* **2,** 2991 (1971).
25. L.-D. Lucas, Maizières-lès-Metz, France, private communication, 1978.
26. B. Gallois and C. H. P. Lupis, *Met. Trans.* **12B,** 549–557 (1981).
27. G. R. Belton, *Met. Trans.* **7B,** 35 (1976).
28. R. H. Fowler and E. A. Guggenheim, *Statistical Thermodynamics*. MacMillan, New York, 1939, p. 430.
29. J. Suzanne, J. P. Coulomb, and M. Bienfait, *Surface Sci.* **44,** 141 (1974).
30. J. C. Shelton, H. R. Patil, and J. M. Blakely, *Surface Sci.* **43,** 493 (1974).
31. J. C. Tracy and P. W. Palmberg, *J. Chem. Phys.* **51,** 4852 (1969).
32. H. Conrad, G. Ertl, J. Koch, and E. E. Latta, *Surface Sci.* **43,** 464 (1974).
33. L. C. Isett and J. M. Blakely, *Surface Sci.* **47,** 645 (1975).
34. L. C. Isett and J. M. Blakely, *J. Vac. Sci. Technol.* **12,** 237 (1975).
35. J. L. Domange, *J. Vac. Sci. Technol.* **9,** 682 (1972).
36. J. Oudar and M. Huber, *J. Cryst. Growth* **31,** 345 (1975).
37. J. L. Domange, J. Oudar, and J. Benard, *Molecular Processes on Solid Surfaces* (E. Drauglis et al., eds.). McGraw-Hill, New York, 1969, p. 353.
38. M. Kostelitz, J. L. Domange, and J. Oudar, *Surface Sci.* **34,** 431 (1973).
39. G. Bernard and C. H. P. Lupis, *Surface Sci.* **42,** 61 (1974).
40. M. Hillert and L.-I. Staffanson, *Acta Chem. Scand.* **24,** 3618 (1970).
41. M. Guttmann, *Met. Trans.* **8A,** 1383 (1977).
42. E. D. Hondros and M. P. Seah, *Met. Trans.* **8A,** 1363 (1977).
43. M. Guttmann, *Surface Sci.* **53,** 213 (1975).
44. G. Metzgar, *Z. Phys. Chem.* **211,** 1 (1953).
45. J. C. Joud, N. Eustathopoulos, A. Bricard, and P. Desré, *J. Chim. Phys.* **71,** 47 (1974).
46. P. Kozakevitch, *Rev. Metallurg.* **46,** 505, 572 (1949).
47. P. P. Evseev and A. F. Filippov, *Izv. V.U.Z. Chem. Met.* **10,** 55 (1967).
48. B. Gallois and C. H. P. Lupis, *Met. Trans.* **12B,** 679–689 (1981).
49. T. J. Whalen, S. M. Kaufman, and M. Humenik, *Trans. Am. Soc. Met.* **55,** 778 (1962).
50. R. Schuhmann, *Acta Met.* **3,** 219 (1955).
51. G. K. Sigworth and J. F. Elliott, *Trans. Met. Soc. AIME* **233,** 257–258 (1965).
52. P. Kozakevitch, S. Chatel, G. Urbain, and M. Sage, *Rev. Met.* **52,** 146 (1955).
53. G. Bernard and C. H. P. Lupis, Carnegie-Mellon University, unpublished results, 1970.
54. C. H. P. Lupis and J. F. Elliott, *Trans. Met. Soc. AIME* **242,** 929 (1968).

XV Statistical Models of Substitutional Metallic Solutions

1. Introduction

In preceding chapters we have often invoked models of statistical thermodynamics in order to justify formalisms or to interpret experimental results. In this and the following chapter we shall not attempt to give a rigorous, exhaustive treatment of the methods of statistical thermodynamics, for which we refer the reader to such classic texts as those of Fowler and Guggenheim [1], Mayer and Mayer [2], Guggenheim [3], Prigogine [4], and Hill [5]. Rather, our purpose now is to show the reader how even a very elementary understanding of the theory of statistical mechanics may be used advantageously to understand, describe, and predict many of the thermodynamic properties of solutions through the development of simple models. We shall confine our interest to properties which have a chemical character (e.g., G^{M}, H^{E}) rather than a dynamic one (e.g., diffusion, viscosity).

To develop a model of a solution, it is usually necessary to introduce the *partition function* of that solution. Let us designate it by Q. It may be defined

$$Q = \sum_j \exp(-E_j/kT) \tag{1}$$

where E_j is the energy of the system in the state j, k is Boltzmann's constant, and T is the temperature; the summation is extended to all possible energy states. If the summation were to be extended to all energy levels, Q would be written

$$Q = \sum_i g_i \exp(-E_i/kT) \tag{2}$$

where g_i is the *degeneracy factor* of the level i, that is, the number of states corresponding to the same energy level E_i.

Once the partition function has been determined, the thermodynamic properties of the solution may be readily obtained. It may be demonstrated that the Helmholtz free energy is simply related to Q by the expression

$$A = -kT \ln Q \tag{3}$$

In addition, for condensed systems (solids and liquids) and at ordinary pressure levels, the Gibbs free energy may be assumed to be equivalent to the Helmholtz free energy, because their difference, the term PV, is usually negligible [6,7]. Thus, for all practical purposes,

$$G \simeq -kT \ln Q \tag{4}$$

Other functions such as the enthalpy, entropy, and volume may be obtained by differentiation.

To calculate Q the maximum term method is very convenient. It allows, in a sum of *very large* numbers, the replacement of the sum by its largest term (even when the largest term is accompanied by several terms which are equal or equivalent). A mathematical treatment of this approximation has been given by Hill [5]. However, it may be readily understood if one reflects on the magnitude of the terms generally involved. A term such as $N!$ with $N = 10^{23}$ is of the order of $10^{10^{24}}$; we may then note, for example, that $100 \times (10^{23}!) \simeq 10^{23}!$ since $100 \times 10^{10^{24}} = 10^{(10^{24}+2)} \simeq 10^{10^{24}}$.

In the development of a model, many assumptions have to be made. A model with too many simplifying assumptions may not represent the data adequately. On the other hand, a model which is too realistic is often mathematically intractable or introduces too many parameters. Thus, for a model to be useful, a compromise is usually necessary.

The energy of a system of atoms is made of many contributions: translational, rotational, vibrational, nuclear, chemical, ... We observe that in calculating a mixing property (the difference in the value of a thermodynamic function after and before the elements of a solution have been mixed), the contributions which have remained unchanged in the process of mixing will cancel out. They may therefore be disregarded from the start. This will be verified occasionally in the sections below.

For condensed systems (as opposed to gaseous systems) the translational contribution may be ignored: the solution itself does not move and the quasi-random movements of its atoms (diffusion) are not sufficiently important. The rotational contribution is to be taken into account only if molecules are involved: for metallic solutions, the "building blocks" are generally atoms instead of molecules, and their rotational contribution is quite negligible. In addition, the electronic contribution is often safely ignored because of the high level of its excitation energy. In most applications, this is also true for the case of the nuclear contribution (due to the existence of the spin). Finally, the energy contributions most likely to be affected by mixing are the chemical and vibrational ones.

The models we shall develop do not distinguish between liquids and disordered solids. This is justified by the following considerations. Unlike ionic solutions, the properties of metallic systems may be interpreted through short-range interactions (Section VI.4.3), and short-range orders in liquids and solids are not very different [8]. The situation is somewhat akin to that of a very crowded room (a room "packed" with people). A person in that room has a certain amount of interaction with his neighbors, and we note that it is relatively difficult for that person to reach another position in the room. If 10% of the people in that room were to leave, a person would still have rather equivalent interactions with his neighbors, but his ability to move through the room would be greatly enhanced. Thus, as long as we confine our study to chemical properties rather than dynamic ones, treating liquids as disordered solids is a justified structural approximation.

2. Ideal Solution

In the model known as the *ideal solution*, the essential assumption is that there is no energy change associated with a rearrangement of the atoms A and B. For example, an arrangement which has all the atoms of the same kind (A or B) segregated together and an arrangement in which the atoms A and B alternate on the lattice sites in any imaginable pattern have exactly the same energy level. It follows that if E_A and E_B represent the energy of n_A moles of pure A and of n_B moles of pure B, the energy of $n_A + n_B$ moles of the solution after mixing remains equal to the energy of the system before mixing: $E_A + E_B$.

Each distinguishable arrangement of the atoms represent a different state. To calculate the unique degeneracy factor (unique because we have assumed that there is only one energy level $E_A + E_B$), we must seek the total number of distinguishable arrangements.

There are $N!$ different ways of distributing N atoms among N lattice sites. This may be seen in the following way. Assume that we have numbered the sites and we proceed to fill them one by one. To fill the first site we have a choice of N atoms. But, for the second site the choice is only among $N - 1$ atoms. For the third, it is $N - 2$, etc. Thus, the total number of arrangements is the product of all possible ways: $N(N - 1)(N - 2) \cdots 1$ or $N!$. However, if a given site is filled with an atom A and another site is also filled with an atom A, a switch between the two atoms leads to an undistinguishable state. If there are N_A atoms, there are also N_A sites filled with A atoms and $N_A!$ ways of redistributing the A atoms among the N_A sites. Similarly, there are $N_B!$ ways of redistributing the B atoms. Consequently, the total number of distinguishable arrangements is $N!/(N_A!N_B!)$. Thus,

$$Q = g\exp(-E/kT) = (N!/N_A!N_B!)\exp[-(E_A + E_B)/kT] \tag{5}$$

and

$$G = A = -kT\ln Q = (E_A + E_B) - kT\ln(N!/N_A!N_B!) \tag{6}$$

To calculate the Helmholtz or Gibbs free energy of mixing, the contributions of the pure components, in this case $E_A + E_B$, must be substracted. Thus,

$$G^M = -kT\ln(N!/N_A!N_B!) \tag{7}$$

Using Sterling's approximation ($\ln x! = x\ln x - x$, for large x) equation (7) becomes

$$\begin{aligned} G^M &= -kT[(N_A + N_B)\ln N - N_A\ln N_A - N_B\ln N_B] + kT(N - N_A - N_B) \\ &= kT[N_A\ln(N_A/N) + N_B\ln(N_B/N)] = RT(n_A\ln X_A + n_B\ln X_B) \end{aligned} \tag{8}$$

where n_A (or n_B) and X_A (or X_B) are respectively the number of moles and the mole fraction of A (or B). For one mole of solution

$$G_m^M = RT(X_A\ln X_A + X_B\ln X_B) \tag{9}$$

We note that

$$S_m^M = -\partial G_m^M/\partial T = -R(X_A\ln X_A + X_B\ln X_B) \tag{10}$$

and

$$H_m^M = G_m^M + TS_m^M = 0 \tag{11}$$

These are familiar results of the ideal solution as defined in Section VI.2 (by stating that the activity of a component is proportional to its mole fraction).

The physical significance of the model should be emphasized. It is a good approximation when the atoms A and B have nearly the same interactions with their neighbors whether in a mixed or pure environment. If the atoms A and B are very dissimilar, the deviations from the model should be considerable. But if the atoms are similar (such as iron and nickel, silver and gold), the deviations should be small, that is, the ideal solution model should be a reasonable approximation.

The results above can be readily extended to the case of a multicomponent system:

$$G^M = -kT\ln(N!/N_A!N_B!N_C!\cdots) \tag{12}$$

or

$$G_m^M = RT(X_A \ln X_A + X_B \ln X_B + X_C \ln X_C + \cdots) \tag{13}$$

with

$$S_m^M = -R(X_A \ln X_A + X_B \ln X_B + X_C \ln X_C + \cdots) \tag{14}$$

and

$$H_m^M = 0 \tag{15}$$

Obviously, real solutions do not behave as ideal ones. However, as seen in Section VI.3, the concept of an ideal solution is so convenient that it has been taken as a reference from which deviations are studied through the formalism of "excess" functions. Consequently, we shall now seek a more sophisticated model in order to understand and predict the excess properties of real solutions.

3. Regular Solution

Among the many statistical models currently in the literature, the *regular solution* model, originally introduced by Hildebrand [9], is probably the most widely used. Its popularity rests in its convenient simplicity. Although it is achieved at the cost of some serious shortcomings, it provides a good qualitative description of many phenomena and a valuable insight in the thermodynamic properties of solutions.

Essentially, the model assumes that the atoms are distributed at random on the sites of a three-dimensional lattice, that there is no vacancy, and that the energy of the system may be calculated as a sum of pairwise interactions. In a binary mixture of N_A atoms A and N_B atoms B, we designate by N_{AA}, N_{BB}, and N_{AB} the number of pairs AA, BB, and AB, associated with atoms which are nearest neighbor to each other. u_{AA}, u_{BB}, and u_{AB} designate the respective energies of these pairs. Z is the coordination number of A and B, that is the number of nearest neighbors to A or B.

Each atom A generates Z pairs of the AA or AB type. Summing on all the atoms A, we have ZN_A pairs that are distributed into $2N_{AA}$ and $1N_{AB}$ pairs:

$$ZN_A = 2N_{AA} + N_{AB} \tag{16}$$

The factor 2 for N_{AA} arises from the fact that each pair AA is counted twice in the summation, once for each atom A. Similarly,

$$ZN_B = 2N_{BB} + N_{AB} \tag{17}$$

As a verification, we note that the sum of equations (16) and (17) leads to

$$\tfrac{1}{2}Z(N_A + N_B) = N_{AA} + N_{AB} + N_{BB} \tag{18}$$

The total number of pairs $\frac{1}{2}ZN$ is indeed equal to the sum of all pairs AA, AB, and BB.

The energy of the lattice is

$$E = N_{AA}u_{AA} + N_{AB}u_{AB} + N_{BB}u_{BB} \tag{19}$$

or, replacing N_{AA} and N_{BB} by their expressions in equations (16) and (17),

$$E = \tfrac{1}{2}ZN_Au_{AA} + \tfrac{1}{2}ZN_Bu_{BB} + N_{AB}[u_{AB} - \tfrac{1}{2}(u_{AA} + u_{BB})] \tag{20}$$

$\frac{1}{2}ZN_A$ is the total number of pairs AA present in the solution of pure A before mixing. Thus, $\frac{1}{2}ZN_A u_{AA}$ is equal to E_A, the energy of N_A atoms A before mixing. Similarly, $E_B = \frac{1}{2}ZN_B u_{BB}$. Introducing the parameter ω_{AB},

$$\omega_{AB} = u_{AB} - \tfrac{1}{2}(u_{AA} + u_{BB}) \tag{21}$$

equation (20) becomes

$$E = E_A + E_B + N_{AB}\omega_{AB} \tag{22}$$

The number of pairs N_{AB} varies according to the distribution of the atoms A and B in the solution; consequently, E varies too. Our assumption of a random solution is incompatible with these considerations, but is nevertheless a convenient simplification. We shall remove this inconsistency in the study of the *quasi-chemical approximation*. However, we retain it here, and thus,

$$g = N!/(N_A!N_B!) \tag{23}$$

To calculate N_{AB} we proceed as follows. In a random arrangement of atoms, the probability of finding an atom A at a given site (i) is equal to the mole fraction of A, X_A. The probability of finding an atom B at the next nearest neighbor site ($i + 1$) is equal to X_B. Consequently, the probability of finding a pair AB at the pair of sites (i, $i + 1$) is $X_A X_B$, and if we do not distinguish between the presence of AB and BA at the sites (i, $i + 1$), this probability is $2X_A X_B$. As N_{AB} must be equal to the total number of pairs $\frac{1}{2}ZN$ multiplied by the probability of finding an AB (or BA) pair among these pairs, the result is

$$N_{AB} = \tfrac{1}{2}ZN2X_A X_B = Z(N_A N_B/N) \tag{24}$$

Therefore,

$$Q = g\exp(-E/kT)$$
$$= \frac{N!}{N_A!N_B!}\exp\left[-\left(E_A + E_B + \frac{ZN_A N_B}{N}\omega_{AB}\right)\Big/kT\right] \tag{25}$$

and

$$G = -kT\ln\frac{N!}{N_A!N_B!} + E_A + E_B + \frac{ZN_A N_B}{N}\omega_{AB} \tag{26}$$

which yields

$$G^M = kT\left(N_A\ln\frac{N_A}{N} + N_B\ln\frac{N_B}{N}\right) + \frac{ZN_A N_B}{N}\omega_{AB} \tag{27}$$

and for one mole of solution

$$G_m^M = RT(X_A\ln X_A + X_B\ln X_B) + Z\omega_{AB}X_A X_B \tag{28}$$

Subtracting the terms corresponding to an ideal solution yields the excess free energy:

$$G_m^E = Z\omega_{AB}X_A X_B \tag{29}$$

We note that

$$H_m^E = Z\omega_{AB}X_A X_B \tag{30}$$

and

$$S_m^E = 0 \tag{31}$$

It is the absence of any excess entropy which is the most characteristic and the most critical feature of the regular solution model. Generally, real solutions have sizable excess entropies. Indeed, the contribution of the term TS^E to G^E is often of the same order of magnitude as H^E (e.g., about $\frac{2}{3}H^E$ for many liquid metallic solutions).

Equations (29) and (30) predict a composition dependence of G^E and H^E which is parabolic and symmetric about $X_A = X_B = 0.5$.

The expressions of the activity coefficients of A and B may also be deduced. Recalling that

$$G_A^E = RT \ln \gamma_A = G_m^E + (1 - X_A)(\partial G_m^E/\partial X_A) \tag{32}$$

yields (with $\beta = 1/kT$)

$$\ln \gamma_A = Z\beta\omega_{AB}X_B^2 \tag{33}$$

Similarly,

$$\ln \gamma_B = Z\beta\omega_{AB}X_A^2 \tag{34}$$

This parabolic dependence entails that the α function

$$\alpha_i = \ln \gamma_i/(1 - X_i)^2 \tag{35}$$

is a constant (at any given temperature) and has the same value for A and B:

$$\alpha_A = \alpha_B = Z\beta\omega_{AB} \tag{36}$$

Let us dwell on the physical significance of the parameter ω_{AB}. It is the difference between the bonding energy of unlike atoms and the arithmetic mean of the bonding energies between like atoms. If the atoms A and B have a greater affinity for each other than for other atoms like themselves, then ω_{AB} is *negative* and equations (29)–(36) show that the excess functions are negative, that is the deviations from ideality are negative. The reverse occurs when like atoms have a greater affinity for each other: ω_{AB} is positive and the deviations from ideality are positive. For $\omega_{AB} = 0$, the solution behaves ideally.

Figure 1 shows the composition dependence of various thermodynamic properties according to the value of the parameter $Z\omega_{AB}$. We note that for a value of $Z\omega_{AB}$ sufficiently large and positive, there is a region where the solution is unstable and segregates into two phases, one rich in A and the other rich in B. The calculations of Section III.4.1 show that instability occurs at the critical temperature $T_C = Z\omega_{AB}/2k$.

The assumptions of the regular solution are too unrealistic. Nevertheless, the model provides useful semiquantitative estimates. The following test of the model illustrates the type of agreement usually encountered.

Equation (34) yields the value of the activity coefficient γ_B at infinite dilution in terms of the parameter $Z\omega_{AB}$:

$$\ln \gamma_B^\infty = Z\beta\omega_{AB} \tag{37}$$

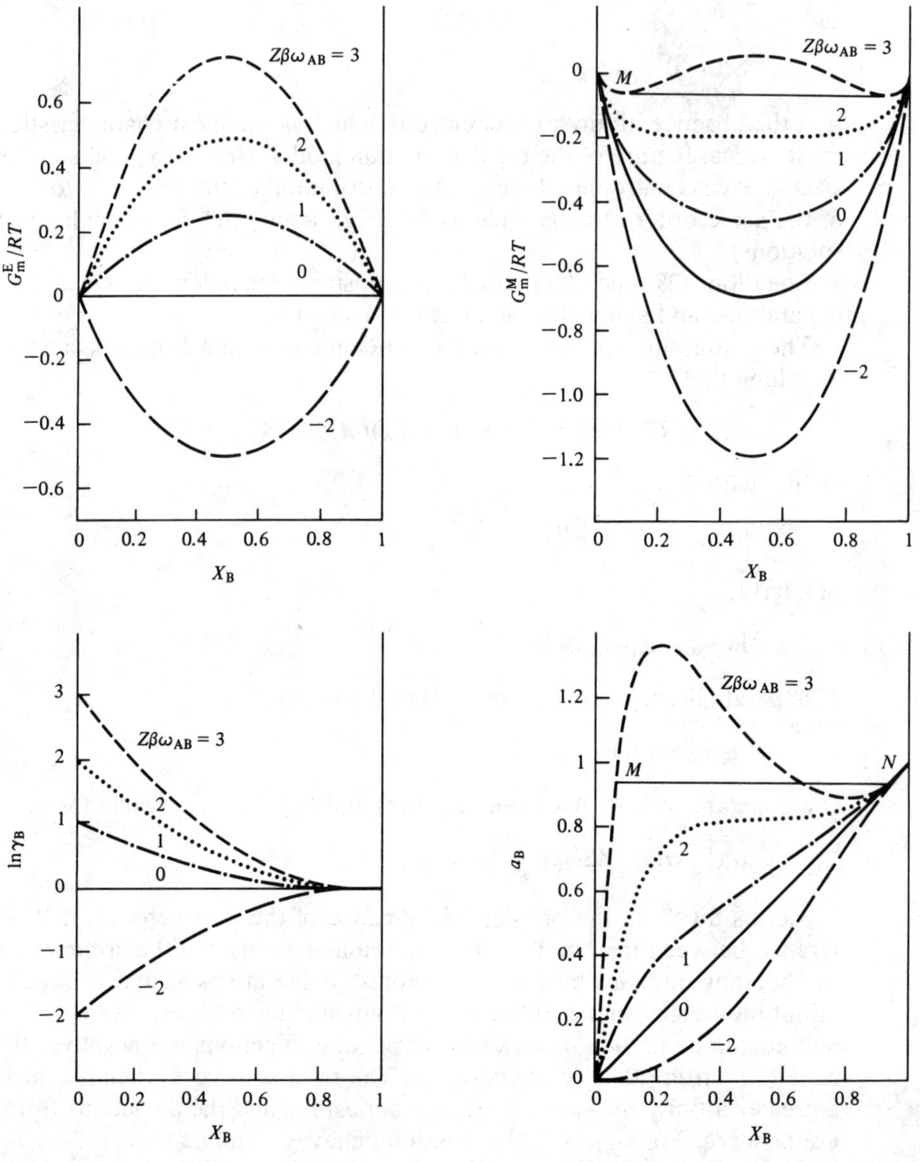

Figure 1. Composition dependence of various thermodynamic functions in the case of a regular solution for different values of the parameter $Z\beta\omega_{AB}$. Note that for $Z\beta\omega_{AB} = 3$ the solution is unstable between the points M and N.

Similarly, the self-interaction coefficient ϵ_B^B is equal to

$$\epsilon_B^B = \left(\frac{\partial \ln \gamma_B}{\partial X_B}\right)_{X_B \to 0} = -2Z\beta\omega_{AB} \tag{38}$$

Consequently, the model predicts the following relationship:

$$\epsilon_B^B = -2 \ln \gamma_B^\infty \tag{39}$$

This result is tested in Fig. 2 and Table 1 for a wide variety of alloys. The

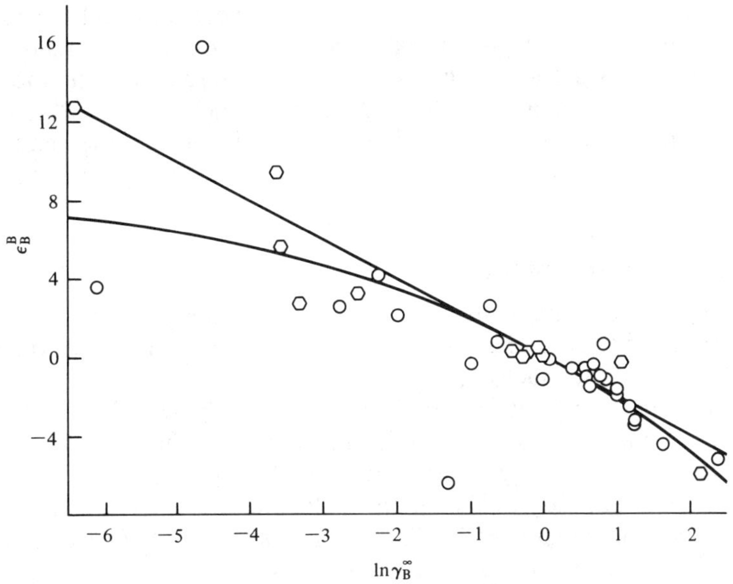

Figure 2. Correlation between the first order interaction coefficient ϵ_B^B and $\ln \gamma_B^\infty$ for various alloys. The straight line is predicted by the regular solution model [equation (39)] and the curved line by the quasi-chemical approximation [equation (69)] for $Z = 10$. The data points are identified in Table 1.

Table 1. Identification of the Experimental Points in Fig. 2

Solvent–solute	$\ln \gamma_B^\infty$	ϵ_B^B	T (°C)	Solvent–solute	$\ln \gamma_B^\infty$	ϵ_B^B	T (°C)
Ag–Al	−2.20	4.2	1000	Fe–Ni	−0.42	0.16	1600
Ag–Pb	0.69	−0.4	1000	Fe–Si	−6.4	12.7	1650
Bi–Cd	0	−1.2	500	Fe–Sn	1.03	−0.3	1600
Bi–Hg	−0.94	−0.3	327	Fe–Ti	−3.30	2.7	1600
Bi–Pb	−0.63	0.7	665	Fe–V	−2.52	3.2	1600
Bi–Sn	0.11	−0.06	330	Pb–Ag	0.40	−0.6	1085
Bi–Te	−1.97	2.1	480	Pb–Bi	−0.72	2.6	500
Bi–Zn	1.25	−3.3	600	Pb–Cd	1.17	−2.6	500
Cd–Bi	−0.16	−6.5	500	Pb–Na	−6.12	3.6	400
Cd–Hg	−2.76	2.5	350	Pb–Sb	−0.22	0.2	500
Cd–Na	−4.61	15.8	395	Pb–Sn	0.83	−1.2	500
Cd–Pb	1.63	−4.5	500	Pb–Te	−0.30	0.1	450
Cd–Sb	1.31	−6.5	500	Pb–Zn	2.40	−5.3	653
Cd–Sn	0.64	−1.6	500	Sn–Cd	0.59	−1.1	500
Cd–Zn	1.03	−1.8	682	Sn–Pb	0.84	0.6	500
Fe–Al	−3.54	5.6	1600	Sn–Te	0.59	−0.5	478
Fe–Co	0.07	0.47	1600	Sn–Zn	0.79	−1.0	437
Fe–Cr	0.00	0.00	1600	Te–Hg	−0.22	0.1	325
Fe–Cu	2.15	−6.0	1600	Te–Sn	1.03	−1.7	414
Fe–Mn	0.26	0	1600	Zn–Cd	1.25	−3.3	682

The values listed are taken from the compilations of Sigworth and Elliott [10] (for the iron alloys) and Dealy and Pehlke [11] (for the other alloys).

agreement between the experimental values and the theoretical curve is fair. Further examination of the properties of the alloys reveals that many of these exhibit sizable excess entropies; this, surprisingly, does not seem to affect in any marked way the agreement with the model, for which $S^{E} = 0$. This apparent discrepancy will be elucidated in Section 5, where we present the central atoms model.

4. Quasi-Chemical Approximation

The *quasi-chemical approximation* was introduced by Guggenheim [3]. The model resembles that of the regular solution. Its advantage lies in a more realistic estimate of the degeneracy factor g.

No rigorous mathematical formula is available for the calculation of g. In the regular solution model, g is taken as identical to the value corresponding to a random distribution. In Guggenheim's text [3] this approximation is called the *zeroth approximation*. Guggenheim's new estimate of g corresponds to the *first approximation*. The model is widely known, however, as the quasi-chemical approximation because it implies an equation which corresponds to a chemical reaction.

4.1. Assumptions of the Model

Let us review the main assumptions inherent in this more sophisticated version of the previous model.

a. The motion of the atoms reduces merely to oscillations about some equilibrium positions. In other words, some sort of lattice structure exists; in the case of liquids, this may be questionable. However, as previously discussed, at temperatures far below the critical temperature the degree of short-range order usually observed tends to support this assumption for the evaluation of the chemical properties of mixing.
b. In the calculation of the energy, only the chemical contribution is taken into account. The vibrational contribution, due to the change in the frequency spectrum of an atom in an environment different from the one it has in the pure component, is neglected. Other types of energy contributions are also neglected but, as discussed in the introductory section, their neglect is likely to be less critical. As a consequence of this assumption, the only source of excess entropy is of a configurational nature; that is, it arises from atomic arrangements which are different from completely random ones. Since that represents a deviation from complete randomness and maximum disorder, the sign of the excess entropy must always be negative. This conclusion is not supported by experimental observations.
c. Only the influence of nearest neighbors is taken into account and pairwise interactions are assumed. Thus, the bonding energy of two atoms is independent of the atoms which surround them.
d. The atoms A and B are of similar size. They occupy the same sort of sites (they are substitutional solutes) and have the same coordination number, which remains independent of composition.

4.2. Derivation

We shall adopt the same notation as for the regular solution and observe that equations (16)–(22) remain unchanged. Thus,

$$Q = \sum_{N_{AB}} g \exp[-(E_A + E_B + N_{AB}\omega_{AB})/kT] \tag{40}$$

and

$$G^M = -kT \ln \sum_{N_{AB}} g \exp(-N_{AB}\omega_{AB}/kT) \tag{41}$$

However, g is now a function of N_{AB}, and N_{AB} is no longer equal to the value we calculated on the assumption of a random distribution. The latter value may be designated by N_{AB}^*. We recall that we obtained

$$N_{AB}^* = ZN_AN_B/N \tag{42}$$

Instead of assuming that we have a random distribution of atoms, we now assume that we have a random distribution of pairs. The resulting expression for g is

$$g_1 = \frac{\frac{1}{2}Z(N_A + N_B)!}{N_{AA}!N_{BB}![(N_{AB}/2)!]^2} \tag{43}$$

The numerator exhibits the total number of pairs while the denominator exhibits the different types of pairs. The factor $[(N_{AB}/2)!]^2$ is equal to $N_{AB}!N_{BA}!$, since, in counting the number of distinguishable arrangements, one must differentiate between the modes of occupation AB and BA of two adjoining lattice sites.

A random distribution of the pairs AA, BB, AB, and BA overestimates the value of the degeneracy factor because the different pairs cannot truly be distributed at random. They do interfere with each other. For instance, Fig. 3 shows that if the pairs AB, BB, and BB are selected for the pairs of sites $(i, i+1)$, $(i+1, i+2)$, and $(i+2, i+3)$, the pair of sites $(i, i+3)$ is necessarily occupied by an AB pair. Rewriting g as

$$g = g_1h \tag{44}$$

where h is a correction term, we now assume that h is independent of the arrangement of the N_A atoms A and N_B atoms B. In other words, h is independent of N_{AB} (but dependent on N_A and N_B).

Figure 3. Pairs of atoms on lattice sites cannot be distributed at random: if the pairs AA, BB, and BB occupy the pairs of sites $(i, i+1)$, $(i+1, i+2)$, $(i+2, i+3)$, the pair of sites $(i, i+3)$ is necessarily occupied by a pair AB.

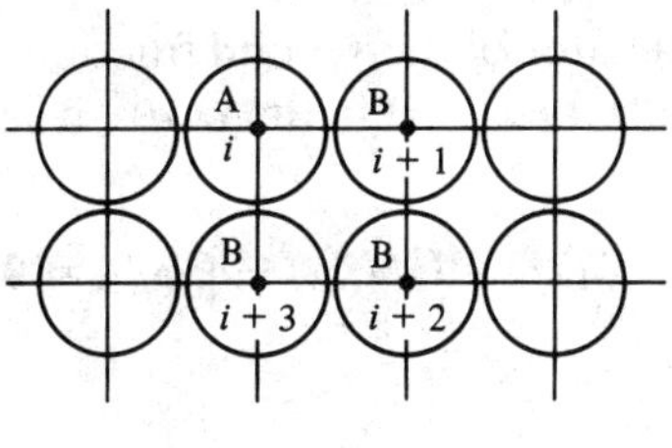

To determine h we note that the value of g is known in the case of a random distribution:

$$g^* = (N_A + N_B)!/N_A!N_B! \tag{45}$$

Since equation (44) should still hold,

$$g^* = h \frac{\frac{1}{2}Z(N_A + N_B)!}{N^*_{AA}!N^*_{BB}![(\frac{1}{2}N^*_{AB})!]^2} \tag{46}$$

where N^*_{AA} and N^*_{BB} are the values of N_{AA} and N_{BB} in a random distribution. h is then immediately deduced:

$$h = \frac{(N_A + N_B)!}{N_A!N_B!} \frac{N^*_{AA}!N^*_{BB}![(\frac{1}{2}N^*_{AB})!]^2}{\frac{1}{2}Z(N_A + N_B)!} \tag{47}$$

The value of N^*_{AB} is known [equation (42)] and the values of N^*_{AA} and N^*_{BB} are obtained through equations (16) and (17):

$$N^*_{AA} = \tfrac{1}{2}ZN_A - \tfrac{1}{2}N^*_{AB} \tag{48}$$

$$N^*_{BB} = \tfrac{1}{2}ZN_B - \tfrac{1}{2}N^*_{AB} \tag{49}$$

Thus, our estimate of the degeneracy factor becomes

$$g = \frac{(N_A + N_B)!}{N_A!N_B!} \frac{(\frac{1}{2}ZN_A - \frac{1}{2}N^*_{AB})!(\frac{1}{2}ZN_B - \frac{1}{2}N^*_{AB})![(\frac{1}{2}N^*_{AB})!]^2}{(\frac{1}{2}ZN_A - \frac{1}{2}N_{AB})!(\frac{1}{2}ZN_B - \frac{1}{2}N_{AB})![(\frac{1}{2}N_{AB})!]^2} \tag{50}$$

To calculate the free energy of mixing

$$G^M = -kT \ln \sum_{N_{AB}} (ge^{-\beta N_{AB}\omega_{AB}}) \tag{51}$$

we replace the sum by its maximum term. This maximum term corresponds to a distribution that for convenience we shall identify by a bar:

$$G^M = -kT \ln(\bar{g}e^{-\beta \bar{N}_{AB}\omega_{AB}}) \tag{52}$$

We note that the maximum term is determined by

$$\frac{\partial}{\partial N_{AB}} (ge^{-\beta N_{AB}\omega_{AB}}) = 0 \tag{53a}$$

or its equivalent

$$\frac{\partial}{\partial N_{AB}} (\ln g - \beta N_{AB}\omega_{AB}) = 0 \tag{53b}$$

Equation (53b) yields the value of $\bar{N}_{AB}$.

In the expression (50) of g only the denominator of the second fraction depends on N_{AB}. Moreover, noting that $d(\ln x!) = d(x \ln x - x) = (\ln x)\, dx$, it is readily verified that

$$\tfrac{1}{2}\ln(\tfrac{1}{2}ZN_A - \tfrac{1}{2}\bar{N}_{AB}) + \tfrac{1}{2}\ln(\tfrac{1}{2}ZN_B - \tfrac{1}{2}\bar{N}_{AB}) - \ln(\tfrac{1}{2}\bar{N}_{AB}) - \beta\omega_{AB} = 0 \tag{54a}$$

or

$$(\tfrac{1}{2}\bar{N}_{AB})^2/(\tfrac{1}{2}ZN_A - \tfrac{1}{2}\bar{N}_{AB})(\tfrac{1}{2}ZN_B - \tfrac{1}{2}\bar{N}_{AB}) = e^{-2\beta\omega_{AB}} \tag{54b}$$

It is readily seen that equation (54b) may be rewritten

$$(\tfrac{1}{2}\overline{N}_{AB})^2/\overline{N}_{AA}\overline{N}_{BB} = e^{-2\beta\omega_{AB}} \tag{55}$$

which corresponds to the chemical equilibrium of the reaction between pairs:

$$AA + BB = AB + BA \tag{56}$$

with the reaction constant

$$k = e^{-\Delta G^\circ/RT} = e^{-(2u_{AB} - u_{AA} - u_{BB})/kT} = e^{-2\beta\omega_{AB}} \tag{57}$$

The origin of the name "quasi-chemical approximation" is in this observation.

It is convenient to introduce the parameter λ:

$$\lambda = e^{2\beta\omega_{AB}} - 1 \tag{58}$$

which has the same sign as ω_{AB}. Equation (54) becomes

$$(\overline{N}_{AB})^2\lambda + \overline{N}_{AB}[Z(N_A + N_B)] - Z^2 N_A N_B = 0 \tag{59}$$

This quadratic expression is easily solved and the result is

$$\overline{N}_{AB} = (ZN/2\lambda)[-1 + (1 + 4X_A X_B \lambda)^{1/2}] \tag{60}$$

The other root of equation (59) yields a negative value for $\overline{N}_{AB}$ and therefore may be readily discarded.

Recalling the series expansion of the square root function,

$$(1 + x)^{1/2} = 1 + \frac{x}{2} - \frac{x^2}{8} + \cdots \tag{61}$$

we note that equation (60) yields

$$\overline{N}_{AB} \simeq (ZN_A N_B/N)(1 - X_A X_B \lambda) \tag{62}$$

As expected, when λ or ω_{AB} is positive (relative repulsion between the atoms A and B), the number of pairs AB is smaller than its value in the case of a random distribution. The converse is true when λ or ω_{AB} is negative. Figure 4 illustrates the percentages of the various pairs at $N_A = N_B$ for various values of λ.

The expression for $\overline{N}_{AB}$ in equation (60) or (62) may now be inserted in equation (50) to obtain $\overline{g}$. G^M is then obtained by equation (52). The calculations are long but straightforward. In particular, it is possible to show that [12]

$$G^E = (N^*_{AB}\omega_{AB}) + RT\left(N^*_{AA}\ln\frac{\overline{N}_{AA}}{N^*_{AA}} + N^*_{AB}\ln\frac{\overline{N}_{AB}}{N^*_{AB}} + N^*_{BB}\ln\frac{\overline{N}_{BB}}{N^*_{BB}}\right) \tag{63}$$

which has the interest of showing the second parenthesis as a correction to the result of the regular solution (first parenthesis).

A closed analytical form of G^E in terms of X_B is not obtainable. However, series expansions with respect to X_B yield [12]

$$G^E/RT = \tfrac{1}{2}ZX_B \ln(1 + \lambda) - \tfrac{1}{2}Z\lambda X_B^2 + \tfrac{1}{2}Z\lambda^2 X_B^3 - Z\lambda^2(\tfrac{1}{4} + \tfrac{5}{3}\lambda)X_B^4 + O(X_B^5) \tag{64}$$

$$\ln \gamma_A = \tfrac{1}{2}Z\lambda X_B^2 - Z\lambda^2 X_B^3 + \tfrac{3}{4}Z\lambda^2(1 + \tfrac{20}{3}\lambda)X_B^4 + O(X_B^5) \tag{65}$$

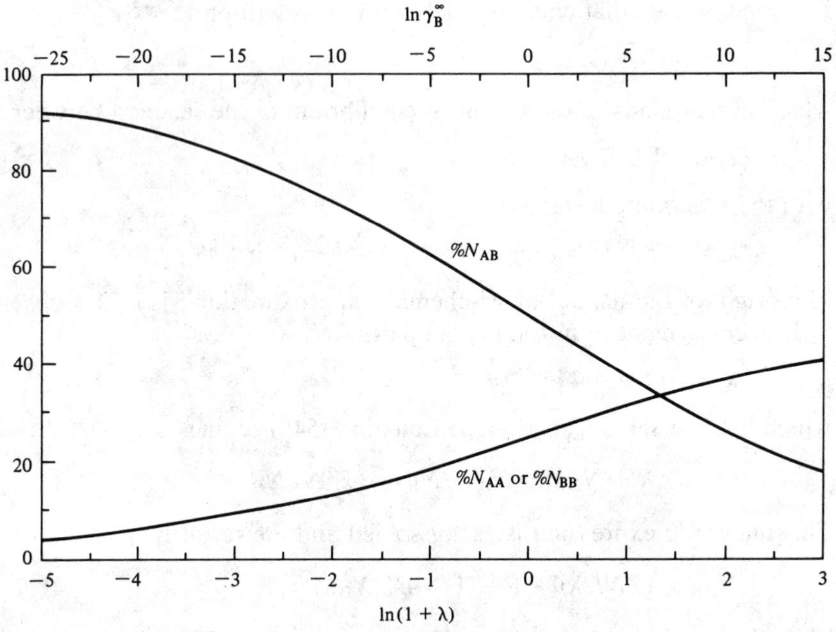

Figure 4. Percentages of pairs in the quasi-chemical model at $X_A = X_B = 0.5$ and calculated for $Z = 10$.

and

$$\ln \gamma_B = \tfrac{1}{2}Z \ln(1 + \lambda) - Z\lambda X_B + \tfrac{1}{2}Z\lambda(1 + 3\lambda)X_B^2 - 2Z\lambda^2(1 + \tfrac{10}{3}\lambda)X_B^3 + O(X_B^4) \quad (66)$$

4.3. Test of the Model and Discussion

To test the model, a procedure analogous to that adopted for the regular solution may be followed. The terms $\ln \gamma_B^\infty$ and ϵ_B^B in equation (66) are readily identified:

$$\ln \gamma_B^\infty = \tfrac{1}{2}Z \ln(1 + \lambda) \quad (67)$$

$$\epsilon_B^B = -Z\lambda \quad (68)$$

Eliminating λ between these two relations, we obtain

$$\ln \gamma_B^\infty = \tfrac{1}{2}Z \ln[1 - (\epsilon_B^B/Z)] \quad (69)$$

Equation (69) is tested in Fig. 2 for a wide variety of alloys. In the calculations a coordination number of 10 was assumed. This is a reasonable average for liquids, and moreover, the curve is only slightly changed by a different value. The agreement between the experimental values and the theoretical curve is satisfactory and better than in the case of the regular solution theory. For small deviations from ideality, the predictions of the quasi-chemical model are equivalent to those of the regular solution model since equation (69) becomes

$$\ln \gamma_B^\infty \simeq \tfrac{1}{2}Z(-\epsilon_B^B/Z) = -\tfrac{1}{2}\epsilon_B^B \quad (70)$$

One of the advantages of the regular solution model is that the values of Z and ω_{AB} are grouped together to form a single parameter. In the quasi-chemical approximation this is no longer so. For liquids, there are very few measurements of Z and generally its value must be estimated. However, in most cases the range of likely values for Z is rather narrow (e.g., from 8 to 12) and, in that range, the expressions for the thermodynamic functions are quite insensitive to the choice of the value of Z. This is because a change in the value of Z can be almost compensated by a change in the value of λ.

When the coordination numbers of the two pure components are different (e.g., $Z = 10$ and $Z = 8$) an average value can be chosen (e.g., $Z = 9$). Alternatively, the range of composition can be divided into three parts: one in which A is the solvent, another in which B is the solvent, and a third in which the results can be interpolated on the basis of the predictions for the dilute regions. Let us consider, for example, the case of the Sn–Cd system at 500°C (Fig. 5). In the region where Sn is the solvent, Z is taken as equal to the coordination of pure tin: $Z = 10$. Similarly, in the region where Cd is the solvent, Z is taken as the coordination number of pure cadmium: $Z = 8$. The values of λ for the two regions are $\frac{6}{5} \times 10^{-1}$ and $\frac{7}{4} \times 10^{-1}$, respectively, and were chosen to provide the best fit to the data. Despite the arbitrariness of these selections, the agreement is excellent (but not necessarily typical). It is interesting to note that the system presents a sizable positive excess entropy [13] which cannot be accounted for by the quasi-chemical model.

Indeed, it was pointed out at the beginning of this section that the excess entropy predicted by the quasi-chemical model is always negative because it arises from an ordering effect (relative to the complete disorder of a random distribution). It is worthwhile to check this result by differentiating the expression of G^E in equation (64) with respect to T. This yields

$$S^E = \tfrac{1}{2}RZ[\lambda - (1 + \lambda)\ln(1 + \lambda)]X_B^2 + O(X_B^3) \tag{71}$$

Figure 5. Test of the quasi-chemical model for the Sn–Cd system at 500°C. ○, experimental values [13]; ———, theoretical curve, from calculations based on equations (65) and (66). The values of λ are chosen to give the best fit.

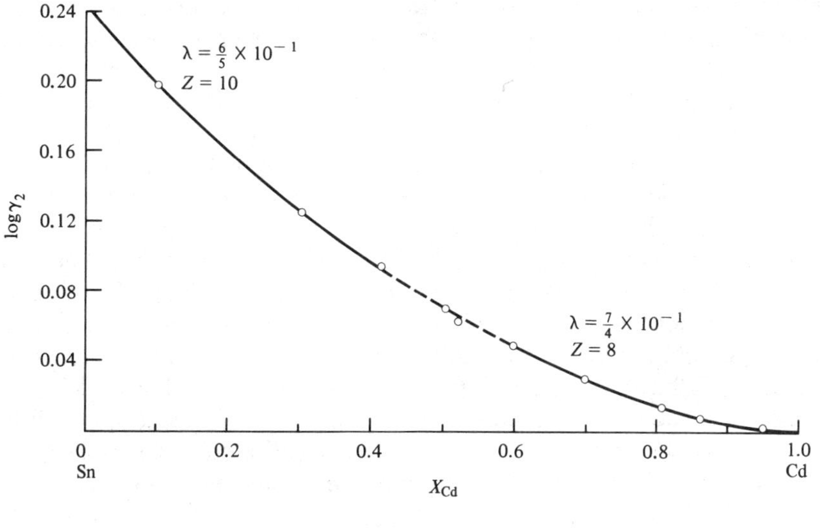

and for small deviations from ideality (λ small)

$$S^{E} \simeq -\tfrac{1}{4}RZ\lambda^2 X_B^2 \tag{72}$$

S^E is seen to be indeed negative. Moreover, we observe that S^E is of the second order with respect to both λ and X_B. In particular,

$$S_B^{E\infty} = 0 \tag{73}$$

since the linear term in the expansion series of S^E is $S_B^{E\infty}X_B$. Thus, the configurational contribution to the excess entropy (the only one taken into account in the quasi-chemical model) is very small for solutions which do not exhibit very strong interactions, and is negligible for very dilute solutions. Experimental excess entropies are generally in disagreement with these predictions of the model.

5. Central Atoms Model

Two assumptions of the quasi-chemical model appear particularly questionable. First, all vibrational contributions to the excess properties have been neglected. Second, the bonding energy between two atoms has been treated as independent of its surroundings. Neither assumption is necessary to the development of the central atoms model.

5.1. General Features of the Model

The central atoms model draws heavily on the cell model of liquids, first introduced by Eyring [14] in 1936 and Eyring and Hirschfelder [15] in 1937. Its present version was introduced in 1965 by Lupis and Elliott [16,17]. At the same time, Mathieu and co-workers [18] developed (independently) the *surrounded atom model,* which is essentially identical to the central atoms model.

For densities of the liquid state at temperatures far below the critical temperature, a certain order may be expected in the distribution of the atoms. Interatomic distances between first neighbors smaller than the atomic diameter are prohibited by large repulsive forces, and distances much larger are statistically unlikely. This introduces a regularity in the spacing of neighboring atoms, with a mean interatomic distance of the order of the atomic diameter. Consequently, each atom may be assumed to be confined to its own cell. The field acting on each atom in its cell is rapidly fluctuating and may be replaced by an average field of spherical symmetry.

The most significant feature of the present model is the replacement of the entity of an atomic pair by the cluster of an atom and its nearest neighbor shell. The partition function of the solution will be described in terms of the probabilities associated with different configurations in the nearest neighbor shell and in terms of the influence of those configurations on the field acting on the central atom. The summation will be extended to all the atoms, each one being considered in turn as a central atom. This is the origin of the name "central atoms model."

Although the assumption of spherical symmetry is not essential, it is extremely convenient and simplifies considerably the model. Indeed, we shall adopt this assumption even in the case of a solid, where it is somewhat more questionable. What it implies is that, for all energy considerations, the configuration of a nearest

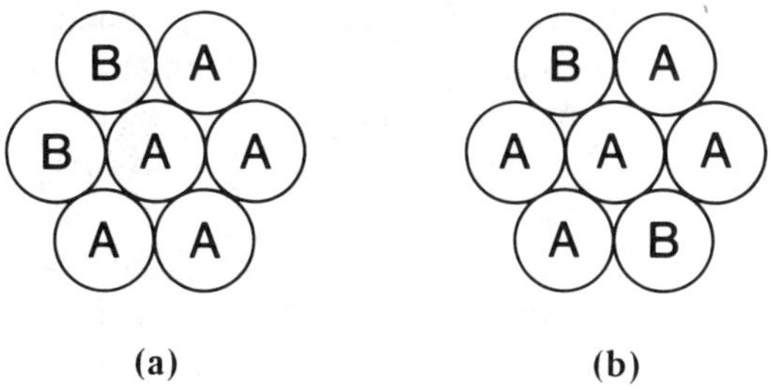

Figure 6. In the central atoms model, the two configurations (**a**) and (**b**) in the nearest neighbor shell of A exert equivalent energy fields on the central atom A but are distinguishable.

neighbor shell is characterized by its composition (number of atoms A and B) rather than by the actual arrangement of the atoms on the lattice sites. For example, a configuration where two atoms B are nearest neighbors in the shell (Fig. 6) exerts on the central atom A an energy field equivalent to the field exerted by a configuration where they are diametrically opposed. It should be observed that this assumption is essentially an averaging technique. The difference in the energies associated with two atoms B in adjacent sites and in nonadjacent sites in the nearest neighbor shell of an atom A is indirectly taken into account when the atoms B are considered, in turn, as central atoms.

In the partition function

$$Q = \sum g \exp(-E/kT) \tag{74}$$

the energy E is that of the ensemble of N_A and N_B atoms A and B. Since it is reasonable to assume that the chemical energy contribution is independent of the other contributions such as the translational and vibrational ones, these may be factored out of the exponential. Q may then be rewritten

$$Q = \sum g q_A^{N_A} q_B^{N_B} e^{-E/kT} \tag{75}$$

E is now the chemical energy of the ensemble and q_A and q_B are *average* partition functions of the atoms A and B which do not include the chemical contribution. If we were numbering the A atoms as $A_1, A_2, \ldots, A_{N_A}$, $q_A^{N_A}$ would represent the product $q_{A_1} q_{A_2} \cdots q_{A_{N_A}}$. In the regular solution model as in the quasi-chemical one, the individual partition functions do not appear because it is assumed that they are unaffected by mixing; hence, they automatically vanish from the thermodynamic excess functions.

5.2. Possible Expressions for the Individual Partition Function q

At some stage in the development of the model, it will be necessary to introduce an assumed dependence of q on the composition of the nearest neighbor shell. This dependence may be considered on an entirely empirical basis, i.e., without justification in terms of the physical parameters represented by q. However, it is preferable to attempt to describe q, even through simple concepts, in order to develop a better interpretation of the model.

Let us consider that an atom is free to move within the small cell of its nearest neighbors and that this cell has a volume v. If the atom is not subject to any net force, the description of its partition function q is an elementary and classical problem in statistical mechanics. The resulting expression is

$$q = (2\pi mkT/h^2)^{3/2} v \tag{76}$$

where m is the mass of the atom and h is Planck's constant. However, since the interaction potential ψ between the atom and its neighbors is a function of its position r in the cell, the probability of observing the central atom in a given element of volume is not uniform. This element of probability is incorporated in the calculations by replacing in equation (76) the cell's volume v by an *effective volume* v_f defined by

$$v_f = \int_{\text{cell}} e^{-[\psi(r) - \psi(0)]/kT} 4\pi r^2 \, dr \tag{77}$$

where the origin $r = 0$ is chosen to correspond to the minimum of ψ (i.e., to the "bottom of the potential well").

Under the simplifying assumption that the potential ψ is parabolic with respect to the distance it may be shown that v_f is approximately equal to

$$v_f = \left[\frac{1}{2\pi kT}\left(\frac{\partial^2\psi}{\partial r^2}\right)\right]^{-3/2} \tag{78}$$

Equation (76) then becomes

$$q = \left(\frac{2\pi m^{1/2} kT}{h}\right)^3 \left(\frac{\partial^2\psi}{\partial r^2}\right)^{-3/2} \tag{79}$$

The same result can be obtained by the consideration of Einstein's model of a crystal in which the atoms are three-dimensional isotropic harmonic oscillators with a frequency ν:

$$\nu = \frac{1}{2\pi m^{1/2}}\left(\frac{\partial^2\psi}{\partial r^2}\right)^{1/2} \tag{80}$$

The characteristic temperature θ is defined as

$$\theta = h\nu/k \tag{81}$$

At temperatures where T is substantially higher than θ, the partition function associated with each mode of vibration may be approximated by $kT/h\nu$; thus, for the three independent modes (e.g., along the three axes x, y, z)

$$q = \left(\frac{kT}{h\nu}\right)^3 = \left(\frac{2\pi m^{1/2} kT}{h}\right)^3 \left(\frac{\partial^2\psi}{\partial r^2}\right)^{-3/2} \tag{82}$$

which is identical to equation (79). This is not surprising, since in both instances we assumed an isotropic interaction potential with a parabolic dependence on the distance.

The proportionality between q, v_f, ν^{-3}, and $(\partial^2\psi/\partial r^2)^{-3/2}$ may be expressed

$$\delta \ln q = \delta \ln v_f = -3\,\delta \ln \nu = -\tfrac{3}{2}\delta \ln(\partial^2\psi/\partial r^2) \tag{83}$$

5.3. Probabilities Associated with Different Configurations and Thermodynamic Functions

In a random solution the probability p_{iB}^{*A} of finding an atom A surrounded in the nearest neighbor shell by i atoms B and $Z - i$ atoms A is

$$p_{iB}^{*A} = C_i^Z X_A^{Z-i} X_B^i \tag{84}$$

where C_Z^i is the combinatorial factor $Z!/(Z-i)!i!$. This may be seen as follows.

If we number the sites in the shell from 1 to Z, the probability of finding an atom B at the lattice site 1 is equal to N_B/N or X_B. Thus the probability of finding atoms B on the lattice sites 1, 2, . . . , i and atoms A on the sites $i+1$, $i+2$, . . . , Z is $X_A^{Z-i}X_B^i$. However, if we do not specify the position of the atoms B on the different sites of the nearest neighbor shell, we must multiply $X_A^{Z-i}X_B^i$ by the number of ways of arranging the i atoms B and $Z - i$ atoms A on the Z lattice sites, which is C_Z^i. Equation (84) is thus verified.

In a nonrandom solution the expression of the probability p_{iB}^A must be corrected from its expression p_{iB}^{*A} in a random solution by a correction factor f_{iB}^A:

$$p_{iB}^A = p_{iB}^{*A} f_{iB}^A / P_A = C_Z^i X_A^{Z-i} X_B^i f_{iB}^A / P_A \tag{85}$$

where P_A is a normalization factor

$$P_A = \sum_{i=0}^{Z} C_Z^i X_A^{Z-i} X_B^i f_{iB}^A \tag{86}$$

such that $\sum_{i=0}^{Z} p_{iB}^A = 1$. It is not necessary at this stage to specify the value of f_{iB}^A. p_{iB}^B has a similar expression:

$$p_{iB}^B = p_{iB}^{*B} f_{iB}^B / P_B = C_Z^i X_A^{Z-i} X_B^Z f_{iB}^B / P_B \tag{87}$$

with

$$P_B = \sum_{i=0}^{Z} C_Z^i X_A^{Z-i} X_B^i f_{iB}^B \tag{88}$$

In the expression of the partition function [equation (75)], the sum may be replaced by its maximum term (Section 1). Consequently,

$$Q = \overline{g}\overline{q}_A^{N_A}\overline{q}_B^{N_B} e^{-\overline{E}/kT} \tag{89}$$

and

$$G \simeq \overline{E} - kT(\ln \overline{g} + \ln \overline{q}_A^{N_A}\overline{q}_B^{N_B}) \tag{90}$$

For convenience, we shall omit in the following equations the bars relating the terms they affect to this maximum term.

Each atom A (or B) surrounded by i atoms B and $Z - i$ atoms A has a potential energy U_{iB}^A (or U_{iB}^B). E may then be expressed as the sum of the energies contributed by each atom considered in turn as a central atom:

$$E = \tfrac{1}{2}X_A \sum_{i=0}^{Z} p_{iB}^A U_{iB}^A + \tfrac{1}{2}X_B \sum_{i=0}^{Z} p_{iB}^B U_{iB}^B \tag{91}$$

p_{iB}^A and p_{iB}^B are the probabilities corresponding to the distribution which yields the maximum term in the partition function's sum (they could have been written more explicitly $\overline{p}_{iB}^{A}$ and $\overline{p}_{iB}^{B}$). The factor of $\frac{1}{2}$ is introduced to avoid counting the interaction energies twice.

Neglecting the contribution of the PV term (generally negligible for solids and liquids), the excess enthalpy corresponding to equation (91) is

$$H_{\text{chem}}^{E} = E - X_A E_A - X_B E_B = E - \tfrac{1}{2} X_A U_{0B}^{A} - \tfrac{1}{2} X_B U_{0B}^{B} \tag{92}$$

Combining equations (91) and (92), we obtain

$$H_{\text{chem}}^{E} = \tfrac{1}{2} X_A \left(\sum_{i=0}^{Z} p_{iB}^{A} U_{iB}^{A} - U_{0B}^{A} \right) + \tfrac{1}{2} X_B \left(\sum_{i=0}^{Z} p_{iB}^{B} U_{iB}^{B} - U_{ZB}^{B} \right) \tag{93}$$

or

$$H_{\text{chem}}^{E} = \tfrac{1}{2} X_A \sum_{i=0}^{Z} p_{iB}^{A} (U_{iB}^{A} - U_{0B}^{A})$$

$$+ \tfrac{1}{2} X_B \sum_{i=0}^{Z} p_{iB}^{B} (U_{iB}^{B} - U_{0B}^{B}) - \tfrac{1}{2} X_B (U_{ZB}^{B} - U_{0B}^{B}) \tag{94}$$

Let

$$\delta U_{iB}^{A} = U_{iB}^{A} - U_{0B}^{A} \quad \text{and} \quad \delta U_{iB}^{B} = U_{iB}^{B} - U_{0B}^{B} \tag{95}$$

Equation (94) becomes

$$H_{\text{chem}}^{E} = \tfrac{1}{2} X_A \sum_{i=0}^{Z} p_{iB}^{A}\, \delta U_{iB}^{A} + \tfrac{1}{2} X_B \sum_{i=0}^{Z} p_{iB}^{B}\, \delta U_{iB}^{B} - \tfrac{1}{2} X_B\, \delta U_{ZB}^{B} \tag{96}$$

δU_{iB}^{A} and δU_{iB}^{B} analyze the perturbation on the potential energy of the central atom created by the substitution of i atoms A by B atoms in a nearest neighbor shell of Z atoms A. This is illustrated in Fig. 7. The reason for not giving a totally symmetric role to the atoms A and B and keeping an environment of all atoms A as a reference configuration is that it is more convenient for the study of dilute solutions.

Equation (90) may be rewritten

$$G = H_{\text{chem}} - TS_{\text{conf}} - TS_{\text{nonconf}} \tag{97}$$

Figure 7. δU_{2B}^{A} is the difference in the potential energy of the central atom A in the configurations (**a**) and (**b**). In other words, the potential energy of the central atom A in the configuration (**a**) is measured relative to the potential energy it would have in a very dilute solution (surrounded by all A atoms).

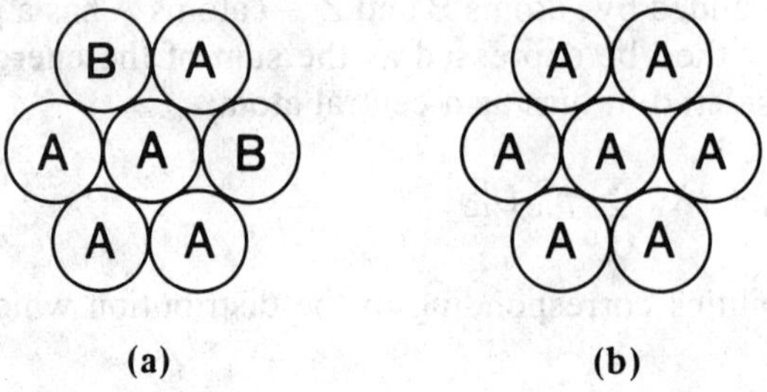

(a) (b)

in which

$$H_{\text{chem}} = E \tag{98}$$

$$S_{\text{conf}} = k \ln g \tag{99}$$

and

$$S_{\text{nonconf}} = k(\ln q_{\text{A}}^{N_{\text{A}}} + \ln q_{\text{B}}^{N_{\text{B}}}) \tag{100}$$

The term H_{chem} has already been analyzed, S_{conf} arises from the number of ways in which the atoms can be distributed among the lattice sites. S_{nonconf} is a contribution arising from the vibrational motion of the atoms in their cells.

Recalling that q_{A} and q_{B} are geometrical averages, S_{nonconf} may be written

$$S_{\text{nonconf}} = R\left(\sum_{i=0}^{Z} X_{\text{A}} p_{i\text{B}}^{\text{A}} \ln q_{i\text{B}}^{\text{A}} + \sum_{i=0}^{Z} X_{\text{B}} p_{i\text{B}}^{\text{B}} \ln q_{i\text{B}}^{\text{B}}\right) \tag{101}$$

and

$$S_{\text{nonconf}}^{\text{E}} = R\left(\sum_{i=0}^{Z} X_{\text{A}} p_{i\text{B}}^{\text{A}} \ln q_{i\text{B}}^{\text{A}} - X_{\text{A}} \ln q_{0\text{B}}^{\text{A}} + \sum_{i=0}^{Z} X_{\text{B}} p_{i\text{B}}^{\text{B}} \ln q_{i\text{B}}^{\text{B}} - X_{\text{B}} \ln q_{Z\text{B}}^{\text{B}}\right) \tag{102}$$

Proceeding as for $H_{\text{chem}}^{\text{E}}$, we obtain

$$S_{\text{nonconf}}^{\text{E}} = R\left(X_{\text{A}} \sum_{i=0}^{Z} p_{i\text{B}}^{\text{A}}\ \delta \ln q_{i\text{B}}^{\text{A}} + X_{\text{B}} \sum_{i=0}^{Z} p_{i\text{B}}^{\text{B}}\ \delta \ln q_{i\text{B}}^{\text{B}} - X_{\text{B}}\ \delta \ln q_{Z\text{B}}^{\text{B}}\right) \tag{103}$$

where

$$\delta \ln q_{i\text{B}}^{\text{A}} = \ln q_{i\text{B}}^{\text{A}} - \ln q_{0\text{B}}^{\text{A}} \quad \text{and} \quad \delta \ln q_{i\text{B}}^{\text{B}} = \ln q_{i\text{B}}^{\text{B}} - \ln q_{0\text{B}}^{\text{B}} \tag{104}$$

The similarity of the expressions of $H_{\text{chem}}^{\text{E}}$ and $S_{\text{nonconf}}^{\text{E}}$ [in equations (96) and (103)] is to be noted. In many cases, it will entail a rough proportionality between the heat of mixing and the excess entropy.

It is easily seen that the expression of the free energy of mixing becomes

$$G^{\text{M}} = -kT \ln g + X_{\text{A}} \sum_{i=0}^{Z} p_{i\text{B}}^{\text{A}}(\tfrac{1}{2}\,\delta U_{i\text{B}}^{\text{A}} - RT\,\delta \ln q_{i\text{B}}^{\text{A}}) + X_{\text{B}} \sum_{i=0}^{Z} p_{i\text{B}}^{\text{B}}(\tfrac{1}{2}\,\delta U_{i\text{B}}^{\text{B}} - RT\,\delta \ln q_{i\text{B}}^{\text{B}}) + X_{\text{B}}(\tfrac{1}{2}\,\delta U_{Z\text{B}}^{\text{B}} - RT\,\delta \ln q_{Z\text{B}}^{\text{B}}) \tag{105}$$

There are $4Z$ parameters in that equation (the $\delta U_{i\text{B}}^{\text{A}}$, $\delta U_{i\text{B}}^{\text{B}}$, $\delta \ln q_{i\text{B}}^{\text{A}}$ and $\delta \ln q_{i\text{B}}^{\text{B}}$), far too many for any practical application. Consequently, additional simplifying assumptions are required.

5.4. *Quasi-Regular Solution*

5.4.1. Assumptions and Derivation

It may be recalled that the regular solution model (Section 3) assumes

a. complete randomness of the atoms,
b. mixing entropy whose source is entirely configurational, and
c. pairwise interactions.

Assumptions a and b entail

$$S^{M}_{conf} = S^{M}_{ideal} = -R(X_A \ln X_A + X_B \ln X_B) \tag{106}$$

and

$$S^{E} = S^{E}_{conf} = 0 \tag{107}$$

As with the regular solution model, the first assumption of complete randomness is now retained in the *quasi-regular solution*. Thus $k \ln g$ is just the ideal entropy of mixing and

$$S^{E}_{conf} = 0 \tag{108}$$

However, the second assumption b is not retained: nonconfigurational contributions (such as vibrational ones) to the total excess entropy are not neglected and

$$S^{E} \neq 0 \tag{109}$$

The second assumption of the quasi-regular solution model is that δU^{A}_{iB}, δU^{B}_{iB}, $\delta \ln q^{A}_{iB}$, and $\delta \ln q^{B}_{iB}$ vary linearly with the number i of atoms B in the nearest neighbor shell:

$$\delta U^{A}_{iB} = i\,\delta U^{A}_{1B}, \qquad \delta U^{B}_{iB} = i\,\delta U^{B}_{1B} \tag{110}$$

$$\delta \ln q^{A}_{iB} = i\,\delta \ln q^{A}_{1B}, \qquad \delta \ln q^{B}_{iB} = i\,\delta \ln q^{B}_{1B} \tag{111}$$

This is equivalent to assuming pairwise interactions, for if we assume that the effect of the nearest neighbors on the central atom is additive, then

$$U^{A}_{iB} = iu_{AB} + (Z - i)u_{AA} \tag{112}$$

where u_{AB} is the fraction of the effect of the nearest neighbor shell on the energy of A due to the presence of B in this shell: that is, the bonding energy between A and B. u_{AA} has a similar meaning. Also,

$$U^{A}_{0B} = Zu_{AA} \tag{113}$$

Consequently,

$$\delta U^{A}_{iB} = i(u_{AB} - u_{AA}) \tag{114}$$

It expresses the fact that the change in the potential of A in the two configurations iB and 0B stems from the replacement of i pairs AA of energy u_{AA} by i pairs of energy u_{AB}. Similarly,

$$\delta U^{B}_{iB} = i(u_{BB} - u_{AB}) \tag{115}$$

Equations (111) for $\delta \ln q_{iB}$ have a similar meaning: other characteristics of the potential well (e.g., its curvature instead of its depth) are also obtained by additivity of the effects of the individual neighbors on the central atom, that is, by the additivity of pairwise interactions.

The assumptions of complete randomness and linearity of the interaction potential with the number of B nearest neighbors permit a relatively easy calculation of the previous thermodynamic formulas. Since the atoms are distributed at random, the term $k \ln g$ becomes identical to the ideal entropy of mixing, $-R(X_A \ln X_A + X_B \ln X_B)$. Moreover, $f_{iB}^{A} = f_{iB}^{B} = 1$ and

$$p_{iB}^{A} = p_{iB}^{B} = C_Z^i X_A^{Z-i} X_B^i \tag{116}$$

H_{chem}^{E} in equation (96) then transforms into

$$\begin{aligned} H_{\text{chem}}^{E} &= \tfrac{1}{2} X_A \sum_{i=0}^{Z} C_Z^i X_A^{Z-i} X_B^i\, i\, \delta U_{1B}^{A} \\ &\quad + \tfrac{1}{2} X_B \sum_{i=0}^{Z} C_Z^i X_A^{Z-i} X_B^i\, i\, \delta U_{1B}^{B} \\ &\quad - \tfrac{1}{2} X_B Z\, \delta U_{1B}^{B} \end{aligned} \tag{117}$$

To simplify this expression we note that

$$(x + y)^z = \sum_{i=0}^{z} C_z^i x^{z-i} y^i \tag{118}$$

and that differentiation with respect to y yields

$$z(x + y)^{z-1} = \sum_{i=0}^{z} i C_z^i x^{z-i} y^{i-1} \tag{119}$$

or

$$yz(x + y)^{z-1} = \sum_{i=0}^{z} i C_z^i x^{z-i} y^i \tag{120}$$

With $x = X_A$, $y = X_B$, and $z = Z$, equation (117) becomes

$$H_{\text{chem}}^{E} = \tfrac{1}{2} X_A Z X_B\, \delta U_{1B}^{A} + \tfrac{1}{2} X_B Z X_B\, \delta U_{1B}^{B} - \tfrac{1}{2} X_B Z\, \delta U_{1B}^{B} \tag{121a}$$

or

$$H_{\text{chem}}^{E} = X_A X_B \tfrac{1}{2}[Z(\delta U_{1B}^{A} - \delta U_{1B}^{B})] \tag{121b}$$

Equations (114) and (115) show that

$$\begin{aligned} \delta U_{1B}^{A} - \delta U_{1B}^{B} &= (u_{AB} - u_{AA}) - (u_{BB} - u_{AB}) \\ &= 2[u_{AB} - \tfrac{1}{2}(u_{AA} + u_{BB})] = 2\omega_{AB} \end{aligned} \tag{122}$$

Consequently,

$$H_{\text{chem}}^{E} = Z \omega_{AB} X_A X_B \tag{123}$$

a result identical to that in the regular solution model.

By a similar procedure, it is easily demonstrated that

$$S_{\text{nonconf}}^{E} = X_A X_B Z(\delta \ln q_{1B}^{A} - \delta \ln q_{1B}^{B}) \tag{124}$$

With

$$\pi_{AB} = \delta \ln q_{1B}^{A} - \delta \ln q_{1B}^{B} \tag{125}$$

equation (124) may be rewritten

$$S_{\text{nonconf}}^{E} = Z\pi_{AB}X_A X_B \tag{126}$$

and

$$G^E = X_A X_B Z(\omega_{AB} - T\pi_{AB}) \tag{127}$$

5.4.2. Applications

Just as for the regular solution model, the quasi-regular solution exhibits a symmetric parabolic dependence of G^E and H^E on the composition. However, the excess entropy of a nonregular solution is not null and its composition dependence is also that of a symmetric parabola.

The quasi-regular solution thus explains the observed discrepancy in the applications of the regular solution model. We recall that the applications of the regular solution model are relatively satisfactory for free energy functions but are inadequate for entropy functions. For instance, we saw that the relationship

$$\epsilon_2^{(2)} = -2 \ln \gamma_2^{\infty} \tag{128}$$

of the regular solution model is semiquantitatively obeyed by systems which exhibit large excess entropies. The same relationship is now predicted by the quasi-regular solution, but without setting S^E equal to zero. Indeed, the quasi-regular solution model predicts that

$$\sigma_2^{(2)} = -2S_2^{E\infty} \tag{129}$$

Unfortunately, entropy data are generally too inaccurate to provide a meaningful test of this new relationship. Nevertheless, agreement with equation (128) is in itself an indication of a reasonable agreement with equation (129).

It is convenient to substitute for the parameter π_{AB} the parameter τ defined as the ratio of ω_{AB} and π_{AB}:

$$\tau = \omega_{AB}/\pi_{AB} \tag{130}$$

G^E becomes

$$G^E = X_A X_B Z\omega_{AB}[1 - (T/\tau)] \tag{131}$$

It is generally observed that as the temperature increases a system tends to become more ideal. τ may then be interpreted as the temperature at which the system would become ideal if its free energy functions were linearly extrapolated. (At extremely high temperatures, the model is obviously inadequate.) Thus, τ should be a positive quantity and larger than the temperatures of investigation: for most metallic solutions, 1500–3500°K is a reasonable order of magnitude for τ.

We also note that equation (131) entails

$$H^E = \tau S^E \tag{132}$$

In the relationship $G^E = H^E - TS^E$, the term TS^E is thus predicted to be of the same sign and order of magnitude as H^E. It is further predicted to be slightly lower than H^E: e.g., $\frac{2}{3}H^E$ if $\tau/T \simeq \frac{3}{2}$. Although these evaluations may be considered somewhat simplistic, they offer in many cases valuable semiquantitative estimates of the entropy or of the temperature dependence of the free energy. They are also an improvement over the predictions of the regular solution model. We note that the latter corresponds to the particular case $\tau = \infty$.

Two assumptions critical to the quasi-regular solution model should be recalled. First, the atoms are randomly distributed. Second, the parameters δU_{iB} and $\delta \ln q_{iB}$ vary linearly with the number i of atoms B in the nearest neighbor shell; we have seen that this is equivalent to assuming pairwise interactions.

5.5. Correlation Between Excess Enthalpy and Entropy

It is interesting to investigate whether a correlation between H^E and S^E similar to that indicated by equation (132) is valid irrespective of the assumptions of the quasi-regular solution.

In the general case, S^M is not simply equal to the sum of the terms we designated by S^M_{conf} and S^M_{nonconf}. It must be obtained from the expression for G^M in equation (97) or (105) by differentiation with respect to T:

$$S^M = -\partial G^M/\partial T \tag{133}$$

and similarly for H^M:

$$H^M = G^M - T\partial G^M/\partial T \tag{134}$$

Equation (97) shows that

$$S^M = S^M_{\text{conf}} + S^E_{\text{nonconf}} + S^E_{\text{corr}} \tag{135}$$

where

$$S^E_{\text{corr}} = -\frac{\partial H^E_{\text{chem}}}{\partial T} + T\frac{\partial S^M_{\text{conf}}}{\partial T} + \frac{\partial S^E_{\text{nonconf}}}{\partial T} \tag{136}$$

Similarly,

$$H^E = H^M = H^E_{\text{chem}} + TS^E_{\text{corr}} \tag{137}$$

From equations (96), (99), and (101) it is readily seen that S^E_{corr} is null when the probabilities p_{iB} are independent of temperature. While this is true in the case of a random distribution, it is not when the factors f_{iB} are different from unity. Nevertheless, S^E_{corr} is of the nature of a correction term and is small when compared to the other terms.

Let us now introduce the parameter τ^A_{iB} defined by the ratio

$$\tau^A_{iB} = \tfrac{1}{2}\,\delta U^A_{iB}/R\,\delta \ln q^A_{iB} \tag{138}$$

τ^A_{iB} measures the correspondence between the change in the potential energy of the central atom A and the change in the logarithm of its partition function when its nearest neighbor shell is changed from a composition of Z atoms A to i atoms B and $Z - i$ atoms A. With the interpretation of $\delta \ln q$ given in Section 5.2, τ^A_{iB} may be rewritten

$$\tau^A_{iB} = -\delta[\psi(r=0)]^A_{iB} \Big/ 3R\,\delta\left[\ln \frac{\partial^2\psi}{\partial r^2}\right]^A_{iB} \tag{139}$$

The numerator measures a change in the depth of the potential well and the denominator the corresponding change in its curvature. Since the distance between the central atom and its nearest neighbor shell is not seriously affected by the composition of the shell, it is intuitively seen that the deeper the well, the more needlelike it will be (i.e., the higher its curvature). It is then reasonable to assume a proportionality between depth and curvature, and that means that τ_{iB}^{A} is independent of i. The same result is achieved in the linear case where the δ parameters are proportional to i (it would be also achieved if they were proportional to i^2 or i^n). Thus, we can write

$$\tau_{iB}^{A} = \tau_A \tag{140}$$

There is a corresponding proportionality constant τ_B for the central atom B. τ_A and τ_B are not expected to be equal. Nevertheless, the variation of τ with the nature of the central atom (A or B) is probably of less importance than the variation of δU and $\delta \ln q$ with the changes in the composition of the atomic shell, so that we shall further assume

$$\tau_A \simeq \tau_B = \tau \tag{141}$$

Comparing the expressions for $H_{\text{chem}}^{\text{E}}$ and $S_{\text{nonconf}}^{\text{E}}$ in equations (96) and (103), we immediately deduce that

$$H_{\text{chem}}^{\text{E}} = \tau S_{\text{nonconf}}^{\text{E}} \tag{142}$$

Equations (135) and (137) then lead to

$$H^{\text{E}} - TS_{\text{corr}}^{\text{E}} = \tau(S^{\text{E}} - S_{\text{conf}}^{\text{E}} - S_{\text{corr}}^{\text{E}}) \tag{143}$$

or

$$H^{\text{E}} = \tau(S^{\text{E}} - S_{\text{conf}}^{\text{E}}) - (\tau - T)S_{\text{corr}}^{\text{E}} \tag{144}$$

As previously observed, the term $S_{\text{corr}}^{\text{E}}$ should be rather small and the term $(\tau - T)S_{\text{corr}}^{\text{E}}$ reasonably negligible in front of the terms τS^{E} and $\tau S_{\text{conf}}^{\text{E}}$. Consequently,

$$H^{\text{E}} \simeq \tau(S^{\text{E}} - S_{\text{conf}}^{\text{E}}) \tag{145}$$

In dilute solutions, $S_{\text{conf}}^{\text{E}}$ is found, through the analysis of the quasi-chemical model, to be second order with respect to the mole fractions of the solutes and also second order with respect to the parameter ω or λ measuring deviations from ideality. The configurational contribution is zero for the zeroth order entropy interaction coefficient, and small (or negligible when deviations from ideality are not very pronounced) for the first order entropy interaction coefficients. Consequently, the functional dependence between zeroth and first order interaction coefficients must approximately be the same for all three thermodynamic functions, free energy, enthalpy, and entropy. This result has already been observed in the particular case of the regular solution model [equations (128) and (129)].

An a priori estimate of τ is desirable. In dilute solutions, it would permit the evaluation of the enthalpy and entropy contributions from free energy data (thus, of the temperature dependence of these data), or, conversely, would permit the evaluation of free energy functions from enthalpy data. As emphasized above, τ is a measure of the correlation between two effects: a change in the depth of the potential well and the corresponding change in its curvature when the com-

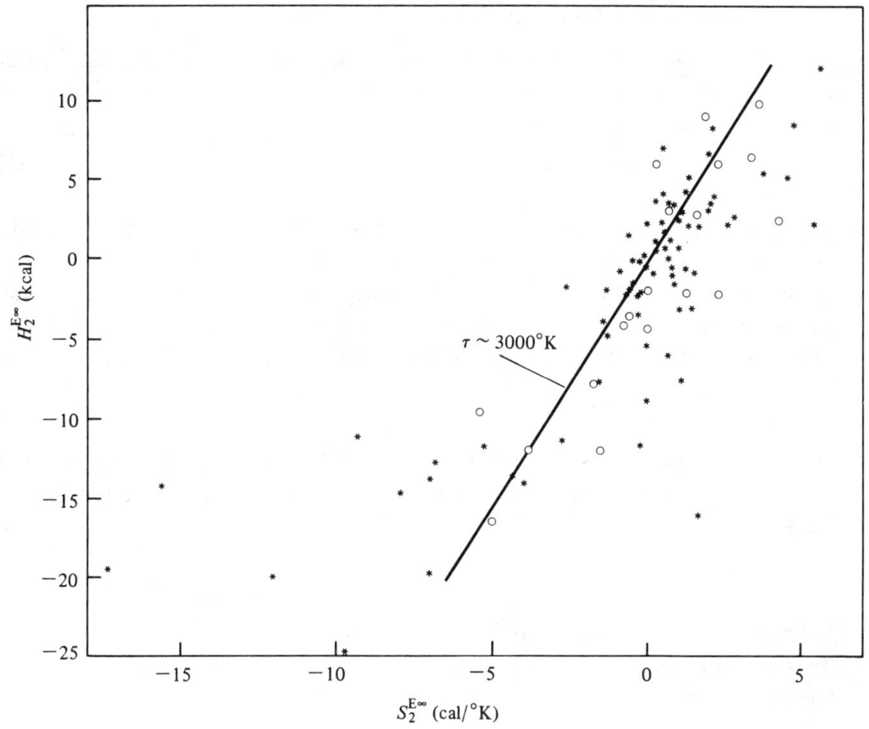

Figure 8. Correlation between excess enthalpy and entropy at infinite dilution of the solute 2 [17]. The data are from the compilation of Hultgren et al. [19]. (*, liquid alloy; ○, solid alloy.)

position of the nearest neighbor shell varies. But is τ strongly dependent on the nature of the atoms involved? More pertinently, assuming that the model is applicable, is τ strongly dependent on the nature of the alloy system considered?

A partial answer to this question may be obtained from Fig. 8, in which $H_2^{E\infty}$ is plotted against $S_2^{E\infty}$ for a wide variety of binary systems—disordered solid alloys or liquid solutions. The zeroth order interaction coefficients were chosen for this test in order to eliminate the contribution of the configurational excess entropy. In spite of a considerable scatter, due certainly to questionable theoretical assumptions but also to large experimental errors (experimental values of S^E are notoriously inaccurate), the correlation seems reasonably established, indicating a generally restricted range of permissible values of τ.

5.6. Assumptions and Discussion

So far in our development of the model we have made a number of simplifying assumptions. If we wish to improve the model and make it more realistic, some of these must be examined or revised.

Two assumptions may be immediately singled out. The first concerns the calculation of the configurational entropy; the second concerns the variation of δU_{iB} and $\delta \ln q_{iB}$ with i, that is, with the composition of the nearest neighbor shell. Let us examine in turn these two assumptions.

5.6.1. Configurational Entropy

Calculating the configurational entropy is equivalent to calculating the degeneracy factor g. In the quasi-regular solution model, the atoms are assumed to be distributed at random and the value of g is

$$g^* = N!/N_A!N_B! \tag{146}$$

If we distinguish between the configurations $\{i\text{B},(Z - i)\text{A}\}$ in the nearest neighbor shell according to the distribution of the i atoms B among the various sites of the shell, the probability α_{iB}^{*A} of each configuration would be the product of the probabilities for each site to be occupied by an A atom (a probability of X_A) or a B atom (a probability of X_B). Hence, the probability α_{iB}^{*A} is equal to

$$\alpha_{iB}^{*A} = X_A^{Z-i}X_B^i \tag{147}$$

Since there are C_Z^i such configurations, the probability that an atom A be surrounded by i atoms B and $Z - i$ atoms A (without specifying which lattice sites they occupy) is

$$p_{iB}^{*A} = C_Z^i\alpha_{iB}^{*A} = C_Z^iX_A^{Z-i}X_B^i \tag{148}$$

When the atoms are not distributed at random α_{iB}^{A} is different from α_{iB}^{*A}. However, if we assume that all the configurations $\{i\text{B},(Z - i)\text{A}\}$ are energetically equivalent, we still have

$$p_{iB}^{A} = C_Z^i\alpha_{iB}^{A} \tag{149}$$

To calculate g we may adopt a method almost identical to that devised by Guggenheim in the quasi-chemical model (Section 4.2). Instead of assuming a random distribution of atoms, or pairs of atoms, we assume that g is proportional to the number of arrangements corresponding to a random distribution of configurations $\{i\text{B},(Z - i)\text{A}\}$:

$$g = hN_A!N_B! \bigg/ \prod_{i=0}^{Z} [(N_A\alpha_{iB}^{A})!]^{C_Z^i} \prod_{i=0}^{Z} [(N_B\alpha_{iB}^{B})!]^{C_Z^i} \tag{150}$$

To determine the constant of proportionality, we note that if the atoms were distributed at random, we would have

$$g^* = h\frac{N_A!N_B!}{\prod_{i=0}^{Z} [(N_A\alpha_{iB}^{*A})!]^{C_Z^i} \prod_{i=0}^{Z} [(N_B\alpha_{iB}^{*B})!]^{C_Z^i}} = \frac{N!}{N_A!N_B!} \tag{151}$$

Consequently,

$$g = \frac{N!}{N_A!N_B!}\,\frac{\prod_{i=0}^{Z} [(N_A\alpha_{iB}^{*A})!]^{C_Z^i} \prod_{i=0}^{Z} [(N_B\alpha_{iB}^{*B})!]^{C_Z^i}}{\prod_{i=0}^{Z} [(N_A\alpha_{iB}^{A})!]^{C_Z^i} \prod_{i=0}^{Z} [(N_B\alpha_{iB}^{B})!]^{C_Z^i}} \tag{152}$$

The maximum term in the expression of the partition function must be obtained by optimization with respect to the unknowns α_{iB}^{A}, α_{iB}^{B}. This is equivalent to a minimization of G^M in equation (105) with respect to the unknowns p_{iB}^{A}, p_{iB}^{B}. These unknowns are not arbitrary. They are subject to the constraints

$$\psi_1 = \sum_{i=0}^{Z} p_{iB}^{A} - 1 = 0 \tag{153a}$$

$$\psi_2 = \sum_{i=0}^{Z} p_{iB}^{B} - 1 = 0 \tag{153b}$$

Moreover, a mass balance on the B atoms, counted as nearest neighbors to the central atoms A and B, yields

$$\psi_3 = N_A \left(\sum_{i=1}^{Z} i p_{iB}^{A} \right) + N_B \left(\sum_{i=1}^{Z} i p_{iB}^{B} \right) - Z N_B = 0 \tag{154}$$

The optimization procedure can be carried out through the technique of Lagrange multipliers, e.g.,

$$\partial[(G^M/RT) + \lambda_1\psi_1 + \lambda_2\psi_2 + \lambda_3\psi_3]/\partial p_{iB}^{J} = 0 \tag{155}$$

where J is A or B. The calculations yield

$$p_{iB}^{J} = C_Z^i X_A^{Z-i} X_B^i e^{i\lambda - \delta\varphi_{iB}^{J}} \Big/ \sum_{i=0}^{Z} C_Z^i X_A^{Z-i} X_B^i e^{i\lambda - \delta\varphi_{iB}^{J}} \tag{156}$$

where λ is the Lagrange multiplier associated with the constraint (154), and the parameter $\delta\varphi_{iB}^{J}$ is defined as equal to

$$\delta\varphi_{iB}^{J} = \frac{\delta U_{iB}^{J}}{2RT} - \delta \ln q_{iB}^{J} \tag{157}$$

The value of λ is obtained by replacing p_{iB}^{J} by its expression (156) in equation (154).

The calculations are readily simplified in the case of pairwise interactions, i.e., when $\delta\varphi_{iB}^{J}$ varies linearly with i:

$$\delta\varphi_{iB}^{J} = i\delta\varphi_{1B}^{J} \tag{158}$$

In particular, equation (156) becomes

$$p_{iB}^{J} = \frac{C_Z^i X_A^{Z-i} X_B e^{i(\lambda - \delta\varphi_{1B}^{J})}}{(X_A + X_B e^{\lambda - \delta\varphi_{1B}^{J}})^Z} \tag{159}$$

and a Taylor series expansion of G^E yields

$$\begin{aligned} G^E/RT = {} & X_B Z(\delta\varphi_{1B}^{A} - \delta\varphi_{1A}^{B}) - X_B^2 Z(e^{(\delta\varphi_{1B}^{A} - \delta\varphi_{1B}^{B})} - 1) \\ & + X_B^3 Z(e^{(\delta\varphi_{1B}^{A} - \delta\varphi_{1B}^{B})} - 1)^2 + \cdots \end{aligned} \tag{160}$$

We note that the solution is found to behave ideally when $\delta\varphi_{1B}^{A} - \delta\varphi_{1B}^{B} = 0$, that is, when $\delta U_{1B}^{A} - \delta U_{1B}^{B} = 0$ and $\delta \ln q_{1B}^{A} - \delta \ln q_{1B}^{B} = 0$. These conditions are equivalent to $\omega_{AB} = 0$ and $\pi_{AB} = 0$, a result already encountered in the quasi-regular solution.

Compared to the treatment leading to the quasi-regular solution, the introduction of a configurational contribution in the excess free energy clearly improves the solution model and makes it more realistic, especially in the case of concentrated solutions. For dilute solutions, the configurational contribution is very small and the predictions of the model are then almost identical to those of the quasi-regular solution.

5.6.2. Variations in a Central Atom's Potential Energy

Further improvements in the model are unlikely to be obtained through refinements in the calculation of the configurational term. Instead, one must question other assumptions of the model. For example, it has been assumed that the coordination number Z is constant and independent of the composition of the nearest neighbor shell and of the nature of the central atom. This is unrealistic if the atoms A and B have very different sizes. It has also been assumed that atomic interactions could be accounted for by a summation of pairwise interactions. This is not a necessary assumption and alternate ones may be explored.

For example, Mathieu et al. [20] have assumed parabolic variations of the potential energy of a central atom J on the number i of B nearest neighbors

$$\delta\varphi^J_{iB} = \left(\frac{i}{Z}\right)^2 \delta\varphi^J_{ZB} \tag{161}$$

or

$$\delta\varphi^J_{iB} = \frac{i}{Z}\left(2 - \frac{i}{Z}\right) \delta\varphi^J_{ZB} \tag{162}$$

In the first parabolic function, the perturbation created by the first atom B in a nearest neighbor shell of all A atoms is smaller than that created by a second atom B, or a third, a fourth, etc. It is larger in the second parabolic function. Such variations have met with some success and the assumption of a parabolic dependence has been retained by Wagner [21] in his simplified treatment of essentially the same model.

A different concept has led Lupis et al. [22] to superpose delta functions to the linear variations of $\delta\varphi^J_{iB}$; e.g.,

$$\delta\varphi^A_{iB} = i\delta\varphi^A_{1B} + \delta_{ij'}\alpha' \tag{163a}$$

$$\delta\varphi^B_{iB} = i\delta\varphi^B_{1B} + \delta_{ij''}\alpha'' \tag{163b}$$

Figure 9. Illustration of the values assumed for $\delta\varphi^A_{iB}$ and $\delta\varphi^B_{iB}$ according to equations (163). $\delta\varphi^A_{1B} = -0.40$, $\delta\varphi^B_{1B} = -0.22$ and the preferred configuration has a stoichiometry AB_2 corresponding to $j' = j'' = 8$ and $\alpha' = \alpha'' = -0.6$.

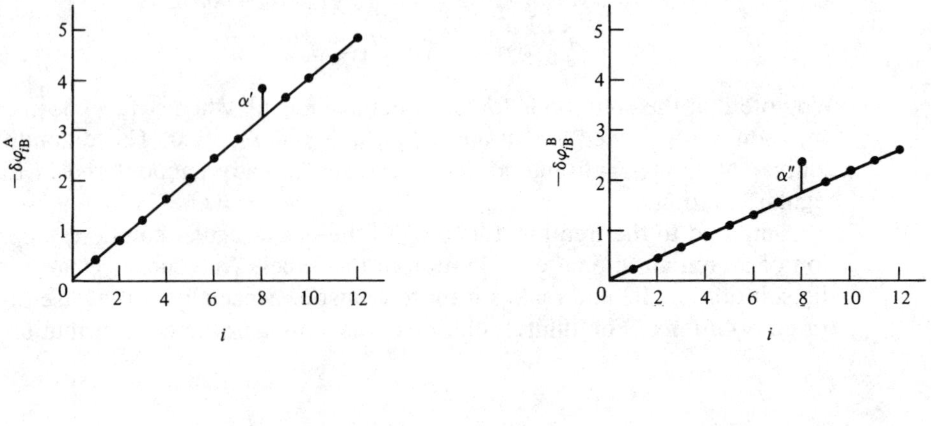

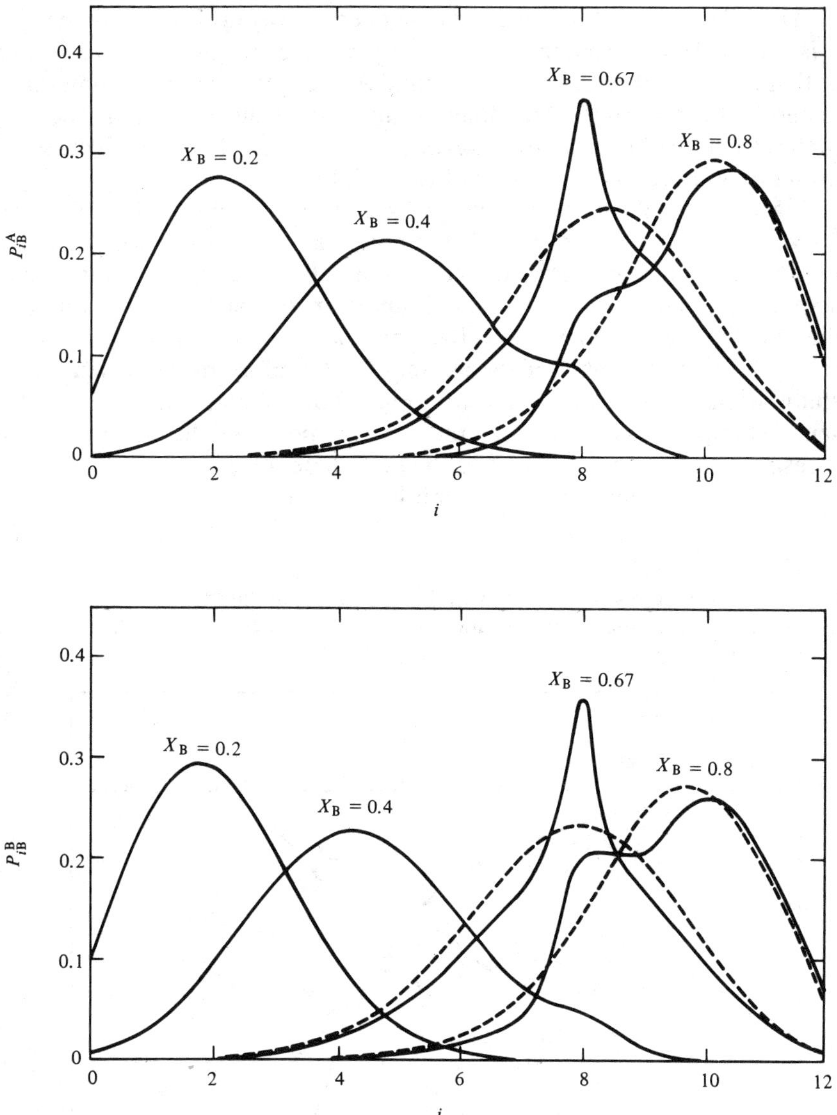

Figure 10. Probabilities for an atom A or B of being surrounded by i atoms B at various concentrations X_B [22]. The dashed lines correspond to the linear variation ($\alpha' = \alpha'' = 0$) and the solid lines to that of the preferred configuration AB_2 (case of Fig. 9).

where $\delta_{ij'}$, $\delta_{ij''}$ are Kronecker's symbols ($\delta_{ij} = 0$ for $i \neq j$ and $\delta_{ij} = 1$ for $i = j$). The significance of such an assumption is that some configuration (corresponding to j' and j'') around a central atom A or B may have a much lower or higher energy because of valence effects, stoichiometric effects, or any other type of effects. As a result, these configurations will have a higher or lower probability of occurrence. The limit case where α' and α'' are infinite and negative implies the existence of stable clusters in the solution. If α' and α'' are infinite and positive, the corresponding configurations are not permitted.

The values of j', j'', α', and α'' are not necessarily arbitrary. For many systems, it is reasonable to assume that the short range order in the liquid is a reflection of that in the solid. Thus, if a compound AB_x exists in the solid state, it is natural to choose for the study of the liquid solution the values of j' and j'' corresponding to the structure of the compound AB_x and the values of α' and α'' corresponding to its stability (i.e., its free energy of formation).

Figures 9–11 illustrate the case of a preferred configuration of the type AB_2 for which $j' = j'' = 8$ and $Z = 12$. We note that, contrary to what may have been expected, the perturbation created by the delta functions is not reflected in the activity coefficient curves by any abrupt change near the composition $X_B = \frac{2}{3}$ of the stoichiometric compound AB_2. The width of the perturbation is somewhat reduced, and the impact of the assumption therefore more evident, through the study of the stability function ψ (see Fig. 10 of Chapter III, corresponding to the same example). It is also clear that the sharpness of the perturbation is an increasing function of the height of the delta function.

Goumiri [23] followed the general lines of this work and investigated the su-

Figure 11. Predicted values of the activity coefficient in the case of Fig. 9 [22]. The dashed lines correspond to the linear variation and the solid lines to that of the preferred configuration AB_2.

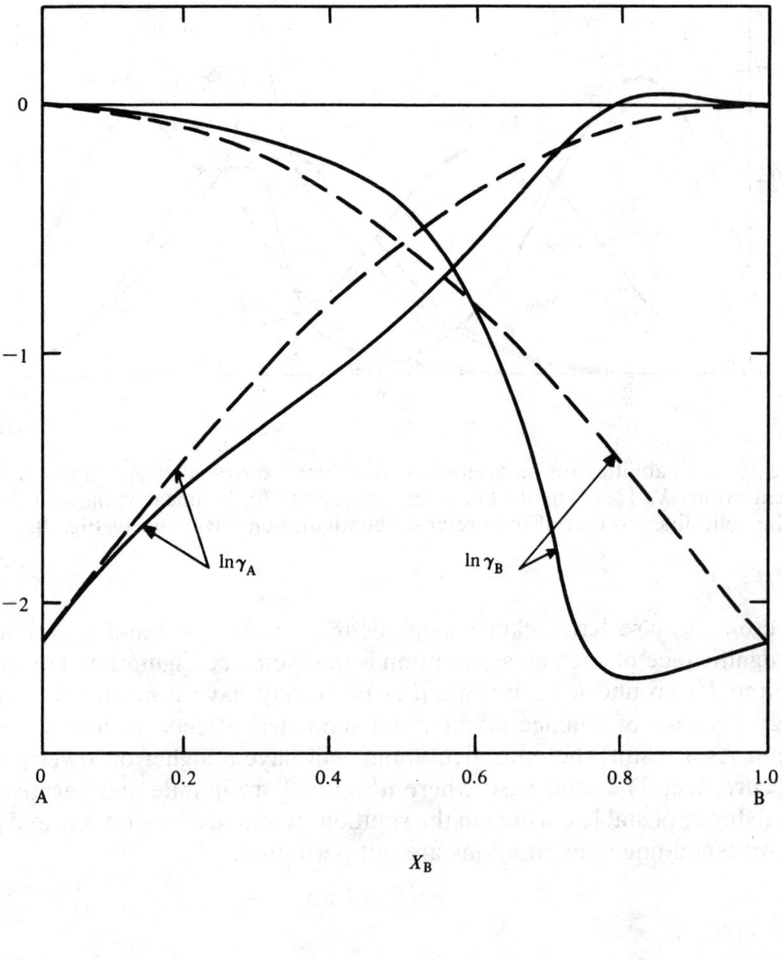

perposition of delta functions to a parabolic variation. His application of the model to systems such as the Fe–Si and Al–Cu binaries gave very encouraging results.

Several such delta functions may be introduced and, of course, ultimately each $\delta\varphi^{J}_{iB}$ may be considered to be a different parameter. The versatility of the model is an advantage, but the introduction of too many parameters makes the model inapplicable and useless. Therefore, a compromise must be sought. The necessary simplifying assumptions will depend upon the system being studied. We note, however, that a model such as that of Miedema [24] (based on changes in electronegativity and electron density) could be adapted to the central atoms model to yield a more rational variation of the potential energy of a central atom.

6. Multicomponent Solutions

All the previous models can be readily extended to the case of multicomponent solutions. Their derivation, however, becomes mathematically more complex and, in many cases can be solved only numerically. For dilute solutions, it is generally possible to express the results in Taylor series and to calculate as many coefficients of these series as may be necessary.

6.1. Regular and Quasi-Regular Solutions

Let us consider the case of a ternary regular solution. With the same assumptions as those described in Section 3, we have now to consider the pairs AA, BB, CC, AB, AC, and BC. Since each atom A generates Z pairs of the AA, AB, and AC types, a summation on all the A atoms yields the following balance on the numbers of pairs:

$$ZN_{\mathrm{A}} = 2N_{\mathrm{AA}} + N_{\mathrm{AB}} + N_{\mathrm{AC}} \tag{164}$$

Similarly,

$$ZN_{\mathrm{B}} = 2N_{\mathrm{BB}} + N_{\mathrm{AB}} + N_{\mathrm{BC}} \tag{165}$$

$$ZN_{\mathrm{C}} = 2N_{\mathrm{CC}} + N_{\mathrm{AC}} + N_{\mathrm{BC}} \tag{166}$$

The energy of the lattice is

$$E = \sum_i \sum_j N_{ij} u_{ij} \tag{167}$$

After rearrangement of its terms, equation (167) becomes

$$E = E_{\mathrm{A}} + E_{\mathrm{B}} + E_{\mathrm{C}} + N_{\mathrm{AB}}\omega_{\mathrm{AB}} + N_{\mathrm{AC}}\omega_{\mathrm{AC}} + N_{\mathrm{BC}}\omega_{\mathrm{BC}} \tag{168}$$

where

$$\omega_{ij} = u_{ij} - \tfrac{1}{2}(u_{ii} + u_{jj}) \tag{169}$$

Since the atoms are distributed at random

$$g = N!/N_{\mathrm{A}}!N_{\mathrm{B}}!N_{\mathrm{C}}! \tag{170}$$

and recalling that

$$Q = ge^{-E/kT} \tag{171}$$

we readily obtain

$$G_m^E = X_A X_B Z\omega_{AB} + X_A X_C Z\omega_{AC} + X_B X_C Z\omega_{BC} \tag{172}$$

It is obvious that in the case of a quasi-regular solution we would find

$$G_m^E = X_A X_B Z\omega_{AB}\left(1 - \frac{T}{\tau_{AB}}\right) + X_A X_C Z\omega_{AC}\left(1 - \frac{T}{\tau_{AC}}\right) + X_B X_C Z\omega_{BC}\left(1 - \frac{T}{\tau_{BC}}\right) \tag{173}$$

If we assume that $\tau_{AB} \simeq \tau_{AC} \simeq \tau_{BC} = \tau$, equation (173) reduces to

$$G_m^E = (X_A X_B Z\omega_{AB} + X_A X_C Z\omega_{AC} + X_B X_C Z\omega_{BC})\left(1 - \frac{T}{\tau}\right) \tag{174}$$

These equations can be easily generalized to the case of a multicomponent solution, e.g.,

$$G_m^E = \sum_i \sum_{j>i} X_i X_j Z\omega_{ij}\left(1 - \frac{T}{\tau_{ij}}\right) \tag{175}$$

This equation shows that the thermodynamic properties of a multicomponent quasi-regular solution may be calculated as sums of the thermodynamic properties of the limiting binary systems.

We recall that the first order free energy interaction coefficient $\epsilon_i^{(j)}$ may be written

$$\epsilon_i^{(j)} = \left(\frac{\partial^2 (G^E/RT)}{\partial X_i\, \partial X_j}\right)_{X_1 \to 1} \tag{176}$$

Consequently, in the regular solution model

$$\epsilon_B^C = Z\beta(\omega_{BC} - \omega_{AB} - \omega_{AC}) = Z[(u_{BC} - u_{AC}) + (u_{AA} - u_{AB})] \tag{177}$$

and in the quasi-regular one

$$\epsilon_B^C = Z\beta\left[\omega_{BC}\left(1 - \frac{T}{\tau_{BC}}\right) - \omega_{AB}\left(1 - \frac{T}{\tau_{AB}}\right) - \omega_{BC}\left(1 - \frac{T}{\tau_{BC}}\right)\right] \tag{178}$$

If the elements B and C have a much stronger affinity for each other than for element A, equation (177) or (178) predicts that ϵ_B^C should be negative. In liquid iron, this is indeed the case, for example, of ϵ_O^{Al} (−433), ϵ_O^{Si} (−15), ϵ_O^{Ti} (−118), and ϵ_O^{V} (−63) [10]. The affinity of Al, Si, Ti, and V for O is evidenced by the large and negative free energy of formation of their oxides. Strong positive values of ϵ_B^C indicates a relative repulsion of an atom B for an atom C.

The values of ϵ_B^C may be estimated from the properties of the binary limiting systems by the equation

$$\epsilon_B^C = (\ln \gamma_C^\infty)_{B-C} - (\ln \gamma_B^\infty)_{A-B} - (\ln \gamma_C^\infty)_{A-C} \tag{179}$$

or

$$\epsilon_B^C = \tfrac{1}{2}[(\epsilon_B^B)_{A-B} + (\epsilon_C^C)_{A-C} - Z_A(\epsilon_C^C/Z_B)_{B-C}] \tag{180}$$

The alternative use of the properties of the solute C in solvent B, or the reverse, in equations (179) and (180) arises from the need to evaluate ω_{BC}, or $\omega_{BC}[1 - (T/\tau_{BC})]$, from the binary solution B–C. Within the framework of the solution models adopted here, there should be no difference in the results. Practically, however, there will be a difference because the binary solutions B–C will rarely be strictly regular or quasi-regular. If the properties of B approximate more closely those of the solvent A than the properties of C do, then it is appropriate to choose, for example, $(\ln \gamma_C^\infty)_{B-C}$ rather than $(\ln \gamma_B^\infty)_{C-B}$ to calculate ω_{BC}. Nevertheless, it is fair to point out that in general such differences are not significant. Usually, only a qualitative or semiquantitative agreement between calculated and experimental values is found [12], or should be sought.

Similar expressions for the enthalpy and entropy first order interaction coefficients may be obtained:

$$\eta_B^C = (H_C^{E\infty})_{B-C} - (H_B^{E\infty})_{A-B} - (H_C^{E\infty})_{A-C} \tag{181}$$

or

$$\eta_B^C = \tfrac{1}{2}[(\eta_B^B)_{A-B} + (\eta_C^C)_{A-C} - Z_A(\eta_C^C/Z_B)_{B-C}] \tag{182}$$

and

$$\sigma_B^C = (S_C^{E\infty})_{B-C} - (S_B^{E\infty})_{A-B} - (S_C^{E\infty})_{A-C} \tag{183}$$

or

$$\sigma_B^C = \tfrac{1}{2}[(\sigma_B^B)_{A-B} + (\sigma_C^C)_{A-C} - Z_A(\sigma_C^C/Z_B)_{B-C}] \tag{184}$$

Equations (181) and (182) are predicted by both the regular and quasi-regular solution models, but equations (183) and (184) are predicted only by the quasi-regular model. Indeed, for the regular model, $S^E = 0$ and σ_B^C is also equal to zero. Table 2 shows two examples of applications of equations (179), (181), and (183) in the cases of the Sn–Zn–Pb and Bi–Ag–Zn liquid solutions at 800°K [17]. On a qualitative basis, the agreement between predicted and experimental values is satisfactory.

Table 2. Predicted and Experimental Values of the Ternary First Order Interaction Coefficients [17]

System A–B–C	Thermodynamic function	Binary interaction coefficients, zeroth order: system B–C	system A–B	system A–C	Ternary interaction coefficients, first order: predicted[a]	experimental
Sn–Pb–Zn	G^E/RT at 800°K	$\ln \gamma_C^\infty = 2.63$	$\ln \gamma_B^\infty = 0.85$	$\ln \gamma_C^\infty = 0.62$	$\epsilon_B^C = 1.2$	$\epsilon_B^C = 2.3$
	H^E (kcal)	$H_C^{E\infty} = 5.75$	$H_B^{E\infty} = 1.36$	$H_C^{E\infty} = 2.22$	$\eta_B^C = 2.2$	$\eta_B^C = 4.1$
	S^E (cal/°K)	$S_C^{E\infty} = 1.97$	$S_B^{E\infty} = 0$	$S_C^{E\infty} = 1.55$	$\eta_B^C = 0.4$	$\eta_B^C = 0.6$
Bi–Ag–Zn	G^E/RT at 800°K	$\ln \gamma_C^\infty = -3.3$	$\ln \gamma_B^\infty = 1.32$	$\ln \gamma_C^\infty = 1.25$	$\epsilon_B^C = -5.9$	$\epsilon_B^C = -2.7$
	H^E (kcal)	$H_C^{E\infty} = -6.5$	$H_B^{E\infty} = 2.94$	$H_C^{E\infty} = 3.68$	$\eta_B^C = -13.1$	$\eta_B^C = -8.8$
	S^E (cal/°K)	$S_C^{E\infty} = -1.5$	$S_B^{E\infty} = 1.06$	$S_C^{E\infty} = 2.10$	$\eta_B^C = -4.7$	$\eta_B^C = -5.6$

[a] Predicted through equations (179), (181), and (183) of the regular solution model.

6.2. Quasi-Chemical Approximations

The development of Guggenheim's quasi-chemical model for the general case of a multicomponent solution follows the same procedure as that outlined above for a binary solution. The calculations are straightforward but somewhat lengthy [12] and need not be reproduced here. We shall, however, briefly describe some of the results obtained.

It is found [12] that the excess Gibbs free energy of the solution may be written

$$G^E = \sum_{i=1}^{m-1} \sum_{j>i}^{m} N_{ij}^{*} \omega_{ij} + RT \sum_{i=1}^{m} \sum_{j \geq i}^{m} N_{ij}^{*} \ln(\overline{N}_{ij}/N_{ij}^{*}) \tag{185}$$

The first sum of this equation yields the results of the regular solution theory; we recall that we have already established that N_{ij}^{*} is equal to ZN_iN_j/N. The second sum accounts for the nonrandomness of the distribution of atoms. Mass balances on the numbers of pairs containing an element i yield

$$2N_{ii} + \sum_{\substack{j=1 \\ i \neq j}}^{m} N_{ij} = ZN_i \tag{186}$$

The $\overline{N}_{ij}$, $\overline{N}_{ii}$, and $\overline{N}_{jj}$ are related by the "quasi-chemical reaction,"

$$\frac{(\frac{1}{2}\overline{N}_{ij})^2}{\overline{N}_{ii}\overline{N}_{jj}} = e^{-2\beta\omega_{ij}} \tag{187}$$

Combining equations (186) and (187) yields the composition dependence of the $\overline{N}_{ij}$ and therefore of G^E. A system of $m(m-1)/2$ equations of the second degree must be solved. An analytic solution may be found in the case of dilute solutions by expressing the results in terms of Taylor series. In particular, the expression

Table 3. Predicted Values of the Free Energy Interaction Coefficients from the Quasi-Chemical Model of Solutions [12]

Interaction coefficient	In terms of quasi-chemical coefficients	In terms of lower order interaction coefficients
1. $\ln \gamma_i^{\infty}$	$\frac{1}{2}Z \ln(1 + \mu_{ii})$	$\frac{1}{2}Z \ln(1 - \epsilon_i^{(i)}/Z)$
2. $\epsilon_i^{(i)}$	$-Z\mu_{ii}$	$Z[1 - (\gamma_i^{\infty})^{2/Z}]$
3. $\rho_i^{(i)}$	$\frac{1}{2}Z\mu_{ii}(1 + 3\mu_{ii})$	$(3/2Z)[\epsilon_i^{(i)}]^2 - \frac{1}{2}(\epsilon_i^{(i)})$
4. $\epsilon_i^{(j)}$	$-Z\mu_{ij}$	$Z\left(1 - \left\{\frac{[1 - (\epsilon_i^{(i)}/Z)]\,[1 - (\epsilon_j^{(j)}/Z)]}{[1 - (\epsilon_i^{(j)}/Z_j)]_{\text{binary } i\text{–}j}}\right\}^{1/2}\right)$
5. $\rho_i^{(i,j)}$	$Z\mu_{ij}(2\mu_{ii} + \mu_{ij} + 1)$	$(\epsilon_i^{(j)}/Z)(2\epsilon_i^{(i)} + \epsilon_i^{(j)} - Z)$
6. $\rho_i^{(j)}$	$Z\mu_{ij}\mu_{jj} + \frac{1}{2}Z(\mu_{ij})^2 + \frac{1}{2}Z\mu_{jj}$	$[(\epsilon_i^{(j)})^2/2Z] + (\epsilon_j^{(j)}/Z)(\epsilon_i^{(j)} - \frac{1}{2}Z)$
7. $\rho_i^{(j,k)}$	$Z(\mu_{ij}\mu_{ik} + \mu_{jk}\mu_{ij} + \mu_{ik}\mu_{jk} + \mu_{jk})$	$(1/Z)(\epsilon_i^{(j)}\epsilon_i^{(k)} + \epsilon_j^{(k)}\epsilon_i^{(j)} + \epsilon_i^{(k)}\epsilon_j^{(k)}) - \epsilon_j^{(k)}$

Note equation (189): $\mu_{ij} = \exp[\beta(\omega_{1i} + \omega_{1j} - \omega_{ij})] - 1$. For $i = j$, $\mu_{ii} = \exp(2\beta\omega_{1i}) - 1$.

for G^E to the third order term with respect to the mole fractions of the solutes is

$$G^E/RT = \sum_{i=2}^{m} ZX_i\beta\omega_{1i} - \tfrac{1}{2}Z\sum_{i=2}^{m}\sum_{j=2}^{m} X_iX_j\mu_{ij} + \tfrac{1}{2}Z\sum_{i=2}^{m} X_i\left(\sum_{j=2}^{m} X_j\mu_{ij}\right)^2 + O(X^4) \tag{188}$$

where

$$\mu_{ij} = \exp[\beta(\omega_{1i} + \omega_{1j} - \omega_{ij})] - 1 \tag{189}$$

and we note that $\mu_{ij} = \mu_{ji}$. The consequent expressions for the zeroth, first, and second order free energy interaction coefficients are summarized in Table 3.

The central atoms model of a binary solution may also be extended to the case of a multicomponent solution. The calculations are straightforward in the case of a linear dependence of the parameters $\delta\varphi$ on the numbers of solute atoms in the nearest neighbor shell and of an estimation of the configuration entropy similar to that outlined in Section 5.6.1. As in the case of the quasi-chemical model, it is convenient to express the results in terms of Taylor series. In particular, G^E is found [22] to be equal to

$$\begin{aligned} G^E/ZRT = {} & \sum_{i=2}^{m} X_i(\delta\varphi^{(1)}_{0,n_i=1,\ldots,0} - \delta\varphi^{(i)}_{0,n_i=1,\ldots,0}) \\ & - \sum_{i=2}^{m}\sum_{j=2}^{m} X_iX_j[\exp(\delta\varphi^{(1)}_{0,n_i=1,\ldots,0} - \delta\varphi^{(j)}_{0,n_i=1,\ldots,0}) - 1] \\ & + \sum_{i=2}^{m} X_i\left\{\sum_{j=2}^{m} X_j[\exp(\delta\varphi^{(1)}_{0,n_i=1,\ldots,0} - \delta\varphi^{(j)}_{0,n_i=1,\ldots,0}) - 1]\right\}^2 \\ & + O(X^4) \end{aligned} \tag{190}$$

7. Conclusions

Many statistical models of solutions may be found in the literature. Only a few fairly representative ones have been presented here. We note that they are based on many simplifying assumptions. For example, only the configurational and vibrational contributions have been considered and our methods of estimating their values are open to question. Depending on the alloy system studied, other contributions (e.g., magnetic) may be important. Moreover, liquid solutions have been regarded as disordered solids, and in the case of disordered solids, we have not considered the effects of stresses. Atoms of different components have been effectively treated as having equal sizes even though they often have significantly different sizes. Bonding energies have been assumed to be constant, although they depend on interatomic distances which are themselves functions of the local environment. Lattice models which take into account lattice deformations have not been effectively developed yet.

In spite of all these shortcomings, simple models such as those presented here are still very useful. The accuracy of the experimental data often does not warrant more than semiquantitative estimates. The development of more accurate models may also signify the introduction of additional parameters, which, from a practical standpoint, is not desirable. In addition, the models may become mathematically intractable. Thus, in general a compromise must be sought.

Problems

1. In Einstein's model, a lattice is considered to be constituted by N independent harmonic oscillators. The vibrational energy levels (in each principal direction x, y, z) differ by $h\nu$ where h is Planck's constant and ν is the frequency of the vibration. Calculate the partition function of the lattice, its associated energy, and its heat capacity C_v. Express the results in terms of the characteristic temperature $\theta = h\nu/k$ (where k is Boltzmann's constant). Determine the limit of C_v for $T \to \infty$.

2. Derive equation (63).

3. **a.** In the quasi-chemical model, develop to the second order term (included) the Taylor series of $\overline{N}_{AB}$ with respect to λ.
b. Also develop to the second order term the Taylor series of G_m^E and S_m^E with respect to λ.

4. Solid A melts at 1200°K with a heat of fusion equal to 3000 cal/mol. The heat capacities of A in the solid and liquid phases are assumed equal. Component B is insoluble in solid A, but its solution in liquid A is assumed to follow the quasi-chemical model with $Z = 12$. At 1100°K, the solubility of B in liquid A is experimentally determined to be equal to $X_B = 0.2$.
a. Estimate γ_B^∞ in the liquid at 1100°K.
b. Estimate the solubility of B at 1150°K.
c. Estimate the slope of the liquidus at 1150°K.

5. **a.** Calculate the value of the Lagrange multiplier λ in the linear case of the central atoms model (Section 5.6.1).
b. Calculate the average number of nearest neighbors B to a central atom B and find the first order term of its Taylor series with respect to X_B.

6. Consider a lattice of N sites occupied by N_A atoms A and N_V vacancies V. Calculate the equilibrium concentration of vacancies following a model similar to that of the regular solution, i.e., assuming random distribution of the atoms and pairwise interactions. Discuss the results. Interpret graphically the parameter ω_{AV} in the case of a one-dimensional lattice.

References

1. R. H. Fowler and E. A. Guggenheim, *Statistical Thermodynamics*. Cambridge University Press, Cambridge, England, 1937.
2. J. E. Mayer and M. G. Mayer, *Statistical Mechanics*. Wiley, New York, 1940.
3. E. A. Guggenheim, *Mixtures*. Oxford University Press, Oxford, 1952.
4. I. Prigogine, *Molecular Theory of Solutions*. North Holland, Amsterdam, 1957.
5. T. L. Hill, *Introduction to Statistical Thermodynamics*. Addison-Wesley, Reading, MA, 1960.
6. C. H. P. Lupis, *Acta Met.* **25,** 751 (1977).
7. C. H. P. Lupis, *Acta Met.* **26,** 211 (1978).

8. J. Frenkel, *Kinetic Theory of Liquids*. Dover, New York, 1965.
9. J. H. Hildebrand and R. L. Scott, *The Solubility of Nonelectrolytes,* 3rd ed. Van Nostrand Reinhold, New York, 1950.
10. G. K. Sigworth and J. F. Elliott, *Met. Sci.* **8,** 298 (1974).
11. J. M. Dealy and R. D. Pehlke, *Trans. Met. Soc. AIME* **227,** 88, 1030 (1963).
12. C. H. P. Lupis and J. F. Elliott, *Acta Met.* **14,** 1019 (1966).
13. J. F. Elliott and J. Chipman, *Trans. Faraday Soc. No. 338* **42**(2), 137 (1951).
14. H. Eyring, *J. Chem. Phys.* **4,** 283 (1936).
15. H. Eyring and J. Hirschfelder, *J. Phys. Chem.* **41,** 249 (1937).
16. C. H. P. Lupis, Sc.D. thesis, M.I.T., 1965.
17. C. H. P. Lupis and J. F. Elliott, *Acta Met.* **15,** 265 (1967).
18. J.-C. Mathieu, F. Durand, and E. Bonnier, *J. Chim. Phys.* **11–12,** 1289 (1965).
19. R. Hultgren, L. R. Orr, P. D. Anderson, and K. K. Kelley, *Selected Values of Thermodynamic Properties of Metals and Alloys*. Wiley, New York, 1963.
20. J.-C. Mathieu, F. Durand, and E. Bonnier, *J. Chim. Phys.* **11–12,** 1297 (1965).
21. C. Wagner, *Acta Met.* **21,** 1297 (1973).
22. C. H. P. Lupis, H. Gaye, and G. Bernard, *Scripta Met.* **4,** 497 (1970).
23. L. Goumiri, Ph.D. thesis, University of Grenoble, France, 1980.
24. A. R. Miedema, *Z. Metallk.* **69,** 183 (1978).

The texts by Hill [5], Prigogine [4], and Guggenheim [3] would be very useful for further study.

XVI Statistical Models of Interstitial Metallic Solutions

1. Introduction

In the previous chapter, we considered that, for a binary phase A–B, a given lattice site can be occupied by either an atom A or B. The components A and B were then said to be substitutional. However, there are many mixtures for which it is not probable that A and B occupy the same kind of site. For example, when carbon is alloyed to iron in the face-centered cubic structure, the carbon atoms, because of their small size, do not replace the iron atoms at their lattice sites but occupy some "holes" in that structure. Carbon is then said to be an *interstitial* solute. All the holes which are large enough to accommodate the carbon atoms may be considered to form another lattice of interstitial sites which is interlocked with the iron lattice. In the case of fcc iron (austenite), these holes are octahedral and form another fcc lattice; its sites are occupied by carbon atoms and vacancies. There are also tetrahedral holes which are smaller in size and, thus, energetically unfavorable. Their presence may be disregarded.

The classical thermodynamics definition of an ideal solution (i.e., the activity of a component i is proportional to its mole fraction X_i) was given in Section VI.2. The formalism of "excess" thermodynamic functions is based upon the properties of that ideal solution. These properties were interpreted in Chapter XV by a model which assumes that the energy of the lattice is independent of any particular distribution of the atoms on identical lattice sites and, therefore, that the atoms are distributed at random on these sites. Such a model does not apply to interstitial solutions and it may be questioned whether the classical definitions of an ideal solution and excess functions should be retained as a framework for the study of interstitial solutions.

It must be acknowledged that in many cases it is difficult to ascertain whether a solute is interstitial or substitutional. For example, in liquids the distinction between interstitial and substitutional sites loses much of its sharpness. It is thus advisable to retain the classical definition of an ideal solution, and the subsequent definitions of excess functions, regardless of the details of a solution's atomic structure.

Adopting a different formalism for interstitial solutions [1] could lead to a proliferation of thermodynamic functions and coefficients quite cumbersome in practice. Generally, in the development of a model, it is appropriate to introduce any function or coordinate which simplifies the calculations. However, it is incumbent upon the developer of that model that he translate the results of the calculations in terms of a common formalism for the sake of the investigators who wish to utilize these results.

It may be added that the definition of an ideal solution belongs to the province of classical thermodynamics. The model of an ideal substitutional solution belongs to the province of statistical thermodynamics, even though it leads to the same thermodynamic equations and is an interpretation of these equations. Although complementary, these two provinces should be clearly distinguished.

We shall now attempt to devise models which can interpret and predict the properties of interstitial solutions. We shall start with the case of binary solutions and then extend our investigation to the case of multicomponent ones.

2. Ideal Interstitial Solution

Let us designate the atoms of the solvent by A, the interstitial atoms by C, the vacancies on the interstitial lattice by V. N_I is the number of interstitial sites and is equal to $N_C + N_V$, the number of C atoms plus the number of vacancies. We shall assume that there are no vacancies on the substitutional lattice. The ratio of the number of interstitial sites N_I to the number of substitutional sites N_S will be designated r:

$$r = N_I/N_S \tag{1}$$

In this case, r is also equal to

$$r = (N_C + N_V)/N_A \tag{2}$$

For an fcc structure r is equal to 1; for a bcc it is equal to 3.

The assumptions of an ideal interstitial solution are similar to those of an ideal substitutional solution. We assume that the energies of the atoms A and C are unaffected by their respective positions on the substitutional and interstitial lattices. Therefore, the energy of the mixture is just

$$E = N_A E_A + N_C E_C \tag{3}$$

Since all the substitutional sites are filled by A atoms, the only source of configurational entropy is in the number of ways of distributing the C atoms among the interstitial sites. We shall assume that this distribution of C atoms (or V vacancies) is random. Consequently,

$$g = N_I!/N_C!N_V! \tag{4}$$

The partition function of the system is equal to

$$Q = \sum g q_A^{N_A} q_C^{N_C} e^{-E/kT} \tag{5}$$

where q_A, q_C are the partition functions of the atoms A and C. Combining equations (3)–(5), we obtain

$$Q = \frac{N_I!}{N_C!N_V!} q_A^{N_A} q_C^{N_C} e^{-(N_A E_A + N_C E_C)/kT} \tag{6}$$

Consequently, the Gibbs free energy of the solution becomes

$$\begin{aligned} G = {} & N_A(E_A - kT \ln q_A) + N_C(E_C - kT \ln q_C) \\ & + kT[N_C \ln(N_C/N_I) + N_V \ln(N_V/N_I)] \end{aligned} \tag{7}$$

In this expression $E_A - kT \ln q_A$ and $E_C - kT \ln q_C$ are the free energies per atom of pure A and pure C in the same structure as that of the solution. For C, this clearly corresponds to a hypothetical state.

Recalling that

$$N_V = N_I - N_C = rN_A - N_C \tag{8}$$

equation (7) becomes

$$\begin{aligned} G = {} & N_A(E_A - kT \ln q_A) + N_C(E_C - kT \ln q_C) \\ & + kT\left[(rN_A - N_C) \ln \frac{rN_A - N_C}{rN_A} + N_C \ln \frac{N_C}{rN_A}\right] \end{aligned} \tag{9}$$

or

$$\begin{aligned} G = {} & N_A(E_A - kT \ln q_A) + N_C(E_C - kT \ln q_C) \\ & + kT\left(rN_A \ln \frac{rN_A - N_C}{rN_A} + N_C \ln \frac{N_C}{rN_A - N_C}\right) \end{aligned} \tag{10}$$

Observing that $kN_i = Rn_i$ (R/k is equal to Avogadro's number), we may rewrite equation (10)

$$G = n_A(E_A - RT \ln q_A) + n_C(E_C - RT \ln q_C - \ln r) + RT\left(n_A r \ln \frac{n_A - (n_C/r)}{n_A} + n_C \ln \frac{n_C}{n_A - (n_C/r)}\right) \tag{11}$$

where E_A and E_C are now the free energies of pure A and C per mole. The function G defined by equation (11) is a homogeneous function of degree 1 with respect to the numbers of moles n_A and n_C (Section II.2). Consequently, it may be easily demonstrated that

$$\overline{G}_A = \mu_A = (E_A - RT \ln q_A) + RTr \ln \frac{X_A - (X_C/r)}{X_A} \tag{12}$$

$$\overline{G}_C = \mu_C = (E_C - RT \ln q_C - \ln r) + RT \ln \frac{X_C}{X_A - (X_C/r)} \tag{13}$$

Recalling that (Section VII.3)

$$\mu_A - \mu_A^{\circ(R)} = RT \ln a_A^{(R)} = RT \ln \gamma_A X_A \tag{14}$$

$$\mu_C - \mu_C^{\circ(H)} = RT \ln a_C^{(H)} = RT \ln \varphi_C X_C \tag{15}$$

Figure 1. Composition dependence of activity functions in an ideal interstitial solution. The standard state of the solvent A is Raoultian, that of the solute C is Henrian. r is the ratio of the number of interstitial sites to the number of substitutional sites. ($r = 1$ for an fcc structure; $r = 3$ for a bcc structure.)

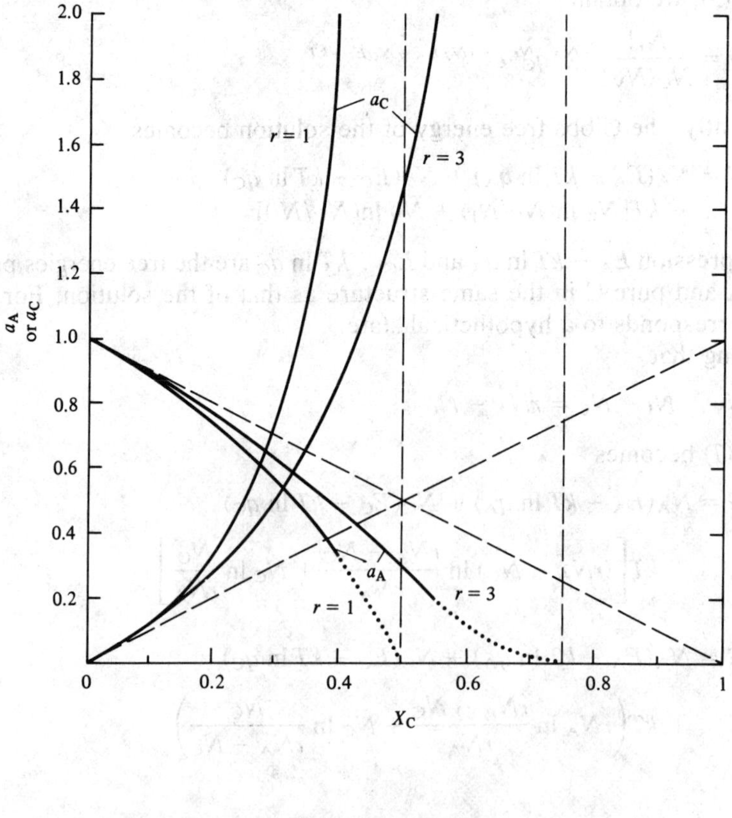

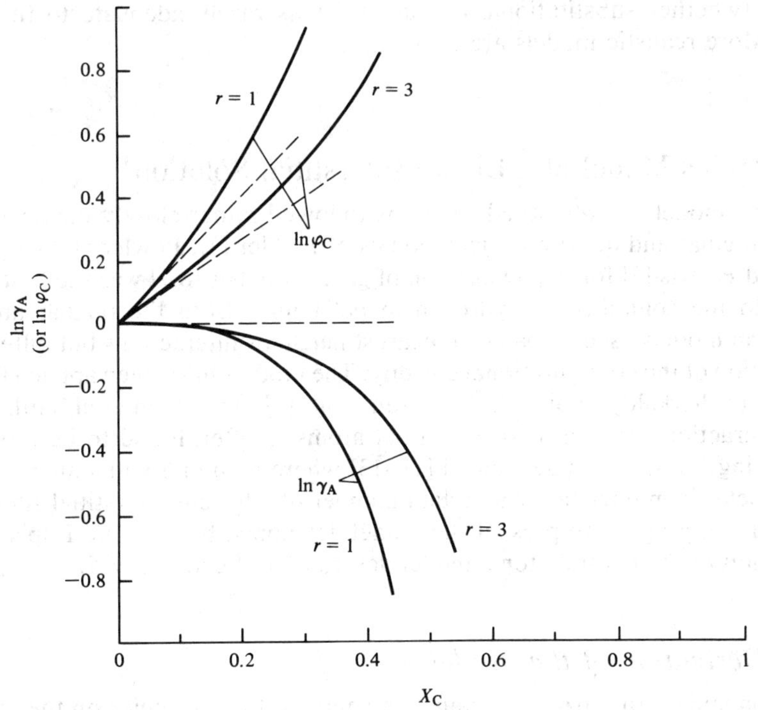

Figure 2. Composition dependence of activity coefficients in an ideal interstitial solution.

we obtain

$$a_A = \left\{\frac{1 - [1 + (1/r)]X_C}{1 - X_C}\right\}^r \tag{16}$$

$$a_C = \frac{X_C}{1 - [1 + (1/r)]X_C} \tag{17}$$

and

$$\ln \gamma_A = r \ln\left[1 - \left(1 + \frac{1}{r}\right)X_C\right] - (1 + r)\ln(1 - X_C) \tag{18}$$

$$\ln \varphi_C = -\ln\left[1 - \left(1 + \frac{1}{r}\right)X_C\right] \tag{19}$$

Figure 1 illustrates the composition dependence of the activities of A and C for the cases $r = 1$ (fcc structure) and $r = 3$ (bcc structure). Figure 2 illustrates the composition dependences of ln γ_A and ln φ_C. We note that

$$\epsilon_C^{(C)} = \left(\frac{d \ln \varphi_C}{dX_C}\right)_{X_C \to 0} = 1 + \frac{1}{r} \tag{20}$$

Let us consider the case of carbon in fcc iron. $r = 1$ and therefore $\epsilon_C^{(C)}$ in the model of an ideal interstitial solution should be equal to 2. At 1000°C, the data yield a value of 8.7 [2]. As in the case of substitutional solutions, the ideal solution model (whether substitutional or interstitial) is rarely adequate to fit accurate data. More realistic models are necessary.

3. Central Atoms Model of a Binary Interstitial Solution

The first models of interstitial solutions followed rather closely the development of the regular and quasi-chemical models of Fowler [3], Fowler and Guggenheim [4], and Peierls [5] for the adsorption of gases. Lacher [6,7] was the first to adapt them to the solubility of hydrogen in palladium. Both the regular and quasi-chemical models assume pairwise nearest neighbor interactions but differ in their evaluation of the configurational entropy. The models have been applied to carbon in iron by Kirkaldy et al. [8], Aaronson et al. [9], and Dunn et al. [10]. Because the interaction between two interstitial atoms is often found to be repulsive, a "blocking" model was developed [11–13] wherein an interstitial atom at a given site blocks from occupancy a certain number of adjacent interstitial sites. In this section, we propose to present the model developed by Foo and Lupis [2] as an extension of the central atoms model described in Section XV.5.

3.1. Derivation of the Model

It is convenient to introduce again as a solute V the vacancies on the interstitial lattice. The vacancies on the substitutional lattice are assumed of negligible importance. For each lattice there are two types of coordination numbers. Z and z will designate respectively the numbers of substitutional and interstitial nearest neighbors to a substitutional site, while Z' and z' will designate respectively the numbers of substitutional and interstitial nearest neighbors to an interstitial site. The ratio r of the number of interstitial sites N_I to the number of substitutional sites N_S may be demonstrated to be also equal to z over Z':

$$r = \frac{N_I}{N_S} = \frac{z}{Z'} \tag{21}$$

For a face-centered cubic lattice (Fig. 3), it may be verified that

$$Z = z' = 12 \tag{22}$$

$$Z' = z = 6 \tag{23}$$

and

$$r = 1 \tag{24}$$

In the calculations pertaining to the model it is convenient to introduce two composition coordinates. The first is the ratio of the number of atoms C or vacancies V to the total number of interstitial sites

$$Y_C = N_C/N_I, \qquad Y_V = N_V/N_I \tag{25}$$

The second is the ratio of N_C to N_V:

$$y_C = N_C/N_V \tag{26}$$

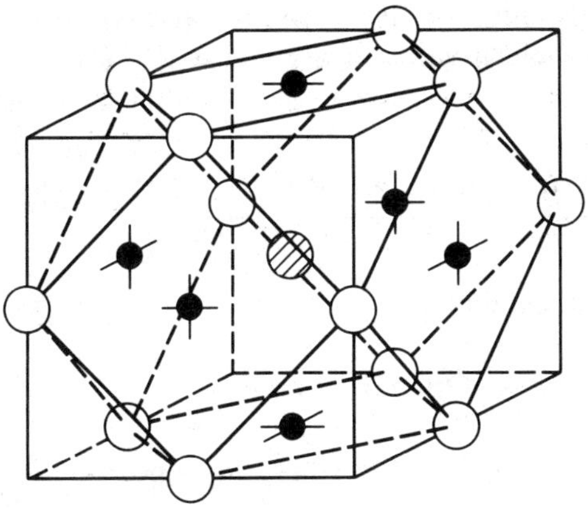

Figure 3. Geometry of the fcc lattice. The dashed circle represents a substitutional central atom. White circles represent substitutional nearest neighbors, black circles interstitial nearest neighbor sites. When the central atom is interstitial, the positions of the atoms are shifted by a translational vector ($\frac{1}{2}$, $\frac{1}{2}$, 0).

These coordinates can be readily related to the mole fraction X_C:

$$Y_C = \frac{X_C}{r(1 - X_C)} \tag{27}$$

$$y_C = \frac{1}{r} \frac{X_C}{1 - X_C[1 + (1/r)]} \tag{28}$$

The derivation of the model follows closely that given in Section XV.5.6.1 for the case of substitutional solutions. Let p_{iC}^{A} be the probability of finding a central atom A surrounded by i atoms C and $z - i$ vacancies V. If the C atoms and V vacancies were assigned to specific sites in the nearest neighbor shell, this probability would be α_{iC}^{A}. There are C_z^i such distributions and, assuming they have equal probabilities,

$$p_{iC}^{A} = C_z^i \alpha_{iC}^{A} = \frac{z!}{i!(z - i)!} \alpha_{iC}^{A} \tag{29}$$

In the case of a random distribution, these probabilities are designated by p_{iC}^{*A} and α_{iC}^{*A}. We note that the probabilities of finding a site occupied by an atom C is N_C/N_I or Y_C. Consequently,

$$p_{iC}^{*A} = C_z^i Y_C^i Y_V^{z-i} \tag{30}$$

Similarly,

$$p_{iC}^{*C} = C_{z'}^i Y_C^i Y_V^{z'-i} = \frac{z'!}{i!(z' - i)!} Y_C^i Y_V^{z'-i} \tag{31}$$

$$p_{iC}^{*V} = C_{z'}^i Y_C^i Y_V^{z'-i} = \frac{z'!}{i!(z' - i)!} Y_C^i Y_V^{z'-i} \tag{32}$$

The partition function of the system is expressed in equation (5). The degeneracy factor g is assumed to be proportional to the number of arrangements corresponding to a random distribution of all the configurations. The coefficient of proportionality is chosen to yield a value of g equal to $N_I!/(N_C!N_V!)$ in the case of a random distribution of the solute atoms C on the interstitial lattice. The consequent expression of g is

$$g = \frac{N_I!}{N_C!N_V!} \frac{\Pi_i[(N_A\alpha_{iC}^{*A})!]^{C_z^i} \Pi_i[(N_C\alpha_{iC}^{*C})!]^{C_{z'}^i} \Pi_i[(N_V\alpha_{iC}^{*V})!]^{C_{z'}^i}}{\Pi_i[(N_A\alpha_{iC}^{A})!]^{C_z^i} \Pi_i[(N_C\alpha_{iC}^{C})!]^{C_{z'}^i} \Pi_i[(N_V\alpha_{iC}^{V})!]^{C_{z'}^i}} \tag{33}$$

The energy of the lattice is calculated as

$$E = \tfrac{1}{2}\left(\sum_{i=0}^{z} N_A p_{iC}^A U_{iC}^A + \sum_{i=0}^{z'} N_C p_{iC}^C U_{iC}^C \right) \tag{34}$$

where U_{iC}^A and U_{iC}^C are the energies of A and C when surrounded by i interstitial atoms C. We also have

$$\ln q_A = \sum_{i=0}^{z} N_A p_{iC}^A \ln q_{iC}^A \tag{35a}$$

$$\ln q_C = \sum_{i=0}^{z'} N_C p_{iC}^C \ln q_{iC}^C \tag{35b}$$

where q_{iC}^A, q_{iC}^C are respectively the individual partition functions of the atoms A and C when surrounded by i interstitial atoms C.

The expressions of g, q_A, q_C, and E in equations (33)–(35) may now be inserted into equation (5) defining the partition function Q. The maximum term in the sum is obtained by optimization with respect to the unknowns p_{iC}^A, p_{iC}^C, and p_{iC}^V under the restrictions provided by mass balances on the C atoms counted as nearest neighbors of the A atoms on the substitutional lattice:

$$\psi_1 = N_A \sum_{i=1}^{z} i p_{iC}^A - Z' N_C = 0 \tag{36a}$$

and counted as nearest neighbors of the C atoms and V vacancies on the interstitial lattice:

$$\psi_2 = N_C \sum_{i=1}^{z'} i p_{iC}^C + N_V \sum_{i=1}^{z'} i p_{iC}^V - z' N_C = 0 \tag{36b}$$

It is convenient to introduce the parameters

$$\varphi_{iC}^A = \tfrac{1}{2}(U_{iC}^A/kT) - \ln q_{iC}^A \tag{37a}$$

$$\varphi_{iC}^C = \tfrac{1}{2}(U_{iC}^C/kT) - \ln q_{iC}^C \tag{37b}$$

Carrying the optimization by the technique of Lagrange multipliers, the following expressions for the probabilities p_{iC} can be obtained:

$$p_{iC}^A = \frac{C_z^i Y_C^i Y_V^{z-i} e^{i\Lambda' - \varphi_{iC}^A}}{\sum_{i=0}^{z} C_z^i Y_C^i Y_V^{z-i} e^{i\Lambda' - \varphi_{iC}^A}} \tag{38a}$$

$$p_{iC}^C = \frac{C_{z'}^i Y_C^i Y_V^{z'-i} e^{i\Lambda - \varphi_{iC}^A}}{\sum_{i=0}^{z'} C_{z'}^i Y_C^i Y_V^{z'-i} e^{i\Lambda - \varphi_{iC}^A}} \tag{38b}$$

$$p_{iC}^{V} = \frac{C_{z'}^{i} Y_C^{i} Y_V^{z'-i} e^{i\Lambda}}{\sum_{i=0}^{z'} C_{z'}^{i} Y_C^{i} Y_V^{z'-i} e^{i\Lambda}} \tag{38c}$$

where Λ and Λ' are two Lagrange multipliers. Their expressions may be obtained by inserting the expressions of p_{iC}^{A}, p_{iC}^{C}, and p_{iC}^{V} above into equations (36).

The calculation of the partition function and of various thermodynamic functions is then straightforward. The parameters of the model are the differences

$$\delta\varphi_{iC}^{A} = \varphi_{iC}^{A} - \varphi_{0C}^{A}, \qquad \delta\varphi_{iC}^{C} = \varphi_{iC}^{C} - \varphi_{0C}^{C} \tag{39}$$

They characterize the change in the state of a central atom (A or C) when i interstitial atoms C are introduced in its nearest neighbor shell originally containing no interstitial atom.

3.2. Linear Variation of a Central Atom's Potential Energy

So far, as stated, the model has $z + z'$ parameters (18 in the case of an fcc lattice) at any given temperature. To reduce that number it is necessary to make an assumption on the dependence of $\delta\varphi_{iC}^{A}$ and $\delta\varphi_{iC}^{C}$ on the number i. The simplest dependence is a linear variation:

$$\delta\varphi_{iC}^{A} = i\delta\varphi_{1C}^{A}, \qquad \delta\varphi_{iC}^{C} = i\delta\varphi_{1C}^{C} \tag{40}$$

It assumes that the effect of i atoms C is equal to i times the effect of one atom C. This result may also be obtained through consideration of pairwise interactions.

With this assumption, it may be shown [2] that the Lagrange multipliers Λ and Λ' are equal to

$$\Lambda = -\ln \tfrac{1}{2}\{[(1 - y_C)^2 + 4y_C \exp(-\delta\varphi_{1C}^{C})]^{1/2} + (1 - y_C)\} \tag{41}$$

$$\Lambda' = \delta\varphi_{1C}^{A} \tag{42}$$

and that the probabilities p_{iC}^{A}, p_{1C}^{C} may be expressed

$$p_{iC}^{A} = \frac{C_z^i y_C^i}{(1 + y_C)^z} \tag{43}$$

$$p_{iC}^{C} = \frac{C_{z'}^{i} y_C^{i} e^{i(\Lambda - \delta\varphi_{1C}^{C})}}{(1 + y_C e^{\Lambda - \delta\varphi_{1C}^{C}})^{z'}} \tag{44}$$

Further calculation (straightforward but rather lengthy) yields the activities of A and C:

$$\ln a_A = -r \ln(1 + y_C) - rz' \ln \frac{1 + y_C e^{\Lambda}}{1 + y_C} \tag{45}$$

$$\ln a_C = \ln y_C + \ln(r\gamma_C^{\infty}) + 2z'\Lambda \tag{46}$$

with

$$\ln(r\gamma_C^{\infty}) = Z'\delta\varphi_{1C}^{A} + \varphi_{0C}^{C} - \varphi_s^{C} \tag{47}$$

where φ_s^{C} represents the state of the atom C when C is pure and in its standard state (e.g., graphite for carbon). Λ is identified in equation (41).

It will prove convenient to introduce the parameter λ_{C-C}:

$$\lambda_{C-C} = 1 - e^{-\delta\varphi_{IC}^{C}} \tag{48}$$

A net repulsion between two adjacent C atoms results in a positive value of λ_{C-C} (and $\delta\varphi_{IC}^{C}$) while a net attraction results in a negative value of λ_{C-C}. For $\lambda_{C-C} = 0$, the equations of the model become those of the ideal interstitial solution. We note that the activity of the interstitial solute C is then proportional to y_C [equations (17) and (28)].

In dilute solutions, equation (41) yields the following approximate expression for the Lagrange parameter Λ:

$$\Lambda = y_C \lambda_{C-C} \tag{49}$$

The expressions of the interaction coefficients ϵ_C^C and ρ_C^C can be deduced from the analytic expression of the activity of C in equation (46). The calculations yield

$$\epsilon_C^C = \left(1 + \frac{1}{r}\right) + \frac{2z'}{r}\lambda_{C-C} \tag{50}$$

$$\rho_C^C = \tfrac{1}{2}\left(1 + \frac{1}{r}\right)^2 + \frac{z'}{r}\lambda_{C-C}\left(2 + \frac{3}{r}\lambda_{C-C}\right) \tag{51}$$

In the case of an fcc lattice, we recall that $r = 1$ and $z' = 12$. In the absence of structural data, it is convenient to assume that liquids have also a close-packed structure corresponding to $r = 1$ and $z' = 12$. With these values, the expressions for ϵ_C^C and ρ_C^C become

$$\epsilon_C^C = 2 + 24\lambda_{C-C} \tag{52}$$

$$\rho_C^C = 2 + 12\lambda_{C-C}(2 + 3\lambda_{C-C}) = \epsilon_C^C + 36(\lambda_{C-C})^2 \tag{53}$$

3.3. Application to the Iron–Carbon System

The model may be tested on the data pertaining to the behavior of carbon in fcc iron. Equations (41) and (46) yield

$$\ln(a_C/y_C) = \ln\gamma_C^{\infty} - 24\ln\tfrac{1}{2}\{[(1 - y_C)^2 + 4y_C e^{-\delta\varphi_{IC}^{C}}]^{1/2} + (1 - y_C)\} \tag{54}$$

Figures 4 and 5 show that this equation represents very satisfactorily the experimental data of Smith [14] and Ban-ya et al. [15,16] on the activity of carbon, and those of Gurry [17], Scheil et al. [18], and Wagman et al. [19] on the solubility of carbon with respect to graphite ($a_C = 1$ for a standard state of carbon corresponding to graphite). The calculated curves are based on the following estimates of $\ln\gamma_C^{\infty}$ and $\delta\varphi_{IC}^{C}$:

$$\ln\gamma_C^{\infty} = -2.1 + (5300/T) \tag{55}$$

$$\delta\varphi_{IC}^{C} = 0.10 + (290/T) \tag{56}$$

They result from a statistical treatment of the data [2]. The temperature dependence of $\ln\gamma_C^{\infty}$ is shown in Fig. 6. The error on $\delta\varphi_{IC}^{C}$ is of the order of 0.03.

Because of experimental difficulties, there are fewer and less accurate data on the activity of carbon in liquid iron (Fig. 7). The data of Dennis and Richardson

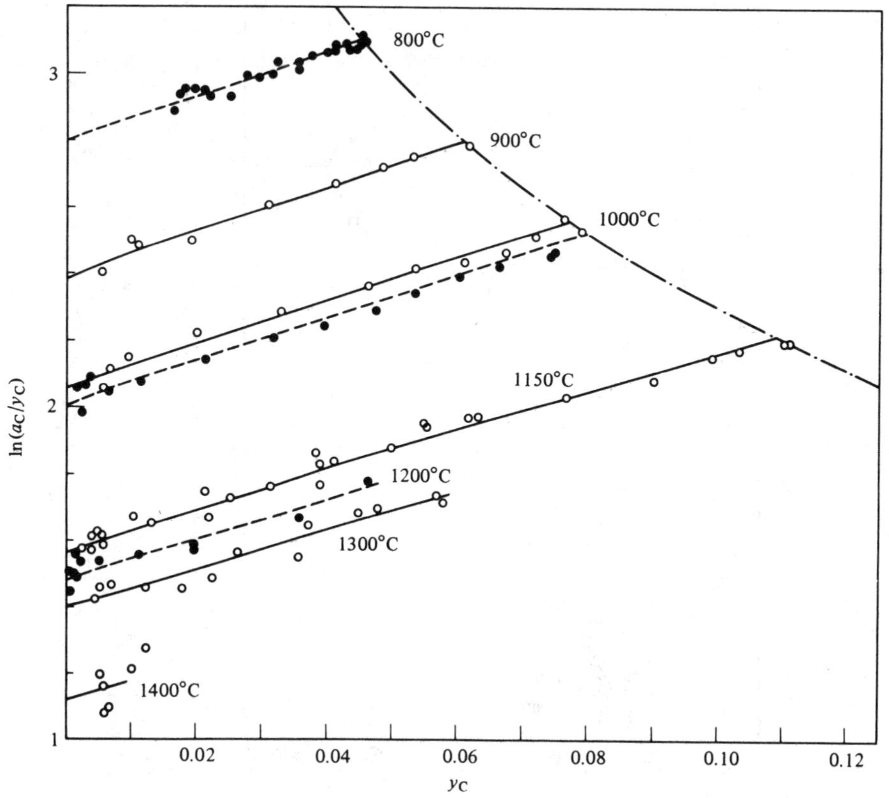

Figure 4. Application of the central atoms model to the activity of carbon in austenite. ○, Ban-ya et al. [15,16]; ●, Smith [14]; –·– graphite saturation.

[20] at 1560°C lead to

$$\ln \gamma_C^{\infty} = -0.37 \pm 0.10 \tag{57}$$

and

$$\delta\varphi_{1C}^{C} = 0.24 \pm 0.06 \tag{58}$$

It is interesting to compare these values with their parallel values in the case of austenite. At 1560°C (1833°K), equation (56) yields $\delta\varphi_{1C}^{C} = 0.26$, a value remarkably close to that directly found for liquid iron. The composition dependences of the activity of carbon are thus nearly identical. Their absolute values are, however, very different since at 1833°K equation (55) yields $\ln \gamma_C^{\infty} = +0.79$. An interpretation of these facts may be proposed along the following lines.

The values of $\ln \gamma_C^{\infty}$ are essentially a measure of the Fe–C interaction while the values of $\delta\varphi_{1C}^{C}$ measure the C–C interaction. The Fe and C atoms are nearest neighbors while two carbon atoms on adjacent interstitial sites are next-to-nearest neighbors and thus further apart. Small changes in the distance between an Fe atom and a C atom are likely to have a large effect on their bonding energy, whereas small changes in the distance between two C atoms will have a much less significant effect. The inclusion of a carbon atom in the rigid lattice of solid

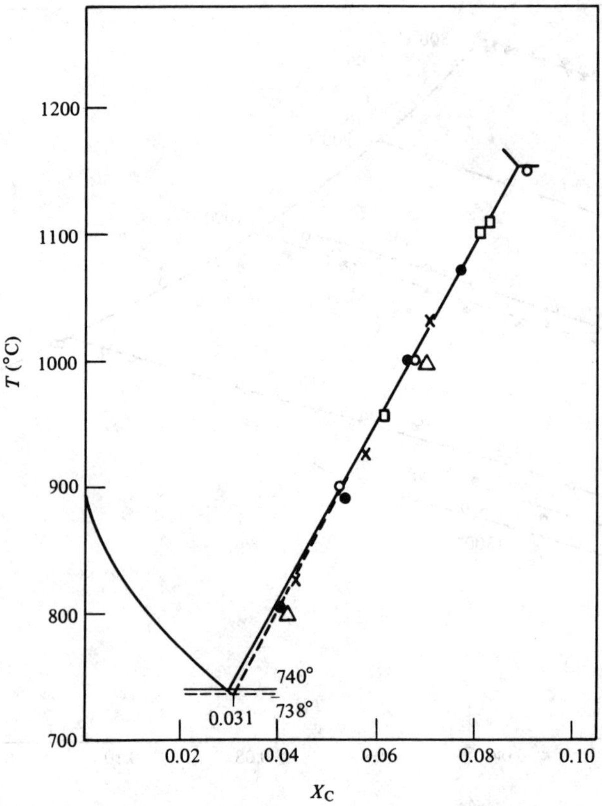

Figure 5. Calculation of the solubility of graphite in austenite by the central atoms model. □, Gurry [17]; ○, Banya et al. [16] (observed); ———, Banya et al. [15] (calculated); △, Smith [14]; ●, Scheil et al. [18]; ×, Wagmann et al. [19]; - - -, calculated by the present study.

austenite is expected to be associated with a sizable strain energy. On the other hand, the structure of the liquid should be much more accommodating to the inclusion of a carbon atom since the iron atoms can be relatively easily pushed away from the carbon atom. Hence, a lower value of γ_C^∞ should indeed be registered in the case of the liquid. On the other hand, the rigidity of the lattice is not likely to affect substantially the distance between two carbon atoms and their interaction energy. Thus, the values of $\delta\varphi_{1C}^C$ should indeed be similar in the cases of austenite and liquid iron.

3.4. Comparison with Other Models

After some rearrangement of the terms, the quasi-chemical model [7] yields the following expression for the activity of the interstitial solute C:

$$\ln a_C/y_C = \ln \gamma_C^\infty - 12 \ln \tfrac{1}{2}\{[(1 - y_C)^2 + 4y_C e^{-u_{CC}/RT}]^{1/2} + (1 - y_C)\} \quad (59)$$

This equation may be compared to equation (54). Identifying the bonding energy between two carbon atoms u_{CC} with $2RT\,\delta\varphi_{1C}^C$, it may be seen through a Taylor

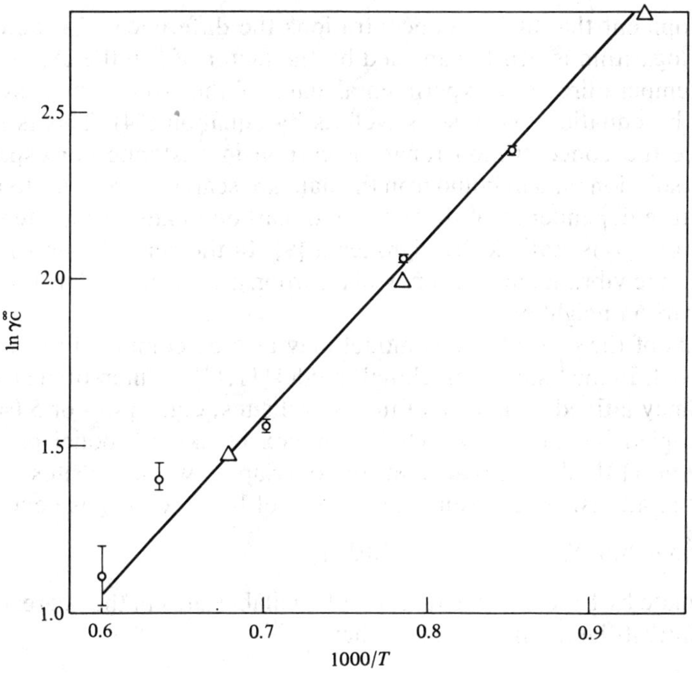

Figure 6. Temperature dependence of the activity coefficient of carbon in austenite at infinite dilution. ○, Ban-ya [15, 16]; Δ, Smith [14].

Figure 7. Application of the central atoms model to the activity of carbon in liquid iron at 1560°C. ———, theory; ○, Dennis and Richardson [20]; ●, graphite saturation.

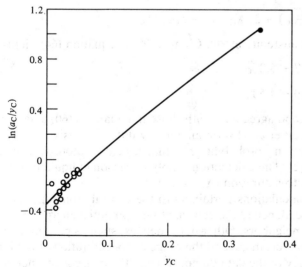

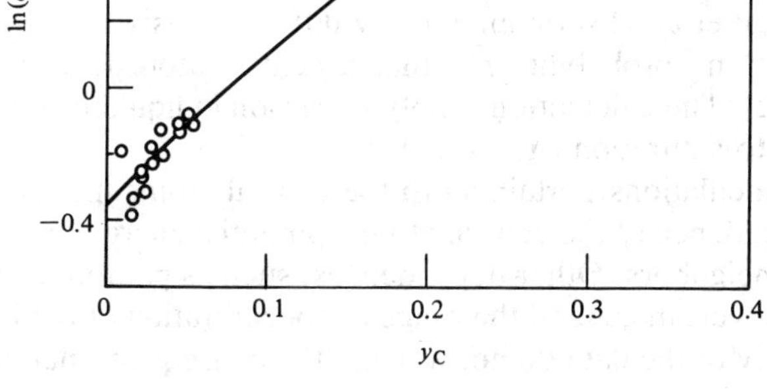

series development that at low concentrations the difference of a factor of 2 in front of the logarithm is nearly canceled by the factor of $\frac{1}{2}$ in the exponential. At any given temperature, the experimental data of the iron–carbon system are represented by equation (59) just as well as by equation (54). This is not unexpected, since the concentration range of carbon in austenite corresponds to a rather dilute solution, and in liquid iron the data are scarce. However, to represent the temperature dependence of the activity of carbon in austenite, a temperature dependence of u_{CC} is statistically significant [9]. In the central atoms model this indicates that the vibrational frequency of a carbon atom is altered by the presence of another carbon neighbor.

The results of the central atoms model may also be compared to those of the blocking model. In the "simple blocking" model [11,12] an interstitial atom blocks from occupancy a fixed number η of interstitial sites, e.g., $\eta = 4$ or 5 (η includes the site occupied by this atom). The "complex blocking" model proposed by McLellan et al. [13] takes into account the overlap between the sites blocked by two interstitial atoms. As a result, the number of blocked sites is decreased:

$$\eta = 5 - 8.67 Y_C, \qquad \eta = 4 - 4.63 Y_C \tag{60}$$

Let us designate by ζ the average number of available sites in the nearest neighbor shell of an interstitial atom. It is clear that

$$\eta + \zeta = z' + 1 \tag{61}$$

Moreover, the product of ζ and Y_C measures the average number of nearest C neighbors:

$$\zeta Y_C = \sum_{i=1}^{z'} i p_{iC}^{C} \tag{62}$$

With the expression of the probability p_{iC}^{C} given in equation (44), equation (62) may be rearranged to yield

$$\zeta = z' e^{\Lambda - \delta\varphi_{1C}^{C}} / [1 - Y_C(1 - e^{\Lambda - \delta\varphi_{1C}^{C}})] \tag{63}$$

In the case of dilute solutions, equations (48), (49), (61), and (63) lead to

$$\eta = (1 + z'\lambda_{C-C}) - 2z'\lambda_{C-C}(1 - \lambda_{C-C}) Y_C \tag{64}$$

Applied to carbon in austenite at 800°C and 1300°C, equation (64) yields

$$(\eta)_{800°C} = 4.7 - 5.2 y_C \tag{65a}$$

$$(\eta)_{1300°C} = 4.0 - 4.5 Y_C \tag{65b}$$

These values are in good agreement with those in equations (60) and which were calculated by McLellan et al. [13] on an entirely different basis.

Figure 8 illustrates the probability p_{iC}^{C} that a central atom C would have i nearest neighbors C [2]. The calculations apply to carbon in liquid iron at 1560°C, from infinite dilution to saturation ($X_C = X_C^*$).

In the preceding calculations pertaining to the central atoms model, we have assumed a linear dependence of the central atom's potential energy on its number of nearest interstitial neighbors. Other dependences, such as parabolic ones, may also be adopted. However, in general the range of concentration of the interstitial solute and the accuracy of the data do not warrant the consequent increase in the

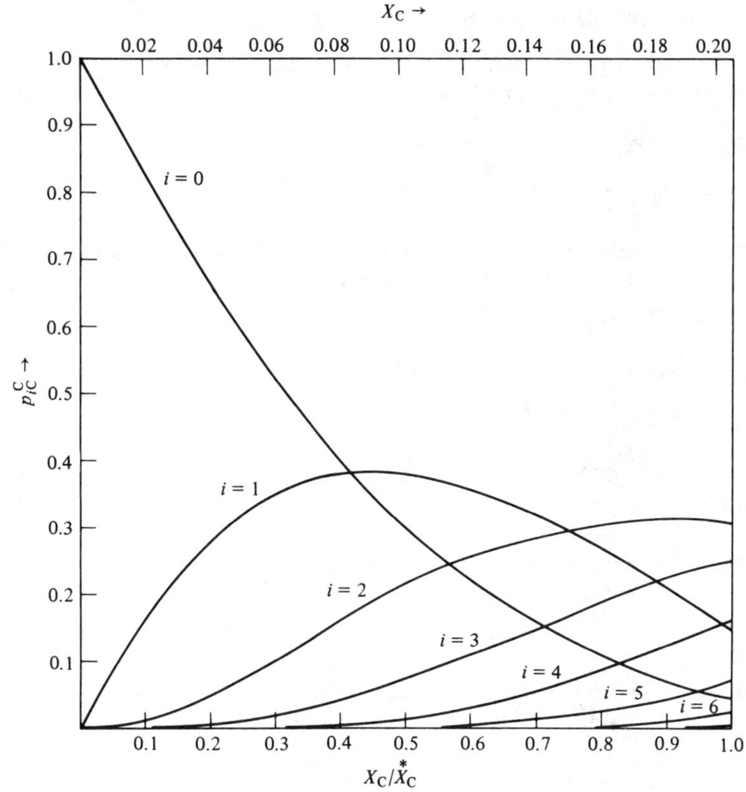

Figure 8. Probability that a carbon atom has i carbon nearest neighbors in liquid iron at 1560°C, from infinite dilution to saturation X_C^* (= 0.205 at 1560°C).

mathematical complexity of the model and do not allow a meaningful test of these different assumptions.

4. Central Atoms Model of an Interstitial Solute in a Multicomponent System

Conceptually, the central atoms model may be readily extended to any multicomponent system of t substitutional solutes and $m - t$ interstitial solutes. The configuration around a central atom J must now be characterized by the numbers i_k of nearest substitutional solutes k ($k = 2$ to t) and the numbers j_l of nearest interstitial solutes l ($l = t + 1$ to m). The vacancies on the interstitial lattice may be treated as an $m + 1$ solute and it may be assumed that there are no vacancies on the substitutional lattice. The model has been fully developed by Foo and Lupis [21] in the case where the potential energy of the central atom J is assumed to be linearly dependent on i_k and j_l. The mathematical treatment of the model is somewhat lengthy and in this text only an outline of its derivation appears warranted. However, the results of the model will be presented at some length as these can be useful in various applications.

4.1. Preliminary Definitions and Parameters

It is again convenient to introduce the coordinates Y and y. Y_u is the ratio of the number of atoms u to the total number of sites to which these u atoms belong:

$$Y_k = N_k/N_S \quad \text{for} \quad u = k \leq t \tag{66a}$$

$$Y_l = N_l/N_I \quad \text{for} \quad u = l > t \tag{66b}$$

where N_S is the total number of substitutional sites and N_I is the total number of interstitial sites. y_u is the ratio of the number of atoms u to the number of solvent atoms or vacant interstitial sites:

$$y_k = N_k/N_1 \quad \text{for} \quad u = k \leq t \tag{67a}$$

$$y_l = N_l/N_{m+1} \quad \text{for} \quad u = l > t \tag{67b}$$

The relations between the coordinates Y_u and y_u and the mole fraction X_u are easily found:

$$Y_k = \frac{X_k}{1 - \sum_{l=t+1}^{m} X_l}, \qquad y_k = \frac{X_k}{1 - \sum_{i=2}^{m} X_i} \tag{68a}$$

and

$$Y_l = \frac{1}{r}\frac{X_l}{1 - \sum_{l=t+1}^{m} X_l}, \qquad y_l = \frac{1}{r}\frac{X_l}{1 - [1 + (1/r)] \sum_{l=t+1}^{m} X_l} \tag{68b}$$

where r is the ratio of N_I to N_S.

The parameters of the model are the differences

$$\delta\varphi^{(J)}_{i_2,\ldots,i_t,j_{t+1},\ldots,j_m} = \varphi^{(J)}_{i_2,\ldots,i_t,j_{t+1},\ldots,j_m} - \varphi^{(J)}_{0_2,\ldots,0_t,0_{t+1},\ldots,0_m} \tag{69}$$

(We recall that $\varphi^{(J)}$ has been defined as $\frac{1}{2}(U^{(J)}/kT) - \ln q^{(J)}$ where $U^{(J)}$ is the energy of the central atom J; $q^{(J)}$ is its atomic partition and is a measure of its vibrational frequency.) The difference $\delta\varphi^{(J)}$ in (69) characterizes the difference in the states of the central atom J, when surrounded by i_k atoms of the substitutional components k and j_l atoms of the interstitial components l, and when surrounded by all solvent atoms on the substitutional lattice and all vacancies on the interstitial lattice. Assuming a linear dependence of $\delta\varphi^{(J)}$ on i_k and j_l, we have

$$\delta\varphi^{(J)}_{i_2,\ldots,i_t,j_{l+1},\ldots,j_m} = \sum_{k=2}^{t} i_k\, \delta\varphi^{(J)}_{0_2,\ldots,1_k,\ldots,0_t,0_{t+1},\ldots,0_m} + \sum_{l=t+1}^{m} j_l\, \delta\varphi^{(J)}_{0_2,\ldots,0_t,0_{t+1},\ldots,1_l,\ldots,0_m} \tag{70a}$$

or, in abbreviated notation,

$$\delta\varphi^{(J)}_{i,j} = \sum_{k=2}^{t} i_k\, \delta\varphi^{(J)}_{1k} + \sum_{l=t+1}^{m} j_l\, \delta\varphi^{(J)}_{1l} \tag{70b}$$

We now let

$$\lambda_{k-u} = 1 - \exp(-\delta\varphi^{(k)}_{1u} + \delta\varphi^{(1)}_{1u}) \tag{71}$$

and

$$\lambda_{l-u} = 1 - \exp(-\delta\varphi_{1u}^{(l)}) \tag{72}$$

It may be demonstrated [21] that

$$\lambda_{J-u} = \lambda_{u-J} \tag{73}$$

where J and u can be either substitutional or interstitial components. A positive value of λ_{J-u} indicates a net repulsion between the atoms J and u whereas a negative value indicates a net attraction.

It is advantageous to define as A_{ij} a recurrent mathematical function:

$$A_{ij} = 2(1 + y_i)/([(1 + y_i)(1 - y_j) + \lambda_{ij}(y_j - y_i)] + \{(1 + y_i)(1 - y_j) + \lambda_{ij}(y_j - y_i)]^2 + 4(1 - \lambda_{ij})y_j(1 + y_i)^2\}^{1/2}) \tag{74}$$

We note that $A_{ij} \neq A_{ji}$ and that for $i = j$

$$A_{ii} = 2/\{(1 - y_i) + [(1 - y_i)^2 + 4(1 - \lambda_{ii})y_i]^{1/2}\} \tag{75}$$

The derivation of the model may proceed in a manner entirely similar to that described in Section 3, i.e., in the case of a binary solution. It is again assumed that the degeneracy factor is proportional to the number of arrangements corresponding to a random distribution of all distinguishable configurations. The maximum term in the partition function is obtained by optimization of the general term with respect to the probabilities of the various configurations under the $2(m - 1)$ constraints provided by the mass balance equations on the substitutional and interstitial lattices. The maximization may be again performed through the technique of Lagrange multipliers. The details of the calculations are given by Foo and Lupis [21]. The results are outlined below.

4.2. *Ternary Solution of Two Substitutional Components and One Interstitial Component*

In the ternary solution 1–2–C, 2 is a substitutional component, C an interstitial one. (With the notation above: $t \equiv 2$, $m \equiv 3 \equiv$ C.) The composition dependence of the activity of C is expressed

$$\ln(a_C/y_C) = \ln(r\gamma_C^\infty) + 2z' \ln A_{CC} + 2Z' \ln A_{2C} \tag{76}$$

This equation may also be transformed into

$$\ln(\gamma_C/\gamma_C^\infty) = -\ln[1 + (1 + r)y_C] + 2z' \ln A_{CC} + 2Z' \ln A_{2C} \tag{77}$$

The first two terms on the right-hand members of equation (76) and (77) analyze only the C–C interactions and are identical to those found in our study of the binary 1–C (Section 3.2). The effect of the substitutional solute 2 is expressed by the term $2Z' \ln A_{2C}$ [where A_{2C} is defined in equation (74)].

A development of the terms in equation (77) in a Taylor series with respect to the mole fraction of X_2 and X_C yields the following expressions for the interaction coefficients:

$$\epsilon_C^{(2)} = 2Z'\lambda_{2-C} \tag{78}$$

$$\rho_C^{(2)} = Z'\lambda_{2-C}^2 \tag{79}$$

$$\rho_C^{(2,C)} = 2Z'\lambda_{2-C}[1 + (\lambda_{2-C}/r)] \tag{80}$$

We note that a negative value of $\epsilon_C^{(2)}$ is interpreted as a net attraction between the components 2 and C and a positive value of $\epsilon_C^{(2)}$ as a net repulsion.

In order to test the model, we note that the influence of a substitutional solute on the solubility or activity coefficient of an interstitial solute C is often measured by equilibrating through a gas phase the activity of C, a_C^b, in a binary alloy 1–C with the activity of C, a_C, in a ternary alloy 1–2–C [22]. We may then write

$$\ln a_C^b = \ln a_C \tag{81}$$

which, through equation (76) becomes

$$\ln y_C^b + 2z' \ln A_{CC}^b = \ln y_C + 2z' \ln A_{CC} + 2Z' \ln A_{2C} \tag{82}$$

If the concentration of C is low, $\ln A_{CC}$ may be approximated by $y_C\lambda_{C-C}$. Equation (82) is thus transformed into

$$\frac{1}{2Z'} \ln \frac{y_C}{y_C^b} + \frac{z'}{Z'} (y_C - y_C^b)\lambda_{C-C} = -\ln A_{2C} \tag{83}$$

Furthermore, it may readily be shown that A_{2C} [defined in equation (74)] becomes

$$(A_{2C})_{y_C \to 0} = \frac{1 + y_2}{1 + y_2(1 - \lambda_{2-C})} = \frac{1}{1 - Y_2\lambda_{2-C}} \tag{84}$$

Thus, equation (83) may be rewritten

$$f \equiv \left(\frac{y_C}{y_C^b}\right)^{1/2Z'} e^{(z'/Z')(y_C - y_C^b)\lambda_{C-C}} - 1 = -Y_2\lambda_{2-C} \tag{85}$$

Figure 9. Effect of manganese on the activity of carbon in austenite at 1000°C. The line is calculated by equation (85) with the value $\lambda_{Mn} = -0.36$. ○, Wada et al. [23]; △, Smith [24].

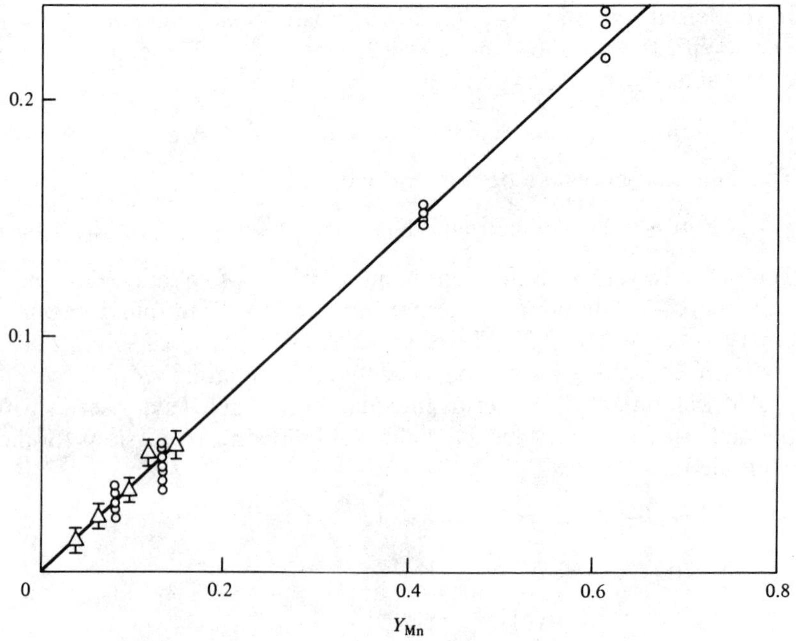

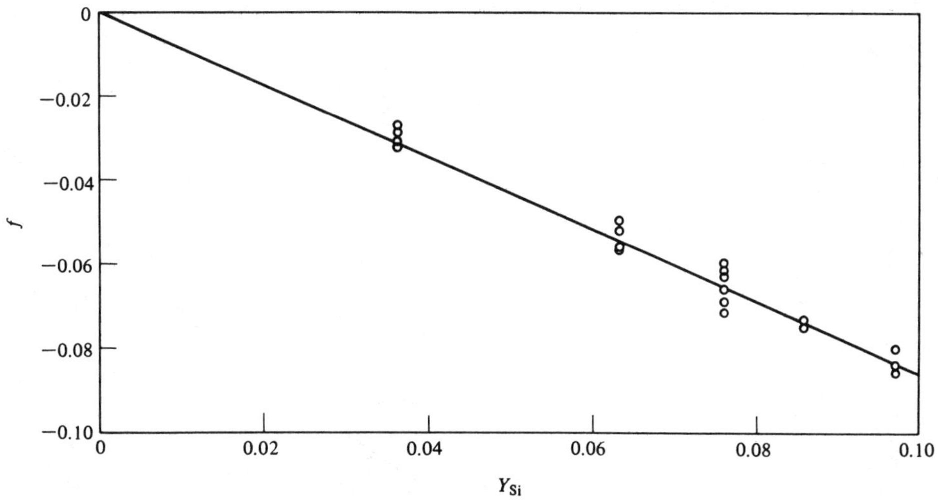

Figure 10. Effect of silicon on the activity of carbon in austenite at 1147°C. The data of Wada et al. [23] are fitted by equation (85) with $\lambda_{Si} = 0.86$.

A plot of the function f vs Y_2 should then result in a straight line. Its slope would equal $-\lambda_{2-C}$.

The application of equation (85) is illustrated in Figs. 9 and 10 for the effects of Mn and Si on the activity of carbon in austenite [23,24]. In liquid iron alloys, the effects of various solutes i on the activity of carbon have usually been determined at relatively dilute concentrations of carbon. Assuming that $Z' = 6$ and $z' = 12$ (as for austenite), the values of λ_{i-C} may be deduced. (For convenience, the notation λ_{i-C} may be abbreviated to λ_i when no confusion is possible.) The results of the dilute solutions may then be extrapolated to high carbon contents through equation (82) in order to predict the effects of these solutes on the solubility of graphite. Figure 11 shows the application of this technique to the effects of Co, Ni, Al, and Si. Additional examples are given in the original article by Foo and Lupis [21]. The agreement between the data and the predictions of the model is generally satisfactory.

Table 1 summarizes the analysis of Foo and Lupis [21] of the interaction coefficients of carbon in alloyed austenite at 1000°C and in liquid iron at 1550°C; the coefficients pertaining to nitrogen in liquid iron at 1600°C are also listed. The table is not meant to provide a test of the model. Rather, the values quoted as reported from the literature are meant to illustrate typical agreement (or disagreement) between different workers or some of the better known coefficients. It is also interesting to observe that, for a given ternary element k and after correction for the temperature difference, the values of λ_{C-k} in austenite and in liquid iron are quite close to each other. The values of λ_{C-k} and λ_{N-k} are also very similar.

4.3. *Ternary Solution of Two Interstitial Solutes*

In the ternary solution 1–C–3, 1 is the solvent, C and 3 are the interstitial solutes. (With the previous notation, $t \equiv 1$, $m \equiv 3$.) In this case, the development of the model does not yield an analytic solution. A Taylor series expansion, however,

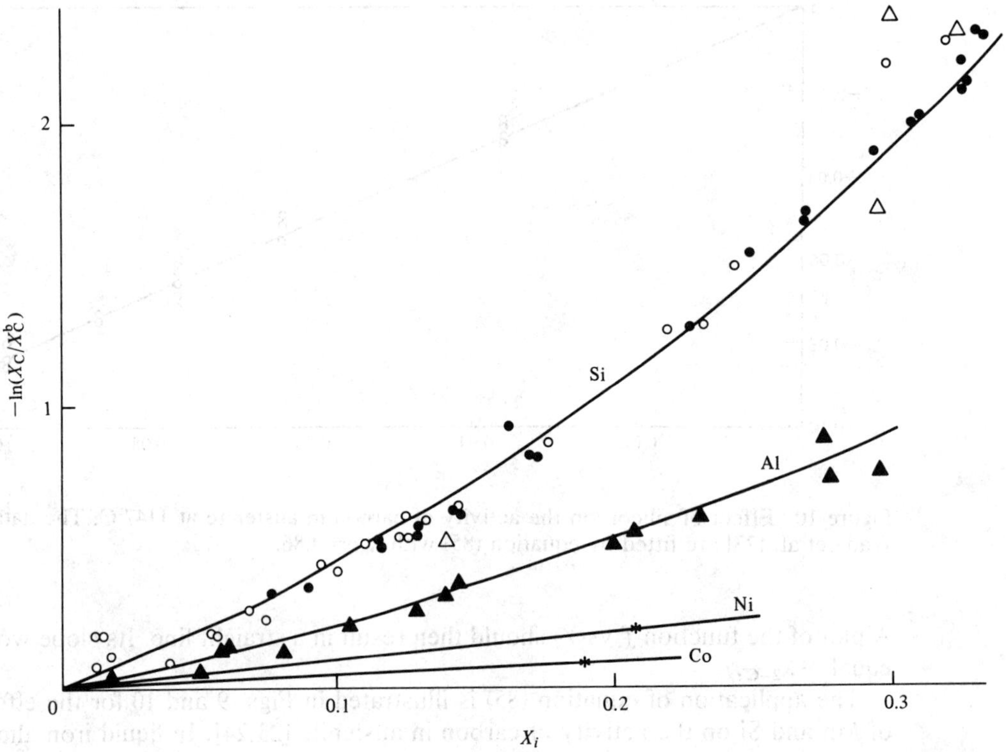

Figure 11. Effects of aluminum, cobalt, nickel, and silicon on the solubility of graphite in liquid iron. The values of the parameters λ_i are deduced from data at low concentrations of carbon. The lines are predicted by the central atoms model. The data on the solubility of graphite are from Refs. [25]–[30]. ○, Ohtani, 1540°C; ●, Chipman et al., 1600°C; △, Chipman et al., 1530°C [Si ($\lambda_{Si} = 0.82$)]; ▲, Mori et al., 1550°C [Al ($\lambda_{Al} = 0.48$)]; *, Turkdogan et al., 1550°C [Ni, Co ($\lambda_{Ni} = 0.20$, $\lambda_{Co} = 0.15$)].

is possible and would yield the values of the activity functions to any desired degree of accuracy. In particular, we find that

$$\epsilon_C^{(3)} = \left(1 + \frac{1}{r}\right) + \frac{2z'}{r}\lambda_{3-C} \tag{86}$$

$$\rho_C^{(3)} = \tfrac{1}{2}\left(1 + \frac{1}{r}\right)^2 + \frac{z'\lambda_{3-C}}{r}\left(2 + \frac{\lambda_{3-C}}{r} + \frac{2}{r}\lambda_{3-3}\right) \tag{87}$$

$$\rho_C^{(C,3)} = \left(1 + \frac{1}{r}\right)^2 + \frac{2z'}{r}\left[\frac{1}{r}\lambda_{3-C}(\lambda_{3-C} + 2\lambda_{C-C}) + (\lambda_{3-C} + \lambda_{C-C})\right] \tag{88}$$

More conveniently, an analytic solution may be found when the solution is highly dilute with respect to one of its solutes, e.g., C. The composition dependence of a_C becomes

$$\ln\left(\frac{a_C}{y_C}\right)_{y_C \to 0} = \ln(r\gamma_C^\infty) + 2z' \ln \frac{1 + y_3 A_{33}}{1 + y_3 A_{33}(1 - \lambda_{3-C})} \tag{89}$$

When the solution is not very dilute in C, the value of a_C may be approximated by the following equation:

$$\ln\left(\frac{a_C}{y_C}\right) = \ln(r\gamma_C^\infty) + 2z'\ln A_{CC} + 2z'\ln\frac{1 + y_3A_{33}}{1 + y_3A_{33}(1 - \lambda_{3\text{-}C})} + r^2\rho_C^{(C,3)}y_Cy_3 \quad (90)$$

The model may be tested by applying it to the solubility of nitrogen in liquid iron–carbon alloys. This solubility has been measured by Schenck et al. [32], Maekawa and Nakagawa [33], Pehlke and Elliott [34], and Gomersall et al. [35]. There is substantial disagreement among the results of these workers. We elect to retain the results of Gomersall et al., which appear to be the most reliable and are in good agreement with those of Schenck et al.

Table 1. Binary and Ternary Interaction Coefficients for Carbon and Nitrogen in Iron Alloys

Temperature (°C)	Interstitial element l	Alloying element k	Parameter λ_{k-l}	Values calculated by the model $\epsilon_l^{(k)}$	$\rho_l^{(k)}$	$\rho_l^{(k,l)}$	Values reported in the literature $\epsilon_l^{(k)}$	$\rho_l^{(k)}$	$\rho_l^{(k,l)}$
1000	C	C	0.28	8.7	11.5				
		Cr	−1	−12	6	0	−10, −10.7, −12		
		Co	0.18	2.2	0.2	2.6	1.6,2.2,2.3		
		Mn	−0.37	−4.3	0.8	−2.8	−4.0		−8
		Ni	0.34	4.1	0.7	5.5	4.1		11.1
		Si	0.93	11.2	5.2	21.5	10.6		39
		V	−4	−48	96	144	−51		
1500	C	C	0.23	7.5	9.4		7.8	13.7	
		Al	0.48	5.8	1.4	8.5	6.7		
		Cr	−0.45	−5.4	1.2	−3.0	−5.1, −7		
							−5.4	1.2	10
							−7.6	1	17
		Co	0.15	1.8	0.13	2.1	1.8,2.9	0	
		Cb	−1.2	−14.4	8.6	2.9	−23	16	
		Cu	0.30	3.6	0.5	4.8	4.2		
		Mn	−0.20	−2.4	0.24	−2			
		Mo	−0.33	−4.0	0.66	−2.6	−3.5, −4.0		
		Ni	0.20	2.4	0.25	2.9	2.4,2.7,2.9		
		Si	0.82	9.8	4	18	9.7	8.8	
							5.6		22
		S	0.70	8.4	3	14.4	12		
							6.4		30
		W	−0.23	−2.8	0.32	−2.1	−2.3		
		V	−0.52	−6.2	1.6	−2.9	−6.2, −8, −16		
1600	N	C	0.21	7.0	8.6		6.15	14.5	
		Cr	−0.80	−9.6	3.8		−9.6, −10.1	3.3	
		Co	0.16	2	0.15		2.7		
		Cb	−2.20	−26.4	2.9		−24.8, −26.3		
		Cu	0.20	2.4	0.2		1		
		Mn	−0.35	−4.2	0.6		−5.2		
		Mo	−0.45	−5.4	1.2		−4.1, −5.0		
		Ni	0.20	2.4	0.2		2.5	0	
		Si	0.50	6.0	1.5		5.9,6	9.7	
		V	−1.78	−21	19		−20.9	25	

From the compilations of Foo et al. [21] and Sigworth et al. [31].

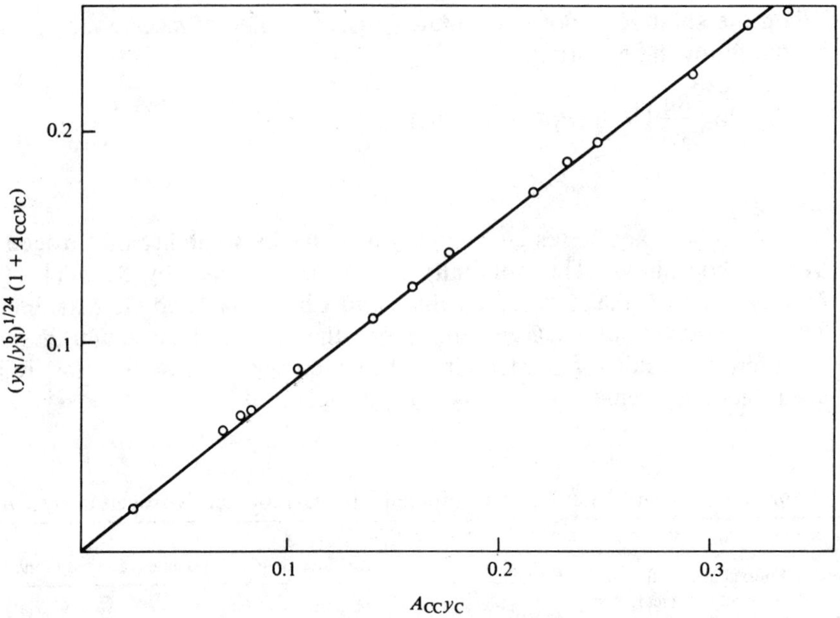

Figure 12. Effect of carbon on the solubility of nitrogen in liquid iron at 1550°C. The data of Gomersall et al. [35] are fitted by equation (92).

Proceeding as in Section 4.2 [equations (81)–(85)], equation (90) (with C→N, 3→C) leads us to

$$\frac{1}{2z'}\ln\frac{y_N}{y_N^b}+\ln\frac{A_{NN}}{A_{NN}^b}=-\ln\frac{1+y_C A_{CC}}{1+y_C A_{CC}(1-\lambda_{N\text{-}C})} \tag{91}$$

where we have neglected the term in $y_N y_C$. Moreover, $\ln(A_{NN}/A_{NN}^b)$ is equivalent to $\lambda_{N-N}(y_N - y_N^b)$ and is also negligible for the small solubilities of nitrogen in iron alloys. Consequently, equation (91) may be transformed into

$$g \equiv \left(\frac{y_N}{y_N^b}\right)^{1/2z'}(1+y_C A_{CC})-1=y_C A_{CC}(1-\lambda_{N\text{-}C}) \tag{92}$$

Plotting g vs $y_C A_{CC}$, where A_{CC} is defined in equation (75), should then result in a straight line of slope equal to $1-\lambda_{N\text{-}C}$. Figure 12 shows that a straight line does indeed fit the data very well. The consequent value of $\lambda_{N\text{-}C}$ is 0.21.

4.4. Quaternary Solution of Two Substitutional and Two Interstitial Components

In the quaternary solution 1–2–C–4, 1 and 2 are the substitutional components, C and 4 the interstitial solutes. An analytic solution of this case can be found only for $y_C \to 0$. It results in the following expression:

$$\ln\left(\frac{a_C}{y_C}\right)_{y_C\to 0}=\ln(r\gamma_C^{\infty})+2Z'\ln\frac{1+y_2A_{42}}{1+y_2A_{42}(1-\lambda_{2\text{-}C})}+2z'\ln\frac{1+y_4A_{44}}{1+y_4A_{44}(1-\lambda_{4\text{-}C})} \tag{93}$$

From this equation, the interaction coefficient $\rho_C^{(2,4)}$ may be deduced:

$$\rho_C^{(2,4)} = 2Z'\lambda_{2-C}[1 + (\lambda_{2-4}/r)] \tag{94}$$

4.5. *Quaternary Solution of Three Substitutional Components and One Interstitial Component*

In this quaternary solution 1–2–3–C, 1, 2, 3 are substitutional components and C is an interstitial solute. An analytic solution of the model is readily found only when the solution is highly dilute in C. The effect of the components 2 and 3 on the activity of C is then simply expressed

$$\ln(a_C/y_C)_{y_C \to 0} = \ln(r\gamma_C^\infty) - 2Z' \ln(1 - Y_2\lambda_{2-C} - Y_3\lambda_{3-C}) \tag{95}$$

From this equation, we may deduce the quaternary interaction coefficient

$$\rho_C^{(2,3)} = 2Z'\lambda_{2-C}\lambda_{3-C} \tag{96}$$

Equation (96) may be generalized to the case of t substitutional components:

$$\ln(a_C/y_C)_{y_C \to 0} = \ln(r\gamma_C^\infty) - 2Z' \ln\left(1 - \sum_{k=2}^{t} Y_k\lambda_{k-C}\right) \tag{97}$$

To test the model, a procedure similar to that followed for the ternary ($t = 2$) may be adopted. The equivalent to equation (85) is

$$f \equiv \left(\frac{y_C}{y_C^b}\right)^{1/2Z'} e^{(z'/Z')(y_C - y_C^b)\lambda_{C-C}} - 1 = -\sum_{k=2}^{t} Y_k\lambda_{k-C} \tag{98}$$

Figure 13. Effect of chromium and nickel on the activity of carbon at 1550°C. The data of Foo and Lupis [22] [○, Fe–C–Ni–Cr; ●, Fe–C–Cr] are fitted by equation (100).

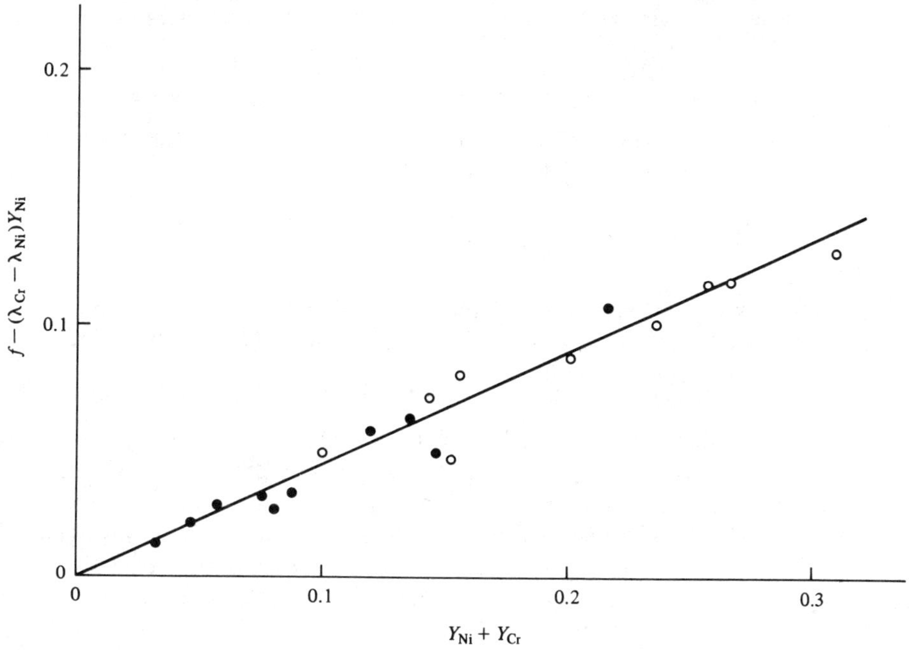

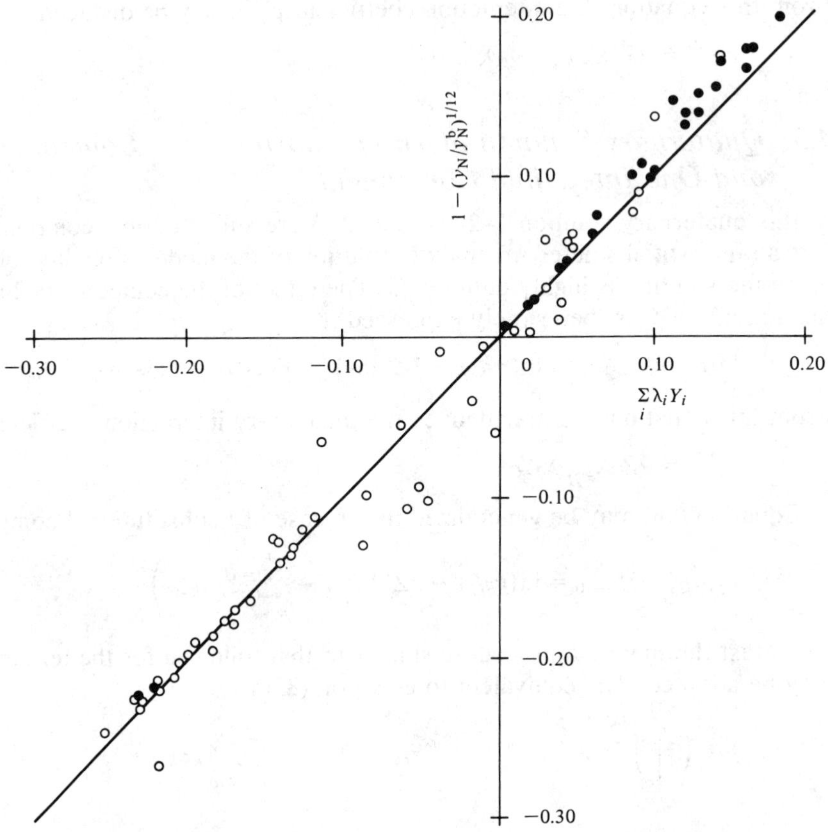

Figure 14. Solubility of nitrogen in quaternary liquid iron alloys at 1600°C. The equality of the ordinates and abscissas represented by the solid line is predicted by equation (99). The data are obtained from the work of Humbert et al. [36] [○, Fe–N–Cr–Ni] and Blossey et al. [37] [●, Fe–N–Co–Ni].

If the concentration of C is very low (such as in the case of nitrogen in liquid iron), equation (98) may be simplified to

$$1-\left(\frac{y_C}{y_C^b}\right)^{1/2Z'} = \sum_{k=2}^{t} Y_k \lambda_{k-C} \tag{99}$$

A few quaternary data obtained by Foo and Lupis [22] on the Fe–C–Cr–Ni system at 1550°C are shown in Fig. 13. Equation (98) has been rearranged into

$$f-(\lambda_{Cr}-\lambda_{Ni})Y_{Ni} = -\lambda_{Cr}(Y_{Cr}+Y_{Ni}) \tag{100}$$

The plot of $f-(\lambda_{Cr}-\lambda_{Ni})Y_{Ni}$ or $f-0.65Y_{Ni}$ vs $(Y_{Cr}+Y_{Ni})$ does result in a straight line with a slope equal to 0.45, which confirms the value of λ_{Cr} obtained through ternary data alone (see Table 1).

Figures 14 and 15 show the application of equation (99) to nitrogen solubilities in liquid iron alloys. The equation predicts that a plot of $1-(y_N/y_N^b)^{1/12}$ vs $\sum \lambda_i Y_i$ should result in a straight line with a slope equal to 1. We see that the correlation is satisfactorily well obeyed.

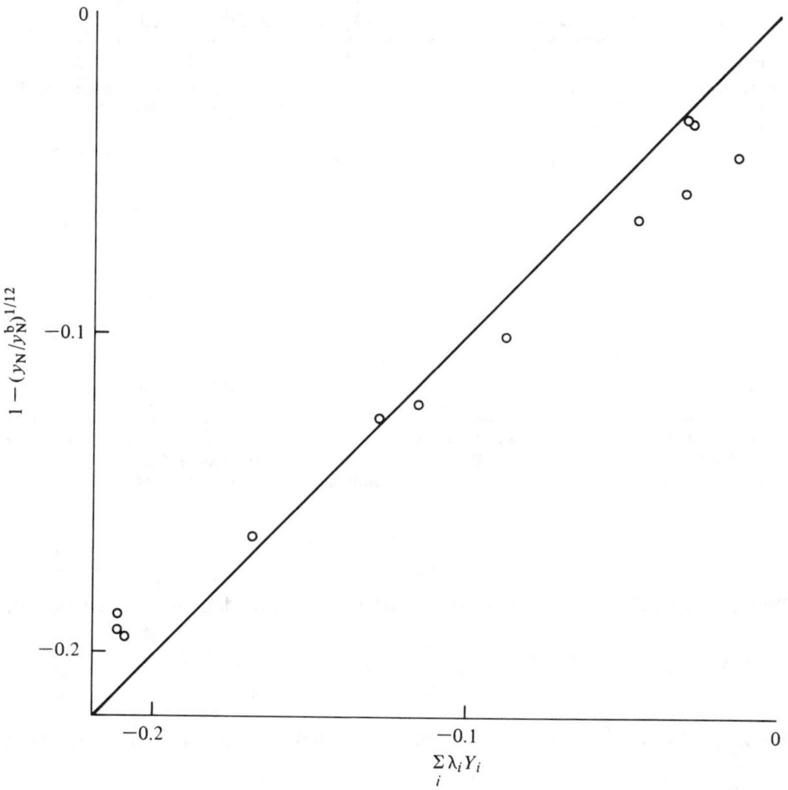

Figure 15. Solubility of nitrogen in multicomponent liquid iron alloys at 1600°C. The equality of the ordinates and abscissas represented by the solid line is predicted by equation (99). The solubility data of Turnock and Pehlke [38] pertain to multicomponent solutions of Fe, Cr, Cb, Mo, Ni, and Si.

4.6. Multicomponent Solutions

The effects of any number of substitutional and interstitial solutes on the activity of an interstitial solute C may be readily calculated in the case of a dilute solution through the concept of interaction coefficients:

$$\ln(\gamma_C/\gamma_C^\infty) = \sum_{i=2}^{m} \epsilon_C^{(i)} X_i + \sum_{i=2}^{m} \rho_C^{(i)} X_i^2 + \sum_{i=2}^{m-1} \sum_{j>i}^{m} \rho_C^{(i,j)} X_i X_j \qquad (101)$$

The expressions of the first and second order interaction coefficients which are predicted by the model may be found in Sections 4.2–4.5 and are listed in Table 2.

We note that a parameter such as λ_{k-C} may be calculated from the properties of the binary solution k–C. Consider, for example, equation (85) for the ternary 1–2–C. If $y_C^{(1)}$ and $y_C^{(2)}$ are the concentration of C in 1 and 2 for the same activity of C (e.g., solubility of carbon with respect to graphite, or solubility of nitrogen at fixed p_{N_2}), then

$$\lambda_{2\text{-}C} = 1 - \left(\frac{y_C^{(2)}}{y_C^{(1)}}\right)^{1/2Z'} \exp\left[\frac{z'}{Z'}(y_C^{(2)} - y_C^{(1)})\lambda_{C-C}\right] \qquad (102)$$

Table 2. Predicted Values of Free Energy Interaction Coefficients

Interaction coefficients	In terms of λ	In terms of other coefficients
$\epsilon_C^{(k)}$	$12\lambda_{k-C}$	—
$\epsilon_C^{(l)}$	$2 + 24\lambda_{l-C}$	—
$\rho_C^{(k)}$	$6\lambda_{k-C}^2$	$\frac{1}{24}[\epsilon_C^{(k)}]^2$
$\rho_C^{(k1,k2)}$	$12\lambda_{k1-C}\lambda_{k2-C}$	$\frac{1}{12}\epsilon_C^{(k1)}\epsilon_C^{(k2)}$
$\rho_C^{(k,l)}$	$12\lambda_{k-C}(1 + \lambda_{k-l})$	$\epsilon_C^{(k)}\left[1 + \frac{\epsilon_l^{(k)}}{12}\right]$
$\rho_C^{(l)}$	$2 + 12\lambda_{l-C}(2 + \lambda_{l-C} + 2\lambda_{l-l})$	$\epsilon_C^{(l)} + \frac{(\epsilon_C^{(l)} - 2)}{48}[\epsilon_C^{(l)} + 2\epsilon_l^{(l)} - 6]$

C is the interstitial element analyzed, l is also an interstitial element (which can be taken as identical to C for studying self-interactions), and k is a substitutional element. The expressions above correspond to the case where $r = 1$, $Z' = 6$, $z' = 12$; for the general case, see equations (78)–(80), (86)–(88), and (94).

If the concentrations of C are low, this expression for $\lambda_{2\text{-C}}$ may be simplified to

$$\lambda_{2\text{-C}} = 1 - \left(\frac{y_C^{(2)}}{y_C^{(1)}}\right)^{1/2Z'} \tag{103}$$

The solubility of C in a multicomponent alloy (at low concentrations of C) may then be estimated by the relatively simple equation

$$(y_C)^{1/2Z'} = \sum_{k=1}^{t} Y_k (y_C^{(k)})^{1/2Z'} \tag{104}$$

For $r = 1$, $y_C \simeq X_C$, $Y_k \simeq X_k$; with $Z' = 6$ (fcc lattice or liquid), equation (104) may be rewritten

$$(X_C)^{1/2Z'} = \sum_{k=1}^{t} X_k (X_C^{(k)})^{1/2Z'} \tag{105}$$

It may be observed that this equation has no adjustable parameters.

5. Conclusions

By and large, the central atoms model is not the only model of multicomponent interstitial solutions. We selected it in this chapter for several reasons. First, its derivation and applications are typical of many other models. Second, it yields very satisfactory results and can be easily adapted to many different kinds of systems (e.g., fused salts [39], nonstoichiometric compounds [40], silicate glasses[1]). Third, many of the assumptions on which we based our derivation can be modified relatively easily. For example, instead of a linear variation of the function $\varphi_{i,j}^{(J)}$ on i and j, a parabolic dependence may be assumed. Other models

[1] The structure of silicate glasses may be approached as a case of an fcc lattice of oxygen anions and vacancies with *two* interstitial lattices of octahedral and tetrahedral sites [41].

are generally less flexible. (A fourth reason for adopting here this model is, of course, that it is the best known to this author!)

The assumption of a parabolic variation of $\varphi_{i,j}^{C}$ on the number i of substitutional atoms in the nearest neighbor shell of C is at the core of a model devised by Wagner [42]. Its derivation is mathematically simple, but from a theoretical standpoint, somewhat less satisfactory than if it were based upon a rigorous statistical model. The resulting equation predicting the activity of C in a ternary 1–2–C contains one adjustable parameter (by contrast, the model above does not contain one). By assigning suitable values to this parameter, good agreement is obtained with experimental data. Chang et al. [43,44] adopted Wagner's treatment but modified it to include two adjustable parameters. Not surprisingly, a better agreement is found. Schwerdtfeger and Pitsch [45] also retained the assumption of a parabolic dependence but included the interactions with second nearest neighbors in their calculations. Their model, like Wagner's, has one adjustable parameter.

The number of parameters in a model is of paramount importance. However, their role and effect are not as easily assessed as it may appear. Let us elaborate on the matter.

It is clear that in the case of a system for which very little accurate information exists, a model with very few parameters is best suited to represent the data and predict properties; e.g., it is certainly unwarranted to fit two data points with a model containing three adjustable parameters! Conversely, it must be realized that a reduction in the number of parameters is generally achieved at the cost of introducing many oversimplifying assumptions.

A model with a single parameter (such as ω_{ij} for the regular or quasi-chemical model, or λ_{i-j} for the central atoms model) gives a good intuitive appreciation of the nature and strength of an important interaction in the solution. In a model with two parameters, different pairs of values for these parameters often yield equivalent fits of the data. Thus, an interpretation of these parameters should not be too narrow but should match their relative accuracy. Also, several models with widely different assumptions may fit equally well the data of a particular system. For example, the activity data of carbon in iron are satisfactorily represented by the quasi-chemical model, the complex blocking model, and the central atoms model (Section 3.5). In such cases, some conclusions must be tentative.

In the development of a "realistic" model of solution, a number of coefficients measuring different atomistic characteristics may appear. If no accurate values can be assigned to these coefficients, they become adjustable parameters. The range of adjustment, however, may occasionally be narrow and the distinction between a coefficient and a parameter is not always easy to make. An additional difficulty in counting the number of parameters is associated with what is assumed known and what is being fitted. Let us take, for example, the application of the central atoms model (Section 4) to a ternary solution 1–2–C where 1 and 2 are substitutional components and C is interstitial. If the concentration range of 2 is not high, the quantity λ_{2-C} will usually play the role of an adjustable parameter. If the concentration range extends from pure 1 to pure 2, the solubility of C in 2 will generally be assumed to be known and will consequently fix the value of λ_{2-C}; this quantity will no longer be regarded as an adjustable parameter to fit the solubility data in the intermediate composition range.

In conclusion, we see that in the evaluation of a model the problem of counting parameters and interpreting their significance must be approached with caution. In the selection or development of a model, a compromise must generally be sought among the mathematical complexity of the model, its number of parameters, and its physical significance.

Problems

1. Calculate the partial molar excess enthalpy and entropy of an interstitial solute in an ideal interstitial solution. Interpret the results.

2. Consider an ideal interstitial solution of components 1–3 where 2 and 3 are interstitial solutes. Derive an expression for the Henrian activity of 2 as a function of composition and calculate the interaction coefficients $\epsilon_2^{(3)}$.

3. Consider the ternary solution A–B–C where A and B are substitutional solutes and C is an interstitial solute. Assume that A and B behave as in a regular solution on the substitutional lattice and that C behaves ideally on the interstitial lattice. Derive expressions for the molar excess Gibbs free energy and the activity coefficients of B and C. Calculate ϵ_B^C.

4. In a ternary solution of two substitutional components A, B, and one interstitial solute C, the probability $p_{i,j}^C$ that an atom C has among its nearest neighbors i atoms B and j atoms C has been calculated by Foo and Lupis [21] in the context of the central atoms model. It was found to be

$$p_{i,j}^C = \frac{Z'!z'!}{i!j!(Z'-i)!(z'-j)!}[y_B A_{CB}(1-\lambda_{B-C})]^i[y_C A_{CC}(1-\lambda_{C-C})]^j/P_C$$

where P_C is a normalization factor such that $\sum_{i,j} p_{i,j}^C = 1$. [A_{CB} and A_{CC} are defined by equations (74) and (75).]

a. Calculate P_C. Give an expression as condensed as possible.

b. Consider the ternary fcc solution of iron, manganese, and carbon at 1000°C, $X_{Mn} = 0.20$, $X_C = 0.04$. Calculate the probability that among its nearest neighbors a carbon atom has one Mn atom and no C atom, two Mn atoms and no C atom.

c. Redo part b after substitution of nickel for manganese.

d. What is the average number of C nearest neighbors to a C atom in cases b and c? What are the average numbers of Mn and Ni nearest neighbors to a C atom?

5. **a.** Compare, in terms of first and second order Gibbs free energy interaction coefficients, the effect of silicon on carbon in liquid iron at 1600°C if silicon were to be treated as an interstitial atom instead of a substitutional one. Assume that experimentally $\epsilon_C^{Si} = 9.8$, $\epsilon_{Si}^{Si} = 12.6$.

b. Calculate the effect of silicon on the solubility of graphite assuming that Si is an interstitial solute, and compare the results with the data in Fig. 11.

References

1. J. Chipman, *Trans. Met. Soc. AIME* **239,** 1332 (1967).
2. E.-H. Foo and C. H. P. Lupis, *Proceedings ICSTIS, Suppl. Trans. Iron Steel Inst. Japan* **11,** 404 (1971).
3. R. H. Fowler, *Proc. Cambridge Phil. Soc.* **32,** 144 (1936).
4. R. H. Fowler and E. A. Guggenheim, *Statistical Thermodynamics*. MacMillan, New York, 1939.
5. R. Peierls, *Proc. Cambridge Phil. Soc.* **32,** 471 (1936).

6. J. R. Lacher, *Proc. Cambridge Phil. Soc.* **33,** 518 (1937).
7. J. R. Lacher, *Proc. Roy. Soc. London Ser. A* **161,** 525 (1937).
8. J. S. Kirkaldy and G. R. Purdy, *Can. J. Phys.* **40,** 202 (1962).
9. H. I. Aaronson, H. A. Domian, and G. M. Pound, *Trans. Met. Soc. AIME* **236,** 753 (1966).
10. W. W. Dunn and R. B. McLellan, *J. Phys. Chem. Solids* **30,** 2631 (1969).
11. E. Scheil, *Arch. Eisenhüttenw.* **22,** 37 (1951).
12. R. Speiser and J. Spretnak, *Trans. Am. Soc. Met.* **47,** 493 (1955).
13. R. B. McLellan, T. L. Garrard, S. J. Horowitz, and J. S. Sprague, *Trans. Met. Soc. AIME* **239,** 528 (1967).
14. R. P. Smith, *J. Am. Chem. Soc.* **68,** 1163 (1946).
15. S. Ban-ya, J. F. Elliott, and J. Chipman, *Trans. Met. Soc. AIME* **245,** 1199 (1969).
16. S. Ban-ya, J. F. Elliott, and J. Chipman, *Met. Trans.* **1,** 1313 (1970).
17. R. W. Gurry, *Trans. Met. Soc. AIME* **150,** 147 (1942).
18. E. Scheil, T. Schmidt, and J. Wüning, *Arch. Eisenhüttenw.* **32,** 251 (1961).
19. D. D. Wagman, J. E. Kilpatrick, W. J. Taylor, K. S. Pitzer, and F. D. Rossini, *J. Res. Nat. Bur. Stand.* **34,** 143 (1945).
20. F. D. Richardson and W. E. Dennis, *Trans. Faraday Soc.* **49,** 171 (1953).
21. E.-H. Foo and C. H. P. Lupis, *Acta Met.* **21,** 1409 (1973).
22. E.-H. Foo and C. H. P. Lupis, *Met. Trans.* **3,** 2125 (1972).
23. T. Wada, H. Wada, J. F. Elliott, and J. Chipman, *Trans. Met. Soc. AIME* **3,** 1657 (1972).
24. R. P. Smith, *J. Am. Chem. Soc.* **70,** 2724 (1948).
25. J. Chipman, R. Alfred, L. Gott, R. Small, D. Wilson, C. Thomson, D. Guernsey, and J. Fulton, *Trans. Am. Soc. Met.* **44,** 1215 (1952).
26. J. Chipman and R. Baschwitz, *Trans. Met. Soc. AIME* **227,** 473 (1963).
27. N. Gokcen and J. Chipman, *Trans. Met. Soc. AIME* **194,** 171 (1952).
28. M. Ohtani, *Sci. Rep. Res. Inst. Tohoku Univ.* **A7,** 487 (1955).
29. T. Mori, K. Fujimara, and H. Kanoshima, *Mem. Fac. Eng. Kyoto Univ.* **25,** 83 (1963).
30. E. T. Turkdogan, R. A. Hancock, S. I. Herlitz, and J. Dentan, *J. Iron Steel Inst.* **183,** 69 (1956).
31. G. K. Sigworth and J. F. Elliott, *Met. Sci.* **8.** 298 (1974).
32. H. Schenck, M. Frohberg, and H. Graf, *Arch. Eisenhüttenw.* **30,** 533 (1959).
33. S. Maekawa and Y. Nakagawa, *Tetsu to Hagane* **46,** 748 (1960).
34. R. Pehlke and J. F. Elliott, *Trans. Met. Soc. AIME* **218,** 1088 (1960).
35. D. W. Gomersall, A. McLean, and R. G. Ward, *Trans. Met. Soc. AIME* **242,** 1309 (1968).
36. J. C. Humbert and J. F. Elliott, *Trans. Met. Soc. AIME* **218,** 1076 (1960).
37. R. J. Blossey and R. D. Pehlke, *Trans. Met. Soc. AIME* **236,** 566 (1966).
38. P. H. Turnock and R. D. Pehlke, *Trans. Met. Soc. AIME* **236,** 1540 (1966).
39. M. Gaune-Escard, J.-C. Mathieu, P. Desré, and Y. Doucet, *J. Chim. Phys.* **70,** 1666 (1973).
40. M. Allibert, Ph.D. thesis, University of Grenoble, France, 1971.
41. C. H. P. Lupis, L. Pargamin, and J. C. Bowker, unpublished results; also L. Pargamin, Ph.D. thesis, Carnegie-Mellon University, Pittsburgh, 1971, and J. C. Bowker, Ph.D. thesis, Carnegie-Mellon University, Pittsburgh, 1979.
42. C. Wagner, *Acta Met.* **21,** 1297 (1973).
43. T. Chiang and Y. A. Chang, *Met. Trans. B* **7,** 453 (1976).
44. R. Y. Lin and Y. A. Chang, *Met. Trans. B* **8,** 293 (1977).
45. K. Schwerdtfeger and W. Pitsch, *Compte Rendu du Congrès Physico-Chimie et Siderurgie,* Versailles, October 1978.

Appendix 1
Units, Useful Constants, and Conversion Factors

Units in the SI System

	SI Base Units	
length	meter	m
mass	kilogram	kg
time	second	sec
electric current	ampere	A
temperature	kelvin	K
luminous intensity	candela	cd
amount of substance	mole	mol
	SI Derived Units	
area	square meter	m^2
volume	cubic meter	m^3
frequency	hertz	Hz, sec^{-1}
density	kilogram per cubic meter	kg/m^3
velocity	meter per second	m/sec
force	newton	N, kg m/sec^2
pressure	pascal	Pa, N/m^2
energy	joule	J, N-m
power	watt	W, J/sec
electric charge	coulomb	C, A sec
potential difference	volt	V, W/A
electric resistance	ohm	Ω, V/A
capacitance	farad	F, A sec/V
magnetic flux	weber	Wb, V sec
inductance	henry	H, V sec/A
magnetic flux density	tesla	T, Wb/m^2

Useful Conversion Factors

Length:
1 ft = 12 in. = 0.3048 m
1 in. = 2.54 cm
1 Å = 10^{-10} m = 10^{-4} μm

Volume:
1 m^3 = 10^6 cm^3 = 35.315 ft^3

Mass:
1 kg = 2.205 lb
1 lb = 453.6 g = 16 oz(avdp) = 14.583 oz(troy)
1 short ton = 2000 lb
1 long ton = 2240 lb
1 metric tonne = 1000 kg = 2204.6 lb

Pressure:
1 atm = 760 mm Hg = 760 Torr
1 atm = 14.696 psi = 1.013 bar
1 atm = 1.013 × 10^5 N/m
1 Pa = 1 N/m

Force:
1 N = 10^5 dyn
1 kg force = 9.80665 N

Force per unit length:
1 dyn/cm = 10^{-3} N/m

Energy:
1 cal = 4.184 J
1 cal = 41.2929 cm^3 atm
1 kcal = 1000 cal = 3.968 Btu
1 eV = 23.053 kcal/mol
1 kW hr = 3.60 × 10^6 J = 860.42 kcal

Temperature:
T(°K) = T(°C) + 273.15
T(°C) = $\frac{5}{9}$[T(°F) − 32]

Useful Constants

Acceleration due to gravity: g = 9.80665 m/sec^2
Avogadro's number: N_A = 6.022 × 10^{23} mol^{-1}
Boltzmann's constant: k = 1.3806 J/°K
Faraday's constant: F = 96,485 C/mol
Gas constant: R = 1.987 cal/°K mol = 82.060 cm^3 atm/°K mol = 8.314 J/°K mol
ln 10 = 2.3026
Planck's constant: h = 6.626 × 10^{-34} J sec
Room temperature: 298°K
Standard conditions (STP): pressure, 1 atm; temperature, 273.15°K or 0°C
Standard volume of ideal gas: 22.414 m^3/kmol

Appendix 2 Atomic Weights and the Periodic Table

The atomic weights listed are based on $C^{12} = 12.0000$, according to the convention adopted by the International Union of Pure and Applied Chemistry in 1961.

The atomic numbers and atomic weights are also shown in the periodic table of the elements.

Atomic Weights

Symbol	*Element*	*Atomic weight*	*Symbol*	*Element*	*Atomic weight*
Ag	Silver	107.870	Dy	Dysprosium	162.50
Al	Aluminum	26.9815	Er	Erbium	167.26
Am^{243}	Americium	243.061	Eu	Europium	151.96
Ar	Argon	39.948	F	Fluorine	18.9984
As	Arsenic	74.9216	Fe	Iron	55.847
Au	Gold	196.967	Ga	Gallium	69.72
B	Boron	10.811	Gd	Gadolinium	157.25
Ba	Barium	137.34	Ge	Germanium	72.59
Be	Beryllium	9.0122	H	Hydrogen	1.00797
Bi	Bismuth	208.980	He	Helium	4.0026
Br	Bromine	79.909	Hf	Hafnium	178.49
C	Carbon	12.01115	Hg	Mercury	200.59
Ca	Calcium	40.08	Ho	Holmium	164.930
Cd	Cadmium	112.40	I	Iodine	126.9044
Ce	Cerium	140.12	In	Indium	114.82
Cl	Chlorine	35.453	Ir	Iridium	192.2
Co	Cobalt	58.9332	K	Potassium	39.102
Cr	Chromium	51.996	Kr	Krypton	83.80
Cs	Cesium	132.905	La	Lanthanum	138.91
Cu	Copper	63.54	Li	Lithium	6.939

(*continued*)

Atomic Weights

Symbol	*Element*	*Atomic weight*	*Symbol*	*Element*	*Atomic weight*
Lu	Lutetium	174.97	S	Sulfur	32.064
Mg	Magnesium	24.312	Sb	Antimony	121.75
Mn	Manganese	54.9380	Sc	Scandium	44.956
Mo	Molybdenum	95.94	Se	Selenium	78.96
N	Nitrogen	14.0067	Si	Silicon	28.086
Na	Sodium	22.9898			
Nb	Niobium	92.906	Sm	Samarium	150.35
Nd	Neodymium	144.24	Sn	Tin	118.69
			Sr	Strontium	87.62
Ne	Neon	20.183	Ta	Tantalum	180.948
Ni	Nickel	58.71	Tb	Terbium	158.924
Np^{237}	Neptunium	237.048			
O	Oxygen	15.9994	Te	Tellurium	127.60
Os	Osmium	190.2	Th	Thorium	232.038
			Ti	Titanium	47.90
			Tl	Thallium	204.37
P	Phosphorus	30.9738	Tm	Thulium	168.934
Pb	Lead	207.19			
Pd	Palladium	106.4	U	Uranium	238.03
Pr	Praseodymium	140.907	V	Vanadium	50.942
Pt	Platinium	195.09	W	Tungsten	183.85
			Xe	Xenon	131.30
Pu^{239}	Plutonium	239.052	Yb	Ytterbium	88.905
Rb	Rubidium	85.47			
Re	Rhenium	186.2	Y	Yttrium	173.04
Rh	Rhodium	102.905	Zn	Zinc	65.37
Ru	Ruthenium	101.07	Zr	Zirconium	91.22

←Metals— — — Nonmetals→

IA	IIA	IIIB	IVB	VB	VIB	VIIB	VIII	VIII	VIII	IB	IIB	IIIA	IVA	VA	VIA	VIIA	0
1 H 1.00797																1 H 1.0079	2 He 4.0026
3 Li 6.939	4 Be 9.012											5 B 10.811	6 C 12.011	7 N 14.007	8 O 15.9994	9 F 18.998	10 Ne 20.183
11 Na 22.990	12 Mg 24.312											13 Al 26.98	14 Si 28.086	15 P 30.97	16 S 32.064	17 Cl 35.453	18 Ar 39.95
19 K 39.102	20 Ca 40.08	21 Sc 44.96	22 Ti 47.90	23 V 50.94	24 Cr 52.00	25 Mn 54.94	26 Fe 55.85	27 Co 58.93	28 Ni 58.71	29 Cu 63.54	30 Zn 65.37	31 Ga 69.72	32 Ge 72.59	33 As 74.92	34 Se 78.96	35 Br 79.91	36 Kr 83.80
37 Rb 85.47	38 Sr 87.62	39 Y 88.91	40 Zr 91.22	41 Nb 92.91	42 Mo 95.94	43 Tc 99	44 Ru 101.07	45 Rh 102.91	46 Pd 106.4	47 Ag 107.87	48 Cd 112.40	49 In 114.82	50 Sn 118.69	51 Sb 121.75	52 Te 127.60	53 I 126.90	54 Xe 131.30
55 Cs 132.90	56 Ba 137.34	57–71 La series*	72 Hf 178.49	73 Ta 180.95	74 W 183.85	75 Re 186.2	76 Os 190.2	77 Ir 192.2	78 Pt 195.1	79 Au 196.97	80 Hg 200.59	81 Tl 204.37	82 Pb 207.19	83 Bi 208.98	84 Po 210	85 At 210	86 Rn 222
87 Fr 223	88 Ra 226	89– Ac series†															

58 Ce 140.12	59 Pr 140.91	60 Nd 144.24	61 Pm 147	62 Sm 150.35	63 Eu 151.96	64 Gd 157.25	65 Tb 158.92	66 Dy 162.50	67 Ho 164.93	68 Er 167.26	69 Tm 168.93	70 Yb 173.04	71 Lu 174.97
90 Th 232.04	91 Pa 231	92 U 238.03	93 Np 237	94 Pu 239	95 Am 241	96 Cm 242	97 Bk 249	98 Cf 252	99 Es 254	100 Fm 253	101 Md	102 No	103 Lw

Appendix 3
Standard Enthalpies and Gibbs Free Energies of Formation at 298°K for Selected Compounds

The values listed below were taken from the following compilations:

D. R. Stull and H. Prophet, *JANAF Thermochemical Tables*, 2nd ed., NSRDS-NBS 37. U.S. GPO, Washington, DC, 1971.

J. F. Elliott and M. Gleiser, *Thermochemistry for Steelmaking*. Addison-Wesley, Reading, MA, 1960, Vol. 1.

C. E. Wicks and F. E. Block, *Thermodynamic Properties of 65 Elements—Their Oxides, Halides, Carbides and Nitrides*. Bur. of Mines, Bull. 605, U.S. GPO, Washington, DC, 1963.

Selected Values of Chemical Thermodynamic Properties, Circular of the National Bureau of Standards 500. U.S. GPO, Washington, DC, 1952.

(*Note*: In the compilation of Elliott and Gleiser the standard state for sulfur at 298°K is the ideal diatomic gas; in the following table, the standard state for sulfur is the rhombic crystal.)

Standard Enthalpies and Gibbs Free Energies of Formation (kcal/mol) at 298°K for Selected Compounds

Compound	$-H^\circ_{f,\,298}$	$-G^\circ_{f,\,298}$
AgCl	30.4	26.2
Ag_2O	7.2	2.5
Ag_2S	7.6	9.5
Al_4C_3	51.6	48.6
AlCl(g)	12.3	18.6
AlF_3	361	342
AlN	76.5	68.6
Al_2O(g)	33.5	42.5
Al_2O_3	400.4	378.1
$Al_2O_3 \cdot H_2O$	471	435
$Al_2O_3 \cdot 3H_2O$	613.7	547.9
Al_2S_3	173	168
$Al_2O_3 \cdot SiO_2$	619.5	584.3
$3Al_2O_3 \cdot 2SiO_2$	1629.8	1539.7
As_2O_3	157	137.7
As_2O_4	176	149
As_2O_5	219	185
As_2S_3	35	
$AuCl_3$	28.3	
Au_2O_3	−19.3	−39.0
B_4C	12.2	12.1
BN	60.0	53.8
B_2O_3	305.4	286.4
B_2S_3	61	57
$BaCO_3$	291	272
$BaCl_2$	205.3	193.3
BaO	133.5	126.5
BaS	103	101
$BaSO_4$	350	323
$BeCl_2$	112.6	102.9
BeO	143.1	136.1
BeS	55.6	55
$BiCl_3$	90.5	76.0
BiO	49.9	43.5
Bi_2O_3	137.9	118.7
CCl_4(l)	33.2	16.3
CCl_4(g)	22.9	12.8
CH_4(g)	17.9	12.1
C_2H_6(g)	20.2	7.9
CO(g)	26.4	32.8
CO_2(g)	94.1	94.3
$COCl_2$(g)	53.3	50.3
COS(g)	34	40.6
CS_2(g)	−27.4	−15.4
CaC_2	14.1	15.3
$CaCO_3$	288.5	269.8
$CaCl_2$	190.2	178.8
CaF_2	293	280.5
Ca_3N_2	105	90
CaO	151.8	144.4
$CaO \cdot Al_2O_3$	560	
$CaO \cdot SiO_2$	379	358
$2CaO \cdot SiO_2$	539	
$3CaO \cdot SiO_2$	688	
Ca_3P_2	112	
CaS	114.3	113
$CaSO_4$	342.4	315.6
$CdCl_2$	93.0	82.7
CdO	61.2	54.1
CdS	34.5	33.6
$CdSO_4$	221	196
CeN	78	70
CeO_2	245	230
Ce_2O_3	435	411
CeS	118	118
Co_2C	−4	−3.3
$CoCl_2$	77.8	67.4
Co_3N	−2	
CoO	57.1	51.5
Co_3O_4	212	184
CoS	20.2	19.8
Co_9S_8	192	192
$CoSO_4$	207.5	182.1
Cr_3C_2	23.2	23.4
Cr_7C_3	98.3	101.0
$Cr_{23}C_6$	44.1	43.5
$CrCl_2$	94.6	85.3
$CrCl_3$	132.5	115.9
CrN	28.2	20.5
Cr_2N	25.6	21.3
CrO_2	139.4	126.9
CrO_3	140	121
Cr_2O_3	272.7	253.2
CuCl	33	28.9
$CuCl_2$	49.2	38.7
CuO	37.3	30.6
Cu_2O	40.7	35.3
CuS	11.6	11.6
Cu_2S	19.5	21
$CuSO_4$	184	158.2
Cs_2O	76	
Cs_2S	87	84
Fe_3C	−6.0	−4.8
$FeCO_3$	178.7	161.1

Standard Enthalpies and Gibbs Free Energies of Formation (*continued*)

Compound	$-H^\circ_{f,\,298}$	$-G^\circ_{f,\,298}$
$FeCl_2$	81.7	72.3
$FeCl_3$	95.7	79.5
Fe_4N	1.1	−1.8
$Fe_{0.95}O$	63.5	58.1
Fe_2O_3	196	177
Fe_3O_4	267	242
FeS	22.4	23.1
FeS_2	12	10.8
$FeSO_4$	220.5	
$2FeO{\cdot}SiO_2$	347.3	
HCl(g)	22.1	22.8
HF(g)	65.1	65.6
H_2O(l)	68.3	56.7
H_2O(g)	57.8	54.6
H_2S(g)	4.8	8.0
H_2SO_4(l)	193.9	
HgCl	31.6	25.3
$HgCl_2$	53.4	42.4
HgO	21.7	14.0
HgS	13.9	11.7
K_2CO_3	273.9	
KCl	104.4	97.7
KF	134.5	127.5
K_2O	86.8	77.0
KOH	101.8	
K_2S	88	85
K_2SO_4	342.7	314.6
$K_2O{\cdot}SiO_2$	365.9	
La_2O_3	458	
La_2S_3	296	293
LiCl	97.7	92.5
Li_2O	142.6	134
MgC_2	−21	−21
Mg_2C_3	−18	−18
$MgCO_3$	266	246
$MgCl_2$	153.4	141.5
Mg_3N_2	110.2	95.9
MgO	143.8	136.1
MgS	84.3	83.4
$MgSO_4$	305.5	280.5
Mg_2Si	18.6	18.4
$MgO{\cdot}SiO_2$	370.2	349.4
$2MgO{\cdot}SiO_2$	520.3	491.9
Mn_7C_3	−19.1	−7.6
$MnCO_3$	213.9	195.4
$MnCl_2$	115.2	105.1
Mn_5N_2	48.2	36.8
MnO	92.1	86.8
MnO_2	124.5	111.4
Mn_2O_3	229	211
Mn_3O_4	331.4	305.9
MnS	48.8	49.8
$MnSO_4$	254.2	228.5
$MnO{\cdot}SiO_2$	302.5	283.3
MoC	−9.7	−5.6
MoC_2	−15.8	−8
Mo_2N	16.6	12
MoO_2	140.8	126.1
MoO_3	178.1	159.7
MoS_2	55.9	54
Mo_2S_3	99	98
NH_3(g)	11.0	3.9
NO(g)	−21.6	−20.7
NO_2(g)	−7.9	−12.2
N_2O(g)	−19.6	−24.9
Na_2CO_3	270.3	250.4
NaCl	98.3	91.9
NaF	136	129
Na_2O	99	90
NaOH	102	
Na_2S	89	86
Na_2SO_4	330.9	302.8
$Na_2O{\cdot}SiO_2$	373.2	350.4
$Na_2O{\cdot}2SiO_2$	590.4	555.5
NbC	33.6	
NbN	56.8	50.0
Nb_2N	61.1	
Nb_2O_4	380.8	354.5
Nb_2O_5	455.2	422.9
Ni_3C	−9.2	−7.6
$NiCO_3$	162.7	146.7
$Ni(CO)_4$(g)	142	
$NiCl_2$	73.0	61.9
Ni_3N	−0.2	
NiO	57.3	50.6
NiS	21.7	21.1
Ni_3S_2	47.7	48
$NiSO_4$	213.0	184.9
PCl_3(l)	76.9	74.5
PCl_3(g)	73.2	68.4
PCl_5(s)	106.5	75.8
PCl_5(g)	95.4	77.6
PO(g)	5.5	12.5
P_4O_{10}	703	644
$POCl_3$(g)	141.5	130.3
$PbCO_3$	167.3	149.7

(*continued*)

Standard Enthalpies and Gibbs Free Energies of Formation (*continued*)

Compound	$-H^\circ_{f,\,298}$	$-G^\circ_{f,\,298}$	Compound	$-H^\circ_{f,\,298}$	$-G^\circ_{f,\,298}$
$PbCl_2$	85.9	75.0	TiC	43.9	43
PbO	52.5	45.2	$TiCl_2$	123	112.2
PbO_2	66.1	52.3	$TiCl_4$(l)	192.2	176.2
Pb_3O_4	175	147	$TiCl_4$(g)	182.4	173.7
PbS	22.3	22.4	TiN	80.5	73.6
$PbSO_4$	219.5	193.9	TiO	124.1	117.1
$PbO{\cdot}SiO_2$	273.7	253.6	TiO_2	225.8	212.6
$2PbO{\cdot}SiO_2$	329.3	302.8	Ti_2O_3	363	343
PdO	20.4		Ti_3O_5	588	554
			TiS	52	51.1
$PtCl$	13	8.5			
$PtCl_2$	29	19.5	Tl_2O	41.9	32.5
$PtCl_3$	43	26.6	Tl_2S	20.8	
$PtCl_4$	53	33			
$Pt(OH)_2$	87.2	68.2	UC_2	42	42
PtS	18	17	UN	80	75
			UO_2	259	247
S_2(g)	−30.8	−19.1	UO_3	292	273
SO_2(g)	70.9	71.7	U_3O_8	854	804
SO_3(g)	94.6	88.7			
			VC	20	20
$SbCl_3$	91.4	77.8	VN	42	35.7
Sb_2O_3	169.9	150.0	VO	100	94
Sb_2S_3	43.5		V_2O_3	296	277
$Sb_2(SO_4)_3$	575		V_2O_4	342	316
SeO_2	55		V_2O_5	373	342
SiC	13.0	12.4	WC	8.4	8.4
$SiCl_4$(l)	153.0	136.9	W_2C	11	11.7
$SiCl_4$(g)	145.7	136.2	WCl_2	36	27
SiF_4(g)	370	360	WCl_4	69	50
SiN_4	178	154.7	WCl_5	82	60
SiO(g)	21.4	27.8	WCl_6	96.9	74
SiO_2	217.7	204.7	W_2N	17.2	
SiS(g)	−16.9	−4.6	WO_2	135	123
SiS_2	43	42	WO_3	200.9	182.5
			W_4O_{11}	741	674
$SnCl_2$	81.0	71.6	WO_4Ca	392.5	
$SnCl_4$(l)	61.4	58.5	WS_2	48	47.5
SnO	68.4	61.4			
SnO(g)	1	6.5	Y_2O_3	455	433
SnO_2	138.8	124.3	YN	71.5	64
SnS	18.4	18			
SnS_2	40	38	$ZnCO_3$	194.2	174.8
			$ZnCl_2$	99.6	88.5
SrS	110	109	Zn_3N_2	5.3	—
			ZnO	83.3	76.1
TaC	38.5	38.1	ZnS	48.5	47.4
Ta_2C	34	33.6	$ZnSO_4$	233.9	208.3
TaN	60	53.2	$2ZnO{\cdot}SiO_2$	391.3	
Ta_2O_5	489	457			
TeO_2	77.7	64.6	ZrC	44.1	43.3
			ZrN	87.3	80.5
Tl_2O	41.9	32.5	ZrO_2	261.5	247.8
Tl_2S	20.8		$ZrO_2{\cdot}SiO_2$	484	456

Appendix 4
JANAF Tables for CO, CO_2, H_2, H_2O, N_2, and O_2

In the late 1950s a project was initiated to establish a data base for the calculations pertaining to rocket propellants. As part of that project, a program was started to critically evaluate and compile consistent tables of thermodynamic properties of propellant combustion products. This program, known as the *JANAF thermochemical tables*, was undertaken by the Dow Chemical Company and sponsored by agencies of the United States Defense Department. The tables found wide utility in technical areas far removed from chemical rocket propulsion, and it was decided to incorporate the second edition of the tables in the National Standard Reference Data System of the National Bureau of Standards.

A complete reference of the JANAF tables may be found on page 48. The tables which follow are extracted from this work.

Table 1. Thermochemical Data for CO (Ideal Gas)[a]

T (°K)	C_p	$S°$	$-(G°-H°_{298})/T$	$H°-H°_{298}$	ΔH_f	$\Delta G°_f$	log K
	(cal/°K mol)				(kcal/mol)		
0	0.000	0.000	∞	−2.072	−27.200	−27.200	∞
100	6.956	39.613	53.401	−1.379	−26.876	−28.741	62.809
200	6.957	44.435	47.851	−0.683	−26.599	−30.718	33.566
298	6.965	47.214	47.214	0.000	−26.417	−32.783	24.029
300	6.965	47.257	47.214	0.013	−26.414	−32.823	23.910
400	7.013	49.265	47.488	0.711	−26.318	−34.975	19.109
500	7.121	50.841	48.006	1.417	−26.296	−37.144	16.235
600	7.276	52.152	48.591	2.137	−26.332	−39.311	14.318
700	7.450	53.287	49.182	2.873	−26.409	−41.468	12.946
800	7.624	54.293	49.759	3.627	−26.514	−43.612	11.914
900	7.786	55.200	50.314	4.397	−26.637	−45.744	11.108
1000	7.931	56.028	50.845	5.183	−26.771	−47.859	10.459
1100	8.057	56.790	51.351	5.983	−26.914	−49.962	9.926
1200	8.168	57.496	51.834	6.794	−27.052	−52.049	9.479
1300	8.263	58.154	52.295	7.616	−27.218	−54.126	9.099
1400	8.346	58.769	52.736	8.446	−27.376	−56.189	8.771
1500	8.417	59.348	53.158	9.285	−27.537	−58.241	8.485
1600	8.480	59.893	53.562	10.130	−27.700	−60.284	8.234
1700	8.535	60.409	53.950	10.980	−27.865	−62.315	8.011
1800	8.583	60.898	54.322	11.836	−28.032	−64.337	7.811
1900	8.626	61.363	54.681	12.697	−28.201	−68.349	7.631
2000	8.664	61.807	55.026	13.561	−28.372	−68.353	7.469
2100	8.698	62.230	55.359	14.430	−28.543	−70.346	7.321
2200	8.728	62.635	55.680	15.301	−28.719	−72.335	7.185
2300	8.756	63.024	55.991	16.175	−28.894	−74.311	7.061
2400	8.781	63.397	56.292	17.052	−29.074	−76.282	6.946
2500	8.804	63.756	56.584	17.931	−29.254	−78.247	6.840
2600	8.825	64.102	56.866	18.813	−29.438	−80.202	6.741
2700	8.844	64.435	57.140	19.696	−29.623	−82.153	6.649
2800	8.863	64.757	57.407	20.582	−29.810	−84.093	6.563
2900	8.879	65.069	57.666	21.469	−30.001	−86.028	6.483
3000	8.895	65.370	57.917	22.357	−30.194	−87.957	6.407
3100	8.910	65.662	58.163	23.248	−30.388	−89.878	6.336
3200	8.924	65.945	58.401	24.139	−30.586	−91.795	6.269
3300	8.937	66.220	58.634	25.032	−30.786	−93.707	6.206
3400	8.949	66.487	58.861	25.927	−30.988	−95.609	6.145
3500	8.961	66.746	59.083	26.822	−31.192	−97.509	6.088
3600	8.973	66.999	59.299	27.719	−31.399	−99.400	6.034
3700	8.984	67.245	59.511	28.617	−31.608	−101.286	5.982
3800	8.994	67.485	59.717	29.516	−31.818	−103.164	5.933
3900	9.004	67.718	59.919	30.416	−32.031	−105.039	5.886
4000	9.014	67.946	60.117	31.316	−32.247	−106.908	5.841
4100	9.024	68.169	60.311	32.218	−32.464	−108.774	5.798
4200	9.033	68.387	60.501	33.121	−32.684	−110.630	5.756
4300	9.042	68.599	60.687	34.025	−32.906	−112.483	5.717
4400	9.051	68.807	60.869	34.930	−33.130	−114.333	5.679
4500	9.059	69.011	61.047	35.835	−33.356	−116.177	5.642

[a] Molecular weight, 28.01055.

Table 2. Thermochemical Data for CO_2 (Ideal Gas)[a]

T (°K)	C_p	$S°$	$-(G° - H°_{298})/T$	$H° - H°_{298}$	ΔH_f	$\Delta G°_f$	log K
	(cal/°K mol)			(kcal/mol)			
0	0.000	0.000	∞	−2.238	−93.965	−93.965	∞
100	6.981	42.758	58.188	−1.543	−93.997	−94.100	205.645
200	7.734	47.769	51.849	−0.816	−94.028	−94.191	102.922
298	8.874	51.072	51.072	0.000	−94.054	−94.265	69.095
300	8.896	51.127	51.072	0.016	−94.055	−94.267	68.670
400	9.877	53.830	51.434	0.958	−94.070	−94.335	51.540
500	10.666	56.122	52.148	1.987	−94.091	−94.399	41.260
600	11.310	58.126	52.981	3.087	−94.124	−94.458	34.405
700	11.846	59.910	53.845	4.245	−94.169	−94.510	29.506
800	12.293	61.522	54.706	5.453	−94.218	−94.556	25.830
900	12.667	62.992	55.546	6.702	−94.270	−94.596	22.970
1000	12.980	64.344	56.359	7.984	−94.321	−94.628	20.680
1100	13.243	65.594	57.143	9.296	−94.371	−94.658	18.806
1200	13.466	66.756	57.896	10.632	−94.419	−94.681	17.243
1300	13.656	67.841	58.620	11.988	−94.469	−94.701	15.920
1400	13.815	68.859	59.315	13.362	−94.515	−94.716	14.785
1500	13.953	69.817	59.984	14.750	−94.562	−94.728	13.801
1600	14.074	70.722	60.627	16.152	−94.607	−94.739	12.940
1700	14.177	71.578	61.246	17.565	−94.650	−94.746	12.180
1800	14.269	72.391	61.843	18.987	−94.696	−94.750	11.504
1900	14.352	73.165	62.418	20.418	−94.742	−94.751	10.898
2000	14.424	73.903	62.974	21.857	−94.788	−94.752	10.353
2100	14.489	74.608	63.512	23.303	−94.834	−94.746	9.860
2200	14.547	75.284	64.031	24.755	−94.885	−94.744	9.411
2300	14.600	75.931	64.535	26.212	−94.936	−94.735	9.001
2400	14.649	76.554	65.023	27.674	−94.991	−94.724	8.625
2500	14.692	77.153	65.496	29.141	−95.048	−94.714	8.280
2600	14.734	77.730	65.956	30.613	−95.107	−94.698	7.960
2700	14.771	78.286	66.402	32.088	−95.170	−94.683	7.664
2800	14.807	78.824	66.836	33.567	−95.235	−94.662	7.388
2900	14.841	79.344	67.259	35.049	−95.305	−94.639	7.132
3000	14.873	79.848	67.670	36.535	−95.377	−94.615	6.892
3100	14.902	80.336	68.071	38.024	−95.451	−94.587	6.668
3200	14.930	80.810	68.461	39.515	−95.530	−94.560	6.458
3300	14.956	81.270	68.843	41.010	−95.611	−94.531	6.260
3400	14.982	81.717	69.215	42.507	−95.696	−94.495	6.074
3500	15.006	82.151	69.578	44.006	−95.784	−94.462	5.898
3600	15.030	82.574	69.933	45.508	−95.874	−94.421	5.732
3700	15.053	82.986	70.280	47.012	−95.968	−94.379	5.574
3800	15.075	83.388	70.620	48.518	−96.064	−94.331	5.425
3900	15.097	83.780	70.953	50.027	−96.162	−94.286	5.283
4000	15.119	84.162	71.278	51.538	−96.263	−94.237	5.149
4100	15.139	84.536	71.597	53.051	−96.367	−94.186	5.020
4200	15.159	84.901	71.909	54.566	−96.473	−94.130	4.898
4300	15.179	85.258	72.216	56.082	−96.583	−94.072	4.781
4400	15.197	85.607	72.516	57.601	−96.694	−94.015	4.670
4500	15.216	85.949	72.811	59.122	−96.807	−93.954	4.563

[a] Molecular weight, 44.00995.

Table 3. Thermochemical Data for H_2 (Reference State—Ideal Gas)[a]

T (°K)	C_p (cal/°K mol)	$S°$ (cal/°K mol)	$-(G°-H°_{298})/T$ (cal/°K mol)	$H°-H°_{298}$ (kcal/mol)	ΔH_f (kcal/mol)	$\Delta G°_f$ (kcal/mol)	log K
0	0.000	0.000	∞	−2.024	0.000	0.000	0.000
100	5.393	24.387	37.035	−1.265	0.000	0.000	0.000
200	6.518	28.520	31.831	−0.662	0.000	0.000	0.000
298	6.892	31.208	31.208	0.000	0.000	0.000	0.000
300	6.894	31.251	31.208	0.013	0.000	0.000	0.000
400	6.975	33.247	31.480	0.707	0.000	0.000	0.000
500	6.993	34.806	31.995	1.406	0.000	0.000	0.000
600	7.009	36.082	32.573	2.106	0.000	0.000	0.000
700	7.036	37.165	33.153	2.808	0.000	0.000	0.000
800	7.087	38.107	33.715	3.514	0.000	0.000	0.000
900	7.148	38.946	34.250	4.226	0.000	0.000	0.000
1000	7.219	39.702	34.758	4.944	0.000	0.000	0.000
1100	7.300	40.394	35.240	5.670	0.000	0.000	0.000
1200	7.390	41.033	35.696	6.404	0.000	0.000	0.000
1300	7.490	41.628	36.130	7.148	0.000	0.000	0.000
1400	7.600	42.187	36.543	7.902	0.000	0.000	0.000
1500	7.720	42.716	36.937	8.668	0.000	0.000	0.000
1600	7.823	43.217	37.314	9.446	0.000	0.000	0.000
1700	7.921	43.695	37.675	10.233	0.000	0.000	0.000
1800	8.016	44.150	38.022	11.030	0.000	0.000	0.000
1900	8.108	44.586	38.356	11.836	0.000	0.000	0.000
2000	8.195	45.004	38.678	12.651	0.000	0.000	0.000
2100	8.279	45.406	38.989	13.475	0.000	0.000	0.000
2200	8.358	45.793	39.290	14.307	0.000	0.000	0.000
2300	8.434	46.166	39.581	15.146	0.000	0.000	0.000
2400	8.506	46.527	39.863	15.993	0.000	0.000	0.000
2500	8.575	46.875	40.136	16.848	0.000	0.000	0.000
2600	8.639	47.213	40.402	17.708	0.000	0.000	0.000
2700	8.700	47.540	40.660	18.575	0.000	0.000	0.000
2800	8.757	47.857	40.912	19.448	0.000	0.000	0.000
2900	8.810	48.166	41.157	20.326	0.000	0.000	0.000
3000	8.859	48.465	41.395	21.210	0.000	0.000	0.000
3100	8.911	48.756	41.628	22.098	0.000	0.000	0.000
3200	8.962	49.040	41.855	22.992	0.000	0.000	0.000
3300	9.012	49.317	42.077	23.891	0.000	0.000	0.000
3400	9.061	49.586	42.294	24.794	0.000	0.000	0.000
3500	9.110	49.850	42.506	25.703	0.000	0.000	0.000
3600	9.158	50.107	42.714	26.616	0.000	0.000	0.000
3700	9.205	50.359	42.917	27.535	0.000	0.000	0.000
3800	9.252	50.605	43.116	28.457	0.000	0.000	0.000
3900	9.297	50.846	43.311	29.385	0.000	0.000	0.000
4000	9.342	51.082	43.502	30.317	0.000	0.000	0.000
4100	9.386	51.313	43.690	31.253	0.000	0.000	0.000
4200	9.429	51.540	43.874	32.194	0.000	0.000	0.000
4300	9.472	51.762	44.055	33.139	0.000	0.000	0.000
4400	9.514	51.980	44.233	34.088	0.000	0.000	0.000
4500	9.555	52.194	44.407	35.042	0.000	0.000	0.000

[a] Molecular weight, 2.016.

Table 4. Thermochemical Data for H_2O (Ideal Gas)[a]

T	C_p	$S°$	$-(G°-H°_{298})/T$	$H°-H°_{298}$	ΔH_f	$\Delta G°_f$	
(°K)	(cal/°K mol)			(kcal/mol)			log K
0	0.000	0.000	∞	−2.367	−57.103	−57.103	∞
100	7.961	36.396	52.202	−1.581	−57.433	−56.557	123.600
200	7.969	41.916	45.837	−0.784	−57.579	−55.635	60.792
298	8.025	45.106	45.106	0.000	−57.798	−54.636	40.048
300	8.027	45.155	45.106	0.015	−57.803	−54.617	39.786
400	8.186	47.484	45.422	0.825	−58.042	−53.519	29.240
500	8.415	49.334	46.026	1.654	−58.277	−52.361	22.886
600	8.676	50.891	46.710	2.509	−58.500	−51.156	18.633
700	8.954	52.249	47.406	3.390	−58.710	−49.915	15.583
800	9.246	53.464	48.089	4.300	−58.905	−48.646	13.289
900	9.547	54.570	48.749	5.240	−59.084	−47.352	11.498
1000	9.851	55.592	49.382	6.209	−59.246	−46.040	10.062
1100	10.152	56.545	49.991	7.210	−59.391	−44.712	8.883
1200	10.444	57.441	50.575	8.240	−59.519	−43.371	7.899
1300	10.723	58.288	51.136	9.298	−59.634	−42.022	7.064
1400	10.987	59.092	51.675	10.384	−59.734	−40.663	6.347
1500	11.233	59.859	52.196	11.495	−59.824	−39.297	5.725
1600	11.462	60.591	52.698	12.630	−59.906	−37.927	5.180
1700	11.674	61.293	53.183	13.787	−59.977	−36.549	4.699
1800	11.869	61.965	53.652	14.964	−60.041	−35.170	4.270
1900	12.048	62.612	54.107	16.160	−60.099	−33.786	3.886
2000	12.214	63.234	54.548	17.373	−60.150	−32.401	3.540
2100	12.366	63.834	54.976	18.602	−60.198	−31.012	3.227
2200	12.505	64.412	55.392	19.846	−60.242	−29.621	2.942
2300	12.634	64.971	55.796	21.103	−60.282	−28.229	2.682
2400	12.753	65.511	56.190	22.372	−60.321	−26.832	2.443
2500	12.863	66.034	56.573	23.653	−60.359	−25.439	2.224
2600	12.965	66.541	56.947	24.945	−60.393	−24.040	2.021
2700	13.059	67.032	57.311	26.246	−60.428	−22.641	1.833
2800	13.146	67.508	57.667	27.556	−60.462	−21.242	1.658
2900	13.228	67.971	58.014	28.875	−60.496	−19.838	1.495
3000	13.304	68.421	58.354	30.201	−60.530	−18.438	1.343
3100	13.374	68.858	58.685	31.535	−60.562	−17.034	1.201
3200	13.441	69.284	59.010	32.876	−60.596	−15.630	1.067
3300	13.503	69.698	59.328	34.223	−60.631	−14.223	0.942
3400	13.562	70.102	59.639	35.577	−60.666	−12.818	0.824
3500	13.617	70.496	59.943	36.936	−60.703	−11.409	0.712
3600	13.669	70.881	60.242	38.300	−60.741	−10.000	0.607
3700	13.718	71.256	60.534	39.669	−60.782	−8.589	0.507
3800	13.764	71.622	60.821	41.043	−60.822	−7.177	0.413
3900	13.808	71.980	61.103	42.422	−60.865	−5.766	0.323
4000	13.850	72.331	61.379	43.805	−60.910	−4.353	0.238
4100	13.890	72.673	61.651	45.192	−60.957	−2.938	0.157
4200	13.957	73.008	61.917	46.583	−61.006	−1.522	0.079
4300	13.963	73.336	62.179	47.977	−61.056	−0.105	0.005
4400	13.997	73.658	62.436	49.375	−61.109	1.311	−0.065
4500	14.030	73.973	62.689	50.777	−61.164	2.729	−0.133

[a] Molecular weight, 18.016.

Table 5. Thermochemical Data for N_2 (Reference State—Ideal Gas)[a]

T (°K)	C_p	$S°$	$-(G°-H°_{298})/T$	$H°-H°_{298}$	ΔH_f	$\Delta G°_f$	log K
	(cal/°K mol)			(kcal/mol)			
0	0.000	0.000	∞	−2.072	0.000	0.000	0.000
100	6.956	38.170	51.957	−1.379	0.000	0.000	0.000
200	6.957	42.992	46.407	−0.683	0.000	0.000	0.000
298	6.961	45.770	45.770	0.000	0.000	0.000	0.000
300	6.961	45.813	45.770	0.013	0.000	0.000	0.000
400	6.990	47.818	46.043	0.710	0.000	0.000	0.000
500	7.069	49.386	46.561	1.413	0.000	0.000	0.000
600	7.196	50.685	47.143	2.125	0.000	0.000	0.000
700	7.350	51.806	47.731	2.853	0.000	0.000	0.000
800	7.512	52.798	48.303	3.596	0.000	0.000	0.000
900	7.670	53.692	48.853	4.355	0.000	0.000	0.000
1000	7.815	54.507	49.378	5.129	0.000	0.000	0.000
1100	7.945	55.258	49.879	5.917	0.000	0.000	0.000
1200	8.061	55.955	50.357	6.718	0.000	0.000	0.000
1300	8.162	56.604	50.813	7.529	0.000	0.000	0.000
1400	8.252	57.212	51.248	8.350	0.000	0.000	0.000
1500	8.330	57.784	51.665	9.179	0.000	0.000	0.000
1600	8.398	58.324	52.065	10.015	0.000	0.000	0.000
1700	8.458	58.835	52.448	10.858	0.000	0.000	0.000
1800	8.512	59.320	52.816	11.707	0.000	0.000	0.000
1900	8.559	59.782	53.171	12.560	0.000	0.000	0.000
2000	8.601	60.222	53.513	13.418	0.000	0.000	0.000
2100	8.638	60.642	53.842	14.280	0.000	0.000	0.000
2200	8.672	61.045	54.160	15.146	0.000	0.000	0.000
2300	8.703	61.431	54.468	16.015	0.000	0.000	0.000
2400	8.731	61.802	54.766	16.886	0.000	0.000	0.000
2500	8.756	62.159	55.055	17.761	0.000	0.000	0.000
2600	8.779	62.503	55.335	18.638	0.000	0.000	0.000
2700	8.800	62.835	55.606	19.517	0.000	0.000	0.000
2800	8.820	63.155	55.870	20.398	0.000	0.000	0.000
2900	8.838	63.465	56.127	21.280	0.000	0.000	0.000
3000	8.855	63.765	56.376	22.165	0.000	0.000	0.000
3100	8.871	64.055	56.619	23.051	0.000	0.000	0.000
3200	8.886	64.337	56.856	23.939	0.000	0.000	0.000
3300	8.900	64.611	57.087	24.829	0.000	0.000	0.000
3400	8.914	64.877	57.312	25.719	0.000	0.000	0.000
3500	8.927	65.135	57.532	26.611	0.000	0.000	0.000
3600	8.939	65.387	57.747	27.505	0.000	0.000	0.000
3700	8.950	65.632	57.957	28.399	0.000	0.000	0.000
3800	8.962	65.871	58.162	29.295	0.000	0.000	0.000
3900	8.972	66.104	58.362	30.191	0.000	0.000	0.000
4000	8.983	66.331	58.559	31.089	0.000	0.000	0.000
4100	8.993	66.553	58.751	31.988	0.000	0.000	0.000
4200	9.002	66.770	58.940	32.888	0.000	0.000	0.000
4300	9.012	66.982	59.124	33.788	0.000	0.000	0.000
4400	9.021	67.189	59.305	34.690	0.000	0.000	0.000
4500	9.030	67.392	59.482	35.593	0.000	0.000	0.000

[a] Molecular weight, 28.0134.

Table 6. Thermochemical Data for O_2 (Reference State—Ideal Gas)[a]

T (°K)	C_p (cal/°K mol)	$S°$ (cal/°K mol)	$-(G°-H°_{298})/T$ (cal/°K mol)	$H°-H°_{298}$ (kcal/mol)	ΔH_f (kcal/mol)	$\Delta G°_f$ (kcal/mol)	log K
0	0.000	0.000	∞	−2.075	0.000	0.000	0.000
100	6.958	41.395	55.205	−1.381	0.000	0.000	0.000
200	6.961	46.218	49.643	−0.685	0.000	0.000	0.000
298	7.020	49.004	49.004	0.000	0.000	0.000	0.000
300	7.023	49.047	49.004	0.013	0.000	0.000	0.000
400	7.196	51.091	49.282	0.724	0.000	0.000	0.000
500	7.431	52.722	49.812	1.455	0.000	0.000	0.000
600	7.670	54.098	50.414	2.210	0.000	0.000	0.000
700	7.883	55.297	51.028	2.988	0.000	0.000	0.000
800	8.063	56.361	51.629	3.786	0.000	0.000	0.000
900	8.212	57.320	52.209	4.600	0.000	0.000	0.000
1000	8.336	58.192	52.765	5.427	0.000	0.000	0.000
1100	8.439	58.991	53.295	6.266	0.000	0.000	0.000
1200	8.527	59.729	53.801	7.114	0.000	0.000	0.000
1300	8.604	60.415	54.283	7.971	0.000	0.000	0.000
1400	8.674	61.055	54.744	8.835	0.000	0.000	0.000
1500	8.738	61.656	55.185	9.706	0.000	0.000	0.000
1600	8.800	62.222	55.608	10.583	0.000	0.000	0.000
1700	8.858	62.757	56.013	11.465	0.000	0.000	0.000
1800	8.916	63.265	56.401	12.354	0.000	0.000	0.000
1900	8.973	63.749	56.776	13.249	0.000	0.000	0.000
2000	9.029	64.210	57.136	14.149	0.000	0.000	0.000
2100	9.084	64.652	57.483	15.054	0.000	0.000	0.000
2200	9.139	65.076	57.819	15.966	0.000	0.000	0.000
2300	9.194	65.483	58.143	16.882	0.000	0.000	0.000
2400	9.248	65.876	58.457	17.804	0.000	0.000	0.000
2500	9.301	66.254	58.762	18.732	0.000	0.000	0.000
2600	9.354	66.620	59.057	19.664	0.000	0.000	0.000
2700	9.405	66.974	59.344	20.602	0.000	0.000	0.000
2800	9.455	67.317	59.622	21.545	0.000	0.000	0.000
2900	9.503	67.650	59.893	22.493	0.000	0.000	0.000
3000	9.551	67.973	60.157	23.446	0.000	0.000	0.000
3100	9.596	68.287	60.415	24.403	0.000	0.000	0.000
3200	9.640	68.592	60.665	25.365	0.000	0.000	0.000
3300	9.682	68.889	60.910	26.331	0.000	0.000	0.000
3400	9.723	69.179	61.149	27.302	0.000	0.000	0.000
3500	9.762	69.461	61.383	28.276	0.000	0.000	0.000
3600	9.799	69.737	61.611	29.254	0.000	0.000	0.000
3700	9.835	70.006	61.834	30.236	0.000	0.000	0.000
3800	9.869	70.269	62.053	31.221	0.000	0.000	0.000
3900	9.901	70.525	62.267	32.209	0.000	0.000	0.000
4000	9.932	70.776	62.476	33.201	0.000	0.000	0.000
4100	9.961	71.022	62.682	34.196	0.000	0.000	0.000
4200	9.988	71.262	62.883	35.193	0.000	0.000	0.000
4300	10.015	71.498	63.081	36.193	0.000	0.000	0.000
4400	10.039	71.728	63.275	37.196	0.000	0.000	0.000
4500	10.062	71.954	63.465	38.201	0.000	0.000	0.000

[a] Molecular weight, 31.9988.

Appendix 5
Interaction Coefficients for Liquid Iron at 1600°C

The values of various interaction coefficients are illustrated in the following tables. The coefficients listed analyze the behaviors of solutes in liquid iron at 1600°C. The definitions of these coefficients, their reciprocal relationships, and their conversion relationships may be found in Tables 1–3 of Chapter IX.

Most of the values listed were taken from a compilation by G. K. Sigworth and J. F. Elliott [*Met. Sci.* **8,** 298 (1974)]. The second order coefficients pertaining to the effects of various solutes on nitrogen were deduced from articles by H. Wada and R. D. Pehlke [*Met. Trans.* **8B,** 679 (1977); **9B,** 441 (1978); **10B,** 409 (1979); **11B,** 51 (1980)]. It should be noted that second order interaction coefficients are difficult to measure with precision and that rather large uncertainties are associated with the quoted values.

Table 1. $\epsilon_i^{(j)}$ for Liquid Iron at 1600°C[a]

	j									
i	Ag	Al	As	Au	B	C	Ca	Ce	Co	Cr
Ag	−19	−8.4				11.5				−2
Al	−8.4	5.6				5.3	−7.5			
As						12.9				
Au										
B					2.5	11.7				
C	11.5	5.3	12.9		11.7	6.9	−15.8		1.8	−5.1
Ca		−7.5				−15.8	0			
Ce										
Co						1.8			0.5	−4.6
Cr	−2					−5.1			−4.6	0.0
Cu						4.1				4.0
Ge						2.1				
H		2.0			3.0	3.8		−1.5	0.4	−0.4
La										
Mg						8				
Mn						−2.7				
Mo						−4.0				0.0
N		−2.6	5.2		5.0	5.86			2.6	−10
Nb						−23.7				
Nd										
Ni						2.9	−10.7		1.9	0.0
O		−433		−6.6	−115	−22				−8.5
P						7.0				−6.3
Pb		2.9				4.1			−0.1	4.4
Pd										
Pt										
Rh										
S		4.4	0.9	0.9	6.8	6.5			0.6	−2.2
Sb										
Se										
Si		7.0			9.5	9.7	−10.7			0.0
Sn						19				3.3
Ta						−17.7				
Te										
Ti										11.9
U		7.1								
V						−16.1				
W						−6.6				
Zr										

Table 1. (*continued*)

j									
Cu	Ge	H	La	Mg	Mn	Mo	N	Nb	Nd
		2.0					−2.6		
							5.2		
		3.0					5.0		
4.1	2.1	3.8		8	−2.7	−4.0	5.86	−23.7	
		−1.5							
		0.4					2.6		
4.0		−0.4				0.0	−10		
6.0		0.0					2.2		
	1.9	2.7							
0.0	2.7	1.0	−17		−0.3	0.2		−1.5	−24
		−17							
		−0.3			0.0		−8.1		
		0.1					−4.9		
2.2					−8.1	−4.9	0.8	−26	
		−1.5					−26	−0.7	
		−24							
		−0.1					1.5		
−3.5		−12			−4.7	0.7	4.0	−54	
6.0		1.9			0.0		6.2		
−7.5					−5.2	−0.7			
		1.8							
		2.5							
−2.3	4.0	1.5			−5.9	0.4	1.4	−5.8	
							3.2		
							1.5		
3.6		3.6			0.5		5.9		
		1.5					2.3		
		−17					−29		
							36		
		−3.6					−105		
		−1.5					−19		
		1.4					−3.4		
							−238		

(*continued*)

Table 1. (*continued*)

	j									
i	Ni	O	P	Pb	Pd	Pt	Rh	S	Sb	Se
Ag										
Al		-433		2.9				4.4		
As								0.9		
Au		-6.6						0.9		
B		-115						6.8		
C	2.9	-22	7.0	4.1				6.5		
Ca	-10.7									
Ce										
Co		1.9		-0.1				0.6		
Cr	0.0	-8.5	-6.3	4.4				-2.2		
Cu		-3.5	6.0	-7.5				-2.4		
Ge								4.0		
H	-0.1	-12	1.9		1.8		2.5	1.5		
La										
Mg										
Mn		-4.7	0.0	-5.2				-5.9		
Mo		0.7		-0.7				0.4		
N	1.5	4.0	6.2					1.4	3.2	1.5
Nb		-54						-5.8		
Nd										
Ni	0.2	1.4	0.0	-4.7				-0.1		
O	1.4	-12.5	9.4		-4.9	1.1	8.1	-17	-13	
P	0.0	9.4	8.4	6.6				4.1		
Pb	-4.7		6.6					-42		
Pd		-4.9			0					
Pt		1.1						4.7		
Rh		8.1								
S	-0.1	-17	4.1	-42	4.7			-3.3	0.7	
Sb		-13						0.7		
Se										
Si	1.2	-15	14.2	6.1				7.8		
Sn		-6.6	5.1	27				-3.3		
Ta								-2.4		
Te										
Ti		-118						-14		
U										
V		-63						-3.3		
W		4.2		-2.3				5.1		
Zr								-20		

Table 1. (*continued*)

j								
Si	Sn	Ta	Te	Ti	U	V	W	Zr
7.0					7.1			
9.5								
9.7	19	−17.7				−16.1	−6.5	
−10.7								
0.0	3.3			11.9				
3.6								
	1.5	−17		−3.6		−1.5	1.4	
0.5								
5.9	2.3	−29	36	−105		−19	−3.4	−238
1.2								
−15	−6.6			−118		−63	4.2	
	5.1							
	27						−2.3	
	−3.3	−2.4		−14			5.1	−20
	7.1							
	−0.3							
				2.7				
					9.4			

[a] $\epsilon_i^j = (\partial \ln \gamma_i / \partial X_j)_{X_1 \to 1} = \epsilon_j^i$.

Table 2. $e_i^{(j)} \times 10^2$ for Liquid Iron at 1600°C[a]

	j									
i	Ag	Al	As	Au	B	C	Ca	Ce	Co	Cr
Ag	−4	−8				22				−1
Al	−1.7	4.5				9.1	−4.7			
As						25				
Au										
B					3.8	22				
C	2.8	4.3	4.3		24	12.4	−9.7		0.76	−2.4
Ca		−7.2				−34	−0.2			
Ce										
Co						2.1			0.22	−2.2
Cr	−0.2					−12			−1.9	−0.03
Cu						6.6				1.8
Ge						3				
H		1.3			5	6		0	0.18	−0.22
La										
Mg						15				
Mn						−7				
Mo						−9.7				−0.03
N		−2.8	1.8		9.4	10.3			1.1	−4.6
Nb						−49				
Nd										
Ni						4.2	−6.7			−0.03
O		−390		−0.5	−260	−45		−3	0.8	−4
P						13				−3
Pb		2.1				6.6			0	2
Pd										
Pt										
Rh										
S		3.5	0.4	0.4	13	11			0.26	−1.1
Sb										
Se										
Si		5.8			20	18	−6.7			−0.03
Sn						37				1.5
Ta						−37				
Te										
Ti										5.5
U		5.9								
V						−34				
W						−15				
Zr										

Table 2. (*continued*)

j									
Cu	Ge	H	La	Mg	Mn	Mo	N	Nb	Nd
		24					−5.8		
							7.7		
		49					7.4		
1.6	0.8	67		7	−1.2	−0.83	8.9	−6	
		−60					3.2		
		−14							
1.6		−33				0.18	−18		
2.3		−24					2.6		
	0.7	41							
0.05	1	0	−2.7		−0.14	0.22		−0.23	−3.8
		−430							
		−31			0		−15		
		−20					−10		
0.9					−3.6	−1.1	0	−6.7	
		−61					−47	0	
		−600							
		−25					1.3		
−1.3		−310	−500		−2.1	0.35	5.7	−14	
2.4		21			0		9.4		
−2.8					−2.3	0			
		20							
		37							
−0.84	1.4	12			−2.6	0.27	1	−1.3	
							4.3		
							1.4		
1.4		64			0.2		9		
		12					27		
		−440					−52		
							60		
		−110					−180		
		−59					−35		
		8.8					−7.2		
							−410		

(*continued*)

Table 2. (*continued*)

	j									
i	Ni	O	P	Pb	Pd	Pt	Rh	S	Sb	Se
Ag										
Al		−660		0.65				3.0		
As								0.37		
Au		−11						0.37		
B		−180						4.8		
C	1.2	−34	5.1	0.79				4.6		
Ca	−4.4									
Ce										
Co		1.8		0.3				0.11		
Cr	0.02	−14	−5.3	0.83				−2.0		
Cu		−6.5	4.4	−0.56				−2.1		
Ge								2.7		
H	0	−19	1.1		0.62		0.63	0.8		
La										
Mg										
Mn		−8.3	−0.35	−0.29				−4.8		
Mo		−0.07		0.23				−0.05		
N	0.63	5	4.5					0.7	0.88	0.6
Nb		−83						−4.7		
Nd										
Ni	0.09	1	−0.35	−0.23				−0.37		
O	0.6	−20	7		−0.9	0.45	1.4	−13.3	−2.3	
P	0.02	13	6.2	1.1				2.8		
Pb	−1.9		4.8					−32		
Pd		−8.4			0.2					
Pt		0.63						3.2		
Rh		11								
S	0	−27	29	−4.6		0.89		−2.8	0.37	
Sb		−20						0.19		
Se										
Si	0.5	−23	11	1				5.6		
Sn		−11	3.6	3.5				−2.8		
Ta								−2.1		
Te										
Ti		−180						−11		
U										
V		−97						−2.8		
W		−5.2		0.05				3.5		
Zr								−16		

Table 2. (*continued*)

j								
Si	Sn	Ta	Te	Ti	U	V	W	Zr
0.56					1.1			
7.8								
8	4.1	−2.1				−7.7	−0.56	
−9.7								
−0.43	0.9			5.9				
2.7								
2.7	0.53	−2		−1.9		−0.74	0.48	
0								
4.7	0.7	−3.6	7	−53		−9.3	−0.15	−63
0.57								
−13.1	−1.11			−60		−30	−0.85	−300
12	1.3							
4.8	5.7						0	
6.3	−0.44	−0.02		−7.2		−1.6	0.97	−5.2
11	1.7					2.5		
5.7	0.16							
				1.3				
					1.3			
4.2						1.5		

[a] $e_i^{(j)} = (\partial \log f_i/\partial \% j)_{\%1\rightarrow 100} = (M_i/M_j)e_j^{(i)} + 0.434 \times 10^{-2}(M_j - M_i)/M_j$.

Table 3. γ_i^{∞} and Changes of Standard States[a] for Liquid Iron at 1600°C

i	γ_i^{∞}	$\Delta G°$ (cal) for M = $\underline{M}$	$\Delta G°$ (cal) for M = $\underline{M}(\%)$
Ag (l)	200	19,700	19,700 − 10.46T
Al (l)	0.029	−15,100 + 1.03T	−15,100 − 6.67T
B (s)	0.022	−15,600 + 0.71T	−15,600 − 5.15T
C (gr)	0.70	5,400 − 3.6T	5,400 − 9.7T
Ca (v)	2240	−9,430 + 20.3T	−9,430 + 11.87T
Co (l)	1.07	240	240 − 9.26T
Cr (l)	1.0	0	− 9.01T
Cr (s)	1.14	4,600 − 2.19T	4,600 − 11.20T
Cu (l)	8.6	8,000	8,000 − 9.41T
$\frac{1}{2}H_2$ (g)	—	—	8,720 + 7.28T
Mn (l)	1.3	976	976 − 9.12T
Mo (l)	1	0	− 10.23T
Mo (s)	1.86	6,600 − 2.29T	6,600 − 12.52T
$\frac{1}{2}N_2$ (g)	—	—	860 + 5.71T
Nb (l)	1.0	0	− 10.2T
Nb (s)	1.4	5,500 − 2.3T	5,500 − 12.5T
Ni (l)	0.66	−5,000 + 1.80T	−5,000 − 7.42T
$\frac{1}{2}O_2$ (g)	—	—	−28,000 − 0.69T
$\frac{1}{2}P_2$ (g)	—	—	−29,200 − 4.6T
Pb (l)	1400	50,800 − 12.7T	50,800 − 25.4T
$\frac{1}{2}S_2$ (g)	—	—	−32,280 + 5.6T
Si (l)	0.0013	−31,430 + 3.64T	−31,430 − 4.12T
Sn (l)	2.8	3,820	3,820 − 10.62T
Ti (l)	0.037	−11,100	−11,000 − 8.85T
Ti (s)	0.038	−7,440 − 1.90T	−7,440 − 10.75T
U (l)	0.027	−13,400	−13,400 − 12.0T
V (l)	0.08	−10,100 + 0.37T	−10,100 − 8.6T
V (s)	0.1	−4,950 − 1.93T	−4,950 − 10.9T
W (l)	1	0	− 11.5T
W (s)	1.2	+7,500 − 3.65T	+7,500 − 15.2T
Zr (l)	0.037	−12,200	−12,200 − 10.13T
Zr (s)	0.043	−8,300 − 1.82T	−8,300 − 11.95T

[a] See Section VIII.5.1.

Table 4. $\eta_i^{(j)}$ (kcal) for Liquid Iron at 1600°C[a]

	j														
i	Al	B	C	Cr	Mn	Mo	N	Nb	Ni	O	S	Si	Ta	Ti	V
Al	14						190			−4550					
B							81.9								
C			15				21.5					37.3			
Cr							−69.8				−40.1				
Mn							−60								
Mo							−25.8								
N	190	81.9	21.5	−69.8	−60	−25.8		−197	4			31.2	−151	−1600	−146
Nb							−197								
Ni							4								
O	−4550									−228					
S				−40.1							61.2				
Si			37.3				31.2					8			
Ta							−151								
Ti							−1600								
V							−146								

[a] $\eta_i^{(j)} = (\partial H_i^E/\partial X_j)_{X_1\to 1} = R\,\partial\epsilon_i^{(j)}/\partial(1/T) = \eta_j^{(i)}$.

Table 5. $h_i^{(j)}$ (cal) for Liquid Iron at 1600°C[a]

	j														
i	Al	B	C	Cr	Mn	Mo	N	Nb	Ni	O	S	Si	Ta	Ti	V
Al	290						7,580			−159,000					
B							3,270								
C			700				860					740			
Cr							−2,780				−698				
Mn							−2,390								
Mo							−1,030								
N	3,930	4230	1000	−750	−610	−150		−1180	38			620	−465	−18,650	−1600
Nb							−7,850								
Ni							160								
O	−94,200									−7,960					
S				−430							1065				
Si			1730				1,240					160			
Ta							−6,000								
Ti							−63,800								
V							−5,820								

[a] $h_i^{(j)} = (\partial \mathcal{H}_i^E/\partial \% j)_{\%1\to 100} = 2.303 \times R\, \partial e_i^{(j)}/\partial(1/T) = (M_i/M_j)h_j^{(i)}$.

Table 6. $\rho_i^{(j)}$ for Liquid Iron at 1600°C[a]

	j								
i	Al	B	C	Cr	P	S	Si	Ti	V
Al	−2.8		0				0		
C			11.6	−0.4	12		8.8		0
Ca	0		0				0		
Cr				0	15		0		
Cu							0		
N		3.7	11.7	5	2.7	0.5	2.8	−15	−1.7
O	8600	−93	−17	−0.6	4.1	−7.4	−7.4	500	−5.5
P				16	−3.1		0		
S	6.9	12	11		6.2	−8.6	14	0	0
Si							−5.5		
Ti				0.8				−1.4	
V							0		−1.6

[a] $\rho_i^{(j)} = \frac{1}{2}(\partial^2 \ln \gamma_i/\partial X_j^2)_{X_1\to 1}$.

Table 7. $r_i^{(j)} \times 10^4$ for Liquid Iron at 1600°C[a]

	j								
i	Al	B	C	Cr	P	S	Si	Ti	V
Al	−10		−40				−6		
C			61	0	41		7		1
Ca	7		120				9		
Cr				0	25		0		
Cu							−3		
N		0	70	2.9	0	0	0	0	0
O	17,000	0	0	0	0	0	0	310	0
P				8	−10		−10		
S	9	74	58		6	−9	17	1	0
Si							−21		
Ti				0				−10	
V							−6		−1

[a] $r_i^{(j)} = \frac{1}{2}[\partial^2 \log f_i/\partial(\% j)^2]_{\%1\to 100}$.

Table 8. $\rho_N^{(i,j)}$ for Liquid Iron at 1600°C[a]

	j						
i	Cr	Mn	Mo	Nb	Ni	Si	Ta
Cr	5	13	12	53	−0.5	−21	57
Mn	13	−18			18		
Mo	12		−13	14	6	−14	
Nb	53		14		−16	−32	
Ni	−0.5	18	6	−16	2	−3.4	33
Si	−21		−14	−32	−3.4	2.8	
Ta	57				33		120

[a] $\rho_N^{(i,j)} = (\partial^2 \ln \gamma_N/\partial X_i \partial X_j)_{X_1 \to 1}$ for $i \neq j$; for $i = j$, $\rho_N^{(i)} = \frac{1}{2}(\partial^2 \ln \gamma_N/\partial X_i^2)_{X_1 \to 1}$.

Table 9. $r_N^{(i,j)} \times 10^4$ for Liquid Iron at 1600°C[a]

	j						
i	Cr	Mn	Mo	Nb	Ni	Si	Ta
Cr	2.9	6.5	1.5	13.6	−0.5	−14.9	5.2
Mn	6.5	−7.9			7.5		
Mo	1.5		−2.4	−1.1	1.6	−3.7	
Nb	13.6		−1.1		−4.1	−8	
Ni	−0.5	7.5	1.6	−4.1	0.7	−3.2	4.5
Si	−14.9		−3.7	−8	−3.2		
Ta	5.2				4.5		2.3

[a] $r_N^{(i,j)} = (\partial^2 \log f_N/\partial \%i\, \partial \%j)_{\%1 \to 100}$ for $i \neq j$; for $i = j$, $r_N^{(i)} = \frac{1}{2}(\partial^2 \log f_N/\partial \%i^2)_{\%1 \to 100}$.

Answers to Problems

Chapter I

1. **a.** 92,210 cal.
b. 709 cal.
c. 77.4°C.
d. 448 cal/sec.

2.
$$\frac{\partial^2 z}{\partial x\,\partial y} = \frac{\partial^2 z}{\partial y\,\partial x} = -\frac{4xy^3}{(x^2+y^4)^2}, \qquad z = -\tfrac{1}{2}\ln\frac{y^2}{x^2+y^4}$$

3. Along the path $y = x$, $Dz_1 = \frac{7}{4}$; along the path $y = x^2$, $Dz_1 = \frac{12}{7}$; dz_1 is not a perfect differential.
dz_2 is a perfect differential. Along both paths, $Dz_2 = 2$. ($z_2 = x^3y + xy^2$.)

4. $C + O_2 = CO_2$; $\Delta H^\circ_{298} = -94.05$ kcal. On the basis of one mole of C, the flue gases consist of 4.14 mol N_2, 0.1 mol O_2, and 1 mol CO_2. The adiabatic flame temperature is determined by

$$4.14(H^\circ_T - H^\circ_{298})_{N_2} + 0.1(H^\circ_T - H^\circ_{298})_{O_2} + 1(H^\circ_T - H^\circ_{298})_{CO_2} = 94.05 \text{ kcal}$$

which yields $T \simeq 2300°K$.

5. **a.**
$$CH_4 + 2O_2 = CO_2 + 2H_2O$$
$$\Delta H^\circ_{298} = -94.05 - 2 \times 57.80 + 17.89$$
$$= -191.76 \text{ kcal.}$$

b.
$$C_2H_6 + \tfrac{7}{2}O_2 = 2CO_2 + 3H_2O$$
$$\Delta H^\circ_{298} = -188.10 - 173.4 + 20.24$$
$$= -341.26 \text{ kcal}$$

c. $(0.83 \times 191.76) + (0.16 \times 341.26) = 213.76$ kcal/mol.

6. **a.** The combustion air contains $1.1(0.83 \times 2 + 0.16 \times \frac{7}{2}) = 2.442$ mol of O_2

and $2.442 \times \frac{79}{21} = 9.186$ mol of N_2. The flue gases are constituted of

CO_2:	$0.83 + 0.32 = 1.15$ mol	(9.1%)
H_2O:	$1.66 + 0.48 = 2.14$ mol	(16.8%)
N_2:	$9.186 + 0.010 = 9.196$ mol	(72.4%)
O_2:	0.222 mol	(1.7%)

The sensible heat in the flue gases is equal to 140.05 kcal.

b. Furnace requirements: 600,000 kcal/hr (for the glass) + 100,000 kcal/hr (for heat losses) = 700,000 kcal/hr. Available heat: 213.76 (from Problem 5c) − 140.05 = 73.71 kcal/mol. Fuel consumption: 9497 mol/hr or 212.8 m^3/hr.

c. Sensible heat in air of combustion: 42.32 kcal. Available heat: 116.03 kcal/mol natural gas. Fuel consumption: 135.2 m^3/hr, or a decrease of 36.5%.

d. Available heat: $73.71 + 117.83r$ kcal/mol of fuel. Consumption: $9497/(1 + 1.60r)$ mol/hr. Cost of fuel: $212.9\alpha/(1 + 1.60r)$. Total cost: $[212.9\alpha/(1 + 1.60r)] + 15\alpha(1 + r)/(1 - r)$. Optimum for $\alpha = 0.477$. $T \simeq 955°K$. Minimum cost: 163.2α/hr.

7. a. Work performed:

$$(-đW) = đq - dE \leq T\,dS - dE; \qquad (-đW)_{T,V} = -(dA)_{T,V}$$

b. Work performed, other than expansion work [1, pp. 104–106]:

$$(-đW') = (-đW) - P\,dV = T\,dS - P\,dV - dE$$

$$(-đW')_{T,P} = (-dG)_{T,P}$$

8. With no other work than expansion forces, $dZ = dE + P\,dV - T\,dS$ [equation (52)] and $dZ = T\,d_iS = đq'$.

9. $(\partial G/\partial T)_V = -S + (\alpha V/\beta)$.

10. $đq = dE + P\,dV$; for a perfect gas, E is a function of T only, and since $C_v = (\partial E/\partial T)_V$, $dE = C_v\,dT$, and $đq = C_v\,dT + P\,dV$. Application of criterion (8) shows that $đq$ is not a perfect differential: $(\partial C_v/\partial V)_T = 0$ and $(\partial P/\partial T)_V = R/V \neq 0$.

$$\frac{đq}{T} = \frac{C_v}{T}dT + \frac{P}{T}dV; \qquad \left(\frac{\partial(C_v/T)}{\partial V}\right)_T = 0; \quad \left(\frac{\partial(P/T)}{\partial T}\right)_V = \left(\frac{\partial(R/V)}{\partial T}\right)_V = 0$$

Thus $đq/T$ is a perfect differential.

11. $dE = T\,dS - P\,dV$; condition $(\partial E/\partial P)_T = 0$ yields $T(\partial S/\partial P)_T - P(\partial V/\partial T)_P = 0$, or $-T(\partial V/\partial T)_P = P(\partial V/\partial P)_T$. Noting that $(\partial P/\partial T)_V = -(\partial V/\partial T)_P/(\partial V/\partial P)_T$ leads to equation (80).

12.
$$(\partial P/\partial T)_V = \frac{P}{T} + \frac{a}{V^2}, \qquad \left(\frac{\partial E}{\partial V}\right)_T = \frac{a}{V^2}, \qquad \beta = \frac{V - b}{PV - a/V + 2ab/V^2}$$

For $a = 0$, $\alpha = (1/T)(1 - b/V)$; $(\partial H/\partial P)_T = b$; $C_P - C_v = R$. With $b = 22$ cm^3, $DH = 53$ cal/mol.

13.
$$DH = \left(\int_{1250}^{1600} C_p\, dT\right)_{P=1\text{ atm}} + \left(\int_{1}^{1000} V(1-\alpha T)\, dP\right)_{T=1600°K}$$

along a path (1250°K, 1 atm)→(1600°K, 1 atm)→(1600°K, 1000 atm).

$$DH = \{5.80 \times 350 + \tfrac{1}{2}(1.98 \times 10^{-3})[(1600)^2 - (1250)^2]\}$$
$$+ \left\{7.31[1 + \alpha(1600 - 1250)](1 - \alpha 1600) \int_1^{1000} [1 - \beta(P-1)]\, dP\right\}$$
$$= 3017 + 163 = 3180 \text{ cal/mol}$$

Similarly, $DS = 2.125 - 0.011 = 2.114$ cal/°K mol. $C_p - C_v = \alpha^2 VT/\beta = 0.798$ and $C_v = 7.477$ cal/°K mol.

14. $\Delta S_{55°K}^{\circ(\alpha\to\beta)} = 2.00$ cal/°K; $\Delta H_{55°K}^{\circ(\alpha\to\beta)} = 110$ cal. $\Delta S_{0°K}^{\circ(\alpha\to\beta)} = 0$ cal/°K; $\Delta H_{0°K}^{\circ(\alpha\to\beta)} = 28$ cal.

15.
$$\left(\frac{\partial\beta}{\partial T}\right)_P = \frac{1}{V^2}\left(\frac{\partial V}{\partial T}\right)_P\left(\frac{\partial V}{\partial P}\right)_T - \frac{1}{V}\frac{\partial^2 V}{\partial P\,\partial T}$$

and with $(\partial V/\partial T)_{P,T\to 0°K} = 0$ [equation (98)],

$$\left(\frac{\partial\beta}{\partial T}\right)_{P,T\to 0°K} = -\frac{1}{V}\left[\frac{\partial}{\partial P}\left(\frac{\partial V}{\partial T}\right)_{P,T\to 0°K}\right] = 0$$

16. $C_v = \frac{12}{5} R\pi^4 (T/\theta_D)^3$.

17.
$$H_T^\circ - H_{298}^\circ = 66.42(T - 298) + 6.85[T^2 - (298)^2]10^{-3}$$
$$+ 16.89(1/T - 1/298)10^5$$
$$= 66.42T + 6.85 \times 10^{-3}T^2 + 16.89 \times 10^5 T^{-1} - 26{,}069$$
$$H_{1000}^\circ - H_{298}^\circ = 48{,}890 \text{ cal/mol}$$
$$S_T^\circ - S_{298}^\circ = 66.42 \ln T/298 + 13.7(T - 298)10^{-3}$$
$$+ 8.445[1/T^2 - 1/(298)^2]10^5$$
$$= 66.42 \ln T + 13.7 \times 10^{-3}T + 8.445 \times 10^{-5}T^{-2} - 391.99$$
$$S_{1000}^\circ - S_{298}^\circ = 81.36 \text{ cal/°K mol}$$

18. **a.** $\Delta H_{298}^\circ = 10{,}514$ cal/mol; $\Delta S_{298}^\circ = 28.42$ cal/°K mol.
b. $P = 3.165 \times 10^{-2}$ atm or 24.05 mm Hg.
c. $T = 294$°K or 21°C.
d.
$$\int_P^1 (\partial G^g/\partial P)_{T=298}\, dP - \int_P^1 (\partial G^l/\partial P)_{T=298}\, dP$$
$$= (G^g - G^l)_{T=298, P=1\text{ atm}}$$

(Fig. P.1) or

$$-RT \ln P + V^l(P - 1) = \Delta H_{298}^\circ - 298\,\Delta S_{298}^\circ$$
$$-1.987 \times 298 \ln P + (\sim 0) = 10{,}514 - 298 \times 28.42$$

yields $\ln P = -3.453$ or $P = 3.165 \times 10^{-3}$ atm.

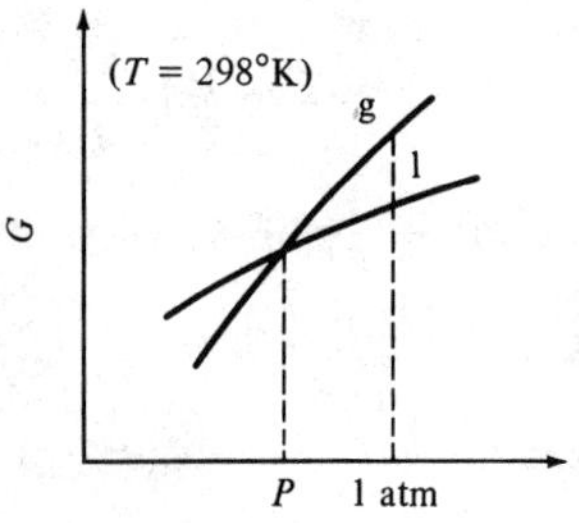

Figure P1. Problem I.18d.

19. The Clausius–Clapeyron equation yields $dP = (\Delta H/\Delta V)(dT/T)$. Integrating, we get

$$P - 1 = -\frac{79.7}{0.09 \times 0.0242} \ln \frac{270.65}{273.15}$$

from which we deduce $P = 337.5$ atm or 4961 psi. (In a well-sharpened skate, the bottom of the blade is in the shape of an inverted U. Thus, the weight of the skater develops large local pressures which melt the ice, form a thin layer of lubricating water, and allow an easy glide.)

20.
$$\ln P \simeq \frac{\Delta H_{\text{vap}}}{R}\left(\frac{1}{T_b} - \frac{1}{T}\right) \simeq \frac{\Delta S_{\text{vap}}}{R}\left(1 - \frac{T_b}{T}\right)$$

Invoking Trouton's rule, $\ln P \simeq 11(1 - 630/298)$, which yields $P \simeq 5 \times 10^{-6}$ atm or 0.004 mm Hg.

21. Using again

$$\ln P \simeq \frac{\Delta H_{\text{vap}}}{R}\left(\frac{1}{T_b} - \frac{1}{T}\right)$$

we have

$$\ln 10^{-4} = \frac{28{,}000}{1.987}\left(\frac{1}{T_b} - \frac{1}{673}\right)$$

which yields $T_b \simeq 1200$°K. (The experimental value is listed as 1180°K.)

22. ΔH_{vap} is not a sensitive function of P, whereas ΔS_{vap} is and cannot be assumed constant.

23. **a.** $V_c = 3b$; $P_c = a/27b^2$; $T_c = 8a/27bR$.
b. $V_c = 128$ cm^3/mol, $P_c = 73.1$ atm, $T_c = 30.8$°C instead of 94.9 cm^3/mol, 72.8 atm, 31.1°C (Fig. 19).

Chapter II

1. $z'_x = e^{y/x}(x + y)[2 - (x + y)(y/x^2)] + y$. $z'_y = e^{y/x}(x + y)[3 + (y/x)] + x$.

2. z_1 is a homogeneous function of degree 0. Therefore, it depends on x and y through the variables $x/(ax + by) = u$ and $y/(ax + by) = v$. Since $au + bv = 1$,

z_1 depends on only a single independent variable. For example, for $a = b = 1$, $z_1 = (1 - 2u)\ \ln[(1 - u)/u] + (1 - 3u + 3u^2)/(1 - 2u + 2u^2)$. z_2 is a homogeneous function of degree 1, and therefore cannot be expressed as a function of only u and v. (Note that v is a function of u, since $v = (1 - au)/b$.)

3. **a.**

$$Y_m(\lambda n_1, \ldots, \lambda n_r) = \frac{Y(\lambda n_1, \ldots, \lambda n_r)}{\lambda n_1 + \cdots + \lambda n_r} = \frac{\lambda Y(n_1, \ldots, n_r)}{\lambda n} = \frac{Y}{n}$$

Therefore, by definition, Y_m is a homogeneous function of degree 0.

b. The definition of a homogeneous function may again be applied.

4. Change variables: $\mu_2(n_1, n_2, n_3) \rightarrow \mu_2(n, n_2/n, n_3/n_1)$.

$$\left(\frac{\partial G}{\partial n_2}\right)_{n_1,n_3} = \mu_2 = \left(\frac{\partial G}{\partial n}\right)_{X_2,X_3/X_1}\left(\frac{\partial n}{\partial n_2}\right)_{n_1,n_3} + \left(\frac{\partial G}{\partial X_2}\right)_{X_3/X_1,n}\left(\frac{\partial X_2}{\partial n_2}\right)_{n_1,n_3} + 0$$

$$= G_{\rm m}\,1 + \frac{1 - X_2}{n}\left(\frac{\partial G}{\partial X_2}\right)_{X_3/X_1,n} = G_{\rm m} + (1 - X_2)\left(\frac{\partial G_{\rm m}}{\partial X_2}\right)_{X_3/X_1}$$

For r components $\mu_2 = G_{\rm m} + (1 - X_2)(\partial G_{\rm m}/\partial X_2)_{X_3/X_1,\ldots,X_r/X_1}$.

5. $dH/dX_2 = (H_2^\circ - H_1^\circ) - 0.1$; $\overline{H}_2 = H_2^\circ + 0.05$.

6. Let Al ≡ 1, Zn ≡ 2. $\mu_2 = \mu_2^\circ + RT \ln X_2 + X_1^2(3150 - 1700X_2)[1 - (T/4000)]$; $\overline{H}_2 = H_2^\circ + X_1^2(3150 - 1700X_2)$; $\overline{S}_2 = S_2^\circ - R \ln X_2 + X_1^2(3150 - 1700X_2)/4000$.

7. $\overline{V}_{\rm C} = 3.73 + 0.15X_{\rm Fe}^2 = 3.86\ {\rm cm^3/mol}$. $D\mu_{\rm C} = \int_1^{2000} \overline{V}_{\rm C}\ dP = 187$ cal/mol. (The effect of β is negligible.)

8. $\overline{Y}_2 = \alpha X_1 X_3(1 - 2X_2) = 0$.

9. $\overline{V}_{\rm A} = 7 - \frac{1}{3} = \frac{20}{3}$; $\overline{V}_{\rm B} = 10 - \frac{1}{3} = \frac{29}{3}$; $\overline{V}_{\rm C} = 12 + \frac{1}{3} = \frac{37}{3}$ (in cm^3/mol).

10. The curve is obtained by forming $G_{\rm m} = X_{\rm Fe}\mu_{\rm Fe} + X_{\rm C}\mu_{\rm C}$. Draw a tangent to that curve from the point $G = G_{\rm C}^{\circ(\rm gr)} = 0$, $X_{\rm C} = 1$. The coordinate of the point of contact yields the solubility of graphite, $X_{\rm C} \simeq 0.21$.

11. **a.** $m = 3$ (Fe, C, argon). $\varphi_{\max} = 3$ (two solid phases and one gas).

b. With pure oxygen, $m = 2$, $\vartheta = 2 - \varphi$, $\varphi_{\max} = 2$. With air, $\varphi_{\max} = 3$.

Chapter III

1. After decomposition $A_{\rm dec} = n'A'(V'_{\rm m}) + n''A''(V''_{\rm m})$, where n' and n'' are the numbers of moles of the fluid with molar volumes $V'_{\rm m}$ and $V''_{\rm m}$. The criterion of stability is $(DA)_{V,T} = (A_{\rm dec} - A)_{V,T} \geq 0$. With the Taylor series

$$A'(V'_{\rm m}) = A(V_{\rm m}) + (\partial A/\partial V_{\rm m})_T(V'_{\rm m} - V_{\rm m}) + \tfrac{1}{2}(\partial^2 A/\partial V_{\rm m}^2)_T(V'_{\rm m} - V_{\rm m})^2 + \cdots$$

and a similar one for A'', this criterion becomes

$$DA = \tfrac{1}{2}(\partial^2 A/\partial V_{\rm m}^2)_T[n'(V'_{\rm m} - V_{\rm m})^2 + n''(V''_{\rm m} - V_{\rm m})^2] + \cdots \geq 0$$

or $(\partial^2 A/\partial V_{\rm m}^2)_T = -(\partial P/\partial V)_T \geq 0$. Consequently, for stability $\beta \geq 0$.

2. The equation of the spinodal (50) may be written $x^2 = (T_c - T)/T_c$. A series development of the miscibility gap's equation (52b) yields $x'^2 + O(x'^4) = 3(T_c - T)/T$. Consequently, $(x'/x)_{T_c \to T} = 3$, or $(AB/PQ)_{T \to T_c} = \sqrt{3}$.

3. From $\partial^2 G_m/\partial X_2^2 = \partial^3 G_m/\partial X_2^3 = 0$, we obtain $X_2^c = 0.398$, $T_c = 610°K$.

4. $y/(T_c - T) \to 1.7 \times 10^{-4}$ mol/°K, or 1.7 at.%/100°K.

5.
$$G_{11} = \frac{X_2}{n}\left(\frac{RT}{X_1} - 2\Omega X_2\right)$$
$$G_{12} = G_{21} = \frac{1}{n}(-RT + 2\Omega X_1 X_2)$$
$$G_{22} = \frac{X_1}{n}\left(\frac{RT}{X_2} - 2\Omega X_1\right)$$

9. $DG/n^\circ = -RT(X_1'^\alpha \ln X_1'^\alpha/X_1^\alpha + X_2'^\alpha \ln X_2'^\alpha/X_2^\alpha) + \Omega^\alpha(X_2'^\alpha - X_2^\alpha)^2$.

Chapter IV

1. $\frac{2}{3}\pi a^3$.

2. **a.** At 273°K, $kT/u^* = 2.31$; $B_2/r^{*3} = -0.8$; $B_2 = -46.7$ Å^3; $\ln(f/P) = -1.254$; $f = 285$.
b. At 1800°K, $kT/u^* = 15.25$; $B_2/r^{*3} = 0.8$; $B_2 = 46.7$ Å^3; $\ln(f/P) = 0.190$; $f = 1210$.

3.
$$\ln\frac{f}{f'} = \int_{P'}^{P} \frac{V}{RT}\,dP = \int_{P'}^{P} \frac{V}{RT}\left(\frac{\partial P}{\partial V}\right)_T dV = \int_{P'}^{P}\left[\frac{1}{RT}\frac{2a}{V^2} - \frac{V}{(V-b)^2}\right] dV$$
$$= -\frac{2a}{RTV} + \frac{b}{V-b} - [\ln(V-b)]_{P'}^{P} \qquad (\text{for } P' \to 0)$$
$$\ln f = -\frac{2a}{RTV} + \frac{b}{V-b} - \ln(V-b) + \ln[P'(V'-b)]$$
$$\ln\frac{f}{P} = -\ln\left[\frac{P(V-b)}{RT}\right] - \frac{2a}{RTV} + \frac{b}{V-b}$$
For $P = 1000$ atm, $T = 273°K$, we find $V = 44.93$ cm^3 and $\ln(f/P) = 0.1759$, $f = 1192$.

4. At 1000 atm and 298°K, $V = 56.05$ cm^3 for pure N_2 and 57.31 cm^3 for pure CH_4. Using the results of Problem 3, $f = 1906$ for pure N_2, $f = 1295$ for pure CH_4. Assuming that the mixture is an ideal solution of imperfect gases, we obtain $f_{N_2} = 1334$ and $f_{CH_4} = 388.5$.

5. Using the results of Problem I.23a we find $V_c = 128.4$ cm^3, $P_c = 45.6$ atm, and $T_c = 190°K$.

6. $p_B = a_B^{(l)}P_B^{(l)} = a_B^{(s)}P_B^{(s)}$; $0.65 \times 2.5 \times 10^{-3}$ atm $= 0.40P_B^{(s)}$ and $P_B^{(s)} = 4.06 \times 10^{-3}$ atm.

7. $\partial\mu_2/\partial X_2 = \partial^2\mu_2/\partial X_2^2 = 0$; $\partial^3\mu_2/\partial X_2^3 \geq 0$. These conditions yield

$$\frac{\partial \ln \gamma_2}{\partial X_2} = -\frac{1}{X_2}; \qquad \frac{\partial^2 \ln \gamma_2}{\partial X_2^2} = \frac{1}{X_2^2}; \qquad \frac{\partial^3 \ln \gamma_2}{\partial X_2^3} \geq -\frac{2}{X_2^3}$$

8. From $a_i = p_i/P_i = \gamma_i X_i$ and $p_1/p_2 = Y_1/Y_2$, we deduce

$$Y_{Ag} = \frac{\gamma_{Ag}X_{Ag}(P_{Ag}/P_{Cu})}{\gamma_{Cu}X_{Cu} + \gamma_{Ag}X_{Ag}(P_{Ag}/P_{Cu})}$$

Using L'Hospital's rule,

$$\left(\frac{dY_{Ag}}{dX_{Ag}}\right)_{X_{Ag}\to 0} = \left(\frac{Y_{Ag}}{X_{Ag}}\right)_{X_{Ag}\to 0} = e^{1.14}\,\frac{3.6 \times 10^{-4}}{8.6 \times 10^{-6}} = 131$$

Similarly,

$$\left(\frac{dY_{Ag}}{dX_{Ag}}\right)_{X_{Ag}\to 1} = \left(\frac{d(1-Y_{Ag})}{d(1-X_{Ag})}\right)_{X_{Ag}\to 1} = \left(\frac{dY_{Cu}}{dX_{Cu}}\right)_{X_{Cu}\to 0}$$

$$= e^{1.14}\,\frac{8.6 \times 10^{-6}}{3.6 \times 10^{-4}} = 0.075$$

9. At $P = 1$ atm, $\ln \gamma_B = -1.8$, $\gamma_B = 0.165$, $a_B = 0.041$. $\ln P_B \simeq (\Delta H_{vap}/R) \times (1/T_b - 1/T)$ and $P_B \simeq 6.1 \times 10^{-8}$ atm. Consequently, at $P = 1$ atm, $p_B = a_B P_B = 0.41 \times 6.1 \times 10^{-8} = 2.5 \times 10^{-9}$ atm.

At $P = 8000$ atm, $(\partial \ln a_B/\partial P)_T = (\overline{V}_B - V_B^\circ)/RT$ and

$$\ln\frac{(a_B)_{8000\text{ atm}}}{(a_B)_{1\text{ atm}}} = \int_1^{8000} \frac{\overline{V}_B - V_B^\circ}{RT}\,dP = \frac{-2.25 \times 7999 \times 0.0242}{1.987 \times 1000} = -0.219$$

Thus $(a_B)_{8000\text{ atm}} = 0.803 \times 0.041 = 0.033$.

With no information on the fugacity of B in the gas phase, we may assume that it behaves as a perfect gas. It follows then that $p_B \simeq 0.033 \times 6.1 \times 10^{-8} = 2.0 \times 10^{-9}$ atm.

Chapter V

1. $T = 417$°K.

2. 22.9% CH_4, 77.1% H_2.

3. 2037°K.

4. $\Delta G^\circ = \Delta H^\circ - T\Delta S^\circ = 29{,}900 - 21.46T$ cal.

5. The reaction taking place may be written $SO_2 + 2H_2S = 2H_2O + 3S(l)$. On the basis of 1 mol SO_2 and 1.5 mol H_2S we obtain

$$K = 5.5 \times 10^3 = \frac{p_{H_2O}^2}{p_{SO_2}p_{H_2S}^2} = \frac{4\lambda^2(2.5 - \lambda)}{(1-\lambda)(1.5 - 2\lambda)}$$

which yields $\lambda = 0.725$. The composition of the flue gases is 15.5% SO_2, 2.8% H_2S, and 81.7% H_2O. The production rate is 31.2 tonnes/hr and the heat generated is 9.6×10^6 kcal/hr (exothermic reaction).

6. $K = 0.664$. On the basis of 1 mol SO_2, x mol SO_3, and the progress variable λ, we have $p_{SO_3}/p_{SO_2} = (x - \lambda)/(1 + \lambda) = 0.664 \times (0.05)^{1/2}$ and $p_{O_2} = 0.05 = \frac{1}{2}\lambda \times 1.2/(1 + x + \lambda/2)$. From these two equations, we deduce $\lambda = 0.0983$ and $x = 0.261$.

7. $p_{H_2O} < 1.8 \times 10^{-8}$ atm. This value corresponds to the formation of TiO.

8. For $MnO + H_2 = Mn + H_2O$, $K = 5.87 \times 10^{-7} = p_{H_2O}/p_{H_2}$. At the rate of 6.425 mol H_2/day, only 37.7×10^{-7} mol MnO/day are transformed. It would take about 1025 years to reduce 100 g of MnO.

9. See "A thermodynamic analysis of the deposition of SnO_2 thin films from the vapor phase," by G. N. Advani, A. Jordan, C. Lupis, and R. Longini [*Thin Solid Films* **62,** 361 (1979)].

10. Two independent reactions: $C + O_2 = CO_2$ and $CO_2 = CO + \frac{1}{2}O_2$. Since all the carbon will be burnt (excess of oxygen), on the basis of 1 mol C, 1.26 mol O_2 (6 mol air), the progress variable λ_1 associated to the first reaction is equal to 1. The second reaction yields $\lambda_2 = 0.14$, from which we deduce the composition of the flue gases: 14.1% CO_2, 2.3% CO, 5.4% O_2, and 78.1% N_2. $DH = -85.84$ kcal/mol C.

11. **a.** $X_B = 0.80$; $X_{BO} = 0.08$. Metallic yield of A, 23.3%; of B, 93.2%.
b. Decrease p_{CO} (sweeping with an inert gas or increasing pumping capacity); increase T.
c. The phase rule shows that all five phases cannot coexist. BO is totally reduced. $X_A = 0.217$. Metallic yield of B, 100%; of A, 27.7%.

12. $p_{CO} = 0.841$; $p_{SiO} = 4.28 \times 10^{-3}$. SiC produced, 1.15 g. C reacted, 1.04 g.

13. Two independent reactions, e.g., $2NH_3 = N_2 + 3H_2$ and $CO_2 + H_2 = CO + H_2O$. On the basis of 1 mol NH_3, $\lambda_1 = 0.15$ and $\lambda_2 = 0.016$. The composition of the gas phase at equilibrium is 21.21% NH_3, 34.85% N_2, 29.82% CO_2, 13.15% H_2, 0.48% CO, and 0.48% H_2O. $DG = -412$ cal.

14. B. **a.** 20.4% O_2, 76.6% N_2, 3% H_2O.
b. Seven unknowns: n_{O_2}, n_{N_2}, n_C, n_{CO}, n_{CO_2}, n_{H_2}, n_{H_2O}. Four mass balances on the elements C, O, N, H. Hence, three independent reactions.
c. For example, $C + O_2 = CO_2$; $CO_2 + C = 2CO$; $C + H_2O = CO + H_2$. $n_{O_2} = 0.204 - \lambda_1 \simeq 0$; $n_{CO_2} = 0.204 - \lambda_2$; $n_{CO} = 2\lambda_2 + \lambda_3$; $n_{H_2} = \lambda_3$; $n_{H_2O} = 0.03 - \lambda_3$; $n_{N_2} = 0.766$; $n_g = 1 + \lambda_2 + \lambda_3$. $p_i = n_i/n_g$.
d. $\lambda_1 = 0.204$, $\lambda_2 = 0.151$, $\lambda_3 = 0.027$.
e. 4.50% CO_2, 27.93% CO, 2.29% H_2, 0.25% H_2O, 65.03% N_2. Carbon consumption: 0.382 mol C. $DH = -12{,}207$ cal/mol air.

C. On the basis of the same reactions $\lambda_1 = 0.25$, $\lambda_2 = 0.179$, $\lambda_3 = 0.0266$. 5.89% CO_2, 31.90% CO, 2.21% H_2, 0.28% H_2O, 59.72% N_2. 0.456 mol C consumed. $DH = -15{,}416$ cal/mol air.

D.

	Case A	Case B	Case C
Carbon consumption (n_C/n_{FE})	4.19	3.99	4.33
Heat evolved (kcal/mol Fe)	155.3	131.9	150.9

15. Let Q represent the point on the line associated with the oxidation reaction of metal M and of abscissa T. The slope of the line CQ can be calculated to be $\Delta S_2^\circ - R\ln(p_{CO}/p_{CO_2})$ where ΔS_2° is the standard entropy change of the reaction $2CO + O_2 = 2CO_2$. This slope is independent of T, and all points on the line CQ correspond to a fixed value of the ratio p_{CO}/p_{CO_2}. The results for point H are similar.

16. **a.** $\lambda_1 = 0.78$ for the reaction $ZnO + CO = Zn(g) + CO_2$ (or $\mathbf{R}_1$); $\lambda = 0.011$ for the reaction $2CO + O_2 = 2CO_2$ (or $\mathbf{R}_2$). The weight of zinc dissociated is 51 g, and the pressure in the chamber is 1.53 atm.

b. Associating $\mathbf{R}_1'$ with the reaction $ZnO + CO = Zn(g) + CO_2$ and $\mathbf{R}_2'$ with $ZnO + C = Zn(g) + CO$, we have

$$\begin{pmatrix}\mathbf{R}_1\\ \mathbf{R}_2\end{pmatrix} = \begin{pmatrix}1 & 0\\ -1 & 1\end{pmatrix}\begin{pmatrix}\mathbf{R}_1'\\ \mathbf{R}_2'\end{pmatrix} \quad \text{and} \quad \begin{pmatrix}\lambda_1'\\ \lambda_2'\end{pmatrix} = \begin{pmatrix}1 & -1\\ 0 & 1\end{pmatrix}\begin{pmatrix}\lambda_1\\ \lambda_2\end{pmatrix}$$

from which we deduce $\lambda_1' = 0.77$ and $\lambda_2' = 0.011$.

Chapter VI

1. Assuming that the composition dependence of $\ln\gamma_B$ is parabolic,

$$\ln\gamma_B = X_A^2(\ln\gamma_B^\infty) \qquad \text{or} \qquad \ln\gamma_B^\infty = 2.302/(0.99)^2 = 2.349$$

Moreover,

$$(\ln\gamma_B^\infty)_{1500°K} = (\ln\gamma_B^\infty)_{1000°K} + \frac{H_B^{E\infty}}{R}\left(\frac{1}{1500} - \frac{1}{1000}\right)$$

$$= 2.349 - \frac{7000}{1.987}\left(\frac{1}{3000}\right) = 1.175$$

and $\ln\gamma_B = (0.98)^2 \times 1.175 = 1.128$, from which $\gamma_B = 3.09$ and $a_B = 0.062$.

2. Equation (76) may be used and yields $\ln\gamma_2 = X_1^2[(\alpha + \frac{1}{2}\beta) + \beta X_2]$,

$$G_m^E/RT = X_1\ln\gamma_1 + X_2\ln\gamma_2 = X_1X_2[(\alpha + \tfrac{1}{2}\beta) + \tfrac{1}{2}\beta X_2]$$

$$G_m^M = RT\{(X_1\ln X_1 + X_2\ln X_2) + X_1X_2[(\alpha + \tfrac{1}{2}\beta) + \tfrac{1}{2}\beta X_2]\}$$

3. **a.** $H_{Cu}^E = X_{Ag}^2(3900 + 3200X_{Cu})$; $S_{Cu}^E = X_{Ag}^2(0.323 + 2.214X_{Cu})$. At $X_{Cu} = 0.5$, $H_{Cu}^E = 1375$ cal, $S_{Cu}^E = 0.358$ cal/°K, $\ln\gamma_{Cu} = 0.306$, $a_{Cu} = 0.679$.

b. From mass balance, $X_{Cu}^{(Ag)} = 0.165$. From the data above, at 1823°K, $\ln\gamma_{Cu}^{(Ag)} = 0.610$ and $a_{Cu}^{(Ag)} = 0.304$. Since $a_{Cu}^{(Ag)} = a_{Cu}^{(Fe)} = \gamma_{Cu}^{(Fe)}X_{Cu}^{(Fe)}$, it follows that $\gamma_{Cu}^{(Fe)} = 10.1$.

4. With $\ln \gamma_1 = \alpha_1 X_2^2$, equation (76) and $G_m^E/RT = X_1 \ln \gamma_1 + X_2 \ln \gamma_2$, the demonstration is straightforward.

Chapter VII

1. From the series development of $(1 + x)^\alpha$

$$(1+x)^\alpha = 1 + \alpha x + \frac{\alpha(\alpha - 1)}{1\cdot 2}x^2 + \cdots + \frac{\alpha(\alpha-1)\cdots(\alpha - n)}{1\cdot 2 \cdots (n+1)}x^{n+1} + \cdots$$

with $\alpha = -1$, $(1 + x)^{-1} = 1 - x + x^2 - x^3 + \cdots$.

By integration, $\ln(1 + x) = x - x^2/2 + x^3/3 - x^4/4 + \cdots$. Consequently, we deduce that $y = 1 + x - \frac{3}{2}x^2 + \frac{11}{6}x^3 - \frac{25}{12}x^4 + \cdots$. At $x = 0.05$, 0.10, and 0.20, we must retain 2, 3, and 4 terms, respectively, for an error smaller than 0.005. For $\Delta y < 0.001$, an additional term must be retained in each case.

2. Proceed as in Section 1.4, identifying in equation (30) the terms in X_2^i to develop the relation between $J_i^{(2)}$ and $J_i^{(1)}$. For the relation between ϕ_i and $J_i^{(1)}$, apply the equation $G_m^E/RT = X_1 \ln \gamma_1 + X_2 \ln \gamma_2$.

3. **b.** $\epsilon_2^{(2)} \simeq 6$.

4. $\epsilon_{Tl}^{Tl} = 4.73$; $\eta_{Tl}^{Tl} = 9.21$ kcal; $\sigma_{Tl}^{Tl} = 2.16$ cal/°K.

5. $G_m^E/RT = X_{Sn}X_{Zn}(X_{Sn} \ln \gamma_{Zn}^\infty + X_{Zn} \ln \gamma_{Sn}^\infty) = X_{Sn}X_{Zn}(0.67X_{Sn} + 1.52X_{Zn})$. At $X_{Sn} = X_{Zn} = 0.5$, $a_{Sn} = 0.59$, $a_{Zn} = 0.73$.

6. $a_{Ag} = 0.178$, $X_{Au} = 0.47$.

7. **a.** See Fig. P.2. If for $T < 1517$°K, $\Delta S_1^\circ = -18.6$ cal/°K, then for 1517°K $< T < 2058$°K, $\Delta S_2^\circ = -20.9$ cal/°K, and for $T > 2058$°K, $\Delta S_3^\circ = -14.6$ cal/°K.

b. $\mu_{O_2}^{\circ(atm)} - \mu_{O_2}^{\circ(mmHg)} = -RT \ln 760 = -13.180T$. $\Delta G^{\circ(mmHg)} = \Delta G^{\circ(atm)} - 6.590T$.

8. a_B (metastable pure liquid) $= 0.52/0.46 = 1.13$.

9. **a.** 0.252% O.

b. %O = 0.128 at $p_{O_2} = 0.21$ and 0.517 at $p_{O_2} = 4$.

c. 97.5 cm³.

10. For the reaction $2\underline{V}(\%) + 3\underline{O}(\%) = V_2O_3(s)$, $\Delta G^\circ = -46{,}960$ cal. If λ represents the quantity of vanadium reacted (in weight percent),

$$(0.25 - \lambda)^2 \left(0.06 - \frac{3 \times 16}{2 \times 50.45}\lambda\right)^3 = 3.3 \times 10^{-6}$$

from which we deduce $\lambda = 0.038$, %V = 0.212, %O = 0.042, and 1.1 lb of V_2O_3.

11. **a.** $e_C^C = 0.20$; $K = 0.041$; $\Delta G^\circ = 7600$ cal.

c. $\gamma_C^\infty = 10$.

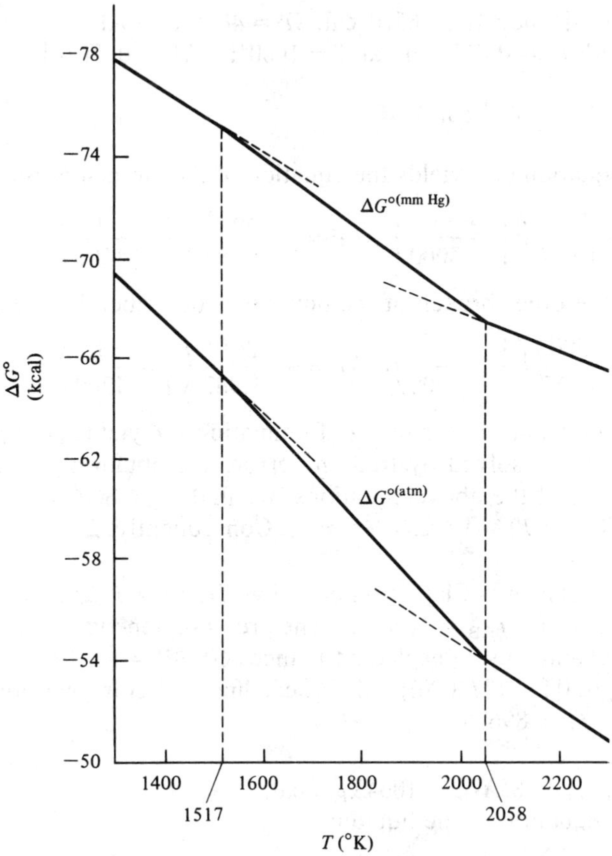

Figure P2. Problem VII.7a.

12. **a.** 5.80×10^{-3}.
b. 5.52×10^{-3}.
c. $e_S^S = -0.038$; $\Delta G° = -22{,}000$ cal.

Chapter VIII

1. Combining equations (16) and (23), we obtain

$$0.78 \exp\left[\frac{3500}{1.987}\left(\frac{1}{1700} - \frac{1}{1500}\right) + \frac{1.1}{1.987}\left(1 - \frac{1500}{1700} + \ln\frac{1500}{1700}\right)\right]$$
$$+ 0.22 \exp\left[\frac{\Delta H^\circ_{f,B}}{1.987}\left(\frac{1}{1700} - \frac{1}{2300}\right)\right] = 1$$

from which we deduce $\Delta H^\circ_{f,B} = 5080$ cal.

2. Assuming that the liquid and solid solutions are regular, equations (37) lead to

$$\ln\frac{0.95}{0.60} + \frac{\Omega^s}{1.987 \times 1200}(0.05)^2 - \frac{\Omega^l}{1.987 \times 1200}(0.40)^2 = \frac{3200}{1.987}\left(\frac{1}{1200} - \frac{1}{1400}\right)$$
$$\ln\frac{0.05}{0.40} + \frac{\Omega^s}{1.987 \times 1200}(0.95)^2 - \frac{\Omega^l}{1.987 \times 1200}(0.60)^2 = \frac{2500}{1.987}\left(\frac{1}{1200} - \frac{1}{1000}\right)$$

from which we deduce $\Omega^s = 6570$ cal, $\Omega^l = 4095$ cal. At $T = 1200°K$, $X_B^l = 0.70$, we find $a_B^l = 0.817$, and at $T = 1600°K$, $X_B^l = 0.70$, $a_B^l = 0.786$.

3. **b.** $\alpha = 1030$ cal, $\beta = 445$ cal.

4. **a.** Applying equation (28) yields the equation of one branch of the liquidus:

$$\ln X_1^l + \frac{500}{1.987}\left(\frac{1}{T} - \frac{1}{3000}\right)(X_2^l)^2 = -\frac{3600}{1.987}\left(\frac{1}{T} - \frac{1}{1500}\right)$$

Similarly, the other branch of the liquidus is described by the equation

$$\ln X_2^l + \frac{500}{1.987}\left(\frac{1}{T} - \frac{1}{3000}\right)(X_1^l)^2 = -\frac{2650}{1.987}\left(\frac{1}{T} - \frac{1}{1200}\right)$$

These equations are linear in $1/T$. Eliminating $1/T$ yields an equation in X_2^l which can be solved by trial and error. We obtain $X_2^l = 0.625$. Replacing in one of the above equations, we find $T_E = 860°K$.

b. $\partial H/\partial P = V(1 - \alpha T) \simeq V$; $\partial \Delta H_f/\partial P \simeq V_f$. Consequently, $\Delta H_{f,A} = 3696$ cal, $\Delta H_{f,B} = 2771$ cal.

Using the Clausius–Clapeyron equation $dT/dP = T\,\Delta V_f/\Delta H_f$, we obtain $T_{f,A} = 1541°K$, $T_{f,B} = 1256°K$. (The pressure dependence of ΔH_f has a small effect and may be neglected.) Since $\partial G^E/\partial P = V^E$, at $P = 10^4$atm, $G^E = X_A X_B 621[1 - (T/3000)]$ cal. Proceeding as before we now obtain $X_2^E = 0.618$, $T_E = 896°K$.

5. **a.** $G_m^{E(l)} = X_{Na}X_{Rb}(1185X_{Na} + 1064X_{Rb})$ cal.

b. Solving the equation of the liquidus,

$$\ln 0.45 + \ln \gamma_{Na}^l = -\frac{621}{1.987}\left(\frac{1}{320} - \frac{1}{371.0}\right)$$

we obtain $\ln \gamma_{Na}^l = 0.664$. With the formalism of the subregular solution and the coefficients determined above, at $X_{Rb} = 0.55$, $\ln \gamma_{Na}^l = 0.558$. Thus, the subregular formalism yields positive deviations from ideality which are too small. The introduction of a positive τ parameter would make this deviation even smaller (the solution reaches ideality when $T \to \tau$, instead of $T \to \infty$). Consequently, the introduction of the A_{22} parameter appears preferable to that of a negative τ parameter.

6. **a.**

$$y = -\frac{(\partial^3 S_m^l/\partial X_2^3)_c}{(\partial^4 G_m^l/\partial X_2^4)_c}(T_c - T)$$

With $y = 0.0585$, $T_c - T = 26°K$,

$$\left(\frac{\partial^3 S_m^l}{\partial X_2^3}\right)_c = -R\left(\frac{1}{X_1^2} - \frac{1}{X_2^2}\right)_c = 11.52 \quad \text{cal/°K},$$

$$\left(\frac{\partial^4 G_m^l}{\partial X_2^4}\right)_c = 2RT\left(\frac{1}{X_1^3} + \frac{1}{X_2^3}\right) + 24A_{22} = 134{,}268 + 24A_{22} \quad \text{cal}$$

We obtain $A_{22} = -5808$ cal.

b. $A_{21} = 7538$ cal; $A_{12} = 8618$ cal.

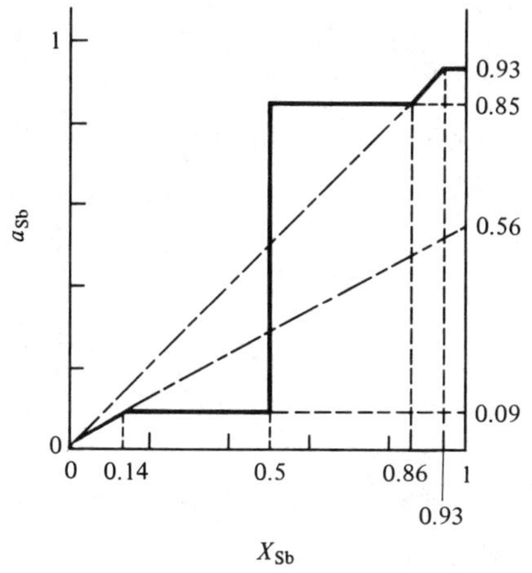

Figure P3. Problem VIII.9b.

7. $G_m^{E(s)} = X_{Co}X_{Cu}(6370X_{Co} + 7850X_{Cu})$ cal. This formalism entails a critical point at $X_{Cu} = 0.573$, $T = 1830°K$. The temperature of the critical point of the solid phase is above that of the liquidus, but that does not violate any rule.

9. **b.** See Fig. P.3. The calculations assume a liquid regular solution.
c. $(G^l)_{X_{Sb}=0.5} = \frac{1}{2}G(\text{GaSb} \text{ or } \beta)$; $\Delta G° = 2G_m^M = 2 \times 1.987 \times \ln 0.5 - 2 \times 1000 \times 0.25 = -3213$ cal.

10. $\ln \gamma_C + \ln X_C = \ln a_C$, from which we deduce $X_C = 0.2067$. Differentiating,

$$\left(\frac{\partial \ln \gamma_C}{\partial(1/T)}\right)_{X_C} d(1/T) + \left[\left(\frac{\partial \ln \gamma_C}{\partial X_C}\right)_T + \frac{1}{X_C}\right] dX_C = 0$$

and

$$\frac{dX_C}{d(1/T)} = -\frac{H_C^E}{R}\frac{X_C}{1 + X_C(\partial \ln \gamma_C/\partial X_C)_T} = -279°K$$

or $dX_C/dT = 0.83 \times 10^{-4}\ °K^{-1}$.

11. $$\frac{dX_B}{dT} = -\frac{\Delta H_{f,A}^\circ}{RT_{f,A}^2} = -\frac{\Delta S_{f,A}^\circ}{R}\frac{1}{T_{f,A}}$$

Chapter IX

1. **a.**

$$\ln \gamma_1 = -\tfrac{1}{2}\epsilon_2^{(2)}X_2^2 - \epsilon_2^{(3)}X_2X_3 - \tfrac{1}{2}\epsilon_3^{(3)}X_3^2 - \tfrac{1}{3}(2\rho_2^{(2)} + \epsilon_2^{(2)})X_2^3$$
$$- (2\rho_3^{(2)} + \epsilon_2^{(2)})X_2^2X_3 - (2\rho_2^{(3)} + \epsilon_3^{(3)})X_2X_3^2 - \tfrac{1}{3}(2\rho_3^{(3)} + \epsilon_3^{(3)})X_3^3$$

b. There is only one quaternary term in the series development of $\ln \gamma_1$ (up to third order) and it is equal to $-2(\rho_2^{(3,4)} + \epsilon_4^{(3)})X_2X_3X_4$.

2. $a_C^{(gr)} = 0.25$; $a_C^{(h)} = 0.68$.

3. $e_S^{Al} = 0.055$; $e_S^C = 0.120$.

4. **a.**
$$d \ln \gamma_{Si} = \left(\frac{\partial \ln \gamma_{Si}}{\partial X_{Si}}\right)_{X_C} dX_{Si} + \left(\frac{\partial \ln \gamma_{Si}}{\partial X_C}\right)_{X_{Si}} dX_C \quad (1)$$

$$\left(\frac{\partial \ln \gamma_{Si}}{\partial X_C}\right)_{a_{Si}} = \left(\frac{\partial \ln \gamma_{Si}}{\partial X_{Si}}\right)_{X_C} \left(\frac{\partial X_{Si}}{\partial X_C}\right)_{a_{Si}} + \left(\frac{\partial \ln \gamma_{Si}}{\partial X_C}\right)_{X_{Si}} \quad (2)$$

Moreover,

$$\left(\frac{\partial \ln \gamma_{Si}}{\partial X_C}\right)_{a_{Si}} = -\left(\frac{\partial \ln X_{Si}}{\partial X_C}\right)_{a_{Si}} \quad (3)$$

and

$$\left(\frac{\partial X_{Si}}{\partial X_C}\right)_{a_{Si}} = X_{Si}\left(\frac{\partial \ln X_{Si}}{\partial X_C}\right)_{a_{Si}} \quad (4)$$

Combining equations (3) and (4) with (2) establishes the desired relationship.

b. $\epsilon_{Si}^{C} \simeq 11$.

5. $\epsilon_i^{(j)} = \theta_i^{(j)} + (1 - v_j)$; $\theta_i^{(j)} = \theta_j^{(i)} + (v_j - v_i)$.

6.
$$-\left(\frac{\partial \ln X_C}{\partial X_{Si}}\right)_{\substack{X_{Si}\to 0\\ a_C = 1}} = \frac{\epsilon_C^i + \rho_C^{C,i} X_C}{1 + \epsilon_C^C X_C + 2\rho_C^C X_C^2} \quad (X_C = 0.205)$$

$$= 0.3\epsilon_C^i + 0.06\rho_C^{C,i}$$

For silicon

$$\rho_C^{C,Si} = \frac{3.55 - 0.3 \times 5.6}{0.085} = 22$$

[see C. H. P. Lupis, *Acta Met.* **16,** 1365 (1968)].

7. $(\%N)_{1873} = 0.22$; $(\%N)_{2073} = 0.16$.

$$(\log \%N)_T = (\log \%N)_{1873} - \left(\frac{\Delta H^\circ + \Sigma h_N^{(i)} \%i}{4.575}\right)\left(\frac{1}{T} - \frac{1}{1873}\right)$$

where ΔH° corresponds to the reaction $\frac{1}{2}N_2(g) = \underline{N}(\%)$ and is equal to $-4.575 d \log K/d(1/T)$ (with $K = \%N$ in pure liquid iron at $p_{N_2} = 1$ atm).

8. Let P be the total cost. The results are shown in Fig. 8.

a. $\alpha < 1.9$: use only C. $P = 1\alpha$.
$\alpha = 1.9$: use C only or B only. P = 1.9.
$\alpha > 1.9$: use B only. P = 1.9.
The use of B *and* C is always more expensive.

b. Same as for part a. D is too expensive.

c. The net cost of D becomes $\alpha/5 - \alpha/5.2 = \alpha/130$.
$\alpha \leq 1.25$: use C and D. $P = 0.991\alpha$. Use 2.6 at.% of D.
$1.25 \leq \alpha \leq 13$: use B and D. $P = (\alpha/13)[1 - \ln(\alpha/13)] + 0.9$. Use $10 \ln(13/\alpha)$ at.% of D.
$\alpha \geq 13$: only B should be used.

d. $\alpha \leq 2.21$: only C should be used. $P = 1.0065\alpha$.
$\alpha \geq 2.21$: only B should be used. $P = 2.23$.

9. **a.** $e_{O}^{Pd} = 0.034$; $e_{O}^{Cu} = -0.34$.
b. $h_{O}^{Pd} \simeq 170$ cal; $h_{O}^{Cu} \simeq -6500$ cal.

c. $$(\log \%O)_T = (\log \%O)_{T'} - \left(\frac{\Delta H^\circ + h_O^{Pd}\%Pd + h_O^{Cu}\%Cu}{4.575}\right)\left(\frac{1}{T} - \frac{1}{T'}\right)$$

where $\Delta H^\circ \simeq -3700$ cal [for the reaction $\frac{1}{2}O_2 = \underline{O}(\%)$].

10. $\ln \varphi_i = 2.3 \log f_i - \ln\{1 + \sum_{j=2}^{m} X_j[(M_j/M_1) - 1]\}$. Differentiating this equation twice with respect to X_j leads to the desired relationship. Note that

$$\frac{\partial \log f_i}{\partial X_j} = \sum_{k=2}^{m} \frac{\partial \log f_i}{\partial \%k}\frac{\partial \%k}{\partial X_j}$$

11. $f_{Si} f_O^2(0.60 - \lambda)\{0.025 - [(2 \times 16)/28.09]\,\lambda\}^2 = 2.27 \times 10^{-5}$. $\lambda = 0.016$; %O $= 0.0068$; %Si $= 0.584$; 0.34 kg of SiO_2/tonne of steel.

12. With a formalism limited to second order coefficients $b_N^i = e_N^i + r_N^{Ni,i}\%Ni + r_N^{Cr,i}\%Cr = e_N^i + 8r_N^{Ni,i} + 18r_N^{Cr,i}$. For Nb, this yields $b_N^{Nb} = -0.067 - 8 \times 4.1 \times 10^{-4} + 18 \times 13.6 \times 10^{-4} = -0.046$, in good agreement with the value of -0.043. Similarly: b_N^{Ta} $18 \times 5.2 \times 10^{-4} = -0.023$, also in good agreement with the value of -0.019.

Chapter X

1. $$\begin{aligned} RT\ln \gamma_2 &= 2X_2(1 - X_2)X_3A_{021} + (1 - 2X_2)X_3^2A_{012} \\ &\quad + 2X_2(1 - X_2)X_1A_{120} + (1 - 2X_2)X_1^2A_{210} \\ &\quad - X_3X_1(2X_3A_{102} + 2X_1A_{201}) + X_1X_3(1 - 2X_2)A_{111} \end{aligned}$$

2. **a.** $RT \ln \gamma_1 = X_2^2\Omega_{12} + X_3^2\Omega_{13} + X_2X_3(\Omega_{12} + \Omega_{13} - \Omega_{23})$.
b. With the formalism of the As

$$\begin{aligned} G_m^E &= X_1X_2(1 - X_3)\Omega_{12} + X_1X_3(1 - X_2)\Omega_{13} + X_2X_3(1 - X_1)\Omega_{23} \\ &= X_1X_2\Omega_{12} + X_1X_3\Omega_{13} + X_2X_3\Omega_{23} - X_1X_2X_3(\Omega_{12} + \Omega_{13} + \Omega_{23}) \end{aligned}$$

The two expressions are identical only in the case where $\Omega_{12} + \Omega_{13} + \Omega_{23} = 0$. Applying equation (27) leads to

$$RT\ln \gamma_1 = X_2^2\Omega_{12} + X_3^2\Omega_{13} - 2X_2X_3\Omega_{23} + 2X_1X_2X_3(\Omega_{12} + \Omega_{13} + \Omega_{23})$$

The difference in the two cases arises from the fact that $X_i + X_j$ is equal to 1 in the binary i–j, but not in the ternary i–j–k.

3. Let Pb = 1, Bi = 2, and Zn = 3. With the formalism of equations (27) and (28) $\ln \gamma_1 = (\ln \gamma_1)_{1-2} + (\ln \gamma_1)_{1-3} + (\ln \gamma_1)_{1-2-3} = -0.301 + 0.122 - 0.059 - 0.238$. Consequently, $\gamma_{Pb} = 0.788$, $a_{Pb} = 0.236$.

4. Applying equation (28c) with Au ≡ 1, Ag ≡ 2, Cu ≡ 3, we obtain

$$(\ln \gamma_{Au})_{X_{Au}\to 0} = X_{Ag}^2(\ln \gamma_{Au}^\infty)_{Ag} + X_{Cu}^2(\ln \gamma_{Au}^\infty)_{Cu}$$

$$- 2X_{Ag}X_{Cu}[X_{Ag}(\ln \gamma_{Cu}^\infty)_{Ag} + X_{Cu}(\ln \gamma_{Ag}^\infty)_{Cu}]$$

from which we deduce $(\gamma_{Au})_{X_{Au}\to 0, X_{Cu}=0.4} = 0.157$.

5. By equation (18), noting that for a subregular solution $G_m^E/RT = X_1X_2(X_1 \ln \gamma_2^\infty + X_2 \ln \gamma_1^\infty)$, we obtain $G_m^E = 846$ cal. The approximations of Kohler and Colinet yield, respectively, 1146 and 1133 cal. The fact that these last two methods give very similar results in this case has no particular significance.

6. In equation (79), let Fe ≡ 1, C ≡ 2, α ≡ liquid, and β ≡ graphite. We deduce

$$\left(\frac{dX_C}{dX_3}\right)_{X_3\to 0} = -X_C\left(\frac{\partial \ln \gamma_C}{\partial X_3}\right)_{X_3\to 0} \Big/ \left(1 + X_C\frac{\partial \ln \gamma_C}{\partial X_C}\right)_{X_3\to 0}$$

The same result would be obtained by differentiating $\ln X_C + \ln \gamma_C = \ln a_C$ = constant:

$$\left(\frac{d \ln X_C}{dX_3}\right) + \left(\frac{\partial \ln \gamma_C}{\partial X_C}\right)_{X_3}\left(\frac{dX_C}{dX_3}\right) + \left(\frac{\partial \ln \gamma_C}{\partial X_3}\right)_{X_C} = 0$$

which after rearrangement yields the desired equation for $X_3 \to 0$. In terms of interaction coefficients,

$$\left(\frac{dX_C}{dX_3}\right)_{X_3\to 0} = -\frac{X_C(\epsilon_C^{(3)} + \rho_C^{(C,3)}X_C)}{1 + \epsilon_C^C X_C + 2\rho_C^C X_C^2}$$

7.

$$\left(\frac{dX_C}{dX_3}\right)_{X_3\to 0} = \left[\frac{3X_C}{1-4X_C} - X_C\left(\frac{\partial \ln \gamma_C}{\partial X_3}\right)_{X_3\to 0}\right] \Big/ \left(1 + X_C\frac{\partial \ln \gamma_C}{\partial X_C}\right)_{X_3\to 0}$$

$$= \left[\frac{3X_C}{1-4X_C} - X_C(\epsilon_C^C + \rho_C^{(C,3)}X_C)\right] \Big/ (1 + \epsilon_C^C X_C + 2\rho_C^C X_C^2)$$

For the direct method, use $\ln a_C + 3 \ln a_{Fe} = \ln a_{Fe_3C} + \ln K$ = constant and the fact that $\partial \ln \gamma_{Fe}/\partial X_i = -(X_C/X_{Fe})(\partial \ln \gamma_C/\partial X_i)$ (with i = C or 3, for $X_3 \to 0$), which is deduced from the Gibbs–Duhem equation.

Chapter XI

2.

$$\frac{dX_3}{dX_2} = -\frac{\partial G_2^M/\partial X_2}{\partial G_2^M/\partial X_3} = -\frac{RT/X_2 + \partial G_2^E/\partial X_2}{\partial G_2^E/\partial X_3}$$

which, using equation (X.26), becomes

$$\frac{dX_3}{dX_2} = -\frac{RT/X_2 + (1-X_2)(\partial^2 G_m^E/\partial X_2^2) - X_3(\partial^2 G_m^E/\partial X_2 \partial X_3)}{(1-X_2)(\partial^2 G_m^E/\partial X_2 \partial X_3) - X_3(\partial^2 G_m^E/\partial X_3^2)}$$

3. **a.** $\partial^2 G_m^E/\partial X_2^2 = -3453$ cal, $\partial^2 G_m^E/\partial X_3^2 = -8079$ cal, $\partial^2 G_m^E/\partial X_2\, \partial X_3 = -13{,}157$ cal.

b. $\psi = 0.0057$.

c. The calculated value of the slope is 0.89.

4. a. From the Maxwell relation, $(\partial\mu_2/\partial n_1)_{n_2,n_3} = (\partial\mu_1/\partial n_2)_{n_1,n_3}$, by changing variables $(n_1, n_2, n_3 \rightarrow n, s, t)$ we obtain

$$\left(\frac{\partial\mu_2}{\partial s}\right)_t = \left(\frac{t}{s}\right)^2\left(\frac{\partial\mu_1}{\partial t}\right)_s$$

which after integration yields the equation we seek. $s = 1$ corresponds to $X_1 = 0$, that is, to the binary 2–3, for which we assumed that the value of $\ln\gamma_2$ was known at any composition (or any t). For additional information, consult N. A. Gokcen [*J. Phys. Chem.* **64,** 401 (1960)].

b. Let $1 - X_1 = u$. The following series expression (Chapter IX)

$$\ln\gamma_1 = -\tfrac{1}{2}\epsilon_2^{(2)}X_2^2 - \epsilon_2^{(3)}X_2X_3 - \tfrac{1}{2}\epsilon_3^{(3)}X_3^2 + \cdots$$

becomes

$$\ln\gamma_1 = -u^2[(\tfrac{1}{2}\epsilon_2^{(2)}(1-t)^2 + \epsilon_2^{(3)}t(1-t) + \tfrac{1}{2}\epsilon_3^{(3)}t^2] + O(u^3)$$

$$= \alpha u^2 + O(u^3)$$

We also note that

$$(dt)_s = \left[d\left(\frac{X_3}{1-X_1}\right)\right]_{X_3/X_1} = \frac{X_3}{X_1}\left[d\left(\frac{X_1}{1-X_1}\right)\right]_s$$

$$= \frac{X_3}{X_1}d\left(\frac{1-u}{u}\right) = -\frac{X_3}{X_1}\frac{du}{u^2} = \frac{-du}{(1-u)(1+t)u}$$

With these remarks, it is now straightforward to establish that

$$\left[\left(\frac{t}{s}\right)^2\left(\frac{\partial\ln\gamma_1}{\partial t}\right)_s\right]_{X_1\rightarrow 1} = -2\alpha(1+t)$$

$$= (1+t)[\epsilon_2^{(2)}(1-t)^2 + 2\epsilon_2^{(3)}t(1-t) + \epsilon_3^{(3)}t^2]$$

5. a. Substitute equations (87) and (88) into equation (83), using the same reference state for the activity of 3 in all three phases. Rearranging the terms and using equation (94a) leads to the desired equation.

b. For $X_2^\beta = 1$, $X_1^\beta = 0$, $\theta_3^{\infty(\beta,B)} = \infty$, we obtain

$$\left(\frac{dX_2}{dX_3}\right)_E = -\frac{X_2^E}{\psi^l}\left(\frac{\partial\ln\gamma_2}{\partial X_3}\right)_l - E_{12}\left(\frac{X_1^E\,\theta_3^{\infty(l,E)}}{X_1^\alpha\,\theta_3^{\infty(\alpha,A)}} - 1\right)\frac{dX_2^{l(\beta)}}{dT}$$

The final result is reached by application of equation (97).

c. For the Fe–C eutectoid, $(dX_C/dX_3)_E = -0.023(\partial\ln\gamma_C/\partial X_3)_{fcc} + 1.7\times10^{-4}(dT/dX_3)_E$. In the case of silicon, $(dT/dX_{Si})_E = 1714°K$ and $(\partial\ln\gamma_C/\partial X_{Si})_{fcc} = \epsilon_C^{Si} + \rho_C^{Si,C}X_C \simeq 14$. Consequently, $(dX_C/dX_{Si})_E \simeq -0.03$.

Chapter XII

1. The temperature of fusion of A_2 remains that of A, $\Delta H_{f,A_2}^\circ = 2\Delta H_{f,A}^\circ$, and $(X_B'/X_B)_{X_B'\rightarrow 0} = 2$. Consequently, in terms of the new species, the equation remains the same: $\lim(dT/dX_B')_{X_B'\rightarrow 0} = -R(T_{f,A_2})^2/\Delta H_{f,A_2}^\circ$.

2. Because of Henry's law, $(\mu_B - RT \ln X_B)_{X_B \to 0} = \mu_B^* + RT \ln \gamma_B^\infty$ has a finite value K. Since $\mu_V = x\mu_A + y\mu_B$ and $X_V = X_B/[y + X_B(1 - x - y)]$, we note that

$$(\mu_V - RT \ln X_V)_{X_V \to 0} = (y\mu_B - RT \ln X_B)_{X_B \to 0} + (x\mu_A)_{X_B \to 0} + RT \ln y$$

$$= (y-1)(\mu_B)_{X_B \to 0} + K + (x\mu_A)_{X_A \to 1} + RT \ln y$$

Since all the terms on the right-hand side of this equation are finite except for $(\mu_B)_{X_B \to 0}$, Henry's law is obeyed by V only in the case where $y = 1$. Similarly,

$$(\mu_U - RT \ln X_U)_{X_U \to 0} = \left(\mu_A - RT \ln \frac{y - X_B(x+y)}{y + X_B(x+y)}\right)_{X_B \to y/(x+y)}$$

For $X_B \to y/(x+y)$, or $X_A \to x/(x+y)$, μ_A tends towards a finite value. However, since the logarithmic term in the right-hand side of the equation tends towards infinity, Henry's law is not obeyed. The only exception occurs for $x = 0$. In that case

$$(\mu_U - RT \ln X_U)_{X_U \to 0} = (\mu_A - RT \ln X_A)_{X_A \to 0} - RT \ln y$$

$$= \mu_A^* + RT \ln \gamma_A^\infty - RT \ln y$$

and Henry's law is obeyed (for $y \neq 0$).

Thus, if both A and B obey Henry's law, U and V both also obey Henry's law only in the case where $x = 0$, $y = 1$. This case corresponds to the original description of the system, $U \equiv A$, $V \equiv B$.

Chapter XIII

1. $\Delta P = 4\sigma/r = 1.2$ mm Hg.

2. $\Delta\rho g h = 2\sigma/r$, where $\Delta\rho$ is the difference in the densities of the liquid alloy and of the atmosphere. $\Delta\rho \simeq \rho_{\text{liquid}} = 2350$ kg/m^3. Consequently, $\sigma = 2350 \times 9.807 \times 0.186 \times 0.4 \times 10^{-3} \times 0.5 = 0.857$ N/m.

3. The increase in vapor pressure is negligible: $(P)_{r=0.1\text{ mm}}/(P)_{r=\infty} = 1.00006$.

4. One side of the square corresponds to an arc of a parabola. Therefore, the equilibrium shape is made of four parabolic arcs (and has a fourfold symmetry).

5. **a.** On the basis of surface free energies before and after separation $W = \sigma_A + \sigma_B - \sigma_{AB}$.

b. If the solid substrate is B and the liquid of the sessile drop is A, application of equation (154) leads to $W = \sigma_{l\text{-}g}(1 + \cos\theta)$.

c. Yes, because surface energies can no longer be taken as equivalent to surface tensions [equation (82)], and the equation measuring W was derived for surface energies.

6. Most metals have surface free energies varying between 1 and 2.5 J/m^2 (1000–2500 erg/cm^2). Water has a much lower surface tension: 0.072 N/m at room temperature (25°C). The interfacial tension σ_{AB} calculated by Girifalco and Good's equation is generally smaller than σ_{metal} by a quantity which is larger than 0.072 N/m. Consequently, in equation (155) k is larger than 1, which corresponds to the case where the solid is completely wetted.

7. **a.** The Helmholtz free energy of the system is a function of the areas of the liquid A–gas, liquid B–gas, and liquid A–liquid B interfaces. Consequently, we may write $dA = (\partial A/\partial s_A)\, ds_A + (\partial A/\partial s_B)\, ds_B + (\partial A/\partial s_{AB}) \times ds_{AB}$. When the area of contact between the liquid lens of B and A expands or contracts, we have $-ds_A = ds_B = ds_{AB}$. Thus, $dA = (\sigma_{AB} + \sigma_B - \sigma_A)\, ds_{AB}$. B spreads on A if the coefficient $C_{A/B} = -(\partial A/\partial s_{AB})$ is positive, i.e., if $\sigma_A - \sigma_B - \sigma_{AB} > 0$.

b. Initially, $C_{A/B} = 72.8 - 28.9 - 35.0 = 8.9$ dyn/cm. After prolonged contact $C_{A/B} = 62.2 - 28.9 - 35.0 = -1.7$ dyn/cm. Consequently, the film of benzene first spreads and then contracts. (For further information, see A. W. Adamson, *Physical Chemistry of Surfaces*. Wiley, New York, 1976, p. 103.)

8. $(\sigma_{Ni-He})_{1200°C} = 2280 - 0.55 \times 140 = 2203$ dyn/cm. From equation (157) we deduce $\sigma_{gb} = 2 \times 2203 \times \cos(78.5) = 860$ dyn/cm (or 0.860 N/m).

9. $\Delta G = -\Delta G_f = -\Delta H_f[1 - (T/T_f)] = -1436 \times (40/273) = -210$ cal/mol (or -880 J/mol). $r_c = -2\sigma V_m/\Delta G = 2 \times 0.022 \times 18 \times 10^{-6}/880 = 9 \times 10^{-10}$ m (or 9 Å).

10. **a.**

$$\frac{d \ln I}{dT} = \frac{1}{kT^2}\left(kT + Q_D + \Delta G^* - T\frac{d\Delta G^*}{dT}\right) = 0$$

where $\Delta G^* = \alpha/(T_f - T)^2$ with $\alpha = 16\pi\sigma^3 V_m^2/3(\Delta S_f)^2$. After rearrangement, we obtain $[(3T - T_f)/(T_f - T)^3]\alpha = kT + Q_D$. By inspection, we see that T_{max}, corresponding to maximum undercooling, is between T_f and $T_f/3$.

b. For gold, $\alpha = 4.83 \times 10^{-14}$. In the equation $(3T - T_f)\alpha/(T_f - T)^3 = kT + Q_D$ at $T = 1076$°K, the left-hand side of the equation is equal to 520×10^{-20} whereas the right-hand side is equal to 8.4×10^{-20}. The agreement is not satisfactory, the calculated maximum rate corresponding to a temperature well below 1076°K. There are several reasons for this discrepancy

On the basis of several classical experiments, it has generally been assumed that, for most metals, nucleation occurs for an undercooling of about $0.2T_f$ (260°K for gold). However, more recent data[1] indicate that the maximum undercooling could be much higher. For Hg, it was observed to be about 90°K, or about $0.38T_f$. Also, many oversimplifying assumptions have been entered in the derivation of the equation for I. The concept of an activation energy for self-diffusion in a liquid is suspect, and the concept of interfacial tension itself becomes questionable for very small particles. We can verify that an undercooling of 260°K corresponds to a critical radius of 11 Å.

[1] J. H. Perepezko and D. H. Rasmussen, *Met. Trans.* **9A**, 1490 (1978).

Chapter XIV

1. From equation (47), we deduce $(\partial\sigma/\partial P)_T = (V/n_1, n_2)$, which for $V^{(s)} = 0$ may be written

$$(\partial\sigma/\partial P)_T = \frac{\Gamma_2(c_1^\beta - c_1^\alpha) - \Gamma_1(c_2^\beta - c_2^\alpha)}{c_1^\alpha c_2^\beta - c_1^\beta c_2^\alpha} = \Gamma_2^{(1)} \frac{(c_1^\beta - c_1^\alpha)V_m^\alpha V_m^\beta}{X_2^\beta - X_2^\alpha}$$

$$= -\Gamma_2^{(1)} V_m^\alpha \left[1 + \frac{X_1^\alpha(V_m^\beta - V_m^\alpha)}{(X_2^\beta - X_2^\alpha)V_m^\alpha}\right]$$

We see that if $V_m^\beta \simeq V_m^\alpha$, $(\partial\sigma/\partial P)_T \simeq -\Gamma_2^{(1)} V_m^\alpha$: σ decreases when P increases if the solute 2 is preferentially adsorbed at the interface (if $\Gamma_2^{(1)} > 0$).

2. **a.** The area per atom is equivalent to a hexagon with an inscribed circle of radius $r = 1.5$ Å. Such a hexagon has an area equal to 7.794 Å^2, from which we deduce $\Gamma_2^{sat} = (6.02 \times 10^{23} \times 7.79 \times 10^{-20})^{-1} = 0.213 \times 10^{-4}$ mol/m^2.
   b. $d\sigma/d \ln X_2 = -RT\Gamma_2^{sat} = -0.175$ N/m and $\sigma \simeq 0.800$ N/m (800 dyn/cm).

3. $$\left(\frac{d\sigma}{dX_2}\right)_{X_2 \to 0} = \left(\frac{d\sigma}{d\theta}\frac{d\theta}{dX_2}\right)_{X_2 \to 0} = -RT\Gamma_1^\circ K \quad \text{and} \quad \nu_2^\infty = K$$

Noting that

$$\left(\frac{d^2\sigma}{dX_2^2}\right) = \frac{d^2\sigma}{d\theta^2}\left(\frac{d\theta}{dX_2}\right)^2 + \frac{d\sigma}{d\theta}\frac{d^2\theta}{dX_2^2} \quad \text{with} \quad \frac{d^2\theta}{dX_2^2} = -\frac{dX_2/d\theta}{(dX_2/d\theta)^3}$$

we deduce

$$(d^2\sigma/dX_2^2)_{X_2 \to 0} = -(1 - 2\alpha)RT\Gamma_1^\circ K^2 + 2(1 - \alpha)RT\Gamma_1^\circ K^2$$

$$= RT\Gamma_1^\circ K^2 = RT\Gamma_1^\circ(\nu_2^\infty)^2$$

and $\xi_2^{(2)} = -\nu_2^\infty(\nu_2^\infty + 1)$.

4. **a.** The perfect solution model yields

$$J_{Pb} = -(d\sigma/dX_{Pb})_{X_{Pb} \to 0} = \Gamma_{Cu}^\circ RT\,[e^{(\sigma_{Cu} - \sigma_{Pb})/\Gamma_{Cu}^\circ RT} - 1]$$

   With $\Gamma_{Cu}^\circ = 0.271 \times 10^{-4}$ mol/m^2, $RT\Gamma_{Cu}^\circ = 0.310$, and $J_{Pb} = 5.39$ N/m mol. $\nu_{Pb}^\infty = J_{Pb}/RT\Gamma_{Cu}^\circ = 17.4$.
   b. The Pb atom is much larger than the Cu atom. Neither the bulk properties nor the surface properties can be expected to obey a perfect solution model. Indeed, large positive deviations from ideality have been observed for the bulk. Repulsion of the Pb atoms by the Cu atoms in the bulk leads to a marked segregation of lead at the surface and, consequently, to a higher value of ν_{Pb}^∞ than that predicted by the perfect solution model.

5. Combining the first equation of state with the Gibbs adsorption equation, we obtain $-d\sigma = RT\Gamma_2\, d \ln a_2 = RT\, d\Gamma_2$, or $d \ln a_2 = d \ln \Gamma_2$. From this we deduce $a_2 = (\Gamma_2^{sat}RT/J_2)\theta$. The second equation of state, $\sigma_1 - \sigma = RT/(A - A_{sat})$, yields $-d\sigma/RT = -dA/(A - A_{sat})^2 = -(1/A)\, d \ln a_2$ or $d \ln a_2 = -A\, dA/(A - A_{sat})^2$. After integration and rearrangement we obtain

$\ln a_2 = \ln[\theta/(1-\theta)] + \theta/(1-\theta) + \ln c$, where c is a constant of integration that can easily be identified as equal to $\Gamma_2^{sat}RT/J_2$. a_2 may then be rewritten $a_2 = (\Gamma_2^{sat}RT/J_2)[\theta/(1-\theta)]e^{\theta/(1-\theta)}$.

6. $$\ln a_2 = \ln\frac{\theta}{1-\theta} + \frac{\theta}{1-\theta} - \frac{2g}{kT} + \ln\frac{\Gamma_2^{sat}RT}{J_2}, \qquad \frac{\sigma_1 - \sigma}{\Gamma_2^{sat}RT} = \frac{\theta}{1-\theta} - \frac{g\theta^2}{kT}$$

The isosteric heat of adsorption $q_{st} = R[\partial \ln a_2/\partial(1/T)]_\theta$ is independent of θ when g is proportional to T.

7. **a.** $J_{1253°K} = 383$ N/m and $J_{1381°K} = 133$ N/m. $q_{st} = \Delta H° = -R[d \ln \nu_O^\infty/d(1/T)]$, assuming that $\Gamma_{Ag}^\circ \simeq$ constant. Therefore, $q_{st} = -31.1$ kcal.
b. Consider Fig. 22. The two isotherms, σ vs X, intersect at a point determined by the equation $\sigma = 0.905 - 383X_O = 0.890 - 133X_O$, which yields $X_O = 6 \times 10^{-5}$ or 9 ppm (by weight).

8. From the Gibbs adsorption equation, we obtain

$$-\frac{1}{RT}(\partial^2\sigma/\partial X_3\partial X_2)_{X_1\to 1} = \xi_2^{(3)} + \nu_3^\infty \epsilon_2^{(3)} + \nu_3^\infty$$

Since $(\partial^2\sigma/\partial X_2\partial X_3)_{X_1\to 1} = (\partial^2\sigma/\partial X_3\partial X_2)_{X_1\to 1}$, we deduce $\xi_2^{(3)} + \nu_3^\infty\epsilon_2^{(3)} + \nu_3^\infty = \xi_3^{(2)} + \nu_2^\infty\epsilon_3^{(2)} + \nu_2^\infty$, which is equivalent to equation (128).

9. **a.** $\nu_2^{*\infty} = \nu_2^\infty$; $\xi_2^{*(3)} = \xi_2^{(3)} + \nu_3^\infty$; $\xi_2^{*(3)} = \xi_3^{*(2)} + \epsilon_2^{(3)}(\nu_2^{*\infty} - \nu_3^{*\infty})$.
b. The value of $\theta_2^{(3)}$ is independent of the choice of k, whereas the value of $\xi_2^{(3)}$ is not. Moreover, $\theta_2^{(3)}$ measures relative changes in the value of ν_2 instead of absolute ones, and that is more convenient, especially in the case of surface-active species. On the other hand, its use in several equations [such as the equivalent to equation (130)] would be more cumbersome.

10. The single-curve feature may be expressed by the equation $(\partial^2\sigma/\partial X_3\,\partial a_2) = 0$, where $2 \equiv$ S, $3 \equiv$ C, and $1 \equiv$ Fe. Considering σ as a function of $\ln a_2$ and $\ln a_3$, we may write

$$\left(\frac{\partial\sigma}{\partial \ln a_2}\right)_{X_3} = \left(\frac{\partial\sigma}{\partial \ln a_2}\right)_{a_3} + \left(\frac{\partial\sigma}{\partial \ln a_3}\right)_{a_2}\frac{(\partial \ln a_3/\partial X_2)_{X_3}}{(\partial \ln a_2/\partial X_2)_{X_3}}$$

or

$$\left(\frac{\partial\sigma}{\partial a_2}\right)_{X_3} = \left(\frac{\partial\sigma}{\partial a_2}\right)_{a_3} + \left(\frac{\partial\sigma}{\partial \ln a_3}\right)_{a_2}\frac{(\partial \ln \varphi_3/\partial X_2)_{X_3}}{(\partial a_2/\partial X_2)_{X_3}}$$

Through the Gibbs adsorption equation

$$-\frac{1}{RT}d\sigma = \left(\overline{\Gamma}_2 - \frac{X_2}{X_1}\overline{\Gamma}_1\right)d\ln a_2 + \left(\overline{\Gamma}_3 - \frac{X_3}{X_1}\overline{\Gamma}_1\right)d\ln a_3$$

we obtain

$$-\frac{1}{RT}\left(\frac{\partial\sigma}{\partial a_2}\right)_{X_3} = \left(\frac{\overline{\Gamma}_2}{X_2\varphi_2} - \frac{\overline{\Gamma}_1}{X_1\varphi_2}\right) + \left(\overline{\Gamma}_3 - \frac{X_3}{X_1}\overline{\Gamma}_1\right)\frac{(\partial \ln \varphi_3/\partial X_2)_{X_3}}{(\partial a_2/\partial X_2)_{X_3}}$$

Differentiation of this equation with respect to X_3, with $\overline{\Gamma}_1 = -\overline{\Gamma}_2 - \overline{\Gamma}_3$ and at $X_1 \to 1$, yields $\xi_2^{(3)} - \nu_2^{\infty}\epsilon_2^{(3)} + \nu_3^{\infty} + \nu_3^{\infty}\epsilon_2^{(3)} = 0$, or $\xi_2^{(3)} = \epsilon_2^{(3)}(\nu_2^{\infty} - \nu_3^{\infty}) - \nu_3^{\infty}$. In the case of the Fe–C–S solution, $\nu_S^{\infty} \gg \nu_C^{\infty}$ (about 4500 vs 0.7 at 1550°C; see Table 2). Thus, $\xi_S^C \simeq \epsilon_S^C \nu_S^{\infty}$. With $\epsilon_S^C = 7$ at 1450°C, $\nu_S^{\infty} = 4500$ at 1550°C and larger at 1450°C, $\xi_S^C \geq 31500$ at 1450°C. (Table 4 lists $\xi_S^C \simeq 30{,}000$ at 1550°C.) The high value of ξ_S^C indicates that carbon in the bulk strongly promotes the segregation of sulfur at the surface.

11. At 1381°K, $\nu_O^{\infty} = 550$. From the perfect solution model (see Fig. 6) we deduce $\nu_{Au}^{\infty} = -0.6$. Equation (132) then yields $\xi_O^{Au} = -551(-0.6 + 0.4\Omega_{Au,O}^{(m)} - 6.8) = 4080 - 220\Omega_{Au,O}^{(m)}$, which is to be compared to an approximate value of 4500. Oxygen has no measurable solubility in gold and consequently no affinity for gold atoms in the bulk. Although the data show no effect of oxygen on the surface of liquid gold, it cannot be deduced that oxygen has no affinity for gold atoms at the surface (a decrease of σ vs X_O is not oberved mainly because $X_O = 0$ for Au). It is indeed possible for $\Omega_{Au}^{(m)}$ to be slightly negative. At $X_{Au} = 0$, the curve X_O^m vs X_O is similar to that shown in Fig. 7; the slope at the origin is equal to 550 (the value of ν_O^{∞}) and the horizontal asymptote has a value of the order of $X_O^{m(sat)} = \frac{1}{4}$. At $X_{Au} = 0.1$, the curve is similar but with a slope at the origin twice as high ($550 + 4500 \times 0.1 = 1000$), and a saturation value slightly lower than the preceding. Thus, the two curves intersect each other.

Chapter XV

1.
$$Q = \left(\sum_{n=0}^{\infty} e^{-nh\nu/kT}\right)^{3N} = \left(\sum_{n=0}^{\infty} e^{-\theta/T^{n}}\right)^{3N} = (1 - e^{-\theta/T})^{-3N}$$

$A = -kT \ln Q = 3NkT \ln(1 - e^{-\theta/T})$ or $3RT \ln(1 - e^{-\theta/T})$ for 1 mol (or when N equals Avogadro's number).

$$S = \left(\frac{\partial A}{\partial T}\right)_V = -3R \ln(1 - e^{-\theta/T}) + 3R \frac{\theta}{T} \frac{e^{-\theta/T}}{1 - e^{-\theta/T}}$$

$$E = A + TS = 3R\theta \frac{e^{-\theta/T}}{1 - e^{-\theta/T}}, \qquad C_v = 3R\left(\frac{\theta}{T}\right)^2 \frac{e^{-\theta/T}}{1 - e^{-\theta/T}}$$

3. a.
$$(1 + x)^{\alpha} = 1 + \alpha x + \frac{\alpha(\alpha - 1)}{1\cdot 2} x^2 + \frac{\alpha(\alpha - 1)(\alpha - 2)}{1\cdot 2\cdot 3} x^3 + \cdots$$

For $\alpha = \frac{1}{2}$, $(1 + x)^{1/2} = 1 + x/2 - x^2/8 + x^3/16 + \cdots$. From equation (60), we deduce $\overline{N}_{AB} = ZX_AX_B(1 - X_AX_B\lambda + 2X_A^2X_B^2\lambda^2)$.

b.
$$\sum N_{ij}^* \ln(\overline{N}_{ij}/N_{ij}^*) = -ZX_A^2X_B^2\lambda^2/4$$

$$G_m^E/RT = \tfrac{1}{2}ZX_AX_B \ln(1 + \lambda) - \tfrac{1}{4}ZX_A^2X_B^2\lambda^2$$

$$= \tfrac{1}{2}ZX_AX_B\lambda(1 - \tfrac{1}{2}\lambda - \tfrac{1}{2}\lambda X_AX_B)$$

$$S_m^E/R = -\tfrac{1}{4}ZX_A^2X_B^2\lambda^2$$

4. a.
$$\ln X_A^l + \ln \gamma_A^l = -\frac{\Delta H_{f,A}}{R}\left(\frac{1}{T} - \frac{1}{T_{f,A}}\right) = -0.114$$

Thus, $\ln \gamma_A^l = -\ln 0.8 - 0.114 = 0.109$. Equation (65) then yields $\lambda = 0.49$ and $\ln \gamma_B^\infty = 2.392$ or $\gamma_B^\infty = 10.9$.

b. $(\lambda)_{1150°K} = 0.464$. $\ln a_A^l = -0.054 = \ln X_A^l + \ln \gamma_A^l$. Through equation (65), we deduce $X_B^l = 0.06$.

c. Equation (VIII.54) yields

$$\frac{dX_A}{dT} = -\frac{X_A}{\psi}\left(\frac{\partial \ln \gamma_A}{\partial T} + \frac{\Delta H^\circ_{f,A}}{RT^2}\right) = 1.103 \times 10^{-3}\ °K^{-1}$$

and $dT/dX_B = -907°K$.

5. a. Equation (154) yields [through equation (120)]

$$e^{\lambda - \delta\varphi_{1B}^B} = 1 + \frac{[1 + 4X_AX_B(e^{\delta\varphi_{1B}^A - \delta\varphi_{1B}^B} - 1)]^{1/2} - 1}{2X_B}$$

b. $$\bar{i} = \sum_{i=0}^{Z} ip_{iB}^B = \frac{ZX_Be^{\lambda - \delta\varphi_{1B}^B}}{X_A + X_Be^{\lambda - \delta\varphi_{1B}^B}} = ZX_Be^{\delta\varphi_{1B}^A - \delta\varphi_{1B}^B} + \cdots$$

6. $$Q = \frac{(N_A + N_V)!}{N_A!N_V!}\exp\left[-\tfrac{1}{2}ZN_{AA}u_{AA} + \tfrac{1}{2}ZN_{VV}u_{VV} + \frac{ZN_AN_V}{N_A + N_V}\frac{\omega_{AV}}{kT}\right]$$

and with $u_{VV} = 0$

$$G = kT\left(N_A \ln \frac{N_A}{N_A + N_V} + N_V \ln \frac{N_V}{N_A + N_V}\right) + \tfrac{1}{2}ZN_{AA}u_{AA} + \frac{ZN_AN_V}{N_A + N_V}\omega_{AV}$$

The perfect crystal (without vacancies) is such that $G^p = \tfrac{1}{2}ZN_Au_{AA}$. Thus,

$$G - G^p = kT\left(N_A \ln\frac{N_A}{N_A + N_V} + N_V \ln\frac{N_V}{N_A + N_V}\right) + \frac{ZN_AN_V}{N_A + N_V}\omega_{AV}$$

The equilibrium concentration is given by $\partial G/\partial N_V = 0$, which yields

$$X_V = e^{-X_A^2 Z\omega_{AV}/RT} \simeq e^{-Z\omega_{AV}/RT}$$

since $X^V \ll 1$ (or $X_A \simeq 1$). The regular solution model is unlikely to be very successful. Indeed, the vibrational frequency of an atom A would be (for example) drastically altered by the presence of an adjacent vacancy. Thus, a vibrational entropy should come into play. The quasi-regular solution model would be more appropriate here. It is also clear that the assumption of randomness is not likely to be respected. In the case of a one-dimensional lattice, the parameter $\omega_{AV} = u_{AV} - \tfrac{1}{2}(u_{AA} + u_{VV})$ may be associated with the following transformation (for $2\omega_{AV}$): ···A–A–A–V–V–V–V···→···A–A–V–A–V–V–V···, or the reaction $\tfrac{1}{2}$AA + $\tfrac{1}{2}$VV = AV.

Chapter XVI

1. $H_C^E = 0$, $S_C^E = R\ \ln\{1 - [1 + (1/r)]X_C\}$. The energies of the atoms A and C (and therefore the energy of the lattice) are unaffected by their respective positions. Consequently, $H_A^E = H_C^E = 0$. However, since the distribution of

the atoms A and C is not random (they cannot exchange sites), the partial molar entropies are not equal to 0.

2. $a_C = X_C/\{1 - X_C[1 + (1/r)] - X_D[1 + (1/r)]\}$; $\epsilon_C^D = 1 + 1/r$.

3. $$G_m^E = \frac{X_A X_B}{1 - X_C} Z\omega_{AB} + RT\left\{ -(1 + r)(1 - X_C)\ln(1 - X_C) + r\left[1 - X_C\left(1 + \frac{1}{r}\right)\right]\ln\left[1 - X_C\left(1 + \frac{1}{r}\right)\right]\right\}$$

$$\frac{G_B^E}{RT} = \ln\gamma_B = \frac{X_A^2}{(1 - X_C)^2}\frac{Z\omega_{AB}}{RT} - (1 + r)\ln(1 - X_C) + r\ln\left[1 - X_C\left(1 + \frac{1}{r}\right)\right]$$

$$\frac{G_C^E}{RT} = \ln\varphi_C = -\ln\left[1 - X_C\left(1 + \frac{1}{r}\right)\right]$$

$$\epsilon_B^C = 0$$

4. **a.** $P_C = [1 + y_B A_{CB}(1 - \lambda_{B-C})]^{Z'}[1 + y_C A_{CC}(1 - \lambda_{C-C})]^{z'}$.
b. In the case of the Fe–Mn–C solution $Z' = 6$, $z' = 12$, $\lambda_{C-C} = 0.28$, $\lambda_{Mn-C} = -0.20$ (see Table 1), $y_{Mn} = 0.263$, $y_C = 0.043$, $A_{CC} = 1.012$, $A_{C,Mn} = 1.002$, $P_C = 7.529$, $p_{1\,Ni,\,0\,C}^C = 0.277$, $p_{2\,Ni,\,0\,C}^C = 0.148$.
c. The case of the Fe-Ni-C solution is similar. The values which differ are $\lambda_{Ni-C} = 0.20, A_{C,Ni} = 1.016, P_C = 4.630, p_{1\,Ni,\,0\,C}^C = 0.277, p_{2\,Ni,\,0\,C}^C = 0.148$.
d. The average number of carbon atoms is $\bar{j} = \sum_{i,j} jp_{i,j}^C = z'y_C A_{CC}(1 - \lambda_{C-C})/[1 + y_C A_{CC}(1 - \lambda_{C-C})] = 0.36$ in both cases b and c. The average number of Mn atoms is $\bar{i} = \sum_{i,j} ip_{i,j}^C = Z'y_{Mn}A_{C,Mn}(1 - \lambda_{C-Mn})/[1 + y_{Mn}A_{C,Mn}(1 - \lambda_{C-Mn})] = 1.44$. Similarly, the average number of Ni atoms is 1.06.

5. **a.** Treating silicon as an interstitial solute, we obtain from equation (86) $\epsilon_C^{Si} = 2 + 24\lambda'_{Si-C} = 9.8$, which yields $\lambda'_{Si-C} = 0.325$. Similarly, $\epsilon_{Si}^{Si} = 12.4$ yields $\lambda'_{Si-Si} = 0.44$. From equations (87) and (88) we then obtain $\rho_C^{Si} = 14.5$ and $\rho_C^{Si,C} = 23.4$.
b. From equation (90), we deduce

$$\begin{aligned}\ln(y_C^t/y_C^b) &= -2z'\ln(A_{CC}^t/A_{CC}^b) - 2z'\ln\frac{1 + y_{Si}A_{SiSi}}{1 + y_{Si}A_{SiSi}(1 - \lambda_{Si-C})} - \rho_C^{C,Si}y_C^t y_{Si}\\ &= -24\ln(A_{CC}^t/A_{CC}^b) - 24\ln\frac{1 + y_{Si}A_{SiSi}}{1 + 0.675y_{Si}A_{SiSi}} - 23.4y_C^t y_{Si}\end{aligned}$$

which may be solved by trial and error (or by computer). We note that for $X_C^b = 0.205$, $y_C^b = 0.347$. For example, for $y_{Si} = 0.20$, we find that $y_C^t = 0.115$. Since $X_C = y_C/(1 + 2y_C + 2y_{Si})$, $X_C^t = 0.070$, and similarly, $X_{Si} = 0.12$. Thus, at $X_{Si} = 0.12$, $-\ln(X_C^t/X_C^b) = 1.07$, which is too high for the data in Fig. 11.

List of Symbols

a_i	activity of solute i
b	as superscript, binary
c	curvature; concentration
d	differential
$đ$	nonperfect differential
e	eigenvalue (p. 78)
$e_i^{(j)}$	first order Gibbs free energy interaction coefficient (wt.% basis)
f	fugacity (p. 494)
f	as subscript, of fusion
f_i	activity coefficient of solute i (wt.% basis)
$f_{i\mathrm{B}}^{\mathrm{A}}$	correction factor for nonrandom probability distribution (p. 455)
g	degeneracy; parameter measuring net interaction forces between adsorbed atoms
g	gas
h	Planck's constant; correction term in quasi-chemical approximation; height of a polyhedron's face
h	as superscript, Henrian standard state (wt.% basis)
$h_i^{(j)}$	first order enthalpy interaction coefficient (wt.% basis)
k	Boltzmann's constant; normalization constant in definition of adsorption coefficient
l	liquid
$l_i^{(j,k)}$	second order enthalpy interaction coefficient (wt.% basis).
m	number of components in phase rule (p. 68) and in solution (see Chapter XVI); mass of atom
m	as subscript, mole; as superscript, relating to monolayer
n	number of moles

p	partial pressure; dimensionless quantity in model of surface-active species (p. 411)
$p_i^{(j,k)}$	second order entropy interaction coefficient (wt.% basis)
p_{iB}^{A}	probability that central atom A is surrounded by i atoms B in its shell of nearest neighbors
q	heat; atomic partition function
r	radius; radius of curvature; macroscopic roughness index (p. 374); ratio of number of interstitial sites to number of substitutional sites
$r_i^{(j,k)}$	second order Gibbs free energy interaction coefficient (wt.% basis)
s	area; entropy density (p. 395)
s	solid
$s_i^{(j)}$	first order entropy interaction coefficient (wt.% basis)
t	triple point; time; tension; number of substutional components (Chapter XVI)
u	ratio of characteristic temperature to temperature (p. 27); energy of bond between two atoms
v	velocity; volume (p. 357)
w	work
x	relative proportion of phase in a system; width of spinodal domain or of miscibility gap
y	asymmetry of spinodal line or of miscibility gap; composition coordinate (see Chapter XVI)
z	coordination number of interstitial atom; composition coordinate (lattice ratio, p. 307)
A	Hemlholtz fre energy
$A^{(\nu)}$	area of polyhedron's face ν (p. 368)
A_{ij}	mathematical function defined on p. 493
B	virial coefficient (p. 100)
C	heat capacity; critical point; interstitial atom
C_p, C_v	heat capacities at constant pressure, at constant volume
C_i^j	combinatorial factor $j!/i!(j-i)!$
D	finite difference; functions for stability conditions (p. 302)
E	energy
E	eutectic; as superscript, excess property
E_{12}	limit of effect of solute 3 on eutectic temperature of binary 1–2
F	force
G	Gibbs free energy
H	enthalpy
H	as superscript, Henrian standard state (mole fraction basis)
$\mathcal{H}_i^E$	excess enthalpy function (p. 185)
I	moment of inertia (p. 26)
J	surface activity
$J_{n_2,\ldots,n_m}^{(i)}$	generalized Gibbs free energy interaction coefficient
K	equilibrium constant of reaction

$K^{(i)}_{n_2, \ldots, n_m}$	generalized entropy interaction coefficient
L	coefficient for analysis of coupled reactions
$L^{(i)}_{n_2, \ldots, n_m}$	generalized enthalpy interaction coefficient
M	atomic weight
M	as superscript, mixing property
N	number of atoms
O	order of remainder in series
P	pressure
P	peritectic
P_i	vapor pressure of pure i
P_{12}	limit of effect of solute 3 on peritectic temperature of binary 1–2
Q	reaction's quotient; partition function
R	universal gas constant; remainder in series
$\mathbf{R}$	vector representing reaction
S	entropy
$\mathcal{S}^{E}_{i}$	excess entropy function (p. 185)
T	temperature
U	compound (Chapter XII); potential energy of central atom
V	volume; compound (Chapter XII)
X	mole fraction
Y	example of homogeneous function (Section II.3) or of solution's thermodynamic property (Section VI.1); composition coordinate (Chapter XVI)
Z	function defined on p. 15; coordination number
α	coefficient of expansion; Darken's stability function (p. 87); function defined for integration of Gibbs–Duhem equation (p. 165); area occupied by 1 mol in surface monolayer (p. 400)
α^{A}_{iC}	probability function (p. 483)
β	coefficient of compressibility; $1/kT$
γ	coefficient for electronic contribution to heat capacity (p. 29)
γ_i	activity coefficient of solute i (Raoultian standard state and mole fraction basis)
Γ	adsorption function; coefficient for pressure dependence of reference state (p. 108)
δ	virtual variation; change in value of central atom's property (p. 456)
δ_{ij}	Kronecker's symbol
Δ	associated with definitions of ΔG, ΔH, ΔS, . . . (e.g., Chapter V)
$\epsilon^{(j)}_i$	first order Gibbs free energy interaction coefficient (mole fraction basis)
η	Lagrange multiplier (p. 369)
$\eta^{(j)}_i$	first order enthalpy interaction coefficient (mole fraction basis)
θ	characteristic temperature; fraction of surface sites occupied by solute; angle of contact (p. 374)
θ_i	activity coefficient (p. 319)

$\theta_i^{(j)}$	interaction coefficient associated with use of lattice ratio (p. 307)
ϑ	variance in phase rule (p. 68)
λ	progress variable of chemical reaction; fraction of solute in given phase (p. 76); parameter in statistical models
$\lambda_i^{(j,k)}$	second order enthalpy interaction coefficient (mole fraction basis)
Λ	Lagrange multiplier
μ_i	chemical potential of element i
ν	stoichiometric coefficient in chemical reaction; frequency; adsorption coefficient
$\xi_i^{(j)}$	adsorption interaction coefficient
π	entropy parameter in statistical model (p. 460)
$\pi_i^{(j,k)}$	second order entropy interaction coefficient (mole fraction basis)
ρ	radius (p. 354)
$\rho_i^{(j,k)}$	second order Gibbs free energy interaction coefficient
σ	surface tension; as superscript, surface excess property (p. 350)
$\sigma_i^{(j)}$	first order entropy interaction coefficient (mole fraction basis)
Σ	geometric surface (Chapter XIII)
τ	parameter linking enthalpy and entropy contributions
φ	number of phases in phase rule (p. 68); dihedral angle (p. 375); dimensionless quantity in model of surface-active species (p. 411)
φ_i	Henrian activity coefficient (mole fraction basis)
ϕ	function for definition of critical point (p. 311); observed contact angle (p. 374)
$\phi_{n_2,\ldots,n_m}$	coefficient in Taylor series development of G^E
ψ	stability function; potential of central atom
ω	parameter of regular solution
Ω	parameter of regular solution
$^\circ$	degree sign; standard state
$\bar{\ }$	above letter, identifies it as partial molar property at constant pressure or reduced adsorption function; also identifies standard state of a compound (Section XII.3.1); below reaction's component identifies standard state of that component as Henrian
$\tilde{\ }$	above letter, identifies it as partial molar property at constant volume
$*$	reference state; random solution
∞	infinity; as superscript, indicates state of infinite dilution
$\rightarrow$	transformation (e.g., $\alpha \rightarrow \beta$: transformation of α into β)
∂	partial derivative
$\sum$	summation
$\prod$	product

Author Index

The name of an author is followed by the numbers of the pages in which his work is cited. If that work is given a reference number in the chapter corresponding to these pages, then that reference number is identified by brackets. For example, Bain E. C., 323, 324 [29], means that pages 323 and 324 cite Bain's work in reference 29 of the chapter corresponding to these pages (in this case, Chapter XI).

Subject Index

D

J

K

L

T